www.RiATV

LINEAR PROGRAMMING

LINEAR PROGRAMMING

James Calvert
William Voxman

UNIVERSITY OF IDAHO

Harcourt Brace Jovanovich, Publishers
and its subsidiary, Academic Press

San Diego   New York   Chicago   Austin   Washington, D.C.

London   Sydney   Tokyo   Toronto

ISBN: 0-15-551027-4

Library of Congress Catalog Card Number: 88-81045

Printed in the United States of America

CONTENTS

SKIP

PREFACE

Two fundamental objectives have guided the development of this textbook: (1) to carefully motivate and explain the basic ideas underlying linear programming and (2) to apply these ideas to a wide variety of mathematical models. To achieve these objectives, we have provided more than the usual number of examples and, in general, favored clarity of presentation over abstraction. Our primary goal has been to write a textbook that allows the student to grasp readily the material presented and to apply the material to a number of challenging problems.

Although linear programming presents many interesting theoretical problems, the focus of an introductory book should be on LP techniques and applications. Consequently, in addition to the various LP algorithms, we have included numerous sections on modeling LP problems. These sections may be omitted without affecting the continuity of the book, but we recommend the study of as many of these sections as possible. They provide excellent practice in learning to set up and solve nontrivial, yet accessible, models.

The book is accompanied by a computer software package, CALIPSO, that can solve fairly large linear and integer programming problems. While CALIPSO is optional, its use will greatly facilitate lengthy calculations and enable the student to solve many LP and IP modeling problems whose hand solution would be impractical. CALIPSO contains tutorials, automatic solution programs, and a modeling program that allows problems to be entered as systems of inequalities. *CALIPSO is available without charge* to institutions that adopt this book and can be obtained by individuals at a nominal charge by mailing the coupon contained in the book.

Chapter 1 of the book introduces a number of problems that can be modeled as LP problems and whose solutions are particularly well suited to techniques developed later in the book. Product-mix, transportation, and scheduling problems are among those discussed in this chapter. An instructor may cover a sampling of these problems initially and refer to these and other problems later.

The underlying geometry of linear programming provides much of the focus of Chapter 2. This material provides insights into the techniques developed throughout the book. In addition, some preliminary algebraic results for developing the simplex method are discussed.

Chapter 3 is the core of the book. It describes in detail the simplex method, one of this century's major mathematical achievements. This chapter features a careful description of the two-phase method as well as a brief introduction to the Big-M method.

The simplex method in a matrix context is the subject of Chapter 4. Material developed in this chapter is used extensively in subsequent chapters, particularly in dealing with the revised simplex method, duality, and sensitivity analysis. We do not assume any linear algebra background on the part of the reader, and we have tried to minimize the level of abstraction as much as possible.

Chapter 5 provides a careful presentation and analysis of the revised simplex method. The advantages of this variation of the simplex method are clearly demonstrated in the complexity analysis developed in this chapter.

Duality is probably the most abstract topic treated in this book. We have, in Chapter 6, given special attention to this concept and illustrated its usefulness. We have also taken extra care to give a thorough and somewhat original presentation of the general concept of duality. The proofs of the major duality results may be found in Section 6.4; readers less theoretically inclined may omit this section with no loss of continuity.

Chapter 7 introduces sensitivity analysis, one of the most important topics in linear programming. A lack of precision characterizes many real world problems, and sensitivity analysis provides a technique for determining how much flexibility there is in dealing with a particular problem. As with the chapter on duality, we have organized the material in such a way that many of the formal proofs of the sensitivity results may be omitted without compromising the continuity of the text or detracting from the applications of the material.

Integer programming is one of the more challenging areas of optimization. In Chapter 8 we present a number of disparate situations where integer programming is put to good use. The three basic techniques for solving IP problems found in this chapter are a branch and bound method, implicit enumeration, and a cutting plane algorithm. You will find CALIPSO especially useful for implementing these techniques.

The transportation problem has played a fundamental role in the development of linear programming. In Chapter 9 we introduce a technique that takes advantage of the special structure of transportation problems.

Chapter 10 treats some of the difficulties encountered in solving large-scale LP problems. Such problems are often the norm in practice, and considerable research has gone into finding efficient ways to deal with them. In particular, we discuss problems with bounded variables and a class of problems for which a clever column generating technique is appropriate.

The exercises in the book range in difficulty from the routine to the reasonably challenging. We have marked with an asterisk those sections and problems that seem particularly difficult or that require more background (e.g., linear algebra) than does the text in general. We have indicated with a dagger (†) those problems that require use of CALIPSO. You will find that the solutions to the odd-numbered exercises are often unusually detailed. We recommend that you spend an appropriate amount of time on an exercise before turning to this appendix to verify your solution or to solve the problem.

The book contains more material than can ordinarily be covered in a one-semester course. This allows for a certain amount of flexibility in the presentation of the material. Most instructors will find it possible to present each section in one 50-minute lecture. The following flow chart describes, with some inconsequential exceptions, the logical dependence of the sections.

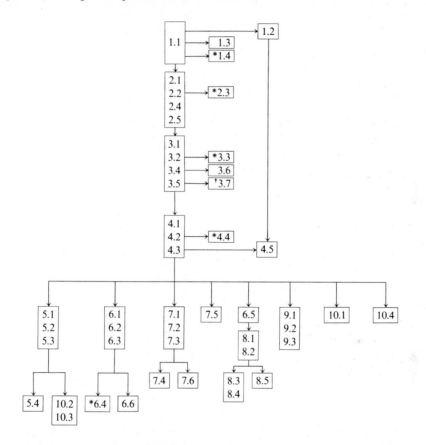

We would like to thank professors Willy Brandal and Sam Stueckle and the many students of theirs and ours who provided valuable suggestions and corrections to the manuscript. I would also like to thank the following reviewers: Ron Barnes, University of Houston; John Birge, University of Michigan; Jerald Dauer, University of Nebraska; J. Rajgopal, University of Pittsburgh; and Mike Todd, Cornell University.

James E. Calvert
William L. Voxman

To obtain a copy of CALIPSO or an Instructor's Manual, call or send a facsimile of this form to:

Harcourt Brace Jovanovich Publishers
College Department
7555 Caldwell Avenue
Chicago, IL 60648
Telephone (312) 647-8822

Please check the appropriate boxes:

☐ I am an instructor using *Linear Programming* by Calvert and Voxman (ISBN 0-15-551027-4).

☐ I want these materials for my personal use. (A nominal charge may be made.)

☐ Send me a CALIPSO disk (ISBN 0-15-551029-0).

☐ Send me an Instructor's Manual (ISBN 0-15-551028-2).

LINEAR PROGRAMMING

Chapter 1

LINEAR PROGRAMMING: AN OVERVIEW AND SAMPLE PROBLEMS

Linear programming is an exciting and relatively new area of mathematics that developed primarily in response to large-scale problems arising during World War II. These problems involved the distribution of materials, troops, aircraft, and so on. Because previously used methods (based essentially on differential calculus) for dealing with such problems proved inadequate, linear programming became the focus of considerable research. A general theory of linear programming began to emerge after the war, in large part through the efforts of George Dantzig, and major industrial applications of linear programming soon became apparent. The advent of computers greatly facilitated the work of Dantzig and others, and now many recognize linear programming as one of the outstanding scientific achievements of this century. In fact, the 1975 Nobel Prize in economics was awarded to T. C. Koopmans and L. V. Kantorovich for their contributions to the theory of optimum allocation of resources, which is at the heart of linear programming. In 1970, it was estimated that linear programming accounted for almost 25% of all computer time. Current applications occur in such diverse areas as transportation of goods, scheduling of plant operations, mixing of various product ingredients, and portfolio management.

This chapter contains a diverse collection of problems that are solvable by linear programming techniques. We focus initially on transcribing the verbal description of a problem into a mathematical form that allows the application of special linear programming techniques. This process of transcription is often called *modeling*, and the mathematical formulation of the problem is often called a *model*. In many ways, modeling is the most difficult aspect of linear programming; therefore, we devote attention to developing the necessary skills throughout this text. Many of the problems we pose in this chapter will be solved later in the text, either as examples or as problems to be solved by you.

1.1 PRODUCT-MIX PROBLEMS

In essence, linear programming problems (LP problems) involve finding the optimal distribution of limited resources, subject to a variety of constraints. Mathematically, this translates into maximizing or minimizing a linear function in several variables

$$z = c_1 x_1 + c_2 x_2 + \cdots + c_n x_n \tag{1}$$

given certain linear constraint inequalities and/or equalities of the form

$$a_1 x_1 + a_2 x_2 + \cdots + a_n x_n \leq b,$$

$$a_1 x_1 + a_2 x_2 + \cdots + a_n x_n \geq b, \tag{2}$$

or $\qquad a_1 x_1 + a_2 x_2 + \cdots + a_n x_n = b$

In (1) and (2), $x_1, x_2, \ldots, x_n$ are variables; $c_1, c_2, \ldots, c_n$ are constants called *cost coefficients*; b is a constant called a *resource value*; and $a_1, a_2, \ldots, a_n$ are constants called *constraint coefficients*. We call z the *objective function*. Thus, for example,

Maximize: $\quad z = 3x_1 - x_2 + 2x_3 \qquad$ (Objective function)

$$\text{Subject to:} \quad \left. \begin{aligned} x_1 - 3x_2 + \tfrac{1}{2}x_3 &\leq 4 \\[4pt] 2x_1 \qquad\quad + 4x_3 &\leq 2 \\[4pt] 3x_1 - x_2 + x_3 &\leq 4 \\[4pt] x_1 \geq 0,\ x_2 \geq 0,\ x_3 &\geq 0 \end{aligned} \right\} \quad \text{(Constraints)} \tag{3}$$

is a typical LP problem. As in (3), we will usually work only with variables that are nonnegative (e.g., $x_1 \geq 0$, $x_2 \geq 0$, $x_3 \geq 0$), though negatively valued variables sometimes arise.

To solve the LP problem (3) we need to find values for $x_1, x_2,$ and x_3 that satisfy all of the constraints and give the largest possible value for the objective function z.

A set of values satisfying all of the constraints of an LP problem is said to form a *feasible solution* to the problem; if the set of values fails to satisfy one or more of the constraints, then we have an *infeasible solution*. In (3), the values $x_1 = \frac{1}{3}$, $x_2 = 3$, $x_3 = \frac{1}{3}$ provide a feasible solution to the problem; the values $x_1 = 1$, $x_2 = 3$, $x_3 = 2$ constitute an infeasible solution because these values fail to satisfy the second constraint. The *feasible region* or *feasible set* for an LP problem is the set of all feasible solutions. An *optimal solution* is a feasible solution that maximizes (or, for a minimization problem, minimizes) the objective function on the feasible region. As you will see in Chapter 3, an optimal solution of (3) is $x_1 = 1, x_2 = 0, x_3 = 0$, and the maximal value of z is 3.

We now consider some situations that we can model as LP problems. We begin with product-mix problems, which, though relatively simple, illustrate many of the basic features of LP problems. In practice, they often arise as building blocks of much larger and more complex problems.

Typically, a product-mix problem involves the manufacture of various products subject to certain production constraints. The goal is to select the optimal mix of these products (taking into account the constraints) that maximizes profits or, perhaps, minimizes losses. In other words, we want to make the best use of the available resources under the imposed restrictions. These restrictions can include limits on the availability of materials needed for the manufacture of the products, insufficient machine capacity, lack of manpower, limited demand for the products, and so on. Our first example leads to a typical small-scale LP problem.

EXAMPLE 1

A small firm produces bookcases and tables. The manufacture of these products involves three basic operations: cutting, assembly, and finishing. Table 1 gives the hours required for each of these activities to produce one bookcase or one table.

Table 1

Item	Cutting time (hr)	Assembly time (hr)	Finishing time (hr)
Bookcase	$\frac{6}{5}$	1	$\frac{3}{2}$
Table	1	$\frac{1}{2}$	2

The profit realized is \$80 for each bookcase sold and \$55 for each table. For simplicity, we assume that the firm can sell all of the items produced. Under this assumption it is clear that if there were no constraints on the manufacture of these products, then the firm should produce an unlimited number of bookcases and tables. In practice, of course, a lack of constraints would be unlikely.

Suppose then that a maximum of 72 work-hours per day is available for cutting, 50 work-hours per day for assembly, and 120 work-hours per day for finishing. Such constraints could be due to a variety of factors: limited personnel, limited machine capacity, and so on.

The firm wants to maximize its profits given these constraints. It must determine the best mix (the appropriate number) of bookcases and tables it should produce to ensure the highest possible return. To achieve this best mix, the firm must decide how most effectively to allocate its resources: the times spent in cutting, assembly, and finishing.

The first step in resolving this problem is to frame it in a manageable form. To this end we let the variable x_1 denote the number of bookcases produced and x_2 the number of tables produced. Since the sale of each bookcase results in a profit of \$80 and the sale of each table in a profit of \$55, we find that the profit from the sale of x_1 bookcases and x_2 tables is

$$z = 80x_1 + 55x_2$$

The function z in the variables x_1 and x_2 is the objective function; it is this function that we want to maximize.

We can describe the constraints in terms of the variables x_1 and x_2 as follows. From Table 1 we see that the cutting time (in hours) required in the manufacture of x_1 bookcases and x_2 tables is

$$\frac{6}{5}x_1 + x_2$$

Since only 72 work-hours per day are available for cutting, we obtain the constraint inequality

$$\frac{6}{5}x_1 + x_2 \le 72$$

In a similar fashion, from Table 1 and the limitation of 50 assembly hours per day, we obtain the constraint inequality

$$x_1 + \frac{1}{2}x_2 \le 50$$

The constraint of 120 finishing hours per day leads to the inequality

$$\frac{3}{2}x_1 + 2x_2 \le 120$$

Finally, since it is impossible to manufacture a negative quantity of bookcases or tables, we have the additional constraints

$$x_1 \ge 0$$
$$x_2 \ge 0$$

Combining this information, we obtain the LP problem

Maximize: $z = 80x_1 + 55x_2$ (Objective function)

Subject to: $\frac{6}{5}x_1 + x_2 \le 72$ (Cutting-time constraint)

$x_1 + \frac{1}{2}x_2 \le 50$ (Assembly-time constraint)

$\frac{3}{2}x_1 + 2x_2 \le 120$ (Finishing-time constraint)

$x_1, x_2 \ge 0$ (Obvious constraints)

As we shall see in Section 2.2, the solution to this problem is $x_1 = 35$, $x_2 = 30$. That is, the values $x_1 = 35$, $x_2 = 30$ result in the greatest possible value ($z = 4450$) of the objective function $z = 80x_1 + 55x_2$ given the various constraints.   ◆

Before taking up another example of the product-mix type, we make a few remarks about what lies ahead. In Example 1 we chose numbers so that the optimal solution would turn out to be integer valued. In general, however, it would be quite possible to obtain a noninteger optimal solution, say $x_1 = 44.6$, $x_2 = 52.3$. Although a firm could conceivably manufacture six-tenths of a bookcase and three-tenths of a table, the market value of these products would be questionable. Thus, we will often impose the additional constraint that x_1 and x_2 be integers. As we shall see, this integer constraint introduces complications. For instance, you might be tempted simply to round off the values 44.6 and 52.3 to obtain an optimal integer solution $x_1 = 44$, $x_2 = 52$. However, rounding off can lead to infeasible solutions or to feasible solutions that are far from optimal integer solutions. Integer programming, which we take up in Chapter 8, deals with the problem of imposing integer requirements on some or all of the variables.

Another question arising in LP programming is that of sensitivity. In Example 1 we assumed that precisely two hours were needed to finish a table. It is natural to inquire to what extent we could change this value without causing a change in the optimal solution: $x_1 = 35$, $x_2 = 30$. We could ask a similar question about the effect of changes in the other work-hour constraints or in the given profit margins. In general, it is useful to know how "sensitive" the solution to an LP problem is to small changes in the original data. We treat problems of this nature in Chapter 7.

We can apply the model underlying product-mix problems to problems that do not actually involve the best mix of manufactured products. In the next example, which is of the type described in Example 1, we determine the best mix of investments subject to certain constraints.

EXAMPLE 2

Suppose you won the state lottery and, after taxes, have $100,000 at your disposal. You decide to place a portion of this money in a savings account and invest the rest in stocks and bonds. You have found that you can expect an average yearly return of 9% on your stock investments, 7% on the bonds, and 6% on your savings account. Because of possible fluctuations in the bond and stock markets, you decide to place at least as much money in your savings account as in stocks and bonds. Moreover, you intend to invest at least twice as much money in the bond market as you do in stocks.

How should you distribute your winnings (under the self-imposed constraints) to maximize your return? In other words, what mix of savings and investments will provide you with the maximal income? To set up this problem we let x_1 denote the amount of money invested in stocks, x_2 the amount invested in bonds, and x_3 the amount placed in a savings account. Then the objective function we want to maximize is

$$z = 0.09x_1 + 0.07x_2 + 0.06x_3$$

The conditions you have set for using your money are

$$x_1 + x_2 \leq x_3$$

$$x_2 \geq 2x_1$$

Since the total amount of disposable winnings is $100,000, we also have the constraint

$$x_1 + x_2 + x_3 = 100,000$$

and, of course, we have the obvious constraints: $x_1 \geq 0$, $x_2 \geq 0$, $x_3 \geq 0$. Thus, we must solve the following LP problem:

Maximize: $z = 0.09x_1 + 0.07x_2 + 0.06x_3$ (Objective function)

Subject to: $x_1 + x_2 + x_3 = 100{,}000$ (Total winnings)

$x_1 + x_2 - x_3 \leq 0$ (Savings vs. stocks and bonds)

$2x_1 - x_2 \leq 0$ (Stocks vs. bonds)

$x_1, x_2, x_3 \geq 0$ (Obvious constraints)

In Chapter 3 we will learn to solve this problem and discover that the solution is $x_1 = \$16{,}666.67$, $x_2 = \$33{,}333.33$, and $x_3 = \$50{,}000$. The maximum return on your investments is $z = \$6{,}833.33$. ◆

To conclude this section we examine a problem that, though strictly speaking is not a product-mix problem, is nevertheless based on similar ideas. The mix in this case is the number of days each of two plants should be operated to minimize costs in the manufacture of certain products.

EXAMPLE 3

A manufacturing firm has two plants P_1 and P_2. Each of these plants is capable of manufacturing four products: *A, B, C,* and *D*. The firm is under contract to supply a distributor with the following number of units of these products:

Product *A*:   1000 units   Product *C*:   900 units

Product *B*:   800 units   Product *D*: 1500 units

Table 2 gives the number of units of these products that plants P_1 and P_2 can produce per day.

Table 2

Product	Plant P_1 (units/day)	Plant P_2 (units/day)
A	200	100
B	60	200
C	90	150
D	130	80

The cost of operating plant P_1 is \$8000 per day, and the cost of operating plant P_2 is \$11,000 per day. The problem is to determine how many days each plant should operate to minimize costs but still meet the demands of the distributor. To set up this problem we let the variable x_1 denote the number of days plant P_1 operates and the variable x_2 the number of days plant P_2 operates. We want to minimize the cost z of operating these plants, where

$$z = 8,000x_1 + 11,000x_2$$

is the objective function.

We determine the constraints as follows. First observe that plant P_1 produces 200 units, and plant P_2 produces 100 units per day of product A. Because the demand for product A is 1000 units, we have the constraint inequality

$$200x_1 + 100x_2 \geq 1000$$

Note that it would not be correct to use an equality in place of the inequality in this constraint. The problem states that we must operate the plants enough days to satisfy the demand; however, to meet all of the constraints simultaneously, it may be necessary to exceed some of the demands. It could be (and in this case it is) impossible to find an optimal schedule that exactly meets all four demands.

In a similar way we obtain the constraint inequalities

$$60x_1 + 200x_2 \geq 800$$
$$90x_1 + 150x_2 \geq 900$$
$$130x_1 + 80x_2 \geq 1500$$

corresponding to the products B, C, and D, respectively. As usual, we also have the obvious constraints $x_1 \geq 0$ and $x_2 \geq 0$. Thus, our LP problem is

Minimize: $z = 8000x_1 + 11,000x_2$

Subject to: $200x_1 + 100x_2 \geq 1000$ (Demand for A constraint)
$\quad\quad\quad\quad 60x_1 + 200x_2 \geq 800$ (Demand for B constraint)
$\quad\quad\quad\quad 90x_1 + 150x_2 \geq 900$ (Demand for C constraint)
$\quad\quad\quad\quad 130x_1 + 80x_2 \geq 1500$ (Demand for D constraint)

$\quad\quad\quad\quad\quad\quad\quad x_1, x_2 \geq 0$ (Obvious constraints)

In Chapter 3 we will solve this problem and find that the solution is $x_1 = 11.132$, $x_2 = 0.660$, and $z = 96,320.75$. We will determine an optimal integer solution in Chapter 8.

Notice that, to meet demands B and D exactly, this solution requires operating the plants for more days than are necessary to meet demands A and C. Thus, as we previously discussed, it would be incorrect to replace the four $\geq$ inequalities with equalities.   ◆

Exercises

We will study methods of solving LP problems in later chapters, and at appropriate times you will be asked to solve many of the following exercises. For now, you need only to set up the problems by introducing variables, forming an objective function, and writing the constraint inequalities or equalities.

1. The Pocatello Potato Processing Plant (4P) produces two kinds of frozen french fries: shoestring and nuggets. The plant's peeling machine can peel a ton of potatoes in one hour. Each of 4P's two slicing machines takes 2.5 hours to slice a ton of shoestring potatoes and 1.5 hours to slice a ton of nuggets. The plant makes $60 per ton of shoestrings sold and $45 per ton of nuggets. Suppose that the machines can be adjusted to produce either type of product. To maximize profit, how many tons of each type should 4P process in an 8-hour work day?

2. Adeline, the cook, wants to mix a cheap granola with a more expensive variety to offer a nutritious breakfast to her dormitory residents. Each pound of Chintz contains 4 units of protein, 4 units of carbohydrate, and 24 units of fat. Each pound of Scrumch contains 12 units of protein, 4 units of carbohydrate, and 8 units of fat. A 50-pound bag of Chintz costs $25, and a 20-pound bag of Scrumch costs $16. Adeline feels that each student should have for breakfast at least 6 units of protein, 4 units of carbohydrates, and 12 units of fat.

   (a) Fill in Table 3 with the conditions of this problem.

   Table 3

Granola	Protein	Carbohydrate	Fat	Cost per pound
Chintz				
Scrumch				
Units needed				

   (b) Set up an LP problem to answer the question: How many pounds of Chintz and how many pounds of Scrumch should Adeline serve each student to provide the minimum dietary requirements at the least cost?

3. Adam Smith has $18,000 to invest in three types of stocks: low-risk, medium-risk, and high-risk. He wants to invest this money in a way that maximizes his total profit subject to three constraints that (theoretically) offer some protection against large losses. He will invest at most $2000 more in low-risk stocks than he invests in medium-risk stocks. He will invest at most $8000 in high-risk stocks, and no more than $14,000 in medium- and high-risk stocks. The expected returns are 7% for low-risk stocks, 9% for medium-risk stocks, and 11% for high-risk stocks. How much money should Adam invest in each type of stock, and how much return can he expect?

4. A farmer owns a 1200-acre farm that produces wheat, potatoes, and peas. Each crop makes a different profit and has different requirements for labor, fertilizer, pesticides, and seeds. The farmer has $36,000 in capital and a large family that can devote 7400 person-hours of labor without pay. Table 4 gives the relevant data *per acre*. The profit from an acre of crop is the difference between the selling price of the crop and the amount spent on fertilizer, pesticide, and seed. How many acres of each crop should the farmer plant to maximize profit?

Table 4

Crop	Labor (hr)	Fertilizer ($)	Pesticide ($)	Seed ($)	Sales ($)
Wheat	6	20	14	2	144
Potatoes	8	12	9	3	125
Peas	3	8	9	1	75

5. A family has a home in the country that has 4 acres of lawn. During the growing season they spend 6 person-hours and $15 per acre to maintain the lawn. They are considering plowing part of the land to grow crops for sale at the local farmers' market and want to estimate how profitable this would be. Table 5 contains their estimates of the labor and investment required and the gross sales expected for each acre. The last row of Table 5 gives limitations on the time (hours) they can spend and the money (dollars) they can invest in such costs as fertilizer, pesticides, and water. Determine how much land they should use for each crop to maximize profit. Any land that is not planted for crops to sell must still be maintained as lawn.

Table 5

Crop	Labor (person-hr)	Expenses ($)	Gross sales ($)
Tomatoes	90	150	750
Beans	18	40	150
Peas	27	30	160
Corn	10	25	50
Carrots	30	55	180
Constraints	300	400	

6. A contractor builds three types of homes: Ranch, Colonial, and Modern. The Ranch requires $24,000 capital and 150 person-days to build and is sold for a profit of $4400. The Colonial requires $30,000 capital, 175 person-days, and is sold for a profit of $5500. The Modern requires $18,000 capital, 100 person-days, and is sold for a profit of $2500. The contractor has $4,800,000, 200 lots, and 34,000 person-days of labor available. To make sure that the neighborhood has a variety of house types, at least 40 Colonial and 30 Modern types must be built. How many of each type should be built to maximize profit?

7. A plant manager is planning a week's production to manufacture five products A, B, C, D, and E. Products can be produced in any combination, except that the plant has already accepted an order for 20 units of product C and 30 units of product D; so at least these amounts of the two products must be made. The manufacture of each of the five products requires time on three machines: M_1, M_2, and M_3. Each machine is available for 80 hours per week. Table 6 gives the processing time (minutes) and the selling price (dollars) for one unit of each product.

The costs of operating the machines are $9 per hour each for machines 1 and 2 and $12 per hour for machine 3. Materials costs are $2 per unit for products A, B, and C and $1 per unit for products D and E.

Table 6

Product	Required time on machine (min)			Selling price ($)
	M_1	M_2	M_3	
A	15	8	6	12.00
B	8	10	9	11.00
C	8	12	10	12.00
D	12	4	12	10.50
E	9	6	0	6.00

(a) Introduce appropriate variables and write the objective function for the profit. The profit is the total selling price of all units, minus the cost of materials used in making the units, and minus the cost of operating the machines. Assume there is no standby cost for a machine not being used.

(b) Write the constraint inequalities.

8. The Mazda Corporation manufactures light bulbs by assembling three components: a base, a glass globe, and a filament. Mazda usually manufactures its own components, though the firm can purchase these components from outside sources when the quantities needed exceed Mazda's production capacity. Mazda has contracted to produce 12,000 bulbs this month. Table 7 gives the costs of manufacturing components inside and purchasing them outside.

Table 7

Component	Inside cost ($)	Outside cost ($)
Base	0.05	0.06
Globe	0.03	0.04
Filament	0.10	0.14

Mazda's plant is organized into three departments. Table 8 gives the time requirements for manufacturing components. The last line of the table indicates Mazda's capacity (hours available for production).

Table 8

Component	Department		
	Cutting (hr)	Shaping (hr)	Assembly (hr)
Base	0.04	0.05	0.06
Globe	0.07	0.03	0.05
Filament	0.06	0.03	0.06
Capacity	1600	1400	1500

How many of each component should be manufactured inside and how many purchased outside to minimize production costs?

9. Suppose that the selling price given for products C and D in Exercise 7 applies only to the units that were already ordered. Units of C in excess of 20 are sold for $9.00, and units of D in excess of 30 are sold for $8.00. Make whatever changes are necessary to the LP problem of Exercise 7 to incorporate this new circumstance.

10. Determine the lowest-cost diet, chosen from the nutrients in Table 9, that supplies the minimum daily requirement of 75 g of protein, 45 mg of vitamin C, and 12 mg of iron.

Table 9

Food	Protein (g/unit)	Vitamin C (mg/unit)	Iron (mg/unit)	Cost ($/unit)
Carrots	0.6	3	0.5	0.05
Peas	0.8	4	0.3	0.08
Apples	0.4	6	0.5	0.10
Bananas	1.2	9	0.5	0.12
Eggs	5.9	0	1.2	0.09

11. By examining the constraint inequalities for the LP problem in Example 2, explain why the solution given for the LP problem in Example 2 is the maximal solution.

12. (a) A farmer can choose from three feeds for his milk cows. Table 10 gives the nutritional requirements and costs. The minimum daily requirements of nutrients A, B, and C are 65, 82, and 70 units, respectively. Determine the mixture of feeds that will supply the minimum nutritional requirements at least cost.

Table 10

Feed	Nutrient			Cost ($/lb)
	A (units/lb)	B (units/lb)	C (units/lb)	
Feed 1	4	7	3	0.10
Feed 2	2	3	4	0.07
Feed 3	5	5	3	0.06

(b) By considering a mixture of 10 lb of Feed 2 and 11 lb of Feed 3 for the farmer's cows, show that Feed 1 will not be used.

13. A paint manufacturer uses three minerals to provide four chemicals required in its paint. The composition of the paint must be at least 4% of chemical A, 3% of chemical B, 30% of chemical C, and 16% of chemical D. Table 11 gives the relevant compositions of the

Table 11

Mineral	Chemical				Cost ($/lb)
	A (%)	B (%)	C (%)	D (%)	
Mineral 1	3	5	35	24	3.50
Mineral 2	7	8	32	12	2.50
Mineral 3	9	1	27	15	3.00

minerals and the unit costs. Because mineral 2 causes an undesirable color when used in excess, no more than 1% of the total mineral content of the paint can be mineral 2. Determine the mixture of minerals that will provide the necessary composition of chemicals at the least cost.

(Suggestion: The least confusing way to analyze a problem involving percentages is to mix a fixed amount of the product, say 100 pounds. Thus, 100 pounds of paint would have to contain 4 pounds of chemical A, 3 pounds of chemical B, 30 pounds of chemical C, and 16 pounds of chemical D.)

1.2 TRANSPORTATION AND ASSIGNMENT PROBLEMS

The transport of goods from designated supply points to various demand points leads to one of the most important kinds of LP problems. In fact, it was work on this type of problem that spurred interest in linear programming and led to some of the most important advances in this field. We can formulate a simple form of the transportation problem as follows.

There are m supply points and n demand points for product P (Figure 1).

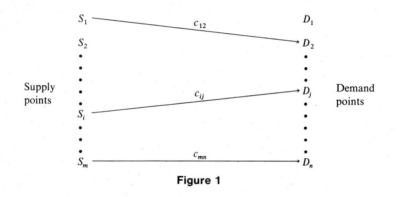

Figure 1

For $1 \leq i \leq m$ and $1 \leq j \leq n$, let c_{ij} denote the cost of shipping a unit of product P from the ith supply point S_i to the jth demand point D_j. Thus, for instance, if the cost of shipping one unit of P from supply point S_2 to demand point D_5 is 300, then $c_{25} = 300$.

For $1 \leq i \leq m$ and $1 \leq j \leq n$, let x_{ij} be the number of units of P shipped from supply point S_i to demand point D_j. Suppose further that, for each i, $1 \leq i \leq m$, s_i units of P are available for shipment from supply point S_i; and for each j, $1 \leq j \leq n$, there is a demand of d_j units of P at demand point D_j.

Our goal is to minimize the cost of shipping product P under the given constraints. We formulate this problem as follows.

$$
\begin{aligned}
\text{Minimize:} \quad z = \quad & c_{11}x_{11} + c_{12}x_{12} + \cdots + c_{1n}x_{1n} \\
& + c_{21}x_{21} + c_{22}x_{22} + \cdots + c_{2n}x_{2n} + \cdots \\
& + c_{m1}x_{m1} + c_{m2}x_{m2} + \cdots + c_{mn}x_{mn}
\end{aligned}
$$

Subject to:
$$x_{11} + x_{12} + \cdots + x_{1n} \leq s_1$$
$$x_{21} + x_{22} + \cdots + x_{2n} \leq s_2 \quad \begin{pmatrix} \text{Supply} \\ \text{constraints} \end{pmatrix}$$
$$\vdots \qquad\qquad \vdots$$
$$x_{m1} + x_{m2} + \cdots + x_{mn} \leq s_m \qquad\qquad (1)$$

$$x_{11} + x_{21} + \cdots + x_{m1} = d_1$$
$$x_{12} + x_{22} + \cdots + x_{m2} = d_2 \quad \begin{pmatrix} \text{Demand} \\ \text{constraints} \end{pmatrix}$$
$$\vdots \qquad\qquad \vdots$$
$$x_{1n} + x_{2n} + \cdots + x_{mn} = d_n \qquad\qquad (2)$$

$$x_{ij} \geq 0 \text{ and integer for } 1 \leq i \leq m, 1 \leq j \leq n$$

Each of the m inequalities in (1) expresses the restriction that no more units can be shipped from a supply point than are stored at that point. (For instance, $x_{21} + x_{22} + \cdots + x_{2n}$ gives the total number of units of P shipped from supply point S_2.) Each of the n equalities in (2) expresses the requirement that the demand must be met exactly. (For instance, $x_{12} + x_{22} + \cdots + x_{m2}$ gives the total number of units of P shipped to destination D_2.) In Chapter 9 we solve transportation problems, taking advantage of the fact that the constraints have a special form: every variable x_{ij} occurs once in an inequality of (1) and once in an equality of (2). In Chapter 9, we also show that we can replace the inequalities of (1) by equalities if we assume that the total supply (sum of the s_i) equals the total demand (sum of the d_i).

EXAMPLE 1

Suppose there are three supply points and five demand points for product P. Table 1 gives the unit costs of shipping product P from a given supply point to a given demand point.

Table 1   Shipping cost per unit

Supply points	D_1	D_2	D_3	D_4	D_5
S_1	60	30	90	10	30
S_2	50	80	110	40	50
S_3	75	40	30	20	40

Suppose that the number of units of this product that are available at each supply point is

Supply point S_1: 1600   Supply point S_2: 900   Supply point S_3: 1300

and the number of units required at each demand point is

Demand point D_1: 700   Demand point D_3: 400   Demand point D_5: 900

Demand point D_2: 1000   Demand point D_4: 800

Combining Table 1 and this information, we obtain the *cost table* for the problem (Table 2).

Table 2 Shipping cost per unit

Supply points	Demand points					Supply
	D_1	D_2	D_3	D_4	D_5	
S_1	60	30	90	10	30	1600
S_2	50	80	110	40	50	900
S_3	75	40	30	20	40	1300
Demand	700	1000	400	800	900	

If for $1 \leq i \leq 3$ and $1 \leq j \leq 5$, we let x_{ij} denote the number of units of P shipped from the supply point S_i to the demand point D_j; then, from the data in Table 2, we obtain the LP problem

$$\text{Minimize:} \quad z = 60x_{11} + 30x_{12} + 90x_{13} + 10x_{14} + 30x_{15}$$
$$+ 50x_{21} + 80x_{22} + 110x_{23} + 40x_{24} + 50x_{25}$$
$$+ 75x_{31} + 40x_{32} + 30x_{33} + 20x_{34} + 40x_{35}$$

$$\text{Subject to:} \quad x_{11} + x_{12} + x_{13} + x_{14} + x_{15} \leq 1600$$
$$x_{21} + x_{22} + x_{23} + x_{24} + x_{25} \leq 900$$
$$x_{31} + x_{32} + x_{33} + x_{34} + x_{35} \leq 1300$$
$$x_{11} + x_{21} + x_{31} = 700$$
$$x_{12} + x_{22} + x_{32} = 1000$$
$$x_{13} + x_{23} + x_{33} = 400$$
$$x_{14} + x_{24} + x_{34} = 800$$
$$x_{15} + x_{25} + x_{35} = 900$$

$$x_{ij} \geq 0 \text{ and integer, } 1 \leq i \leq 3, 1 \leq j \leq 5 \qquad \blacklozenge$$

When we learn to solve these problems, we shall discover that the solution is $x_{11} = 0$, $x_{12} = 100$, $x_{13} = 0$, $x_{14} = 800$, $x_{15} = 700$, $x_{21} = 700$, $x_{22} = 0$, $x_{23} = 0$, $x_{24} = 0$, $x_{25} = 200$, $x_{31} = 0$, $x_{32} = 900$, $x_{33} = 400$, $x_{34} = 0$, and $x_{35} = 0$. The minimum shipping cost is $125,000

We can apply the transportation model to problems that have nothing to do with the transport of goods. In the next example we see how to set up a certain scheduling problem as a transportation problem.

EXAMPLE 2

A firm that manufactures product P wants to establish an optimal production schedule for the months January, February, March, and April, based on the data in Table 3.

Table 3

Month	Monthly demand	Normal production capacity	Overtime production capacity	Unit storage costs	Normal unit costs	Overtime additional unit costs
Jan	400	450	200	4	30	10
Feb	700	700	0	4	25	10
Mar	600	900	100	4	25	10
Apr	800	500	100	4	35	10

The firm wants to meet the monthly demands at a minimum cost. To set this problem up as a transportation problem, we consider each month to be a supply point, distinguishing between normal and overtime production during each month. Thus, our "supply periods" and the respective available "supplies" are

S_1: January, normal production   450

S_2: January, overtime production   200

S_3: February, normal production   700

S_4: March, normal production   900

S_5: March, overtime production   100

S_6: April, normal production   500

S_7: April, overtime production   100

Consider each month also to be a demand point. Thus, we have

D_1: January   400 D_3: March   600

D_2: February   700 D_4: April   800

Observe that the total production capacity (normal + overtime) is 2950 units, whereas the total demand is 2500 units; thus there is an adequate supply to meet the demand. Table 4 is the cost table associated with this problem.

If for $1 \le i \le 7$, $1 \le j \le 4$, we let x_{ij} be the number of units "supplied" by S_i to meet the demand D_j, then the LP problem is (with $\infty = 10{,}000$)

$$
\begin{aligned}
\text{Minimize:} \quad z = \ & 30x_{11} + 34x_{12} + 38x_{13} + 42x_{14} \\
& + 40x_{21} + 44x_{22} + 48x_{23} + 52x_{24} \\
& + 10{,}000x_{31} + 25x_{32} + 29x_{33} + 33x_{34} \\
& + 10{,}000x_{41} + 10{,}000x_{42} + 25x_{43} + 29x_{44} \\
& + 10{,}000x_{51} + 10{,}000x_{52} + 35x_{53} + 39x_{54} \\
& + 10{,}000x_{61} + 10{,}000x_{62} + 10{,}000x_{63} + 35x_{64} \\
& + 10{,}000x_{71} + 10{,}000x_{72} + 10{,}000x_{73} + 45x_{74}
\end{aligned}
$$

Subject to:
$$x_{11} + x_{12} + x_{13} + x_{14} \leq 450$$
$$x_{21} + x_{22} + x_{23} + x_{24} \leq 200$$
$$x_{31} + x_{32} + x_{33} + x_{34} \leq 700$$
$$x_{41} + x_{42} + x_{43} + x_{44} \leq 900$$
$$x_{51} + x_{52} + x_{53} + x_{54} \leq 100$$
$$x_{61} + x_{62} + x_{63} + x_{64} \leq 500$$
$$x_{71} + x_{72} + x_{73} + x_{74} \leq 100$$
$$x_{11} + x_{21} + x_{31} + x_{41} + x_{51} + x_{61} + x_{71} = 400$$
$$x_{12} + x_{22} + x_{32} + x_{42} + x_{52} + x_{62} + x_{72} = 700$$
$$x_{13} + x_{23} + x_{33} + x_{43} + x_{53} + x_{63} + x_{73} = 600$$
$$x_{14} + x_{24} + x_{34} + x_{44} + x_{54} + x_{64} + x_{74} = 800$$

$$x_{ij} \geq 0 \text{ and integer, } 1 \leq i \leq 7, 1 \leq j \leq 4$$

Table 4

Supply points	Demand points				Supply
	D_1	D_2	D_3	D_4	
S_1	30	34^a	38	42	450
S_2	40^b	44	48	52	200
S_3	∞^c	25	29	33	700
S_4	∞	∞	25	29	900
S_5	∞	∞	35	39	100
S_6	∞	∞	∞	35	500
S_7	∞	∞	∞	45	100
Demand	400	700	600	800	

[a] The unit cost of P produced in January to meet the demand in February. This cost is \$34 = [normal January unit production cost (30)] + [one month unit storage cost (4)].

[b] The overtime cost for production in January (and to meet the January demand). This cost is \$40 = [normal January unit production cost (30)] + [additional January unit overtime cost (10)].

[c] There is no February production to meet the January demand; ∞ represents a number large enough that the variable x_{31} associated with S_3 and D_1 will take on the value 0 for the optimal solution.

When we learn to solve these problems, we shall discover that the solution is $x_{11} = 400$, $x_{32} = 700$, $x_{43} = 600$, $x_{44} = 300$, $x_{64} = 500$, and the other variables are 0. The minimum value of z is \$70,700.   ◆

In Example 2, we could have avoided the use of large cost coefficients to force some of the variables to be zero in the optimal solution by simply omitting those variables from the model. We elected not to do so because omitting some variables would have destroyed the special structure of transportation problems; (1) and (2) show this structure. Since we will develop a more efficient method of solving transportation problems in Chapter 9, we prefer to include a variable to describe the transportation of goods from every source to every destination.

The *assignment problem* is closely related to the transportation problem. Suppose there are n tasks $t_1, t_2, \ldots, t_n$ to be performed, and there are n persons $p_1, p_2, \ldots, p_n$ capable of accomplishing each of these tasks. For each i, $1 \le i \le n$, and for each j, $1 \le j \le n$, the cost of assigning person p_i to task t_j is c_{ij}. We want to assign a person to a task in such a way that we assign no person to more than one task and so that the total assignment cost is a minimum.

To set up this problem we introduce the variables x_{ij} ($1 \le i, j \le n$), where

$$x_{ij} = \begin{cases} 1 & \text{if we assign the } i\text{th person } p_i \text{ to the } j\text{th task } t_j \\ 0 & \text{otherwise} \end{cases}$$

Note that since we assign every person p_i to exactly one task, it follows that for each *fixed* integer i, $x_{ij} = 1$ for precisely one integer j. Thus, for each i, $1 \le i \le n$, we have the constraint

$$x_{i1} + x_{i2} + \cdots + x_{in} = 1$$

Similarly, since only one person is assigned to a task t_j, it follows that for each *fixed* integer j, $x_{ij} = 1$ for precisely one integer i. Thus, for each j, $1 \le j \le n$, we have the constraint

$$x_{1j} + x_{2j} + \cdots + x_{nj} = 1$$

With these observations in mind we see that the LP formulation of the assignment problem is

$$
\begin{aligned}
\text{Minimize:} \quad z = \ & c_{11}x_{11} + c_{12}x_{12} + \cdots + c_{1n}x_{1n} \\
& + c_{21}x_{21} + c_{22}x_{22} + \cdots + c_{2n}x_{2n} + \cdots \\
& + c_{n1}x_{n1} + c_{n2}x_{n2} + \cdots + c_{nn}x_{nn}
\end{aligned}
$$

$$
\begin{aligned}
\text{Subject to:} \quad & x_{11} + x_{12} + \cdots + x_{1n} = 1 \\
& x_{21} + x_{22} + \cdots + x_{2n} = 1 \\
& \quad \vdots \\
& x_{n1} + x_{n2} + \cdots + x_{nn} = 1 \\
& x_{11} + x_{21} + \cdots + x_{n1} = 1 \\
& x_{12} + x_{22} + \cdots + x_{n2} = 1 \\
& \quad \vdots \\
& x_{1n} + x_{2n} + \cdots + x_{nn} = 1 \\
& x_{ij} = 0 \text{ or } 1, \ 1 \le i, j \le n
\end{aligned}
\tag{3}
$$

Using summation ($\sum$) notation, we can express (3) more compactly as

$$\text{Minimize:} \quad \sum_{j=1}^{n} \sum_{i=1}^{n} c_{ij} x_{ij}$$

$$\text{Subject to:} \quad \sum_{j=1}^{n} x_{ij} = 1; \qquad 1 \leq i \leq n$$

$$\sum_{i=1}^{n} x_{ij} = 1; \qquad 1 \leq j \leq n \tag{4}$$

$$x_{ij} = 0 \text{ or } 1, 1 \leq i,j \leq n$$

We can view (and solve) the assignment problem as a special case of the transportation problem. There are more efficient ways (e.g., the Hungarian method) to solve this problem that utilize the special nature of the constraints; however, we do not present these methods in this book.

Exercises

You can only set up the following exercises as transportation or assignment problems now. As we develop techniques for solving LP problems in later chapters, you will be asked to solve some of these exercises. In the spirit of the statement following Example 2, you should formulate LP models for these exercises that have the general form of a transportation problem, as given in (1) and (2), or the general form of an assignment problem, as given in (3).

1. The Del Valle Ketchup Company has ketchup plants in Davis, Fresno, and Merced and has warehouses in San Francisco, Stockton, Los Angeles, and San Diego to store the ketchup for later distribution to stores. The Davis and Fresno plants can each produce 50,000 bottles of ketchup per week, and the Merced plant can produce 30,000 bottles per week. The warehouses in San Francisco, Stockton, Los Angeles, and San Diego require 35,000, 25,000, 40,000, and 30,000 bottles per week, respectively. Table 5 gives shipping costs in dollars per thousand bottles.

Table 5

	Warehouse			
Plant	San Francisco	Stockton	Los Angeles	San Diego
Davis	10	6	12	15
Fresno	8	4	8	10
Merced	11	8	5	6

Find a shipping schedule that meets the demands with the least shipping cost.

2. Find a minimum-cost shipping schedule for the transportation costs given in Table 6.
3. Find a minimum-cost shipping schedule for the transportation costs given in Table 7. The symbol ∞ indicates that there is no shipping route from the source to the supply.
4. John is concerned by a long-term weather report predicting a hotter than normal summer. He will be working at a forest fire lookout in a remote site and wants to be sure

Table 6

Supply points	Demand points					Supply
	D_1	D_2	D_3	D_4	D_5	
S_1	1	3	12	15	9	100
S_2	5	4	9	7	10	120
S_3	7	6	20	10	11	90
Demand	60	80	70	60	40	

Table 7

Supply points	Demand points					Supply
	D_1	D_2	D_3	D_4	D_5	
S_1	4	∞	6	12	21	250
S_2	5	5	12	6	∞	100
S_3	10	7	15	12	10	300
S_4	∞	8	10	10	12	150
Demand	100	200	250	150	100	

of having an adequate beer supply. He can haul 12 cases when driving to the site in June at a cost of $8 per case. A friend will visit him in July and can bring up to 20 cases at $9.50 per case. Another friend with a smaller car can bring 8 cases in July and another 8 cases in August, at $9.00 per case. John estimates he will need 10 cases in June, 12 cases in July, and 15 cases in August to cope with the increasing heat. How should he schedule the deliveries to keep up with his thirst at the least cost?

5. Harvard is a small community in Idaho that draws its water supply from the Palouse River. The Palouse is a small river with a flow that fluctuates with the seasons. To meet its higher demand for water in the summer and fall, Harvard needs to store water in a reservoir during periods of high runoff. There is a cost associated with this storage, and there also are treatment costs associated with pumping water from the Palouse River and treating it for human consumption. No storage cost is incurred for water that is used in the same quarter in which it is drawn from the river. Table 8 contains estimates of the supply, demand, and cost for each quarter of a year. Water is measured in acre-feet, storage cost is in dollars per acre-foot per quarter, and treatment costs are in dollars per acre-foot.

Table 8

Season	Demand (acre-ft)	Supply (acre-ft)	Storage cost ($)	Treatment cost ($)
Jan–Mar	150	300	10	25
Apr–Jun	200	900	10	40
Jul–Sep	600	220	10	30
Oct–Dec	400	60	10	35

Set up a transportation model to design a storage program that minimizes storage and treatment costs. Assume that water gathered in a quarter must be stored an average of 3 months for use in the next quarter, 6 months for use two quarters hence, and so forth.

6. (a) A small construction firm has arranged to borrow up to $40,000 over the next year to finance the expansion of its business. The bank has agreed to provide up to $10,000 of this money on the first day of each quarter of the year. The quarterly interest rate is 4%. The bank charges a loan initiation fee on each installment of 3% if the amount borrowed is $4000 or less and 5% on the amount exceeding $4000. (Thus, if $3000 is borrowed, the initiation fee is $150; and if $9000 is borrowed, the initiation fee is $350.) The firm is not required to borrow all of the available money. The firm estimates it will need $6,000, $9,000, $8,000, and $15,000 for the first, second, third, and fourth quarters, respectively. Set up a transportation model to minimize fees and interest paid at the end of the year.

   (b) What is the solution to (a)? It is fairly easy to find the solution without knowing how to solve transportation problems in general.

7. A caterer must supply 95 party napkins on Monday, 75 on Tuesday, 108 on Wednesday, 90 on Thursday, and 120 on Friday. New napkins cost 35 cents each. Soiled napkins can be laundered by a 1-day service at a cost of 8 cents per napkin, a 2-day service at 5 cents per napkin, a 3-day service at 3 cents per napkin, or a 4-day service at 3 cents per napkin. At the end of the week all napkins are thrown away. Find the cheapest schedule to supply the napkins. (Suggestion: Start by defining the supply and demand points and forming a cost table. Fill in the cost table completely by assigning a very high cost to any demand that cannot be met from a supply point.)

8. Suppose the napkin laundry of Exercise 7 discontinues its 1-day service. How should the caterer alter the napkin supply plan?

   In Exercises 9–13, you should set up the problems as assignment problems, in the form given in (3). To accomplish this for some of these problems, you will need to make a clever choice of cost coefficients. Since there are so many inequalities to write, you may prefer to use the summation notation (4). You can assume that the values of c_{ij} have been defined as in the cost tables given with each exercise.

9. Four people (p_1, p_2, p_3, p_4) are to be assigned to four tasks (t_1, t_2, t_3, t_4). Table 9 gives the cost of assigning the ith person to the jth task. Find an assignment that minimizes cost.

Table 9

	t_1	t_2	t_3	t_4
p_1	8	10	6	9
p_2	10	8	8	11
p_3	12	9	10	9
p_4	10	10	9	14

10. Suppose that person p_3 cannot perform task 4 but that other assignment costs are the same as in Exercise 9. How should the work force be assigned now?

11. For the situation described in Exercise 9, suppose that person p_3 calls in sick, making it necessary to reassign the work force; thus only three of the tasks can be completed. Determine a way of assigning the three people available that omits one task and results in the least cost. (Suggestion: If person p_3 were considered capable of doing any task at no cost, the most costly task would be assigned to that person.)

12. Suppose that the supervisor planning the assignment schedule of Exercise 9 decides that task 4 will not be undertaken and wants to lay off one of the employees for the day. Which employee should be laid off, and how should the remaining employees be assigned?

13. An airline offers six vacation trips as a bonus for six of its employees. To distribute the impact of providing free space on its flights, only one employee can go to each vacation site. The employees are asked to rank their vacation preferences from 1 to 6, where 1 represents the most preferred. Table 10 gives their preferences. Find a plan that minimizes the total dissatisfaction of the six people.

Table 10

	t_1	t_2	t_3	t_4	t_5	t_6
p_1	6	5	4	3	2	1
p_2	1	2	3	4	5	6
p_3	5	1	2	6	4	3
p_4	5	1	2	6	4	3
p_5	6	4	2	3	5	1
p_6	1	3	4	5	2	6

1.3 GENERAL COMMENTS REGARDING LP MODELS

Linear programming has proven to be a remarkably useful method for modeling a broad spectrum of problems. Despite its versatility, however, it does suffer from the limitations of any mathematical model. Like all models, linear programming represents a balance (or compromise) between the often incompatible virtues of simplicity and completeness. For a model to be useful, it must be *tractable*; that is, it must be sufficiently simple to be solvable with available computing equipment. Generally speaking, the simpler the model the greater its tractability; however, the price paid for simplicity may be a low correlation between reality and the results of the model.

As a potential decision maker, you will have to make delicate (and sometimes difficult) choices in selecting the amount of information or the variables you need to incorporate into your models. Generally, you should try to ignore factors that have minimal effect on the results of your model, while using those factors that enable you to attain a good approximation of the reality you are modeling. Approximating reality takes a variety of forms. To achieve tractability you may need to decrease the number of variables either by eliminating some or by regarding a constellation of variables as a single variable. You may also need to alter variables or use an average to represent change. For instance, in Example 2 of Section 1.4 we will use discrete monthly data to obtain an approximate representation of continuous stream flows into a reservoir. Figure 1 abstracts the general modeling process.

Ideally, the conclusions derived from a model should, when properly interpreted, coincide with the results derived directly from the real system. The extent to which this ideal is attained depends on many factors, from the concept of the model itself to the quality of the information available for use in the model.

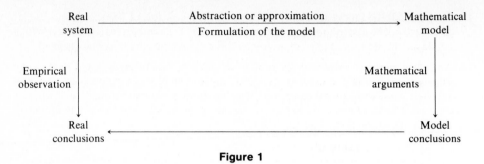

Figure 1

Although used in diverse situations, LP models have a number of elements in common. They all involve "decision variables" (the variables x_i or x_{ij} in the previous examples). These variables are often referred to as *activity variables* because they correspond to the various activities that define the system—activities such as the production of bookcases, the days a plant operates, and the shipping of goods from supply point S_i to demand point D_j.

The principal characteristic of decision or activity variables is that they are controllable; usually you can use this attribute to identify these variables. To clarify your thinking in dealing with the decision variables, it is helpful to list and label them; for instance, you might have

$$x_3 = \text{number of green cars manufactured}$$

or

$$x_{25} = \text{amount of loans made from financial}$$
$$\text{branch office } \#2 \text{ to corporate client } \#5$$

Once you have identified the controllable decision variables, you must also quantify them. The first step is to determine the objective function, which is a function of the decision variables. This function measures the effectiveness of the system and is the function you want to maximize or minimize. It is important to note that, in practice, various objective functions may be appropriate to a situation. For instance, you may want to view a problem from the standpoint of maximizing profits; on the other hand, it may be equally appropriate to look at the same problem with an eye to minimizing costs. This does not imply that the two approaches are necessarily equivalent because the available information and the constraints of the problem may lead to different optimal solutions for each case. Obviously, it is important to determine whether the problem involves maximizing or minimizing the objective function. Since in this text we usually express problems in terms of maximizing an objective function, you should be especially sensitive to those problems whose objective functions are to be minimized.

There are a number of parameters over which you usually have little control. These include the coefficients in the objective function (often called the *cost coefficients*), the coefficients of the decision variables, and the resource values. In practice, you may not know the exact values of these parameters, and a major problem in linear programming is to see to what extent the parameters can vary without producing changes in the optimal solution. Sensitivity analysis (discussed in Chapter 7) deals with problems of this sort.

Once you have determined the decision variables and the objective function, you need to delineate the various constraints. These constraints are defined in terms of the decision variables and the given resource limitations. In determining the constraints you should be aware of a number of pitfalls.

1. **Omitting constraints.** If you happen to omit a needed constraint, you probably will not obtain the correct optimal solution. In fact, in some cases, omission of a constraint results in arbitrarily large values of the objective function. For instance, if we ignore the third constraint in the LP problem,

$$\text{Maximize:} \quad z = 4x_1 + 2x_2$$
$$\text{Subject to:} \quad -x_1 + x_2 \leq 4$$
$$x_1 - 2x_2 \leq 2 \qquad (1)$$
$$2x_1 + 3x_3 \leq 8$$
$$x_1, x_2 \geq 0$$

then it is easy to see that any solution of (1) of the form (k, k) is a feasible solution; hence, for increasing values of k, the objective function becomes arbitrarily large. Thus, without this constraint there is no optimal solution to this problem.

2. **Introducing contradictory constraints.** You must avoid constraints that contradict one another. Any pair of contradictory constraints results in a situation in which there is no feasible solution and, hence, no optimal solution. For instance, the LP problem

$$\text{Minimize:} \quad z = 3x_1 - x_2$$
$$\text{Subject to:} \quad x_1 + x_2 \leq 3$$
$$2x_1 + x_2 \leq 5$$
$$x_1 + 2x_2 \geq 7$$
$$x_1, x_2 \geq 0$$

has no feasible solutions because no solution can satisfy simultaneously the first and third constraints. In practice, contradictory constraints usually mean you have made an error in constructing your model. Thus, such constraints can provide a check on your work: They may indicate mathematical or conceptual (or just careless) errors.

3. **Using inconsistent units of measurement.** A common error is to use different units of measurement in the same constraint. In particular, you should make sure that the units used for the left-hand side of a constraint inequality or equality coincide with those used on the right-hand side. Substantial errors result if, for example, you use ounces on the left-hand side of the constraint and pounds on the right-hand side.

4. Introducing redundant constraints. In the LP problem

$$\text{Maximize:} \quad z = x_1 - x_2$$

$$\text{Subject to:} \quad x_1 - x_2 \leq 3$$
$$2x_1 + x_2 \leq 12$$
$$x_1 + x_2 \leq 2$$

$$x_1, x_2 \geq 0$$

the constraint

$$2x_1 + x_2 \leq 12 \tag{2}$$

is *redundant* (not needed): Any solution satisfying the constraint

$$x_1 + x_2 \leq 2$$

automatically satisfies (2); so (2) does not add a restriction not already imposed by the other constraints. It is not incorrect to include the constraint (2), but it adds unnecessary computations to solving the problem. Whenever they are detected, redundant constraints should be eliminated from the LP model; however, redundancy is usually not a serious problem because it does not cause errors in the solution. Moreover, in large problems it is rarely worth the effort to locate redundant constraints; the calculating effort needed to find such constraints may far exceed that needed to solve the problem with the redundant constraints included.

In summary, to build an LP model you should carefully read and digest the given information. Your first task is to determine what is (or what needs to be) asked. You should identify or define the decision variables (those variables over which you have some degree of control). Having determined the decision variables, you are in a position to describe the objective function and state the various constraints pertinent to the model. During this process you should try to eliminate the conditions that have minimal effect on the system and to define clearly those factors that have a significant bearing on the final results, results that should accurately reflect the system you are modeling.

*1.4 DYNAMIC PLANNING

Many practical problems involve activities that change with time. Such problems are often called *dynamic* or *multiperiod* planning problems. In this section we consider two examples of these problems, one involving production scheduling and the other resource allocation.

We first consider the problem of determining an optimum production schedule when there is an uneven demand for the goods produced. Scheduling is a major concern of many firms, and linear programming provides a powerful tool for resolving problems of this kind. In fact, large-scale LP problems are perhaps most frequently encountered in connection with production scheduling. Scheduling problems commonly involve a sequence of product-mix problems that are linked in some prescribed way.

If demand for a product varies considerably over time, and if no inventory is kept, then production must vary to meet the demand. This means, though, that to minimize production expenses it may be necessary to continually adjust the number of employees. This results in what are called *hiring–firing costs*. To reduce these costs it may be advisable to try to even out the production levels while maintaining an inventory sufficient to meet customer demands. Such a policy, however, introduces storage costs. Our immediate objective is to find a balance between storage costs and hiring–firing costs that minimizes the total cost. Example 1 illustrates this type of problem.

EXAMPLE 1

The Merry Mannequin Manufacturing Company (3M) operates a profitable sideline making model horses for carousels. 3M knows from experience that the demand for horses fluctuates with the seasons and is planning production based on the quarterly predictions given in Table 1.

Table 1   Predicted demand for horses

Jan–Mar	Apr–Jun	Jul–Sep	Oct–Dec
500	900	400	800

One obvious way to meet the demand is to produce exactly 500 horses during January–March, 900 during April–June, 400 during July–September, and 800 during October–December. This would eliminate storage costs but could lead to hiring–firing costs. Workers that were needed to produce 900 horses during the second quarter of the year might be laid off in the third quarter when demand is slack, only to be rehired in the fourth quarter.

To reduce the hiring–firing costs the manufacturer could opt to use a production schedule such as that given in Table 2.

Table 2   Production schedule

Time period	Horses produced
Jan–Mar	700
Apr–Jun	700
Jul–Sep	600
Oct–Dec	600

This schedule would increase storage costs but would tend to decrease hiring–firing costs.

The optimal schedule depends on the storage costs and the hiring–firing costs incurred when production levels are changed. To determine this schedule we proceed as follows.

Let T_1, T_2, T_3, and T_4 denote the periods January–March, April–June, July–September, and October–December, respectively; and let T_0 denote the October–December period preceding T_1 (T_0 is useful for considering situations involving carryover of horses or people from the previous year). For $i = 0, 1, 2, 3, 4$, let P_i denote the number of horses produced in period T_i; and let I_i denote the number of horses stored during this same time period. We include I_0 to allow for inventory left over from the previous year. You will soon see why we need P_0. We assume that it costs \$2 to store a horse for each period T_i. For simplicity, we also assume that the average number of horses stored during a period T_i is the number in storage at the end of the period. In particular, if we suppose that the number of horses produced annually is equal to the annual demand, then $I_4 = 0$.

Next, we take into consideration the hiring–firing costs. Suppose that on the average it costs \$8 for each additional horse added to the production schedule from one period to the next. For simplicity, we assume that the production of horses is proportional to the employee-hours worked. With this assumption, we can interpret the \$8 cost of making an additional horse as the hiring cost because the actual hiring cost per employee would be a fixed multiple of this cost per horse. Suppose further that on the average it costs \$3 for each decrease of one in the number of horses produced from one period to the next. We can interpret this cost as proportional to the firing cost. In this example, for simplicity, we assume these constants of proportionality are 1.

Finally, for $i = 1, 2, 3, 4$, let U_i denote the increase in the number of horses produced from period T_{i-1} to period T_i; and let D_i denote the decrease in the number of horses produced from period T_{i-1} to period T_i. With this notation, the sum

$$8U_1 + 8U_2 + 8U_3 + 8U_4 + 3D_1 + 3D_2 + 3D_3 + 3D_4$$

gives the total hiring–firing cost. It follows that the total yearly storage cost plus the hiring–firing cost is

$$
\begin{aligned}
z = \quad & 8U_1 + 8U_2 + 8U_3 + 8U_4 \\
& + 3D_1 + 3D_2 + 3D_3 + 3D_4 \\
& + 2I_1 + 2I_2 + 2I_3 + 2I_4
\end{aligned}
\tag{1}
$$

and it is this sum we want to minimize.

What are the constraints in this problem? We must meet the required demand during each period. This means that the production during the current period plus the carry-over inventory from the previous period must be equal to the demand during the current period plus the inventory that is left over at the end of the current period. That is,

$$
\begin{bmatrix}
\text{Carry-over} \\
\text{inventory} \\
\text{from the} \\
\text{previous} \\
\text{period}
\end{bmatrix}
+
\begin{bmatrix}
\text{Production} \\
\text{during the} \\
\text{current} \\
\text{period}
\end{bmatrix}
=
\begin{bmatrix}
\text{Demand} \\
\text{during the} \\
\text{current} \\
\text{period}
\end{bmatrix}
+
\begin{bmatrix}
\text{Leftover} \\
\text{inventory} \\
\text{at end of} \\
\text{current} \\
\text{period}
\end{bmatrix}
$$

In terms of the T_i it follows that

$$\begin{bmatrix} \text{Inventory} \\ \text{at end of} \\ \text{period } T_{i-1} \end{bmatrix} + \begin{bmatrix} \text{Production} \\ \text{during} \\ \text{period } T_i \end{bmatrix} = \begin{bmatrix} \text{Demand} \\ \text{during} \\ \text{period } T_i \end{bmatrix} + \begin{bmatrix} \text{Inventory} \\ \text{at end of} \\ \text{period } T_i \end{bmatrix}$$

If we assume that the beginning and ending inventories are 0 (that is, $I_0 = 0$ and $I_4 = 0$), we obtain the constraints

$$0 + P_1 = 500 + I_1$$
$$I_1 + P_2 = 900 + I_2$$
$$I_2 + P_3 = 400 + I_3 \qquad (2)$$
$$I_3 + P_4 = 800 + 0$$

Observe that the change in production from period T_{i-1} to period T_i is $U_i - D_i$ (note also that for every i either U_i or D_i is 0). We can also express this change in production as $P_i - P_{i-1}$. If we assume that no employees were left on the staff from the last period of the previous year (T_0), then $P_0 = 0$, $D_1 = 0$, $U_1 = P_1$, and we have the constraints

$$U_1 \qquad = P_1$$
$$U_2 - D_2 = P_2 - P_1$$
$$U_3 - D_3 = P_3 - P_2 \qquad (3)$$
$$U_4 - D_4 = P_4 - P_3$$

From (1), (2), (3), and the assumptions that $D_1 = 0$ and $I_4 = 0$, we obtain the LP problem

Minimize: $z = 8U_1 + 8U_2 + 8U_3 + 8U_4 + 3D_2 + 3D_3$
$\qquad\qquad + 3D_4 + 2I_1 + 2I_2 + 2I_3$

Subject to: $U_1 \qquad = P_1$
$\qquad\qquad U_2 - D_2 = P_2 - P_1$
$\qquad\qquad U_3 - D_3 = P_3 - P_2$
$\qquad\qquad U_4 - D_4 = P_4 - P_3$
$\qquad\qquad\qquad P_1 = 500 + I_1 \qquad (4)$
$\qquad\qquad I_1 + P_2 = 900 + I_2$
$\qquad\qquad I_2 + P_3 = 400 + I_3$
$\qquad\qquad I_3 + P_4 = 800$

$$U_i \geq 0, 1 \leq i \leq 4; D_i \geq 0, 2 \leq i \leq 4$$
$$I_i \geq 0, 1 \leq i \leq 3; P_i \geq 0, 1 \leq i \leq 4$$

In later chapters you will learn to solve this problem; the solution is $D_3 = 100$, $I_1 = 200$, $I_3 = 200$, $P_1 = 700$, $P_2 = 700$, $P_3 = 4600$, $P_4 = 600$, $U_1 = 700$, and the value of the other variables is 0. The minimum of z is \$6700. You can find various modifications of this problem in the exercises.   ◆

The problems we have considered thus far have been of a *discrete* nature. For instance, we asked how to maximize profit for a company that had two choices: manufacture either tables or chairs. We also asked how to minimize shipping costs between a finite number of supply points and a finite number of destination points. In a discrete problem, the solution can be given as a finite list of values: Manufacture 10 tables and 12 chairs, or ship 12 units from supply point S_1 to demand point D_1 and 15 from supply point S_2 to demand point D_2. In fact, linear programming methods can only solve problems of a discrete nature. Even so, it is possible to construct models of a discrete nature that approximate solutions to problems of a *continuous* nature. For example, consider the problem of setting the rate of release of water from a dam built to generate electrical power and to provide storage for irrigation water. Water flows into the reservoir continuously as a function of time; it is not dumped into the reservoir in a series of discrete events. Nevertheless, we can approximate the real rate of inflow by considering only the amount of water that flows in each month or, if a month is not a fine enough measure, then each week, or each day, or each hour. We can approximate the outflow by giving the number of units of water to be released over a period of time (month, week, or hour) even though the actual answer to the problem of regulating the dam requires that we specify the amount to be released at every instant of time. The next example gives a discrete model for a real situation that varies continuously with time.

EXAMPLE 2

The Bountiful Power Administration (BPA) operates two dams on the Squawfish River. The High Mountain Sheep Dam (HMSD) produces hydroelectric power and conserves water for irrigation of farmland downstream. After passing through the turbines, water can be diverted to an irrigation project or left to flow down the stream. The Valley Goat Dam (VGD) is downstream from the farm land and is used only for hydroelectric power generation. Each month, at least one unit of water must be left to flow downstream from each dam to maintain the river's population of squawfish. The reservoir behind HMSD has a capacity of 10 units of water. It contains 5 units of water at the beginning of the year (January) and must contain 5 units of water at the end of the year (December). To stay within the maximum load for the turbines at HMSD, the maximum amount of water that can be released in any month is 7 units; the turbines at VGD can handle far more water than the Bountiful River is expected to carry. Table 1 gives the return in millions of dollars per unit of water (M\$/unit of water) for each unit of water released from HMSD for hydroelectric power and irrigation, estimated inflow to the HMSD reservoir for each month, constraints on the amount of water to be used for irrigation, and the return (M\$/unit of water) for hydroelectric power at VGD from water that passed through HMSD.

Table 1

	Jan	Feb	Mar	Apr	May	Jun	Jul	Aug	Sep	Oct	Nov	Dec
Inflow to HMSD (units of water)	2	2	3	4	3	2	2	1	2	3	3	2
Hydropower profit from HMSD (M$/unit of water)	1.6	1.7	1.8	1.9	2.0	2.0	2.0	1.9	1.8	1.7	1.6	1.5
Irrigation profit (M$/unit of water)	1.0	1.2	1.8	2.0	2.2	2.2	2.5	2.2	1.8	1.4	1.1	1.0
Hydropower profit from VGD (M$/unit of water)	1.4	1.3	1.5	1.9	2.3	2.1	1.8	1.8	2.1	1.8	1.3	1.3
Minimum need for irrigation (units of water)	0	0	0	0.5	1.0	2.5	2.8	1.2	0.5	0.5	0	0
Maximum needed for irrigation (units of water)	0.5	0.2	0.1	1.8	4.0	5.0	5.4	2.5	0.8	0.7	0.5	0.2

Let the numbers 1 through 12 correspond to the months January through December; and for $1 \leq i \leq 12$, let x_i be the amount of water released through the turbines in month i and y_i be the amount diverted for irrigation in month i. We consider first the constraint that the HMSD reservoir can hold no more than 10 units of water. In any month, the amount of water stored is

$$5 + \text{(total units that flowed in)} - \text{(total units that flowed out)}$$

For instance, in March the amount of water stored is

$$5 + (2 + 2 + 3) - (x_1 + x_2 + x_3) \tag{5}$$

There is always a nagging doubt, when forming an expression such as (5), about whether to include the 3 units of inflow for March and the 3 units of outflow for March. The trick is not to consider March to be a period of extended time but rather as a discrete entity; think of all March events as happening at the same instant rather than as spread over a period of time. The following 12 constraints express the fact that the reservoir capacity is 10 units.

$7 - x_1 \leq 10$	(Jan)	$23 - x_1 - \cdots - x_7 \leq 10$ (Jul)
$9 - x_1 - x_2 \leq 10$	(Feb)	$24 - x_1 - \cdots - x_8 \leq 10$ (Aug)
$12 - x_1 - x_2 - x_3 \leq 10$	(Mar)	$26 - x_1 - \cdots - x_9 \leq 10$ (Sep)
$16 - x_1 - \cdots - x_4 \leq 10$	(Apr)	$29 - x_1 - \cdots - x_{10} \leq 10$ (Oct)
$19 - x_1 - \cdots - x_5 \leq 10$	(May)	$32 - x_1 - \cdots - x_{11} \leq 10$ (Nov)
$21 - x_1 - \cdots - x_6 \leq 10$	(Jun)	$34 - x_1 - \cdots - x_{12} \leq 10$ (Dec)

The requirement that at least 5 units remain in the reservoir at the end of the year is given by

$$34 - x_1 - \cdots - x_{12} \geq 5$$

The requirement that at least one unit of water flows downstream each month is given by the 12 constraints

$$y_i \leq x_i - 1, \qquad 1 \leq i \leq 12$$

The limitation of 7 units per month through the turbines is given by the 12 constraints

$$x_i \leq 7, \qquad 1 \leq i \leq 12$$

The minimum and maximum uses for irrigation are given by the 24 constraints

$$y_1 \geq 0; \quad y_1 \leq 0.5; \quad y_2 \geq 0; \quad y_2 \leq 0.2; \quad y_3 \geq 0; \quad y_3 \leq 0.1;$$
$$y_4 \geq 0.5; \quad y_4 \leq 1.8; \quad y_5 \geq 1.0; \quad y_5 \leq 4.0; \quad y_6 \geq 2.5; \quad y_6 \leq 5.0;$$
$$y_7 \geq 2.8; \quad y_7 \leq 5.4; \quad y_8 \geq 1.2; \quad y_8 \leq 2.5; \quad y_9 \geq 0.5; \quad y_9 \leq 0.8;$$
$$y_{10} \geq 0.5; \quad y_{10} \leq 0.7; \quad y_{11} \geq 0; \quad y_{11} \leq 0.5; \quad y_{12} \geq 0; \quad y_{12} \leq 0.2$$

And, finally, all variables are nonnegative,

$$x_i \geq 0, \quad y_i \geq 0, \quad \text{for} \quad 1 \leq i \leq 12$$

The objective function that we want to maximize is

$$
\begin{aligned}
z = \quad & 1.6x_1 + 1.7x_2 + 1.8x_3 + 1.9x_4 + 2.0x_5 + 2.0x_6 \\
& + 2.0x_7 + 1.9x_8 + 1.8x_9 + 1.7x_{10} + 1.6x_{11} + 1.5x_{12} \\
& + 1.0y_1 + 1.2y_2 + 1.8y_3 + 2.0y_4 + 2.2y_5 + 2.2y_6 \\
& + 2.5y_7 + 2.2y_8 + 1.8y_9 + 1.4y_{10} + 1.1y_{11} + 1.0y_{12} \\
& + 1.4(x_1 - y_1) + 1.3(x_2 - y_2) + 1.5(x_3 - y_3) \\
& + 1.9(x_4 - y_4) + 2.3(x_5 - y_5) + 2.1(x_6 - y_6) \\
& + 1.8(x_7 - y_7) + 1.8(x_8 - y_8) + 2.1(x_9 - y_9) \\
& + 1.8(x_{10} - y_{10}) + 1.3(x_{11} - y_{11}) + 1.3(x_{12} - y_{12})
\end{aligned}
$$

In Chapter 3, you will learn to solve this problem, and with the aid of CALIPSO find that the maximal profit is M\$113.92 and that the solution is

$$
\begin{array}{llll}
x_1 = 1.0 & y_1 = 0.0 & x_7 = 6.4 & y_7 = 5.4 \\
x_2 = 1.1 & y_2 = 0.1 & x_8 = 2.2 & y_8 = 1.2 \\
x_3 = 1.1 & y_3 = 0.1 & x_9 = 1.5 & y_9 = 0.5 \\
x_4 = 3.0 & y_4 = 1.8 & x_{10} = 1.5 & y_{10} = 0.5
\end{array}
$$

$$x_5 = 5.9 \quad y_5 = 1.0 \qquad x_{11} = 1.0 \quad y_{11} = 0.0$$
$$x_6 = 3.5 \quad y_6 = 2.5 \qquad x_{12} = 1.0 \quad y_{12} = 0.0 \qquad \blacklozenge$$

Exercises

1. Suppose the 3M company of Example 1 began the year with an inventory of 300 horses and wants to end the year with an inventory of 500 horses for sale in the next year. Make whatever changes are needed to LP problem (4) to model this new situation.

2. In Example 1 we are not told how many workers the 3M company employs at any time. The size of the work force is measured in terms of productivity (e.g., the number of workers needed to produce 100 horses). Suppose the 3M company ended the previous year with a work force capable of making 100 horses in a quarter. It wants to meet the production goals given in Example 1 in such a way that its work force is of a size capable of making 200 horses in the fourth quarter of the current year. Make whatever changes are needed to LP problem (4) to model this situation.

3. How many decision variables would be required to change Example 1 to model the 3M company's annual horse production on a monthly rather than a quarterly basis?

4. The 3M company would like to see if planning quarterly production over a 2-year period rather than a 1-year period would reduce their costs. Assuming that the demands for the second year will repeat those for the first year and that storage and hiring–firing costs do not change, formulate the LP problem.

5. The Puffy Bicycle Company is planning its 1989 production schedule by quarters (winter, spring, summer, and fall). Puffy has such a large plant that the only limit on production capacity is the size of the labor force. At the beginning of the winter quarter, Puffy has a labor force capable of producing 4000 bikes per quarter. Puffy expects the bicycle market to decline in 1990 and wants to reduce its labor force to a size capable of producing 3000 bikes per quarter during the fall quarter of 1989.

   Puffy expects to sell 4000, 5000, 3000, and 2500 bikes during the winter, spring, summer, and fall quarters, respectively. At the beginning of the winter quarter, Puffy has 750 bikes in stock and wants to end the year with 500 bikes in stock. Puffy can increase the work force at a cost of $25 per bike and reduce the work force at a cost of $10 per bike. Storage charges for bikes made for future delivery are $5 for each period a bike is stored.
   (a) Write the production constraints for each quarter using the relation:

   $$\text{beginning inventory} + \text{production} = \text{demand} + \text{ending inventory}$$

   (b) Write the hiring–firing constraints. The one for the first period would be

   $$U_1 - D_1 = P_1 - 4000$$

   (c) Write the objective function to minimize storage and hiring–firing costs.

6. Suppose the Puffy Bicycle Company wants to consider using overtime payments to workers to augment the plan described in Exercise 5 rather than hiring so many new workers. Suppose that regular pay adds a cost of $10 per bike and overtime pay adds $15 per bike, and that, in each period, overtime pay cannot exceed 20% of regular pay. Formulate the LP problem to minimize the sum of storage costs, hiring–firing costs, and overtime costs.

7. Suppose that the circumstances of the Puffy Company are as described in Exercise 5 except that the availability of parts limits quarterly production to 4000 bikes during the winter and spring quarters. Formulate the LP model for this problem.

8. The National Business Machine Company (NBM) assembles computers from components imported from various countries. NBM has a regular work force of 20 people and hires temporary workers when it has more orders than the regular workers can assemble. All new workers must undergo a 1-month training period. Temporary workers are hired at the beginning of a month and laid off at the end of a month.

   New workers are trained by one of the regular employees. One regular employee can train five new workers. During training, all six of these employees are nonproductive.

   Each worker (regular or temporary) can assemble 100 computers per month. A temporary worker is paid $1600 per month. The salary of the regular workers is a fixed cost that is not relevant to this study because regular workers are not laid off. When workers are laid off at the end of a month, they receive termination payments of $400.

   NBM has received an order for 24,000 computers, all of which must be delivered within 6 months. The buyer would actually prefer to have the computers sooner and has stipulated that at least 1000 must be delivered by the end of the first month, at least 4000 (including those delivered in the first month) by the end of the second month, at least 8000 by the end of the third month, at least 12,000 by the end of the fourth month, at least 15,000 by the end of the fifth month, and all 24,000 by the end of the sixth month. There is no penalty for delivering computers sooner than scheduled.

   We introduce the following variables for $1 \le i \le 6$.

   H_i: number of temporary workers hired at start of month i

   F_i: number of temporary workers fired at end of month i

   T_i: number of workers engaged in training (new or old)

   W_i: number of producing workers (not engaged in training)

   z_i: cost of paying and firing temporary workers

   You may need to introduce more variables.

   The cost function and the constraints describing the first month are

   Cost: $z_1 = 1600H_1 + 400F_1$

   Train the new workers: $T_1 = H_1 + (\frac{1}{5})H_1$

   Hire enough temporaries: $H_1 + 20 = T_1 + W_1$

   Deliver at least 1000 computers: $100W_1 \ge 1000$

   Complete the model by writing cost functions and constraints for the other five months.

9. For the solution given for Example 2, explain why only the minimum allowable amount of water is used for irrigation in May.

10. For the solution given for Example 2, verify that exactly 5 units of water are left in the reservoir behind HMSD at the end of the year.

11. Which constraints in Example 2 are redundant? That is, which constraints can be omitted from the problem because their validity is implied by other constraints?

12. In the solution given for Example 2, how much water flows into VGD in May?

13. The State Lottery (SL) offers a prize of $500,000. Actually the lottery does not plan to pay the winner that much cash; instead, SL uses an investment plan that pays the winner $50,000 per year for 10 years. Because SL is run by the state, it is required to invest in the State Credit Union or in securities sold to finance state projects. At the present time, two securities are available for purchase. Security 1 pays interest of 8% per year and matures at the end of 6 years; at maturity, the purchaser receives the principal amount plus 10% of the principal. Security 2 pays interest of 6% per year plus 25% when it matures at the end of 7 years. The State Credit Union pays 5% interest on funds on deposit for the previous year.

Molly, the SL linear programmer, has started to develop an LP problem by introducing the following variables:

x_1: the amount invested in Security 1

x_2: the amount invested in Security 2

b_i: the amount put in savings at the beginning of year i, $1 \leq i \leq 9$

On the day a winner is declared, Molly plans to pay the winner the first $50,000 installment and invest a sum of money in the two securities and the savings account to provide enough to pay nine more annual installments of $50,000. The savings account will be used to hold all idle cash; it must have a zero balance after the tenth payment.

Molly began her LP problem as follows:

Minimize: $z = x_1 + x_2 + b_1 + 50{,}000$

Subject to: $0.08x_1 + 0.06x_2 + 1.05b_1 - b_2 = 50{,}000$ (Year 2)
$\qquad\qquad 0.08x_1 + 0.06x_2 + 1.05b_2 - b_3 = 50{,}000$ (Year 3)

Unfortunately (for SL), Molly won the lottery and quit her job. Her partial solution was left on her desk. Complete it.

CHECKLIST: CHAPTER 1

DEFINITIONS

CONCEPTS AND RESULTS

Chapter 2

GEOMETRIC AND ALGEBRAIC PRELIMINARIES

2.1 HALF-SPACES, HYPERPLANES, AND CONVEX SETS

Although eventually we will use an algebraic procedure (the simplex method) to solve LP problems, it is instructive now to study the underlying geometry of linear programming. In particular, we will consider some basic geometric ideas that enable us to see where optimal solutions of LP problems occur. For certain simple situations (such as the one described in Example 1 of Section 1.1), we can use these geometric insights to solve LP problems.

To begin, we define the *Euclidean plane R^2* to be the set of all ordered pairs of real numbers; that is,

$$R^2 = \{(x_1, x_2) \mid x_1 \text{ and } x_2 \text{ are real numbers}\}$$

Geometrically, we represent R^2 as in Figure 1, in which, for instance, the ordered pair $(3, -\frac{5}{2})$ corresponds to the indicated point.

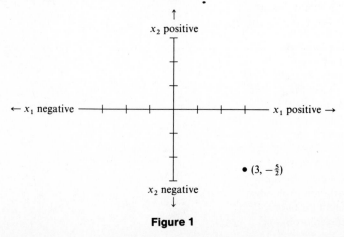

Figure 1

We recall that the *graph* in R^2 of any equation (or inequality) in the variables x_1 and x_2 is the set of all points (x_1, x_2) satisfying the given equation (or inequality).

EXAMPLE 1

The graph in R^2 of an equation of the form

$$a_1 x_1 + a_2 x_2 = c$$

(where a_1, a_2, and c are constants) is a straight line. For instance, the graph in R^2 of the equation

$$2x_1 - 3x_2 = 6 \tag{1}$$

is the line indicated in Figure 2.

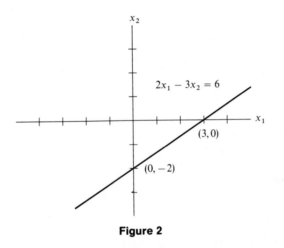

Figure 2

To draw the line in Figure 2, we need two points: We first set $x_1 = 0$ in (1) and solve for x_2, and then set $x_2 = 0$ and solve for x_1, thus obtaining the points at which the line crosses the coordinate axes (these points are usually called *intercepts*). We have labeled the intercepts in this figure.   ◆

Sometimes an equation of a line has a zero coefficient for either x_1 or x_2. For instance, the equation

$$4x_1 = -2 \tag{2}$$

would be interpreted in R^2 as

$$4x_1 + 0x_2 = -2$$

Thus, x_2 could be any number, and $x_1 = -\frac{2}{4} = -\frac{1}{2}$. Figure 3 gives the graph of (2).

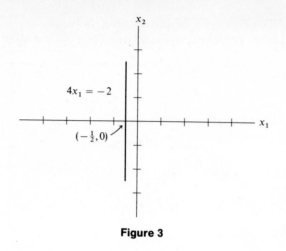

Figure 3

EXAMPLE 2

The graph in R^2 of the inequality

$$a_1 x_1 + a_2 x_2 \leq c$$

is the set of all points in R^2 lying on the line $a_1 x_1 + a_2 x_2 = c$ together with all points lying to one side of this line. For instance, the shaded region in Figure 4 is the graph of the inequality

$$2x_1 - 3x_2 \leq 6 \tag{3}$$

All points on one side of the line $2x_1 - 3x_2 = 6$ satisfy the inequality $2x_1 - 3x_2 < 6$, and all points on the other side of the line $2x_1 - 3x_2 = 6$ satisfy the inequality $2x_1 - 3x_2 > 6$. To determine on which side of the line this region lies, consider a point, say $(0,0)$, not lying on the line but satisfying the inequality (3); the side of the line containing this point is the one corresponding to the inequality.

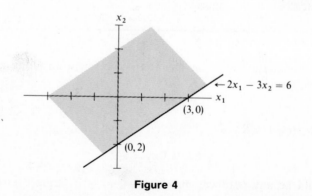

Figure 4

EXAMPLE 3

The shaded region in Figure 5 is the graph in R^2 of the inequality

$$4x_1 \geq -2$$

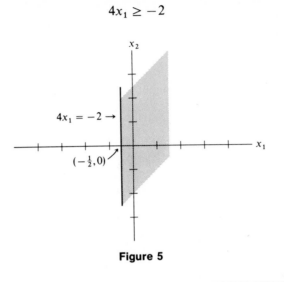

Figure 5   ◆

Subsets of the plane such as those given in Figures 4 and 5 are called half-spaces.

Definition.   A *half-space* in R^2 is the set of all points in R^2 satisfying an inequality of the form

$$a_1 x_1 + a_2 x_2 \leq c$$

or an inequality of the form

$$a_1 x_1 + a_2 x_2 \geq c$$

where at least one of the constants a_1 or a_2 is nonzero.

Next we consider 3-dimensional Euclidean space, R^3. We define R^3 to be the family of all ordered triples; that is,

$$R^3 = \{(x_1, x_2, x_3) \,|\, x_1, x_2, x_3 \text{ are real numbers}\}$$

Geometrically, we represent R^3 as in Figure 6, in which, for instance, the ordered triple $(-2, 3, 4)$ corresponds to the indicated point.

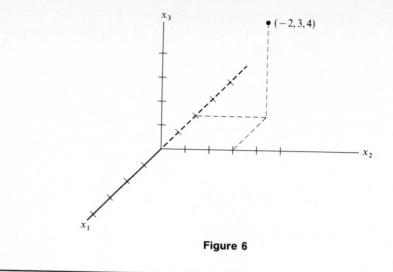

Figure 6

EXAMPLE 4

The graph in R^3 of any equation of the form

$$a_1 x_1 + a_2 x_2 + a_3 x_3 = c$$

is a plane. For instance, the graph in R^3 of the equation

$$-4x_1 + 6x_2 + 3x_3 = 12 \tag{4}$$

is the plane indicated in Figure 7. Since any three points in R^3 (not lying on a line) determine exactly one plane in R^3, an effective way to plot a plane is to determine three points that lie on it by assigning values to two of the variables and solving (4) for the value of the third variable. When the coefficients of all the variables are

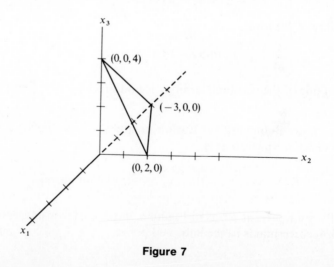

Figure 7

nonzero, we can assign zero to each pair of variables and solve the resulting equation for the other variable. Thus, the graph of the plane (4) passes through the points $(0, 2, 0)$, $(-3, 0, 0)$, and $(0, 0, 4)$. ◆

EXAMPLE 5

The graph in R^3 of the inequality

$$a_1 x_1 + a_2 x_2 + a_3 x_3 \leq c$$

or of the inequality

$$a_1 x_1 + a_2 x_2 + a_3 x_3 \geq c$$

is the set of all points in R^3 lying in the plane

$$a_1 x_1 + a_2 x_2 + a_3 x_3 = c$$

together with all points in R^3 lying to one side of this plane. For instance, the region on one side of the plane in Figure 8 is the graph in R^3 of the inequality

$$-4x_1 + 6x_2 + 3x_3 > 12 \qquad (5)$$

As was the case for R^2, to determine on which side of the plane $-4x_1 + 6x_2 + 3x_3 = 12$ this region lies, consider a point, say $(0, 0, 0)$, not lying in this plane but satisfying the inequality (5); the side of the plane not containing this point is the one corresponding to the inequality.

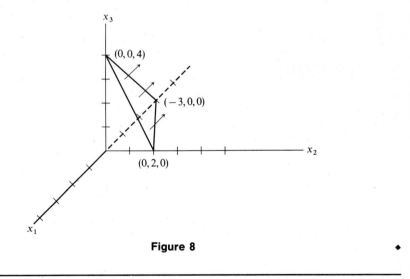

Figure 8 ◆

The region in Example 5 defined by $-4x_1 + 6x_2 + 3x_3 \geq 12$ is called a half-space in R^3.

Definition.   A *half-space* in R^3 is the set of all points in R^3 satisfying an inequality of the form

$$a_1x_1 + a_2x_2 + a_3x_3 \leq c$$

or an inequality of the form

$$a_1x_1 + a_2x_2 + a_3x_3 \geq c$$

where at least one of the constants a_1, a_2, a_3 is nonzero.

We can generalize these ideas to arbitrary Euclidean spaces, R^n, where for $n = 1, 2, \ldots,$

$$R^n = \{(x_1, x_2, \ldots, x_n) \mid x_1, x_2, \ldots, x_n \text{ are real numbers}\}$$

Definition.   A *hyperplane* in R^n is the set of points in R^n satisfying an equality of the form

$$a_1x_1 + a_2x_2 + \cdots + a_nx_n = c$$

where at least one of the constants $a_1, a_2, \ldots, a_n$ is nonzero.

Note that a hyperplane in R^2 is a line, and a hyperplane in R^3 is a plane.

Definition.   A *half-space* in R^n is the set of points in R^n satisfying an inequality of the form

$$a_1x_1 + a_2x_2 + \cdots + a_nx_n \leq c$$

or an inequality of the form

$$a_1x_1 + a_2x_2 + \cdots + a_nx_n \geq c$$

where at least one of the constants $a_1, a_2, \ldots, a_n$ is nonzero.

EXAMPLE 6

The set of points in R^5 satisfying

$$3x_1 + \frac{1}{2}x_2 - x_3 + x_4 + \frac{2}{3}x_5 = -9$$

is a hyperplane in R^5, and the set of points in R^5 satisfying the inequality

$$3x_1 + \frac{1}{2}x_2 - x_3 + x_4 + \frac{2}{3}x_5 \geq -9$$

is a half-space in R^5.   ◆

Next we briefly describe the idea of convexity, which plays a major role in linear programming.

Definition. A subset K of R^n is *convex* if K is empty or is a single point, or if for each two distinct points **p** and **q** in K, the line segment connecting **p** and **q** lies entirely in K.

EXAMPLE 7

The sets in Figure 9 are convex.

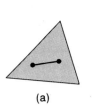 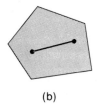 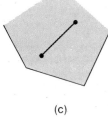

| (a) | (b) | (c) | (d) |

Figure 9

The sets in Figure 10 are not convex.

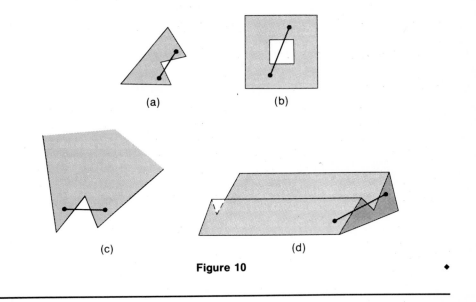

(a)   (b)

(c)   (d)

Figure 10   ◆

In Theorem 1 we show that half-spaces are convex. This seems obvious in the lower-dimensional cases; however, the formal proof requires a bit of work. We begin by noting that if

$$\mathbf{p} = (p_1, p_2, \ldots, p_n) \qquad \text{and} \qquad \mathbf{q} = (q_1, q_2, \ldots, q_n)$$

are points in R^n, then the *line segment* joining $\mathbf{p}$ and $\mathbf{q}$ consists of all points of the form

$$(1 - t)\mathbf{p} + t\mathbf{q}, \qquad 0 \le t \le 1 \tag{6}$$

where

$$(1 - t)\mathbf{p} + t\mathbf{q} = (1 - t)(p_1, p_2, \ldots, p_n) + t(q_1, q_2, \ldots, q_n)$$
$$= [(1 - t)p_1 + tq_1, (1 - t)p_2 + tq_2, \ldots, (1 - t)p_n + tq_n]$$

Observe in (6) that if $t = 0$, then

$$(1 - t)\mathbf{p} + t\mathbf{q} = \mathbf{p}$$

and if $t = 1$, then

$$(1 - t)\mathbf{p} + t\mathbf{q} = \mathbf{q}$$

EXAMPLE 8

The line segment in R^2 joining the points $\mathbf{p} = (3, 6)$ and $\mathbf{q} = (-4, 5)$ is the set of points

$$(1 - t)(3, 6) + t(-4, 5) = [(1 - t)3 - 4t, (1 - t)6 + 5t]$$
$$= (3 - 7t, 6 - t), \qquad 0 \le t \le 1$$

Figure 11 gives the graph of this line.

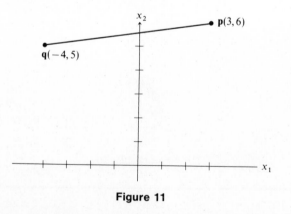

Figure 11

♦

Observe in Example 8 that if $t = 0$, then

$$(1 - t)(3, 6) + t(-4, 5) = (3, 6) = \mathbf{p}$$

and if $t = 1$, then

$$(1 - t)(3, 6) + t(-4, 5) = (-4, 5) = \mathbf{q}$$

Note also that as t moves from 0 to 1, the corresponding points on the line segment move proportionally from point $\mathbf{p}$ to point $\mathbf{q}$. For example, setting $t = \frac{1}{3}$ gives the point

$$\mathbf{r} = \frac{2}{3}(3, 6) + \frac{1}{3}(-4, 5) = \left(\frac{2}{3}, \frac{17}{3}\right)$$

which is $\frac{1}{3}$ of the way from $\mathbf{p}$ to $\mathbf{q}$, as shown in Figure 12.

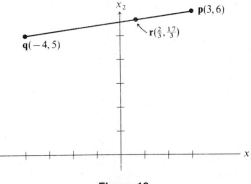

Figure 12

Theorem 1. A half-space H in R^n that is defined either by the inequality

$$a_1 x_1 + a_2 x_2 + \cdots + a_n x_n \leq c \tag{7}$$

or the inequality

$$a_1 x_1 + a_2 x_2 + \cdots + a_n x_n \geq c \tag{8}$$

is convex.

PROOF. We establish this result for the half-space H defined by the inequality (7). A similar argument holds for half-spaces defined by (8) (see Exercise 24). Suppose that the points $\mathbf{p} = (p_1, p_2, \ldots, p_n)$ and $\mathbf{q} = (q_1, q_2, \ldots, q_n)$ lie in H; that is, these points satisfy (7). To show that the line segment connecting these points lies entirely in H, it suffices to show that for each t, $0 \leq t \leq 1$, the point $(1 - t)\mathbf{p} + t\mathbf{q}$ also satisfies (7). Since

$$(1 - t)\mathbf{p} + t\mathbf{q} = [(1 - t)p_1 + tq_1, (1 - t)p_2 + tq_2, \ldots, (1 - t)p_n + tq_n]$$

we must show that

$$a_1[(1-t)p_1 + tq_1] + a_2[(1-t)p_2 + tq_2] + \cdots + a_n[(1-t)p_n + tq_n] \le c \quad (9)$$

First, observe that since the points **p** and **q** satisfy (7), we have

$$a_1 p_1 + a_2 p_2 + \cdots + a_n p_n \le c$$

and

$$a_1 q_1 + a_2 q_2 + \cdots + a_n q_n \le c$$

To show that inequality (9) is valid, we note that

$$a_1[(1-t)p_1 + tq_1] + a_2[(1-t)p_2 + tq_2] + \cdots + a_n[(1-t)p_n + tq_n]$$
$$= (1-t)(a_1 p_1 + a_2 p_2 + \cdots + a_n p_n) + t(a_1 q_1 + a_2 q_2 + \cdots + a_n q_n)$$
$$\le (1-t)c + tc = c$$

and this concludes the proof. ∎

The following result is also useful in linear programming.

Theorem 2. If $K_1, K_2, \ldots, K_r$ are convex subsets of R^n, then the intersection of these sets, $K = K_1 \cap K_2 \cap \cdots \cap K_r$ is also convex.

PROOF. If K is empty or consists of a single point, then K is convex by definition. Suppose then that K consists of more than one point, and let **p** and **q** be any two distinct points in K. Since both **p** and **q** are in each convex set K_i, $1 \le i \le r$, the line segment L connecting **p** and **q** also lies entirely in each K_i. Therefore, L lies in the intersection K of these sets, and we conclude that K is convex. ∎

The next result is an immediate consequence of Theorems 1 and 2.

Theorem 3. A hyperplane M in R^n defined by

$$a_1 x_1 + a_2 x_2 + \cdots + a_n x_n = c$$

is convex.

PROOF. M is the intersection of the convex half-spaces

$$a_1 x_1 + a_2 x_2 + \cdots + a_n x_n \le c$$

and

$$a_1 x_1 + a_2 x_2 + \cdots + a_n x_n \ge c$$

By Theorem 2, this intersection is convex. ∎

We will be especially interested in corner points of convex sets. We define corner points in terms of interior points of line segments, where an *interior point* of the line segment

$$L = \{\mathbf{x} \,|\, \mathbf{x} = (1 - t)\mathbf{p} + t\mathbf{q}, \qquad 0 \le t \le 1\}$$

joining the distinct points **p** and **q** is any point **x** corresponding to a *t* that is not 0 or 1. For example, the point **r** of Figure 12 is an interior point.

Definition.   A point **q** is a *corner point* (or an *extreme point*) of a convex set *K* if **q** is not an interior point of any line segment contained in *K*.

EXAMPLE 9

The points $\mathbf{q}_1, \mathbf{q}_2, \mathbf{q}_3, \mathbf{q}_4, \mathbf{q}_5$ are the corner points of the convex set in Figure 13.

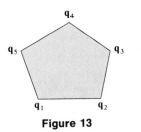

Figure 13

The points $\mathbf{q}_1, \mathbf{q}_2, \ldots, \mathbf{q}_8$ are the corner points of the convex set in Figure 14.

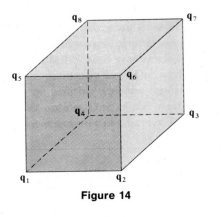

Figure 14   ◆

As you will see in the next section, corner points are of special significance because optimal solutions of LP problems occur at these points.

Exercises

1. Draw the graphs in R^2 of the lines
   (a) $3x_1 + 2x_2 = 6$
   (b) $-2x_1 + 4x_2 = 12$
   (c) $x_1 = 4$
   (d) $2x_2 = 4$

2. Draw the graphs in R^2 of the lines
   (a) $x_1 = x_2$
   (b) $2x_1 = -3x_2 + 6$
   (c) $x_1 + x_2 = 2$
   (d) $-2x_2 = 3$

3. Draw the graphs in R^2 of the half-spaces
   (a) $3x_1 + 2x_2 \le 6$
   (b) $-2x_1 + 4x_2 \ge 12$
   (c) $x_1 \ge 4$
   (d) $2x_2 \le 4$

4. Draw the graphs in R^2 of the half-spaces
   (a) $x_1 \le x_2$
   (b) $2x_1 \ge -3x_2 + 6$
   (c) $x_1 + x_2 \le 2$
   (d) $-2x_2 \ge 3$

5. Draw the graphs in R^3 of the planes
   (a) $x_1 + x_2 + x_3 = 3$
   (b) $x_1 - x_2 + x_3 = 3$
   (c) $2x_1 + 3x_2 + 4x_3 = 12$
   (d) $-2x_1 + 3x_2 + x_3 = 6$

6. Draw the graphs in R^3 of the planes
   (a) $4x_1 + 2x_2 - x_3 = 6$
   (b) $-2x_1 + 2x_2 - 3x_3 = 12$
   (c) $-x_1 + 3x_2 - x_3 = 5$
   (d) $x_1 + x_2 = x_3 + 4$

7. Draw the graphs in R^3 of the half-spaces
   (a) $x_1 + x_2 + x_3 \le 3$
   (b) $x_1 - x_2 + x_3 \ge 3$
   (c) $2x_1 + 3x_2 + 4x_3 \ge 12$
   (d) $-2x_1 + 3x_2 + x_3 \le 6$

8. Draw the graphs in R^3 of the half-spaces
   (a) $4x_1 + 2x_2 - x_3 \ge 6$
   (b) $-2x_1 + 2x_2 - 3x_3 \ge 12$
   (c) $-x_1 + 3x_2 - x_3 \le 5$
   (d) $x_1 + x_2 \le x_3 + 4$

9. Which of the following expressions define hyperplanes?
   (a) $2.5x_1 - 3.2x_2 + 4x_3 - 2.1x_5 + 6.3x_6 = 12$
   (b) $3x_1^2 + 2x_2 + x_3 + x_4 = 6$
   (c) $x_1 + x_2 = 1$

10. Which of the following expressions define hyperplanes?
   (a) $2x_1 + 2x_2 + 3x_3 + 4x_4 \le 5$
   (b) $2x_1 + 3x_3 = x_2 - x_4 + 3$
   (c) $x_1 + x_2 + x_3 + x_4 = 0$

11. Which of the following expressions define half-spaces?
   (a) $x_1 + x_2 - x_3 - x_4 \le 0$
   (b) $x_1 + x_2 \le x_3 + x_4$
   (c) $x_1 = x_2 + x_3$
   (d) $x_1 x_2 \le 1$

12. Which of the following expressions define half-spaces?
   (a) $x_1 + x_2^3 + 2x_3 \ge 9$
   (b) $x_1 \le 6 + 1/x_3$
   (c) $x_1 - 3x_2 \ge 0$

13. Which of the following sets are convex?

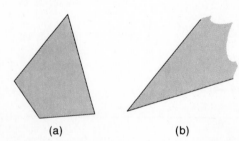

(a)   (b)

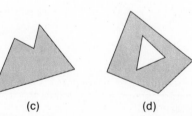

(c)   (d)

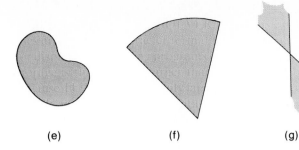
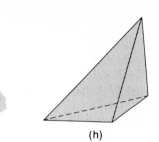

(e) (f) (g) (h)

14. Which of the following sets are convex?

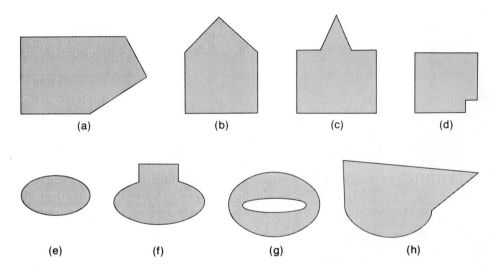

(a) (b) (c) (d)

(e) (f) (g) (h)

15. (a) Is a line segment convex?
(b) Do the sides of a triangle form a convex set?
(c) Is R^3 convex?

16. (a) Is an eggshell convex?
(b) Is a baseball convex?

17. Let $\mathbf{p} = (1, 2)$ and $\mathbf{q} = (2, 4)$ be two points in R^2. Find the coordinates of the midpoint of the line segment joining $\mathbf{p}$ to $\mathbf{q}$.

18. Let $\mathbf{p} = (1, 2, 3)$ and $\mathbf{q} = (2, 4, -1)$ be two points in R^3. Find the coordinates of the midpoint of the line segment joining $\mathbf{p}$ to $\mathbf{q}$.

19. For the points $\mathbf{p}$ and $\mathbf{q}$ of Exercise 17, find the coordinates of the point that is $\frac{1}{4}$ of the way from $\mathbf{p}$ to $\mathbf{q}$.

20. For the points $\mathbf{p}$ and $\mathbf{q}$ of Exercise 18, find the coordinates of the point that is $\frac{1}{3}$ of the way from $\mathbf{p}$ to $\mathbf{q}$.

21. Let $\mathbf{p}$ and $\mathbf{q}$ be points of R^n. Show that the line segment from $\mathbf{p}$ to $\mathbf{q}$ can be defined as the set of points $t_1\mathbf{p} + t_2\mathbf{q}$, where $t_1 + t_2 = 1, 0 \le t_1 \le 1, 0 \le t_2 \le 1$.

22. Consider a convex set consisting of a triangle and its interior. Label its corner points $\mathbf{p}, \mathbf{q}, \mathbf{r}$, and locate the points $\mathbf{x}$ and $\mathbf{s}$ as in Figure 15.

Write $\mathbf{x} = (1 - t)\mathbf{s} + t\mathbf{p}$ and $\mathbf{s} = (1 - k)\mathbf{q} + k\mathbf{r}$ where $0 \le t, k \le 1$. Show that $\mathbf{x} = t_1\mathbf{p} + t_2\mathbf{q} + t_3\mathbf{r}$, where $t_1 + t_2 + t_3 = 1$ and $t_i \ge 0, 1 \le i \le 3$.

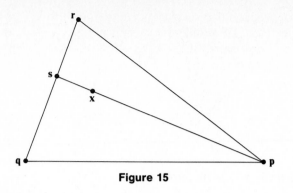

Figure 15

23. A point $\mathbf{p}$ in R^n is a *convex combination* of the points $\mathbf{p}_1, \mathbf{p}_2, \ldots, \mathbf{p}_k$ in R^n if for some real numbers $t_1, t_2, \ldots, t_k$ that satisfy

$$t_1 + t_2 + \cdots + t_k = 1 \qquad \text{and} \qquad t_i \geq 0, \qquad 1 \leq i \leq k$$

we can write

$$\mathbf{p} = t_1\mathbf{p}_1 + t_2\mathbf{p}_2 + \cdots + t_k\mathbf{p}_k$$

For instance, $\frac{1}{2}\mathbf{p}_1 + \frac{1}{2}\mathbf{p}_2$ and $\frac{1}{4}\mathbf{p}_1 + \frac{3}{4}\mathbf{p}_2$ are two convex combinations of $\mathbf{p}_1$ and $\mathbf{p}_2$. Show that the set of all convex combinations of $\mathbf{p}_1, \mathbf{p}_2, \ldots, \mathbf{p}_k$ is a convex set.

24. Prove Theorem 1 for inequalities of type (8).

25. Give an example of two convex sets whose union is not convex.

2.2 FEASIBLE SETS AND GEOMETRIC SOLUTIONS OF LP PROBLEMS

In Example 1 of Section 1.1 we considered an enterprise that manufactured tables and bookcases and determined that the LP problem to maximize profit was

$$\text{Maximize:} \quad z = 80x_1 + 55x_2$$

$$\text{Subject to:} \quad \frac{6}{5}x_1 + \quad x_2 \leq \quad 72$$

$$x_1 + \frac{1}{2}x_2 \leq \quad 50 \tag{1}$$

$$\frac{3}{2}x_1 + 2x_2 \leq 120$$

$$x_1, x_2 \geq 0$$

Recall that the optimal solution to such a problem is a feasible solution [an ordered pair (x_1, x_2) where x_1 and x_2 satisfy all of the constraints of (1)] that maximizes the objective function on the feasible set. (The feasible set or region for this problem is the set of all ordered pairs that satisfy the given constraints.) In general, recall that the *feasible region* of an LP problem in n variables $x_1, x_2, \ldots, x_n$ is the set of all n-tuples $(s_1, s_2, \ldots, s_n)$ that satisfy all of the constraints of the problem when we set $x_1 = s_1, x_2 = s_2, \ldots,$ and $x_n = s_n$.

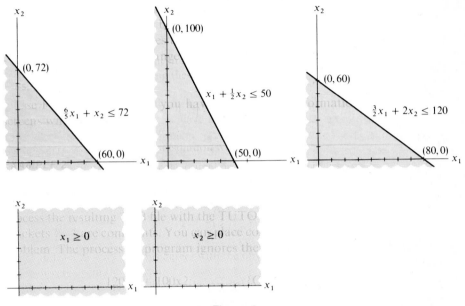

Figure 1

Note that each of the constraints in the LP problem (1) defines a half-space. Figure 1 illustrates these half-spaces.

The feasible set consists of all points common to each of these half-spaces; in other words, the feasible set consists of all points in the intersection of these half-spaces. The shaded area in each drawing in Figure 2 indicates this set of points.

Observe that the feasible region in Figure 2 is convex. This illustrates the following general result.

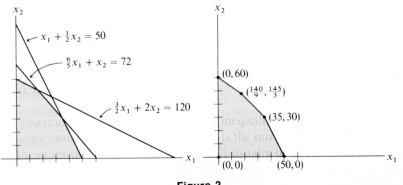

Figure 2

Theorem 1.   The feasible set in R^n corresponding to any number of constraints of the types

$$a_1 x_1 + a_2 x_2 + \cdots + a_n x_n \leq b$$

$$a_1 x_1 + a_2 x_2 + \cdots + a_n x_n = b \qquad\qquad (2)$$

$$a_1 x_1 + a_2 x_2 + \cdots + a_n x_n \geq b$$

$$x_1, x_2, \ldots, x_n \geq 0$$

is convex.

PROOF.   The inequality constraints of type (2) define half-spaces, and the equality constraints define hyperplanes. By Theorems 1 and 3 of Section 2.1 these half-spaces and hyperplanes are convex sets. Since the feasible set is the intersection of these convex sets, it follows from Theorem 2 of Section 2.1 that the feasible set is convex.   ■

EXAMPLE 1

Figure 3 shows the graph of the convex set determined by the following system of inequalities.

$$x_1 + 2x_2 \leq 6$$

$$x_1 - x_2 \geq 2$$

$$x_1 - x_2 \leq 4$$

$$x_2 \leq 1$$

$$x_1, x_2 \geq 0$$

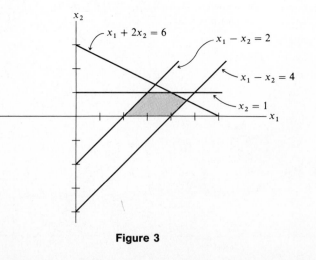

Figure 3   ◆

The following result is one of the most important in linear programming. We defer its proof to Section 2.3.

Theorem 2.   (Corner Point Theorem) Consider the LP problem

$$\text{Maximize (or Minimize):} \quad z = c_1 x_1 + c_2 x_2 + \cdots + c_n x_n$$

Subject to a system of inequalities of the types:

$$
\begin{aligned}
a_1 x_1 + a_2 x_2 + \cdots + a_n x_n &\leq b \\
a_1 x_1 + a_2 x_2 + \cdots + a_n x_n &= b \\
a_1 x_1 + a_2 x_2 + \cdots + a_n x_n &\geq b
\end{aligned}
\tag{3}
$$

$$x_1, x_2, \ldots, x_n \geq 0$$

(a) If the feasible region of (3) is bounded, then an optimal solution is attained at a corner point of this feasible region.
(b) If the feasible region is unbounded, then an optimal solution may not exist; however, if an optimal solution exists, it is attained at a corner point of (3).

To see that this result is plausible, we return to LP problem (1). In this problem, our goal is to find a point (x_1, x_2) in the feasible region shown in Figure 2 that maximizes the objective function

$$z = 80x_1 + 55x_2 \tag{4}$$

Note that for each fixed value of z, the graph of (4) is a straight line. Moreover, by changing the value of z we obtain a family of parallel lines in R^2. A line that is obtained from an objective function such as (4) by assigning a value to z is called a *level line*. Figure 4 illustrates the level lines corresponding to the values $z = 1100$, $z = 2200$, and $z = 4400$.

Observe in Figure 4 that increasing the value of z causes the level line to move away from the origin. The maximum value of the objective function occurs for the value z^* of z for which the level line

$$z^* = 80x_1 + 55x_2$$

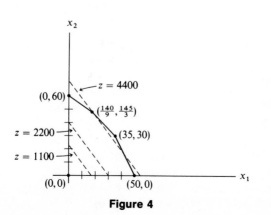

Figure 4

has moved as far from the origin as possible while still intersecting the feasible set. From Figure 4 it is apparent that the largest such z corresponds to the line that intersects the feasible set in the corner point $(35, 30)$. We conclude that $x_1 = 35$ and $x_2 = 30$ is the optimal solution to the problem, and that $z^* = (80)(35) + (55)(30) = 4450$ is the maximal value of z.

It follows from the Corner Point Theorem that we can also solve LP problem (1) by listing all of the corner points of the feasible region (shown in Figure 2) and calculating the value of the objective function (4) at each corner point. By Theorem 2, the values of x_1 and x_2, obtained in this way, that produce the largest value of the objective function form the maximal solution to (1). We have tabulated these values in Table 1 and conclude again that $x_1 = 35$ and $x_2 = 30$ is the maximal solution.

Table 1

x_1	x_2	$z = 80x_1 + 55x_2$
0	0	0
0	60	3300
140/9	145/3	35125/9
35	30	4450
50	0	4000

The next example illustrates Theorem 2 in the case of LP problems with three variables.

EXAMPLE 2

$$\text{Maximize:} \quad z = x_1 + 2x_2 + 2x_3$$

$$\text{Subject to:} \quad 117x_1 + 78x_2 - 117x_3 \leq 780$$

$$2x_2 - \quad x_3 \geq 6$$

$$\begin{array}{ll} x_1 \leq 4 & x_1 \geq 1 \\ x_2 \leq 13 & x_2 \geq 5 \\ x_3 \leq 8 & x_3 \geq 4 \\ x_i \geq 0, & 1 \leq i \leq 3 \end{array} \qquad (5)$$

Observe that the graphs corresponding to the constraints in LP problem (5) are half-spaces. Figure 5 illustrates the convex region that is the intersection of these half-spaces.

Figure 5 also shows the graph of the plane resulting from the objective function when we set $z = 2$. The planes obtained as graphs of objective functions by setting z equal to a constant z^* are called *level planes*. The level planes are parallel. The arrows by the level plane in Figure 5 indicate the direction in which the level planes for the objective function of (5) move as z increases; that is, as z increases, the level planes move through the feasible region shown in Figure 5. You can see that the corner point $(4, 13, 8)$ is the feasible point for which z is largest and that this point is the maximal solution to LP problem (5).

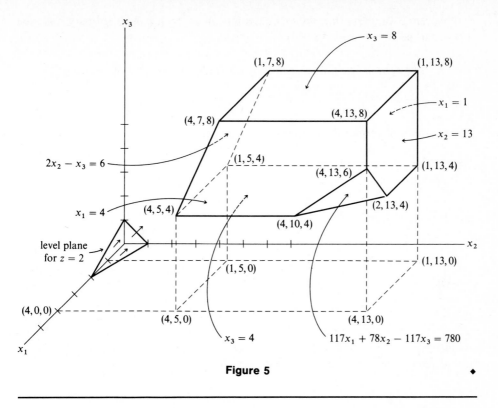

Figure 5   ◆

The importance of Theorem 2 is it shows that, to solve an LP problem, it suffices to consider only the corner points of the feasible region. It should be clear that a finite number of constraints results in a finite number of corner points. Theoretically then, as we have indicated, we can check the values of z at each of these points to determine the maximum (or minimum) value of the objective function. In practice, however, the difficulties with this approach are often insurmountable. One problem lies in determining which points are corner points. For instance, in Figure 2 we saw that each corner point was the intersection of two lines determined by the constraints. Note, however, in Figure 2, that the intersection of the lines $x_1 + \frac{1}{2}x_2 = 50$ and $\frac{3}{2}x_1 + 2x_2 = 120$ does not produce a corner point of the feasible region of LP problem (1), even though these lines correspond to two of the constraints. You can imagine that an LP problem with many variables and constraints has an astronomical number of possible corner points. In such problems, determining which combinations of constraints actually produce corner points is, for all practical purposes, impossible, even with the aid of a high-speed computer. For instance, a problem involving 10 variables and 50 constraints has 10,272,000,000 potential corner points, and LP problems with hundreds of variables and constraints are not uncommon. Fortunately, a procedure known as the *simplex method* usually significantly reduces the number of corner points checked, thus yielding optimal solutions to large problems in a reasonable time. In the next chapter we will examine this method, which is at the heart of linear programming.

Theorem 2 suggests that an optimal solution of an LP problem may not exist. There can be two reasons for this:

1. The feasible region is empty; that is, there are no feasible solutions.

2. The feasible region is unbounded.

The next two examples illustrate these situations.

EXAMPLE 3

The feasible region of the LP problem

$$\text{Maximize:} \quad z = x_1 + x_2$$
$$\text{Subject to:} \quad x_1 - 2x_2 \geq 6$$
$$2x_1 + x_2 \leq 4 \tag{6}$$
$$x_1, x_2 \geq 0$$

is empty (Figure 6). No point (x_1, x_2) satisfies all of the constraints of (6); thus, there are no feasible solutions and, in particular, no optimal solution.

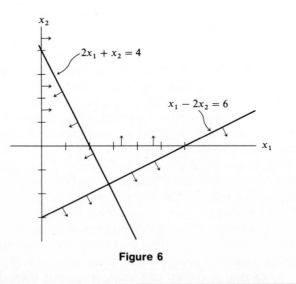

Figure 6 ◆

EXAMPLE 4

Figure 7 gives the feasible region of the LP problem

$$\text{Maximize:} \quad z = x_1 + x_2$$

$$\text{Subject to:} \qquad x_1 - x_2 \le 5$$
$$-2x_1 + x_2 \le 4 \qquad\qquad (7)$$
$$x_1, x_2 \ge 0$$

Although this region is convex, it is unbounded.

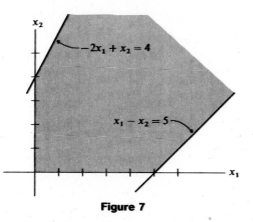

Figure 7

Figure 8 indicates level lines corresponding to three fixed values of the objective function z of (7).

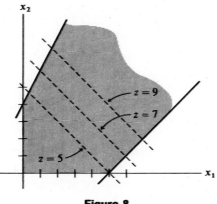

Figure 8

Note that for any value of $z \ge 0$, the line $z = x_1 + x_2$ intersects the feasible region; hence, there are feasible points (x_1, x_2) that give arbitrarily large values of z. It follows that this LP problem has no optimal solution.   ◆

However, an LP problem with an unbounded feasible region can have an optimal solution. Example 5 illustrates this possibility.

EXAMPLE 5

Figure 9 gives the (unbounded) feasible region of the LP problem

$$\text{Maximize:} \quad z = -2x_1 + x_2$$
$$\text{Subject to:} \quad -3x_1 + x_2 \le 1$$
$$-x_1 + x_2 \le 3$$
$$-x_1 + 2x_2 \le 2 \tag{8}$$
$$x_1, x_2 \ge 0$$

The dashed line in Figure 9 is a typical level line for the objective function. The arrows indicate the direction the dashed line moves as z increases. We see that the optimal solution occurs at the corner point $(1, 4)$.

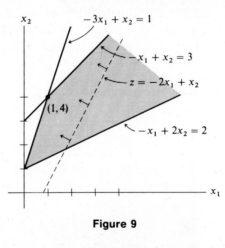

Figure 9   ◆

We note the possibility that there is more than one optimal solution to an LP problem. This can occur if the level lines (or more generally, the level hyperplanes) are parallel to one of the lines (hyperplanes) defining the boundary of the feasible region. Exercises 13 and 14 illustrate this.

We conclude this section with a brief introduction to sensitivity analysis, which is an important aspect of linear programming that we will investigate extensively in Chapter 7. Sensitivity analysis seeks ranges over which the constants of an LP problem can vary without changing the optimal solution (the optimal values of the decision variables $x_1, x_2, \ldots$). For instance, the cost coefficients c_i in an objective function

$$z = c_1 x_1 + c_2 x_2 + \cdots + c_n x_n$$

often represent material costs or selling prices in an application. Such quantities may fluctuate over time, and it is of considerable interest to the manager of an enterprise to know when changes to these coefficients become large enough to warrant a change in the operation of the enterprise. The next example interprets changing values of c_i geometrically.

EXAMPLE 6

From Figure 4 we found that the optimal solution of the prototype problem (1) is $x_1 = 35$, $x_2 = 30$. The objective function of that problem is

$$z = 80x_1 + 55x_2$$

To perform a sensitivity analysis, we write this objective function as

$$z = c_1x_1 + c_2x_2 \tag{9}$$

and seek a range of values over which c_1 and c_2 can vary from their present values of 80 and 55 without changing the optimal solution $x_1 = 35$, $x_2 = 30$. In Figure 10, which is derived from Figure 4, the dashed line represents the graph of the level line of $z = 80x_1 + 55x_2$ that passes through the optimal point $(35, 30)$. The arrows indicate the region that is swept out by rotating this level line about $(35, 30)$ in a clockwise direction until the line labeled L_1 is reached, and in a counterclockwise direction until the line labeled L_2 is reached. These rotations correspond to changes in the values of c_1 and c_2 in (9).

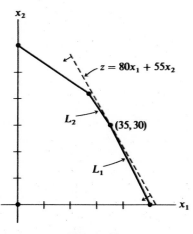

Figure 10

We can find numerical limits on c_1 and c_2 by comparing the slopes of L_1 and L_2 to the slopes of level lines of (9). The line L_1 corresponds to the equation

$$x_1 + \frac{1}{2}x_2 = 50 \tag{10}$$

derived from the second inequality of (1). You may recall from studying analytic geometry that we can find the slope of a line such as (10) by solving the equation for

x_2. Doing so, we obtain

$$x_2 = -2x_1 + 100$$

The slope of (10) is the coefficient of x_1, -2. Since the line L_2 corresponds to the equation

$$\frac{3}{2}x_1 + 2x_2 = 120 \tag{11}$$

derived from the third constraint of (1), we solve (11) for x_2 to obtain

$$x_2 = -\frac{3}{4}x_1 + 60 \tag{12}$$

and conclude that the slope of L_2 is $-\frac{3}{4}$. In a similar way, we find the slope of the level lines of (9) by solving (9) for x_2 to obtain

$$x_2 = -\frac{c_1}{c_2}x_1 + \frac{z}{c_2} \tag{13}$$

The algebraic condition that the level lines of (9) first meet the feasible region shown in Figure 10 is

$$-2 \leq -\frac{c_1}{c_2} \leq -\frac{3}{4}$$

or, equivalently,

$$\frac{3}{4} \leq \frac{c_1}{c_2} \leq 2 \tag{14}$$

Thus, as long as c_1 and c_2 satisfy (14), the optimal solution to LP problem (1) is $x_1 = 35$, $x_2 = 30$. ◆

___

Exercises

In Exercises 1–10, sketch the feasible region of the given LP problem, find the coordinates of all corner points of the feasible region, and determine the optimal solution (or show that one does not exist) by graphing level lines of the objective function.

1. Maximize: $z = 120x_1 + 100x_2$

Subject to: $x_1 + x_2 \leq 4$

$5x_1 + 3x_2 \leq 15$

$x_1, x_2 \geq 0$

2. Maximize: $z = 4x_1 + 6x_2$

Subject to: $x_1 + 3x_2 \leq 9$

$2x_1 + 3x_2 \leq 12$

$5x_1 + 2x_2 \leq 15$

$x_1, x_2 \geq 0$

3. Minimize: $z = -6x_1 + 2x_2$

   Subject to: $-3x_1 + x_2 \leq 6$
   $$3x_1 + 5x_2 \geq 15$$
   $$x_1, x_2 \geq 0$$

4. Maximize: $z = 2x_1 + 3x_2$

   Subject to: $x_1 + x_2 \leq 3$
   $$3x_1 + x_2 \leq 6$$
   $$x_1 + 2x_2 \leq 5$$
   $$x_1, x_2 \geq 0$$

5. Maximize: $z = 7x_1 + 5x_2$

   Subject to: $x_1 - x_2 \leq 2$
   $$x_1 + x_2 \geq 3$$
   $$2x_1 + 3x_2 \leq 12$$
   $$2x_1 - x_2 \geq 1$$
   $$x_1, x_2 \geq 0$$

6. Minimize: $z = 7x_1 + 5x_2$

   Subject to the same constraints as in Exercise 5.

7. Maximize: $z = 5x_1 + 2x_2$

   Subject to: $2x_1 - x_2 \geq 2$
   $$x_1 + 3x_2 \geq 5$$
   $$x_1 + 3x_2 \leq 9$$
   $$x_1 + x_2 \leq 6$$
   $$x_1, x_2 \geq 0$$

8. Maximize: $z = -2x_1 + x_2$

   Subject to: $4x_1 - 9x_2 \leq 3$
   $$-3x_1 + 2x_2 \leq 6$$
   $$-x_1 + 3x_2 \leq 3$$
   $$x_1, x_2 \geq 0$$

9. Minimize: $z = -4x_1 + 7x_2$

   Subject to: $x_1 + x_2 \geq 3$
   $$-x_1 + x_2 \leq 3$$
   $$2x_1 + x_2 \leq 8$$
   $$x_1, x_2 \geq 0$$

10. Maximize: $z = x_1 + x_2 + x_3$

    Subject to: $x_1 + x_3 \leq 4$
    $$2x_2 - x_3 \leq 0$$
    $$x_i \geq 0, \qquad 1 \leq i \leq 3$$

11. For each of the preceding exercises that is listed below, calculate the value of the objective function at each corner point of the feasible region, and determine the optimal value by using Theorem 2.
    (a) Exercise 1   (b) Exercise 3   (c) Exercise 5

12. Follow the same instructions as in Exercise 11.
    (a) Exercise 2   (b) Exercise 4   (c) Exercise 6

13. Graph the feasible region and objective function for the following LP problem. Explain why the problem has more than one maximal solution.

$$\text{Maximize:} \quad z = 2x_1 + 2x_2$$
$$\text{Subject to:} \quad x_1 + 2x_2 \leq 4$$
$$x_1 + x_2 \leq 3$$
$$x_1 - x_2 \geq 1$$
$$x_1, x_2 \geq 0$$

14. The LP problem of Exercise 13 not only has more than one solution, it has infinitely many solutions. Is it possible for an LP problem in two variables to have more than one solution but not infinitely many solutions?

15. Calculate the value of the objective function for (5) at each of the ten corner points of the feasible region (see Figure 5). Find the maximal and minimal solutions by using Theorem 2.

16. (a) Draw four lines that intersect in six points.
    (b) Draw four lines that intersect in five points.
    (c) What is the maximum number of points in which four lines can intersect?

In Exercises 17–20, perform a sensitivity analysis on the cost coefficients of the LP problem, as we did in Example 6.

17. Exercise 1.   18. Exercise 2.   19. Exercise 5.   20. Exercise 4.

21. Show that the largest possible number of points of intersection of five lines in R^2 is 10. (You could count the number of pairs of lines as follows: There are five ways to choose the first line in a pair; and, having chosen the first line, there are four ways to choose the second. In this manner, $5 \cdot 4 = 20$ pairs are formed. Why, then, are there only ten possible points of intersection?)

22. Show that the largest possible number of points of intersection of n lines in R^2 is $n(n-1)/2$.

23. Show that the largest possible number of points of intersection of five planes in R^3 is 10. (You could count the number of ways of choosing three planes from a set of five planes as follows: There are five ways of choosing the first plane, then four ways of choosing the second plane, and then three ways of choosing the third plane. In this manner, $5 \cdot 4 \cdot 3 = 60$ triples are formed. Show that each set of three planes occurs in six triples formed in this manner.)

24. Show that the maximum possible number of points of intersection of n planes in R^3 is $n(n-1)(n-2)/6$.

25. The problem of determining the maximum possible number of points of intersection of n hyperplanes in R^k is equivalent to the combinatorial problem of determining how many subsets of k elements can be formed from a set of n elements. This number is given by the binomial coefficient,

$$\binom{n}{k} = \frac{n!}{k!(n-k)!}$$

For example,

$$\binom{5}{3} = \frac{5!}{3!(5-3)!} = \frac{5 \cdot 4 \cdot 3 \cdot 2 \cdot 1}{(3 \cdot 2 \cdot 1)(2 \cdot 1)} = \frac{5 \cdot 4}{2 \cdot 1} = 10$$

Compute the following binomial coefficients.

(a) $\binom{7}{4}$ (b) $\binom{10}{5}$ (c) $\binom{20}{10}$ (d) $\binom{16}{6}$

---

*2.3 PROOF OF THE CORNER POINT THEOREM

Our goal in this section is to establish the Corner Point Theorem, which, as we have seen, enables us to limit our search for an optimal solution of an LP problem to the corner points of the feasible region. For convenience, we restate the result.

Theorem 1. (Corner Point Theorem) Consider the LP problem

$$\text{Maximize (or Minimize):} \quad z = c_1 x_1 + c_2 x_2 + \cdots + c_n x_n$$

Subject to a system of inequalities of the types:

$$a_1 x_1 + a_2 x_2 + \cdots + a_n x_n \leq b$$
$$a_1 x_1 + a_2 x_2 + \cdots + a_n x_n = b \qquad (1)$$
$$a_1 x_1 + a_2 x_2 + \cdots + a_n x_n \geq b$$

$$x_1, x_2, \ldots, x_n \geq 0$$

(a) If the feasible region of (1) is bounded, then an optimal solution is attained at a corner point of this feasible region.
(b) If the feasible region is unbounded, then an optimal solution may not exist; however, if an optimal solution exists, it is attained at a corner point of (1).

To prove Theorem 1 we need some additional notation and some preliminary results. First, we adopt functional notation to describe the objective function. We denote

$$z = c_1 x_1 + c_2 x_2 + \cdots + c_n x_n$$

by the function $f: R^n \to R^1$ defined by

$$f(x_1, x_2, \ldots, x_n) = c_1 x_1 + c_2 x_2 + \cdots + c_n x_n \qquad (2)$$

Note that if $\mathbf{p} = (p_1, p_2, \ldots, p_n)$ is a point in R^n, then

$$f(\mathbf{p}) = f(p_1, p_2, \ldots, p_n) = c_1 p_1 + c_2 p_2 + \cdots + c_n p_n$$

Definition. A *polyhedron* is the intersection of a finite number of half-spaces and/or hyperplanes. Points that lie in the polyhedron and on one or more of the half-spaces or hyperplanes defining the polyhedron are called *boundary points*. Points that lie in the polyhedron but are not boundary points are called *interior points*.

Note that it follows from Theorem 1, Section 2.2, that a polyhedron is convex, and that the feasible region of an LP problem is a polyhedron.

EXAMPLE 1

Figure 1 depicts the polyhedron that is the intersection of the half-spaces

$$x_1 + x_2 \leq 4$$
$$5x_1 + 3x_2 \leq 15$$
$$3x_1 + x_2 \geq 6$$

$$x_1, x_2 \geq 0$$

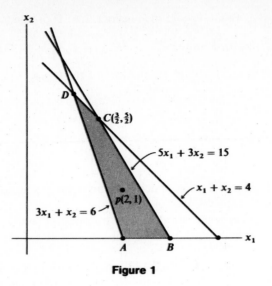

Figure 1

The points A, B, C, and D in Figure 1 are corner points. The points on the segments from A to B, B to C, C to D, and D to A are boundary points. The point P with coordinates $(2, 1)$ is an example of an interior point.   ◆

EXAMPLE 2

Let $\mathbf{p} = (1, 2, 3)$ and $\mathbf{q} = (2, 4, 6)$. Then,

$$2\mathbf{p} = (2, 4, 6), \qquad \frac{1}{2}\mathbf{q} = (1, 2, 3)$$

$$\mathbf{p} + \mathbf{q} = (3, 6, 9), \qquad 2\mathbf{p} + 3\mathbf{q} = (8, 16, 24)$$ ◆

Theorem 2.   Suppose that $f: R^n \to R^1$ is defined by (2). If $\mathbf{p} = (p_1, p_2, \ldots, p_n)$ and $\mathbf{q} = (q_1, q_2, \ldots, q_n)$ are two points in R^n, and if

$$\mathbf{r} = (1 - t)\mathbf{p} + t\mathbf{q}$$

is any point on the line segment joining $\mathbf{p}$ and $\mathbf{q}$, then $f(\mathbf{r})$ is between $f(\mathbf{p})$ and $f(\mathbf{q})$. That is, if $f(\mathbf{p}) \leq f(\mathbf{q})$, then $f(\mathbf{p}) \leq f(\mathbf{r}) \leq f(\mathbf{q})$; or, if $f(\mathbf{q}) \leq f(\mathbf{p})$, then $f(\mathbf{q}) \leq f(\mathbf{r}) \leq f(\mathbf{p})$.

PROOF.   Suppose that $f(\mathbf{p}) \leq f(\mathbf{q})$. (You can prove the case $f(\mathbf{q}) \leq f(\mathbf{p})$ by interchanging $\mathbf{p}$ and $\mathbf{q}$ in the following argument.) We first observe that

$$f(\mathbf{r}) = (1 - t)f(\mathbf{p}) + tf(\mathbf{q}) \tag{3}$$

To see this, note that

$$f(\mathbf{r}) = f[(1 - t)\mathbf{p} + t\mathbf{q}]$$
$$+ f[(1 - t)p_1 + tq_1, (1 - t)p_2 + tq_2, \ldots, (1 - t)p_n + tq_n]$$
$$= c_1[(1 - t)p_1 + tq_1] + c_2[(1 - t)p_2 + tq_2]$$
$$+ \cdots + c_n[(1 - t)p_n + tq_n]$$
$$= (1 - t)(c_1p_1 + c_2p_2 + \cdots + c_np_n)$$
$$+ t(c_1q_1 + c_2q_2 + \cdots + c_nq_n)$$
$$= (1 - t)f(\mathbf{p}) + tf(\mathbf{q})$$

Since $f(\mathbf{p}) \le f(\mathbf{q})$, we have from (3) and this calculation

$$f(\mathbf{p}) = (1 - t)f(\mathbf{p}) + tf(\mathbf{p})$$
$$\le (1 - t)f(\mathbf{p}) + tf(\mathbf{q}) = f(\mathbf{r})$$

and

$$f(\mathbf{r}) = (1 - t)f(\mathbf{p}) + tf(\mathbf{q})$$
$$\le (1 - t)f(\mathbf{q}) + tf(\mathbf{q})$$
$$= f(\mathbf{q})$$

from which it follows that

$$f(\mathbf{p}) \le f(\mathbf{r}) \le f(\mathbf{q}) \qquad \blacksquare$$

We will first establish the Corner Point Theorem in the case that $n = 2$ and then give a proof that is valid for all n but only for bounded regions; however, for this proof, we will have to assume a result (Theorem 3) that will not be proven.

PROOF OF THEOREM 1 ($n = 2$). Let K be the feasible region for (1).

(a) Suppose K is bounded.

Let $f(x_1, x_2) = c_1x_1 + c_2x_2$ denote the objective function, and suppose that the corner points of the feasible region K are $\mathbf{k}_1, \mathbf{k}_2, \ldots, \mathbf{k}_m$. We assume (without loss of generality) that these corner points have been labeled so that

$$f(\mathbf{k}_1) \le f(\mathbf{k}_i) \le f(\mathbf{k}_m), \qquad 2 \le i \le m - 1 \tag{4}$$

Let $\mathbf{p}$ be an arbitrary point in the feasible region K. We will show that

$$f(\mathbf{k}_1) \le f(\mathbf{p}) \le f(\mathbf{k}_m) \tag{5}$$

Since $\mathbf{p}$ is an arbitrary point, it follows from (5) that the maximal value of $z = f(x_1, x_2)$ occurs at the corner point $\mathbf{k}_m$, and the minimal value occurs at the corner point $\mathbf{k}_1$.

There are two cases to consider: (i) $\mathbf{p}$ is on the boundary of K and (ii) $\mathbf{p}$ is an interior point of K.

CASE I.   If $\mathbf{p}$ is a corner point, then $\mathbf{p} = \mathbf{k}_i$ for some i. By (4),

$$f(\mathbf{k}_1) \le f(\mathbf{p}) \le f(\mathbf{k}_m)$$

as we need to show. If $\mathbf{p}$ is on the boundary but is not a corner point, then $\mathbf{p}$ is an interior point of a line segment joining a corner point $\mathbf{k}_u$ and a corner point $\mathbf{k}_v$. If we label these corner points so that $f(\mathbf{k}_u) \le f(\mathbf{k}_v)$, then, by Theorem 2,

$$f(\mathbf{k}_u) \le f(\mathbf{p}) \le f(\mathbf{k}_v)$$

By (4), it follows that

$$f(\mathbf{k}_1) \le f(\mathbf{k}_u) \le f(\mathbf{p}) \le f(\mathbf{k}_v) \le f(\mathbf{k}_m)$$

as we needed to show.

CASE II.   Suppose that $\mathbf{p}$ is an interior point (i.e., does not lie on the boundary) of K. Let L be the line segment joining $\mathbf{k}_1$ to $\mathbf{p}$ and meeting the boundary of K at the point $\mathbf{q}$. (This is where we use the assumption that K is bounded because, otherwise, L need not meet the boundary at a point $\mathbf{q}$.) Line segment L is shown in Figure 2.

Suppose (without loss of generality) that $f(\mathbf{k}_u) \le f(\mathbf{k}_v)$; then by Theorem 2

$$f(\mathbf{k}_u) \le f(\mathbf{q}) \le f(\mathbf{k}_v)$$

By (4), $f(\mathbf{k}_1) \le f(\mathbf{k}_u)$, and so $f(\mathbf{k}_1) \le f(\mathbf{q})$. It follows by Theorem 2 that

$$f(\mathbf{k}_1) \le f(\mathbf{p}) \le f(\mathbf{q})$$

and hence that

$$f(\mathbf{k}_1) \le f(\mathbf{p}) \le f(\mathbf{q}) \le f(\mathbf{k}_v) \le f(\mathbf{k}_m)$$

Thus,

$$f(\mathbf{k}_1) \le f(\mathbf{p}) \le f(\mathbf{k}_m)$$

and this concludes the proof of Part (a).

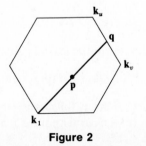

Figure 2

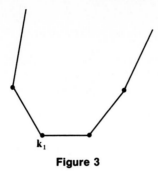

Figure 3

(b) Suppose K is unbounded. Suppose the feasible region K is as in Figure 3.

Assume that we want to maximize the objective function f. (In Exercise 2 you are asked to modify the following proof when we want to minimize the objective function.)

As before, suppose the corner points are labeled so that

$$f(\mathbf{k}_1) \le f(\mathbf{k}_i) \le f(\mathbf{k}_m), \qquad 2 \le i \le m - 1$$

Let $\mathbf{r}$ be a point of K at which an optimal solution of (1) exists. Suppose $\mathbf{r}$ is not a corner point of K (if it is a corner point, there is nothing more to prove). If $\mathbf{r}$ lies on the boundary of K, let $\mathbf{q}$ be a point on the boundary of K so that $\mathbf{r}$ lies between $\mathbf{q}$ and a corner point $\mathbf{k}_u$ of K as indicated in Figure 4a. If $\mathbf{r}$ lies in the interior of K, choose $\mathbf{q}$ in K so that $\mathbf{r}$ lies between $\mathbf{q}$ and a corner point $\mathbf{k}_u$ of K as indicated in Figure 4b.

We show that in either case $f(\mathbf{r}) = f(\mathbf{k}_u)$. From this it follows from (4) that $f(\mathbf{k}_1) \le f(\mathbf{r}) \le f(\mathbf{k}_m)$; and thus, since $f(\mathbf{r})$ is maximal, it must be the case that $f(\mathbf{r}) = f(\mathbf{k}_m)$.

By (3),

$$f(\mathbf{r}) = (1 - t)f(\mathbf{k}_u) + tf(\mathbf{q}) \tag{6}$$

where $0 < t < 1$ (since $\mathbf{r} \ne \mathbf{k}_u$ and $\mathbf{r} \ne \mathbf{q}$). Moreover, by Theorem 2, either

$$f(\mathbf{q}) \le f(\mathbf{r}) \le f(\mathbf{k}_u) \tag{7}$$

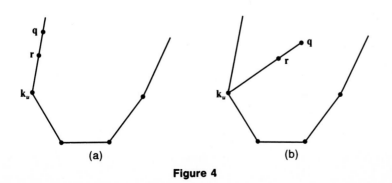

(a)   (b)

Figure 4

or

$$f(\mathbf{k}_u) \leq f(\mathbf{r}) \leq f(\mathbf{q}) \tag{8}$$

If (7) holds, then, since $f(\mathbf{r})$ is a maximal value of the objective function, it follows that $f(\mathbf{r}) = f(\mathbf{k}_u) = f(\mathbf{k}_m)$, and we are done. If (8) holds, then $f(\mathbf{r}) = f(\mathbf{q})$. Substituting $f(\mathbf{r})$ for $f(\mathbf{q})$ in (6), we obtain

$$f(\mathbf{r}) = (1 - t)f(\mathbf{k}_u) + tf(\mathbf{r}) \tag{9}$$

Rearranging (9) by moving $tf(\mathbf{r})$ to the left side and factoring gives

$$(1 - t)f(\mathbf{r}) = (1 - t)f(\mathbf{k}_u), \qquad \text{for some } t, \quad 0 < t < 1$$

Since $1 - t \neq 0$, we have $f(\mathbf{r}) = f(\mathbf{k}_u)$. Thus, in this case as well, the maximum occurs at a corner point, and this completes the proof of Theorem 1 (for $n = 2$). ∎

If $n \geq 2$, we can base a proof of Theorem 1, for the case in which the feasible region is bounded, on Theorem 3. We will not prove Theorem 1 for unbounded feasible regions, nor will we prove Theorem 3. Proofs of these theorems can be found in Tiel [14]. First we give a definition.

Definition.   Let $\mathbf{k}_1, \mathbf{k}_2, \ldots, \mathbf{k}_m$ be points in R^n. A *convex combination* of $\mathbf{k}_1, \mathbf{k}_2, \ldots, \mathbf{k}_m$ is any point $\mathbf{p}$ that can be written as

$$\mathbf{p} = a_1\mathbf{k}_1 + a_2\mathbf{k}_2 + \cdots + a_m\mathbf{k}_m$$

where $a_1, a_2, \ldots, a_m$ are nonnegative numbers such that

$$a_1 + a_2 + \cdots + a_m = 1$$

EXAMPLE 3

The point $(\frac{21}{8}, \frac{7}{4})$ is a convex combination of the points $(1, 2)$, $(0, 1)$, $(3, 1)$, and $(4, 2)$ because

$$\left(\frac{21}{8}, \frac{7}{4}\right) = \frac{1}{4}(1, 2) + \frac{1}{8}(0, 1) + \frac{1}{8}(3, 1) + \frac{1}{2}(4, 2) \qquad \blacklozenge$$

Theorem 3.   Suppose that K is a bounded polyhedron with corner points $\mathbf{k}_1, \mathbf{k}_2, \ldots, \mathbf{k}_m$. Then any point $\mathbf{p}$ in K is a convex combination of $\mathbf{k}_1, \mathbf{k}_2, \ldots, \mathbf{k}_m$.

In Exercise 11 you are asked to establish the following result.

Theorem 4.   Suppose that $f: R^n \to R^1$ is defined by

$$f(x_1, x_2, \ldots, x_n) = c_1x_1 + c_2x_2 + \cdots + c_nx_n$$

and let $\mathbf{k}_1, \mathbf{k}_2, \ldots, \mathbf{k}_m$ be any m points in R^n. Then for any constants $b_1, b_2, \ldots, b_m$,

$$f(b_1 \mathbf{k}_1 + b_2 \mathbf{k}_2 + \cdots + b_m \mathbf{k}_m) = b_1 f(\mathbf{k}_1) + b_2 f(\mathbf{k}_2) + \cdots + b_m f(\mathbf{k}_m)$$

We now use Theorems 3 and 4 to establish Theorem 1.

PROOF OF THEOREM 1 (FOR A BOUNDED FEASIBLE REGION K IN R^n). Suppose that $\mathbf{k}_1, \mathbf{k}_2, \ldots, \mathbf{k}_m$ are the corner points of the feasible region K of (1), and, as before, assume that these points have been labeled so that

$$f(\mathbf{k}_1) \le f(\mathbf{k}_i) \le f(\mathbf{k}_m)$$

for each i. Let $\mathbf{p}$ be any point in K. Then by Theorem 3 there are nonnegative constants $a_1, a_2, \ldots, a_m$ whose sum is 1, and so that

$$\mathbf{p} = a_1 \mathbf{k}_1 + a_2 \mathbf{k}_2 + \cdots + a_m \mathbf{k}_m$$

By Theorem 4,

$$\begin{aligned} f(\mathbf{p}) &= f(a_1 \mathbf{k}_1 + a_2 \mathbf{k}_2 + \cdots + a_m \mathbf{k}_m) \\ &= a_1 f(\mathbf{k}_1) + a_2 f(\mathbf{k}_2) + \cdots + a_m f(\mathbf{k}_m) \end{aligned} \qquad (10)$$

Since $a_1 + a_2 + \cdots + a_m = 1$, it follows from (4) and (10) that

$$\begin{aligned} f(\mathbf{k}_1) &= (a_1 + a_2 + \cdots + a_m) f(\mathbf{k}_1) \\ &= a_1 f(\mathbf{k}_1) + a_2 f(\mathbf{k}_1) + \cdots + a_m f(\mathbf{k}_1) \\ &\le a_1 f(\mathbf{k}_1) + a_2 f(\mathbf{k}_2) + \cdots + a_m f(\mathbf{k}_m) = f(\mathbf{p}) \end{aligned}$$

and

$$\begin{aligned} f(\mathbf{p}) &= a_1 f(\mathbf{k}_1) + a_2 f(\mathbf{k}_2) + \cdots + a_m f(\mathbf{k}_m) \\ &\le a_1 f(\mathbf{k}_m) + a_2 f(\mathbf{k}_m) + \cdots + a_m f(\mathbf{k}_m) \\ &= (a_1 + a_2 + \cdots + a_m) f(\mathbf{k}_m) = f(\mathbf{k}_m) \end{aligned}$$

from which it follows that

$$f(\mathbf{k}_1) \le f(\mathbf{p}) \le f(\mathbf{k}_m) \qquad (11)$$

for any point $\mathbf{p}$ in K. From (11), we conclude that the objective function $z = f(x_1, x_2, \ldots, x_n)$ takes on a maximum value at the corner point $\mathbf{k}_m$ and a minimum value at the corner point $\mathbf{k}_1$. ∎

Exercises

1. Suppose an LP problem with objective function

$$z = f(x_1, x_2) = 2x_1 + 3x_2$$

has the feasible region given in Figure 5.

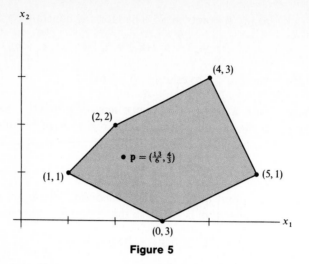

Figure 5

(a) Identify the points $\mathbf{k}_1$ and $\mathbf{k}_m$ (here $m = 5$) used in the proof of Theorem 1 for the case $n = 2$.

(b) For the point $\mathbf{p}$ in Figure 5, find the coordinates of the point labeled $\mathbf{q}$ in Figure 2 that was used in the proof of Theorem 1.

(c) By evaluating the objective function at the various points, verify that

$$f(\mathbf{k}_1) \leq f(\mathbf{p}) \leq f(\mathbf{q}) \leq f(\mathbf{k}_5)$$

2. Modify the proof of Theorem 1 (part b) for the problem of minimizing the objective function.

3. Let $f(x_1, x_2, x_3) = 2x_1 - x_2 + 3x_3$, $\mathbf{p}_1 = (1, 2, 3)$, $\mathbf{p}_2 = (2, 2, 4)$, $\mathbf{p}_3 = (1, 0, 2)$, $b_1 = 2$, $b_2 = 4$, and $b_3 = 5$.
   Verify that

$$f(b_1\mathbf{p}_1 + b_2\mathbf{p}_2 + b_3\mathbf{p}_3) = b_1 f(\mathbf{p}_1) + b_2 f(\mathbf{p}_2) + b_3 f(\mathbf{p}_3)$$

by evaluating both sides of the equation.

Exercises 4–9 illustrate the proof of Theorem 3 for some special cases.

4. Let $\mathbf{k}_1, \mathbf{k}_2$, and $\mathbf{k}_3$ be the vertices of a triangle T in R^2 and let $\mathbf{p}$ be a point in the interior of T. Join $\mathbf{k}_1$ to $\mathbf{p}$ by a line segment, and extend this segment to intersect the boundary of the triangle in point $\mathbf{q}$, as shown in Figure 6.

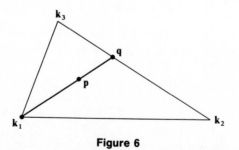

Figure 6

Observe that $\mathbf{q} = (1 - t)\mathbf{k}_2 + t\mathbf{k}_3$ for some t, $0 \leq t \leq 1$, and that $\mathbf{p} = (1 - s)\mathbf{k}_1 + s\mathbf{q}$ for some s, $0 \leq s \leq 1$. Show that $\mathbf{p} = t_1\mathbf{k}_1 + t_2\mathbf{k}_2 + t_3\mathbf{k}_3$ for some $t_1, t_2, t_3 \geq 0$, and $t_1 + t_2 + t_3 = 1$.

5. Regard the points k_1, k_2, k_3 as vertices of a triangle in R^2 and p as a point in the interior or on the boundary of the triangle. Express (if possible) p as a convex combination of k_1, k_2, and k_3.
   (a) $p = (\frac{3}{2}, \frac{5}{2})$, $k_1 = (1, 2)$, $k_2 = (2, 3)$, and $k_3 = (0, 4)$.
   (b) $p = (2, 2)$, $k_1 = (1, 1)$, $k_2 = (3, 4)$, and $k_3 = (4, 0)$.
   (c) $p = (\frac{5}{2}, \frac{5}{2})$, $k_1 = (1, 2)$, $k_2 = (3, 4)$, and $k_3 = (1, 1)$.

6. Let k_1, k_2, k_3, and k_4 be the vertices of a quadrilateral in R^2. Join k_2 to k_4 by a segment as shown in Figure 7.

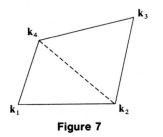

Figure 7

Let p be any point in the interior of the quadrilateral. Then p is an interior point of one of the triangles or is on the segment from k_2 to k_4. Use Exercise 4 to conclude that

$$p = t_1 k_1 + t_2 k_2 + t_3 k_3 + t_4 k_4$$

for some $t_i \geq 0$, $1 \leq i \leq 4$ and $t_1 + t_2 + t_3 + t_4 = 1$.

7. The points $(1, 1), (1, 2), (2, 2)$, and $(2, 1)$ are corner points of a rectangle in R^2.
   (a) Express $(\frac{3}{2}, \frac{3}{2})$ as a convex combination of these points.
   (b) Express $(\frac{5}{4}, \frac{9}{8})$ as a convex combination of these points.

8. The five-sided polyhedron K shown in Figure 8 has been subdivided into triangles. Observe that any point in K can be written as a convex combination of three or fewer corner points of K. Conclude that any point in K can be written as a convex combination of the five corner points.

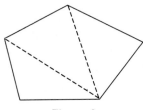

Figure 8

9. Use the ideas developed in Exercises 4–8 to prove Theorem 3 for bounded polyhedra in R^2.

10. Consider the polyhedron with corner points $(1, 1), (2, 3), (2, 5)$, and $(1, 4)$. Observe that $p = (\frac{5}{4}, \frac{13}{4})$ can be expressed as the convex combination

$$p = \frac{1}{4}(1, 1) + \frac{1}{8}(2, 3) + \frac{1}{8}(2, 5) + \frac{1}{2}(1, 4)$$

Find a different way of writing p as a convex combination of these four corner points. This example illustrates that there can be more than one way of representing a point as a convex combination of corner points of a polyhedron.

11. Prove Theorem 4. (Suggestion: For $1 \leq i \leq m$, write

$$\mathbf{k}_i = (x_{i1}, x_{i2}, \ldots, x_{in})$$

    and substitute this expression for each occurrence of $\mathbf{k}_i$.)

12. Let $f(\mathbf{x}) = c_1 x_1 + c_2 x_2 + \cdots + c_n x_n$ be the objective function for an LP problem whose feasible set is a polyhedron K. Show that if $\mathbf{p}$ and $\mathbf{q}$ are two points in K and $f(\mathbf{p}) = f(\mathbf{q})$, then $f(\mathbf{r}) = f(\mathbf{p}) = f(\mathbf{q})$ for all points $\mathbf{r}$ on the line segment joining $\mathbf{p}$ to $\mathbf{q}$.

13. This problem is for students who have studied calculus. Recall that a necessary condition for a differentiable function to assume a maximum or minimum at an interior point $\mathbf{p}$ of its domain is that its partial derivatives at $\mathbf{p}$ are all zero. For an objective function

$$f(x_1, x_2, \ldots, x_n) = c_1 x_1 + c_2 x_2 + \cdots + c_n x_n$$

    show that this can occur only when $c_i = 0$, $1 \leq i \leq n$. Conclude that an objective function for an LP problem cannot assume its optimal value at an interior point unless the function is identically zero.

2.4 ALGEBRAIC PRELIMINARIES

Thus far we have used a geometric approach to solve certain LP problems. We have observed, however, that this procedure is limited to problems with two or three variables. The *simplex method*, which we describe in the next chapter, is essentially algebraic in nature and is more efficient than its geometric counterpart. In this section we discuss briefly some underlying algebraic ideas used in the simplex method.

Recall that a *linear equation* in n variables $x_1, x_2, \ldots, x_n$ is an equation of the form

$$a_1 x_1 + a_2 x_2 + \cdots + a_n x_n = b \qquad (1)$$

where the coefficients b and a_i, $1 \leq i \leq n$, are constants. A solution to (1) is an n-tuple $(s_1, s_2, \ldots, s_n)$ of real numbers with the property that

$$a_1 s_1 + a_2 s_2 + \cdots + a_n s_n = b$$

EXAMPLE 1

The equation

$$-2x_1 + x_2 + \frac{1}{2}x_3 = 7 \qquad (2)$$

is a linear equation in the variables x_1, x_2, x_3. The 3-tuple $(1, 6, 6)$ is a solution of (2) since $(-2)(1) + (1)(6) + (\frac{1}{2})6 = 7$. Note that $(0, 2, 10)$ is also a solution of (2); in fact, there is an infinite number of solutions to (2). ◆

Definition. A *system of linear equations* is a family of m linear equations in n variables:

$$a_{11}x_1 + a_{12}x_2 + \cdots + a_{1n}x_n = b_1$$
$$a_{21}x_1 + a_{22}x_2 + \cdots + a_{2n}x_n = b_2$$
$$\vdots$$
$$a_{m1}x_1 + a_{m2}x_2 + \cdots + a_{mn}x_n = b_m$$

(3)

A solution to (3) is an n-tuple

$$(s_1, s_2, \ldots, s_n)$$

of real numbers with the property that if

$$x_1 = s_1, \qquad x_2 = s_2, \qquad \ldots, \qquad x_n = s_n$$

then all of the m equations of (3) are valid.

EXAMPLE 2

The 3-tuple $(2, -3, -4)$ is a solution of the system of linear equations (two equations in three variables)

$$x_1 - 2x_2 + x_3 = 4$$
$$-x_1 + x_2 - 2x_3 = 3$$

(4)

since

$$x_1 = 2, \qquad x_2 = -3 \qquad \text{and} \qquad x_3 = -4$$

satisfies both of these equations. Note that the 3-tuple $(11, 0, -7)$ is also a solution of (4); and, in fact, it is easy to see that there is an infinite number of solutions of (4). The 3-tuple $(2, 3, 8)$ is not a solution of (4) because, though this 3-tuple satisfies the first equation of (4), it is not a solution of the second equation. ◆

In developing the simplex method we will be interested especially in systems of linear equations of the following form.

Definition. A system of linear equations with m equations and n variables

$$a_{11}x_1 + a_{12}x_2 + \cdots + a_{1n}x_n = b_1$$
$$a_{21}x_1 + a_{22}x_2 + \cdots + a_{2n}x_n = b_2$$
$$\vdots$$
$$a_{m1}x_1 + a_{m2}x_2 + \cdots + a_{mn}x_n = b_m$$

(5)

is in *canonical form* (or is a *canonical* system of linear equations) if

(a) $m \leq n$ and

(b) there are m "distinguished" variables, called *basic variables*, such that each basic variable x_i has coefficient 1 in exactly one equation of (5) and coefficient 0 in the other equations, and such that each equation in (5) contains exactly one basic variable.

EXAMPLE 3

The systems of linear equations (6), (7), and (8) are in canonical form with the indicated basic variables:

$$\left.\begin{array}{l} 2x_1 + 0x_2 + x_3 + 3x_4 + 0x_5 = 6 \\ -x_1 + x_2 + 0x_3 + 0x_4 + 0x_5 = 2 \\ 3x_1 + 0x_2 + 0x_3 + 5x_4 + x_5 = -3 \end{array}\right\} \quad \text{(Basic variables: } x_2, x_3, x_5) \quad (6)$$

$$\left.\begin{array}{l} x_1 - 3x_2 + 0x_3 = \dfrac{3}{2} \\ 0x_1 + 2x_2 + x_3 = 0 \end{array}\right\} \quad \text{(Basic variables: } x_1, x_2) \quad (7)$$

$$\left.\begin{array}{l} x_1 - 2x_2 + x_3 + 0x_4 = 6 \\ 0x_1 + 3x_2 + 0x_3 + x_4 = 4 \end{array}\right\} \quad \text{(Basic variables: either } x_1, x_4 \text{ or } x_3, x_4) \quad (8)$$

For our purposes, the significance of a canonical system of linear equations is that we can readily obtain a solution to it. If we set the nonbasic variables in the canonical system to zero, then the right-hand side of the system provides the solution values for the basic variables. For instance, if in (6) we set the nonbasic variables x_1 and x_4 equal to 0, then the values $x_2 = 2$, $x_3 = 6$, and $x_5 = -3$ give us the solution $(0, 2, 6, 0, -3)$ to (6). Similarly, by setting the nonbasic variable x_2 in (7) equal to 0, we find from the right-hand side of (7) that $(\frac{3}{2}, 0, 0)$ is a solution to (7). In (8) we have a choice of the solutions $(6, 0, 0, 4)$ or $(0, 0, 6, 4)$. Solutions obtained in this manner are called basic solutions.

Definition. A *basic solution* of a canonical system of linear equations is any solution obtained by

(a) setting each nonbasic variable equal to zero, and

(b) setting each basic variable equal to the right-hand side of the equation in which the basic variable appears with coefficient 1.

Definition. Two systems of linear equations in n variables are *equivalent* if they have the same set of solutions.

As we will see, the simplex method involves a systematic consideration of successive, equivalent systems of linear equations, each of which is in canonical form. The corresponding basic solutions provide increasingly favorable values of the objective function, leading eventually to an optimal solution (unless the feasible solutions are unbounded).

You are asked to show in Exercise 15 that applying the following two *row operations* to a given system of linear equations yields an equivalent system of equations, that is, one with the same solution set as the original system of equations.

1. Multiply any equation in (3) by a nonzero constant k; that is, replace an equation

$$a_{s1}x_1 + a_{s2}x_2 + \cdots + a_{sn}x_n = b_s$$

with the equation

$$ka_{s1} + ka_{s2}x_2 + \cdots + ka_{sn}x_n = kb_s$$

2. Add a multiple k of one equation

$$a_{r1}x_1 + a_{r2}x_2 + \cdots + a_{rn}x_n = b_r$$

to another equation

$$a_{s1}x_1 + a_{s2}x_2 + \cdots + a_{sn}x_n = b_s \tag{9}$$

and replace (9) with the resulting equation

$$(ka_{r1} + a_{s1})x_1 + (ka_{r2} + a_{s2})x_2 + \cdots + (ka_{rn} + a_{sn})x_n = kb_r + b_s$$

EXAMPLE 4

We use row operations to show that the systems of linear equations

$$
\begin{aligned}
x_1 - 2x_2 + x_3 + \tfrac{1}{2}x_4 &= 6 \\
-2x_1 + x_2 - 3x_3 &= 9 \\
2x_1 + 2x_2 + 4x_3 - 2x_4 &= -5
\end{aligned}
\tag{10}
$$

and

$$
\begin{aligned}
x_1 + \tfrac{5}{3}x_3 &= -\tfrac{73}{6} \\
x_2 + \tfrac{1}{3}x_3 &= -\tfrac{46}{3} \\
x_4 &= -25
\end{aligned}
\tag{11}
$$

are equivalent. Note that (11) is in canonical form with basic variables x_1, x_2, and x_4. We describe in detail the sequence of row operations used in transforming (10) into (11). We make x_1 a basic variable in the first equation by multiplying the first equation of (10) by 2 and adding the result to the second equation of (10), and then multiplying the first equation of (10) by -2 and adding the result to the third equation. This gives us the equivalent system of linear equations

$$
\begin{aligned}
x_1 - 2x_2 + x_3 + \frac{1}{2}x_4 &= 6 \\
- 3x_2 - x_3 + x_4 &= 21 \\
6x_2 + 2x_3 - 3x_4 &= -17
\end{aligned}
\tag{12}
$$

with basic variable x_1. Next we make x_2 a basic variable for the second equation of (12). Multiplying this equation by $-\frac{1}{3}$, we obtain the equivalent system

$$
\begin{aligned}
x_1 - 2x_2 + x_3 + \frac{1}{2}x_4 &= 6 \\
x_2 + \frac{1}{3}x_3 - \frac{1}{3}x_4 &= -7 \\
6x_2 + 2x_3 - 3x_4 &= -17
\end{aligned}
\tag{13}
$$

Multiplying the second equation of (13) by 2 and adding the result to the first equation, and then multiplying the second equation of (13) by -6 and adding the result to the third equation, gives the equivalent system of linear equations

$$
\begin{aligned}
x_1 + \frac{5}{3}x_3 - \frac{1}{6}x_4 &= -8 \\
x_2 + \frac{1}{3}x_3 - \frac{1}{3}x_4 &= -7 \\
-x_4 &= 25
\end{aligned}
\tag{14}
$$

for which x_1 and x_2 are basic variables.

Finally, we convert x_4 to a basic variable in (14) by first multiplying the third equation by -1 to obtain

$$
\begin{aligned}
x_1 + \frac{5}{3}x_3 - \frac{1}{6}x_4 &= -8 \\
x_2 + \frac{1}{3}x_3 - \frac{1}{3}x_4 &= -7 \\
x_4 &= -25
\end{aligned}
\tag{15}
$$

We then multiply the third equation of (15) by $\frac{1}{6}$ and add the result to the first equation of (15), and then multiply the third equation by $\frac{1}{3}$ and add the result to the second equation of (15). This gives us the canonical system (11).   ◆

In Example 4 we converted the system of equations (10) into an equivalent canonical system with basic variables x_1, x_2, and x_4. To do this we made repeated use of the following steps:

1. Select an equation E and a variable in E whose coefficient a is not zero. (This is the variable that becomes basic in E.)

2. Multiply the equation E by $\frac{1}{a}$, and replace E with the resulting equation.

3. Add appropriate multiples of the equation resulting from step 2 to the remaining equations so that the coefficients of the basic variable in these equations are 0.

We often refer to steps 1, 2, and 3 as a *pivoting operation*, and we call the original coefficient a of the potential basic variable the *pivot element*.

Matrix notation provides a convenient way to carry out pivot operations. We discuss matrices in more detail in the next chapter; for our present purpose we simply recall that a matrix is a rectangular array of numbers. Since pivot operations involve only the coefficients of the variables, we can replace a system of equations with matrices comprised of these coefficients. The position of the entries in the matrix corresponds to the position of the coefficient in the linear system. To illustrate this idea, we use matrices to parallel the steps employed in Example 4. In each of the following matrices, the circled entry represents the pivot element.

We begin by forming a matrix from the coefficients of the original system of equations (10), with a column at the right that contains the right-hand sides of (10).

$$\begin{bmatrix} \boxed{1} & -2 & 1 & \frac{1}{2} & 6 \\ -2 & 1 & -3 & 0 & 9 \\ 2 & 2 & 4 & -2 & -5 \end{bmatrix}$$

Pivoting on the circled entry, we obtain the matrix corresponding to (15):

$$\begin{bmatrix} 1 & -2 & 1 & \frac{1}{2} & 6 \\ 0 & -3 & -1 & 1 & 12 \\ 0 & 6 & 2 & -3 & -17 \end{bmatrix} \tag{16}$$

Next, we pivot on the circled entry in (16) to obtain the matrix corresponding to (14):

$$\begin{bmatrix} 1 & 0 & \frac{5}{3} & -\frac{1}{6} & -8 \\ 0 & 1 & \frac{1}{3} & -\frac{1}{3} & -7 \\ 0 & 0 & 0 & -1 & 25 \end{bmatrix} \tag{17}$$

Finally, pivoting on the circled entry in (17) yields the canonical matrix

$$\begin{bmatrix} 1 & 0 & \frac{5}{3} & 0 & -\frac{73}{6} \\ 0 & 1 & \frac{1}{3} & 0 & -\frac{46}{3} \\ 0 & 0 & 0 & 1 & 25 \end{bmatrix}$$

which corresponds to the canonical system of linear equations (11).

EXAMPLE 5

We use matrix notation to find a system of linear equations that is equivalent to (10) and is in canonical form with basic variables x_2, x_3, and x_4. In each of the following matrices we circle the pivot elements. We begin with the matrix

$$\begin{bmatrix} 1 & -2 & 1 & \frac{1}{2} & 6 \\ -2 & ① & -3 & 0 & 9 \\ 2 & 2 & 4 & -2 & -5 \end{bmatrix} \tag{18}$$

Pivoting on the circled entry we obtain

$$\begin{bmatrix} -3 & 0 & ⊖5 & \frac{1}{2} & 24 \\ -2 & 1 & -3 & 0 & 9 \\ 6 & 0 & 10 & -2 & -23 \end{bmatrix} \tag{19}$$

and we see that x_2 has become a basic variable. Pivoting on the circled entry in (19) we obtain

$$\begin{bmatrix} \frac{3}{5} & 0 & 1 & -\frac{1}{10} & -\frac{24}{5} \\ -\frac{1}{5} & 1 & 0 & -\frac{3}{10} & -\frac{27}{5} \\ 0 & 0 & 0 & ⊖1 & 25 \end{bmatrix} \tag{20}$$

As a result of this pivoting operation, x_3 has become a basic variable. (Note that we could also have made x_3 a basic variable by pivoting on the entry 10 in the third row and third column of (19) instead of pivoting on the -5.)

Pivoting on the circled entry in (20), we obtain the canonical matrix

$$\begin{bmatrix} \frac{3}{5} & 0 & 1 & 0 & -\frac{73}{10} \\ -\frac{1}{5} & 1 & 0 & 0 & -\frac{129}{10} \\ 0 & 0 & 0 & 1 & -25 \end{bmatrix} \tag{21}$$

with basic variables x_2, x_3, and x_4. We see from (21) that the 4-tuple $(0, -\frac{129}{10}, -\frac{73}{10}, -25)$ is a solution of (10). ♦

As you are asked to show in Exercise 18, there are systems of linear equations that cannot be converted into canonical form.

We point out that it will be important for you to "read" certain matrices as systems of equations. For instance, you should be able to interpret the second row of (21) as the equation

$$-\frac{1}{5}x_1 + x_2 = -\frac{129}{10}$$

and the third row as the equation

$$x_4 = -25$$

In the sequel we will work with linear equations and with linear inequalities. A *linear inequality* in n variables is an expression of either the form

$$a_1 x_1 + a_2 x_2 + \cdots + a_n x_n \leq b \qquad (22)$$

or the form

$$a_1 x_1 + a_2 x_2 + \cdots + a_n x_n \geq b \qquad (23)$$

A *solution* to (22), or (23), is an n-tuple

$$(s_1, s_2, \ldots, s_n)$$

of real numbers with the property that

$$a_1 s_1 + a_2 s_2 + \cdots + a_n s_n \leq b$$

or

$$a_1 s_1 + a_2 s_2 + \cdots + a_n s_n \geq b$$

EXAMPLE 6

The expression

$$3x_1 - 2x_2 + 4x_3 - 2x_4 \leq 6 \qquad (24)$$

is a linear inequality in four variables. The 4-tuple

$$(1, 4, -2, -3)$$

is a solution of (24) since

$$3(1) - 2(4) + 4(-2) - 2(-3) \leq 6 \qquad \blacklozenge$$

A *system of linear inequalities* is a family of m linear inequalities in n variables. An n-tuple $(s_1, s_2, \ldots, s_n)$ is a *solution* to such a system if it satisfies each of the inequalities.

EXAMPLE 7

The 3-tuple $(1, -2, 4)$ is a solution to the system of linear inequalities

$$
\begin{aligned}
3x_1 - 2x_2 + x_3 &\leq 14 \\
x_1 + x_2 - x_3 &\geq -16 \\
x_2 + 3x_3 &\leq 10
\end{aligned}
$$

since $x_1 = 1$, $x_2 = -2$, $x_3 = 4$ satisfies each of the inequalities. $\qquad \blacklozenge$

Exercises

1. Which of the following equations are not linear and why?
   (a) $2x_1 + 5x_2 - 3x_3 = 5$
   (b) $3x^2 + 5x + 7y - 4t = 2$
   (c) $4xy + 2x - 5y = 3$

2. Which of the following equations are not linear and why?
   (a) $3/x + 2/y = 9$
   (b) $\pi x - y = 3$
   (c) $x_1 + 3\sqrt{x_2} + x_3 = 6$

3. Convert each of the following equations or inequalities to a form having all the variables on the left, all constants on the right, and each variable occurring only once.
   (a) $2x_1 + 3x_2 + 5 \le 4x_3 - 5x_4$
   (b) $2x - 3y = 5x - y + 3$
   (c) $x + y \le 2x - 3y$

4. Convert each of the following equations or inequalities to a form having a positive constant term on the right.
   (a) $2x_1 - 3x_2 \le -5$
   (b) $x_1 - 3x_2 + x_3 = -9$
   (c) $2x_1 - 2x_2 - 3x_3 > -3$

5. Find two solutions to each of the following:
   (a) $x + 2y - z \le 9$
   (b) $3x_1 + 5x_2 + 2x_3 = 30$

6. Find two solutions to each of the following:
   (a) $x_1 + 3x_2 - 2x_3 = 4$
   (b) $-x_1 + 3x_2 - 4x_3 + 5x_4 = 6$

7. Verify that $(2, 1, -3)$ is a solution to the system

$$x_1 + 3x_2 - 2x_3 = 11$$
$$-3x_1 + 5x_2 - 4x_3 = 11$$

8. Find two sets of values of (x_1, x_2, x_3) that satisfy

$$x_1 + x_2 + \ x_3 \le 6$$
$$2x_1 - x_2 - 4x_3 \le 2$$
$$x_i \ge 0, \quad 1 \le i \le 3$$

9. Rearrange the following system into canonical form, and identify the basic variables.

$$x_1 + 2x_2 = 20 - x_4 - x_5$$
$$2x_1 - 10 - x_2 = -x_3 + x_1 - x_6$$
$$-x_1 + 4x_2 - 3x_4 + x_7 = -2x_3 + 40$$

10. Rearrange the following system into canonical form, and identify the basic variables.

$$10 = x_1 + 3x_2 - 2x_3$$
$$x_5 + 2x_3 = 5x_4 + 2x_2 + 9$$
$$2x_2 - 2x_3 - 4 = -x_6 - 9x_4$$

11. Which of the following systems are in canonical form? For those that are in canonical form, identify the basic variables. Explain why the others are not in canonical form.

   (a) $x_1 - 2x_2 - 3x_3 - 2x_4 = \ 3$

   $\qquad x_1 - \ x_2 + 2x_3 + \ x_4 = 11$

   (b) $x - 2y + \ z - 3w - 2t = 1$

   $\qquad x - \ y + 2z + \ w + \ t = 2$

   (c) $x_1 + x_3 + x_5 = 1$

   $\qquad x_4 + x_3 + x_5 = 2$

   $\qquad x_2 + x_3 + x_6 = 3$

12. Follow the instructions given in Exercise 11 for the systems:

(a) $\quad x_1 + 2x_2 + 3x_3 + x_5 = 1$

$\quad 2x_1 + 3x_2 - x_3 + x_4 = 4$

$\quad x_1 - x_2 + x_3 + x_6 = 5$

$\quad 2x_1 + x_2 - 3x_3 + x_7 = 9$

(b) $x_1 + 2x_3 + 4x_6 = 9$

$\quad x_2 + x_4 + 3x_6 = 6$

$\quad x_3 + x_5 - 2x_6 = 4$

(c) $x_1 + x_2 + x_3 = 4$

$\quad x_1 - x_2 + x_4 = 5$

$\quad x_1 + 2x_2 = 7$

13. For each system from Exercise 11 that is in canonical form, find a basic solution.

14. For each system in Exercise 12 that is in canonical form, find a basic solution.

15. Show that the following two systems are equivalent by explaining why each solution to System 1 is a solution to System 2, and why each solution to System 2 is a solution to System 1.

System 1:

$$a_{11}x_1 + a_{12}x_2 + a_{13}x_3 = b_1$$
$$a_{21}x_1 + a_{22}x_2 + a_{23}x_3 = b_2$$
$$a_{31}x_1 + a_{32}x_2 + a_{33}x_3 = b_3$$

System 2:

$$a_{11}x_1 + a_{12}x_2 + a_{13}x_3 = b_1$$
$$(ka_{11} + a_{21})x_1 + (ka_{12} + a_{22})x_2 + (ka_{13} + a_{23})x_3 = kb_1 + b_2$$
$$a_{31}x_1 + a_{32}x_2 + a_{33}x_3 = b_3$$

16. For each of the following systems of linear equations, form a matrix representation of the system, perform pivots to reduce the matrix to canonical form, and write the system of equations corresponding to the canonical form.

(a) $x_1 + x_2 + 2x_3 + x_4 = 6$

$\quad 3x_2 + x_3 + 8x_4 = 3$

(b) $\quad\quad 3x_2 - 3x_3 = 15$

$\quad x_1 + x_2 + x_3 = 0$

$\quad 3x_1 + 5x_2 - 6x_3 = 4$

17. Follow the instructions given in Exercise 16 for the systems

(a) $\quad x_1 + x_2 - 2x_3 + 3x_4 = 2$

$\quad -2x_1 + x_3 = 2$

(b) $\quad x_1 + 2x_2 = 3$

$\quad -3x_2 + 4x_3 = 5$

$\quad 2x_1 + 2x_2 - x_3 = 9$

18. A system of linear equations is *inconsistent* if the system has no solution. The following system is obviously inconsistent:

$$x_1 + x_2 = 1$$
$$x_1 + x_2 = 2$$

Although it is not immediately obvious, the following system is also inconsistent:

$$x_1 + x_2 - 2x_3 = 7$$
$$3x_2 - x_3 = 15$$
$$-2x_1 + x_2 + 3x_3 = 2$$

Attempt to reduce this system to canonical form by pivoting. Describe in general how one detects inconsistency when a system is being reduced to canonical form.

19. A system of linear equations is *redundant* if one or more equations of the system is a linear combination of other equations of the system. The following system,

$$x_1 + x_2 + x_3 = 1$$
$$2x_1 + 3x_2 - x_3 = 2$$
$$4x_1 + 5x_2 + x_3 = 4$$

is redundant because the third equation equals 2 × (first equation) + (second equation). The following system is also redundant:

$$x_1 + 2x_2 \quad\quad + x_4 = 20$$
$$2x_1 + x_2 + x_3 \quad\quad = 10$$
$$-x_1 + 4x_2 - 2x_3 + 3x_4 = 40$$

Reduce this system to canonical form by pivoting. Describe in general how one detects redundancy when a system is reduced to canonical form.

2.5 SLACK VARIABLES AND CONSTRAINTS IN CANONICAL FORM

We will now convert the constraint inequalities of an LP problem into equalities in such a way that we can work with canonical systems of linear equations. We will see that each corner point of a feasible region corresponds to such a system; moreover, the basic solutions of these systems provide us with the values of the objective function at the various corner points.

To illustrate these ideas we begin with the LP problem

Maximize: $z = 3x_1 + 2x_2$

Subject to: $2x_1 - x_2 \leq 1$
$-3x_1 + 4x_2 \leq 13$ (1)
$x_1 + x_2 \leq 5$

$x_1, x_2 \geq 0$

Figure 1 shows the lines that make up the boundary of the (shaded) feasible region for this problem.

To apply the simplex method to this problem we first introduce *slack variables* into the inequalities in (1). We use these variables to convert the constraint

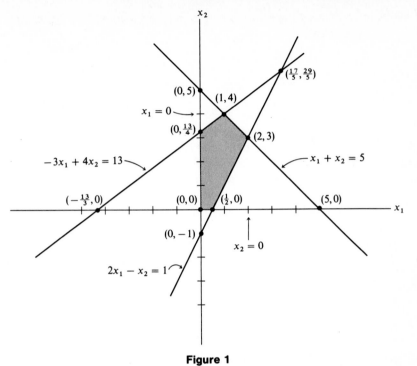

Figure 1

inequalities of (1) to equalities by, as the name suggests, taking up the slack in each inequality. For example, corresponding to the first inequality in (1),

$$2x_1 - x_2 \le 1 \tag{2}$$

we define a slack variable x_3 by

$$x_3 = 1 - (2x_1 - x_2) \tag{3}$$

or, equivalently,

$$2x_1 - x_2 + x_3 = 1$$

Note that it follows from (2) and (3) that $x_3 \ge 0$.

Similarly, corresponding to the second inequality in (1),

$$-3x_1 + 4x_2 \le 13 \tag{4}$$

we introduce the slack variable x_4 defined by

$$x_4 = 13 - (-3x_1 + 4x_2) \tag{5}$$

or, equivalently,

$$-3x_1 + 4x_2 + x_4 = 13$$

Note that it follows from (4) and (5) that $x_4 \ge 0$.

Finally, introducing a slack variable x_5 to the third inequality in (1), we obtain the system of linear equations

$$
\begin{aligned}
2x_1 - x_2 + x_3 \qquad\qquad &= 1 \\
-3x_1 + 4x_2 \qquad + x_4 \qquad &= 13 \\
x_1 + x_2 \qquad\qquad + x_5 &= 5
\end{aligned}
\tag{6}
$$

where x_3, x_4, x_5 are (nonnegative) slack variables. In this context, we sometimes refer to the original variables x_1 and x_2 as *decision variables*.

Observe that the system of linear equations (6) is in canonical form and that the slack variables are the basic variables. Since x_1 and x_2 are nonbasic variables in (6), it follows that

$$(0, 0, 1, 13, 5)$$

is the basic solution of (6); geometrically, this solution corresponds to the point $(0, 0)$ in Figure 1 (since $x_1 = 0, x_2 = 0$). This point is a corner point of the feasible region of the LP problem (1); and as you will see, it serves as the initial corner point in applying the simplex method.

In the remainder of this section we will see that the remaining corner points of the feasible region (as well as other intersection points of the lines defining the feasible region) correspond to various canonical systems of linear equations that are equivalent to (6). These canonical systems correspond in turn to particular selections of basic variables. For instance, the corner point $(2, 3)$ corresponds to the choice of x_1, x_2, and x_4 as basic variables. To see this, we apply appropriate pivot operations to (6) so that these variables become basic.

Pivoting on the coefficient 2 of x_1 in the first equation of (6), we obtain the equivalent system of linear equations

$$
\begin{aligned}
x_1 - \frac{1}{2}x_2 + \frac{1}{2}x_3 \qquad\qquad &= \frac{1}{2} \\
\frac{5}{2}x_2 + \frac{3}{2}x_3 + x_4 \qquad &= \frac{29}{2} \\
\frac{3}{2}x_2 - \frac{1}{2}x_3 \qquad + x_5 &= \frac{9}{2}
\end{aligned}
\tag{7}
$$

and we see that x_1 is now a basic variable.

Next, we pivot on the coefficient $\frac{3}{2}$ of x_2 in the third equation of (7) to obtain the equivalent canonical system of linear equations

$$
\begin{aligned}
x_1 \qquad + \frac{1}{3}x_3 \qquad + \frac{1}{3}x_5 &= 2 \\
\frac{7}{3}x_3 + x_4 - \frac{5}{3}x_5 &= 7 \\
x_2 - \frac{1}{3}x_3 \qquad + \frac{2}{3}x_5 &= 3
\end{aligned}
$$

with basic variables x_1, x_2, x_4. The corresponding basic solution is

$$(2, 3, 0, 7, 0)$$

and this solution is represented geometrically by the corner point $(2, 3)$ of the feasible region.

Each choice of basic variables corresponds geometrically to an intersection point of two or more lines defined by the constraints of (1); however, for some choices, these points lie outside the feasible region. For instance, suppose we were to select x_1, x_3, and x_5 as basic variables. To calculate the corresponding canonical system of equations, we pivot on the coefficient -3 of x_1 in (6) to obtain

$$
\frac{5}{3}x_2 + x_3 + \frac{2}{3}x_4 \qquad\qquad = \frac{29}{3}
$$
$$
x_1 - \frac{4}{3}x_2 \qquad\quad - \frac{1}{3}x_4 \qquad = -\frac{13}{3} \qquad (8)
$$
$$
\frac{7}{3}x_2 \qquad\quad + \frac{1}{3}x_4 + x_5 = \frac{28}{3}
$$

The resulting basic solution of (8) is

$$\left(-\frac{13}{3}, 0, \frac{29}{3}, 0, \frac{28}{3}\right)$$

and this solution corresponds to the point $(-\frac{13}{3}, 0)$ in Figure 1, which does not lie in the feasible region.

Table 1 shows the relationships between the various choices of basic variables and the points of intersection in Figure 1. The table also lists the values of the objective function z that correspond to the corner points. Note that for each choice of basic variables we can determine the corresponding pair of intersecting lines by setting the nonbasic variables to 0 in (6).

From Table 1 we obtain Figure 2.

Table 1

Basic variables	Intersecting pairs of lines	Points of intersection	Feasible point	Value of z
x_3, x_4, x_5	$x_1 = 0$; $x_2 = 0$	$(0, 0)$	Yes	0
x_1, x_4, x_5	$x_2 = 0$; $2x_1 - x_2 = 1$	$(\frac{1}{2}, 0)$	Yes	$\frac{3}{2}$
x_1, x_4, x_2	$2x_1 - x_2 = 1$; $x_1 + x_2 = 5$	$(2, 3)$	Yes	12
x_1, x_3, x_2	$-3x_1 + 4x_2 = 13$; $x_1 + x_2 = 5$	$(1, 4)$	Yes	11
x_1, x_3, x_5	$x_2 = 0$; $-3x_1 + 4x_2 = 13$	$(-\frac{13}{3}, 0)$	No	-13
x_1, x_3, x_4	$x_2 = 0$; $x_1 + x_2 = 5$	$(5, 0)$	No	15
x_2, x_3, x_4	$x_1 = 0$; $x_1 + x_2 = 5$	$(0, 5)$	No	10
x_2, x_3, x_5	$x_1 = 0$; $-3x_1 + 4x_2 = 13$	$(0, \frac{13}{4})$	Yes	$\frac{13}{2}$
x_2, x_4, x_5	$x_1 = 0$; $2x_1 - x_2 = 1$	$(0, -1)$	No	-2
x_2, x_1, x_5	$-3x_1 + 4x_2 = 13$; $2x_1 - x_2 = 1$	$(\frac{17}{5}, \frac{29}{5})$	No	$\frac{109}{5}$

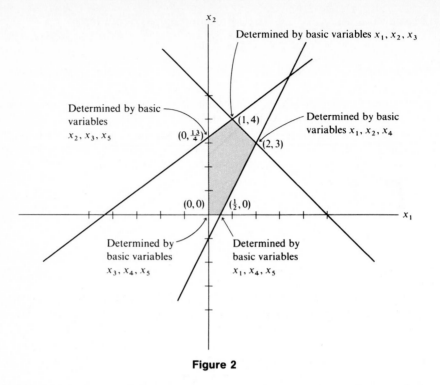

Figure 2

It is theoretically possible to make up a table like Table 1 for any LP problem. From such a table we could determine by inspection the optimal value for the objective function z. However, in practice, the calculations needed to produce these tables are extraordinarily time-consuming. In Exercise 10 you will have a chance to investigate the magnitude of such a task.

The simplex method is a procedure for choosing a succession of adjacent corner points (by selecting appropriate sets of basic variables) that leads to an optimal solution. This procedure usually requires far fewer calculations than does enumerating all possibilities, as in Table 1. We say "usually" because there are examples in which the simplex method checks all corner points, though such examples occur rarely, if ever, in practical problems.

As we shall see, the simplex method enables us to solve (6) by choosing sets of basic variables in the order

$$x_3, x_4, x_5; \qquad x_1, x_4, x_5; \qquad x_1, x_2, x_4$$

Note that these sets produce a succession of adjacent corner points along the boundary of the feasible region in the manner shown in Figure 3. Moreover, this "movement" along the boundary produces *increasingly favorable* values of the objective function. The optimal solution is attained at the point corresponding to the basic variables x_1, x_2, x_4.

Observe also from Table 1 that as we move from one corner point of the feasible region to an adjacent corner point, we change precisely one basic variable. That is,

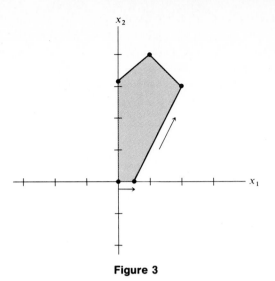

Figure 3

one current basic variable becomes nonbasic, and a current nonbasic variable becomes basic. At each stage of this process, the variable that becomes basic is called the *entering basic variable*, and the variable that becomes nonbasic is called the *departing basic variable*. For instance, as we move from the initial corner point $(0, 0)$ (which corresponds to the basic variables x_3, x_4, x_5) to the adjacent corner point $(0, \frac{13}{4})$ (which corresponds to the basic variables x_3, x_2, x_5), x_2 is the entering basic variable, and x_4 is the departing basic variable. Similarly, as we move from the corner point $(0, \frac{13}{4})$ to the adjacent corner point $(1, 4)$ (which corresponds to the basic variables x_3, x_2, x_1), x_1 is the entering basic variable, and x_5 is the departing basic variable.

Thus, the problem now becomes: How do we determine entering and departing basic variables that cause us to move from one corner point of the feasible region to an appropriate adjacent corner point in such a way that there is an improvement in the value of the objective function. The next section provides the solution to this problem, which is the essence of the simplex method.

To apply the simplex method to an LP problem, the problem must be stated in the form we give in the next definition.

Definition.   An LP problem

$$\text{Maximize:} \quad z = c_1 x_1 + c_2 x_2 + \cdots + c_n x_n$$

$$\text{Subject to:} \quad a_{11}x_1 + a_{12}x_2 + \cdots + a_{1n}x_n = b_1$$
$$a_{21}x_1 + a_{22}x_2 + \cdots + a_{2n}x_n = b_2$$
$$\vdots$$
$$a_{m1}x_1 + a_{m2}x_2 + \cdots + a_{mn}x_n = b_m$$
$$x_i \geq 0, \, 1 \leq i \leq n$$

is in *canonical form* if its system of constraint equations is in canonical form (see Section 2.4).

In the previous discussion of LP problem (1), we showed how to use slack variables to put a problem in canonical form when (a) the original problem is a maximization problem, (b) all constraints are of the $\le$ type, and (c) all decision variables are nonnegative. In Section 3.4 we will learn to put problems having $=$ or $\ge$ constraints in canonical form. The next three examples illustrate ways of coping with minimization problems and with problems that allow some decision variables to be negative.

EXAMPLE 1

We cannot put the LP problem

$$\begin{aligned} \text{Minimize:} \quad & z = 3x_1 - 2x_2 \\ \text{Subject to:} \quad & 2x_1 - 4x_2 \le 9 \\ & 3x_1 + 6x_2 \le 27 \\ & x_1, x_2 \ge 0 \end{aligned} \tag{9}$$

in canonical form by adding slack variables because the objective function is to be minimized. To circumvent this problem we convert (9) to a maximization problem by introducing

$$z' = -z = -3x_1 + 2x_2$$

Since the smallest value of z corresponds to the largest value of z', the LP problem

$$\begin{aligned} \text{Maximize:} \quad & z' = -3x_1 + 2x_2 \\ \text{Subject to:} \quad & 2x_1 - 4x_2 \le 9 \\ & 3x_1 + 6x_2 \le 27 \\ & x_1, x_2 \ge 0 \end{aligned}$$

has the same solution as (9), and the value of z' at this solution is the negative of the value of z at the minimal solution to (9). ◆

EXAMPLE 2

We cannot put the LP problem

$$\begin{aligned} \text{Maximize:} \quad & z = 3x_1 - 2x_2 + 4x_3 \\ \text{Subject to:} \quad & 5x_1 - 2x_2 + 3x_3 \le 6 \\ & 2x_1 + 3x_2 - 5x_3 \le 9 \\ & x_1, x_2 \le 0; \quad x_3 \ge 0 \end{aligned} \tag{10}$$

in canonical form by adding slack variables because x_1 and x_2 are not nonnegative. To circumvent this problem, we let $x_1' = -x_1$ and $x_2' = -x_2$ and substitute in (10) to obtain the LP problem

$$\text{Maximize:} \quad z = -3x_1' + 2x_2' + 4x_3$$
$$\text{Subject to:} \quad -5x_1' + 2x_2' + 3x_3 \le 6$$
$$-2x_1' - 3x_2' - 5x_3 \le 9 \tag{11}$$
$$x_1' \ge 0, \quad x_2' \ge 0, \quad x_3 \ge 0$$

We can put (11) in canonical form by adding slack variables. We can then obtain the solution to LP problem (10) from the one for problem (11) by using the relations $x_1 = -x_1'$ and $x_2 = -x_2'$. ◆

EXAMPLE 3

We cannot put the LP problem

$$\text{Maximize:} \quad z = 4x_1 - 2x_2$$
$$\text{Subject to:} \quad 3x_1 - 2x_2 \le 7$$
$$-5x_1 + 3x_2 \le 12 \tag{12}$$
$$x_1 \ge 0; \quad x_2 \quad \text{unrestricted}$$

in canonical form by adding slack variables because x_2 is unrestricted (it can be positive, zero, or negative). To circumvent this problem, we take advantage of the fact that we can write any real number as the difference of two nonnegative numbers (in infinitely many ways). Thus, we let $x_2 = u_2 - v_2$ where u_2 and v_2 are two new nonnegative variables. Substituting for x_2 in (12), we obtain the LP problem

$$\text{Maximize:} \quad z = 4x_1 - 2u_2 + 2v_2$$
$$\text{Subject to:} \quad 3x_1 - 2u_2 + 2v_2 \le 7$$
$$-5x_1 - 3u_2 + 3v_2 \le 12 \tag{13}$$
$$x_1 \ge 0, \quad u_2 \ge 0, \quad v_2 \ge 0$$

We can put (13) in canonical form by adding slack variables. We can then obtain the solution to (12) from the solution to (13) by using the relation $x_2 = u_2 - v_2$. ◆

We conclude this section with the following example, variations of which are found in later sections.

EXAMPLE 4

The Carter Nut Company (CNC) supplies a variety of nut mixtures for sale to companies that package and resell them. Its most expensive product is Bridge Mix, a mixture of peanuts, almonds, and cashews. Bridge Mix contains no more than 25% peanuts and no less than 40% cashews. There is no limitation on the percentage of almonds. The current selling price of Bridge Mix is $0.80 per pound. CNC estimates that it can process up to 1000 pounds of Bridge Mix this month. Table 2 gives the amounts of inventory and cost of the nuts to be used.

Table 2

Type	Cost ($ per lb)	Pounds available
Peanuts	0.20	400
Almonds	0.35	250
Cashews	0.50	200

How much Bridge Mix should CNC produce this month to maximize profit?

We can express all conditions of the problem in terms of three variables x_1, x_2, and x_3 representing the number of pounds of peanuts, almonds, and cashews, respectively, contained in the quantity of Bridge Mix produced. The amount of Bridge Mix produced is

$$x_1 + x_2 + x_3$$

The condition that no more than 1000 pounds can be produced yields the constraint

$$x_1 + x_2 + x_3 \le 1000$$

The condition that no more than 25% of the product is peanuts yields the constraint

$$x_1 \le 0.25(x_1 + x_2 + x_3) \tag{14}$$

The condition that at least 40% is cashews yields the constraint

$$x_3 \ge 0.40(x_1 + x_2 + x_3) \tag{15}$$

The limits on availability of the types of nuts yield

$$x_1 \le 400; \qquad x_2 \le 250; \qquad x_3 \le 200$$

The profit is the difference between the selling price and the cost of the nuts:

$$z = 0.80(x_1 + x_2 + x_3) - 0.20x_1 - 0.35x_2 - 0.50x_3 \tag{16}$$

Expressions (14), (15), and (16) are not in the form we are using to express LP problems. We must combine coefficients of like variables in each expression and move all variables to the left side of inequalities. Making these changes results in the LP problem:

Maximize: $z = 0.60x_1 + 0.45x_2 + 0.30x_3$

$$
\begin{aligned}
\text{Subject to:} \quad 0.75x_1 - 0.25x_2 - 0.25x_3 &\le 0 \\
0.40x_1 + 0.40x_2 - 0.60x_3 &\le 0 \\
x_1 + x_2 + x_3 &\le 1000 \\
x_1 &\le 400 \\
x_2 &\le 250 \\
x_3 &\le 200
\end{aligned}
\tag{17}
$$

$$x_i \ge 0, \qquad 1 \le i \le 3$$

Adding slack variables to (17) yields the canonical system of linear equations

Maximize: $z = 0.60x_1 + 0.45x_2 + 0.30x_3$

$$
\begin{aligned}
\text{Subject to:} \quad 0.75x_1 - 0.25x_2 - 0.25x_3 + s_1 \quad\quad\quad\quad\quad\quad &= \quad 0 \\
0.40x_1 + 0.40x_2 - 0.60x_3 \quad + s_2 \quad\quad\quad\quad &= \quad 0 \\
x_1 + \quad x_2 + \quad x_3 \quad\quad + s_3 \quad\quad &= 1000 \\
x_1 \quad\quad\quad\quad\quad\quad\quad + s_4 \quad &= \ 400 \\
x_2 \quad\quad\quad\quad\quad\quad + s_5 \quad &= \ 250 \\
x_3 \quad\quad\quad\quad + s_6 &= \ 200
\end{aligned}
$$
(18)

$$x_i \geq 0, \quad 1 \leq i \leq 3; \quad s_i \geq 0, \quad 1 \leq i \leq 6$$

The slack variables $s_1, s_2, s_3, s_4, s_5, s_6$ form a basic feasible solution from which the simplex method can start. We will solve (18) in Section 3.2. ◆

Exercises

1. By making an appropriate pivot in (6) verify the entries in the row of Table 1 that contains the point of intersection $(0, \frac{13}{4})$.

2. By making an appropriate pivot in (8), verify the entries in the row of Table 1 that contains the point of intersection $(5, 0)$.

3. Consider the following LP problem and Figure 4.

$$
\begin{aligned}
\text{Maximize:} \quad z &= 120x_1 + 100x_2 \\
\text{Subject to:} \quad x_1 + \ x_2 &\leq \ 4 \\
5x_1 + 3x_2 &\leq 15 \\
x_1 - \ x_2 &\leq \ 2 \\
x_1, x_2 &\geq 0
\end{aligned}
$$

The letters in the figure label the points of intersection of the five lines corresponding to the five constraints of this LP problem.

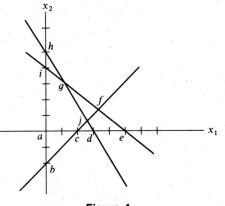

Figure 4

(a) Locate the feasible region and list the letters of its five corner points.

(b) Evaluate the objective function at each of the five corner points.

(c) Identify two paths along the boundary of the feasible region that begin at the origin and extend to the point yielding a maximum value of the objective function. Each successive point in each of your paths should produce increasing values of the objective function.

4. Consider the LP problem

$$\text{Maximize:} \quad z = 10x_1 - 2x_2$$
$$\text{Subject to:} \quad 2x_1 - x_2 \le 4$$
$$x_1 - x_2 \le 1$$
$$x_1, x_2 \ge 0$$

(a) Draw the feasible region, and find the coordinates of the corner points of the feasible region.

(b) Evaluate the objective function at the corner points.

(c) Find a point in the feasible region at which the objective function z attains a larger value than at any of the corner points. (The lesson is that we need to build a mechanism into the simplex algorithm to detect an unbounded situation.)

5. Suppose the nonconvex polygon in Figure 5 defines a feasible region. Evaluate the objective function $z = x_1 + x_2$ at each corner point. Is it possible to progress from the corner point $(3, 2)$ to the corner point yielding the maximal value of z by passing through corner points of steadily increasing z value? (The lesson is that when the feasible domain is not convex, it is not necessarily possible to reach an optimal value by traversing the boundary through steadily increasing values of the objective function.)

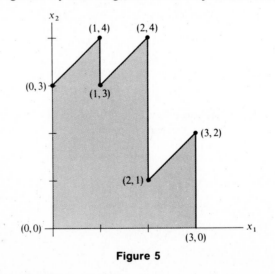

Figure 5

6. The feasible region for the constraints

$$x_1 + x_2 + x_3 \le 1$$
$$x_1 + x_2 + 4x_3 \le 2$$
$$x_i \ge 0, 1 \le i \le 3$$

is the intersection of the two tetrahedrons shown in Figure 6.

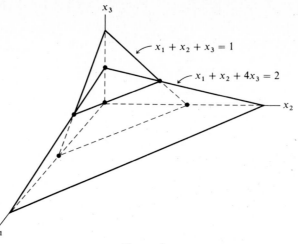

Figure 6

Compute the value of the objective function $x_1 - x_2 + x_3$ at the six corner points of this feasible set. Observe that, starting at the origin, you can reach the point that yields the maximum by progressing through corner points that give steadily increasing values of the objective function.

7. Consider the problem

$$\text{Maximize:} \quad z = 120x_1 + 100x_2$$
$$\text{Subject to:} \quad x_1 + x_2 \le 4$$
$$5x_1 + 3x_2 \le 15$$
$$x_1, x_2 \ge 0$$

Adding slack variables yields the problem

$$\text{Maximize:} \quad z = 120x_1 + 100x_2$$
$$\text{Subject to:} \quad x_1 + x_2 + s = 4$$
$$5x_1 + 3x_2 + t = 15$$
$$x_1, x_2, s, t \ge 0$$

There are six points of intersection of the lines

$$x_1 + x_2 = 4, \qquad 5x_1 + 3x_2 = 15, \qquad x_1 = 0, \qquad \text{and} \qquad x_2 = 0$$

The point $x_1 = 0$, $x_2 = 0$ corresponds to the basic solution with $s = 4$ and $t = 15$ in the preceding equations and also to the intersection of the lines $x_1 = 0$ and $x_2 = 0$. Perform five pivots of the system to discover the other five points of intersection, and plot these points on an x_1–x_2 graph. Which of the six points are feasible?

8. For each of the following systems, there is only one choice of pivot that will make x_1 a basic variable without introducing a negative number on the right-hand side of an equation. What is the choice?

(a) $2x_1 + 3x_2 + 4x_3 + s_1 = 5$
$3x_1 + x_2 - 2x_3 + s_2 = 2$
$4x_1 - x_2 + 3x_3 + s_3 = 3$

(b) $-x_1 + 2x_2 - 3x_3 + s_1 = 1$
$2x_1 - x_2 + x_3 + s_2 = 2$
$-3x_1 + 2x_2 + 2x_3 + s_3 = 2$

9. Can you state a general principle for making the choice described in Exercise 8? Specifically, on which coefficient of x_1 would you pivot if $b_1, b_2, b_3 \geq 0$ and the right-hand sides are to remain nonnegative in the following case:

$$a_{11}x_1 + a_{12}x_2 + a_{13}x_3 + s_1 = b_1$$

$$a_{21}x_1 + a_{22}x_2 + a_{23}x_3 + s_2 = b_2$$

$$a_{31}x_1 + a_{32}x_2 + a_{33}x_3 + s_3 = b_3$$

10. The purpose of this exercise is to study the magnitude of the calculations involved in solving LP problems by exhaustive analysis of the points of intersection of the hyperplanes bounding the feasible set (as was done in producing Table 1). Since binomial coefficients are useful in this and many other counting problems that we will consider, we review their definition. For positive integers k and m with $k \geq m$,

$$\binom{k}{m} = \frac{k!}{m!(k-m)!}$$

gives the number of subsets of m elements that can be formed from a set of k elements. For example,

$$\binom{5}{2} = \frac{5!}{2!3!} = 10$$

(a) Compute $\binom{6}{4}$ and $\binom{7}{3}$.

(b) Let $S = \{a, b, c, d\}$. Form the $\binom{4}{2}$ two-element subsets of S.

(c) Use a binomial coefficient to find the number of points of intersection of the lines determining the feasible set of the LP problem of Exercise 7.

(d) In Exercise 6 we considered five planes that determined six corner points of the feasible region. Since a point of R^3 is determined by the intersection of three planes, there are $\binom{5}{3} = 10$ possible points of intersection of the five planes. There can be fewer than ten points of intersection because sometimes three planes do not intersect in a point. Show that there are nine points of intersection of the five planes given in Figure 6.

Parts (c) and (d) illustrate the maximum number of points in which several lines or several planes can intersect. The general result is that a system of linear inequalities with m constraints and n nonnegative decision variables can define a feasible region with as many as

$$\binom{n+m}{m}$$

corner points.

(e) Consider a system of linear inequalities with six nonnegative decision variables and seven constraint inequalities. How many possible points of intersection of the hyperplanes bounding the feasible region are there?

(f) Show that there are 184,756 possible points of intersection of a system with ten nonnegative decision variables and ten constraints.

In Exercises 11–16 use the techniques illustrated in Examples 1, 2, and 3 to form equivalent LP problems that you can put into canonical form by adding slack variables.

11. Minimize: $z = 2x_1 - 3x_2$

Subject to: $5x_1 - 2x_2 \leq 3$

$-2x_1 + 3x_2 \leq 5$

$x_1 \geq 0, \quad x_2 \leq 0$

12. Minimize: $z = 2x_1 - 3x_2 + 5x_3$

Subject to: $6x_1 - 2x_2 + 3x_3 \leq 5$

$5x_1 + 2x_2 - 4x_3 \leq 10$

$x_1 \geq 0, \quad x_2 \leq 0, \quad x_3 \geq 0$

13. Maximize: $z = 2x_1 - 3x_2$

Subject to: $-x_1 + 2x_2 \leq 5$

$x_1 - 3x_2 \leq 6$

$x_1 \geq 0, \quad x_2 \quad \text{unrestricted}$

14. Maximize: $z = -2x_1 + 3x_2 - x_3$

Subject to: $2x_1 - 3x_2 + 5x_3 \leq 7$

$-x_1 + 2x_2 - x_3 \leq 5$

$x_1, x_2 \quad \text{unrestricted}; \quad x_3 \geq 0$

15. Minimize: $z = 2x_1 - 3x_2$

Subject to: $2x_1 - 5x_2 \leq 12$

$-3x_1 + 2x_2 \leq 15$

$x_1 \leq 0; \quad x_2 \quad \text{unrestricted}$

16. Minimize: $z = -5x_1 + 3x_2 - 6x_3$

Subject to: $-3x_1 + 2x_2 - x_3 \leq 6$

$2x_1 - 5x_2 + 3x_3 \leq 9$

$x_1, x_2 \quad \text{unrestricted}; \quad x_3 \leq 0$

CHECKLIST: CHAPTER 2

DEFINITIONS

CONCEPTS AND RESULTS

Bounded, unbounded, empty feasible region

Geometric introduction to sensitivity analysis

Geometric relation between basic variables and points of intersection

Theorem 1 (Feasible Regions Are Convex).

The feasible set in R^n corresponding to any number of constraints of the types

$$a_1 x_1 + a_2 x_2 + \cdots + a_n x_n \leq b$$

$$a_1 x_1 + a_2 x_2 + \cdots + a_n x_n = b$$

$$a_1 x_1 + a_2 x_2 + \cdots + a_n x_n \geq b$$

$$x_1, x_2, \ldots, x_n \geq 0$$

is convex.

Theorem 2 (Corner Point Theorem).

If an optimal solution of the following LP problem exists,

$$\text{Maximize (or minimize)} \quad z = c_1 x_1 + c_2 x_2 + \cdots + c_n x_n$$

Subject to a system of inequalities of the types:

$$a_1 x_1 + a_2 x_2 + \cdots + a_n x_n \leq b$$

$$a_1 x_1 + a_2 x_2 + \cdots + a_n x_n = b$$

$$a_1 x_1 + a_2 x_2 + \cdots + a_n x_n \geq b$$

$$x_1, x_2, \ldots, x_n \geq 0$$

then there is an optimal solution at a corner point of the feasible region.

*Theorem 2.

Suppose that $f: R^n \to R^1$ is defined by (2). If $\mathbf{p} = (p_1, p_2, \ldots, p_n)$ and $\mathbf{q} = (q_1, q_2, \ldots, q_n)$ are two points in R^n, and if

$$\mathbf{r} = (1 - t)\mathbf{p} + t\mathbf{q}$$

is any point on the line segment joining $\mathbf{p}$ and $\mathbf{q}$, then $f(\mathbf{r})$ is between $f(\mathbf{p})$ and $f(\mathbf{q})$. That is, if $f(\mathbf{p}) \leq f(\mathbf{q})$, then $f(\mathbf{p}) \leq f(\mathbf{r}) \leq f(\mathbf{q})$; or if $f(\mathbf{q}) \leq f(\mathbf{p})$, then $f(\mathbf{q}) \leq f(\mathbf{r}) \leq f(\mathbf{p})$.

*Theorem 3.

Suppose that K is a bounded polyhedron with corner points $\mathbf{k}_1, \mathbf{k}_2, \ldots, \mathbf{k}_m$. Then any point $\mathbf{p}$ in K is a convex combination of $\mathbf{k}_1, \mathbf{k}_2, \ldots, \mathbf{k}_m$.

*Theorem 4.

Suppose that $f: R^n \to R^1$ is defined by

$$f(x_1, x_2, \ldots, x_n) = c_1 x_1 + c_2 x_2 + \cdots + c_n x_n$$

and let $\mathbf{k}_1, \mathbf{k}_2, \ldots, \mathbf{k}_m$ be any m points in R^n. Then for any constants $b_1, b_2, \ldots, b_m$,

$$f(b_1 \mathbf{k}_1 + b_2 \mathbf{k}_2 + \cdots + b_m \mathbf{k}_m) = b_1 f(\mathbf{k}_1) + b_2 f(\mathbf{k}_2) + \cdots + b_m f(\mathbf{k}_m)$$

Chapter 3

THE SIMPLEX METHOD

3.1 SELECTION OF ENTERING AND DEPARTING BASIC VARIABLES

In the previous chapter we saw how introducing slack variables gives us an initial basic feasible solution for some LP problems. We also saw the necessity for developing rules for choosing the entering and departing basic variables so that we can easily move from an already examined corner point to an appropriate adjacent corner point (one that yields an improved value in the objective function) *without* leaving the feasible region.

In this section we continue to restrict our attention to LP problems of the form

$$
\begin{aligned}
\text{Maximize:} \quad & z = c_1 x_1 + c_2 x_2 + \cdots + c_n x_n \\
\text{Subject to:} \quad & a_{11} x_1 + a_{12} x_2 + \cdots + a_{1n} x_n \leq b_1 \\
& a_{21} x_1 + a_{22} x_2 + \cdots + a_{2n} x_n \leq b_2 \\
& \quad \vdots \\
& a_{m1} x_1 + a_{m2} x_2 + \cdots + a_{mn} x_n \leq b_m \\
& x_i \geq 0, \qquad 1 \leq i \leq n
\end{aligned}
\tag{1}
$$

where $b_i \geq 0$ for each i, $1 \leq i \leq m$. As before, we introduce slack variables x_{n+1}, $x_{n+2}, \ldots, x_{n+m}$ to obtain the canonical system of linear equations

$$
\begin{aligned}
a_{11} x_1 + a_{12} x_2 + \cdots + a_{1n} x_n + x_{n+1} & && = b_1 \\
a_{21} x_1 + a_{22} x_2 + \cdots + a_{2n} x_n && + x_{n+2} & = b_2 \\
\quad \vdots & \\
a_{m1} x_1 + a_{m2} x_2 + \cdots + a_{mn} x_n && + x_{n+m} & = b_m
\end{aligned}
\tag{2}
$$

The slack variables are the basic variables for (2) and the $(m + n)$-tuple $(0, 0, \ldots, 0, b_1, b_2, \ldots, b_m)$ is the basic feasible solution of (2) that serves as a starting point for the simplex method. Geometrically, this solution corresponds to the origin (of R^n) where the objective function z takes on the value

$$c_1 \cdot 0 + c_2 \cdot 0 + \cdots + c_n \cdot 0 = 0$$

We now want either to move from the origin to an adjacent corner point of the feasible region where the value of z increases, or to determine that there is no such adjacent point, in which case the current feasible solution provides a maximal value for z. To do this we need to consider the choice of entering and departing basic variables.

Choosing the Entering Basic Variable

To motivate the selection of the entering basic variable we return to the example given in Section 2.5

$$\text{Maximize:} \quad z = 3x_1 + 2x_2$$

$$
\begin{aligned}
\text{Subject to:} \quad 2x_1 - x_2 + x_3 \qquad\qquad &= 1 \\
-3x_1 + 4x_2 \qquad + x_4 \qquad &= 13 \\
x_1 + x_2 \qquad\qquad + x_5 &= 5 \\
x_i \geq 0, \qquad 1 \leq i \leq 5 &
\end{aligned}
\tag{3}
$$

where the variables x_3, x_4, and x_5 are slack variables. Initially, the variables x_1 and x_2 are nonbasic, and hence $z = 3x_1 + 2x_2 = 0$. Observe now that if either of the current basic variables x_1 and x_2 were to become basic and take on a positive value, then the value of the objective function z would increase. Therefore, we can choose either x_1 or x_2 as the entering basic variable. (We also observe that if, say, the coefficient of x_1 were negative and if x_1 were to become basic and take on a positive value, then the value of z would decrease.)

In Section 3.3 and in Chapter 5 we will examine a variety of rules for selecting the entering basic variable, each of which has certain advantages. For the present we will make the "greediest" choice by selecting the variable with largest positive coefficient in the objective function (thus, in our current example, we select x_1). A unit increase in this variable yields a greater increase than does a unit increase in any of the other variables.

These considerations lead us to adopt the following rule for selecting an entering basic variable.

GREEDY ENTERING BASIC VARIABLE RULE

If the objective function z for a maximization problem is written in terms of the current nonbasic variables (i.e., if in the current expression defining z, all of the cost coefficients of the current basic variables are 0), then the *largest positive cost coefficient* determines the entering basic variable.

You will see presently how to write c ojective functions in terms of current nonbasic variables.

The Greedy Entering Basic Variable ule appears reasonable because, as we observed, allowing the nonbasic variable with the largest, positive cost coefficient to become basic yields the greatest *unit* increase in the value of z. For instance, if x_1, x_2, and x_3 are the current nonbasic variables, and if

$$z = 3x_1 - 2x_2 + 6x_4$$

then a unit increase in x_4 results in a greater favorable change in z than does a unit increase in x_1; a unit increase in x_2 yields an unwanted decrease in z. You should realize, however, that the actual amount you can increase x_1 or x_4 depends on the given feasible region. In fact, for some feasible regions you actually may be able to increase x_1 more than x_4 without leaving the feasible region; thus, choosing x_4 as the entering basic variable may not necessarily lead to the greatest increase in z (see Exercises 14 and 15). In general, you can designate *any* nonbasic variable with a positive cost coefficient as the entering basic variable, though the Greedy Entering Basic Variable Rule selects the one having the largest cost coefficient. Note, however, that nonbasic variables with negative cost coefficients (such as x_2 in $z = 3x_1 - 2x_2 + 6x_4$) cannot be considered to be entering basic variables because any increase in them causes a decrease in z.

EXAMPLE 1

If an objective function z is written in the form

$$z = 0x_1 - 2x_2 + 4x_3 + 0x_4 + 2x_5 - 6x_6 + 0x_7$$

where x_1, x_4, and x_7 are the current basic variables, then the Greedy Entering Basic Variable Rule selects x_3 as the entering basic variable because a unit increase in x_3 results in greater positive change in z than does a unit increase in any of the other nonbasic variables. Note that x_5 also could serve as an entering basic variable because increasing x_5 also produces an increase in z. The variables x_2 and x_6 cannot serve as entering basic variables because increasing them decreases z.   ◆

EXAMPLE 2

If an objective function z is written in the form

$$z = 0x_1 - 2x_2 - 4x_3 + 0x_4 + 0x_5 - \frac{1}{2}x_6 + 0x_7$$

where x_1, x_5, x_7 are the current basic variables, then the current value of z must be maximal because allowing any of the nonbasic variables x_2, x_3, x_4, x_6 to become positive would not increase the value of z.   ◆

Example 2 illustrates one of the rules for ending the simplex procedure.

OPTIMALITY CRITERION

If the objective function z is written in terms of the nonbasic variables and if the coefficients of the nonbasic variables are all negative or zero, then the current value of z is the maximal value of the objective function, and an optimal solution has been attained.

Choosing the Departing Basic Variable

We return to LP problem (3) to understand the choice of a departing basic variable. In that problem we saw that the initial entering basic variable was x_1. We base the selection of the departing basic variable on the following two considerations:

We want to increase the entering basic variable x_1 as much as possible (thereby increasing z as much as currently possible).

We do not want to leave the feasible region.

In other words, we want the greatest possible increase of x_1 without causing any of the other basic variables (currently x_3, x_4, x_5) to become negative. Note that it follows from the definition of slack variables that we remain in the feasible region as long as the variables x_3, x_4, x_5 are nonnegative. Thus, since $x_2 = 0$ in the basic solution, we see from (3) that x_1 can be as large as the following restrictions

$$0 \le x_3 = 1 - 2x_1 \tag{4}$$

$$0 \le x_4 = 13 + 3x_1 \tag{5}$$

$$0 \le x_5 = 5 - x_1 \tag{6}$$

allow. Ignoring x_3, x_4, x_5 in (4)–(6) and moving the x_1 terms to the left side of the inequalities, we have

$$2x_1 \le 1 \tag{4'}$$

$$-3x_1 \le 13, \quad \text{which is true for any} \quad x_1 \ge 0 \tag{5'}$$

$$x_1 \le 5 \tag{6'}$$

Inequality (4') is the most restrictive of these constraints; thus, it follows that the largest allowable positive value for x_1 is $x_1 = \frac{1}{2}$. From (4), we see that setting $x_3 = 0$ makes $x_1 = \frac{1}{2}$, and therefore x_3 becomes the departing basic variable.

We observe that the departing basic variable is selected by forming ratios of the resource values of (3) to the *positive* coefficients of x_1. Table 1 illustrates this procedure.

Table 1

Coefficients of x_1	Resource values	Ratios
2	1	$\frac{1}{2}$
-3	13	Skip
1	5	$\frac{5}{1}$

DEPARTING BASIC VARIABLE RULE

Assume that the entering basic variable has been chosen. For every equation for which the coefficient of the entering basic variable is positive, form the ratio of the resource value to that coefficient. Select the equation that produces the smallest of these ratios (if there is a tie among the equations for smallest, choose any one of these equations). The basic variable for the selected equation is the departing basic variable.

In our present problem, the original basic variables are x_3, x_4, and x_5. The Greedy Entering Basic Variable Rule has selected x_1 as the entering basic variable, and the Departing Basic Variable Rule has selected x_3 as the departing basic variable. To convert the system to an equivalent system with basic variables x_3, x_4, and x_5, we pivot on $2x_1$ in the first equation of (3), obtaining

$$x_1 - \frac{1}{2}x_2 + \frac{1}{2}x_3 \qquad\qquad = \frac{1}{2}$$

$$\frac{5}{2}x_2 + \frac{3}{2}x_3 + x_4 \qquad = \frac{29}{2} \qquad\qquad (7)$$

$$\frac{3}{2}x_2 - \frac{1}{2}x_3 \qquad + x_5 = \frac{9}{2}$$

$$x_i \geq 0, \qquad 1 \leq i \leq 5$$

From (7) we obtain the basic feasible solution

$$\left(\frac{1}{2}, 0, 0, \frac{29}{2}, \frac{9}{2}\right)$$

The current value of the objective function is

$$z = 3x_1 + 2x_2 = 3\left(\frac{1}{2}\right) + 2(0) = \frac{3}{2} \qquad\qquad (8)$$

Observe that, geometrically, we have moved from the corner point $(0,0)$ to the corner point $(\frac{1}{2}, 0)$.

To determine if $z = \frac{3}{2}$ is the optimum value of the objective function, we rewrite (8) in terms of the current nonbasic variables. From (7) we see that

$$x_1 = \frac{1}{2} + \frac{1}{2}x_2 - \frac{1}{2}x_3$$

and it follows that

$$z = 3x_1 + 2x_2 = 3\left(\frac{1}{2} + \frac{1}{2}x_2 - \frac{1}{2}x_3\right) + 2x_2$$

$$= \frac{3}{2} + \frac{7}{2}x_2 - \frac{3}{2}x_3 \tag{9}$$

It is apparent from (9) that we can increase z by allowing x_2 to become a basic variable because the coefficient of x_2 is the largest (in this case, the only) positive coefficient of the nonbasic variables. Thus, x_2 is the new entering basic variable.

Before determining the new departing basic variable, we note that we could also obtain (9) by augmenting (3) with the equation

$$z - 3x_1 - 2x_2 = 0$$

This gives us

$$
\begin{aligned}
\boxed{2x_1} - \ x_2 + x_3 & = 1 \\
-3x_1 + 4x_2 \qquad + x_4 & = 13 \\
x_1 + \ x_2 \qquad\qquad + x_5 & = 5 \\
z - 3x_1 - 2x_2 \qquad\qquad & = 0
\end{aligned}
\tag{10}
$$

Pivoting as before on the circled entry in (10), we have

$$
\begin{aligned}
x_1 - \frac{1}{2}x_2 + \frac{1}{2}x_3 \qquad\qquad & = \frac{1}{2} \\[4pt]
\frac{5}{2}x_2 + \frac{3}{2}x_3 + x_4 \qquad & = \frac{29}{2} \\[4pt]
\frac{3}{2}x_2 - \frac{1}{2}x_3 \qquad + x_5 & = \frac{9}{2} \\[4pt]
z \qquad - \frac{7}{2}x_2 + \frac{3}{2}x_3 \qquad\qquad & = \frac{3}{2}
\end{aligned}
\tag{11}
$$

$$x_i \geq 0, \qquad 1 \leq i \leq 5$$

The last equation in (11) is the same as (9). Note that in the formulations given by (10) and (11), we choose the new entering basic variable to be the variable that corresponds to the *largest negative* coefficient appearing in the last equations of these systems.

To find the new departing basic variable we use the Departing Basic Variable Rule. The ratios are formed from the coefficients of x_2 and the resource values in (11), as in Table 2. Notice that you must not use the coefficient of x_2 in the equation for the objective function [the last equation in (11)] in forming these ratios. The choice of departing basic variables involves only the constraint equations.

Table 2

Coefficients of x_2	Resource values	Ratios
$-\frac{1}{2}$	$\frac{1}{2}$	Skip
$\frac{5}{2}$	$\frac{29}{2}$	$\frac{29}{5}$
$\frac{3}{2}$	$\frac{9}{2}$	$\frac{9}{3}$

Since $\frac{9}{3}$ is the smallest of the ratios in Table 2, the departing basic variable is the basic variable from the third equation in (11), that is, x_5. The next pivot is on the coefficient of the entering basic variable x_2 in the equation corresponding to the departing basic variable. Thus, we pivot on the coefficient $\frac{3}{2}$ of x_2 in the third equation of (11) to obtain the new canonical system of equations

$$
\begin{aligned}
x_1 \quad +\frac{1}{3}x_3 \quad +\frac{1}{3}x_5 &= 2 \\[2mm]
\frac{7}{3}x_3 + x_4 - \frac{5}{3}x_5 &= 7 \\[2mm]
x_2 - \frac{1}{3}x_3 \quad +\frac{2}{3}x_5 &= 3 \\[2mm]
z \quad +\frac{1}{3}x_3 \quad +\frac{7}{3}x_5 &= 12
\end{aligned}
\tag{12}
$$

Note that in the last equation of (12) we have written z in terms of the current nonbasic variables x_3 and x_5. That is, we have

$$
z = 12 - \frac{1}{3}x_3 - \frac{7}{3}x_5 \tag{13}
$$

It follows from (13) that the current value of z is 12; furthermore, it is clear that we cannot improve this value by allowing either x_3 or x_5 to become basic. Thus, by the Optimality Criterion, we have reached an optimal solution to (3):

$$
(2, 3, 0, 7, 0)
$$

This corresponds to the corner point $(2, 3)$ in Figure 1 of Section 2.5.

During this procedure we considered only three corner points: $(0, 0)$, $(\frac{1}{2}, 0)$ and $(2, 3)$ to obtain the maximal value of the objective function z. Since there are ten possible points of intersection of the five lines that determine the feasible region, the simplex method has found the solution to this problem very efficiently.

EXAMPLE 3

As we saw in Chapter 2, LP problems may have unbounded feasible solutions. To illustrate how the simplex method detects this kind of situation we consider the

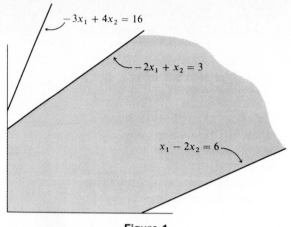

Figure 1

problem

$$\text{Maximize:} \quad z = x_1 + 2x_2$$

$$\text{Subject to:} \quad -2x_1 + x_2 \leq 3$$
$$-3x_1 + 4x_2 \leq 16 \tag{14}$$
$$x_1 - 2x_2 \leq 6$$

$$x_1, x_2 \geq 0$$

Figure 1 shows the feasible region for (14).

Introducing slack variables in (14) and including the objective function as an equation at the bottom, we have the system

$$-2x_1 + \text{\textcircled{x_2}} + x_3 = 3$$
$$-3x_1 + 4x_2 + x_4 = 16$$
$$x_1 - 2x_2 + x_5 = 6 \tag{15}$$
$$z - x_1 - 2x_2 = 0$$

$$x_i \geq 0, \quad 1 \leq i \leq 5$$

Applying the Greedy Entering Basic Variable Rule to (15), we see that x_2 is the initial entering basic variable. Since $\frac{3}{1}$ is the smallest ratio formed in applying the Departing Basic Variable Rule, we pivot on x_2 in the first equation of (15) to obtain

$$-2x_1 + x_2 + x_3 = 3$$
$$\text{\textcircled{$5x_1$}} - 4x_3 + x_4 = 4$$
$$-3x_1 + 2x_3 + x_5 = 12 \tag{16}$$
$$z - 5x_1 + 2x_3 = 6$$

From (16), we see that x_1 is the new entering basic variable and that x_4 is the new departing basic variable (why?). Pivoting on x_1 in the second equation of (16), we have

$$
\begin{aligned}
x_2 - \frac{3}{5}x_3 + \frac{2}{5}x_4 \qquad &= \frac{23}{5} \\
x_1 \qquad - \frac{4}{5}x_3 + \frac{1}{5}x_4 \qquad &= \frac{4}{5} \\
- \frac{2}{5}x_3 + \frac{3}{5}x_4 + x_5 &= \frac{72}{5} \\
z \qquad - 2x_3 + \ x_4 \qquad &= 10
\end{aligned}
\tag{17}
$$

It follows from (17) that x_3 is the entering basic variable. Keeping in mind that the variable x_4 is nonbasic (and hence equal to 0 in the basic solution), we have from (17)

$$
x_2 = \frac{23}{5} + \frac{3}{5}x_3
$$

$$
x_1 = \frac{4}{5} + \frac{4}{5}x_3
$$

$$
x_5 = \frac{72}{5} + \frac{2}{5}x_3
$$

$$
z = 10 + 2x_3
$$

These equations show that the variables x_2, x_1, and x_5 are positive for any non-negative value for x_3. Therefore, we can increase the entering basic variable x_3 as much as we want and still remain in the feasible region. But increasing x_3 leads to a corresponding increase in $z = 10 + 2x_3$ [the last equation in (17)]. Thus, we can obtain arbitrarily large values for z; in other words, the objective function is unbounded on the feasible region. ◆

UNBOUNDED TERMINATION RULE

Suppose the objective function is written in terms of the current nonbasic variables. If there is an entering basic variable whose coefficient in each constraint is nonpositive, then the objective function is unbounded on the feasible region.

Such a situation indicates the termination of the simplex method, with the conclusion that the objective function can take on arbitrarily large values; hence, though there are feasible solutions, there is no optimal solution.

Summary of the Simplex Method

For LP problems of the form

$$\text{Maximize:} \quad z = c_1 x_1 + c_2 x_2 + \cdots + c_n x_n$$

$$\text{Subject to:} \quad a_{11} x_1 + a_{12} x_2 + \cdots + a_{1n} x_n \leq b_1$$

$$a_{21} x_1 + a_{22} x_2 + \cdots + a_{2n} x_n \leq b_2 \tag{18}$$

$$\vdots$$

$$a_{m1} x_1 + a_{m2} x_2 + \cdots + a_{mn} x_n \leq b_m$$

$$x_i \geq 0, \quad 1 \leq i \leq n$$

where $b_i \geq 0$, $1 \leq i \leq m$, the simplex method consists of the following steps:

1. Introduce slack variables to the constraints of (18) to obtain the canonical system of linear equations,

$$a_{11} x_1 + a_{12} x_2 + \cdots + a_{1n} x_n + x_{n+1} \qquad\qquad = b_1$$

$$a_{21} x_1 + a_{22} x_2 + \cdots + a_{2n} x_n \qquad + x_{n+2} \qquad = b_2 \tag{19}$$

$$\vdots$$

$$a_{m1} x_1 + a_{m2} x_2 + \cdots + a_{mn} x_n \qquad\qquad\qquad + x_{n+m} = b_m$$

with basic variables $x_{n+1}, x_{n+2}, \ldots, x_{n+m}$

2. Express the objective function as an equation of the form

$$z - c_1 x_1 - \cdots - c_n x_n - c_{n+1} x_{n+1} - \cdots - c_{n+m} x_{n+m} = 0 \tag{20}$$

where the coefficients of all nonbasic variables are 0 (initially, $c_{n+1} = c_{n+2} = \cdots = c_{n+m} = 0$).

Append (20) to (19), as the last equation, so that you can pivot the objective function along with the constraints (this ensures that the objective function is expressed in terms of the current nonbasic variables). In the sequel, we often write the objective function in the form of (20) so that we can treat it similarly to constraint equations in pivot operations. Since the signs of all c_i have been changed in forming (20), we modify the Greedy Entering Basic Variable Rule to select the most *negative* cost coefficient in (20) rather than the largest positive cost coefficient, as was done before when z was expressed as in (18). The statement in step 3 incorporates this change.

3. Determine the entering basic variable using the *Greedy Entering Basic Variable Rule*: If the objective function z for a maximization problem is written in the form of (20) and in terms of the current nonbasic variables (i.e., if in the current expression that defines z all of the cost coefficients of the current basic variables are 0), then the *most negative cost coefficient* determines the entering basic variable. If all cost coefficients in (20) are nonnegative, the process terminates; an optimal solution has been found.

4. Determine the departing basic variable using the *Departing Basic Variable Rule*: For each equation for which the coefficient of the entering basic variable is positive, form the ratio of the resource value to that coefficient. Select the equation producing the smallest of these ratios (if there is a tie among equations for smallest, choose any of these equations). The basic variable for the equation selected is the departing basic variable.

   If all coefficients of the entering basic variable are nonpositive, the objective function is unbounded on the feasible region, and therefore there is no optimal solution.

5. Pivot on the coefficient of the entering basic variable in the equation for which the departing variable is basic. Also include the objective function in the pivoting operation so that it is expressed in terms of the new nonbasic variable.

6. Return to step 3, and continue the process until it terminates in step 3 at a maximal solution, or in step 4 (indicating that the objective function is unbounded).

In the next section we formalize the steps of the simplex method with the aid of the simplex tableau. These tableaus will simplify our work; moreover, they are easily adaptable to the computer and will enable us to deal with LP problems that have many variables and constraints.

Exercises

1. Consider the following expressions as objective functions to be maximized. Assume they are written in terms of the current nonbasic variables of an LP problem. Choose the next entering basic variable used by the simplex algorithm.
   (a) $5x_1 + 2x_2 + 3x_3 - x_4$
   (b) $x_1 - 2x_2 - 3x_3$
   (c) $2x_1 + 2x_2 + x_3$

2. Follow the instructions given in Exercise 1 for the objective functions
   (a) $7x_1 - 2x_2 + 3x_3 + 16x_4 - x_5 + 3x_6$
   (b) $-x_1 + 3x_2 + 2x_3 + 4x_4 - x_5 + x_6 + 3x_7$
   (c) $3x_1 + 4x_2 - 2x_3 + 5x_4 - 18x_5 + x_6 + 4x_7$

3. For the system of linear equations

$$3x_1 + 2x_2 - 4x_3 + 6x_4 + x_5 = 5$$
$$2x_1 - 3x_2 + 2x_3 + 5x_4 + x_6 = 6$$
$$x_1 + x_2 - x_3 + x_4 + x_7 = 3$$
$$4x_1 \qquad + 3x_3 + 2x_4 + x_8 = 2$$

choose the departing basic variable when the entering basic variable is (a) x_1, (b) x_2, (c) x_3, (d) x_4.

4. For the system of linear equations

$$2x_1 - 4x_2 - 3x_3 + 5x_4 + x_5 = 7$$
$$x_1 + 3x_2 - 3x_3 + 4x_4 + x_6 = 8$$
$$-x_1 + 2x_2 - x_3 - x_4 + x_7 = 6$$
$$3x_1 \qquad - 2x_3 + 3x_4 + x_8 = 8$$

choose the departing basic variable when the entering basic variable is (a) x_1, (b) x_2, (c) x_3, (d) x_4.

5. Sometimes the natural formulation of an LP problem leads to a linear system that is in canonical form with an identified basis but has an objective function involving some basic variables. The LP problem

$$\text{Maximize:} \quad z = 2x_1 - x_2 + x_3 + 2x_4$$
$$\text{Subject to:} \quad 3x_1 + 2x_2 + x_3 \qquad = 5$$
$$-2x_1 + x_2 \qquad + x_4 = 6$$
$$x_i \geq 0, \qquad 1 \leq i \leq 4$$

is such an example. Write an equivalent objective function that involves only nonbasic variables.

6. Follow the instructions of Exercise 5 for

$$\text{Maximize:} \quad z = x_1 + x_4 - x_6$$
$$\text{Subject to:} \quad x_1 + 2x_2 + x_3 \qquad - x_5 \qquad = 2$$
$$-2x_1 \qquad + x_4 + 2x_5 \qquad = 4$$
$$3x_1 \qquad + x_5 + x_6 = 3$$
$$x_i \geq 0, 1 \leq i \leq 6$$

7. Solve the following problems by the simplex method.

(a) Maximize: $z = 120x_1 + 100x_2$

Subject to: $x_1 + x_2 \leq 4$
$5x_1 + 3x_2 \leq 15$

$x_1, x_2 \geq 0$

(b) Maximize: $z = 6x_1 + 7x_2$

Subject to: $2x_1 + 3x_2 \leq 400$
$x_1 + x_2 \leq 150$

$x_1, x_2 \geq 0$

(c) Maximize: $z = 10x_1 + 12x_2 + 15x_3$

Subject to: $x_1 - 2x_2 \qquad \leq 6$
$3x_1 \qquad + x_3 \leq 9$
$x_2 + 3x_3 \leq 12$

$x_i \geq 0, 1 \leq i \leq 3$

(d) Maximize: $z = 2x_1 + x_2$

Subject to: $-2x_1 + x_2 \leq 2$
$x_1 - 2x_2 \leq 1$

$x_1, x_2, x_3 \geq 0$

8. Solve the following problems by the simplex method.

(a) Maximize: $z = 120x_1 + 100x_2$

Subject to: $2x_1 + 2x_2 \leq 8$
$5x_1 + 3x_2 \leq 15$
$6x_1 + 2x_2 \leq 12$

$x_1, x_2 \geq 0$

(b) Maximize: $z = 8x_1 + 3x_2$

Subject to: $2x_1 + 3x_2 \leq 5$
$4x_1 + 3x_2 \leq 12$

$x_1, x_2 \geq 0$

(c) Maximize: $z = 2x_1 + 3x_2 + 4x_3$

Subject to:
$$x_1 + x_2 + x_3 \le 8$$
$$2x_1 + 3x_2 + 4x_3 \le 32$$
$$x_1 + 5x_2 + 2x_3 \le 28$$
$$x_i \ge 0, 1 \le i \le 3$$

(d) Maximize: $z = x_1 + 2x_2 + x_3$

Subject to:
$$2x_1 - x_2 + 2x_3 \le 2$$
$$3x_1 + 3x_2 - 2x_3 \le 6$$
$$x_i \ge 0, 1 \le i \le 3$$

In Exercises 9 and 10 you are given LP problems that have feasible solutions making the objective function arbitrarily large. Write equations that show that z can be made arbitrarily large, as we did in Example 3.

9.
$$x_1 - 2x_2 + 3x_3 \qquad\qquad = 4$$
$$- 3x_2 + 4x_3 \qquad + x_5 = 6$$
$$- x_2 - 2x_3 + x_4 \qquad = 3$$
$$z \qquad - 4x_2 + 3x_3 \qquad\qquad = 14$$

10.
$$- x_1 + 2x_2 - 3x_3 + x_4 \qquad\qquad = 6$$
$$- 3x_1 + x_2 - 3x_3 \qquad + x_5 \qquad = 3$$
$$- 5x_1 - 3x_2 + x_3 \qquad\qquad + x_6 = 5$$
$$z - 2x_1 - x_2 + 2x_3 \qquad\qquad\qquad = 36$$

11. A furniture manufacturer produces chairs and sofas, which are sold at prices of $160 and $140, respectively. The manufacturing process requires the following person-hours of labor per item:

	Carpentry	Upholstery	Finishing
Sofa	6	2	1
Chair	3	6	1

Each week there are at most 240 hours of carpentry time, 180 hours of upholstery time, and 45 hours of finishing time available. How many sofas and how many chairs should the manufacturer produce each week to maximize total sales?

12. A developer builds two types of homes. The first requires $24,000 in materials and 160 worker-days to build and is sold for a profit of $4200; the second requires $36,000 in materials and 200 worker-days to build and is sold for a profit of $5000. The developer owns 180 lots, has 30,000 days of worker time available, and can invest $6,000,000 in materials. How many of each type house should be built to maximize profits?

13. This exercise suggests an alternative derivation of the Departing Basic Variable Rule. Consider the canonical system of linear equations

$$a_{11}x_1 + a_{12}x_2 + a_{13}x_3 + x_4 \qquad\qquad = b_1$$
$$a_{21}x_1 + a_{22}x_2 + a_{23}x_3 \qquad + x_5 \qquad = b_2 \qquad\qquad (21)$$
$$a_{31}x_1 + a_{32}x_2 + a_{33}x_3 \qquad\qquad + x_6 = b_3$$

For simplicity, suppose that a_{ij} and b_i are positive for $1 \le i \le 3$ and $1 \le j \le 3$. Suppose that we have determined that x_1 is to be the entering basic variable and want to determine which of x_4, x_5, or x_6 is to be the departing basic variable.

(a) Pivot on a_{11} in (21). Show that the new resource values are

$$b_1' = \frac{b_1}{a_{11}}$$

$$b_2' = b_2 - a_{21}\left(\frac{b_1}{a_{11}}\right) = a_{21}\left(\frac{b_2}{a_{21}} - \frac{b_1}{a_{11}}\right)$$

$$b_3' = b_3 - a_{31}\left(\frac{b_1}{a_{11}}\right) = a_{31}\left(\frac{b_3}{a_{31}} - \frac{b_1}{a_{11}}\right)$$

Notice that b_2' and b_3' are nonnegative if and only if b_1/a_{11} is less than or equal to both b_2/a_{21} and b_3/a_{31}.

(b) Perform a pivot on a_{21} in (21), and write equations for the new resource values similar to those produced in (a). What condition on b_2/a_{21} guarantees that the new resource values are nonnegative?

(c) Either by performing a pivot on a_{31} in (21) or by inspection of the result of (a) and (b), state the condition on b_3/a_{31} that guarantees that a pivot on a_{31} will produce new resource values that are nonnegative.

(d) Explain how you can derive the Departing Basic Variable Rule from these relations.

14. Consider the LP problem

$$\text{Maximize:} \quad z = 5x_1 + 6x_2$$
$$\text{Subject to:} \quad 2x_1 + 3x_2 \le 4$$
$$2x_1 + x_2 \le 15$$
$$x_1, x_2 \ge 0$$

The Greedy Entering Basic Variable Rule selects x_2 as the first entering basic variable. Verify that selecting x_1 will yield a larger increase in z.

15. Consider the LP problem

$$\text{Maximize:} \quad z = 100x_1 + 90x_2$$
$$\text{Subject to:} \quad x_1 + x_2 \le 4$$
$$5x_1 + 3x_2 \le 15$$
$$x_1, x_2 \ge 0$$

The Greedy Entering Basic Variable Rule selects x_1 as the first entering basic variable. Verify that selecting x_2 will yield a larger increase in z.

Exercise 16 introduces a way of discovering that an LP problem has only one solution, and Exercise 17 introduces a way of discovering that an LP problem has more than one solution. We continue this discussion in Exercises 12 and 13 of Section 3.2.

16. You can show that the solution to LP problem (3) is unique by rewriting the optimal system of equations (12) for this problem as

$$x_1 = 2 - \frac{1}{3}x_3 - \frac{1}{3}x_5$$

$$x_4 = 7 - \frac{7}{3}x_3 + \frac{5}{3}x_5$$

$$x_2 = 3 + \frac{1}{3}x_3 - \frac{2}{3}x_5 \qquad (12')$$

$$z = 12 - \frac{1}{3}x_3 - \frac{7}{3}x_5$$

Explain each of the following claims.

(a) The last equation in (12') shows that any feasible solution for which $z = 12$ has $x_3 = 0$ and $x_5 = 0$. Since 12 is also the maximal value of z, it follows that any optimal solution has $x_3 = 0$ and $x_5 = 0$.

(b) The first three equations in (12′) show that the only feasible solution with $x_3 = 0$ and $x_5 = 0$ is $x_1 = 2$, $x_4 = 7$, and $x_2 = 3$.

(c) From (a) and (b) it follows that the solution to LP problem (3) is unique.

17. We introduce slack variables to the LP problem

$$\text{Maximize:} \quad z = 60x_1 + 35x_2 + 20x_3$$

$$\text{Subject to:} \quad 8x_1 + 6x_2 + \quad x_3 \leq 48$$

$$4x_1 + 2x_2 + \frac{3}{2}x_3 \leq 20$$

$$2x_1 + \frac{3}{2}x_2 + \frac{1}{2}x_3 \leq 8$$

$$x_i \geq 0, \quad 1 \leq i \leq 3$$

to obtain the canonical system of equations

$$8x_1 + 6x_2 + \quad x_3 + x_4 \qquad\qquad = 48$$

$$4x_1 + 2x_2 + \frac{3}{2}x_3 \qquad + x_5 \qquad = 20$$

$$2x_1 + \frac{3}{2}x_2 + \frac{1}{2}x_3 \qquad\qquad + x_6 = 8$$

$$z - 60x_1 - 35x_2 - 20x_3 \qquad\qquad\qquad = 0$$

By performing two pivots of the simplex method, you can reduce this system to the optimal system of equations

$$-2x_2 \qquad + x_4 + 2x_5 - 8x_6 = 24$$

$$-2x_2 + x_3 \qquad + 2x_5 - 4x_6 = 8$$

$$x_1 + \frac{5}{4}x_2 \qquad - \frac{1}{2}x_5 + \frac{3}{2}x_6 = 2$$

$$z \qquad\qquad + 10x_5 + 10x_6 = 280$$

Explain each of the following claims.

(a) The variables x_5 and x_6 are 0 in any solution for which $z = 280$. Since $z = 280$ is the maximal value of z, it follows that $x_5 = 0$ and $x_6 = 0$ in any maximal solution.

(b) If we set $x_5 = 0$ and $x_6 = 0$, then

$$x_4 = 24 + 2x_2$$

$$x_3 = 8 + 2x_2$$

$$x_1 = 2 - \frac{5}{4}x_2$$

Thus, whenever

$$0 \leq x_2 \leq \frac{8}{5}$$

we have a feasible solution for which $z = 280$.

(c) This problem has infinitely many feasible solutions that yield the maximal value $z = 280$.

3.2 SIMPLEX TABLEAUS

Simplex tableaus provide a convenient method for carrying out the steps of the simplex method. To see how this is done we retrace (via tableaus) the steps we followed in the previous section in solving the LP problem

$$\text{Maximize:} \quad z = 3x_1 + 2x_2$$

$$\text{Subject to:} \quad \begin{aligned} 2x_1 - x_2 &\le 1 \\ -3x_1 + 4x_2 &\le 13 \\ x_1 + x_2 &\le 5 \end{aligned} \tag{1}$$

$$x_1, x_2 \ge 0$$

Introducing slack variables to (1), we obtain the LP problem in canonical form,

$$\text{Maximize:} \quad z = 3x_1 + 2x_2$$

$$\text{Subject to:} \quad \begin{aligned} 2x_1 - x_2 + x_3 \quad\quad\quad &= 1 \\ -3x_1 + 4x_2 \quad + x_4 \quad\quad &= 13 \\ x_1 + x_2 \quad\quad\quad + x_5 &= 5 \end{aligned} \tag{2}$$

$$x_i \ge 0, \quad 1 \le i \le 5$$

We now construct an array (tableau) consisting of the coefficients that appear in the constraints of (2). We label each column of the tableau with the appropriate variable and list the current basic variables to the left of their corresponding rows:

	x_1	x_2	x_3	x_4	x_5	
x_3	2	-1	1	0	0	1
x_4	-3	4	0	1	0	13
x_5	1	1	0	0	1	5

$$\tag{3}$$

In the previous section we found it convenient to work simultaneously with the objective function

$$z = 3x_1 + 2x_2 \tag{4}$$

rewritten as

$$z - 3x_1 - 2x_2 + 0x_3 + 0x_4 + 0x_5 = 0 \tag{5}$$

If we augment (3) with a column corresponding to z and a row corresponding to (5), we obtain the initial simplex tableau

	z	x_1	x_2	x_3	x_4	x_5	
x_3	0	2	-1	1	0	0	1
x_4	0	-3	4	0	1	0	13
x_5	0	1	1	0	0	1	5
	1	-3	-2	0	0	0	0

(6)

We refer to the bottom row of such a tableau (excluding the current z value in the lower right-hand corner of the tableau) as the *objective row*. It is important to note that entries appearing in the objective row are opposite in sign to the cost coefficients of the objective function z. We call the right-hand column of (6), excluding the value of z, the *resource column*.

We now determine the entering basic variable from (6) by locating the most *negative* entry in the objective row (which results from the largest positive cost coefficient of z). From the objective row of this tableau we see that x_1 is the initial entering basic variable. We call the column headed by the entering basic variable the *pivot column*.

By the Departing Basic Variable Rule of Section 3.1, the initial departing basic variable corresponds to the smallest ratio of entries in the resource column of (6) to *positive* entries in the column headed by the entering basic variable x_1. Since the minimum of $\{\frac{1}{2}, \frac{5}{1}\}$ is $\frac{1}{2}$, it follows that x_3 is the departing basic variable. We refer to the row corresponding to the departing basic variable as the *pivot row* and to the entry at the intersection of the pivot row and the pivot column as the *pivot element*. We circle this entry in Tableau 1, where we also use arrows to denote the entering and departing basic variables.

Tableau 1

	z	x_1 $\downarrow$	x_2	x_3	x_4	x_5	
$\leftarrow x_3$	0	②	-1	1	0	0	1
x_4	0	-3	4	0	1	0	13
x_5	0	1	1	0	0	1	5
	1	-3	-2	0	0	0	0

Pivoting on the circled entry in Tableau 1 we obtain Tableau 2, the second simplex tableau:

Tableau 2

	z	x_1	x_2 $\downarrow$	x_3	x_4	x_5	
x_1	0	1	$-\frac{1}{2}$	$\frac{1}{2}$	0	0	$\frac{1}{2}$
x_4	0	0	$\frac{5}{2}$	$\frac{3}{2}$	1	0	$\frac{29}{2}$
$\leftarrow x_5$	0	0	③ $\frac{3}{2}$	$-\frac{1}{2}$	0	1	$\frac{9}{2}$
	1	0	$-\frac{7}{2}$	$\frac{3}{2}$	0	0	$\frac{3}{2}$

Note that Tableau 2 corresponds precisely to the canonical system of equations (11) in Section 3.1. As indicated previously, you should be able to "read off" these equations from the information given in Tableau 2; for instance, the second row of this tableau represents the equation

$$\frac{5}{2}x_2 + \frac{3}{2}x_3 + x_4 = \frac{29}{2}$$

and the last row represents the equation

$$z - \frac{7}{2}x_2 + \frac{3}{2}x_3 = \frac{3}{2}$$

or, equivalently,

$$z = \frac{7}{2}x_2 - \frac{3}{2}x_3 + \frac{3}{2}$$

where the objective function z is written in terms of the current nonbasic variables x_2 and x_3. Since x_2 and x_3 are equal to 0 in the basic solution, we have $z = \frac{3}{2}$ at this stage of the simplex method; this is the value that appears in the lower right-hand corner of the simplex tableau.

There is just one negative entry in the objective row of Tableau 2; hence, x_2 is the new entering basic variable. Checking the ratio of entries in the resource column of this tableau to positive entries in the column headed by the entering basic variable x_2, we see that x_5 is the departing basic variable; thus, the pivot element is the entry circled. Pivoting on this entry, we obtain Tableau 3, the third simplex tableau:

Tableau 3

	z	x_1	x_2	x_3	x_4	x_5	
x_1	0	1	0	$\frac{1}{3}$	0	$\frac{1}{3}$	2
x_4	0	0	0	$\frac{7}{3}$	1	$-\frac{5}{3}$	7
x_2	0	0	1	$-\frac{1}{3}$	0	$\frac{2}{3}$	3
	1	0	0	$\frac{1}{3}$	0	$\frac{7}{3}$	12

Observe that Tableau 3 corresponds precisely to the canonical system of equations (12) in Section 3.1.

All of the entries in the objective row of Tableau 3 are nonnegative; therefore, by the Optimality Criterion, the optimal solution is

$$(2, 3, 0, 7, 0)$$

and the maximal value of the objective function given by this tableau is $z = 12$.

You may have observed that the column headed by z did not change during this procedure. In fact, it should be clear from the nature of the pivot operations that this column always remains unchanged. For this reason, the z column is customarily omitted from the tableaus. We will follow this convention from now on, but we emphasize once again the need for you to be able to translate the objective row (with or without the z column) into the equation it represents.

In the next example we apply the simplex method to the unbounded LP problem given in Example 3 of Section 3.1.

EXAMPLE 1

From (15) of Section 3.1 we obtain Tableau 1, the initial simplex tableau,

Tableau 1

	x_1	x_2 ↓	x_3	x_4	x_5	
← x_3	-2	①	1	0	0	3
x_4	-3	4	0	1	0	16
x_5	1	-2	0	0	1	6
	-1	-2	0	0	0	0

The circled entry indicates the pivot element, and the arrows indicate the entering and departing basic variables. Pivoting on the circled entry yields Tableau 2,

Tableau 2

	x_1 ↓	x_2	x_3	x_4	x_5	
x_2	-2	1	1	0	0	3
← x_4	⑤	0	-4	1	0	4
x_5	-3	0	2	0	1	12
	-5	0	2	0	0	6

which corresponds to the canonical system of equations (16) in Section 3.1. Pivoting on the circled entry in Tableau 2 gives us Tableau 3, the third simplex tableau,

Tableau 3

	x_1	x_2	x_3	x_4	x_5	
x_2	0	1	$-\frac{3}{5}$	$\frac{2}{5}$	0	$\frac{23}{5}$
x_1	1	0	$-\frac{4}{5}$	$\frac{1}{5}$	0	$\frac{4}{5}$
x_5	0	0	$-\frac{2}{5}$	$\frac{3}{5}$	1	$\frac{72}{5}$
	0	0	-2	1	0	10

The objective row of Tableau 3 indicates that x_3 is the entering basic variable. However, there is no departing basic variable corresponding to this variable because all entries in the column headed by x_3 are nonpositive. This situation illustrates the second way in which the simplex algorithm terminates: by finding that there are feasible solutions that make the objective function arbitrarily large. To see why you can conclude this from Tableau 3, see (17) in Example 3 of Section 3.1. ◆

Definition. The *reduced cost coefficients* are the coefficients of the objective function corresponding to a tableau obtained by applying the simplex method.

EXAMPLE 2

The objective function corresponding to Tableau 3 of Example 1 is

$$z = 0x_1 + 0x_2 + 2x_3 - x_4 + 0x_5$$

The corresponding reduced cost coefficients are

$$0, 0, 2, -1, 0 \qquad \qquad \blacklozenge$$

The next example shows a close connection between the Departing Basic Variable Rule and the Unbounded Termination Rule.

EXAMPLE 3

In the tableau

	x_1	x_2	x_3	x_4	x_5	x_6	x_7	
x_2	a_{11}	1	a_{13}	a_{14}	0	a_{16}	0	b_1
x_7	a_{21}	0	a_{23}	a_{24}	0	a_{26}	1	b_2
x_5	a_{31}	0	a_{33}	a_{34}	1	a_{36}	0	b_3
	$-t_1$	0	$-t_3$	$-t_4$	0	$-t_6$	0	f

suppose that $b_1, b_2, b_3 \geq 0$ and that $t_3 > 0$ (so that $-t_3 < 0$). Further suppose that x_3 has been selected as the entering basic variable and that we are interested in continuing the simplex method by either selecting a departing basic variable or determining that the method terminates because the objective function is unbounded on the feasible region. Because x_3 has been selected as the entering basic variable, the other nonbasic variables x_1, x_4, and x_6 continue to have value 0 in the basic solution after x_3 enters. We write the system of equations corresponding to the tableau with x_1, x_4, and x_6 equal to 0.

$$
\begin{aligned}
x_2 + a_{13}x_3 &= b_1 \qquad \text{or} \qquad x_2 = b_1 - a_{13}x_3 \\
x_7 + a_{23}x_3 &= b_2 \qquad \text{or} \qquad x_7 = b_2 - a_{23}x_3 \\
x_5 + a_{33}x_3 &= b_3 \qquad \text{or} \qquad x_5 = b_3 - a_{33}x_3 \\
z - t_3x_3 &= f \qquad \text{or} \qquad z = f + t_3x_3
\end{aligned}
\qquad (7)
$$

CASE 1. If all of the coefficients a_{13}, a_{23}, and a_{33} are negative, then we can assign any nonnegative value to x_3 and obtain a solution to the problem by computing values of x_2, x_7, and x_5 from (7) and setting x_1, x_4, and x_6 equal to 0. In this case, we can make z arbitrarily large.

CASE 2. If any of the coefficients a_{13}, a_{23}, a_{33} are positive, then the values we can assign to x_3 to produce a feasible solution are bounded. To see this we use the fact that all variables in the feasible region are nonnegative to conclude from (7) that

$$x_2 = b_1 - a_{13}x_3 \geq 0$$
$$x_7 = b_2 - a_{23}x_3 \geq 0$$
$$x_5 = b_3 - a_{33}x_3 \geq 0 \qquad\qquad (8)$$
$$z = f + t_3 x_3$$

Any positive coefficient a_{13}, a_{23}, or a_{33} produces an upper bound on the amount that x_3 can increase in the corresponding inequality of (8).

CASE 3. All of a_{13}, a_{23}, a_{33} are 0. This case never occurs in a tableau derived by the simplex method unless the coefficients of x_3 are all zero in the original tableau. Ordinarily, we do not include a variable in an LP problem if its coefficient in every constraint is 0.   ♦

In Section 2.2 we observed that some LP problems have infinitely many optimal solutions (all yielding the same value of the objective function). Geometrically speaking, this situation occurs when the hyperplane determined by the objective function is parallel to one of the hyperplanes bounding the feasible region. In the next example we show how to discover the existence of alternative optimal solutions by examining the final tableau of the simplex method.

EXAMPLE 4

We apply the simplex method to the LP problem

$$\text{Maximize:} \quad z = 100x_1 + 100x_2$$
$$\text{Subject to:} \quad 2x_1 + 2x_2 \leq 8$$
$$5x_1 + 3x_2 \leq 15$$
$$x_1, x_2 \geq 0$$

to obtain Tableaus 1–3.

Tableau 1

	$\downarrow$ x_1	x_2	x_3	x_4	
x_3	2	2	1	0	8
$\leftarrow x_4$	⑤	3	0	1	15
	-100	-100	0	0	0

Tableau 2

	x_1	x_2	x_3	x_4	
$\leftarrow x_3$	0	$\frac{4}{5}$	1	$-\frac{2}{5}$	2
x_1	1	$\frac{3}{5}$	0	$\frac{1}{5}$	3
	0	-40	0	20	300

Tableau 3

	x_1	x_2	x_3	x_4	
x_2	0	1	$\frac{5}{4}$	$-\frac{1}{2}$	$\frac{5}{2}$
x_1	1	0	$-\frac{3}{4}$	$\frac{1}{2}$	$\frac{3}{2}$
	0	0	50	0	400

Because all entries in its objective row are nonnegative, Tableau 3 gives the optimal solution

$$x_1 = \frac{3}{2}, \qquad x_2 = \frac{5}{2}; \qquad z = 400$$

Notice that the entry in the objective row of Tableau 3 in the column labeled by the *nonbasic* variable x_4 is 0. Such a 0 in a column corresponding to a nonbasic variable indicates that there are alternative optimal solutions. To find a corner point yielding an alternative optimal solution, we can perform a pivot on any positive entry in this column. In this case, we pivot on the $\frac{1}{2}$ in row x_1 and column x_4 to obtain Tableau 4, another optimal tableau,

Tableau 4

	x_1	x_2	x_3	x_4	
x_2	1	1	$\frac{1}{2}$	0	4
x_4	2	0	$-\frac{3}{2}$	1	3
	0	0	50	0	400

Tableau 4 gives the alternative optimal solution

$$x_2 = 4, \qquad x_4 = 3; \qquad z = 400$$

See Exercises 12–15 for a continuation of this discussion. Exercise 13 concerns the situation in which all entries in a column corresponding to a nonbasic variable with a 0 reduced cost coefficient are nonpositive.   ◆

EXAMPLE 5

In Example 1 of Section 2.5, we formulated a product-mix problem for the Carter Nut Company. Tableau 1 shows the problem in tableau form.

Tableau 1

	x_1 ↓	x_2	x_3	s_1	s_2	s_3	s_4	s_5	s_6	
← s_1	(0.75)	−0.25	−0.25	1	0	0	0	0	0	0
s_2	0.40	0.40	−0.60	0	1	0	0	0	0	0
s_3	1	1	1	0	0	1	0	0	0	1000
s_4	1	0	0	0	0	0	1	0	0	400
s_5	0	1	0	0	0	0	0	1	0	250
s_6	0	0	1	0	0	0	0	0	1	200
	−0.60	−0.45	−0.30	0	0	0	0	0	0	0

The simplex method yields Tableaus 2–4 (entries are displayed only to two decimal places)

Tableau 2

	x_1	x_2 ↓	x_3	s_1	s_2	s_3	s_4	s_5	s_6	
x_1	1	−0.33	−0.33	1.33	0	0	0	0	0	0
← s_2	0	(0.53)	−0.47	−0.53	1	0	0	0	0	0
s_3	0	1.33	1.33	−1.33	0	1	0	0	0	1000
s_4	0	0.33	0.33	−1.33	0	0	1	0	0	400
s_5	0	1	0	0	0	0	0	1	0	250
s_6	0	0	1	0	0	0	0	0	1	200
	0	−0.65	−0.50	0.80	0	0	0	0	0	0

Tableau 3

	x_1	x_2	x_3 ↓	s_1	s_2	s_3	s_4	s_5	s_6	
x_1	1	0	−0.63	1	0.63	0	0	0	0	0
x_2	0	1	−0.88	−1	1.88	0	0	0	0	0
s_3	0	0	2.50	0	−2.50	1	0	0	0	1000
s_4	0	0	0.63	−1	−0.63	0	1	0	0	400
s_5	0	0	0.88	1	−1.88	0	0	1	0	250
← s_6	0	0	(1)	0	0	0	0	0	1	200
	0	0	−1.07	0.15	1.22	0	0	0	0	0

Tableau 4

	x_1	x_2	x_3	s_1	s_2	s_3	s_4	s_5	s_6	
x_1	1	0	0	1	0.63	0	0	0	0.63	125
x_2	0	1	0	−1	1.88	0	0	0	0.88	175
s_3	0	0	0	0	−2.50	1	0	0	−2.50	500
s_4	0	0	0	−1	−0.63	0	1	0	−0.63	275
s_5	0	0	0	1	−1.88	0	0	1	−0.88	75
x_3	0	0	1	0	0	0	0	0	1	200
	0	0	0	0.15	1.22	0	0	0	1.07	213.75

Tableau 4 is optimal and yields the following production plan for the Carter Nut Company (CNC):

$$x_1 = 125 \text{ pounds of peanuts}$$

$$x_2 = 175 \text{ pounds of almonds}$$

$$x_3 = 200 \text{ pounds of cashews}$$

$$\text{Profit} = \$213.75$$

As is often the case, the final values of the slack variables contain useful information.

$s_3 = 500$ means that CNC should not use 500 pounds of available production capacity.

$s_4 = 275$ means that 275 pounds of peanuts were surplus.

$s_5 = 75$ means that 75 pounds of almonds were surplus.

$s_6 = 0$ means that all the cashews were used.   ◆

Exercises

1. Consider the LP problem:

$$\text{Maximize:} \quad z = 100x_1 + 120x_2$$

$$\text{Subject to:} \quad x_1 + x_2 \leq 4$$

$$3x_1 + 5x_2 \leq 15$$

$$x_1, x_2 \geq 0$$

(a) Verify that the following tableau results from adding slack variables s_1 and s_2 to this problem.

	x_1	x_2	s_1	s_2	
s_1	1	1	1	0	4
s_2	3	5	0	1	15
	-100	-120	0	0	0

(b) The following tableau was obtained from the preceding tableau by performing one pivot. Verify that this is so.

	x_1	x_2	s_1	s_2	
s_1	$\frac{2}{5}$	0	1	$-\frac{1}{5}$	1
x_2	$\frac{3}{5}$	1	0	$\frac{1}{5}$	3
	-28	0	0	24	360

(c) Write the system of equations corresponding to the tableau in (b). Include the objective function.

(d) What are the values of the current basic variables from (b) or (c) in the basic solution?

(e) Evaluate the objective function obtained in (c) and the original objective function at the values of the basic variables obtained in (c). You should, of course, get the same answer.

(f) Verify that the following tableau results from the one in part (b) after one pivot. Notice that the tableau represents the maximal solution.

	x_1	x_2	s_1	s_2	
x_1	1	0	$\frac{5}{2}$	$-\frac{1}{2}$	$\frac{5}{2}$
x_2	0	1	$-\frac{3}{2}$	$\frac{1}{2}$	$\frac{3}{2}$
	0	0	70	10	430

(g) Write the system of equations corresponding to the tableau in part (f). Evaluate the objective function as it appears in part (f) and as it appears in part (a) at the final basic solution. The values should, of course, be the same.

2. Consider the LP problem

$$\text{Maximize:} \quad z = 320x_1 + 240x_2$$
$$\text{Subject to:} \quad 5x_1 + 3x_2 \leq 12$$
$$4x_1 + 6x_2 \leq 24$$
$$x_1, x_2 \geq 0$$

(a) Verify that the following tableau results from adding slack variables s_1 and s_2 to this problem.

	x_1	x_2	s_1	s_2	
s_1	5	3	1	0	12
s_2	4	6	0	1	24
	-320	-240	0	0	0

(b) The next tableau was obtained from the preceding tableau by performing one pivot. Verify that this is so.

	x_1	x_2	s_1	s_2	
x_1	1	$\frac{3}{5}$	$\frac{1}{5}$	0	$\frac{12}{5}$
s_2	0	$\frac{18}{5}$	$-\frac{4}{5}$	1	$\frac{72}{5}$
	0	-48	64	0	768

(c) Write the system of equations corresponding to the tableau in (b). Include the objective function.

(d) What are the values of the current basic variables from (b) or (c) in the basic solution?

(e) Evaluate the objective function obtained in (c) and the original objective function at the values of the basic variables obtained in (c). You should, of course, get the same answer.

(f) Verify that the next tableau results from the one in part (b) after one pivot. Notice that the tableau represents the maximal solution.

	x_1	x_2	s_1	s_2	
x_1	1	0	$\frac{1}{3}$	$-\frac{1}{6}$	0
x_2	0	1	$-\frac{2}{9}$	$\frac{5}{18}$	4
	0	0	$\frac{160}{3}$	$\frac{40}{3}$	960

(g) Write the system of equations corresponding to the tableau in (f). Evaluate the objective function as it appears in (f) and as it appears in (a) at the final basic solution. The values should, of course, be the same.

In Exercises 3–10 form the appropriate simplex tableau and solve by the simplex method.

3. Maximize: $z = 10x_1 + 6x_2 - 8x_3$

Subject to: $5x_1 - 2x_2 + 6x_3 \le 20$

$10x_1 + 4x_2 - 6x_3 \le 30$

$x_i \ge 0, \quad 1 \le i \le 3$

4. Minimize: $z = -5x_1 - 14x_2 + 7x_3 + 9x_4$

Subject to: $x_1 - 2x_2 + x_3 + x_4 = 5$

$-x_1 - 3x_2 \qquad + 5x_4 \le 10$

$2x_1 + 2x_2 \qquad + 4x_4 \le 10$

$x_i \ge 0, \quad 1 \le i \le 4$

Be careful! You need to use x_3 as a basic variable in Exercise 4, even though it also occurs in the objective function.

5. Maximize: $z = 2x_1 + 3x_2 + 3x_3$

   Subject to: $3x_1 + 2x_2 \qquad \le 60$
$$-x_1 + x_2 + 4x_3 \le 10$$
$$2x_1 - 2x_2 + 5x_3 \le 5$$
$$x_i \ge 0, \quad 1 \le i \le 3$$

6. Maximize: $z = x_3 - x_4$

   Subject to: $x_1 \qquad\quad -3x_4 + x_5 = 1$
$$x_2 \qquad + 6x_4 - 5x_5 = 6$$
$$x_3 - 3x_4 + 2x_5 = 5$$
$$x_i \ge 0, \quad 1 \le i \le 5$$

Something is wrong with the objective function.

7. Minimize: $z = 3x_1 - 2x_2 - 5x_3$

   Subject to: $-3x_1 + 2x_2 + 6x_3 \le 9$
$$x_1 + 2x_2 - x_3 \le 6$$
$$-2x_1 + 4x_2 + 2x_3 \le 5$$
$$x_i \ge 0, \quad 1 \le i \le 3$$

8. Minimize: $z = -2x_1 + 3x_2 - 4x_3$

   Subject to: $x_1 - 2x_2 + 4x_3 \le 15$
$$2x_1 - x_2 + 5x_3 \le 10$$
$$x_i \ge 0, \quad 1 \le i \le 3$$

9. Maximize: $z = 2x_1 - x_2$

   Subject to: $2x_1 - 3x_2 \le 12$
$$-x_1 + 3x_2 \le 8$$
$$x_1 \le 0, \quad x_2 \text{ unrestricted}$$

10. Maximize: $z = 3x_1 - 2x_2 + x_3$

   Subject to: $3x_1 - 2x_2 + x_3 \le 15$
$$-x_1 + 2x_2 + 3x_3 \le 12$$
$$x_1, x_2 \le 0; \quad x_3 \text{ unrestricted}$$

11. Demonstrate that the LP problem of Example 1 is unbounded by carrying out the following steps.
(a) Write the system of equations corresponding to Tableau 3 of Example 1. Also write the objective function corresponding to the objective row.
(b) Set the nonbasic variable x_4 equal to zero in each of the equations obtained in (a).
(c) Solve the equations obtained in (b) for the basic variables and for z.
(d) Observe that you can obtain a solution to the problem from the equations of (c) for any positive value of x_3. Conclude that you can make z as large as desired.

Exercises 12–15 continue the discussion of LP problems that have infinitely many optimal solutions which we began in Exercises 16 and 17 of Section 3.1.

12. Draw the graph of the feasible set for the LP problem in Example 4. Use the graph to explain why this problem has alternative optimal solutions at two corner points of the feasible set.

13. Consider the LP problem

$$\text{Maximize:} \quad z = 120x_1 + 100x_2 - 105x_3$$
$$\text{Subject to:} \quad 2x_1 + 2x_2 - 2x_3 \le 8$$
$$5x_1 + 3x_2 - \frac{7}{2}x_3 \le 15$$
$$x_i \ge 0, \quad 1 \le i \le 3$$

(a) Solve this problem by the simplex method.
(b) Observe that x_3 is nonbasic in the optimal tableau and has zero reduced cost. Notice that all entries in the column for x_3 are negative, and, therefore, that you cannot move to another feasible corner point by pivoting.
(c) Explain why this problem has infinitely many optimal solutions.

14. The LP problem in Exercise 17 of Section 3.1 has the initial tableau

	x_1	x_2	x_3	x_4	x_5	x_6	
x_4	8	6	1	1	0	0	48
x_5	4	2	$\frac{3}{2}$	0	1	0	20
x_6	2	$\frac{3}{2}$	$\frac{1}{2}$	0	0	1	8
	-60	-35	-20	0	0	0	0

(a) Use the simplex method with the Greedy Entering Basic Variable Rule to show that the final tableau is

	x_1	x_2	x_3	x_4	x_5	x_6	
x_4	0	-2	0	1	2	-8	24
x_3	0	-2	1	0	2	-4	8
x_1	1	$\frac{5}{4}$	0	0	$-\frac{1}{2}$	$\frac{3}{2}$	2
	0	0	0	0	10	10	280

(b) Notice that the nonbasic variable x_2 has zero reduced cost in the final tableau. The Departing Variable Rule selects $\frac{5}{4}$ as the pivot element. Pivot on this element to obtain the next tableau and observe that the objective function value does not change.

	x_1	x_2	x_3	x_4	x_5	x_6	
x_4	$\frac{8}{5}$	0	0	1	$\frac{6}{5}$	$-\frac{28}{5}$	$\frac{136}{5}$
x_3	$\frac{8}{5}$	0	1	0	$\frac{6}{5}$	$-\frac{8}{5}$	$\frac{56}{5}$
x_2	$\frac{4}{5}$	1	0	0	$-\frac{2}{5}$	$\frac{6}{5}$	$\frac{8}{5}$
	0	0	0	0	10	10	280

(c) The tableau in part (a) shows that the corner point $(x_1, x_2, x_3) = (2, 0, 8)$ gives an optimal solution, and the tableau in part (b) shows that the corner point $(x_1, x_2, x_3) = (0, \frac{8}{5}, \frac{56}{5})$ gives another optimal solution. Explain why

$$(x_1, x_2, x_3) = \left(2 - 2\lambda, \frac{8}{5}\lambda, 8 + \frac{16}{5}\lambda\right)$$

gives an optimal solution for $0 \le \lambda \le 1$.

15. Describe a general test that you can perform on any optimal tableau to determine whether or not the problem has more than one optimal solution.

    The following exercises should be done using the computer program CALIPSO. In the sequel, exercises that should be done on a computer have the symbol † before the problem number. You should do Exercise 16 before attempting other exercises.

†**16.** Turn to Appendix A and read Sections A.1 and A.2. Perform all of the exercises described in A.2.

Use the TUTOR program to solve the following problems.

†**17.** Maximize: $z = 3x_1 + x_2 + 3x_3$

Subject to: $2x_1 + x_2 + x_3 \leq 2$

$x_1 + 2x_2 + 3x_3 \leq 5$

$2x_1 + 2x_2 + x_3 \leq 6$

$x_i \geq 0, \quad 1 \leq i \leq 3$

†**18.** Solve Exercise 8 from Section 3.1.

†**19.** Maximize: $z = \dfrac{2}{3}x_1 + 2x_2 + \dfrac{1}{2}x_3 + x_4$

Subject to: $x_1 + 3x_2 \qquad + x_4 \leq 4$

$x_1 + \dfrac{1}{2}x_2 \qquad\qquad \leq \dfrac{3}{2}$

$x_2 + 4x_3 + x_4 \leq 3$

$x_i \geq 0, \quad 1 \leq i \leq 4$

†**20.** Maximize: $z = 3x_1 - 2x_2 + x_3 + 4x_4$

Subject to: $2x_1 - x_2 + 3x_3 + x_4 \leq 12$

$x_1 + 2x_2 + 4x_3 + 5x_4 \leq 42$

$3x_1 + x_2 - x_3 + 4x_4 \leq 24$

$x_i \geq 0, \quad 1 \leq i \leq 4$

†**21.** Maximize: $z = -10x_1 + 32x_2 + 48x_3 + 54x_4$

Subject to: $2x_1 + 3x_2 + 5x_3 + x_4 \leq 24$

$5x_1 + 2x_2 + x_3 + 3x_4 \leq 32$

$8x_1 + 5x_2 + 6x_3 + 10x_4 \leq 64$

$3x_1 + 6x_2 + 9x_3 + 12x_4 \leq 81$

$x_i \geq 0, \quad 1 \leq i \leq 4$

†**22.** Maximize: $z = 120x_1 - 32x_2 + 48x_3 + 64x_4$

Subject to: $3x_1 + 2x_2 - x_3 + 4x_4 \leq 42$

$-x_1 + 2x_2 + 3x_3 + 12x_4 \leq 36$

$4x_1 + 3x_2 + 5x_3 + 21x_4 \leq 45$

$8x_1 + 3x_2 + 4x_3 + 5x_4 \leq 28$

$6x_1 + 2x_2 + 4x_3 + x_4 \leq 14$

$x_i \geq 0, \quad 1 \leq i \leq 4$

†**23.** Use the simplex method to show that the following problem is unbounded.

Minimize: $z = 13x_2 - 6x_3 + x_4$

Subject to: $-2x_1 + 6x_2 + 2x_3 - 3x_4 \leq 20$

$-4x_1 + 7x_2 + x_3 - x_4 \leq 10$

$-5x_2 + 3x_3 - x_4 \leq 60$

$x_i \geq 0, \quad 1 \leq i \leq 4$

†**24.** Solve Exercise 1 of Section 1.1.

*3.3 DEGENERACY, CYCLING, AND CONVERGENCE

In the LP problems we have considered thus far, each step of the simplex method has led to an improvement in the value of the objective function. In the next example we see that this does not always happen, though we still are able to obtain an optimal solution to the problem.

EXAMPLE 1

We consider the LP problem

$$\text{Maximize:} \quad z = 3x_1 + 4x_2$$

$$\text{Subject to:} \quad \begin{aligned} -x_1 + x_2 &\le 3 \\ -x_1 + 2x_2 &\le 10 \\ x_1 + 4x_2 &\le 32 \\ 2x_1 + 3x_2 &\le 29 \\ x_1 - 2x_2 &\le 4 \end{aligned} \tag{1}$$

$$x_1, x_2 \ge 0$$

Introducing slack variables gives us the LP problem in canonical form with basic variables x_3, x_4, x_5, x_6, x_7.

$$\text{Maximize:} \quad z = 3x_1 + 4x_2$$

$$\text{Subject to:} \quad \begin{aligned} -x_1 + x_2 + x_3 &= 3 \\ -x_1 + 2x_2 + x_4 &= 10 \\ x_1 + 4x_2 + x_5 &= 32 \\ 2x_1 + 3x_2 + x_6 &= 29 \\ x_1 - 2x_2 + x_7 &= 4 \end{aligned} \tag{2}$$

$$x_1, x_2 \ge 0$$

Figure 1 illustrates the feasible region for (1).

Note that the intersection of *any pair* of the lines

$$\begin{aligned} -x_1 + x_2 &= 3 \\ -x_1 + 2x_2 &= 10 \\ x_1 + 4x_2 &= 32 \\ 2x_1 + 3x_2 &= 29 \end{aligned}$$

determines the corner point (4, 7). Applying the simplex method to (2) results first in

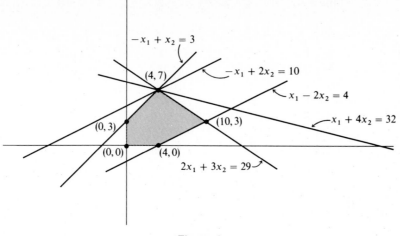

Figure 1

Tableau 1, which begins (geometrically) at the origin, determined by the lines $x_1 = 0$ and $x_2 = 0$.

Tableau 1

	x_1	x_2	x_3	x_4	x_5	x_6	x_7	
x_3	-1	①	1	0	0	0	0	3
x_4	-1	2	0	1	0	0	0	10
x_5	1	4	0	0	1	0	0	32
x_6	2	3	0	0	0	1	0	29
x_7	1	-2	0	0	0	0	1	4
	-3	-4	0	0	0	0	0	0

Tableau 2 represents (geometrically) a move from the origin to corner point $(0, 3)$ determined by the lines $x_1 = 0$ and $-x_1 + x_2 = 3$. At this stage, note that the entering basic variable is x_1 and that the departing basic variable can be x_4, x_5, or x_6.

Tableau 2

	x_1	x_2	x_3	x_4	x_5	x_6	x_7	
x_2	-1	1	1	0	0	0	0	3
x_4	①	0	-2	1	0	0	0	4
x_5	5	0	-4	0	1	0	0	20
x_6	5	0	-3	0	0	1	0	20
x_7	-1	0	2	0	0	0	1	10
	-7	0	4	0	0	0	0	12

Selecting x_4 as the departing basic variable, we obtain Tableau 3.

Tableau 3

	x_1	x_2	x_3	x_4	x_5	x_6	x_7	
x_2	0	1	-1	1	0	0	0	7
x_1	1	0	-2	1	0	0	0	4
x_5	0	0	6	-5	1	0	0	0
x_6	0	0	⑦	-5	0	1	0	0
x_7	0	0	0	1	0	0	1	14
	0	0	-10	7	0	0	0	40

Tableau 3 represents (geometrically) a move from corner point $(0,3)$ to corner point $(4,7)$. Since in Tableau 3, $x_3 = 0$ and $x_4 = 0$, the lines $-x_1 + x_2 = 3$ and $-x_1 + 2x_2 = 10$ are the lines determining the corner point $(4,7)$. The entering basic variable is now x_3, and again we have a choice of departing basic variables: x_5 or x_6. If we select x_6 as the departing basic variable, we obtain Tableau 4.

Tableau 4

	x_1	x_2	x_3	x_4	x_5	x_6	x_7	
x_2	0	1	0	$\frac{2}{7}$	0	$\frac{1}{7}$	0	7
x_1	1	0	0	$-\frac{3}{7}$	0	$\frac{2}{7}$	0	4
x_5	0	0	0	$-\frac{5}{7}$	1	$-\frac{6}{7}$	0	0
x_3	0	0	1	$-\frac{5}{7}$	0	$\frac{1}{7}$	0	0
x_7	0	0	0	①	0	0	1	14
	0	0	0	$-\frac{1}{7}$	0	$\frac{10}{7}$	0	40

Tableau 4 represents (geometrically) remaining at $(4,7)$, though now we view this point as the intersection of lines $-x_1 + 2x_2 = 10$ and $2x_1 + 3x_2 = 29$ (since $x_4 = 0$ and $x_6 = 0$).

At this point in the simplex method we have encountered a troublesome phenomenon: There has been no improvement in the objective function! As you will soon see, this means the simplex method could be caught in an endless loop. Geometrically, we have not moved from one corner point to an adjacent one but instead remain at the same corner point and simply regard it as the intersection of a different pair of lines.

In applying the simplex method, we may encounter this situation whenever there is a "tie" in selecting the new departing basic variable. It is easy to show (see Exercise 4) that such a tie leads to a resource entry of 0. If, in this case, we carry out one more step of the simplex method, then the Departing Basic Variable Rule may select the row having this 0 resource entry as the pivot row. Consequently, on pivoting we do not improve the current z value because we only add 0 to the current value of the objective function.

We say that *degeneracy* occurs if a basic variable becomes 0. In most instances, degeneracy is not of particular significance because subsequent applications of the simplex method usually result in an optimal solution. In our example, pivoting on the circled entry in Tableau 4 yields Tableau 5.

Tableau 5

	x_1	x_2	x_3	x_4	x_5	x_6	x_7	
x_2	0	1	0	0	0	$\frac{1}{7}$	$-\frac{2}{7}$	3
x_1	1	0	0	0	0	$\frac{2}{7}$	$\frac{3}{7}$	10
x_5	0	0	0	0	1	$-\frac{6}{7}$	$\frac{5}{7}$	10
x_3	0	0	1	0	0	$\frac{1}{7}$	$\frac{5}{7}$	10
x_4	0	0	0	1	0	0	1	14
	0	0	0	0	0	$\frac{10}{7}$	$\frac{1}{7}$	42

Since the objective row of Tableau 5 has no negative entries, it follows that we have obtained an optimal solution $(10, 3, 10, 14, 10, 0, 0)$ of (2), which corresponds geometrically to the corner point $(10, 3)$.   ◆

In extremely rare cases, degeneracy can lead to *cycling*. This occurs if, in the case of ties, we make a series of unfortunate choices of departing basic variables leading to a tableau coinciding with a previously obtained tableau. If we continue to make the same cycle of choices of departing basic variables, we create an infinite loop of tableaus and never attain an optimal solution. This possibility is of particular concern in programming a computer to solve LP problems because computer programs are usually written to make the same choice in cases of ties. The next example, which is due to Beale [1], illustrates this possibility.

EXAMPLE 2

We consider the following LP problem in canonical form.

Maximize: $z = 10x_1 - 57x_2 - 9x_3 - 24x_4$

Subject to: $\frac{1}{2}x_1 - \frac{11}{2}x_2 - \frac{5}{2}x_3 + 9x_4 + x_5 \qquad\qquad = 0$

$\frac{1}{2}x_1 - \frac{3}{2}x_2 - \frac{1}{2}x_3 + x_4 \qquad + x_6 \qquad = 0$

$x_1 \qquad\qquad\qquad\qquad\qquad + x_7 = 1$

$x_i \geq 0, \qquad 1 \leq i \leq 7$

The simplex method yields the following sequence of tableaus (Tableaus 1–7). Note that Tableau 1, the initial tableau, is degenerate because the *basic* variables x_5 and x_6 are 0.

Tableau 1

	x_1	x_2	x_3	x_4	x_5	x_6	x_7	
x_5	$(\frac{1}{2})$	$-\frac{11}{2}$	$-\frac{5}{2}$	9	1	0	0	0
x_6	$\frac{1}{2}$	$-\frac{3}{2}$	$-\frac{1}{2}$	1	0	1	0	0
x_7	1	0	0	0	0	0	1	1
	-10	57	9	24	0	0	0	0

Tableau 2

	x_1	x_2	x_3	x_4	x_5	x_6	x_7	
x_1	1	-11	-5	18	2	0	0	0
x_6	0	(4)	2	-8	-1	1	0	0
x_7	0	11	5	-18	-2	0	1	1
	0	-53	-41	204	20	0	0	0

Tableau 3

	x_1	x_2	x_3	x_4	x_5	x_6	x_7	
x_1	1	0	$(\frac{1}{2})$	-4	$-\frac{3}{4}$	$\frac{11}{4}$	0	0
x_2	0	1	$\frac{1}{2}$	-2	$-\frac{1}{4}$	$\frac{1}{4}$	0	0
x_7	0	0	$-\frac{1}{2}$	4	$\frac{3}{4}$	$-\frac{11}{4}$	1	1
	0	0	$-\frac{29}{2}$	98	$\frac{27}{4}$	$\frac{53}{4}$	0	0

Tableau 4

	x_1	x_2	x_3	x_4	x_5	x_6	x_7	
x_3	2	0	1	-8	$-\frac{3}{2}$	$\frac{11}{2}$	0	0
x_2	-1	1	0	(2)	$\frac{1}{2}$	$-\frac{5}{2}$	0	0
x_7	1	0	0	0	0	0	1	1
	29	0	0	-18	-15	93	0	0

Tableau 5

	x_1	x_2	x_3	x_4	x_5	x_6	x_7	
x_3	-2	4	1	0	$(\frac{1}{2})$	$-\frac{9}{2}$	0	0
x_4	$-\frac{1}{2}$	$\frac{1}{2}$	0	1	$\frac{1}{4}$	$-\frac{5}{4}$	0	0
x_7	1	0	0	0	0	0	1	1
	20	9	0	0	$-\frac{21}{2}$	$\frac{141}{2}$	0	0

Tableau 6

	x_1	x_2	x_3	x_4	x_5	x_6	x_7	
x_5	-4	8	2	0	1	-9	0	0
x_4	$\frac{1}{2}$	$-\frac{3}{2}$	$-\frac{1}{2}$	1	0	①	0	0
x_7	1	0	0	0	0	0	1	1
	-22	93	21	0	0	-24	0	0

Tableau 7

	x_1	x_2	x_3	x_4	x_5	x_6	x_7	
x_5	$\frac{1}{2}$	$-\frac{11}{2}$	$-\frac{5}{2}$	9	1	0	0	0
x_6	$\frac{1}{2}$	$-\frac{3}{2}$	$-\frac{1}{2}$	1	0	1	0	0
x_7	1	0	0	0	0	0	1	1
	-10	57	9	24	0	0	0	0

Observe that Tableau 7 coincides with Tableau 1; so no progress has been made toward finding an optimal solution (which, as we shall see in Example 3, is $(1, 0, 1, 0, 2, 0, 0)$ with $z = 1$). Note, moreover, that if we were to continue to use the same choices for departing basic variables, we would cycle forever.   ◆

Cycling is not as serious a problem as it might appear. It occurs very rarely in practice; so rarely, in fact, that most commercial computer codes ignore this possibility.

In 1952 Charnes [4] developed a perturbation technique that eliminates cycling. Charnes based his method essentially on slightly moving (perturbing) the hyperplanes that define the feasible regions so that no basic variable becomes 0. In 1977, Bland [3] showed that the following simple method for choosing entering and departing basic variables prevents cycling.

BLAND'S RULE

1. For the entering basic variable: Of all negative coefficients in the objective row, choose the one with smallest subscript.
2. For the departing basic variable: When there is a tie between one or more ratios computed using the ordinary Departing Basic Variable Rule, choose the candidate for departing basic variable that has the smallest subscript.

Bland showed that if these choices are made, then the simplex method always terminates. In Exercise 8, you are asked to fill in the details of a proof showing that cycling cannot occur when Bland's Rule is used.

EXAMPLE 3

In examining the tableaus in Example 2, we see that we followed Bland's Rule until we reached Tableau 6. According to Bland's Rule x_1 (rather than x_6) is the entering basic variable in Tableau 6; x_4 remains the departing basic variable. Pivoting on the entry $\frac{1}{2}$ in Tableau 6 in Example 2, we obtain Tableau 7',

Tableau 7'

	x_1	x_2	x_3	x_4	x_5	x_6	x_7	
x_5	0	−4	−2	8	1	−1	0	0
x_1	1	−3	−1	2	0	2	0	0
x_7	0	3	①	−2	0	−2	1	1
	0	27	−1	44	0	20	0	0

Continuing with the simplex method, we obtain Tableau 8,

Tableau 8

	x_1	x_2	x_3	x_4	x_5	x_6	x_7	
x_5	0	2	0	4	1	−5	2	2
x_1	1	0	0	0	0	0	1	1
x_3	0	3	1	−2	0	−2	1	1
	0	30	0	42	0	18	1	1

Thus $(1, 0, 1, 0, 2, 0, 0)$ is the optimal solution of the problem of Example 2, and the maximal value of the objective function is $z = 1$. ◆

It is not difficult to see that the simplex method converges if degeneracy does not occur. Theorem 1 states this more carefully.

Theorem 1. Suppose that an LP problem in canonical form has the property that degeneracy never occurs during execution of the simplex method. Then the simplex method must terminate, either by producing an optimal solution or by reaching a tableau that shows that the objective function is unbounded on the feasible region.

PROOF. There are a finite number of corner points of the feasible region of an LP problem. Thus, if we demonstrate that (in the absence of degeneracy) the simplex method actually moves us from one corner point to another corner point with each pivot and that we never return to a corner point previously examined, then we have proven the theorem.

Our proof is based on the fact that (in the absence of degeneracy) pivoting at an entry selected by the simplex method always leads to an improvement in the value of the objective function. At any stage of the simplex method in which the tableau has not reached a terminal form and in which degeneracy does not occur, the rules for selecting a pivot element produce (a) a value $q < 0$ from the objective row, (b) a resource value $b > 0$ (in the pivot row), and (c) a pivot element $a > 0$. If the value of the objective function before the pivot is z_0, then by the nature of the pivot operation, the value after the pivot is $z_0 - qb/a$, a positive increase.

Thus, every pivot performed by the simplex method results (in the absence of degeneracy) in an increase in the value of the objective function. To conclude the proof of the theorem, we observe that in the absence of degeneracy every pivot results (geometrically) in a move from one corner point of the feasible region to another. This is clear because we can always calculate the value of the objective function by evaluating the original expression

$$z = c_1 x_1 + c_2 x_2 + \cdots + c_n x_n \tag{3}$$

at the current corner point $(x_1, x_2, \ldots, x_n)$, where the values of $x_1, x_2, \ldots, x_n$ are taken from any of the tableaus. Therefore, if pivoting one tableau to another leads to a change in the value of the objective function, the coordinates of the corner point must have changed.

Only one question remains unanswered to complete the proof: Why cannot the simplex method cycle (i.e., return to a corner point that was previously examined)? It cannot do so because the value of the objective function *increases* at every step. Since (3) shows that we can calculate the value of the objective function from the coordinates of the corner point, the method never returns to a previously examined corner point.   ∎

EXAMPLE 4

As an example of the proof of Theorem 1 consider the LP problem

Maximize: $z = 2x_1 + 3x_2 + 3x_3$

Subject to:
$$3x_1 + 2x_2 \quad\quad + x_4 \quad\quad\quad\quad\quad = 60$$
$$-x_1 + x_2 + 4x_3 \quad\quad + x_5 \quad\quad = 10$$
$$2x_1 - 2x_2 + 5x_3 \quad\quad\quad\quad + x_6 = 50$$
$$x_i \geq 0, \quad 1 \leq i \leq 6$$

The initial tableau (Tableau 1) is

Tableau 1

	x_1	x_2	x_3	x_4	x_5	x_6	
x_4	3	2	0	1	0	0	60
x_5	−1	①	4	0	1	0	10
x_6	2	−2	5	0	0	1	50
	−2	−3	−3	0	0	0	0

The corner point corresponding to Tableau 1 is $(0, 0, 0)$ and

$$z = 2(0) + 3(0) + 3(0) = 0$$

Pivoting on the circled entry in Tableau 1 yields Tableau 2,

Tableau 2

	x_1	x_2	x_3	x_4	x_5	x_6	
x_4	⑤	0	-8	1	-2	0	40
x_2	-1	1	4	0	1	0	10
x_6	0	0	13	0	2	1	70
	-5	0	9	0	3	0	30

The corner point corresponding to Tableau 2 is $(0, 10, 0)$ and

$$z = 2(0) + 3(10) + 3(0) = 30$$

It is, of course, obvious that we moved from one corner point to another in going from Tableau 1 to 2. To comprehend the argument given in proof of Theorem 1, the point to notice is that knowing that z increased is enough to conclude that a move between corner points occurred. We now pivot on the circled entry to obtain Tableau 3,

Tableau 3

	x_1	x_2	x_3	x_4	x_5	x_6	
x_1	1	0	$-\frac{8}{5}$	$\frac{1}{5}$	$-\frac{2}{5}$	0	8
x_2	0	1	$\frac{12}{5}$	$\frac{1}{5}$	$\frac{3}{5}$	0	18
x_6	0	0	13	0	2	1	70
	0	0	1	1	1	0	70

The corner point corresponding to Tableau 3 is $(8, 18, 0)$ and

$$z = 2(8) + 3(18) + 3(0) = 70$$

The simplex method now terminates, having found a maximal solution.   ◆

Exercises

1. The LP problem

$$\text{Maximize:} \quad z = 2x_1 - x_2 + 8x_3$$

$$\text{Subject to:} \quad 2x_3 \le 1$$
$$2x_1 - 4x_2 + 6x_3 \le 3$$
$$-x_1 + 3x_2 + 4x_3 \le 2$$
$$x_i \ge 0, \quad 1 \le i \le 3$$

becomes degenerate after one pivot of the simplex method; however, it does have a maximal solution. Form the canonical tableau, and solve the problem.

2. The LP problem

$$\text{Maximize:} \quad z = 2x_1 + \frac{3}{2}x_3$$

$$\text{Subject to:} \quad x_1 - x_2 \qquad \leq 2$$
$$2x_1 \qquad + x_3 \leq 4$$
$$x_1 + x_2 + x_3 \leq 3$$
$$x_i \geq 0, \qquad 1 \leq i \leq 3$$

becomes degenerate after one pivot of the simplex method; however, it does have a maximal solution. Form the canonical tableau, and solve the problem.

†3. The LP problem

$$\text{Maximize:} \quad z = \frac{3}{4}x_1 - 20x_2 + \frac{1}{2}x_3 - 6x_4$$

$$\text{Subject to:} \quad \frac{1}{4}x_1 - 8x_2 - x_3 + 9x_4 \leq 0$$

$$\frac{1}{2}x_1 - 12x_2 - \frac{1}{2}x_3 + 3x_4 \leq 0$$

$$x_3 \qquad \leq 1$$

$$x_i \geq 0, \qquad 1 \leq i \leq 4$$

is one of the earliest examples of an LP problem that cycles under some sequence of pivots of the simplex method. It is due to E. M. L. Beale [1]. Form the initial tableau and perform the following pivots. Since so many calculations are involved, you probably want to use the TUTOR program of CALIPSO.

(a) Solve the problem with the simplex method by using Bland's Rule to break ties for entering and departing basic variables.

(b) Perform the following sequence of pivots. Note that all choices of pivot element are legitimate choices for the simplex method.

Pivot	Column	Row
1	1	1
2	2	2
3	3	1
4	4	2
5	5	1
6	6	2

After the sixth pivot, you should be back at the initial tableau.

4. In this exercise you are asked to show that a basic variable always becomes zero when a tie occurs in selecting the departing variable by the Departing Basic Variable Rule. The "skeleton" given in Tableau 1 has dots or blanks that replace entries that are not of immediate interest.

Tableau 1

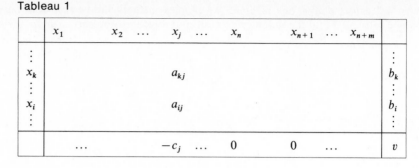

Suppose that a_{kj} and a_{ij} are positive and that

$$\frac{b_k}{a_{kj}} = \frac{b_i}{a_{ij}}$$

Suppose further that this common value is the smallest of the ratios formed for the Departing Basic Variable Rule.

(a) If a_{ij} is chosen as the pivot element, show that the new resource value for row k is

$$b_k^* = b_k - a_{kj}\left(\frac{b_i}{a_{ij}}\right) \qquad (4)$$

(b) Show that $b_k^* = 0$.

5. In proving Theorem 1 we argued that the objective function increases in value at each pivot of the simplex method provided degeneracy does not occur. We could also extract this conclusion from (4). Using the fact that the objective row is pivoted in the same way as a constraint row, show how to interpret (4) as

$$v^* = v + c_j\left(\frac{b_i}{a_{ij}}\right)$$

and conclude that $v^* > v$ (in the absence of degeneracy).

*6. We could have done the proof of Theorem 1 somewhat differently by observing that the tableaus generated by the simplex method depend uniquely on the sets of basic variables associated with each tableau. That is, if $T_1, T_2, \ldots, T_t$ is a sequence of tableaus generated by the simplex method, and if two of these tableaus, say T and T^*, have the same set of basic variables, then $T = T^*$. Here is an outline of a proof of this fact. Fill in the missing details.

(a) Let B be the set of subscripts of basic variables for both T and T^*. Then the tableau for T describes a system of linear equations that we can write in the form

$$x_i + \sum_{j \notin B} a_{ij}x_j = b_i, \qquad \text{for each} \quad i \in B$$

$$z = v + \sum_{j \notin B} c_j x_j \qquad (5)$$

(b) Select any $k \notin B$ and any number y. A solution of (5) is given by

$$x_k = y, \quad x_j = 0, \qquad \text{for} \quad j \notin B \quad \text{and} \quad j \neq k$$

$$x_i = b_i - a_{ik}y, \qquad \text{for each} \quad i \in B \qquad (6)$$

$$z = v + c_k y$$

Verify that this is a solution by substitution in (5).

(c) Since T^* has the same set B of basic variables as T, we can write the system of linear equations corresponding to T^* as

$$x_i + \sum_{j \notin B} a_{ij}^* x_j = b_i^*, \qquad \text{for each} \quad i \in B \tag{7}$$

$$z = v^* + \sum_{j \notin B} c_j^* x_j$$

Since all of the tableaus generated by the simplex method have the same set of solutions, (6) is also a solution of (7). Substitute (6) into (7) and obtain

$$b_i - a_{ik} y = b_i^* - a_{ik}^* y, \qquad \text{for each} \quad i \in B$$

$$v + c_k y = v^* + c_k^* y$$

(d) Rewrite the identities from (c) as

$$b_i - b_i^* = (a_{ik} - a_{ik}^*) y, \qquad \text{for each} \quad i \in B$$

$$v - v^* = (c_k - c_k^*) y$$

Now recall that y was an arbitrary number. Note that y occurs on the right-hand side of these equations but not on the left. The only way this is possible is for both sides to be zero. Thus,

$$b_i = b_i^* \qquad \text{and} \qquad a_{ik} = a_{ik}^*, \qquad \text{for each} \quad i \in B$$

$$v = v^* \qquad \text{and} \qquad c_k = c_k^*$$

Now recall that k was also arbitrary to conclude that $T = T^*$.

7. Use the technique given in Examples 1 and 2 to show that an LP problem has an unbounded objective function on the feasible region if the following tableau occurs during application of the simplex method.

	x_1	x_2	x_3	x_4	x_5	x_6	
x_4	-3	2	0	1	0	0	60
x_5	-1	1	4	0	1	0	10
x_6	-2	-2	5	0	0	1	50
	-2	-3	-3	0	0	0	0

*8. Carry out the following steps, which lead to a proof that Bland's Rule keeps a degenerate problem from cycling. (These steps are adapted from a proof given in Chvatal [5].)
   (a) Suppose that Bland's Rule is used to select entering and departing basic variables when applying the simplex method to a problem, but that nevertheless there is a degenerate tableau T_0 that occurs twice. That is, there exists a sequence of degenerate tableaus:

$$T_0, T_1, \ldots, T_k = T_0$$

Explain why we prove that Bland's Rule works if we can derive a contradiction from this assumption.
   (b) We call a variable *fickle* if it is nonbasic in some of the tableaus $T_0, T_1, \ldots, T_k = T_0$ and basic in others. Among all the fickle variables, let x_t have the largest subscript. In the sequence $T_0, T_1, \ldots, T_k = T_0$, there is a tableau T with x_t departing (basic in T but

nonbasic in the next tableau) and with some other fickle variable x_s entering (nonbasic in T but basic in the next tableau). Further along in the sequence

$$T_0, T_1, \ldots, T_k, T_0, T_1, \ldots, T_k$$

there must be a tableau T^* with x_t entering. Let B be the set of subscripts of basic variables for tableau T, and let v be the value of the objective function for tableau T. Explain why we can write the system of equations corresponding to T as

$$x_i = b_i - \sum_{j \notin B} a_{ij} x_j, \qquad i \in B$$
$$z = v + \sum_{j \notin B} c_j x_j \tag{8}$$

(c) Explain why the objective function value is v for both tableau T and tableau T^*. If we let c_j^* denote the cost coefficients for T^*, explain why the objective row of T^* can be written as

$$z = v + \sum_{j=1}^{n+m} c_j^* x_j \tag{9}$$

(Hint: $c_j^* = 0$ whenever x_j is a basic variable for T^*.)

(d) Explain why (9) must be satisfied by every solution of the system of linear equations (8).

(e) For any real number y, show that the following equations satisfy (8):

$$x_s = y$$
$$x_j = 0, \qquad j \notin B \quad \text{but} \quad j \neq s$$
$$x_i = b_i - a_{is} y, \qquad i \in B$$
$$z = v + c_s y$$

(f) Conclude from (d) and (e) that

$$v + c_s y = v + c_s^* y + \sum_{i \in B} c_i^* (b_i - a_{is} y)$$

Simplify this expression to obtain

$$\left(c_s - c_s^* + \sum_{i \in B} c_i^* a_{is} \right) y = \sum_{i \in B} c_i^* b_i \tag{10}$$

(g) Use the fact that y is an arbitrary real number to explain why (10) implies that

$$c_s - c_s^* + \sum_{i \in B} c_i^* a_{is} = 0 \tag{11}$$

(h) Use the fact that x_s is the entering basic variable in T but x_s is not the entering basic variable in T^* (even though $s < t$) to conclude that $c_s > 0$ and $c_s^* \leq 0$. Conclude from (11) that

$$c_r^* a_{rs} < 0, \qquad \text{for some} \quad r \in B$$

(i) Use the fact that $r \in B$ and $c_r^* \neq 0$ to conclude that x_r is fickle.

(j) Use the fact that $x_r \neq x_t$ to conclude that $r < t$. Use the fact that x_t is the departing basic variable in T to conclude that $a_{ts} > 0$. Thus, $c_t^* a_{ts} > 0$, $r < t$, and x_r is not entering in T^*. Conclude that $c_r^* \leq 0$.

(k) Use (h) and (j) to conclude that $a_{rs} > 0$. Note that $b_r = 0$. Explain why this is a contradiction of the fact that we picked x_t as the departing basic variable in tableau T.

3.4 THE TWO-PHASE METHOD

Thus far we have dealt with LP problems of the form

$$\text{Maximize:} \quad z = c_1 x_1 + c_2 x_2 + \cdots + c_n x_n$$

$$\text{Subject to:} \quad a_{11} x_1 + a_{12} x_2 + \cdots + a_{1n} x_n \leq b_1$$

$$a_{21} x_1 + a_{22} x_2 + \cdots + a_{2n} x_n \leq b_2$$

$$\vdots$$

$$a_{m1} x_1 + a_{m2} x_2 + \cdots + a_{mn} x_n \leq b_m$$

$$x_i \geq 0, \quad 1 \leq i \leq n$$

(1)

where $b_i \geq 0, 1 \leq i \leq m$. In this section we begin the study of LP problems for which the resource values b_i are not necessarily nonnegative, and for which some or all of the inequalities $\leq$ in (1) are replaced by $\geq$ or $=$.

We first observe that in some cases we can convert a given LP problem into an equivalent problem of the form (1). For example, the LP problem

$$\text{Minimize:} \quad z = 2x_1 - 4x_2$$

$$\text{Subject to:} \quad 2x_1 + 3x_2 \leq 5$$

$$x_1 - 3x_2 \geq -2$$

$$x_1, x_2 \geq 0$$

(2)

is equivalent to the LP problem

$$\text{Maximize:} \quad z = -2x_1 + 4x_2$$

$$\text{Subject to:} \quad 2x_1 + 3x_2 \leq 5$$

$$-x_1 + 3x_2 \leq 2$$

$$x_1, x_2 \geq 0$$

(3)

which has the form (1). The optimal solution of (3), which we can obtain from our work so far, is also the optimal solution of (2). Consider, however, the LP problem:

$$\text{Maximize:} \quad z = 2x_1 + 3x_2$$

$$\text{Subject to:} \quad -4x_1 + 3x_2 \leq 12$$

$$2x_1 + x_2 \leq 6$$

$$x_1 + x_2 \geq 3$$

$$5x_1 + x_2 \geq 4$$

$$x_1, x_2 \geq 0$$

(4)

Although multiplication of the third and fourth inequalities of (4) by -1 changes the direction of the inequality signs, it also results in negative resource values.

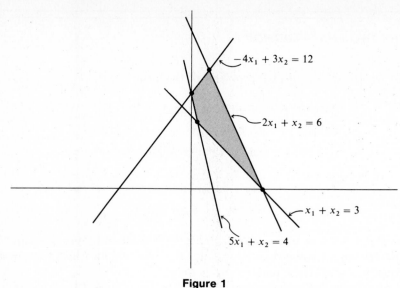

$-4x_1 + 3x_2 = 12$

$2x_1 + x_2 = 6$

$x_1 + x_2 = 3$

$5x_1 + x_2 = 4$

Figure 1

Because of these negative values, we cannot use our usual choice of nonbasic variables (x_1 and x_2) to initiate the simplex method, since setting $x_1 = 0$ and $x_2 = 0$ results in slack variables with negative values. Figure 1 shows the feasible region of (4); note that the origin is not a feasible corner point of this region.

The problem now is to find an initial basic feasible solution of (4) that will allow us to initiate the simplex method. In this particular example, we could use Figure 1 to find such a solution. However, when there are more than two decision variables, it is usually impractical, if not impossible, to find a solution from a sketch of the feasible region.

There are a number of ways to circumvent this problem. In this and the next section we present examples of the most commonly used method: the two-phase method. In Section 3.6 we consider another method called the Big-M method.

Common to both of these methods is the introduction of "artificial variables" in addition to slack variables. We use LP problem (4) to illustrate this idea.

Introducing (nonnegative) slack variables x_3, x_4, x_5, x_6 to (4), we have

$$
\begin{aligned}
\text{Maximize:} \quad & z = 2x_1 + 3x_2 \\
\text{Subject to:} \quad & -4x_1 + 3x_2 + x_3 && = 12 \\
& 2x_1 + x_2 && + x_4 && = 6 \\
& x_1 + x_2 && - x_5 && = 3 \\
& 5x_1 + x_2 && - x_6 && = 4 \\
& x_i \geq 0, \quad 1 \leq i \leq 6
\end{aligned}
\tag{5}
$$

(In this context, for fairly obvious reasons, we sometimes refer to the slack variables x_5 and x_6 as *surplus variables*.) We note that at this stage the slack variables $x_3, x_4,$ x_5, x_6 cannot be basic variables because assigning the value 0 to the other variables

x_1 and x_2 would result in the infeasible solution

$$(0, 0, 12, 6, -3, -4)$$

We apply the simplex method to a different but related LP problem to find an initial basic feasible solution to (5). To this end, we introduce nonnegative *artificial* variables y_1 and y_2 to those constraints in (5) for which the slack variables appear with a negative sign. This results in the LP problem

$$\begin{aligned}
\text{Maximize:} \quad & z = 2x_1 + 3x_2 \\
\text{Subject to:} \quad & -4x_1 + 3x_2 + x_3 && = 12 \\
& 2x_1 + x_2 + x_4 && = 6 \\
& x_1 + x_2 - x_5 + y_1 && = 3 \\
& 5x_1 + x_2 - x_6 + y_2 && = 4 \\
& x_i \geq 0, 1 \leq i \leq 6; \quad y_i \geq 0, 1 \leq i \leq 2
\end{aligned} \tag{6}$$

You should note that LP problems (5) and (6) are not equivalent. Nevertheless, if $(t_1, t_2, \ldots, t_6, 0, 0)$ is a basic feasible solution to (6) for which the artificial variables y_1 and y_2 are two of the four nonbasic variables, then $(t_1, t_2, \ldots, t_6)$ is a basic feasible solution to (5).

Clearly, the 8-tuple

$$(x_1, x_2, x_3, x_4, x_5, x_6, y_1, y_2) = (0, 0, 12, 6, 0, 0, 3, 4) \tag{7}$$

is a basic feasible solution to (6). The fundamental idea of the two-phase method is to find a way of passing from a solution like (7) to another basic feasible solution of (6) of the form

$$(x_1, x_2, x_3, x_4, x_5, x_6, y_1, y_2) = (t_1, t_2, \ldots, t_6, 0, 0) \tag{8}$$

in which the artificial variables are nonbasic and have value 0. Then, as we have just observed,

$$(x_1, x_2, x_3, x_4, x_5, x_6) = (t_1, t_2, \ldots, t_6) \tag{9}$$

is a basic feasible solution to (6). (We sometimes say the artificial variables have been "driven out" when they have value zero and are nonbasic.) The second phase (phase 2) of the two-phase method uses the basic feasible solution (9) to LP problem (5) as a starting point for the simplex method.

In summary, as the name suggests, the two-phase method consists of

PHASE 1, which drives out the artificial variables by finding a basic feasible solution for which the artificial variables have value zero and are nonbasic, and

PHASE 2, which starts from this basic feasible solution and produces an optimal solution.

We illustrate the method by continuing the discussion of LP problem (6). To accomplish phase 1 we need to find a basic feasible solution of (6) that does not involve the artificial variables y_1 and y_2. To do this we consider an LP problem that minimizes the sum of these variables subject to constraints (6). Since the variables y_1 and y_2 are both nonnegative, it follows that if the minimal value of their sum is 0, then both of these variables must also be 0, and we have succeeded in driving these variables to 0. Thus, we want to minimize $y_1 + y_2$ or, equivalently, maximize $-y_1 - y_2$. This leads us to consider the *auxiliary* LP problem

$$
\begin{aligned}
\text{Maximize:} \quad & w = -y_1 - y_2 \\
\text{Subject to:} \quad & -4x_1 + 3x_2 + x_3 && = 12 \\
& 2x_1 + x_2 \phantom{{}+x_2} + x_4 && = 6 \\
& x_1 + x_2 \phantom{{}+x_4} - x_5 \phantom{{}+} + y_1 && = 3 \\
& 5x_1 + x_2 \phantom{{}+x_4+x_5} - x_6 \phantom{{}+} + y_2 && = 4
\end{aligned}
\tag{10}
$$

$$
x_i \geq 0,\, 1 \leq i \leq 6; \qquad y_i \geq 0,\, 1 \leq i \leq 2
$$

The basic variables in (10) are x_3, x_4, y_1, and y_2. Note that (10) is not quite in canonical form because the objective function is not written in terms of nonbasic variables. To rewrite $w = -y_1 - y_2$ in terms of the nonbasic variables x_1, x_2, x_5, x_6, we rearrange the third and fourth constraints of (10) as

$$
\begin{aligned}
-y_1 &= x_1 + x_2 - x_5 - 3 \\
-y_2 &= 5x_1 + x_2 - x_6 - 4
\end{aligned}
$$

Adding these equations, we have

$$
w = -y_1 - y_2 = 6x_1 + 2x_2 - x_5 - x_6 - 7
$$

Thus, an equivalent system to (10) in canonical form is

$$
\begin{aligned}
\text{Maximize:} \quad & w = 6x_1 + 2x_2 - x_5 - x_6 - 7 \\
\text{Subject to:} \quad & -4x_1 + 3x_2 + x_3 && = 12 \\
& 2x_1 + x_2 \phantom{{}+x_2} + x_4 && = 6 \\
& x_1 + x_2 \phantom{{}+x_4} - x_5 \phantom{{}+} + y_1 && = 3 \\
& 5x_1 + x_2 \phantom{{}+x_4+x_5} - x_6 \phantom{{}+} + y_2 && = 4
\end{aligned}
\tag{11}
$$

$$
x_i \geq 0,\, 1 \leq i \leq 6; \qquad y_i \geq 0,\, 1 \leq i \leq 2
$$

In Tableau 1, the tableau form of (11), we include a second objective row corresponding to the objective function z in the original problem, where z is also written in terms of the current nonbasic variables x_1, x_2, x_5, x_6. This row plays no part in phase 1 but will be useful in initiating phase 2. We include this row in the tableau to pivot on during phase 1, so that at the end of phase 1 we have z expressed in terms of the nonbasic variables. Application of the simplex method yields Tableaus 2 and 3. We always determine entering basic variables by examining the (w) objective row.

Tableau 1

	x_1	x_2	x_3	x_4	x_5	x_6	y_1	y_2	
x_3	-4	3	1	0	0	0	0	0	12
x_4	2	1	0	1	0	0	0	0	6
y_1	1	1	0	0	-1	0	1	0	3
y_2	⑤	1	0	0	0	-1	0	1	4
(z)	-2	-3	0	0	0	0	0	0	0
(w)	-6	-2	0	0	1	1	0	0	-7

Tableau 2

	x_1	x_2	x_3	x_4	x_5	x_6	y_1	y_2	
x_3	0	$\frac{19}{5}$	1	0	0	$-\frac{4}{5}$	0	$\frac{4}{5}$	$\frac{76}{5}$
x_4	0	$\frac{3}{5}$	0	1	0	$\frac{2}{5}$	0	$-\frac{2}{5}$	$\frac{22}{5}$
y_1	0	$\left(\frac{4}{5}\right)$	0	0	-1	$\frac{1}{5}$	1	$-\frac{1}{5}$	$\frac{11}{5}$
x_1	1	$\frac{1}{5}$	0	0	0	$-\frac{1}{5}$	0	$\frac{1}{5}$	$\frac{4}{5}$
(z)	0	$-\frac{13}{5}$	0	0	0	$-\frac{2}{5}$	0	$\frac{2}{5}$	$\frac{8}{5}$
(w)	0	$-\frac{4}{5}$	0	0	1	$-\frac{1}{5}$	0	$\frac{6}{5}$	$-\frac{11}{5}$

Tableau 3

	x_1	x_2	x_3	x_4	x_5	x_6	y_1	y_2	
x_3	0	0	1	0	$\frac{19}{4}$	$-\frac{7}{4}$	$-\frac{19}{4}$	$\frac{7}{4}$	$\frac{19}{4}$
x_4	0	0	0	1	$\frac{3}{4}$	$\frac{1}{4}$	$-\frac{3}{4}$	$-\frac{1}{4}$	$\frac{11}{4}$
x_2	0	1	0	0	$-\frac{5}{4}$	$\frac{1}{4}$	$\frac{5}{4}$	$-\frac{1}{4}$	$\frac{11}{4}$
x_1	1	0	0	0	$\frac{1}{4}$	$-\frac{1}{4}$	$-\frac{1}{4}$	$\frac{1}{4}$	$\frac{1}{4}$
(z)	0	0	0	0	$-\frac{13}{4}$	$\frac{1}{4}$	$\frac{13}{4}$	$-\frac{1}{4}$	$\frac{35}{4}$
(w)	0	0	0	0	0	0	1	1	0

Since all the entries in the (w) objective row of Tableau 3 are nonnegative, we find that the maximum value of $w = -y_1 - y_2$ (or equivalently, the minimum value of $y_1 + y_2$) is 0. Therefore, we have achieved our goals of driving the artificial variables y_1 and y_2 to 0 and of obtaining a basic feasible solution to the original problem:

$$(x_1, x_2, x_3, x_4, x_5, x_6) = \left(\frac{1}{4}, \frac{11}{4}, \frac{19}{4}, \frac{11}{4}, 0, 0\right)$$

We are now ready for phase 2, during which we will use the result of phase 1 to optimize the original objective function $z = 2x_1 + 3x_2$. Note that Tableau 3 contains the equivalent expression for z in terms of the current nonbasic variables in the row labeled (z). We now eliminate the two artificial variable columns and the (w)

objective row from Tableau 3 to obtain Tableau 4,

Tableau 4

	x_1	x_2	x_3	x_4	x_5	x_6	
x_3	0	0	1	0	$\frac{19}{4}$	$-\frac{7}{4}$	$\frac{19}{4}$
x_4	0	0	0	1	$\frac{3}{4}$	$\frac{1}{4}$	$\frac{11}{4}$
x_2	0	1	0	0	$-\frac{5}{4}$	$\frac{1}{4}$	$\frac{11}{4}$
x_1	1	0	0	0	$\left(\frac{1}{4}\right)$	$-\frac{1}{4}$	$\frac{1}{4}$
(z)	0	0	0	0	$-\frac{13}{4}$	$\frac{1}{4}$	$\frac{35}{4}$

Applying the simplex method to Tableau 4, we obtain Tableaus 5–7,

Tableau 5

	x_1	x_2	x_3	x_4	x_5	x_6	
x_3	-19	0	1	0	0	$\left(3\right)$	0
x_4	-3	0	0	1	0	1	2
x_2	5	1	0	0	0	-1	4
x_5	4	0	0	0	1	-1	1
(z)	13	0	0	0	0	-3	12

Tableau 6

	x_1	x_2	x_3	x_4	x_5	x_6	
x_6	$-\frac{19}{3}$	0	$\frac{1}{3}$	0	0	1	0
x_4	$\left(\frac{10}{3}\right)$	0	$-\frac{1}{3}$	1	0	0	2
x_2	$-\frac{4}{3}$	1	$\frac{1}{3}$	0	0	0	4
x_5	$-\frac{7}{3}$	0	$\frac{1}{3}$	0	1	0	1
(z)	-6	0	1	0	0	0	12

Tableau 7

	x_1	x_2	x_3	x_4	x_5	x_6	
x_6	0	0	$-\frac{3}{10}$	$\frac{19}{10}$	0	1	$\frac{19}{5}$
x_1	1	0	$-\frac{1}{10}$	$\frac{3}{10}$	0	0	$\frac{3}{5}$
x_2	0	1	$\frac{1}{5}$	$\frac{2}{5}$	0	0	$\frac{24}{5}$
x_5	0	0	$\frac{1}{10}$	$\frac{7}{10}$	1	0	$\frac{12}{5}$
(z)	0	0	$\frac{2}{5}$	$\frac{9}{5}$	0	0	$\frac{78}{5}$

Tableau 7 is optimal and shows that the maximal solution of (6) is

$$(x_1, x_2, x_3, x_4) = \left(\frac{3}{5}, \frac{24}{5}, 0, 0\right)$$

$$z = \frac{78}{5}$$

As the next example illustrates, we also add artificial variables to equality constraints.

EXAMPLE 1

Introducing a slack and an artificial variable to the $\geq$ constraint and an artificial variable to the $=$ constraint of the LP problem

$$\text{Maximize:} \quad z = 80x_1 + 60x_2 + 42x_3$$
$$\text{Subject to:} \quad 2x_1 + 3x_2 + x_3 \leq 12$$
$$5x_1 + 6x_2 + 3x_3 \geq 15$$
$$2x_1 - 3x_2 + x_3 = 8$$
$$x_i \geq 0, \quad 1 \leq i \leq 3$$

we have

$$\text{Maximize:} \quad z = 80x_1 + 60x_2 + 42x_3$$
$$\text{Subject to:} \quad 2x_1 + 3x_2 + x_3 + x_4 = 12$$
$$5x_1 + 6x_2 + 3x_3 - x_5 + y_1 = 15 \quad (12)$$
$$2x_1 - 3x_2 + x_3 + y_2 = 8$$
$$x_i \geq 0, 1 \leq i \leq 5; \quad y_1, y_2 \geq 0$$

To express $w = -y_1 - y_2$ in terms of the nonbasic variables, we solve the second and third equations of (12) for $-y_1$ and $-y_2$, obtaining

$$-y_1 = 5x_1 + 6x_2 + 3x_3 - x_5 - 15$$
$$-y_2 = 2x_1 - 3x_2 + x_3 - 8$$

Adding these equations, we have

$$w = -y_1 - y_2 = 7x_1 + 3x_2 + 4x_3 - x_5 - 23$$

The *auxiliary* problem tableaus are Tableaus 1–3,

Tableau 1

	x_1	x_2	x_3	x_4	x_5	y_1	y_2	
x_4	2	3	1	1	0	0	0	12
y_1	⑤	6	3	0	-1	1	0	15
y_2	2	-3	1	0	0	0	1	8
(z)	-80	-60	-42	0	0	0	0	0
(w)	-7	-3	-4	0	1	0	0	-23

Tableau 2

	x_1	x_2	x_3	x_4	x_5	y_1	y_2	
x_4	0	$\frac{3}{5}$	$-\frac{1}{5}$	1	$\frac{2}{5}$	$-\frac{2}{5}$	0	6
x_1	1	$\frac{6}{5}$	$\frac{3}{5}$	0	$-\frac{1}{5}$	$\frac{1}{5}$	0	3
y_2	0	$-\frac{27}{5}$	$-\frac{1}{5}$	0	$\left(\frac{2}{5}\right)$	$-\frac{2}{5}$	1	2
(z)	0	36	6	0	-16	16	0	240
(w)	0	$\frac{27}{5}$	$\frac{1}{5}$	0	$-\frac{2}{5}$	$\frac{7}{5}$	0	-2

Tableau 3

	x_1	x_2	x_3	x_4	x_5	y_1	y_2	
x_4	0	6	0	1	0	0	-1	4
x_1	1	$-\frac{3}{2}$	$\frac{1}{2}$	0	0	0	$\frac{1}{2}$	4
x_5	0	$-\frac{27}{2}$	$-\frac{1}{2}$	0	1	-1	$\frac{5}{2}$	5
(z)	0	-180	-2	0	0	0	40	320
(w)	0	0	0	0	0	1	1	0

Since the auxiliary objective function has value 0 in Tableau 3, we have discovered an initial basic feasible solution to the problem. Omitting the artificial variable columns and the objective function, we have Tableau 4

Tableau 4

	x_1	x_2	x_3	x_4	x_5	
x_4	0	(6)	0	1	0	4
x_1	1	$-\frac{3}{2}$	$\frac{1}{2}$	0	0	4
x_5	0	$-\frac{27}{2}$	$-\frac{1}{2}$	0	1	5
(z)	0	-180	-2	0	0	320

Applying the simplex method to Tableau 4 gives Tableaus 5 and 6,

Tableau 5

	x_1	x_2	x_3	x_4	x_5	
x_2	0	1	0	$\frac{1}{6}$	0	$\frac{2}{3}$
x_1	1	0	$\left(\frac{1}{2}\right)$	$\frac{1}{4}$	0	5
x_5	0	0	$-\frac{1}{2}$	$\frac{9}{4}$	1	14
(z)	0	0	-2	30	0	440

Tableau 6

	x_1	x_2	x_3	x_4	x_5	
x_2	0	1	0	$\frac{1}{6}$	0	$\frac{2}{3}$
x_3	2	0	1	$\frac{1}{2}$	0	10
x_5	1	0	0	$\frac{5}{2}$	1	19
(z)	4	0	0	31	0	460

Tableau 6 is optimal and gives the solution

$$(x_1, x_2, x_3, x_4, x_5) = (0, \tfrac{2}{3}, 10, 0, 19)$$

and the maximum objective function value $z = 460$.   ◆

The next example shows how the two-phase method detects an infeasible problem.

EXAMPLE 2

Introducing slack and artificial variables to the constraints of the LP problem

$$\text{Maximize:} \quad z = -2x_1 + 3x_2$$
$$\text{Subject to:} \quad -x_1 + x_2 \geq 3$$
$$3x_1 + x_2 \leq 5$$
$$x_1 + x_2 \geq 6$$
$$x_1, x_2 \geq 0$$

we have

$$\text{Maximize:} \quad z = -2x_1 + 3x_2$$

$$\begin{aligned}
\text{Subject to:} \quad -x_1 + x_2 - x_3 \qquad\qquad\quad + y_1 \qquad &= 3 \\
3x_1 + x_2 \qquad + x_4 \qquad\qquad\quad &= 5 \qquad (13)\\
x_1 + x_2 \qquad\qquad - x_5 \quad + y_2 &= 6
\end{aligned}$$

$$x_i \geq 0, 1 \leq i \leq 5; \qquad y_i \geq 0, 1 \leq i \leq 2$$

We could express $w = -y_1 - y_2$ in terms of the nonbasic variables, as we did in Example 1; however, we suggest an alternative for this step, which you may prefer. We first form Tableau 0 as the auxiliary problem tableau, using $w = -y_1 - y_2$ as the auxiliary objective function.

Tableau 0

	x_1	x_2	x_3	x_4	x_5	y_1	y_2	
y_1	-1	1	-1	0	0	①	0	3
x_4	3	1	0	1	0	0	0	5
y_2	1	1	0	0	-1	0	①	6
(z)	2	-3	0	0	0	0	0	0
(w)	0	0	0	0	0	1	1	0

We express the (w) objective row in terms of the nonbasic variables by performing two pivots on the circled entries in Tableau 0. In this way we obtain Tableau 1, which is in the canonical form for starting the simplex method.

Tableau 1

	x_1	x_2	x_3	x_4	x_5	y_1	y_2	
y_1	-1	①	-1	0	0	1	0	3
x_4	3	1	0	1	0	0	0	5
y_2	1	1	0	0	-1	0	1	6
(z)	2	-3	0	0	0	0	0	0
(w)	0	-2	1	0	1	0	0	-9

Applying the simplex method, we obtain Tableaus 2 and 3,

Tableau 2

	x_1	x_2	x_3	x_4	x_5	y_1	y_2	
x_2	-1	1	-1	0	0	1	0	3
x_4	4	0	①	1	0	-1	0	2
y_2	2	0	1	0	-1	-1	1	3
(z)	-1	0	-3	0	0	3	0	9
(w)	-2	0	-1	0	1	2	0	-3

Tableau 3

	x_1	x_2	x_3	x_4	x_5	y_1	y_2	
x_2	3	1	0	1	0	0	0	5
x_3	4	0	1	1	0	-1	0	2
y_2	-2	0	0	-1	-1	0	1	1
(z)	11	0	0	3	0	0	0	15
(w)	2	0	0	1	1	1	0	-1

Observe that all of the (w) objective row entries are nonnegative, yet

$$w = -y_1 - y_2 = -1 \neq 0$$

Thus, the minimal value of $y_1 + y_2$ is 1, and we have not succeeded in driving the artificial variables y_1 and y_2 to 0. ◆

If, as in Example 2, we cannot drive the artificial variables to 0, then the feasible region of the original problem must be empty. This fact follows from Theorem 1 in the next section.

Summary of the Two-Phase Method

This summary is incomplete because it does not consider the problem that occasionally arises when an artificial variable remains in the basis with value 0 at the end of phase 1. We deal with this problem in the next section.

PHASE 1

STEP 1. Form the auxiliary problem.

a. For each $\leq$ constraint, add a nonnegative slack variable s to obtain an equation of the form

$$a_1 x_1 + a_2 x_2 + \cdots + a_n x_n + s = b$$

Use a different variable (or a different subscript) for each slack variable that you add.

b. For each $=$ constraint, add a nonnegative artificial variable y to obtain an equation of the form

$$a_1 x_1 + a_2 x_2 + \cdots + a_n x_n + y = b$$

Use a different variable (or a different subscript) for each artificial variable that you add.

c. For each $\geq$ constraint, subtract a nonnegative slack variable s, and add a nonnegative artificial variable y to obtain an equation of the form

$$a_1 x_1 + a_2 x_2 + \cdots + a_n x_n - s + y = b$$

Use a different variable (or a different subscript) for each slack and artificial variable that you add.

d. Form the auxiliary objective function

$$w = -y_1 - y_2 - \cdots - y_k$$

where $y_1, y_2, \ldots, y_k$ are all of the artificial variables. Since $y_1, y_2, \ldots, y_k$ are basic variables of the auxiliary problem, you must write each y_i in terms of the decision variables of the original problem and the slack variables before starting the simplex method for the auxiliary problem.

   You can write the variables y_i in terms of the decision and slack variables by solving for each y_i in the one constraint equation in which it occurs (see Example 1). Alternatively, you can form the tableau with artificial variables in the auxiliary objective function and pivot on each artificial variable before starting the simplex method (see Example 2).

STEP 2. Find a feasible solution.

a. Form the auxiliary tableau, including objective rows for w and z.

b. Apply the simplex method to the auxiliary tableau using the (w) objective row to determine the entering basic variable. Also perform pivots on the (z) objective row so that it is continually expressed in terms of the nonbasic variables. If the simplex method finds a solution with the maximum value of w equal to 0, then the final tableau gives a feasible solution to the problem. (See the next section to find out what to do if an artificial variable remains in the basis.) If the maximum value of w is negative, then the problem is infeasible.

PHASE 2

If you found a feasible solution in phase 1, remove the columns corresponding to artificial variables and the artificial objective row from the tableau. Start the simplex method using the (z) objective row to find the maximal solution in the usual way.

Exercises

In Exercises 1–4, add slack and artificial variables to each of the systems to put them in canonical form for solution by the two-phase method. Form the initial tableaus with rows labeled by the initial basic variables and the auxiliary objective row expressed in terms of nonbasic variables. It is not necessary to solve the problems.

1. Maximize: $z = -2x_1 - 2x_2 + 5x_3$

Subject to: $3x_1 + 2x_2 - 4x_3 = 7$

$x_1 - x_2 + 3x_3 = 2$

$x_i \geq 0, \quad 1 \leq i \leq 3$

2. Maximize: $z = 3x_1 - x_3$

Subject to: $-x_1 + 2x_2 + x_3 \leq 6$

$3x_1 - x_2 + x_3 = 3$

$x_i \geq 0, \quad 1 \leq i \leq 3$

3. Minimize: $z = x_1 + 2x_2 - x_3$

Subject to: $3x_1 - x_2 + 4x_3 \geq 5$

$-x_1 + 2x_2 + 5x_3 \geq 6$

$2x_1 + x_2 - x_3 \geq 2$

$x_i \geq 0, \quad 1 \leq i \leq 3$

4. Maximize: $z = a + 2b + 3c + 4d$

Subject to: $a + c - 4d = 2$

$b - c + 3d = 9$

$a + b - 2c - 3d = 21$

$a, b, c, d \geq 0$

In Exercises 5–12, use phase 1 of the two-phase method to find a solution with nonnegative variables to the systems of linear equations and inequalities.

5. $x_1 - x_2 = 1$

$2x_1 + x_2 - x_3 = 3$

6. $x_1 + x_2 + 2x_3 = 13$

$2x_1 - 3x_2 + 6x_3 = 1$

7. $x_1 + 2x_2 - 3x_3 + 4x_4 = 6$

$3x_1 - 2x_2 + x_3 + 5x_4 = 8$

$2x_1 + 3x_2 - x_3 + 3x_4 = 12$

8. $-2x_1 + 3x_2 + x_3 - 2x_4 \leq 10$

$3x_1 - 2x_2 + 3x_3 + 5x_4 = 42$

9. $3x_1 + 2x_2 \geq 4$

$2x_1 - x_2 \geq 2$

10. $4x_1 + 6x_2 \geq 5$

$6x_1 + 3x_2 \geq 8$

11. $5x_1 + 2x_2 + 3x_3 \geq 7$

$6x_1 + 2x_2 + 8x_3 \geq 24$

12. $2x_1 + 3x_2 + 5x_3 \geq 15$

$-x_1 + 2x_2 + 4x_3 \geq 9$

$x_1 + x_2 + x_3 = 4$

Solve Exercises 13–20 by the two-phase method.

13. Maximize: $z = 30x_1 - 40x_2$

Subject to: $x_1 + x_2 \leq 5$
$3x_1 - 2x_2 \geq 12$

$x_1, x_2 \geq 0$

14. Minimize: $z = x_1 + 5x_2 + 3x_3$

Subject to: $x_1 + x_2 + x_3 \geq 20$
$2x_1 + x_2 \qquad \geq 10$

$x_i \geq 0, \qquad 1 \leq i \leq 3$

15. Minimize: $z = 13x_1 + 4x_2$

Subject to: $x_2 \geq -2x_1 + 11$
$x_2 \leq -x_1 + 10$
$x_2 \leq -\dfrac{1}{3}x_1 + 6$
$x_2 \geq -\dfrac{1}{4}x_1 + 4$

$x_1, x_2 \geq 0$

16. Maximize: $z = 20x_1 + 32x_2 + 40x_3$

Subject to: $3x_1 + 2x_2 + 4x_3 \leq 12$
$2x_1 + x_2 + 3x_2 > 4$

$x_i \geq 0, \qquad 1 \leq i \leq 3$

17. Maximize: $z = 30x_1 + 32x_2 - 4x_3$

Subject to: $2x_1 - 3x_2 + 4x_3 = 15$
$3x_1 + 2x_2 + x_3 \geq 12$
$x_1 + x_2 + x_3 \leq 24$

$x_i \geq 0, \qquad 1 \leq i \leq 3$

18. Solve Exercise 12 if $x_1 \leq 0, x_2 \geq 0, x_3 \geq 0$.

19. Solve Exercise 13 if $x_1 \geq 0$ and x_2 unrestricted.

20. Do Exercise 6 from Section 1.1.

21. In Tableau 2 of Example 2, the Greedy Entering Basic Variable Rule would have selected the 4 in the column labeled by x_1 as the pivot element. We chose a different pivot because it avoided the introduction of fractions. Verify that the same solution results from pivoting on the 4.

In Exercises 22–25, use phase 1 to demonstrate that the problems are infeasible.

22. Maximize: $z = 2x_1 + 3x_2$

Subject to: $x_1 + 4x_2 \leq 4$
$x_1 - 2x_2 \leq 3$
$x_1 - x_2 \geq 4$

$x_1, x_2 \geq 0$

23. Maximize: $z = 3x_1 + 2x_2$

Subject to: $3x_1 + 4x_2 \leq 5$
$2x_1 + x_2 \geq 2$
$-x_1 + x_2 \geq 1$

$x_1, x_2 \geq 0$

24. Maximize: $z = x_1 + x_2 + x_3$

Subject to: $x_1 + 2x_2 + 3x_3 \leq 27$
$2x_1 + x_2 + x_3 \geq 54$
$x_1 - x_2 + x_3 = 9$

$x_i \geq 0, \qquad 1 \leq i \leq 3$

25. Minimize: $z = 2x_1 - 3x_2 - 5x_3$

Subject to: $3x_1 + 4x_2 + x_3 \leq 5$
$-2x_1 + 3x_2 - 4x_3 \geq 7$
$5x_1 + x_2 + 5x_3 \geq 2$

$x_i \geq 0, \qquad 1 \leq i \leq 3$

26. A small college has $30 million in endowed scholarship funds to invest. The rules of the endowment require that the money be divided among treasury notes, bonds, and stocks, with at least 10% of the money in each type of investment. Furthermore, the college must invest at least half the money in treasury notes and bonds, with the amount invested in bonds not to exceed twice the amount invested in treasury notes. The annual yields for the investments are 7% for treasury notes, 8% for bonds, and 9% for stocks. How should the college allocate the money among the various investments to produce the largest return?

27. The Big Spud Company has potato processing plants in Idaho Falls and Pocatello and potato fields near Burley and Twin Falls. The cost of shipping a ton of potatoes from Burley to Idaho Falls is $6; from Burley to Pocatello, $3; from Twin Falls to Idaho Falls, $9; and from Twin Falls to Pocatello, $5. Suppose the Idaho Falls plant needs 25 tons and the Pocatello plant needs 30 tons. Suppose the Burley field can provide up to 45 tons and the Twin Falls field can provide up to 40 tons. What is the most economical way to ship the potatoes?

You should use the TUTOR program for Exercises 28–34. You are advised not to begin using the automatic-solution programs for the two-phase method until told to do so because it is important to build your skill with the mechanical aspects of the simplex method. The tutorials allow you to make all the decisions and help you out only with the arithmetic. Note that computer programs cannot handle subscripts. The best way to enter a subscripted variable like x_1 is x1.

†28. You can use the computer programs to express the auxiliary objective function in terms of the nonbasic variables by the following technique. The tableau that follows is the same as Tableau 1 of LP problem (11), except that the auxiliary objective function has been entered as $w = -y_1 - y_2$.

	x_1	x_2	x_3	x_4	x_5	x_6	y_1	y_2	
x_3	-4	3	1	0	0	0	0	0	12
x_4	2	1	0	1	0	0	0	0	6
y_1	1	1	0	0	-1	0	1	0	3
y_2	5	1	0	0	0	-1	0	1	4
(z)	-2	-3	0	0	0	0	0	0	0
(w)	0	0	0	0	0	0	1	1	-7

Enter this tableau in the computer. Pivot first on column y_1 and row y_1 and then on column y_2 and row y_2. Observe that the tableau is now the same as Tableau 1.

Since you now have entered this tableau, you may want to practice your skill with the program by performing the pivots that were done in the text.

†29. Use the two-phase method to solve the LP problems that were given earlier in this set of exercises:
(a) Exercise 13   (b) Exercise 15   (c) Exercise 17   (d) Exercise 19

†30. Use the two-phase method to solve the LP problems that were given earlier in this set of exercises:
(a) Exercise 2   (b) Exercise 4

†31. Solve the product-mix problem given in Exercise 3 of Section 1.1.

†32. Solve the assignment problem given as Exercise 9 in Section 1.2.

33. Demonstrate that the following two LP problems have the same optimal objective function values by solving both problems. (These problems are examples of *duality*, a topic we consider in Chapter 6.)

(a) Maximize: $z = 13x_1 - 2x_2 + 5x_3$

Subject to: $2x_1 - 6x_2 + 7x_3 \leq 100$

$x_1 + 9x_2 - 8x_3 \leq 150$

$x_i \geq 0, \quad 1 \leq i \leq 3$

(b) Minimize: $z = 100x_1 + 150x_2$

Subject to: $2x_1 + x_2 \geq 13$

$-6x_1 + 9x_2 \geq -2$

$7x_1 - 8x_2 \geq 5$

$x_1, x_2 \geq 0$

(What are you going to do about the negative resource value in the second equation?)

†34. Solve by the two-phase method.

Minimize: $z = 5a_1 + 6a_2 + 2a_3 + 4b_1 + 3b_2 + 7b_3 + 2c_1 + 9c_2 + 8c_3$

Subject to: $a_1 + a_2 + a_3 \leq 8$

$b_1 + b_2 + b_3 \leq 9$

$c_1 + c_2 + c_3 \leq 6$

$a_1 + b_1 + c_1 = 6$

$a_2 + b_2 + c_2 = 5$

$a_3 + b_3 + c_3 = 9$

$a_i, b_i, c_i \geq 0, \quad 1 \leq i \leq 3$

(This is a good time to emphasize a point you may have overlooked in reading the computer instructions in Appendix A: You do not have to type 0 values in the tableaus. They will be inserted automatically the first time you pivot.)

3.5   THE TWO-PHASE METHOD: SOME COMPLICATIONS

In this section we examine some difficulties that arise in applying the two-phase method. We also show that a general LP problem has a feasible solution if and only if all artificial variables have value 0 at the end of phase 1.

A complication that can arise in phase 1 is that though we may drive the artificial variables to 0, one or more of these variables may remain as basic variables. The next example illustrates this possibility.

EXAMPLE 1

Introducing artificial variables to the LP problem

$$\text{Maximize:} \quad z = -3x_1 - x_2 + x_3$$

$$\text{Subject to:} \quad 11x_2 + 6x_3 \leq 2$$

$$x_1 - \frac{3}{2}x_2 - 2x_3 = 3$$

$$x_1 - x_2 - x_3 = 3$$

$$x_i \geq 0, \quad 1 \leq i \leq 3$$

we have

$$\text{Maximize:} \quad z = -3x_1 - x_2 + x_3$$

$$\text{Subject to:} \quad 11x_2 + 6x_3 + x_4 \qquad\qquad = 2$$

$$x_1 - \frac{3}{2}x_2 - 2x_3 \quad + y_1 \qquad = 3$$

$$x_1 - x_2 - x_3 \qquad\qquad + y_2 = 3$$

$$x_i \geq 0, 1 \leq i \leq 3; \qquad y_i \geq 0, 1 \leq i \leq 2$$

Since $w = -y_1 - y_2 = 2x_1 - (\frac{5}{2})x_2 - 3x_3 - 6$, the initial auxiliary tableau (Tableau 1) is

Tableau 1

	x_1	x_2	x_3	x_4	y_1	y_2	
x_4	0	11	6	1	0	0	2
y_1	1	$-\frac{3}{2}$	-2	0	1	0	3
y_2	①	-1	-1	0	0	1	3
(z)	3	1	-1	0	0	0	0
(w)	-2	$\frac{5}{2}$	3	0	0	0	-6

The simplex method yields Tableau 2,

Tableau 2

	x_1	x_2	x_3	x_4	y_1	y_2	
x_4	0	11	6	1	0	0	2
y_1	0	$-\frac{1}{2}$	-1	0	1	-1	0
x_1	1	-1	-1	0	0	1	3
(z)	0	4	2	0	0	-3	-9
(w)	0	$\frac{1}{2}$	1	0	0	2	0

In Tableau 2 we see that we have driven $y_1 + y_2$ to 0, but y_1 remains as a basic variable. Therefore, we have not achieved our goal of finding a *basic* feasible solution to the *original* problem (there are only two decision or slack variables in the basis; there should be three). Notice, however, that we could drive y_1 from the basis by making one more pivot on any nonzero entry in the row labeled y_1 that is in a column headed by a nonartificial variable; thus, in this case, we could pivot on either $-\frac{1}{2}$ or -1. (Note that pivoting on a negative value does not cause a resource value to become negative because the resource value for the row labeled y_1 is 0.)

Pivoting on $-\frac{1}{2}$ gives Tableau 3,

Tableau 3

	x_1	x_2	x_3	x_4	y_1	y_2	
x_4	0	0	-16	1	22	-22	2
x_2	0	1	2	0	-2	2	0
x_1	1	0	1	0	-2	3	3
(z)	0	0	-6	0	8	-11	-9
(w)	0	0	0	0	1	1	0

◆

In Example 1, we were able to drive the artificial variable y_1 out of the basis because there was a nonzero entry in its row that occurred in a column headed by a nonartificial variable (a variable of the sort we need to fill out the basis). If all of the entries under nonartificial variables are zero, the constraint of the original problem corresponding to this row of zeros is *redundant*. A redundant row can be written as a sum of multiples of other rows and thus could be omitted from the set of constraints without changing the feasible region. Example 2 illustrates redundancy.

EXAMPLE 2

Introducing artificial variables into the LP problem

$$\text{Minimize:}\quad z = 4x_1 + 15x_2$$
$$\text{Subject to:}\quad 2x_1 + 5x_2 - x_3 \qquad\qquad = 12$$
$$2x_1 + 3x_2 \qquad - x_4 = 10 \qquad\qquad (1)$$
$$6x_1 + 11x_2 - x_3 - 2x_4 = 32$$
$$x_i \geq 0, \qquad 1 \leq i \leq 4$$

we obtain the constraints

$$2x_1 + 5x_2 - x_3 \qquad + y_1 \qquad\qquad = 12$$
$$2x_1 + 3x_2 \qquad - x_4 \qquad + y_2 \qquad = 10$$
$$6x_1 + 11x_2 - x_3 - 2x_4 \qquad\qquad + y_3 = 32$$
$$x_i \geq 0, \qquad 1 \leq i \leq 4$$

Since

$$w = -y_1 - y_2 - y_3 = 10x_1 + 19x_2 - 2x_3 - 3x_4 - 54$$

the initial auxiliary tableau (Tableau 1) is

Tableau 1

	x_1	x_2	x_3	x_4	y_1	y_2	y_3	
y_1	2	⑤	-1	0	1	0	0	12
y_2	2	3	0	-1	0	1	0	10
y_3	6	11	-1	-2	0	0	1	32
(z)	4	15	0	0	0	0	0	0
(w)	-10	-19	2	3	0	0	0	-54

Applying the phase 1 method gives Tableaus 2 and 3,

Tableau 2

	x_1	x_2	x_3	x_4	y_1	y_2	y_3	
x_2	$\frac{2}{5}$	1	$-\frac{1}{5}$	0	$\frac{1}{5}$	0	0	$\frac{12}{5}$
y_2	$\frac{4}{5}$	0	$\frac{3}{5}$	-1	$-\frac{3}{5}$	1	0	$\frac{14}{5}$
y_3	$\left(\frac{8}{5}\right)$	0	$\frac{6}{5}$	-2	$-\frac{11}{5}$	0	1	$\frac{28}{5}$
(z)	-2	0	3	0	-3	0	0	-36
(w)	$-\frac{12}{5}$	0	$-\frac{9}{5}$	3	$\frac{19}{5}$	0	0	$-\frac{42}{5}$

Tableau 3

	x_1	x_2	x_3	x_4	y_1	y_2	y_3	
x_2	0	1	$-\frac{1}{2}$	$\frac{1}{2}$	$\frac{3}{4}$	0	$-\frac{1}{4}$	1
y_2	0	0	0	0	$\frac{1}{2}$	1	$-\frac{1}{2}$	0
x_1	1	0	$\frac{3}{4}$	$-\frac{5}{4}$	$-\frac{11}{8}$	0	$\frac{5}{8}$	$\frac{7}{2}$
(z)	0	0	$\frac{9}{2}$	$-\frac{5}{2}$	$-\frac{23}{4}$	0	$\frac{5}{4}$	-29
(w)	0	0	0	0	$\frac{1}{2}$	0	$\frac{3}{2}$	0

In Tableau 3 we see that we have driven the sum $y_1 + y_2 + y_3$ to zero, but y_2 remains as a basic variable. This time it is not possible to remove y_2 from the basis by pivoting because only zeros occur in entries of the y_2 row corresponding to nonartificial variables. In this example, the second equation of (1) is redundant; we can obtain this constraint by subtracting half of the first constraint of (1) from the third constraint and dividing the result by 2. Actually, redundancy causes no trouble in proceeding to phase 2 because the simplex method never selects an entry from a row of zeros as a pivot. Thus, we can ignore this row. We obtain Tableau 4 from Tableau 3 by omitting the artificial variables and the (w) objective function. (We could also omit the row of zeros, but we do not choose to do so because we want to emphasize that leaving the row in place does not interfere with phase 2. In an implementation of the two-phase method on a computer, it would be wasteful of computer time to remove such a row.)

Tableau 4

	x_1	x_2	x_3	x_4	
x_2	0	1	$-\frac{1}{2}$	$\frac{1}{2}$	1
y_2	0	0	0	0	0
x_1	1	0	$\frac{3}{4}$	$-\frac{5}{4}$	$\frac{7}{2}$
(z)	0	0	$\frac{9}{2}$	$-\frac{5}{2}$	-29

Applying the simplex method to Tableau 4 gives Tableau 5,

Tableau 5

	x_1	x_2	x_3	x_4	
x_4	0	2	-1	1	2
y_2	0	0	0	0	0
x_1	1	$\frac{5}{2}$	$-\frac{1}{2}$	0	6
(z)	0	5	2	0	-24

So the optimal solution to (1) is

$$(x_1, x_2, x_3, x_4) = (6, 0, 0, 2)$$

$$z = 24$$

◆

Before giving a general statement of the two-phase method, we note that by rearranging the order of the constraints, if necessary, we can write a general LP problem in the form

Maximize: $z = c_1 x_1 + c_2 x_2 + \cdots + c_n x_n$

Subject to:

$$
\begin{aligned}
a_{11}x_1 + \quad a_{12}x_2 + \cdots + \quad a_{1n}x_n \le b_1 \\
a_{21}x_1 + \quad a_{22}x_2 + \cdots + \quad a_{1n}x_n \le b_2 \\
\vdots \\
a_{r1}x_1 + \quad a_{r2}x_2 + \cdots + \quad a_{rn}x_n \le b_r
\end{aligned}
\tag{2}
$$

$$
\begin{aligned}
a_{r+1,1}x_1 + a_{r+1,2}x_2 + \cdots + a_{r+1,n}x_n = b_{r+1} \\
a_{r+2,1}x_1 + a_{r+2,2}x_2 + \cdots + a_{r+2,n}x_n = b_{r+2} \\
\vdots \\
a_{t1}x_1 + \quad a_{t2}x_2 + \cdots + \quad a_{tn}x_n = b_t
\end{aligned}
\tag{3}
$$

$$
\begin{aligned}
a_{t+1,1}x_1 + a_{t+1,2}x_2 + \cdots + a_{t+1,n}x_n \ge b_{t+1} \\
a_{t+2,1}x_1 + a_{t+2,2}x_2 + \cdots + a_{t+2,n}x_n \ge b_{t+2} \\
\vdots \\
a_{m1}x_1 + \quad a_{m2}x_2 + \cdots + \quad a_{mn}x_n \ge b_m
\end{aligned}
\tag{4}
$$

$$x_j \ge 0, \qquad 1 \le j \le n$$

Systems such as (2)–(4) are cumbersome to write. In the sequel we write them in the more compact sigma (Σ) notation

$$\sum_{j=1}^{n} a_{ij}x_j \leq b_i, \qquad 1 \leq i \leq r \tag{5}$$

$$\sum_{j=1}^{n} a_{ij}x_j = b_i, \qquad r+1 \leq i \leq t \tag{6}$$

$$\sum_{j=1}^{n} a_{ij}x_j \geq b_i, \qquad t+1 \leq i \leq m \tag{7}$$

$$x_j \geq 0, \qquad 1 \leq j \leq n$$

Expressions (5), (6), (7) have the same meaning as expressions (2), (3), (4), respectively. The notation enables us to omit some of the types of constraints. For instance, setting $t = r$ would indicate that there are no numbers i satisfying $r + 1 \leq i \leq t$ and, consequently, no equality constraints of type (6).

Summary of the Two-Phase Method

PHASE 1

STEP 1. Form the auxiliary problem.

a. For each constraint of type (5), add a nonnegative slack variable to obtain the constraints

$$\sum_{j=1}^{n} a_{ij}x_j + s_i = b_i, \qquad 1 \leq i \leq r \tag{8}$$

b. For each constraint of type (6), add a nonnegative artificial variable to obtain the constraints

$$\sum_{j=1}^{n} a_{ij}x_j + y_i = b_i, \qquad r+1 \leq i \leq t \tag{9}$$

c. For each constraint of type (7), subtract a nonnegative slack variable and add a nonnegative artificial variable to obtain the constraints

$$\sum_{j=1}^{n} a_{ij}x_j - s_i + y_i = b_i, \qquad t+1 \leq i \leq m \tag{10}$$

d. Form the auxiliary objective function

$$w = \sum_{j=r+1}^{m} (-y_i) = -y_{r+1} - y_{r+2} - \cdots - y_m \tag{11}$$

Since the artificial variables (y_i, $r+1 \leq i \leq m$) are basic variables in the initial auxiliary problem, we need to rewrite w in terms of nonbasic variables. These nonbasic variables are $x_j (1 \leq j \leq n)$ and $s_i (t+1 \leq i \leq m)$. To rewrite w we solve each equation from (9) and (10) for y_i and substitute in (11).

STEP 2. Find a feasible solution.

a. Form the auxiliary tableau including objective rows for w and z.

b. Apply the simplex method to the auxiliary tableau using the (w) objective function to determine the entering basic variable. Also, perform pivots on the (z) objective function so that it is continually expressed in terms of the nonbasic variables. If the maximum value of the (w) objective function is 0, then we have found a feasible solution to the problem. If the maximum value of (w) is negative, the problem is infeasible.

STEP 3. Remove any artificial variables remaining in the basis after step 2.

If a feasible solution was found in step 2, it could happen that some artificial variables remain in the basis at value 0. Examples 1 and 2 illustrated this possibility. There are two cases to consider:

CASE 1. The row corresponding to the artificial variable that remains in the basis contains a nonzero entry (call it a) in a column corresponding to a decision or slack variable. Pivoting on a replaces the artificial variable by a nonartificial variable. Note that it does not matter if a happens to be negative. Since the resource value for the row corresponding to the artificial variable is 0, no negative resource values will result from the pivot. Example 1 illustrates this case.

CASE 2. The row that corresponds to the artificial variable that remains in the basis contains only zero entries in columns corresponding to decision or slack variables. In this case, the constraint in the original problem corresponding to this row is redundant (i.e., the constraint is a linear combination of other constraints). We can omit this row from further consideration without changing the feasible set, as Example 2 illustrates.

PHASE 2

If phase 1 found a feasible solution, then we eliminate from the tableau the row corresponding to the artificial objective function and perform the simplex method on the resulting tableau.

Theorem 1 gives the theoretical basis for phase 1 of the two-phase method.

Theorem 1. A general LP problem has a feasible solution if and only if all of the artificial variables have value 0 at the end of phase 1.

PROOF. We show that the systems of constraints (5), (6), (7) define the same set of feasible solutions as the system of constraints corresponding to the final tableau of phase 1 when phase 1 has terminated with all artificial variables having a zero value.

PART 1. Assume that the set of constraints (5), (6), (7) has a feasible solution $(x_1^*, x_2^*, \ldots, x_n^*)$. Express (5), (6), (7) as a system of linear equalities by adding slack variables (but not artificial variables), thus obtaining

$$\sum_{j=1}^{n} a_{ij}x_j + s_i = b_i, \qquad 1 \le i \le r \tag{5'}$$

$$\sum_{j=1}^{n} a_{ij}x_j = b_i, \qquad r+1 \le i \le t \tag{6'}$$

$$\sum_{j=1}^{n} a_{ij}x_j - s_i = b_i, \qquad t+1 \le i \le m \tag{7'}$$

All variables are nonnegative.

Solve equations (5') and (7') for the s_i in terms of the feasible solution x_i^*, $1 \le i \le n$ to obtain a set of values of the slack variables,

$$\{s_i^* \mid 1 \le i \le r, \quad t+1 \le i \le m\}$$

for which the x_i^* and s_i^* form a feasible solution to (5'), (6'), (7'). Obtain a feasible solution to (8), (9), (10) by setting all of the artificial variables equal to 0. Therefore, since the artificial variables are nonnegative, the minimum value of the auxiliary objective function (w) is zero, and phase 1 must produce a solution with artificial variables 0.

PART 2. Suppose phase 1 produces a solution with all artificial variables 0. Since all of the tableaus produced by the simplex method are equivalent (in the sense that they define the same solution sets), the solution found by phase 1 is also a solution of (8), (9), (10). Omitting all of the slack variables in this solution provides a feasible solution to (5), (6), (7). ∎

Exercises

1. Each of the following tableaus resulted from performing some steps of phase 1. Describe the next operation that is to be performed to continue phase 1. If the operation is a pivot, specify the element to be pivoted on. (The y_i are artificial variables.)

(a)

	x_1	x_2	s_1	y_1	y_2	y_3	
y_1	0	$\frac{5}{2}$	0	1	0	0	5
y_3	0	$-\frac{23}{6}$	0	0	$-\frac{1}{3}$	1	0
x_1	1	$-\frac{5}{12}$	0	0	$-\frac{1}{6}$	0	2
s_1	0	$-\frac{13}{12}$	1	0	$\frac{1}{6}$	0	1
(w)	0	$-\frac{7}{2}$	0	0	1	0	-2

(b)

	x_1	x_2	s_1	y_1	y_2	y_3	
y_1	0	$\frac{5}{2}$	0	1	0	0	0
y_3	0	$-\frac{23}{6}$	0	0	$-\frac{1}{3}$	1	0
x_1	1	$-\frac{5}{12}$	0	0	$-\frac{1}{6}$	0	2
s_1	0	$-\frac{13}{12}$	1	0	$\frac{1}{6}$	0	1
(w)	0	$\frac{7}{2}$	0	0	1	0	0

2. Follow the instructions from Exercise 1 for the following tableaus.

(a)

	x_1	x_2	s_1	y_1	y_2	y_3	
x_1	1	0	1	1	0	9	0
x_2	0	1	2	0	0	23	0
y_2	0	0	-3	0	1	0	0
(w)	0	0	0	0	0	2	0

(b)

	x_1	x_2	s_1	y_1	y_2	y_3	
y_1	0	0	0	1	0	9	0
x_2	0	1	0	0	$-\frac{1}{3}$	23	0
x_1	1	0	0	0	$-\frac{1}{6}$	0	2
(w)	0	0	0	0	1	2	0

3. Use phase 1 to show that the constraints of the following LP problem are redundant:

$$\text{Maximize:} \quad z = -x_1 - 3x_2 + 2x_3$$
$$\text{Subject to:} \quad x_1 - 2x_2 + 3x_3 \qquad = 1$$
$$2x_2 - 4x_3 + 2x_4 = 2$$
$$3x_1 - 2x_2 + x_3 + 4x_4 = 7$$
$$x_i \geq 0, \qquad 1 \leq i \leq 4$$

4. Use phase 1 to show that the constraints of the following LP problem are redundant:

$$\text{Minimize:} \quad z = -3x_1 + 2x_2 - 2x_3 - 2x_4$$
$$\text{Subject to:} \quad -x_1 + 3x_2 - x_3 + 2x_4 = 1$$
$$8x_2 - 3x_3 + 5x_4 = 4$$
$$2x_1 + 2x_2 - x_3 + x_4 = 2$$
$$x_i \geq 0, \qquad 1 \leq i \leq 4$$

5. Use phase 1 to demonstrate that the following LP problem is infeasible:

$$\text{Minimize:} \quad z = x_1 + x_2 + x_3$$
$$\text{Subject to:} \quad x_1 + 2x_2 - x_3 \leq 1$$
$$2x_1 \qquad - x_3 \geq 4$$
$$2x_1 - x_2 + x_3 = 4$$
$$x_i \geq 0, \qquad 1 \leq i \leq 3$$

6. Use phase 1 to demonstrate that the following LP problem is infeasible:

$$\text{Maximize:} \quad z = 2x_1 + 3x_2$$
$$\text{Subject to:} \quad x_1 + 4x_2 \leq 4$$
$$x_1 - 2x_2 \leq 4$$
$$x_1 - x_2 \geq 6$$
$$x_1, x_2 \geq 0$$

7. Write the systems of inequalities without Σ notation.
   (a) $\sum_{j=1}^{3}(i+j)x_j \leq i,$ $1 \leq i \leq 3$
   (b) $\sum_{j=1}^{3}(a_{ij}+j)y_j \leq 1,$ $1 \leq i \leq 2$
   (c) $\sum_{j=1}^{4}(-1)^{i+j}x_j \geq i,$ $1 \leq i \leq 3$

8. Write the systems of equations without Σ notation.
   (a) $\sum_{j=1}^{3} a_{ij}x_j + s_i = b_i,$ $1 \leq i \leq 2$
   (b) $\sum_{j=1}^{4} x_j + \sum_{j=1}^{3} iy_j + s_i = b_i,$ $1 \leq i \leq 2$
   (c) $\sum_{j=1}^{3} a_{ij}x_j = s_i - y_i + b_i,$ $1 \leq i \leq 3$

9. By appropriately assigning a_{ij} and b_i, write the following systems using one Σ symbol.
   (a) $x_1 + 3x_2 - 4x_3 = 3$
          $x_1 \qquad + 2x_3 = 5$

   (b) $2x_1 - 3x_2 + 5x_3 \leq 3$
         $-3x_1 + \ x_2 - 5x_3 \geq 1$
         $x_1 \qquad + \ x_3 \leq 1$

10. By appropriately assigning a_{ij} and b_i, express the constraints of Example 1 in Σ notation.

11. Explain why the following tableau could not appear during the execution of phase 1.

	x_1	x_2	s_1	s_2	y_1	y_2	
x_1	1	2	-2	3	0	-2	2
y_1	0	1	-3	1	1	1	1
(w)	0	0	-2	-1	0	3	-1

12. The phase 1 method is an efficient *algorithmic* way to find a basic feasible solution to an LP problem. It is crucially important that it be an algorithmic method because computers usually do the calculations. That does not mean that an algorithmic method is the fastest way for humans to solve some (especially small) problems. For example, it is easy to find a basic solution to

$$x_1 + \ x_2 \qquad = 1$$
$$2x_1 + 3x_2 + x_3 = 3$$

by multiplying the first equation by -2 and adding the result to the second equation. In a similar way, find basic solutions to Exercises 1 and 2 of Section 3.4.

13. Suppose an LP problem has eight decision variables and three constraints of type (5), five constraints of type (6), and six constraints of type (7). How many total variables does the phase 1 problem have?

3.6   THE BIG-M METHOD

The Big-M method provides another approach to solving LP problems with artificial variables. To illustrate this method we return to the LP problem (4) given in Section 3.4.

$$\text{Maximize:} \quad z = 2x_1 + 3x_2$$

$$\text{Subject to:} \quad \begin{aligned} -4x_1 + 3x_2 &\le 12 \\ 2x_1 + x_2 &\le 6 \\ x_1 + x_2 &\ge 3 \\ 5x_1 + x_2 &\ge 4 \end{aligned} \tag{1}$$

$$x_1, x_2 \ge 0$$

Introducing slack and artificial variables to (1), we have (as before) the LP problem

$$\text{Maximize:} \quad z = 2x_1 + 3x_2$$

$$\text{Subject to:} \quad \begin{aligned} -4x_1 + 3x_2 + x_3 &&&&&&= 12 \\ 2x_1 + x_2 && + x_4 &&&&= 6 \\ x_1 + x_2 &&&- x_5 &+ y_1 &&= 3 \\ 5x_1 + x_2 &&&&- x_6 &+ y_2 &= 4 \end{aligned} \tag{2}$$

$$x_i \ge 0,\ 1 \le i \le 6; \qquad y_i \ge 0,\ 1 \le i \le 2$$

As with the two-phase method, we want to find a solution to (2) for which the artificial variables have been "driven out." The strategy of the Big-M method, however, is quite different from that of the two-phase method in that we will treat all variables (decision, slack, and artificial) equally. We drive the artificial variables to zero by giving them an "unfavorable" status in the objective function.

To begin, we select a large positive number M (thus the name, the Big-M method) and form a new objective function

$$z' = 2x_1 + 3x_2 - M(y_1 + y_2)$$

which we attempt to maximize. Observe that since y_1 and y_2 are nonnegative, and since we chose M to be a very large positive number, we obtain a maximum for z' only if both y_1 and y_2 are 0. The difficulty of determining how large M must be to accomplish this is a weakness of the Big-M method.

To form a canonical tableau for problem (2), we must express the objective function z' in a form that does not explicitly depend on y_1 and y_2 because we want to use these variables as our initial basic variables. To rewrite z' we use the third and fourth equations of (2) to obtain

$$y_1 + y_2 = -6x_1 - 2x_2 + x_5 + x_6 + 7$$

Substituting this expression in the objective function for (2) gives

$$z' = 2x_1 + 3x_2 - M(-6x_1 - 2x_2 + x_5 + x_6 + 7)$$

$$= (2 + 6M)x_1 + (3 + 2M)x_2 - Mx_5 - Mx_6 - 7M$$

The corresponding tableau (Tableau 1) is

Tableau 1

	x_1	x_2	x_3	x_4	x_5	x_6	y_1	y_2	
x_3	-4	3	1	0	0	0	0	0	12
x_4	2	1	0	1	0	0	0	0	6
y_1	1	1	0	0	-1	0	1	0	3
y_2	5	1	0	0	0	-1	0	1	4
(z')	$-(2+6M)$	$-(3+2M)$	0	0	M	M	0	0	$-7M$

Setting $M = 1000$ gives Tableau 2, to which we can apply the simplex method,

Tableau 2

	x_1	x_2	x_3	x_4	x_5	x_6	y_1	y_2	
x_3	-4	3	1	0	0	0	0	0	12
x_4	2	1	0	1	0	0	0	0	6
y_1	1	1	0	0	-1	0	1	0	3
y_2	⑤	1	0	0	0	-1	0	1	4
(z)	-6002	-2003	0	0	1000	1000	0	0	-7000

The simplex method eventually produces the optimal Tableau 3.

Tableau 3

	x_1	x_2	x_3	x_4	x_5	x_6	y_1	y_2	
x_5	0	0	0.1	0.7	1	0	-1	0	2.4
x_6	0	0	-0.3	1.9	0	1	0	-1	3.8
x_2	0	1	0.2	0.4	0	0	0	0	4.8
x_1	1	0	-0.1	0.3	0	0	0	0	0.6
(z)	0	0	0.4	1.8	0	0	1000	1000	15.6

Tableau 3 shows that we have accomplished our goal of driving the artificial variables to 0. The optimal solution is

$$(x_1, x_2, x_3, x_4, x_5, x_6, y_1, y_2) = (0.6, 4.8, 0, 0, 2.4, 3.8, 0, 0)$$

$$z' = 15.6$$

It follows that the optimal solution to (1) is

$$(x_1, x_2) = (0.6, 4.8)$$

$$z = 15.6$$

We can also use artificial variables to deal with equality constraints, as Example 1 illustrates.

EXAMPLE 1

Introducing slack variables to the LP problem

$$\text{Maximize:} \quad z = -x_1 + x_2$$

$$\text{Subject to:} \quad x_1 + 2x_2 \geq 6$$
$$2x_1 + x_2 \leq 6 \tag{3}$$
$$4x_1 + x_2 = 4$$

$$x_1, x_2 \geq 0$$

gives us

$$\text{Maximize:} \quad z = -x_1 + x_2$$

$$\text{Subject to:} \quad x_1 + 2x_2 - x_3 \quad\quad = 6$$
$$2x_1 + x_2 \quad\quad + x_4 = 6 \tag{4}$$
$$4x_1 + x_2 \quad\quad\quad = 4$$

$$x_i \geq 0, \quad 1 \leq i \leq 4$$

To obtain an initial set of basic variables, we add nonnegative artificial variables to the first and third constraints of (4) to obtain the canonical system of linear equations (5) with basic variables x_4, y_1, y_2. As before, we introduce a large number M to the original objective function.

$$\text{Maximize:} \quad z' = -x_1 + x_2 - M(y_1 + y_2)$$

$$\text{Subject to:} \quad x_1 + 2x_2 - x_3 \quad\quad + y_1 \quad\quad = 6$$
$$2x_1 + x_2 \quad\quad + x_4 \quad\quad = 6 \tag{5}$$
$$4x_1 + x_2 \quad\quad\quad + y_2 = 4$$

$$x_i \geq 0, 1 \leq i \leq 4; \quad y_i \geq 0, 1 \leq i \leq 2$$

To rewrite the objective function of (5) in terms of the current nonbasic variables, we use the first and third equations of (5) to obtain

$$y_1 + y_2 = 10 - (5x_1 - 3x_2 + x_3)$$

and substituting the right-hand side of this equation in

$$z' = -x_1 + x_2 - M(y_1 + y_2)$$

results in the LP problem

$$\text{Maximize:} \quad z' = -10M + (-1 + 5M)x_1 + (1 + 3M)x_2 - Mx_3$$

$$\text{Subject to:} \quad x_1 + 2x_2 - x_3 \quad\quad + y_1 \quad\quad = 6$$
$$2x_1 + x_2 \quad\quad + x_4 \quad\quad = 6$$
$$4x_1 + x_2 \quad\quad\quad + y_2 = 4$$

$$x_i \geq 0, 1 \leq i \leq 4; \quad y_i \geq 0, 1 \leq i \leq 2$$

Setting $M = 500$, we obtain the initial Tableau 1.

Tableau 1

	x_1	x_2	x_3	x_4	y_1	y_2	
y_1	1	2	-1	0	1	0	6
x_4	2	1	0	1	0	0	6
y_2	④	1	0	0	0	1	4
	-2499	-1501	500	0	0	0	-5000

The simplex method eventually produces the optimal Tableau 2.

Tableau 2

	x_1	x_2	x_3	x_4	y_1	y_2	
x_2	4	1	0	0	0	1	4
x_4	-2	0	0	1	0	-1	2
x_3	7	0	1	0	-1	2	2
	5	0	0	0	500	501	4

In Tableau 2 we see that we have driven the artificial variables to 0 and obtained the optimal solution

$$(x_1, x_2, x_3, x_4, y_1, y_2) = (0, 4, 2, 2, 0, 0)$$

$$z = 4$$

It follows that the optimal solution to the original LP problem (3) is

$$(x_1, x_2) = (0, 4)$$

$$z = 4 \qquad \blacklozenge$$

The next example illustrates how the Big-M method detects an empty feasible region.

EXAMPLE 2

Introducing slack and artificial variables to the LP problem

$$\text{Maximize:} \quad z = x_1 - 2x_2$$

$$\text{Subject to:} \quad 2x_1 + 3x_2 \le 6$$

$$-x_1 + x_2 \le 2 \qquad (6)$$

$$x_1 + 2x_2 \ge 10$$

$$x_1, x_2 \ge 0$$

gives the constraints

$$\begin{aligned}
2x_1 + 3x_2 + x_3 \qquad\qquad\qquad &= 6 \\
-x_1 + x_2 \qquad + x_4 \qquad\qquad &= 2 \\
x_1 + 2x_2 \qquad\qquad - x_5 + y_1 &= 10
\end{aligned} \qquad (7)$$

From the third equation of (7) we have

$$y_1 = 10 - x_1 - 2x_2 + x_5$$

and we can form the Big-M objective function

$$z' = x_1 - 2x_2 - My_1 = (1 + M)x_1 + (-2 + M)x_2 - Mx_5 - 10M$$

The LP problem in Big-M form is

$$\begin{aligned}
\text{Maximize:}\quad & z' = (1 + M)x_1 + (-2 + M)x_2 - Mx_5 - 10M \\
\text{Subject to:}\quad & 2x_1 + 3x_2 + x_3 \qquad\qquad\qquad = 6 \\
& -x_1 + x_2 \qquad + x_4 \qquad\qquad = 2 \\
& x_1 + 2x_2 \qquad\qquad - x_5 + y_1 = 10 \\
& x_i \geq 0,\ 1 \leq i \leq 5; \qquad y_1 \geq 0
\end{aligned}$$

The corresponding initial tableau (Tableau 1) is

Tableau 1

	x_1	x_2	x_3	x_4	x_5	y_1	
x_3	2	3	1	0	0	0	6
x_4	-1	1	0	1	0	0	2
y_1	1	2	0	0	-1	1	10
	$-1 - M$	$2 - M$	0	0	M	0	$-10M$

Pivoting on the circled entry, we obtain Tableau 2,

Tableau 2

	x_1	x_2	x_3	x_4	x_5	y_1	
x_1	1	$\frac{3}{2}$	$\frac{1}{2}$	0	0	0	3
x_4	0	$\frac{5}{2}$	$\frac{1}{2}$	1	0	0	5
y_1	0	$\frac{1}{2}$	$-\frac{1}{2}$	0	-1	1	7
	0	$(7 + M)/2$	$(1 + M)/2$	0	M	0	$3 - 7M$

Note that all terms in the objective row of Tableau 2 are positive, but the artificial variable y_1 is still a basic variable and *has a nonzero value*. We conclude that of all feasible solutions to (7), the smallest value y_1 can have is 7. Since there is no *feasible* solution of (7) with $y_1 = 0$, there is no feasible solution of (6). To see that this is indeed the case, you are asked in Exercise 6 to describe geometrically the original constraints of this problem. ◆

A serious drawback of the Big-*M* method is its inherent propensity for numerical roundoff error. This property virtually eliminates the Big-*M* method from consideration as a practical computer algorithm for large problems. Because the method works by making some entries very large with respect to others, in a lengthy computer calculation this usually causes serious roundoff errors in the many additions and subtractions involved in the simplex method. Although the Big-*M* method is occasionally useful theoretically, the two-phase method is almost always used for computer solutions. Even for small hand-calculations, the two-phase method usually requires fewer calculations, especially when unnecessary calculations of the pivoted entries in artificial variable columns are avoided.

Exercises

1. Solve the following exercises from Section 3.4 by the Big-*M* method.
   (a) Exercise 1   (b) Exercise 3   (c) Exercise 13

2. Solve the following exercises from Section 3.4 by the Big-*M* method.
   (a) Exercise 2   (b) Exercise 4   (c) Exercise 14

3. Example 1 of Section 3.5 required an additional step in the phase 1 procedure to eliminate from the basis an artificial variable with final value 0. Apply the Big-*M* method to this problem. How does the Big-*M* method handle the difficulty?

4. Answer the questions from Exercise 3 for Example 2 of Section 3.5.

5. Solve Exercise 23 of Section 3.4 with the Big-*M* method.

6. Describe geometrically the feasible set for Example 2 (of this section) to verify that it is empty.

7. Use the Big-*M* method to show that Exercise 5 of Section 3.5 is infeasible.

8. Use the Big-*M* method to show that Exercise 6 of Section 3.5 is infeasible.

   You can solve Exercises 9–13 with the TUTOR program if you do not use so large a value of *M* that the four-digit limit on computed fractions is exceeded; $M = 100$ works for these problems.

†9. Use the Big-*M* method to solve the following LP problems from Section 3.4.
    (a) Exercise 15   (b) Exercise 17

†10. Use the Big-*M* method to solve Exercise 16 in Section 3.4.

†11. Use the Big-*M* method to solve Exercise 33 (b) of Section 3.4.

†12. Use the Big-*M* method to solve Exercise 34 of Section 3.4.

†13. Use the Big-*M* method to solve Exercise 3 of Section 3.5. How does the Big-*M* method detect redundancy?

†3.7 LP MODELS

In this section we develop several practical LP models and introduce you to some of the automatic-solution programs of CALIPSO. We begin with two introductory examples.

EXAMPLE 1

We start with a slightly modified version of Example 1 of Section 2.5. The Carter Nut Company (CNC) supplies a variety of nut mixtures for sale to companies that package and resell them. CNC's most expensive product is Bridge Mix, a mixture of peanuts, almonds, and cashews. Bridge Mix contains no more than 25% peanuts and no less than 40% cashews. There is no limitation on the percentage of almonds. The current selling price of Bridge Mix is $.80 per pound. Table 1 gives the amounts of nuts available for purchase and the cost of the nuts.

Table 1   Supply and cost of nuts this month

Variable	Type	Cost per pound ($)	Pounds available
b_{11}	Peanuts	0.20	400
b_{12}	Almonds	0.35	250
b_{13}	Cashews	0.50	150

How much Bridge Mix should CNC produce this month to maximize profit?

The formulation of this problem is very similar to the one given in Example 4 of Section 2.5. We have put double subscripts on the variables to assist in further development of the problem in Example 2 below. We have also used the letter e for some of the slack variables. The new formulation is

$$\text{Maximize:} \quad z_1 = 0.60b_{11} + 0.45b_{12} + 0.30b_{13}$$

$$
\begin{aligned}
\text{Subject to:} \quad & 0.75b_{11} - 0.25b_{12} - 0.25b_{13} + s_{11} && = 0 \\
& 0.40b_{11} + 0.40b_{12} - 0.60b_{13} \quad + s_{12} && = 0 \\
& b_{11} \qquad\qquad\qquad\qquad\qquad + e_{11} && = 400 \qquad (1) \\
& \qquad b_{12} \qquad\qquad\qquad\qquad + e_{12} && = 250 \\
& \qquad\qquad b_{13} \qquad\qquad\qquad + e_{13} && = 150
\end{aligned}
$$

All variables are nonnegative.

We are going to solve this problem using the automatic solution programs of CALIPSO. You should read Section A.3 of Appendix A and work through the steps described below on a computer.

STEP 1. From the CALIPSO menu, select TUTOR. From the TUTOR menu, select Tableau Editor.

STEP 2. The screen that appears is the same as the one you have previously used to enter and pivot tableaus; however, this time you will only enter the tableau and store it in a disk file.

Enter the following tableau, which corresponds to system (1). You do not need to type the zeros when entering the tableau on the computer because they will be automatically inserted in blank spaces. Because you cannot type subscripts with TUTOR, you must type them in line with the letter (i.e., b_{11} is entered as b11).

	b_{11}	b_{12}	b_{13}	s_{11}	s_{12}	e_{11}	e_{12}	e_{13}	RHS
s_{11}	0.75	−0.25	−0.25	1	0	0	0	0	0
s_{12}	0.40	0.40	−0.60	0	1	0	0	0	0
e_{11}	1	0	0	0	0	1	0	0	400
e_{12}	0	1	0	0	0	0	1	0	250
e_{13}	0	0	1	0	0	0	0	1	150
OBJ	−0.60	−0.45	−0.30	0	0	0	0	0	0

When the tableau has been entered, store (save) it (using the F7 key) with the name NUT1. Then press the Esc key to get back to the TUTOR menu.

STEP 3. Enter the file name NUT1 in the space provided, and select the automatic-solution program. The menu that appears offers you a variety of choices to display and print information. Since we only want to display the solution on the screen, you can press ENTER without making any choices. The following solution should appear.

Number of pivots to find a maximum = 3

Maximum value of the objective function = 160.31250

Basic Variables	Value
b_{11}	93.75000
b_{12}	131.25000
e_{11}	306.25000
e_{12}	118.75000
b_{13}	150.00000

(2)

Solution (2) shows that the availability of cashews is the most restrictive constraint ($b_{13} = 150$). There is a surplus of 306.25 pounds of peanuts ($e_{11} = 306.25$) and 118.75 pounds of almonds ($e_{12} = 118.75$).

STEP 4. When you have finished viewing the solution, press ENTER. The next menu allows you to print the solution.

EXAMPLE 2

The Carter Nut Company decides to plan its production for the next two months, rather than for just one month. CNC projects that nuts will be more expensive next month because it will be later in the season. It also predicts that it can sell mixed nuts for $0.90 per pound next month. CNC has already accepted orders totaling 200 pounds that must be delivered in the first month. Table 2 gives the predicted cost and availability of nuts for the second month.

Table 2　Predicted supply and cost of nuts next month

Variable	Type	Cost per pound ($)	Pounds available
b_{21}	Peanuts	0.22	200
b_{22}	Almonds	0.40	75
b_{23}	Cashews	0.72	300

CNC has enough storage to allow the purchase of nuts at this month's prices for mixing next month. It is willing to neglect the cost of storage and investment of money in nuts. How should CNC plan its production for two months to maximize its profit from Bridge Mix?

The slack variables e_{11}, e_{12}, e_{13} contain the amounts of the three kinds of nuts that were not purchased in the first month. To allow CNC to purchase some of these excess nuts to store for the next month's production, we replace e_{1i} by $e_{1i} + p_{1i}$ ($1 \leq i \leq 3$) where p_{1i} represents the quantity of nuts to be stored for use next month. With this change in LP problem (1), we can link the model for the first month to that for the second month by using the following constraints.

$$b_{11} + b_{12} + b_{13} \geq 200 \qquad \text{(Fill first month's orders)}$$

$$b_{21} + p_{11} \leq 0.25(b_{21} + p_{11} + b_{22} + p_{12} + b_{23} + p_{13}) \qquad \text{(At most 25\% peanuts)}$$

$$b_{23} + p_{13} \geq 0.40(b_{21} + p_{11} + b_{22} + p_{12} + b_{23} + p_{13}) \qquad \text{(At least 40\% cashews)}$$

$$b_{21} \leq 200 \qquad \text{(Available peanuts)}$$

$$b_{22} \leq 75 \qquad \text{(Available almonds)}$$

$$b_{23} \leq 300 \qquad \text{(Available cashews)}$$

The objective function for the 2-month period is

$$z = 0.80(b_{11} + b_{12} + b_{13}) \qquad \text{(Selling price in first month)}$$

$$-0.20b_{11} - 0.35b_{12} - 0.50b_{13} \qquad \text{(Cost of nuts in first month)}$$

$$-0.20p_{11} - 0.35p_{12} - 0.50p_{13} \qquad \text{(Cost of stored nuts)}$$

$$+0.90(b_{21} + p_{11} + b_{22} + p_{12} + b_{23} + p_{13}) \qquad \text{(Selling price in second month)}$$

$$-0.22b_{21} - 0.40b_{22} - 0.72b_{23} \qquad \text{(Cost of nuts in second month)}$$

The LP problem for the 2-month period is

Maximize: $z = 0.60b_{11} + 0.45b_{12} + 0.30b_{13} + 0.68b_{21} + 0.50b_{22} + 0.18b_{23}$
$$+ 0.70p_{11} + 0.55p_{12} + 0.40p_{13}$$

Subject to:

$$
\begin{aligned}
b_{11} + b_{12} + b_{13} - t &= 200 \\
0.75b_{11} - 0.25b_{12} - 0.25b_{13} + s_{11} &= 0 \\
0.40b_{11} + 0.40b_{12} - 0.60b_{13} + s_{12} &= 0 \\
b_{11} + e_{11} + p_{11} &= 400 \\
b_{12} + e_{12} + p_{12} &= 250 \\
b_{13} + e_{13} + p_{13} &= 150 \qquad (3) \\
0.75b_{21} - 0.25b_{22} - 0.25b_{23} + 0.75p_{11} - 0.25p_{12} - 0.25p_{13} + s_{21} &= 0 \\
0.40b_{21} + 0.40b_{22} - 0.60b_{23} + 0.40p_{11} + 0.40p_{12} - 0.60p_{13} + s_{22} &= 0 \\
b_{21} - p_{11} + e_{21} &= 200 \\
b_{22} - p_{12} + e_{22} &= 75 \\
b_{23} - p_{13} + e_{23} &= 300
\end{aligned}
$$

All variables are nonnegative.

The slack variable t of the first constraint in (3) has a negative coefficient; so we cannot use t as an initial basic variable. To enter the tableau corresponding to (3) for solution by SOLVER, you leave the row label for this constraint blank. The blank row label causes the automatic-solution program to add an artificial variable in this row and begin processing with phase 1. Note that you do not add any artificial variables to the input tableau. The program adds an artificial variable for each row that is not labeled with a basic variable. In this example, the program adds one artificial variable for the first constraint of (3).

You should enter the tableau for (3) into the computer and solve it. The solution as reported by the program is

Maximum value of the objective function = 449.83333

Basic Variables	Value
b_{11}	50.00000
b_{12}	70.00000
b_{13}	80.00000
b_{22}	75.00000
b_{23}	300.00000
p_{11}	208.33333
p_{12}	180.00000
p_{13}	70.00000
e_{11}	141.66667
e_{21}	200.00000
s_{22}	36.66667

In Exercise 1 you are asked to interpret this solution.   ◆

Example 2 is a common type of LP model called a *multiperiod model.* It is customary for enterprises to organize their activities over a unit of time such as a day, a week, or a year while maintaining a long-range plan that spans many of their time units. The statement of CNC's problem suggested that its production schedule is based on a time unit of one month. Even so, it is still interested in linking these basic units to take advantage of such things as surplus inventory. The linking was easily established in this case. It is

(beginning inventory) + (purchases) = (ending inventory) + (sales)

The constraint on peanuts in the second month, from (3), can be viewed in this form as

$$p_{11} + 200 = e_{21} + b_{21}$$

Two other linking equations of the same form exist for the other kinds of nuts. It is a simple matter to extend the linear model for CNC to additional monthly periods. Notice that the first six constraints of (3) correspond to the first month, and the next five constraints correspond to the second month. The constraints in these two groups have a similar form except that the second set has no minimum production requirement (though it would have been natural to impose one). Extending the model to cover more periods is a simple matter of adding new variables and more groups of constraints similar to the two we have. You are asked to make such an extension in Exercise 2.

Financial problems often involve linking a series of short-term investments into a plan that maximizes yield over a long period. Example 3 illustrates this kind of problem.

EXAMPLE 3

A man wants to establish a fund to finance his granddaughter's college education. She is now 8 years old and is expected to begin college 10 years from now. The grandfather wants to calculate the amount of money he must set aside now and to determine a sequence of investments from a limited selection of choices so that his granddaughter will have $20,000 available to her in each of the 4 years starting 10 years from now. He has identified three investments that appear attractive:

1. He can purchase a municipal bond at 52% of its value; the bond will be redeemable in 12 years.

2. He can purchase a government security that pays 6% interest per year and is redeemable for its face value in 10 years.

3. He can deposit money in a savings account at 4% interest per year.

Find an investment schedule that will minimize the present cost to the grandfather. Introduce the following variables,

b_1 = face value of the bond

b_2 = face value of the security

d_i = amount deposited in the savings account at the start of year i for
$i = 0, 1, \ldots, 12$ (We let d_0 denote the original deposit.)

Except for years in which the principal is received from investments in the bond or the security, the cash income consists of a 6% return from the security and a 4% return on the previous year's savings. For the years prior to the start of college, this cash will be redeposited in savings according to the formula

$$d_i = 0.06b_2 + 1.04d_{i-1}$$

We write the LP problem in the form that enables you to enter it with the inequality editor, EDITOR, of CALIPSO. You can read about EDITOR in Appendix A. It creates a tableau file that you can use to solve the problem with

SOLVER. We have printed the problem without subscripts as you would enter it with EDITOR.

Note the following conventions for entering inequalities with EDITOR (these and other conventions are more elaborately described in Appendix A).

a. EDITOR identifies an expression as an objective function if the expression does not contain one of the symbols $\leq$, $\geq$, or $=$.

b. If an objective function is preceded by MIN, EDITOR multiplies it by -1, effectively changing the LP problem to a maximization problem.

c. You must mark the end of each objective and constraint expression with a semicolon.

d. EDITOR ignores phrases enclosed in curly brackets; such phrases are used as comments to document your problem.

e. EDITOR assumes that all variables are nonnegative; therefore, you should not include constraints that make the variables ≥ 0.

$$\text{MIN } .52b1 + b2 + d0;$$

$.06b2 + 1.04d0 - d1 = 0;$ {Year 1}

$.06b2 + 1.04d1 - d2 = 0;$ {Year 2}

$.06b2 + 1.04d2 - d3 = 0;$ {Year 3}

$.06b2 + 1.04d3 - d4 = 0;$ {Year 4}

$.06b2 + 1.04d4 - d5 = 0;$ {Year 5}

$.06b2 + 1.04d5 - d6 = 0;$ {Year 6}

$.06b2 + 1.04d6 - d7 = 0;$ {Year 7}

$.06b2 + 1.04d7 - d8 = 0;$ {Year 8}

$.06b2 + 1.04d8 - d9 = 0;$ {Year 9}

$1.06b2 + 1.04d9 - d10 = 20000;$ {Start college. Security matures}

$1.04d10 - d11 = 20000;$ {Year 11}

$b1 + 1.04d11 - d12 = 20000;$ {Bond matures}

$1.04d12 = 20000;$ {Graduation year}

None of the constraints for this problem contain a variable that we can use initially in the basis. If you examine the tableau created by EDITOR, you will see that the rows are not labeled with a basis variable. Omission of these row labels indicates to the automatic solution program of SOLVER that it must add artificial variables and solve the problem with the two-phase method. You never introduce artificial variables into a problem to be solved with automatic solution programs; the programs add them.

If you choose not to use EDITOR, you can solve the problem with CALIPSO in the following way:

1. Enter the problem in tableau format using the tableau editor that you reach from the TUTOR menu. Enter a basic variable as a row label when a constraint has a suitable basic variable; leave the row label blank when it does not.

2. Save the tableau in a disk file, and return to the CALIPSO menu.

3. Find the solution by selecting SOLVER.

The solution found by CALIPSO is to purchase a bond worth $39,230.77 and a security worth $22,803.73 for an initial investment of $43,203.73. Nothing should be deposited initially in the savings account.   ◆

Exercises

1. Answer the following questions about the solution to Example 2.
   (a) How many pounds of peanuts, almonds, and cashews were available but unused?
   (b) How much profit did CNC sacrifice in the first month of operation under the plan of Example 2 that it would have made under the plan of Example 1?

2. Suppose that CNC wants to extend its manufacturing plan for Bridge Mix to a 3-month period. The profits and costs for the first two months are as in Examples 1 and 2. The projected selling price of Bridge Mix in the third month is $.92 per pound, and Table 3 gives the projected cost of nuts and their availability.

   Table 3

Type	Cost per pound ($)	Pounds available
Peanuts	0.24	150
Almonds	0.42	100
Cashews	0.56	180

In addition to the 200 pounds of Bridge Mix that must be delivered in the first month, suppose CNC has accepted advance orders for 100 pounds in the second month and 75 pounds in the third month. Set up the LP problem for CNC to maximize profit over the 3-month period.

3. The Whitney Iron and Steel Co. (WISC) manufactures various sizes of wire rope. They have three machines A, B, and C that weave wire rope of small, medium, and large diameter. Table 4 gives the diameter that can be woven, speed of operation, hours of availability per week, and labor cost of each machine.

   To fill standing orders, WISC must produce 200,000 ft of $\frac{1}{4}$-in. rope, 115,000 ft of $\frac{3}{8}$-in. rope, and 110,000 ft of $\frac{1}{2}$-in. rope this week. WISC can produce more than these minimum amounts of rope in these sizes; however, because of limited storage capacity the length of rope of all sizes that can be produced is 600,000 ft. The selling price per foot of rope (excluding labor costs) is $0.015 for $\frac{1}{4}$ in., $0.018 for $\frac{3}{8}$ in., and $0.020 for $\frac{1}{2}$ in. WISC wants

Table 4

Machine	Diameter (in.)	Speed (ft/min)	Availability (hr/week)	Labor ($/hour)
A	$\frac{3}{16}-\frac{3}{8}$	200	30	10
B	$\frac{5}{16}-\frac{1}{2}$	150	35	14
C	$\frac{3}{8}-\frac{3}{4}$	125	40	16

to maximize its profit measured as the difference between income from sales of rope and labor costs to produce the rope.

Rope of $\frac{1}{4}$-in. diameter can be made only on machine A, rope of $\frac{3}{8}$-in. diameter on any machine, and rope of $\frac{1}{2}$-in. diameter on machine B or C. We introduce five variables to describe the production of each type of rope by each machine.

$$A_1 = \text{thousands of feet of } \frac{1}{4} \text{ in. produced on machine A}$$

$$A_2 = \text{thousands of feet of } \frac{3}{8} \text{ in. produced on machine A}$$

$$B_2 = \text{thousands of feet of } \frac{3}{8} \text{ in. produced on machine B}$$

$$B_3 = \text{thousands of feet of } \frac{1}{2} \text{ in. produced on machine B}$$

$$C_2 = \text{thousands of feet of } \frac{3}{8} \text{ in. produced on machine C}$$

$$C_3 = \text{thousands of feet of } \frac{1}{2} \text{ in. produced on machine C}$$

The machine capacity constraint for each machine must be expressed in hours of weaving time for 1000 ft. For machine C this would be

$$1000 \text{ divided by } \left(125 \frac{\text{ft}}{\text{min}} \right)\left(60 \frac{\text{min}}{\text{hr}} \right) = \frac{2}{15} \text{hour}$$

Show that the corresponding hours to weave 1000 ft of rope for machines A and B are $\frac{1}{12}$ and $\frac{1}{9}$.

The constraint for machine A is

$$\frac{1}{12}A_1 + \frac{1}{12}A_2 \leq 30$$

Two other constraints express limitations on machine time, another expresses the limitation on storage, and three more express the minimum production requirements. Formulate the LP problem to maximize profit.

†4. Solve Exercise 3 with CALIPSO. Note that CALIPSO's editor accepts fractions. Entering fractions rather than the decimal approximations to them increases the accuracy of calculations.

†**5.** Solve the following product-mix problems from Section 1.1.
    (a) Exercise 5   (b) Exercise 7   (c) Exercise 9   (d) Exercise 13

†**6.** Solve the following product-mix problems from Section 1.1.
    (a) Exercise 8   (b) Exercise 10   (c) Exercise 12

†**7.** Solve the following transportation problems from Section 1.2.
    (a) Exercise 1   (b) Exercise 3   (c) Exercise 5   (d) Exercise 7

†**8.** Solve the following transportation problems from Section 1.2.
    (a) Exercise 2   (b) Exercise 4   (c) Exercise 6   (d) Exercise 8

†**9.** Solve the following assignment problems from Section 1.2.
    (a) Exercise 9   (b) Exercise 11   (c) Exercise 13

†**10.** Solve the following transportation problems from Section 1.2.
    (a) Exercise 2   (b) Exercise 4   (c) Exercise 6

†**11.** Solve the following hiring–firing problems from Section 1.4.
    (a) Exercise 1   (b) Exercise 5   (c) Exercise 7

†**12.** Solve the following assignment problems from Section 1.2.
    (a) Exercise 10   (b) Exercise 12

†**13.** Solve the multiperiod problem given in Exercise 13 of Section 1.4.

†**14.** Solve the following hiring–firing problems from Section 1.4.
    (a) Exercise 2   (b) Exercise 4   (c) Exercise 6   (d) Exercise 8

CHECKLIST: CHAPTER 3

DEFINITIONS

CONCEPTS AND RESULTS

Theorem 1, Section 3.3

Suppose that an LP problem in canonical form has the property that degeneracy never occurs during execution of the simplex method. Then the simplex method must terminate, either by producing an optimal solution or by reaching a tableau that shows that the objective function is unbounded on the feasible region.

Bland's Rule

1. For the Entering Basic Variable: Of all negative coefficients in the objective row, choose the one with smallest subscript.

2. For the Departing Basic Variable: When there is a tie between one or more ratios computed using the ordinary Departing Basic Variable Rule, choose the candidate for departing variable that has the smallest subscript.

Theorem 1, Section 3.5

A general LP problem has a feasible solution if and only if all of the artificial variables have value 0 at the end of phase 1.

Chapter 4

THE SIMPLEX METHOD IN MATRIX NOTATION

4.1 MATRICES

In the remainder of this text, we will make frequent use of matrices and matrix notation. The first two sections of this chapter provide the fundamental matrix results needed for our purposes; for a more complete exposition of this and other matrix-related material, see any introductory algebra book such as Roman [12].

Definition. An $m \times n$ *matrix* is a rectangular array of numbers consisting of m rows and n columns. If $m = n$, the matrix is said to be a *square matrix*. The numbers in a matrix are called the *entries* of the matrix.

EXAMPLE 1

The matrix

$$\begin{bmatrix} 1 & -3 & \frac{1}{2} \\ 0 & 2 & 6 \end{bmatrix}$$

is a 2 × 3 matrix, because it has two rows and three columns. ◆

EXAMPLE 2

The matrix

$$[3 \quad -9 \quad \tfrac{2}{3} \quad 0.63]$$

is a 1 × 4 matrix. A matrix consisting of one row is called a *row matrix*. ◆

EXAMPLE 3

The matrix

$$\begin{bmatrix} 4 \\ 0 \\ 7 \end{bmatrix}$$

is a 3×1 matrix. A matrix consisting of one column is called a *column matrix*.   ◆

We usually use a capital letter to denote a matrix and the same lower-case letter, subscripted by the row and column number, to denote the entries of the matrix. For example,

$$A = \begin{bmatrix} a_{11} & a_{12} & \cdots & a_{1n} \\ a_{21} & a_{22} & \cdots & a_{2n} \\ \vdots & & & \\ a_{m1} & a_{m2} & \cdots & a_{mn} \end{bmatrix}$$

is a typical $m \times n$ matrix. We often shorten this notation to

$$A = [a_{ij}]_{m \times n}$$

where it is understood that $1 \le i \le m$ and $1 \le j \le n$.

We now define the operations of addition, subtraction, matrix multiplication, and scalar multiplication.

Definition.   (Matrix Addition and Subtraction.)
If

$$A = [a_{ij}]_{m \times n} \qquad B = [b_{ij}]_{m \times n}$$

then

$$A + B = [a_{ij} + b_{ij}]_{m \times n}$$

and

$$A - B = [a_{ij} - b_{ij}]_{m \times n}$$

Note that matrix addition and subtraction are defined only between matrices of the same size (i.e., the same number of rows and the same number of columns).

EXAMPLE 4

If

$$A = \begin{bmatrix} 3 & -1 & 4 & 0 \\ 2 & \frac{1}{2} & 16 & 1 \\ 0 & 1 & 0 & 4 \end{bmatrix}, \qquad B = \begin{bmatrix} -3 & -7 & -\frac{3}{4} & 10 \\ 1 & 1 & 2 & 3 \\ -2 & 4 & 0 & 6 \end{bmatrix}$$

then

$$A + B = \begin{bmatrix} 0 & -8 & \frac{13}{4} & 10 \\ 3 & \frac{3}{2} & 18 & 4 \\ -2 & 5 & 0 & 10 \end{bmatrix}, \qquad A - B = \begin{bmatrix} 6 & 6 & \frac{19}{4} & -10 \\ 1 & -\frac{1}{2} & 14 & -2 \\ 2 & -3 & 0 & -2 \end{bmatrix} \qquad \bullet$$

Definition. (Scalar Multiplication). If k is a number and

$$A = [a_{ij}]_{m \times n}$$

is a matrix, then

$$kA = [ka_{ij}]_{m \times n}$$

EXAMPLE 5

If $k = -5$ and

$$A = \begin{bmatrix} 3 & -1 & 2 \\ 4 & 6 & -17 \end{bmatrix}$$

then

$$-5A = \begin{bmatrix} -15 & 5 & -10 \\ -20 & -30 & 85 \end{bmatrix} \qquad \bullet$$

Matrix multiplication is somewhat more complicated than the other operations and, at first glance, appears to be quite contrived. As you will see, however, the definition of matrix multiplication proves to be natural and useful.

Definition. (Matrix Multiplication). If

$$A = [a_{ij}]_{m \times q}, \qquad B = [b_{ij}]_{q \times n}$$

then the matrix product AB is the $m \times n$ matrix

$$C = [c_{ij}]_{m \times n}$$

where, for $1 \leq i \leq m$ and $1 \leq j \leq n$,

$$c_{ij} = \sum_{k=1}^{q} a_{ik}b_{kj} = a_{i1}b_{1j} + a_{i2}b_{2j} + \cdots + a_{iq}b_{qj} \qquad (1)$$

Note that this definition applies only to matrices A and B with the property that the *number of columns* of A is equal to the *number of rows* of B. The product C has the same number of rows as A and the same number of columns as B. Equation (1) appears to be a formidable formula; in practice, however, it is quite easy to form. To obtain the entry c_{ij} we "multiply" the ith row of A by the jth column of B. We accomplish this by multiplying each entry in the ith row of A by the corresponding entry in the jth column of B and then forming the sum of all these products.

The following example should help clarify matters.

EXAMPLE 6

If

$$A = \begin{bmatrix} 1 & -3 & 0 \\ 2 & 4 & 6 \end{bmatrix}, \qquad B = \begin{bmatrix} 3 & -2 & 1 & 7 \\ -6 & -4 & 5 & 4 \\ 3 & 0 & -2 & 5 \end{bmatrix}$$

then

$$C = AB = \begin{bmatrix} c_{11} & c_{12} & c_{13} & c_{14} \\ c_{21} & c_{22} & c_{23} & c_{24} \end{bmatrix} = \begin{bmatrix} 21 & 10 & -14 & -5 \\ 0 & -20 & 10 & 60 \end{bmatrix}$$

To calculate the entry c_{11} in C, we multiply the first row of A and the first column of B to obtain

$$c_{11} = (1)(3) + (-3)(-6) + (0)(3) = 21$$

Similarly, to determine the entry c_{23} we multiply the second column of A and the third column of B to obtain

$$c_{23} = (2)(1) + (4)(5) + (6)(-2) = 10 \qquad \blacklozenge$$

The following notation and observations will be useful in the sequel. Suppose that A is an $m \times n$ matrix. If we let $A_1, A_2, \ldots, A_m$ denote the m columns of A, then we can write

$$A = [A_1 \mid A_2 \mid \cdots \mid A_m]$$

Suppose that A is an $m \times q$ matrix and

$$B = [B_1 \mid B_2 \mid \cdots \mid B_n]$$

is a $q \times n$ matrix with columns $B_1, B_2, \ldots, B_n$. Then it follows easily from the definition of multiplication that

$$AB = [AB_1 \mid AB_2 \mid \cdots \mid AB_n]$$

In other words, the ith column of the matrix product AB is the product of A and the ith column of B.

EXAMPLE 7

If

$$A = \begin{bmatrix} 1 & 3 & 4 \\ -2 & 0 & 6 \\ 4 & 1 & 7 \\ 5 & -6 & -2 \end{bmatrix}, \quad B = \begin{bmatrix} 2 & -3 & 4 & 0 & 1 \\ 4 & 0 & 1 & 6 & 8 \\ -5 & 4 & 2 & 3 & 0 \end{bmatrix}$$

then the fourth column of the product AB is

$$AB_4 = \begin{bmatrix} 1 & 3 & 4 \\ -2 & 0 & 6 \\ 4 & 1 & 7 \\ 5 & -6 & -2 \end{bmatrix} \begin{bmatrix} 0 \\ 6 \\ 3 \end{bmatrix} = \begin{bmatrix} 30 \\ 18 \\ 27 \\ -42 \end{bmatrix}$$

At times it is convenient to view a matrix as being composed of a number of "blocks." For instance, we can view the matrix

$$A = \begin{bmatrix} 1 & 2 & 2 & 7 & 6 \\ -1 & 3 & 4 & 0 & 1 \\ 1 & 4 & 5 & 6 & 7 \end{bmatrix}$$

in the form

$$A = \begin{bmatrix} 1 & 2 & 2 & 7 & 6 \\ -1 & 3 & 4 & 0 & 1 \\ 1 & 4 & 5 & 6 & 7 \end{bmatrix}$$

In this case, A consists of the four blocks

$$\begin{bmatrix} 1 & 2 \\ -1 & 3 \end{bmatrix} \quad [1 \quad 4] \quad \begin{bmatrix} 2 & 7 & 6 \\ 4 & 0 & 1 \end{bmatrix} \quad [5 \quad 6 \quad 7]$$

The following result is an immediate consequence of the definition of matrix multiplication. Suppose that the $m \times q$ matrix A and the $q \times n$ matrix B are decomposed into the blocks

$$A_{m \times q} = \left[\begin{array}{c|c} C_{p \times r} & D_{p \times s} \\ \hline E_{t \times r} & F_{t \times s} \end{array} \right]_{m \times q}, \qquad B_{q \times n} = \left[\begin{array}{c|c} K_{r \times u} & L_{r \times v} \\ \hline M_{s \times u} & N_{s \times v} \end{array} \right]_{q \times n}$$

where $m = p + t$, $q = r + s$, $n = u + v$. Then

$$(AB)_{m \times n} = \left[\begin{array}{c|c} (CK + DM)_{p \times u} & (CL + DN)_{p \times v} \\ \hline (EK + FM)_{t \times u} & (EL + FN)_{t \times v} \end{array} \right] \tag{2}$$

Note that to carry out the multiplication in (2), we have treated the blocks as we would treat matrix entries.

EXAMPLE 8

Suppose that

$$A = \left[\begin{array}{cc|ccc} 1 & 3 & 0 & 1 & 5 \\ 2 & 4 & 1 & -1 & 6 \\ \hline 6 & 2 & 2 & 0 & 2 \end{array} \right]$$

and

$$B = \left[\begin{array}{ccc|ccccc} 4 & 6 & 3 & 1 & 5 & 6 & 2 & 1 \\ 1 & 8 & 7 & 3 & -1 & 2 & 4 & 5 \\ \hline 1 & -1 & 2 & 4 & 3 & 1 & 6 & 2 \\ 4 & 0 & 1 & 5 & 0 & -2 & 3 & 1 \\ 2 & 1 & 3 & 2 & 0 & 0 & 1 & 2 \end{array} \right]$$

Then, in the notation preceding this example, we have

$$C = \left[\begin{array}{cc} 1 & 3 \\ 2 & 4 \end{array} \right] \qquad D = \left[\begin{array}{ccc} 0 & 1 & 5 \\ 1 & -1 & 6 \end{array} \right]$$

$$E = \left[\begin{array}{cc} 6 & 2 \end{array} \right] \qquad F = \left[\begin{array}{ccc} 2 & 0 & 2 \end{array} \right]$$

$$K = \left[\begin{array}{ccc} 4 & 6 & 3 \\ 1 & 8 & 7 \end{array} \right] \qquad L = \left[\begin{array}{ccccc} 1 & 5 & 6 & 2 & 1 \\ 3 & -1 & 2 & 4 & 5 \end{array} \right]$$

$$M = \left[\begin{array}{ccc} 1 & -1 & 2 \\ 4 & 0 & 1 \\ 2 & 1 & 3 \end{array} \right] \qquad N = \left[\begin{array}{ccccc} 4 & 3 & 1 & 6 & 2 \\ 5 & 0 & -2 & 3 & 1 \\ 2 & 0 & 0 & 1 & 2 \end{array} \right]$$

You may verify that

$$(CK + DM) = \left[\begin{array}{ccc} 21 & 35 & 40 \\ 21 & 49 & 53 \end{array} \right] \qquad (CL + DN) = \left[\begin{array}{ccccc} 25 & 2 & 10 & 22 & 27 \\ 25 & 9 & 23 & 29 & 35 \end{array} \right]$$

$$(EK + FM) = \left[\begin{array}{ccc} 32 & 52 & 42 \end{array} \right] \qquad (EL + FN) = \left[\begin{array}{ccccc} 24 & 34 & 42 & 34 & 24 \end{array} \right]$$

and that

$$AB = \begin{bmatrix} (CK + DM) & (CL + DN) \\ \hline (EK + FM) & (EL + FN) \end{bmatrix}$$

$$= \begin{bmatrix} 21 & 35 & 40 & 25 & 2 & 10 & 22 & 27 \\ 21 & 49 & 53 & 25 & 9 & 23 & 29 & 35 \\ \hline 32 & 52 & 42 & 24 & 34 & 42 & 34 & 24 \end{bmatrix}$$

♦

Definition. For any positive integer n, the $n \times n$ *identity matrix* is the matrix with 1's on the diagonal from upper left to lower right and 0's elsewhere.

For instance,

$$\begin{bmatrix} 1 & 0 \\ 0 & 1 \end{bmatrix}$$

is the 2×2 identity matrix, and

$$\begin{bmatrix} 1 & 0 & 0 \\ 0 & 1 & 0 \\ 0 & 0 & 1 \end{bmatrix}$$

is the 3×3 identity matrix.

It should be clear from the definition of matrix multiplication that, in general, $AB \neq BA$; in fact, if A is a 2×4 matrix and B a 4×3 matrix, then the product AB exists, but the product BA does not. Thus, matrix multiplication is not commutative. Nevertheless, many of the other properties of real numbers do carry over to matrices. In Exercise 20 you are asked to verify the properties given in the following theorem for 2×2 matrices.

Theorem 1. Assume that the sizes of the matrices A, B, and C and the identity matrix I are such that the following operations are defined, and let α and β be real numbers. Then

(a) $A(BC) = (AB)C$

(b) $A(B + C) = AB + AC$

(c) $A + (B + C) = (A + B) + C$

(d) $\alpha(A + B) = \alpha A + \alpha B$

(e) $(\alpha + \beta)A = \alpha A + \beta A$

(f) $\alpha(AB) = (\alpha A)B = A(\alpha B)$

(g) $IA = AI = A$

We will use the inverse of a matrix many times in the sequel.

Definition.   An *inverse* of a square matrix A is a matrix B with the property that $AB = I = BA$. If a square matrix A has an inverse, then it is said to be an *invertible* matrix. Note that if B is an inverse of A, then A is an inverse of B.

You can find proofs of the following two useful facts about square matrices in most linear algebra books.

An invertible matrix has only one inverse. We often denote the inverse of A by A^{-1}.

If a matrix B satisfies $AB = I$, then $B = A^{-1}$ is the inverse of A; that is, it is not necessary to verify that $BA = I$. Similarly, if $BA = I$, then $B = A^{-1}$.

EXAMPLE 9

You may verify that

$$B = \begin{bmatrix} \frac{3}{5} & -\frac{1}{5} & -\frac{4}{5} \\ -\frac{1}{5} & \frac{2}{5} & \frac{3}{5} \\ -\frac{2}{5} & \frac{4}{5} & \frac{1}{5} \end{bmatrix}$$

is the inverse of the matrix

$$A = \begin{bmatrix} 2 & 3 & -1 \\ 1 & 1 & 1 \\ 0 & 2 & -1 \end{bmatrix}$$

by simply multiplying these matrices together and seeing that the products AB and BA are both equal to the 3×3 identity matrix I.   ◆

EXAMPLE 10

The matrix

$$A = \begin{bmatrix} 1 & 4 \\ -2 & -8 \end{bmatrix}$$

does not have an inverse. To see this, suppose to the contrary that there is a matrix

$$B = \begin{bmatrix} b_{11} & b_{12} \\ b_{21} & b_{22} \end{bmatrix}$$

such that

$$AB = \begin{bmatrix} 1 & 4 \\ -2 & -8 \end{bmatrix}\begin{bmatrix} b_{11} & b_{12} \\ b_{21} & b_{22} \end{bmatrix} = \begin{bmatrix} 1 & 0 \\ 0 & 1 \end{bmatrix}$$

Then we would have

$$b_{11} + 4b_{21} = 1$$

$$b_{12} + 4b_{22} = 0$$

$$-2b_{11} - 8b_{21} = 0$$

$$-2b_{12} - 8b_{22} = 1$$

Note, however, that there are no values of b_{11} and b_{21} that satisfy both the first and third equations (similarly, there are no values b_{12} and b_{22} that satisfy the second and fourth equations); thus, the inverse of the matrix A does not exist. ◆

The next theorem and its corollary will be used repeatedly in the sequel.

Theorem 2. If A and B are invertible $n \times n$ matrices, then AB is invertible and $(AB)^{-1} = B^{-1}A^{-1}$.

PROOF. Observe that by Theorem 1

$$(AB)(B^{-1}A^{-1}) = A(BB^{-1})A^{-1} = AIA^{-1} = AA^{-1} = I$$

and

$$(B^{-1}A^{-1})(AB) = B^{-1}(A^{-1}A)B = B^{-1}IB = B^{-1}B = I$$ ∎

Corollary. If $A_1, A_2, \ldots, A_k$ are invertible $n \times n$ matrices, then $A_1 A_2 \cdots A_k$ is invertible and $(A_1 A_2 \cdots A_k)^{-1} = A_k^{-1} \cdots A_2^{-1} A_1^{-1}$.

PROOF. The proof follows by repeated application of Theorem 2.   ∎

You have probably observed that we have not yet indicated how to find the inverse of a matrix (we take up this topic in the next section). We conclude here with an operation that we will use many times.

Definition. The *transpose* of an $m \times n$ matrix A is the $n \times m$ matrix that results from interchanging the rows and columns of A. The transpose is denoted A^T.

Note that if

$$A = [a_{ij}]_{m \times n}$$

then

$$A^T = [a_{ji}]_{n \times m}$$

EXAMPLE 11

If

$$A = \begin{bmatrix} 1 & 2 & 2 & 7 & 6 \\ -1 & 3 & 4 & 0 & 1 \\ 1 & 4 & 5 & 6 & 7 \end{bmatrix}, \quad \text{then} \quad A^\mathsf{T} = \begin{bmatrix} 1 & -1 & 1 \\ 2 & 3 & 4 \\ 2 & 4 & 5 \\ 7 & 0 & 6 \\ 6 & 1 & 7 \end{bmatrix} \qquad \blacklozenge$$

We will often use the following theorem. You are asked to verify these properties in the exercises.

Theorem 3.   If A and B are $m \times q$ matrices, C is a $q \times n$ matrix, and α is any real number, then

   (a) $(A^\mathsf{T})^\mathsf{T} = A$

   (b) $(A + B)^\mathsf{T} = A^\mathsf{T} + B^\mathsf{T}$

   (c) $(A - B)^\mathsf{T} = A^\mathsf{T} - B^\mathsf{T}$

   (d) $(\alpha A)^\mathsf{T} = \alpha A^\mathsf{T}$

   (e) $(AC)^\mathsf{T} = C^\mathsf{T} A^\mathsf{T}$

EXAMPLE 12

Consider the matrices A and B from Example 6. We form the product $B^\mathsf{T} A^\mathsf{T}$.

$$B^\mathsf{T} A^\mathsf{T} = \begin{bmatrix} 3 & -6 & 3 \\ -2 & -4 & 0 \\ 1 & 5 & -2 \\ 7 & 4 & 5 \end{bmatrix} \begin{bmatrix} 1 & 2 \\ -3 & 4 \\ 0 & 6 \end{bmatrix} = \begin{bmatrix} 21 & 0 \\ 10 & -20 \\ -14 & 10 \\ -5 & 60 \end{bmatrix}$$

By comparing this result with the product AB obtained in Example 6, you will see that $(AB)^\mathsf{T} = B^\mathsf{T} A^\mathsf{T}$. $\qquad \blacklozenge$

Exercises

  1. Let

$$A = \begin{bmatrix} 1 & 3 & -1 \\ 2 & 1 & 0 \end{bmatrix} \quad \text{and} \quad B = \begin{bmatrix} 4 & 1 & 3 \\ -1 & 2 & 1 \end{bmatrix}$$

   Find: (a) $A + B$ (b) $A - B$ (c) $2A$ (d) $2A + 3B$

2. Let $\alpha = 3$, $\beta = -3$,

$$A = \begin{bmatrix} 1 & 2 \\ 3 & 1 \\ -1 & 4 \end{bmatrix} \quad \text{and} \quad B = \begin{bmatrix} 0 & 2 \\ 4 & 1 \\ -1 & 3 \end{bmatrix}$$

Find: (a) $A + B$ (b) $B - A$ (c) αA (d) $(\alpha + \beta)A$ (e) $\alpha A + \beta B$
(f) $\alpha A + \beta A$

3. Let

$$A = \begin{bmatrix} 1 & 3 & -1 \\ 2 & 1 & 0 \end{bmatrix}, \quad B = \begin{bmatrix} 4 & -1 \\ 1 & 2 \\ 3 & 1 \end{bmatrix}, \quad \text{and} \quad C = \begin{bmatrix} 1 & -1 & 3 \\ 0 & 1 & 0 \\ 1 & 3 & -1 \end{bmatrix}$$

Compute: (a) AB (b) AC (c) CB (d) CC

4. Let

$$A = \begin{bmatrix} 1 & -1 & 3 \\ 2 & 1 & 2 \\ 3 & 0 & 1 \end{bmatrix}, \quad B = \begin{bmatrix} 1 & 2 & 1 & -1 \\ 0 & 1 & 0 & 0 \\ 1 & -1 & 0 & 1 \end{bmatrix}, \quad C = \begin{bmatrix} 2 \\ 1 \\ 3 \end{bmatrix}, \quad D = \begin{bmatrix} 1 & 2 & -1 \end{bmatrix}$$

Compute: (a) AB (b) DA (c) DB (d) DC (e) CD

In Exercises 5 and 6, verify that $AB \neq BA$ by computing AB and BA.

5.
$$A = \begin{bmatrix} 1 & 2 \\ 3 & 4 \end{bmatrix} \quad B = \begin{bmatrix} 2 & 1 \\ 3 & 4 \end{bmatrix}$$

6.
$$A = \begin{bmatrix} 1 & 2 & 0 \\ -1 & 3 & 1 \\ 1 & 2 & 1 \end{bmatrix} \quad B = \begin{bmatrix} 2 & 1 & 1 \\ 0 & 1 & 0 \\ 3 & 0 & 1 \end{bmatrix}$$

7. Let

$$A = \begin{bmatrix} -1 \\ 2 \\ 1 \\ 3 \end{bmatrix} \quad \text{and} \quad B = \begin{bmatrix} 3 & 1 & 2 & -1 \end{bmatrix}$$

Compute: (a) AB (b) BA

8. Let

$$A = \begin{bmatrix} 2 & 1 \\ -1 & 3 \end{bmatrix}$$

Compute: (a) $A^2 (A^2 = AA)$ (b) $A^3 (A^3 = AAA)$.

In Exercises 9–12, perform the products using block multiplication, and observe that you can obtain the same answer by multiplying in the usual way.

9.
$$\left[\begin{array}{ccc|cc} 1 & 0 & -1 & 1 & -1 \\ 2 & 1 & 0 & 2 & 1 \\ \hline -1 & 1 & 1 & 1 & 3 \\ 2 & 0 & 1 & -1 & 1 \end{array}\right] \left[\begin{array}{cc|cc} 1 & -1 & 1 & -1 \\ 2 & 0 & 2 & 1 \\ 3 & 1 & 2 & 4 \\ \hline -2 & 1 & 0 & 2 \\ 0 & 1 & 1 & -1 \end{array}\right]$$

10.
$$\begin{bmatrix} 1 & 1 & 2 & -1 & | & 1 \\ 0 & 1 & 1 & 0 & | & 3 \\ \hline -1 & 1 & 3 & 1 & | & 4 \\ 2 & 1 & -1 & 2 & | & 1 \end{bmatrix} \begin{bmatrix} 2 & -1 & | & 1 & -1 \\ 1 & 0 & | & -1 & 1 \\ \hline 3 & 1 & | & 4 & 0 \\ -1 & 2 & | & 1 & 0 \\ \hline 4 & 2 & | & 2 & 0 \end{bmatrix}$$

11.
$$\begin{bmatrix} 1 & 1 & | & 2 & 0 & | & 1 \\ -1 & 1 & | & 1 & 1 & | & 1 \end{bmatrix} \begin{bmatrix} 1 & 2 & 0 \\ \hline 1 & -1 & 1 \\ \hline 1 & 2 & 2 \\ 0 & 1 & 1 \\ \hline 2 & 0 & 1 \end{bmatrix}$$

12.
$$\begin{bmatrix} 1 & 2 & | & 0 & 1 & | & 4 \\ 2 & 0 & | & 1 & -1 & | & 3 \end{bmatrix} \begin{bmatrix} 4 & 1 & | & 1 & 2 & 3 \\ 0 & 1 & | & 0 & 2 & 1 \\ \hline 1 & 2 & | & 1 & -1 & 2 \\ 1 & 0 & | & 2 & 0 & 1 \\ \hline 1 & 1 & | & 3 & 1 & 0 \end{bmatrix}$$

In Exercises 13 and 14, form a system of equations that must be satisfied by the inverse matrix (as in Example 10), and solve the system to find the inverse.

13. $\begin{bmatrix} 1 & 2 \\ -1 & 1 \end{bmatrix}$ **14.** $\begin{bmatrix} 2 & 0 \\ 3 & 1 \end{bmatrix}$

15. Find the transpose of each of the following matrices:

$$A = \begin{bmatrix} 1 & 2 \\ 0 & 1 \end{bmatrix} \qquad B = \begin{bmatrix} 1 & 2 \\ 1 & 0 \\ 3 & 1 \end{bmatrix} \qquad C = \begin{bmatrix} 1 & 2 & 1 \\ 0 & 1 & 1 \\ 1 & 3 & 1 \end{bmatrix}$$

$$D = \begin{bmatrix} 1 & 2 & 1 & 3 \end{bmatrix} \qquad E = \begin{bmatrix} 2 \\ 1 \\ -1 \\ 4 \end{bmatrix} \qquad F = \begin{bmatrix} 1 & 2 & 3 & 4 & 5 \\ 6 & 7 & 8 & 9 & 10 \end{bmatrix}$$

16. For the matrices A and B of Exercise 2, verify that
(a) $(A + B)^T = A^T + B^T$ (c) $(A^T)^T = A$
(b) $(9A)^T = 9A^T$ (d) $(A - B)^T = A^T - B^T$

17. For the matrices A and B of Exercise 1, verify that
(a) $(A + B)^T = A^T + B^T$ (c) $(A^T)^T = A$
(b) $(5A)^T = 5A^T$ (d) $(A - B)^T = A^T - B^T$

18. For the matrices A, B, C, and D of Exercise 4, verify that
(a) $(AB)^T = B^T A^T$ (c) $(CA)^T = A^T C^T$
(b) $(CD)^T = D^T C^T$ (d) $(DC)^T = C^T D^T$

19. For the matrices A, B, and C of Exercise 3, verify that
(a) $(AB)^T = B^T A^T$ (c) $(CA)^T = A^T C^T$
(b) $(CB)^T = B^T C^T$

20. Prove Theorem 1 for 2×2 matrices.

21. Show that if A is invertible and $\alpha \neq 0$, then αA is invertible.

22. True or false? If A and B are invertible matrices, then $A + B$ is invertible.

23. A square matrix is *symmetric* if $A = A^T$. Show that BB^T is symmetric for any matrix B.

24. Show that $B + B^T$ is symmetric for any square matrix B.

4.2 ELEMENTARY MATRICES, MATRIX INVERSES, AND PIVOTING

We now develop a method for calculating the inverse of an invertible matrix. In addition, we derive some formulas related to pivoting that will be useful in developing the revised simplex method.

In Section 2.4 we introduced two row operations for use in the simplex method: multiplying a row by a constant and adding a multiple of one row to another row. There is a third row operation that enables you to interchange two rows of a matrix; even though this operation is not used in the simplex method, we include it in our discussion for completeness.

Definition. An *elementary matrix* is a matrix that results from performing a single row operation on an identity matrix.

EXAMPLE 1

We give an example of each of the three types of elementary matrices that correspond to the three row operations:

$$E_1 = \begin{bmatrix} 1 & 0 & 0 \\ 0 & 2 & 0 \\ 0 & 0 & 1 \end{bmatrix} \qquad \text{(We multiplied the second row of } I_3 \text{ by 2.)}$$

$$E_2 = \begin{bmatrix} 1 & 0 & 0 \\ 0 & 1 & 0 \\ 2 & 0 & 1 \end{bmatrix} \qquad \text{(We added 2 times the first row of } I_3 \text{ to the third row of } I_3 \text{.)}$$

$$E_3 = \begin{bmatrix} 0 & 1 & 0 \\ 1 & 0 & 0 \\ 0 & 0 & 1 \end{bmatrix} \qquad \text{(We interchanged the first and second rows of } I_3 \text{.)} \qquad \blacklozenge$$

The following result illustrates the importance of elementary matrices.

Theorem 1. Let E be an elementary matrix corresponding to a particular row operation on I_m, and let A be an $m \times n$ matrix. Then the product EA is the matrix that results from applying this same row operation to A.

Rather than provide a formal proof of Theorem 1, we illustrate its correctness with examples of each type of elementary matrix.

EXAMPLE 2

Let E_1, E_2, and E_3 be the matrices used in Example 1, and let

$$A = \begin{bmatrix} 2 & -1 & 4 & 3 \\ 1 & 2 & -4 & 1 \\ 3 & 2 & 1 & -1 \end{bmatrix}$$

To multiply the second row of A by 2, we premultiply A by E_1 to obtain

$$E_1 A = \begin{bmatrix} 1 & 0 & 0 \\ 0 & 2 & 0 \\ 0 & 0 & 1 \end{bmatrix} \begin{bmatrix} 2 & -1 & 4 & 3 \\ 1 & 2 & -4 & 1 \\ 3 & 2 & 1 & -1 \end{bmatrix} = \begin{bmatrix} 2 & -1 & 4 & 3 \\ 2 & 4 & -8 & 2 \\ 3 & 2 & 1 & -1 \end{bmatrix}$$

To replace the third row of A with the sum of the third row of A and 2 times the first row of A, we premultiply A by E_2 to obtain

$$E_2 A = \begin{bmatrix} 1 & 0 & 0 \\ 0 & 1 & 0 \\ 2 & 0 & 1 \end{bmatrix} \begin{bmatrix} 2 & -1 & 4 & 3 \\ 1 & 2 & -4 & 1 \\ 3 & 2 & 1 & -1 \end{bmatrix} = \begin{bmatrix} 2 & -1 & 4 & 3 \\ 1 & 2 & -4 & 1 \\ 7 & 0 & 9 & 5 \end{bmatrix}$$

To interchange the first and second rows of A, we premultiply A by E_3 to obtain

$$E_3 A = \begin{bmatrix} 0 & 1 & 0 \\ 1 & 0 & 0 \\ 0 & 0 & 1 \end{bmatrix} \begin{bmatrix} 2 & -1 & 4 & 3 \\ 1 & 2 & -4 & 1 \\ 3 & 2 & 1 & -1 \end{bmatrix} = \begin{bmatrix} 1 & 2 & -4 & 1 \\ 2 & -1 & 4 & 3 \\ 3 & 2 & 1 & -1 \end{bmatrix} \quad \blacklozenge$$

Elementary matrices are invertible. Table 1 shows how to construct their inverses.

Table 1

Row operation on I that produces E	Row operation on I that produces E^{-1}
Multiply row i by $c \neq 0$	Multiply row i by $1/c$
Interchange rows i and j	Interchange rows i and j
Add c times row i to row j	Add $-c$ times row i to row j

EXAMPLE 3

We illustrate the construction of the inverse of an elementary matrix with the matrices E_1, E_2, and E_3 used in Example 1.

$$E_1 E_1^{-1} = \begin{bmatrix} 1 & 0 & 0 \\ 0 & 2 & 0 \\ 0 & 0 & 1 \end{bmatrix} \begin{bmatrix} 1 & 0 & 0 \\ 0 & \frac{1}{2} & 0 \\ 0 & 0 & 1 \end{bmatrix} = \begin{bmatrix} 1 & 0 & 0 \\ 0 & 1 & 0 \\ 0 & 0 & 1 \end{bmatrix}$$

$$E_2 E_2^{-1} = \begin{bmatrix} 1 & 0 & 0 \\ 0 & 1 & 0 \\ 2 & 0 & 1 \end{bmatrix} \begin{bmatrix} 1 & 0 & 0 \\ 0 & 1 & 0 \\ -2 & 0 & 1 \end{bmatrix} = \begin{bmatrix} 1 & 0 & 0 \\ 0 & 1 & 0 \\ 0 & 0 & 1 \end{bmatrix}$$

$$E_3 E_3^{-1} = \begin{bmatrix} 0 & 1 & 0 \\ 1 & 0 & 0 \\ 0 & 0 & 1 \end{bmatrix} \begin{bmatrix} 0 & 1 & 0 \\ 1 & 0 & 0 \\ 0 & 0 & 1 \end{bmatrix} = \begin{bmatrix} 1 & 0 & 0 \\ 0 & 1 & 0 \\ 0 & 0 & 1 \end{bmatrix}$$

♦

Definition.   A matrix A is *row equivalent* to a matrix B, if B can be obtained from A by a sequence of row operations.

We can describe row equivalence as premultiplication by a sequence of elementary matrices. Performing a single row operation on A is equivalent to premultiplying A by some elementary matrix E_1, obtaining $E_1 A$. Performing a second row operation is equivalent to premultiplying $E_1 A$ by another elementary matrix E_2, obtaining $E_2 E_1 A$. If we perform k row operations to get B from A, we can write the result as

$$B = E_k \cdots E_2 E_1 A$$

The next example shows that the pivoting process of the simplex method produces a row-equivalent matrix. It also shows how to perform pivoting operations by premultiplying by a sequence of elementary matrices.

EXAMPLE 4

Let A be the 3×4 matrix given in Example 2. We want to pivot on the entry 4 in the third column of A to obtain an equivalent matrix B. First, we multiply row 1 of A by $\frac{1}{4}$ using the elementary matrix

$$E_1 = \begin{bmatrix} \frac{1}{4} & 0 & 0 \\ 0 & 1 & 0 \\ 0 & 0 & 1 \end{bmatrix}$$

to obtain

$$E_1 A = \begin{bmatrix} \frac{1}{4} & 0 & 0 \\ 0 & 1 & 0 \\ 0 & 0 & 1 \end{bmatrix} \begin{bmatrix} 2 & -1 & 4 & 3 \\ 1 & 2 & -4 & 1 \\ 3 & 2 & 1 & -1 \end{bmatrix} = \begin{bmatrix} \frac{1}{2} & -\frac{1}{4} & 1 & \frac{3}{4} \\ 1 & 2 & -4 & 1 \\ 3 & 2 & 1 & -1 \end{bmatrix}$$

Next, to add 4 times row 1 to row 2, we premultiply E_1A by the elementary matrix

$$E_2 = \begin{bmatrix} 1 & 0 & 0 \\ 4 & 1 & 0 \\ 0 & 0 & 1 \end{bmatrix}$$

to obtain

$$E_2E_1A = \begin{bmatrix} 1 & 0 & 0 \\ 4 & 1 & 0 \\ 0 & 0 & 1 \end{bmatrix} \begin{bmatrix} \frac{1}{2} & -\frac{1}{4} & 1 & \frac{3}{4} \\ 1 & 2 & -4 & 1 \\ 3 & 2 & 1 & -1 \end{bmatrix} = \begin{bmatrix} \frac{1}{2} & -\frac{1}{4} & 1 & \frac{3}{4} \\ 3 & 1 & 0 & 4 \\ 3 & 2 & 1 & -1 \end{bmatrix}$$

Finally, to add -1 times the first row of E_2E_1A to the third row, we premultiply E_2E_1A by the elementary matrix

$$E_3 = \begin{bmatrix} 1 & 0 & 0 \\ 0 & 1 & 0 \\ -1 & 0 & 1 \end{bmatrix}$$

to obtain

$$E_3E_2E_1A = \begin{bmatrix} 1 & 0 & 0 \\ 0 & 1 & 0 \\ -1 & 0 & 1 \end{bmatrix} \begin{bmatrix} \frac{1}{2} & -\frac{1}{4} & 1 & \frac{3}{4} \\ 3 & 1 & 0 & 4 \\ 3 & 2 & 1 & -1 \end{bmatrix} = \begin{bmatrix} \frac{1}{2} & -\frac{1}{4} & 1 & \frac{3}{4} \\ 3 & 1 & 0 & 4 \\ \frac{5}{2} & \frac{9}{4} & 0 & -\frac{7}{4} \end{bmatrix} \quad \blacklozenge$$

The next result enables us to determine if an $n \times n$ matrix A is invertible; as you will see, the proof of this result also provides a means for calculating the inverse of A.

Theorem 2.   If A is an $n \times n$ matrix that is row equivalent to the identity matrix I_n, then A is invertible.

PROOF.   Let $E_1, E_2, \ldots, E_k$ be a sequence of elementary matrices that reduces A to I_n. Thus,

$$E_k \cdots E_2 E_1 A = I_n \tag{1}$$

We premultiply both sides of (1) successively by $E_k^{-1}, E_{k-1}^{-1}, \ldots, E_2^{-1}, E_1^{-1}$, to obtain

$$E_{k-1} \cdots E_2 E_1 A = E_k^{-1} I_n$$

$$E_{k-2} \cdots E_2 E_1 A = E_{k-1}^{-1} E_k^{-1} I_n$$

$$\vdots$$

$$A = E_1^{-1} E_2^{-1} \cdots E_{k-1}^{-1} E_k^{-1} I_n \tag{2}$$

Since in (2) we have expressed A as a product of invertible matrices, we conclude from the corollary to Theorem 2 of Section 4.1 that A is invertible. ∎

Next note that by inserting parentheses in (1) we can view $E_k \cdots E_2 E_1 A$ as the product of two matrices: $E_k \cdots E_2 E_1$ and A, i.e.,

$$(E_k \cdots E_2 E_1)A = I_n$$

and it follows that

$$A^{-1} = E_k \cdots E_2 E_1 = (E_k \cdots E_2 E_1)I_n \qquad (3)$$

Equation (3) shows that we can compute the inverse of A by applying to I_n the same elementary matrices used to reduce A to I_n.

EXAMPLE 5

We illustrate the use of formula (3) by computing the inverse of

$$A = \begin{bmatrix} 4 & -2 & 3 \\ 8 & -3 & 5 \\ 7 & -2 & 4 \end{bmatrix}$$

We apply to I_3 the same elementary matrices needed to reduce A to I_3. Rather than making two separate calculations (first applying a sequence of elementary matrices to A and then applying the same sequence to I_3), we form a single 3×6 matrix A' by adjoining I_3 to A,

$$A' = \begin{bmatrix} 4 & -2 & 3 & 1 & 0 & 0 \\ 8 & -3 & 5 & 0 & 1 & 0 \\ 7 & -2 & 4 & 0 & 0 & 1 \end{bmatrix}$$

and perform the row operations once on A'.

Usually it is easier to perform the row operations directly on the A' matrix than to premultiply it by elementary matrices. The elementary matrices are used mainly as theoretical tools to understand concepts. Nevertheless, in the following calculation we have indicated the elementary matrix used in each step of this procedure. You can easily see which row operation we used on A' by studying the corresponding elementary matrix (Table 2). We have reduced A to I_3, and we see that the matrix to which I_3 was reduced by this operation is

$$A^{-1} = \begin{bmatrix} -2 & 2 & -1 \\ 3 & -5 & 4 \\ 5 & -6 & 4 \end{bmatrix}$$

Table 2

Elementary matrices	Row equivalences of A'

$$\begin{bmatrix} 4 & -2 & 3 & 1 & 0 & 0 \\ 8 & -3 & 5 & 0 & 1 & 0 \\ 7 & -2 & 4 & 0 & 0 & 1 \end{bmatrix}$$

$$\begin{bmatrix} \tfrac{1}{4} & 0 & 0 \\ 0 & 1 & 0 \\ 0 & 0 & 1 \end{bmatrix} \qquad \begin{bmatrix} 1 & -\tfrac{1}{2} & \tfrac{3}{4} & \tfrac{1}{4} & 0 & 0 \\ 8 & -3 & 5 & 0 & 1 & 0 \\ 7 & -2 & 4 & 0 & 0 & 1 \end{bmatrix}$$

$$\begin{bmatrix} 1 & 0 & 0 \\ -8 & 1 & 0 \\ 0 & 0 & 1 \end{bmatrix} \qquad \begin{bmatrix} 1 & -\tfrac{1}{2} & \tfrac{3}{4} & \tfrac{1}{4} & 0 & 0 \\ 0 & 1 & -1 & -2 & 1 & 0 \\ 7 & -2 & 4 & 0 & 0 & 1 \end{bmatrix}$$

$$\begin{bmatrix} 1 & 0 & 0 \\ 0 & 1 & 0 \\ -7 & 0 & 1 \end{bmatrix} \qquad \begin{bmatrix} 1 & -\tfrac{1}{2} & \tfrac{3}{4} & \tfrac{1}{4} & 0 & 0 \\ 0 & 1 & -1 & -2 & 1 & 0 \\ 0 & \tfrac{3}{2} & -\tfrac{5}{4} & -\tfrac{7}{4} & 0 & 1 \end{bmatrix}$$

$$\begin{bmatrix} 1 & \tfrac{1}{2} & 0 \\ 0 & 1 & 0 \\ 0 & 0 & 1 \end{bmatrix} \qquad \begin{bmatrix} 1 & 0 & \tfrac{1}{4} & -\tfrac{3}{4} & \tfrac{1}{2} & 0 \\ 0 & 1 & -1 & -2 & 1 & 0 \\ 0 & \tfrac{3}{2} & -\tfrac{5}{4} & -\tfrac{7}{4} & 0 & 1 \end{bmatrix}$$

$$\begin{bmatrix} 1 & 0 & 0 \\ 0 & 1 & 0 \\ 0 & -\tfrac{3}{2} & 1 \end{bmatrix} \qquad \begin{bmatrix} 1 & 0 & \tfrac{1}{4} & -\tfrac{3}{4} & \tfrac{1}{2} & 0 \\ 0 & 1 & -1 & -2 & 1 & 0 \\ 0 & 0 & \tfrac{1}{4} & \tfrac{5}{4} & -\tfrac{3}{2} & 1 \end{bmatrix}$$

$$\begin{bmatrix} 1 & 0 & 0 \\ 0 & 1 & 0 \\ 0 & 0 & 4 \end{bmatrix} \qquad \begin{bmatrix} 1 & 0 & \tfrac{1}{4} & -\tfrac{3}{4} & \tfrac{1}{2} & 0 \\ 0 & 1 & -1 & -2 & 1 & 0 \\ 0 & 0 & 1 & 5 & -6 & 4 \end{bmatrix}$$

$$\begin{bmatrix} 1 & 0 & -\tfrac{1}{4} \\ 0 & 1 & 0 \\ 0 & 0 & 1 \end{bmatrix} \qquad \begin{bmatrix} 1 & 0 & 0 & -2 & 2 & -1 \\ 0 & 1 & -1 & -2 & 1 & 0 \\ 0 & 0 & 1 & 5 & -6 & 4 \end{bmatrix}$$

$$\begin{bmatrix} 1 & 0 & 0 \\ 0 & 1 & 1 \\ 0 & 0 & 1 \end{bmatrix} \qquad \begin{bmatrix} 1 & 0 & 0 & -2 & 2 & -1 \\ 0 & 1 & 0 & 3 & -5 & 4 \\ 0 & 0 & 1 & 5 & -6 & 4 \end{bmatrix}$$

◆

We now consider the use of elementary matrices to express the pivoting operation.

Definition. A *pivoting matrix* is a matrix formed by multiplying together the sequence of elementary matrices used to carry out the pivoting operation.

Note that, since elementary matrices are invertible, it follows by the corollary to Theorem 2 of Section 4.1 that pivoting matrices are invertible.

As we will see, we can accomplish the desired pivot by premultiplying by the corresponding pivoting matrix. The next example illustrates this idea. (Also recall Example 4.)

EXAMPLE 6

The first three row operations performed in Example 5 accomplish a pivot on the entry in row 1 and column 1 of A'. The product of the three elementary matrices corresponding to the row operations is

$$P = \begin{bmatrix} \frac{1}{4} & 0 & 0 \\ -2 & 1 & 0 \\ -\frac{7}{4} & 0 & 1 \end{bmatrix} \tag{4}$$

We can obtain the matrix P by multiplying the three elementary matrices together or by simply observing from Table 2 that I_3 has been reduced to P. We could have performed the pivot on A' in one step by calculating PA'.   ◆

Theorem 3 gives the general form of a pivoting matrix.

Theorem 3. Let

$$A = \begin{bmatrix} a_{11} & a_{12} & \cdots & a_{1j} & \cdots & a_{1n} \\ \vdots & & & & & \\ a_{i1} & a_{i2} & \cdots & a_{ij} & \cdots & a_{in} \\ \vdots & & & & & \\ a_{m1} & a_{m2} & \cdots & a_{mj} & \cdots & a_{mn} \end{bmatrix}$$

and suppose that the entry a_{ij} in row i and column j of A is nonzero. Let A^* be the matrix resulting from A by pivoting on a_{ij}, and let

$$L = \begin{bmatrix} -a_{1j}/a_{ij} \\ -a_{2j}/a_{ij} \\ \vdots \\ 1/a_{ij} \\ \vdots \\ -a_{mj}/a_{ij} \end{bmatrix} \leftarrow \text{row } i$$

be formed from the j column of A. Let P be the matrix formed by replacing the i column of the $m \times m$ identity matrix I_m with the column matrix L. Then $A^* = PA$.

Rather than presenting a formal proof of Theorem 3, we illustrate its application with the following example.

EXAMPLE 7

Consider the matrix

$$A = \begin{bmatrix} 1 & -11 & -5 & 18 & 2 & 0 & 0 \\ 0 & 4 & 2 & -8 & -1 & 1 & 0 \\ 0 & 11 & 5 & -18 & -2 & 0 & 1 \end{bmatrix}$$

To pivot on entry 2 in the third column of A, we let

$$L = \begin{bmatrix} \frac{5}{2} \\ \frac{1}{2} \\ -\frac{5}{2} \end{bmatrix}$$

and

$$P = \begin{bmatrix} 1 & \frac{5}{2} & 0 \\ 0 & \frac{1}{2} & 0 \\ 0 & -\frac{5}{2} & 1 \end{bmatrix}$$

then

$$PA = \begin{bmatrix} 1 & -1 & 0 & -2 & -\frac{1}{2} & \frac{5}{2} & 0 \\ 0 & 2 & 1 & -4 & -\frac{1}{2} & \frac{1}{2} & 0 \\ 0 & 1 & 0 & 2 & \frac{1}{2} & -\frac{5}{2} & 1 \end{bmatrix} \tag{5}$$

gives us the matrix resulting from the pivot operation.

In the revised simplex method, we do not always need all of the entries in a matrix obtained by pivoting. Notice that we could obtain any column of PA by multiplying the corresponding column of A by P. For instance, we can compute the third column of PA from the third column of A by

$$P \begin{bmatrix} -5 \\ 2 \\ 5 \end{bmatrix} = \begin{bmatrix} 1 & \frac{5}{2} & 0 \\ 0 & \frac{1}{2} & 0 \\ 0 & -\frac{5}{2} & 1 \end{bmatrix} \begin{bmatrix} -5 \\ 2 \\ 5 \end{bmatrix} = \begin{bmatrix} 0 \\ 1 \\ 0 \end{bmatrix}$$

We could obtain the result of pivoting on columns 4, 6, and 7 of A by calculating

$$P \begin{bmatrix} 18 & 0 & 0 \\ -8 & 1 & 0 \\ -18 & 0 & 1 \end{bmatrix} = \begin{bmatrix} -2 & \frac{5}{2} & 0 \\ -4 & \frac{1}{2} & 0 \\ 2 & -\frac{5}{2} & 1 \end{bmatrix} \qquad \blacklozenge$$

You are asked to prove the following corollary to Theorem 3 in Exercise 19.

Corollary.   If A^* is obtained from A through a sequence of pivots, then there is an invertible matrix D such that

$$A^* = DA$$

In Chapter 5, we apply Theorem 3 and its corollary to develop the revised simplex method. At that time, it will be important to notice that the pivoting matrices that are used to describe the pivots performed in the course of the simplex method always have a right-most column consisting of all zeros except for a 1 in the last row. The following theorem, which you are asked to prove in Exercise 21, states this property exactly.

Theorem 4.   Let A be an $m \times n$ matrix. Suppose that a sequence of pivots is performed on A and that none of these pivots is performed on an entry from the last row of a matrix. Let D be the invertible matrix described in the corollary to Theorem 3 such that

$$A^* = DA$$

Then column m of D has a 1 in row m and 0's elsewhere. That is, D has the form

$$D = \begin{bmatrix} d_{11} & d_{12} & \cdots & d_{1,m-1} & 0 \\ \vdots & & & & \\ d_{j1} & d_{j2} & \cdots & d_{j,m-1} & 0 \\ \vdots & & & & \\ d_{m1} & d_{m2} & \cdots & d_{m,m-1} & 1 \end{bmatrix}$$

In Chapter 5, we will use the next theorem to rearrange the columns of certain tableaus.

Theorem 5.   Suppose that A^* is derived from A through a sequence of pivot operations, and let D be the pivoting matrix such that

$$A^* = DA$$

Suppose that S is obtained from A by rearranging some columns of A, and that S^* is obtained from A^* by rearranging columns of A^* in the same way that A was rearranged to obtain S. Then

$$S^* = DS$$

You are asked to prove this theorem in Exercise 20.

EXAMPLE 8

Let A and P be the matrices introduced in Example 7. We illustrate Theorem 5 with $D = P$ and $A^* = DA$.

$$S = \begin{bmatrix} -11 & 1 & -5 & 18 & 0 & 2 & 0 \\ 4 & 0 & 2 & -8 & 1 & -1 & 0 \\ 11 & 0 & 5 & -18 & 0 & -2 & 1 \end{bmatrix}$$

$$S^* = \begin{bmatrix} -1 & 1 & 0 & -2 & \frac{5}{2} & -\frac{1}{2} & 0 \\ 2 & 0 & 1 & -4 & \frac{1}{2} & -\frac{1}{2} & 0 \\ 1 & 0 & 0 & 2 & -\frac{5}{2} & \frac{1}{2} & 1 \end{bmatrix}$$

You may verify that $S^* = DS$. ◆

Exercises

1. Which of the following are elementary matrices?

(a) $\begin{bmatrix} 1 & 0 \\ 0 & 3 \end{bmatrix}$
  (b) $\begin{bmatrix} 1 & 0 \\ 3 & 1 \end{bmatrix}$
  (c) $\begin{bmatrix} 0 & 1 \\ 1 & 0 \end{bmatrix}$

(d) $\begin{bmatrix} 0 & 2 \\ 2 & 0 \end{bmatrix}$
  (e) $\begin{bmatrix} 1 & 0 & 0 \\ 0 & 1 & 0 \\ 1 & 1 & 1 \end{bmatrix}$
  (f) $\begin{bmatrix} 1 & 0 & 0 & 0 \\ 0 & 1 & 0 & 0 \\ 0 & 0 & 0 & 1 \end{bmatrix}$

In Exercises 2–6 find elementary matrices E that satisfy the given equations.

2. $E \begin{bmatrix} 1 & 2 & 1 & -1 \\ 3 & 0 & 1 & 0 \\ 1 & 2 & -2 & 1 \end{bmatrix} = \begin{bmatrix} 1 & 2 & 1 & -1 \\ 4 & 2 & 2 & -1 \\ 1 & 2 & -2 & 1 \end{bmatrix}$

3. $E \begin{bmatrix} 1 & 2 & 1 & -1 \\ 3 & 0 & 1 & 0 \\ 1 & 2 & -2 & 1 \end{bmatrix} = \begin{bmatrix} 1 & 2 & -2 & 1 \\ 3 & 0 & 1 & 0 \\ 1 & 2 & 1 & -1 \end{bmatrix}$

4. $E \begin{bmatrix} 1 & 2 & 1 & -1 \\ 3 & 0 & 1 & 0 \\ 1 & 2 & -2 & 1 \end{bmatrix} = \begin{bmatrix} 1 & 2 & 1 & -1 \\ 3 & 0 & 1 & 0 \\ -1 & -2 & -4 & 3 \end{bmatrix}$

5. $E \begin{bmatrix} 1 & 2 & 3 & -1 \\ 2 & 1 & 0 & 3 \\ 1 & 1 & 2 & 2 \\ 4 & -1 & 2 & 1 \end{bmatrix} = \begin{bmatrix} 9 & 0 & 7 & 1 \\ 2 & 1 & 0 & 3 \\ 1 & 1 & 2 & 2 \\ 4 & -1 & 2 & 1 \end{bmatrix}$

6. $E \begin{bmatrix} -2 & 1 & 3 & 5 \\ 6 & 1 & 2 & 3 \\ 7 & 1 & 0 & 5 \\ 4 & -1 & 2 & 1 \end{bmatrix} = \begin{bmatrix} -2 & 1 & 3 & 5 \\ 6 & 1 & 2 & 3 \\ 7 & 1 & 0 & 5 \\ 16 & 1 & 6 & 7 \end{bmatrix}$

7. Let

$$A = \begin{bmatrix} 2 & 3 & 0 & 1 \\ 1 & 0 & 2 & 2 \\ -1 & 1 & -2 & 2 \end{bmatrix}$$

Perform the following row operations on A by premultiplying A by a suitable elementary matrix:
(a) Multiply the second row by 3.
(b) Add 2 times row 1 to row 2.
(c) Subtract row 2 from row 1.

8. Let

$$A = \begin{bmatrix} 2 & 5 & -1 & 2 & 1 \\ 1 & 0 & 3 & 1 & -1 \\ 2 & 1 & 0 & 2 & 3 \\ 5 & 0 & 1 & 1 & -1 \end{bmatrix}$$

Perform the following row operations on A by premultiplying A by a suitable elementary matrix:
(a) Multiply the third row by 23.
(b) Add 4 times row 3 to row 1.
(c) Subtract row 1 from row 4.

In Exercises 9–13 use the method of Example 5 to find the inverses of the following matrices. You do not need to use elementary matrices.

9. $\begin{bmatrix} 2 & 1 \\ -1 & 1 \end{bmatrix}$ **10.** $\begin{bmatrix} 1 & 2 & 1 \\ 1 & -1 & 0 \\ 2 & 1 & 2 \end{bmatrix}$ **11.** $\begin{bmatrix} 1 & 1 \\ -1 & 1 \end{bmatrix}$

12. $\begin{bmatrix} 1 & 2 & 4 \\ 4 & 1 & 2 \\ 3 & 2 & 1 \end{bmatrix}$ **13.** $\begin{bmatrix} 1 & 0 & 0 \\ 1 & 2 & 0 \\ 1 & 2 & 3 \end{bmatrix}$

†**14.** Use the TUTOR program to find the inverse of

$$\begin{bmatrix} 1 & 2 & -2 & -1 & 3 \\ -\frac{1}{4} & 1 & 2 & 4 & -\frac{1}{2} \\ 2 & 0 & 1 & 0 & 0 \\ 3 & 1 & -2 & 4 & 8 \\ 4 & 5 & -1 & 2 & 3 \end{bmatrix}$$

In Exercises 15–18, use Theorem 3 to find the pivoting matrix P that can be used to perform a pivot on the circled entry of the given matrices.

15. (a) $\begin{bmatrix} 1 & 2 \\ -1 & ① \end{bmatrix}$
(b) $\begin{bmatrix} 1 & 2 & 2 & 2 \\ 2 & 2 & -6 & 4 \\ -1 & 1 & ① & 3 \end{bmatrix}$

16. (a) $\begin{bmatrix} 2 & 1 & 3 \\ ④ & 2 & 1 \\ 1 & -1 & 2 \end{bmatrix}$
(b) $\begin{bmatrix} 2 & 1 & 3 & 1 \\ 4 & 2 & 1 & 2 \\ 1 & -1 & 2 & ③ \end{bmatrix}$

17. (a) $\begin{bmatrix} 2 & ①{2} & 3 \\ 1 & 4 & -1 \\ 2 & \frac{3}{2} & 1 \end{bmatrix}$
(b) $\begin{bmatrix} 1 & 2 & -3 & 1 \\ 4 & 1 & 2 & 1 \\ -2 & 1 & ② & 3 \end{bmatrix}$

18. $\begin{bmatrix} 1 & 2 & -2 & -1 & 3 \\ -\frac{1}{4} & 1 & 2 & 4 & \boxed{-\frac{1}{2}} \\ 2 & 0 & 1 & 0 & 0 \\ 3 & 1 & -2 & 4 & 8 \\ 4 & 5 & -1 & 2 & 3 \end{bmatrix}$ **19.** Prove the corollary to Theorem 3.

20. Prove Theorem 5. Suggestion: Distribute the multiplications over columns, as in

$$DA = [DA_1 | DA_2 | \cdots | DA_n]$$

21. Prove Theorem 4.

4.3   MATRIX REPRESENTATION OF THE SIMPLEX METHOD: PART 1

In this section we examine matrix relationships between tableaus generated by the simplex method. In particular, we identify an invertible matrix B (known as the basis matrix) that we can use as an alternative to pivoting in calculating the simplex tableaus. In the next chapter we use the inverse of the basis matrix to develop the revised simplex method, a more efficient implementation of the simplex method. We will also use the inverse of the basis matrix to develop some important ideas concerning sensitivity analysis (see Chapter 7).

In matrix notation, the LP problem (which we assume to be in canonical form)

$$
\begin{aligned}
\text{Maximize:} \quad & z = c_1 x_1 + c_2 x_2 + \cdots + c_n x_n + z_0 \\
\text{Subject to:} \quad & a_{11} x_1 + a_{12} x_2 + \cdots + a_{1n} x_n = b_1 \\
& a_{21} x_1 + a_{22} x_2 + \cdots + a_{2n} x_n = b_2 \\
& \quad \vdots \\
& a_{m1} x_1 + a_{m2} x_2 + \cdots + a_{mn} x_n = b_m \\
& x_i \geq 0, \quad 1 \leq i \leq n
\end{aligned}
\tag{1}
$$

becomes

$$
\begin{aligned}
\text{Maximize:} \quad & z = \mathbf{c}\mathbf{x} + z_0 \\
\text{Subject to:} \quad & A\mathbf{x} = \mathbf{b} \\
& \mathbf{x} \geq \mathbf{0}
\end{aligned}
$$

where

$\mathbf{c} = [c_1 \quad c_2 \quad \cdots \quad c_n]$ is the *cost matrix*;

z_0 is a constant;

$\mathbf{x} = \begin{bmatrix} x_1 \\ x_2 \\ \vdots \\ x_n \end{bmatrix}$ is the *variable matrix*;

$$A = \begin{bmatrix} a_{11} & a_{12} & \cdots & a_{1n} \\ a_{21} & a_{22} & \cdots & a_{2n} \\ \vdots & & & \\ a_{m1} & a_{m2} & \cdots & a_{mn} \end{bmatrix} \text{ is the } \textit{constraint matrix; and}$$

$$\mathbf{b} = \begin{bmatrix} b_1 \\ b_2 \\ \vdots \\ b_m \end{bmatrix} \text{ is the } \textit{resource matrix.}$$

Notice that we have used a row representation for the cost matrix and a column representation for the variable and resource matrices. This convention simplifies the matrix formulas to be developed.

With this notation the tableau associated with (1) becomes

$$\left[\begin{array}{c|c} A & \mathbf{b} \\ \hline -\mathbf{c} & z_0 \end{array} \right] \tag{2}$$

EXAMPLE 1

Consider the LP problem

$$\text{Maximize:} \quad z = 100x_1 + 120x_2$$

$$\text{Subject to:} \quad \begin{aligned} 2x_1 + 2x_2 + x_3 \quad\quad &= 8 \\ 3x_1 + 5x_2 \quad\quad + x_4 &= 15 \end{aligned}$$

$$x_i \geq 0, \quad 1 \leq i \leq 4$$

The matrices associated with this problem are

$$A = \begin{bmatrix} 2 & 2 & 1 & 0 \\ 3 & 5 & 0 & 1 \end{bmatrix}$$

$$\mathbf{x} = \begin{bmatrix} x_1 \\ x_2 \\ x_3 \\ x_4 \end{bmatrix} \quad \mathbf{b} = \begin{bmatrix} 8 \\ 15 \end{bmatrix}$$

$$\mathbf{c} = \begin{bmatrix} 100 & 120 & 0 & 0 \end{bmatrix}$$

$$z_0 = 0$$

The tableau, written in the notation (2), is

$$\left[\begin{array}{c|c} A & \mathbf{b} \\ \hline -\mathbf{c} & 0 \end{array} \right] \tag{3}$$

◆

Expression (2) does not completely describe the tableau corresponding to LP problem (1) because it does not specify the basic variables. To create a useful notation for describing the basic variables, we make the following definition.

Definition. Let m and n be positive integers with $m \leq n$. An *injection* h of $\{1, 2, \ldots, m\}$ into $\{1, 2, \ldots, n\}$ is a sequence $h(1), h(2), \ldots, h(m)$ of distinct integers in the set $\{1, 2, \ldots, n\}$.

EXAMPLE 2

Let $m = 3$ and $n = 7$. We define three injections h_1, h_2, and h_3 of $\{1, 2, 3\}$ into $\{1, 2, 3, 4, 5, 6, 7\}$, by

$$h_1(1) = 1, \qquad h_1(2) = 2, \qquad h_1(3) = 3$$
$$h_2(1) = 7, \qquad h_2(2) = 3, \qquad h_2(3) = 4$$
$$h_3(1) = 2, \qquad h_3(2) = 1, \qquad h_3(3) = 3$$

There are many other injections of $\{1, 2, 3\}$ into $\{1, 2, 3, 5, 4, 6, 7\}$. In fact, to define an injection h we have seven choices for $h(1)$; once we choose $h(1)$, we have six choices for $h(2)$; and once we choose $h(1)$ and $h(2)$, we have five choices for $h(3)$. Thus, there are $7 \cdot 6 \cdot 5 = 210$ ways of defining h.   ◆

Definition. Let

$$\left[\begin{array}{c|c} A^* & \mathbf{b}^* \\ \hline -\mathbf{c}^* & z^* \end{array} \right]$$

be any tableau derived from tableau (2) by the simplex method. Let $h: \{1, 2, \ldots, m\} \to \{1, 2, \ldots, n\}$ be an injection such that for $1 \leq i \leq m$, $h(i)$ is the subscript of the basic variable corresponding to row i (i.e., $x_{h(i)}$ is the basic variable for row i). The *basic variable matrix* is

$$\mathbf{x}_B = \begin{bmatrix} x_{h(1)} \\ x_{h(2)} \\ \vdots \\ x_{h(m)} \end{bmatrix}$$

Be careful not to confuse the variable matrix $\mathbf{x}$ with the basic variable matrix $\mathbf{x}_B$. The m entries of $\mathbf{x}_B$ form a subset of the n entries of $\mathbf{x}$.

Definition. The *basic cost matrix* associated with the basic variable matrix $\mathbf{x}_B$ is

$$\mathbf{c}_B = \begin{bmatrix} c_{h(1)} & c_{h(2)} & \cdots & c_{h(m)} \end{bmatrix}$$

where $c_{h(1)}, c_{h(2)}, \ldots, c_{h(m)}$ are the coefficients of the basis variables $x_{h(1)}, x_{h(2)}, \ldots, x_{h(m)}$ in the original objective function $z = c_1 x_1 + c_2 x_2 + \cdots + c_n x_n$.

It is important to notice that we form c_B from the *original* objective function for the LP problem described in (1) and *not* from the objective function for some tableau derived by applying the simplex method.

EXAMPLE 3

We interpret the definitions of the basic variable matrix and basic cost matrix for the LP problem

$$\text{Maximize:} \quad z = 6x_1 + 8x_2 + x_3$$

$$\text{Subject to:} \quad 3x_1 + 5x_2 + 3x_3 + x_4 \qquad\qquad = 20$$
$$x_1 + 3x_2 + 2x_3 \qquad + x_5 \qquad = 9$$
$$6x_1 + 2x_2 + 5x_3 \qquad\qquad + x_6 = 30$$

$$x_i \geq 0, \qquad 1 \leq i \leq 6$$

Applying the simplex method gives the sequence of Tableaus 1–4.

	x_1	x_2	x_3	x_4	x_5	x_6		
x_4	3	5	3	1	0	0	20	
x_5	1	③	2	0	1	0	9	Tableau 1
x_6	6	2	5	0	0	1	30	
	-6	-8	-1	0	0	0	0	
x_4	$\textcircled{$\frac{4}{3}$}$	0	$-\frac{1}{3}$	1	$-\frac{5}{3}$	0	5	
x_2	$\frac{1}{3}$	1	$\frac{2}{3}$	0	$\frac{1}{3}$	0	3	Tableau 2
x_6	$\frac{16}{3}$	0	$\frac{11}{3}$	0	$-\frac{2}{3}$	1	24	
	$-\frac{10}{3}$	0	$\frac{13}{3}$	0	$\frac{8}{3}$	0	24	
x_1	1	0	$-\frac{1}{4}$	$\frac{3}{4}$	$-\frac{5}{4}$	0	$\frac{15}{4}$	
x_2	0	1	$\frac{3}{4}$	$-\frac{1}{4}$	$\frac{3}{4}$	0	$\frac{7}{4}$	Tableau 3
x_6	0	0	5	-4	⑥	1	4	
	0	0	$\frac{7}{2}$	$\frac{5}{2}$	$-\frac{3}{2}$	0	$\frac{73}{2}$	
x_1	1	0	$\frac{19}{24}$	$-\frac{1}{12}$	0	$\frac{5}{24}$	$\frac{55}{12}$	
x_2	0	1	$\frac{1}{8}$	$\frac{1}{4}$	0	$-\frac{1}{8}$	$\frac{5}{4}$	Tableau 4
x_5	0	0	$\frac{5}{6}$	$-\frac{2}{3}$	1	$\frac{1}{6}$	$\frac{2}{3}$	
	0	0	$\frac{19}{4}$	$\frac{3}{2}$	0	$\frac{1}{4}$	$\frac{75}{2}$	

The injection used to define the basic variable and cost matrix for Tableau 2 is

$$h(1) = 4, \qquad h(2) = 2, \qquad \text{and} \qquad h(3) = 6$$

The basic variable matrix for this tableau is

$$\mathbf{x_B} = \begin{bmatrix} x_4 \\ x_2 \\ x_6 \end{bmatrix}$$

and its basic cost matrix is

$$\mathbf{c_B} = \begin{bmatrix} c_4 & c_2 & c_6 \end{bmatrix} = \begin{bmatrix} 0 & 8 & 0 \end{bmatrix}$$

The injection used to define the basic variable and cost matrix for Tableau 4 is

$$h(1) = 1, \qquad h(2) = 2, \qquad \text{and} \qquad h(3) = 5$$

The basic variable matrix for this tableau is

$$\mathbf{x_B} = \begin{bmatrix} x_1 \\ x_2 \\ x_5 \end{bmatrix}$$

and its basic cost matrix is

$$\mathbf{c_B} = \begin{bmatrix} c_1 & c_2 & c_5 \end{bmatrix} = \begin{bmatrix} 6 & 8 & 0 \end{bmatrix} \qquad \blacklozenge$$

Definition. A *unit column* is an $m \times 1$ column matrix consisting of one 1 and $m - 1$ 0's. For $1 \le j \le m$, we define the unit column $\mathbf{e}_j$ by

$$\mathbf{e}_j = \begin{bmatrix} 0 \\ 0 \\ \vdots \\ 1 \\ \vdots \\ 0 \end{bmatrix} \quad \leftarrow \text{row } j$$

We now introduce the important concept of the *basis matrix B* and its inverse B^{-1}. Suppose that

$$T = \begin{bmatrix} A & \mathbf{b} \\ \hline -\mathbf{c} & z_0 \end{bmatrix}$$

is the initial tableau written in canonical form for an LP problem with m constraints. Then T contains m unit columns that correspond to the initial basic solution. Tableau 1 of Example 3 is an example of such a matrix T. Note that in that example the unit columns form the identity matrix located in the right-hand portion of the A matrix. In general, however, the m unit columns can appear anywhere in the A matrix.

Let

$$T^* = \begin{bmatrix} A^* & \mathbf{b}^* \\ \hline -\mathbf{c}^* & z^* \end{bmatrix}$$

be any tableau obtained from T by applying the simplex method, and let

$$\mathbf{x_B} = \begin{bmatrix} x_{h(1)} \\ x_{h(2)} \\ \vdots \\ x_{h(m)} \end{bmatrix}$$

be the basic variable matrix corresponding to T^*.

Since the simplex method always produces tableaus in canonical form, T^* also contains m unit columns corresponding to a basic solution.

The *basis matrix B for T^** is the matrix formed from columns of T corresponding to the variables heading the unit columns of T^*; the order of the columns of B is the same as the order of the basic variables in T^*. More precisely, if $h(1), \ldots, h(m)$ is the injection corresponding to the basis for T^*, then

$$B = [A_{h(1)} \mid A_{h(2)} \quad \cdots \mid A_{h(m)}]$$

where $A_{h(i)}$ is the $h(i)$ column of A.

EXAMPLE 4

The basis matrix for Tableau 2 of Example 3 is

$$B = \begin{bmatrix} 1 & 5 & 0 \\ 0 & 3 & 0 \\ 0 & 2 & 1 \end{bmatrix}$$

The basis matrix for Tableau 4 of Example 3 is

$$B = \begin{bmatrix} 3 & 5 & 0 \\ 1 & 3 & 1 \\ 6 & 2 & 0 \end{bmatrix}$$

◆

The following theorem (which you are asked to prove in Exercise 12) shows that the basis matrix is invertible.

Theorem 1. Suppose that T and T^* are defined as before and that B is the basis matrix for T^*. Then B is an invertible matrix.

It is a simple matter to find the B^{-1} matrix for any tableau T^* derived from a tableau T by the simplex method. It follows from Theorem 5 and the proof of Theorem 2 of Section 4.2 that the first column of B^{-1} is the column in T^* headed by the variable corresponding to the unit column $\mathbf{e}_1$ of T; the second column of B^{-1} is the column in T^* headed by the variable corresponding to the unit column $\mathbf{e}_2$ of T;

and so forth. When the unit columns in T are not located so as to form an identity matrix (as they are in Tableau 1 of Example 3), you must exercise some care in assembling the columns of B^{-1} in the correct order (see Exercises 7, 8, and 9).

EXAMPLE 5

Consider the B matrix given for Tableau 4 in Example 4. The identity submatrix of Tableau 1 of Example 3 has been transformed by a sequence of pivots into a corresponding submatrix of Tableau 4. By Theorem 2 of Section 4.2 this submatrix must be B^{-1}; that is,

$$B^{-1} = \begin{bmatrix} -\frac{1}{12} & 0 & \frac{5}{24} \\ \frac{1}{4} & 0 & -\frac{1}{8} \\ -\frac{2}{3} & 1 & \frac{1}{6} \end{bmatrix}$$

Similarly, the inverse of the basis matrix for Tableau 2 of Example 3 is

$$B^{-1} = \begin{bmatrix} 1 & -\frac{5}{3} & 0 \\ 0 & \frac{1}{3} & 0 \\ 0 & -\frac{2}{3} & 1 \end{bmatrix}$$

◆

The next theorem gives the fundamental matrix formulas relating a tableau produced by the simplex method to the initial tableau. We will prove this theorem in Section 4.4.

Theorem 2. Suppose that

$$T = \left[\begin{array}{c|c} A & \mathbf{b} \\ \hline -\mathbf{c} & z_0 \end{array} \right]$$

is the tableau for an LP problem in canonical form where A contains the unit columns $\mathbf{e}_1, \mathbf{e}_2, \ldots, \mathbf{e}_m$ in some order. Let T^* be any tableau derived from T by one or more pivots of the simplex method, and let B be the basis matrix for T^*. Denote the submatrices of T^* by

$$T^* = \left[\begin{array}{c|c} A^* & \mathbf{b}^* \\ \hline -\mathbf{c}^* & z^* \end{array} \right]$$

Then

$$A^* = B^{-1}A$$

$$\mathbf{b}^* = B^{-1}\mathbf{b}$$

$$-\mathbf{c}^* = \mathbf{c}_B B^{-1}A - \mathbf{c}$$

$$z^* = \mathbf{c}_B B^{-1}\mathbf{b} + z_0$$

10. Solve by the two-phase method, and identify the final B^{-1} matrix.

$$\text{Maximize:} \quad z = x_1 - x_2 - 3x_3$$

$$\text{Subject to:} \quad x_1 + 2x_2 - x_3 \le 1$$
$$2x_1 + x_2 + x_3 \ge 2$$
$$x_1 + x_2 + x_3 = 1$$
$$x_i \ge 0, \quad 1 \le i \le 3$$

11. Solve by the two-phase method, and identify the B^{-1} matrix.

$$\text{Minimize:} \quad z = 2x_1 + x_2 + x_3$$

$$\text{Subject to:} \quad 3x_1 + 2x_2 - x_3 \ge 3$$
$$2x_1 - x_2 + 2x_3 \ge 2$$
$$x_i \ge 0, \quad 1 \le i \le 3$$

12. Prove Theorem 1.

13. Let A be an $m \times n$ matrix for an LP problem in canonical form.
    (a) Show that there are $n!/(n-m)! = n(n-1)\cdots(n-m+1)$ injections of $\{1, 2, \ldots, m\}$ into $\{1, 2, \ldots, n\}$.
    (b) Show that there are

$$\frac{n!}{m!(n-m)!}$$

    different subsets of $\{x_1, x_2, \ldots, x_n\}$ that could be sets of basic variables.
    (c) In Exercise 6 of Section 3.3 we outlined a proof of the fact that there is only one simplex tableau for each set of basic variables. However, (a) and (b) show that there are more injections than sets of basic variables. How can this be?

---

*4.4 MATRIX REPRESENTATION OF THE SIMPLEX METHOD: PART 2

We now develop the matrix formulas for the simplex method, with the particular goal of proving Theorem 2 of the previous section.

We continue to consider the general, canonical-form LP problem (1) of Section 4.3. We write the initial tableau for the problem as

$$T_0 = \left[\begin{array}{c|c} A_0 & \mathbf{b}_0 \\ \hline -\mathbf{c}_0 & z_0 \end{array} \right]$$

As we saw in Section 4.2, we can perform a pivot on T_0 by premultiplying T_0 by a pivot matrix P_1. We express this pivot by

$$P_1 T_0 = P_1 \left[\begin{array}{c|c} A_0 & \mathbf{b}_0 \\ \hline -\mathbf{c}_0 & z_0 \end{array} \right] = \left[\begin{array}{c|c} A_1 & \mathbf{b}_1 \\ \hline -\mathbf{c}_1 & z_1 \end{array} \right] = T_1$$

A second pivot of the simplex method (using matrix P_2) results in a new tableau T_2 defined by

$$T_2 = P_2 T_1 = P_2 P_1 T_0$$

Applying the simplex method r times in this way, we obtain the sequence of tableaus $T_0, T_1, \ldots, T_r$ satisfying

$$T_r = P_r T_{r-1} = P_r P_{r-1} T_{r-2} = \cdots = P_r P_{r-1} \cdots P_1 T_0$$

Theorem 3 of Section 4.2 gives the general form of a pivoting matrix. A pivoting matrix that performs a pivot on the entry in row i and column j is obtained by replacing the ith column of an $m \times m$ identity matrix with the column L given in Theorem 3. Since the simplex method never pivots on an entry in the bottom row (the objective row) of a tableau, it follows from Theorem 4 of Section 4.2 that the last column of a corresponding pivoting matrix is never replaced by a column L. Thus, the last column of a pivoting matrix used in a step of the simplex method is e_m, a unit column, we summarize and further develop these ideas in the following theorem.

Theorem 1.   Let T_0 be the initial tableau for an LP problem in canonical form. Let $T_0, T_1, \ldots, T_r$ be the sequence of tableaus resulting from r pivots by the simplex method; let $P_1, P_2, \ldots, P_r$ be the pivoting matrices that produce these tableaus; and let

$$D = P_r P_{r-1} \cdots P_2 P_1$$

be the product of these r matrices. Then,

(i) $T_r = D T_0$

(ii) D can be partitioned as

$$D = \left[\begin{array}{c|c} R & 0 \\ \hline s & 1 \end{array} \right]$$

where R is an $m \times m$ matrix, s a $1 \times m$ matrix, 0 an $m \times 1$ matrix all of whose entries are 0, and 1 a 1×1 matrix whose entry is 1.

PROOF.   Part (i) follows immediately from the remarks that precede Theorem 1. To prove part (ii) it suffices to show that the product of any two $m \times m$ matrices whose last columns are e_m has a last column e_m. Let U and V be two matrices whose last column is e_m, and let $W = UV$. The entry in row m and column m of W is

$$w_{mm} = u_{m1} v_{1m} + u_{m2} v_{2m} + \cdots + u_{mm} v_{mm}$$

$$= u_{m1} \cdot 0 + u_{m2} \cdot 0 + \cdots + 1 \cdot 1 = 1$$

The entry in row i and column m, for $1 \leq i < m$, is

$$w_{im} = u_{i1} v_{1m} + u_{i2} v_{2m} + \cdots + u_{im} v_{mm}$$

$$= u_{i1} \cdot 0 + u_{i2} \cdot 0 + \cdots + 0 \cdot 1 = 0$$

Therefore, the only nonzero entry in column m of W is in row m and column m; hence, column m is $\mathbf{e}_m$. ∎

EXAMPLE 1

Applying the simplex method to the LP problem

$$\text{Maximize:} \quad z = 125x_1 + 100x_2$$
$$\text{Subject to:} \quad x_1 + x_2 \leq 4$$
$$5x_1 + 3x_2 \leq 15$$
$$x_1, x_2 \geq 0$$

results in the following sequence of tableaus.

	x_1	x_2	x_3	x_4		
x_3	1	1	1	0	4	
x_4	⑤	3	0	1	15	(Tableau 1)
	-125	-100	0	0	0	
x_3	0	②∕₅	1	$-\frac{1}{5}$	1	
x_1	1	$\frac{3}{5}$	0	$\frac{1}{5}$	3	(Tableau 2)
	0	-25	0	25	375	
x_2	0	1	$\frac{5}{2}$	$-\frac{1}{2}$	$\frac{5}{2}$	
x_1	1	0	$-\frac{3}{2}$	$\frac{1}{2}$	$\frac{3}{2}$	(Tableau 3)
	0	0	$\frac{125}{2}$	$\frac{25}{2}$	$\frac{875}{2}$	

We can use the formula in Theorem 3 of Section 4.2 to obtain the pivoting matrix used to derive Tableau 2 from Tableau 1. This matrix is

$$P_1 = \begin{bmatrix} 1 & -\frac{1}{5} & 0 \\ 0 & \frac{1}{5} & 0 \\ 0 & 25 & 1 \end{bmatrix}$$

You should verify directly that we can obtain Tableau 2 from Tableau 1 by multiplying Tableau 1 by P_1.

The pivoting matrix for obtaining Tableau 3 from Tableau 2 is

$$P_2 = \begin{bmatrix} \frac{5}{2} & 0 & 0 \\ -\frac{3}{2} & 1 & 0 \\ \frac{125}{2} & 0 & 1 \end{bmatrix}$$

and the matrix for obtaining Tableau 3 from Tableau 1 is

$$D = P_2 P_1 = \begin{bmatrix} \frac{5}{2} & -\frac{1}{2} & 0 \\ -\frac{3}{2} & \frac{1}{2} & 0 \\ \frac{125}{2} & \frac{25}{2} & 1 \end{bmatrix}$$

         ◆

The next theorem gives the general form of D.

Theorem 2. Suppose that T is the tableau for an LP problem in canonical form; that is,

$$T = \left[\begin{array}{c|c} A & \mathbf{b} \\ \hline -\mathbf{c} & z_0 \end{array} \right]$$

where A contains the unit columns $\mathbf{e}_1, \mathbf{e}_2, \ldots, \mathbf{e}_m$ in some order. Let T^* be any tableau obtained from T by one or more pivots of the simplex method. Let D be the matrix for which

$$T^* = DT$$

Then,

$$D = \left[\begin{array}{c|c} B^{-1} & \mathbf{0} \\ \hline \mathbf{c_B} B^{-1} & 1 \end{array} \right] \tag{1}$$

where B is the basis matrix.

PROOF. Since T^* was obtained from T by the simplex method, it also has unit columns $\mathbf{e}_1, \mathbf{e}_2, \ldots, \mathbf{e}_m$ in some order. Rearranging these unit columns so that they appear, in order, at the left we have

$$S^* = \left[\begin{array}{c|c|c} I & Q^* & \mathbf{b}^* \\ \hline \mathbf{0} & \mathbf{q}^* & z^* \end{array} \right]$$

If we rearrange the columns of T in the same way we rearranged the columns of T^*, we obtain

$$S = \left[\begin{array}{c|c|c} B & Q & \mathbf{b} \\ \hline \mathbf{r} & \mathbf{q} & z_0 \end{array} \right] \tag{2}$$

It follows from Theorem 5 of Section 4.2 that $S^* = DS$ where D is defined by (1).

Let h be the injection that denotes the basic variables for S^*; that is, $x_{h(1)}$ is the basic variable for row 1, $x_{h(2)}$ is the basic variable for row 2, and so forth. Then the tableaus S and S^*, with the relevant columns and rows labeled with basic variables, are

	$x_{h(1)}$	$x_{h(2)}$	$\cdots$	$x_{h(m)}$	$\cdots$		
$\cdots$	b_{11}	b_{12}	$\cdots$	b_{1m}	$\cdots$	b_1	
$\cdots$	b_{21}	b_{22}	$\cdots$	b_{2m}	$\cdots$	b_2	(Tableau S)
$\cdots$	b_{m1}	b_{m2}	$\cdots$	b_{mm}	$\cdots$	b_m	
	$-c_{h(1)}$	$-c_{h(2)}$	$\cdots$	$-c_{h(m)}$	$\cdots$	z_0	
$x_{h(1)}$	1	0	$\cdots$	0	$\cdots$	b_1^*	
$x_{h(2)}$	0	1	$\cdots$	0	$\cdots$	b_2^*	(Tableau S^*)
$x_{h(m)}$	0	0	$\cdots$	1	$\cdots$	b_m^*	
	0	0	$\cdots$	0	$\cdots$	z^*	

We can identify two of the submatrices in (2) by inspecting Tableau S. The submatrix B of (2) is the basis matrix corresponding to S^* and (T^*) because it is correspondingly located above the identity submatrix of S^*. The submatrix $\mathbf{r}$ of (2) is $-\mathbf{c}_B$. The submatrices Q and $\mathbf{q}$ of (2) will be of no further interest, and we do not attempt to describe them. Thus,

$$S = \left[\begin{array}{c|c|c} B & Q & \mathbf{b} \\ \hline -\mathbf{c}_B & \mathbf{q} & z_0 \end{array}\right] \tag{3}$$

Using (3) and the expression for D from Theorem 1, part (ii), we have

$$DS = \left[\begin{array}{c|c} R & \mathbf{0} \\ \hline \mathbf{s} & 1 \end{array}\right]\left[\begin{array}{c|c|c} B & Q & \mathbf{b} \\ \hline -\mathbf{c}_B & \mathbf{q} & z_0 \end{array}\right] = S^* = \left[\begin{array}{c|c|c} I & Q^* & \mathbf{b}^* \\ \hline \mathbf{0} & \mathbf{q}^* & z^* \end{array}\right] \tag{4}$$

To prove the theorem, we only need to carry out the multiplication in (4) for the left-hand upper and lower blocks; that is to say, we need only the following part of equality (4):

$$\left[\begin{array}{c|c} R & \mathbf{0} \\ \hline \mathbf{s} & 1 \end{array}\right]\left[\begin{array}{c} B \\ \hline -\mathbf{c}_B \end{array}\right] = \left[\begin{array}{c} I \\ \hline \mathbf{0} \end{array}\right] \tag{5}$$

Multiplying these blocks, we obtain

$$\left[\begin{array}{c} RB + \mathbf{0} \\ \hline \mathbf{s}B - \mathbf{c}_B \end{array}\right] = \left[\begin{array}{c} I \\ \hline \mathbf{0} \end{array}\right]$$

Hence,

$$RB = I; \quad \text{so} \quad R = B^{-1}$$

$$\mathbf{s}B - \mathbf{c}_B = \mathbf{0}; \quad \text{so} \quad \mathbf{s}B = \mathbf{c}_B \quad \text{or} \quad \mathbf{s} = \mathbf{c}_B B^{-1}$$

Thus,

$$D = \left[\begin{array}{c|c} B^{-1} & \mathbf{0} \\ \hline \mathbf{c_B}B^{-1} & 1 \end{array}\right]$$

∎

EXAMPLE 2

Consider the expression for D obtained in Example 1. It is clear that the upper left-hand block of D is the B^{-1} matrix for Tableau 3 of Example 1. We need only verify that the lower left-hand block is $\mathbf{c_B}B^{-1}$. Here is the calculation:

$$\mathbf{c_B}B^{-1} = [100 \quad 125]\left[\begin{array}{cc} \frac{5}{2} & -\frac{1}{2} \\ -\frac{3}{2} & \frac{1}{2} \end{array}\right] = [\frac{125}{2} \quad \frac{25}{2}]$$

◆

We now prove Theorem 2 from Section 4.3. It gives the fundamental matrix formulas relating a tableau produced by the simplex method to the initial tableau. For convenience we state the theorem again.

Theorem 3.   Suppose that T is the tableau for an LP problem in canonical form; that is,

$$T = \left[\begin{array}{c|c} A & \mathbf{b} \\ \hline -\mathbf{c} & z_0 \end{array}\right]$$

where A contains the unit columns $\mathbf{e_1}, \mathbf{e_2}, \ldots, \mathbf{e_m}$ in some order. Let T^* be any tableau derived from T by one or more applications of the simplex method. Denote the submatrices by

$$T^* = \left[\begin{array}{c|c} A^* & \mathbf{b}^* \\ \hline -\mathbf{c}^* & z^* \end{array}\right]$$

Then,

$$A^* = B^{-1}A$$

$$\mathbf{b}^* = B^{-1}\mathbf{b}$$

$$-\mathbf{c}^* = \mathbf{c_B}B^{-1}A - \mathbf{c}$$

$$z^* = \mathbf{c_B}B^{-1}\mathbf{b} + z_0$$

PROOF.   Let D be the matrix introduced in the proof of Theorem 3, for which

$$DT = T^*$$

Using the form developed for D in Theorem 2, we can write

$$DT = \left[\begin{array}{c|c} B^{-1} & \mathbf{0} \\ \hline \mathbf{c_B}B^{-1} & 1 \end{array}\right]\left[\begin{array}{c|c} A & \mathbf{b} \\ \hline -\mathbf{c} & z_0 \end{array}\right] = T^* = \left[\begin{array}{c|c} A^* & \mathbf{b}^* \\ \hline -\mathbf{c}^* & z^* \end{array}\right]$$

Taking the block products gives

$$\left[\begin{array}{c|c} B^{-1}A & B^{-1}\mathbf{b} \\ \hline \mathbf{c_B}B^{-1}A - \mathbf{c} & \mathbf{c_B}B^{-1}\mathbf{b} + z_0 \end{array}\right] = \left[\begin{array}{c|c} A^* & \mathbf{b}^* \\ \hline -\mathbf{c}^* & z^* \end{array}\right]$$

Equating corresponding blocks gives the desired formulas.   ∎

We will usually use the formulas given in Theorem 3 for A^* and $\mathbf{c}^*$ to compute a single column of A^* or a single element of $\mathbf{c}^*$. The formulas in the following corollary are more convenient for this application.

Corollary. For any j, $1 \leq j \leq n$, let A_j^* be the jth column of A^* and c_j^* be the jth entry in $\mathbf{c}^*$. Then,

$$A_j^* = B^{-1}A_j$$

and

$$-c_j^* = \mathbf{c_B}B^{-1}A_j - c_j$$

PROOF. The formulas follow immediately from those in Theorem 2 by equating the jth columns of the matrices.   ∎

You may want to review Example 5 of Section 4.3, which illustrates the use of the preceding formulas on a simple case. We illustrate the use of the formulas on the following example, in which the unit columns of the initial A matrix do not occur in their natural order so that an identity matrix is not a submatrix of A.

EXAMPLE 3

$$\text{Maximize:} \quad z = -13x_3 + 6x_5 - 2x_7 + 100$$

$$\text{Subject to:} \quad \begin{aligned} -2x_1 \quad\quad\quad + 6x_3 \quad\quad\quad + 2x_5 + x_6 - 3x_7 &= 20 \\ -4x_1 + x_2 + 7x_3 \quad\quad\quad + x_5 \quad\quad - x_7 &= 10 \\ -5x_3 + x_4 + 3x_5 \quad\quad - x_7 &= 60 \end{aligned}$$

$$x_i \geq 0, \quad 1 \leq i \leq 7$$

Tableaus 1–4 solve this problem by the simplex method:

Tableau 1

	x_1	x_2	x_3	x_4	x_5	x_6	x_7	
x_6	-2	0	6	0	2	1	-3	20
x_2	-4	1	7	0	①	0	-1	10
x_4	0	0	-5	1	3	0	-1	60
	0	0	13	0	-6	0	2	100

Column 2 of B^{-1} ↓ (at x_2), Column 3 of B^{-1} ↓ (at x_4), Column 1 of B^{-1} ↓ (at x_6)

Tableau 2

x_6	⑥	-2	-8	0	0	1	-1	0
x_5	-4	1	7	0	1	0	-1	10
x_4	12	-3	-26	1	0	0	2	30
	-24	6	55	0	0	0	-4	160

Tableau 3

x_1	1	$-\frac{1}{3}$	$-\frac{4}{3}$	0	0	$\frac{1}{6}$	$-\frac{1}{6}$	0
x_5	0	$-\frac{1}{3}$	$\frac{5}{3}$	0	1	$\frac{2}{3}$	$-\frac{5}{3}$	10
x_4	0	1	-10	1	0	-2	④	30
	0	-2	23	0	0	4	-8	160

Tableau 4

x_1	1	$-\frac{7}{24}$	$-\frac{7}{3}$	$\frac{1}{24}$	0	$\frac{1}{12}$	0	$\frac{5}{4}$
x_5	0	$\frac{1}{12}$	$-\frac{5}{2}$	$\frac{5}{12}$	1	$-\frac{1}{6}$	0	$\frac{45}{2}$
x_7	0	$\frac{1}{4}$	$-\frac{5}{2}$	$\frac{1}{4}$	0	$-\frac{1}{2}$	1	$\frac{15}{2}$
	0	0	3	2	0	0	0	220

Notice that in every tableau the B^{-1} matrix is located in the same columns (the unit columns x_6, x_2, and x_4 in the original tableau). On the other hand, the columns of the B matrix change. In fact, one column of the B matrix departs and another enters at each step. The B matrix for each tableau is the matrix formed from Tableau 1 by choosing columns corresponding to the unit columns of a given tableau

Table 1

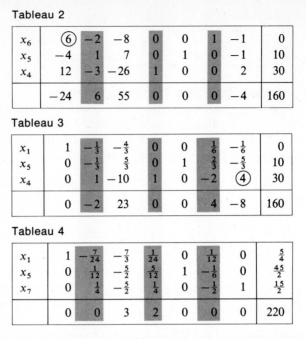

Tableau number	B matrix	B^{-1} matrix
1	$\begin{array}{ccc} x_6 & x_2 & x_4 \\ \begin{bmatrix} 1 & 0 & 0 \\ 0 & 1 & 0 \\ 0 & 0 & 1 \end{bmatrix} \end{array}$	$\begin{array}{ccc} x_6 & x_2 & x_4 \\ \begin{bmatrix} 1 & 0 & 0 \\ 0 & 1 & 0 \\ 0 & 0 & 1 \end{bmatrix} \end{array}$
2	$\begin{array}{ccc} x_6 & x_5 & x_4 \\ \begin{bmatrix} 1 & 2 & 0 \\ 0 & 1 & 0 \\ 0 & 3 & 1 \end{bmatrix} \end{array}$	$\begin{array}{ccc} x_6 & x_2 & x_4 \\ \begin{bmatrix} 1 & -2 & 0 \\ 0 & 1 & 0 \\ 0 & -3 & 1 \end{bmatrix} \end{array}$
3	$\begin{array}{ccc} x_1 & x_5 & x_4 \\ \begin{bmatrix} -2 & 2 & 0 \\ -4 & 1 & 0 \\ 0 & 3 & 1 \end{bmatrix} \end{array}$	$\begin{array}{ccc} x_6 & x_2 & x_4 \\ \begin{bmatrix} \frac{1}{6} & -\frac{1}{3} & 0 \\ \frac{2}{3} & -\frac{1}{3} & 0 \\ -2 & 1 & 1 \end{bmatrix} \end{array}$
4	$\begin{array}{ccc} x_1 & x_5 & x_7 \\ \begin{bmatrix} -2 & 2 & -3 \\ -4 & 1 & -1 \\ 0 & 3 & -1 \end{bmatrix} \end{array}$	$\begin{array}{ccc} x_6 & x_2 & x_4 \\ \begin{bmatrix} \frac{1}{12} & -\frac{7}{24} & \frac{1}{24} \\ -\frac{1}{6} & \frac{1}{12} & \frac{5}{12} \\ -\frac{1}{2} & \frac{1}{4} & \frac{1}{4} \end{bmatrix} \end{array}$

(taken in the order $e_1, e_2, \ldots, e_m$). Table 1 shows the B and B^{-1} matrices for the four tableaus.

Since we always find the columns of the B^{-1} matrix in the same columns in every tableau, we can compute the next B^{-1} matrix by premultiplying the current B^{-1} matrix by the pivoting matrix. For example, the pivoting matrix for obtaining Tableau 3 from Tableau 2 is

$$P_2 = \begin{bmatrix} \frac{1}{6} & 0 & 0 \\ \frac{4}{6} & 1 & 0 \\ -\frac{12}{6} & 0 & 1 \end{bmatrix}$$

and the B^{-1} matrix for Tableau 2 is

$$B^{-1} = \begin{bmatrix} 1 & -2 & 0 \\ 0 & 1 & 0 \\ 0 & -3 & 1 \end{bmatrix}$$

Thus,

$$P_2 B^{-1} = \begin{bmatrix} \frac{1}{6} & 0 & 0 \\ \frac{4}{6} & 1 & 0 \\ -\frac{12}{6} & 0 & 1 \end{bmatrix} \begin{bmatrix} 1 & -2 & 0 \\ 0 & 1 & 0 \\ 0 & -3 & 1 \end{bmatrix} = \begin{bmatrix} \frac{1}{6} & -\frac{1}{3} & 0 \\ \frac{2}{3} & -\frac{1}{3} & 0 \\ -2 & 1 & 1 \end{bmatrix}$$

produces the B^{-1} matrix for Tableau 3.   ◆

Exercises

In Exercises 1–6, for the given tableaus, form the D matrix defined in Theorems 1 and 2.

1. LP problem: Example 1 of Section 3.2
   Initial tableau: Tableau 1
   Final tableau: Tableau 3

2. LP problem: Example 4 of Section 3.2
   Initial tableau: Tableau 1
   Final tableau: Tableau 4

3. LP problem: Example 1 of Section 3.3
   Initial tableau: Tableau 1
   Final tableau: Tableau 4

4. LP problem: Example 2 of Section 3.3
   Initial tableau: Tableau 1
   Final tableau: Tableau 4

5. Prototype problem (11) of Section 3.4
   (a) Initial tableau: Tableau 1
       Final tableau: Tableau 3
   (b) Initial tableau: Tableau 4
       Final tableau: Tableau 7

6. Phase 1 problem: Example 1 of Section 3.4
   Initial tableau: Tableau 1
   Final tableau: Tableau 3

7. Give the sizes of all submatrices that occur in expression (4) of this section.

8. Explain how expression (5) is derived from expression (4) of this section.

9. The LP problem

$$\text{Maximize:} \quad z = 2x_2 - 3x_4 + 4x_5 - 2x_7 + 54$$

$$\text{Subject to:} \quad \begin{aligned} x_1 + 2x_2 \quad\quad\quad - 2x_4 - 3x_5 \quad\quad\quad + 4x_7 &= 10 \\ -x_2 \quad\quad\quad + 4x_4 + \ x_5 + x_6 + 6x_7 &= 12 \\ 3x_2 + \ x_3 + \ x_4 + 2x_5 \quad\quad\quad + \ x_7 &= 24 \end{aligned}$$

$$x_i \geq 0, \quad 1 \leq i \leq 7$$

is solved by the sequence of tableaus:

	x_1	x_2	x_3	x_4	x_5	x_6	x_7		
x_1	1	2	0	-2	-3	0	4	10	
x_6	0	-1	0	④	1	1	6	12	(Tableau T)
x_3	0	3	1	1	2	0	1	24	
	0	-2	0	3	-4	0	2	54	
x_1	1	$\frac{3}{2}$	0	0	$-\frac{5}{2}$	$\frac{1}{2}$	7	16	
x_4	0	$-\frac{1}{4}$	0	1	$\frac{1}{4}$	$\frac{1}{4}$	$\frac{3}{2}$	3	
x_3	0	$\frac{13}{4}$	1	0	$\frac{7}{4}$	$-\frac{1}{4}$	$-\frac{1}{2}$	21	
	0	$-\frac{5}{4}$	0	0	$-\frac{19}{4}$	$-\frac{3}{4}$	$-\frac{5}{2}$	45	
x_1	1	-1	0	10	0	3	22	46	
x_5	0	-1	0	4	1	1	6	12	
x_3	0	⑤	1	-7	0	-2	-11	0	
	0	-6	0	19	0	4	26	102	
x_1	1	0	$\frac{1}{5}$	$\frac{43}{5}$	0	$\frac{13}{5}$	$\frac{99}{5}$	46	
x_5	0	0	$\frac{1}{5}$	$\frac{13}{5}$	1	$\frac{3}{5}$	$\frac{19}{5}$	12	(Tableau T^*)
x_2	0	1	$\frac{1}{5}$	$-\frac{7}{5}$	0	$-\frac{2}{5}$	$-\frac{11}{5}$	0	
	0	0	$\frac{6}{5}$	$\frac{53}{5}$	0	$\frac{8}{5}$	$\frac{64}{5}$	102	

(a) Identify x_B and c_B for Tableau T^*.

(b) Identify the submatrices B^{-1} of Tableau T^* and B of Tableau T.

(c) Rearrange the columns of Tableaus T and T^* so that Tableau T is in the form of S given in expression (2). Identify the submatrices named B, Q, $\mathbf{b}$, $\mathbf{r}$, $\mathbf{q}$, and z_0 from expression (2).

(d) Form the pivoting matrix D, and verify relation (4).

10. Suppose that another variable x_3 is added to the LP problem of Example 1 to obtain the problem:

$$\text{Maximize:} \quad z = 125x_1 + 100x_2 + 50x_3$$

$$\text{Subject to:} \quad \begin{aligned} x_1 + \ x_2 + \ x_3 &\leq 4 \\ 5x_1 + 3x_2 + 2x_3 &\leq 15 \end{aligned}$$

$$x_i \geq 0, \quad 1 \leq i \leq 3$$

Use the formulas of Theorem 3 and Tableaus 1 and 2 of Example 1 to find the solution to this problem. Do not solve the problem by starting the solution over again.

11. What is the solution to Exercise 10 if the cost coefficient of x_3 is changed from 50 to 90?

4.5 LP MODELS

We introduced the transportation problem in Section 1.2, and here we use the simplex method and the CALIPSO software to solve such problems. We also introduce the transshipment problem, an extension of the transportation problem. In Chapter 9 we will consider the *transportation method*, which is an alternative way of implementing the simplex method for these problems. Even though the transportation method that we present in Chapter 9 provides a more efficient way for solving transportation problems, software is not always readily available for implementing that transportation method. We can use the method described in this section to solve transportation problems with any software capable of solving general LP problems; and if the problem is not too large, we can find the solution in a reasonable amount of computer time.

The next example introduces a simple transportation problem, which we will solve with CALIPSO.

EXAMPLE 1

The Del Valle Ketchup Company has canneries in Sacramento and Eugene and warehouses in Boise, Butte, and Elko. Del Valle moves its ketchup by truck from the canneries to the warehouses. Table 1 gives the cost of moving a truckload of ketchup from each cannery to each warehouse, the supply of ketchup (in truckloads) at each cannery, and the demand for ketchup (in truckloads) at each warehouse.

Table 1

Cannery	Boise 1	Butte 2	Elko 3	Supply (truckloads)
	Warehouse			
Sacramento 1	200	300	450	45
Eugene 2	200	250	300	30
Demand (truckloads)	30	20	25	

If we denote canneries with subscripts 1 and 2 and warehouses with subscr pts 1, 2, and 3, as indicated in Table 1, and if we let x_{ij} denote the number of truc loads moved from cannery i to warehouse j, then we can formulate this problem as the LP problem

$$
\begin{aligned}
\text{Minimize:} \quad z = \quad & 200x_{11} + 300x_{12} + 450x_{13} \\
+ \ & 200x_{21} + 250x_{22} + 300x_{23}
\end{aligned}
$$

$$
\begin{aligned}
\text{Subject to:} \quad x_{11} + x_{12} + x_{13} \quad & \quad \leq 45 \\
x_{21} + x_{22} + x_{23} & \leq 30 \\
x_{11} \quad\quad + x_{21} \quad\quad & = 30 \\
x_{12} \quad\quad + x_{22} \quad & = 20 \\
x_{13} \quad\quad + x_{23} & = 25 \\
x_{ij} \geq 0; \quad 1 \leq i \leq 2, \quad 1 \leq j \leq 3 &
\end{aligned}
\tag{1}
$$

It can be shown that all transportation problems have integer optimal solutions (i.e., all the values of x_{ij} in the optimal solution are integers). We do not prove this result because doing so requires a greater knowledge of linear algebra than we presume. Essentially, this result is a consequence of the special form of the coefficient matrix of the constraints in (1). To see this form better, we give the coefficient matrix as

$$
A = \begin{bmatrix}
1 & 1 & 1 & 0 & 0 & 0 \\
0 & 0 & 0 & 1 & 1 & 1 \\
1 & 0 & 0 & 1 & 0 & 0 \\
0 & 1 & 0 & 0 & 1 & 0 \\
0 & 0 & 1 & 0 & 0 & 1
\end{bmatrix}
$$

Notice that A has exactly two 1's in each column. All transportation problems have a coefficient matrix of this form. ♦

We can use the inequality-generation features of CALIPSO's MODELER to enter LP problem (1). You can read about these features in Appendix A. We recommend that you use the following format to solve the exercises in this section because it facilitates altering the values of &m and &n and using the data to generate inequalities corresponding to any transportation problem.

Format for solving a transportation problem with CALIPSO

```
&m:= the number of supply points;
&n:= the number of demand points;
assign(&cij, i = 1 to &m, j = 1 to &n, insert the cost coefficient matrix);
assign(&si, i = 1 to &m, insert list of supplies);
assign(&dj, j = 1 to &n, insert list of demands);
for i = 1 to &m do add(j = 1 to &n, −&cij*xij); {cost function}
for i = 1 to &m do add(j = 1 to &n, xij) <= &si;
for j = 1 to &n do add(i = 1 to &m, xij) = &dj;
```

EXAMPLE 2

Introducing the data for (1) in this format, we have

&m:= 2;

&n:= 3;

assign(&cij, i = 1 to &m, j = 1 to &n, 200, 300, 450, 200, 250, 300);

assign(&si, i = 1 to &m, 45, 30);

assign(&dj, j = 1 to &n, 30, 20, 25);

for i = 1 to &m do add(j = 1 to &n, −&cij∗xij); {cost function}

for i = 1 to &m do add(j = 1 to &n, xij) <= &si;

for j = 1 to &n do add(i = 1 to &m, xij) = &dj;

If you direct **MODELER** to create a file of type RSM, then you can solve the problem with the automatic-solution program of **SOLVER**. The solution is

$$(x_{11}, x_{12}, x_{13}, x_{21}, x_{22}, x_{23}) = (30, 15, 0, 0, 5, 25)$$

and the minimum cost is $z = \$19{,}250$. Figure 1 depicts this solution.

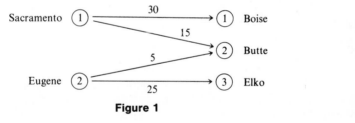

Figure 1   ◆

The Transshipment Problem

The transshipment problem is a variation of the transportation problem; and, as you will see, we can solve such problems as transportation problems.

Suppose the trucking company that hauls Del Valle's ketchup offers alternative routes for moving ketchup from the canneries to the warehouses. These alternative routes involve trucking the ketchup to various *junction points*, at which the ketchup is switched to another truck or other means of conveyance. These junction points could be any of Del Valle's canneries or warehouses, or they could be depots owned by the trucking company in the cities Weed, Redmond, and Moscow. A scenario of this sort is called a *transshipment* problem because it involves shipping goods to an intermediate point and then transferring them to another conveyance before shipping them further toward their destination.

In fact, it may be more efficient to ship goods through an intermediate point than directly to the final destination; for instance, in the Del Valle example, it may be more economical to ship the ketchup from Sacramento to Weed and then to Boise, than to ship directly from Sacramento to Boise. Table 2 gives the costs of the various shipping alternatives available to Del Valle. The cities are designated in Table 2 by numbers: 1, Sacramento; 2, Eugene; 3, Weed; 4, Redmond; 5, Moscow; 6, Boise; 7, Butte; and 8, Elko.

Table 2   Costs of various shipping alternatives

From \ To		Cannery		Junction			Warehouse			Supply (truckloads)
		1	2	3	4	5	6	7	8	
Cannery	1	0	100	120	210	250	250	300	450	45
	2	100	0	100	130	170	200	150	300	30
Junction	3	130	110	0	50	150	100	75	120	
	4	220	130	50	0	200	110	150	100	
	5	250	170	150	200	0	120	100	150	
Warehouse	6	250	200	100	110	120	0	100	85	
	7	300	250	75	150	100	25	0	75	
	8	450	300	120	100	150	85	100	0	
Demand (truckloads)							30	20	25	

Table 2 is reminiscent of the cost tables we used in Section 1.2 to represent transportation problems, except that some of its rows lack supply amounts, and some of its columns lack demand amounts. We first form an LP problem that models the Del Valle transshipment problem and then form a transportation problem from whose solution we can derive the solution of the transshipment problem.

EXAMPLE 3

For $1 \leq i \leq 8$, $1 \leq j \leq 8$, and $i \neq j$, let x_{ij} denote the number of truckloads of ketchup shipped from city i to city j, and let c_{ij} denote the costs given in Table 2. The cost function to be minimized is

$$\text{Minimize:} \quad z = \sum_{\substack{i=1 \\ i \neq j}}^{8} \sum_{j=1}^{8} c_{ij} x_{ij} \tag{2}$$

We describe the dual function of a cannery as a source of ketchup and as a junction point for transshipment by

$$(\text{amount shipped out}) - (\text{amount shipped in}) = \text{supply}$$

For the two canneries of Table 2, this balance gives the constraints

$$(x_{12} + x_{13} + x_{14} + x_{15} + x_{16} + x_{17} + x_{18})$$
$$-(x_{21} + x_{31} + x_{41} + x_{51} + x_{61} + x_{71} + x_{81}) = 45 \qquad (3)$$
$$(x_{21} + x_{23} + x_{24} + x_{25} + x_{26} + x_{27} + x_{28})$$
$$-(x_{12} + x_{32} + x_{42} + x_{52} + x_{62} + x_{72} + x_{82}) = 30$$

We can express constraints (3) more compactly by using summation notation:

$$\sum_{k=2}^{8} x_{1k} - \sum_{k=2}^{8} x_{k1} = 45$$
$$\sum_{\substack{k=1 \\ k \neq 2}}^{8} x_{2k} - \sum_{\substack{k=1 \\ k \neq 2}}^{8} x_{k2} = 30 \qquad (4)$$

We describe the condition that cities 3, 4, and 5 are junctions by

(amount shipped out) − (amount shipped in) = 0

For the three junctions in Table 2, this balance gives the constraints

$$\sum_{\substack{k=1 \\ k \neq i}}^{8} x_{ik} - \sum_{\substack{k=1 \\ k \neq i}}^{8} x_{ki} = 0, \qquad \text{for} \quad 3 \leq i \leq 5 \qquad (5)$$

We describe the condition that cities 6, 7, and 8 serve as destinations as well as warehouses by

(amount shipped in) − (amount shipped out) = demand

For the three warehouses in Table 2 this balance gives

$$\sum_{\substack{k=1 \\ k \neq 6}}^{8} x_{k6} - \sum_{\substack{k=1 \\ k \neq 6}}^{8} x_{6k} = 30 \qquad (6)$$

$$\sum_{\substack{k=1 \\ k \neq 7}}^{8} x_{k7} - \sum_{\substack{k=1 \\ k \neq 7}}^{8} x_{7k} = 20 \qquad (7)$$

$$\sum_{\substack{k=1 \\ k \neq 8}}^{8} x_{k8} - \sum_{\substack{k=1 \\ k \neq 8}}^{8} x_{8k} = 25 \qquad (8)$$

In Exercise 15 you are asked to show that the solution to this problem is

$$x_{13} = 45; \; x_{27} = 30; \; x_{36} = 20; \; x_{76} = 10; \; x_{38} = 25; \; z^* = \$15{,}150$$

Figure 2 depicts this solution.

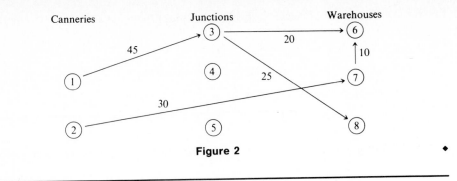

Figure 2

Notice that in the LP problem in Example 3 each variable occurs in exactly two constraints, once with a coefficient of 1 and once with a coefficient of -1. This special form implies that the transshipment problem always has integer solutions. (Just as for the transportation problem, we do not prove this result because the proof requires a considerable knowledge of linear algebra.) Since all solutions are integers, we can use the simplex method rather than an integer programming method to solve transshipment problems.

Notice that equations (4) and (5) differ in structure only in that the right-hand side of each equation in (5) is 0. If we regard a junction as being a source with supply 0, then there is no need to distinguish between sources and junctions. (With a transparent change of focus, we could also view a junction as a demand point with demand 0.) Thus, by allowing some supplies and demands to be 0, we can state the transshipment problem more simply as follows.

Suppose there are m supply points $P_1, P_2, \ldots, P_m$ and n demand points $P_{m+1}, P_{m+2}, \ldots, P_{m+n}$. For $1 \leq i \leq m$, supply point P_i has s_i units of the product available; for $1 \leq j \leq n$, demand point P_{m+j} requires d_{m+j} units of the product. To ensure that the problem has a solution, we must assume that the total demand does not exceed the total supply. For the transshipment problem, it is convenient to assume that the total demand equals the total supply; that is,

$$\sum_{j=1}^{n} d_{m+j} = \sum_{i=1}^{m} s_i \tag{9}$$

(In Chapter 9 we learn that we can always replace a transportation or transshipment problem by an equivalent problem with total supply equal to total demand.) For $1 \leq r, t \leq m + n$ and $r \neq t$, suppose there is one, and only one, route for shipping product from P_r to P_t (thus we assume that any point can serve as a junction point). Finally, let c_{rt} be the cost of shipping a unit of product from P_r to P_t. We do *not* assume that $c_{rt} = c_{tr}$; thus, the cost of shipping from P_r to P_t can be different from the cost of shipping from P_t to P_r. If it is impossible to ship product between points P_r and P_t, we set $c_{rt} = \infty$ (an exhorbitant shipping cost). Figure 3 depicts such a set of transportation routes for a situation with two supply points P_1 and P_2 and three demand points P_3, P_4, and P_5. The double arrows indicate that we may distinguish the direction in which shipping occurs between two points for the purpose of assessing a shipping cost.

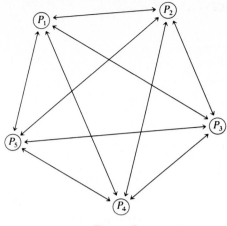

Figure 3

The *LP transshipment problem* is

Minimize: $z = \displaystyle\sum_{\substack{i=1 \\ i \neq j}}^{m+n} \sum_{j=1}^{m+n} c_{ij}x_{ij}$

Subject to: $\displaystyle\sum_{\substack{k=1 \\ k \neq i}}^{m+n} x_{ik} - \sum_{\substack{k=1 \\ k \neq i}}^{m+n} x_{ki} = s_i, \qquad 1 \leq i \leq m$ (10)

$\displaystyle\sum_{\substack{k=1 \\ k \neq m+j}}^{m+n} x_{k,m+j} - \sum_{\substack{k=1 \\ k \neq m+j}}^{m+n} x_{m+j,k} = d_{m+j}, \qquad 1 \leq j \leq n$

$x_{rt} \geq 0; \qquad 1 \leq r,t \leq m+n; \qquad r \neq t$

The next example gives a format that you can use to enter transshipment problems with MODELER.

EXAMPLE 4

The constraints in (10) require forming sums that omit the x_{ii} terms. Without doing a lot of typing, it is difficult to express such sums using the ADD function of MODELER. To avoid this difficulty, we write the first constraint of (10) as

$$\sum_{\substack{k=1 \\ k \neq i}}^{m+n} x_{ik} - \sum_{\substack{k=1 \\ k \neq i}}^{m+n} x_{ki} = \sum_{k=1}^{m+n} (x_{ik} - x_{ki}) = s_i$$

We write the second constraint in a similar way. The format we can use with MODELER is the following.

Format for MODELER

&m:= *number of supply points*
&r:= *number of supply points* +1
&n:= *number of demand points*
&t:= *total number of points* (&m + &n);
assign(&cij, i = 1 to &t, j = 1 to &t, *insert the matrix of cost coefficients*);
for i = 1 to &t do add(j = 1 to &t, −&cij∗xij);
for i = 1 to &m do add(k = 1 to &t, xik − xki) = &si;
for j = &r to &t do add(k = 1 to &t, xkj − xjk) = &dj;   ◆

We want to pose a transportation problem from whose solution we can derive a solution to LP problem (10). To this end, notice that problem (10) differs from the general model for transportation problems by the absence of variables x_{ii} and by the presence of minus signs. If we let

$$s = \sum_{j=1}^{n} d_{m+j} = \sum_{i=1}^{m} s_i$$

then, since the total amount shipped into any point P_i cannot exceed the total supply, we have

$$\sum_{\substack{k=1 \\ k \neq i}}^{m+n} x_{ki} \leq s, \qquad 1 \leq i \leq m \tag{11}$$

Similarly, since the total amount shipped out of any point P_j cannot exceed s, we have

$$\sum_{\substack{k=1 \\ k \neq m+j}}^{m+n} x_{m+j,k} \leq s, \qquad 1 \leq j \leq n \tag{12}$$

We introduce slack variables into (11) and (12) to obtain

$$\sum_{\substack{k=1 \\ k \neq i}}^{m+n} x_{ki} + x_{ii} = s \tag{13}$$

$$\sum_{\substack{k=1 \\ k \neq m+j}}^{m+n} x_{m+j,k} + x_{m+j,m+j} = s \tag{14}$$

We simplify (13) and (14) by moving x_{ii} and $x_{m+j,m+j}$ into the summation to obtain

$$\sum_{k=1}^{m+n} x_{ki} = s \tag{15}$$

$$\sum_{k=1}^{m+n} x_{m+j,k} = s \tag{16}$$

We substitute (15) and (16) into the first and second constraints of (10) to obtain

$$\sum_{k=1}^{m+n} x_{ik} = s + s_i, \qquad 1 \le i \le m \tag{17}$$

$$\sum_{k=1}^{m+n} x_{k,m+j} = s + d_{m+j}, \qquad 1 \le j \le n \tag{18}$$

Thus, we can write problem (10) as

$$\text{Minimize:} \quad z = \sum_{\substack{i=1 \\ i \ne j}}^{m+n} \sum_{j=1}^{m+n} c_{ij} x_{ij} \tag{19}$$

$$\text{Subject to:} \quad \sum_{k=1}^{m+n} x_{ki} = s, \qquad 1 \le i \le m$$

$$\sum_{k=1}^{m+n} x_{k,m+j} = s + d_{m+j}, \qquad 1 \le j \le n$$

$$\sum_{k=1}^{m+n} x_{ik} = s + s_i, \qquad 1 \le i \le m$$

$$\sum_{k=1}^{m+n} x_{m+j,k} = s, \qquad 1 \le j \le n$$

$$x_{ij} \ge 0; \qquad 1 \le i \le m, \quad 1 \le j \le n$$

We can more easily recognize (19) as a transportation problem by introducing

$$\delta_i = s, \qquad 1 \le i \le m$$
$$\delta_{m+j} = s + d_{m+j}, \qquad 1 \le j \le n$$
$$\sigma_i = s + s_i, \qquad 1 \le i \le m$$
$$\sigma_{m+j} = s, \qquad 1 \le j \le n$$

With these substitutions (19) becomes

$$\text{Minimize:} \quad z = \sum_{\substack{i=1 \\ i \ne j}}^{m+n} \sum_{j=1}^{m+n} c_{ij} x_{ij} \tag{20}$$

$$\text{Subject to:} \quad \sum_{k=1}^{m+n} x_{ki} = \delta_i, \qquad 1 \le i \le m+n$$

$$\sum_{k=1}^{m+n} x_{ik} = \sigma_i, \qquad 1 \le i \le m+n$$

$$x_{ij} \ge 0; \qquad 1 \le i \le m, \quad 1 \le j \le n$$

which is in the usual form of a transportation problem.

EXAMPLE 5

For the Del Valle problem, we regard the junction points P_1, P_2, P_3 as supply points with an initial supply of 0. Table 3 gives the cost table corresponding to problem (19).

Table 3   Costs of shipping

From \ To		Cannery 1	Cannery 2	Junction 3	Junction 4	Junction 5	Warehouse 6	Warehouse 7	Warehouse 8	Supply (truckloads)
Cannery	1	0	100	120	210	250	250	300	450	120
	2	100	0	100	130	170	200	150	300	105
Junction	3	130	110	0	50	150	100	75	120	75
	4	220	130	45	0	200	110	150	100	75
	5	250	170	150	220	0	120	100	150	75
Warehouse	6	250	200	100	110	120	0	100	85	75
	7	300	250	75	150	100	25	0	75	75
	8	450	300	120	100	150	85	100	0	75
Demand (truckloads)		75	75	75	75	75	105	95	100	

In Exercise 16 you are asked to use CALIPSO to obtain the solution

$$x_{13} = 45, \quad x_{27} = 30, \quad x_{33} = 30, \quad x_{36} = 20, \quad x_{38} = 25, \quad x_{76} = 10, \quad x_{77} = 65$$

and $x_{ii} = 75$ for all i, $1 \le i \le 8$, except $i = 3$ and 7. From (13) with $i = 3$, we see that $x_{33} = 30$ indicates that $75 - 30 = 45$ truckloads were shipped into city 3. Thus, city 3 (Weed) is a junction point through which 45 truckloads were shipped. Similarly, $x_{77} = 65$ indicates that warehouse 7 serves as a junction point for transshipping $75 - 65 = 10$ truckloads. Because the other values of x_{ii} are all 75, (13) and (14) show that no ketchup was hauled through other junction cities. See Figure 2, which depicts this solution.   ◆

The purpose of restructuring LP transshipment problem (10) as a transportation problem (19) was to set the stage for using a special method of solving transportation problems to solve transshipment problems. (We give this special method in Chapter 9.) Notice, however, that we pay a heavy price in converting a transshipment problem to a transportation problem. The transshipment problem (10) has $(m - 1)(n - 1)$ variables and $m + n$ constraints; the transportation problem (19) has mn variables and $2(m + n)$ constraints. As you will see in Chapter 6, the time spent in solving an LP problem with the simplex method depends mainly on the number of constraints and increases rapidly as the number of constraints increases.

Even though the special transportation method we study in Chapter 9 can be a more efficient way of solving these problems, much of the advantage can be lost in dealing with the additional constraints imposed by the transportation model of the transshipment problem.

Exercises

In Exercises 1–8 you are asked to solve problems from Section 1.2. You can use the transportation format we gave in that section to enter these problems with MODELER and solve them with SOLVER.

†1. Exercise 1   †2. Exercise 2   †3. Exercise 3   †4. Exercise 4   †5. Exercise 5

†6. Exercise 6   †7. Exercise 7   †8. Exercise 8

In Exercises 9–13 you are asked to solve some assignment problems from Section 1.2. Since the assignment problem is a special case of the transportation problem, you can use the transportation format to enter these problems with MODELER and solve them with SOLVER.

†9. Exercise 9   †10. Exercise 10   †11. Exercise 11   †12. Exercise 12

†13. Exercise 13

†14. Solve transportation problem (1).

†15. Solve the transshipment problem given in Example 3.

†16. Solve the transportation problem given in Example 5.

†17. Suppose the rail freight charge per ton between seven cities is given by Table 4. An entry of ∞ indicates that no rail service is available between the two locations. In a computer solution, ∞ would be replaced by a large enough value to ensure that the optimal solution does not select such a route.

Table 4   Rail freight charge per ton

From \ To	A	B	C	D	E	F	G
A	0	12	20	18	45	35	∞
B	18	0	24	∞	16	24	32
C	22	12	0	24	22	26	19
D	∞	20	20	0	12	18	20
E	36	32	24	27	0	24	28
F	27	28	32	42	36	0	∞
G	∞	32	24	25	27	42	0

Goods can be sent through intermediate cities at a cost equal to the sum of the freight costs for each leg of the journey. Suppose 100, 80, and 90 tons are to be shipped from cities A, B, and C, respectively. Suppose 75, 110, and 65 tons are to be received at cities E, F, and G, respectively. Find the minimum cost shipping schedule.

†18. Suppose a company has a large supply of wheat at city A and needs to ship 20, 45, 30, and 15 tons to cities B, D, E, and F, respectively. Using the rail routes and shipping costs given in Table 4, find the minimum cost shipping schedule.

†**19.** An airline offers direct flights from San Francisco or Los Angeles to London or Paris. The airline also offers the option of making these flights by changing planes at Chicago, New York, or Anchorage. The fare for a flight requiring a change of plane is the sum of the fares between various cities. Table 5 gives the fares. Find the cheapest routes from the origins to the destinations. The various cities are designated by numbers: 1, Los Angeles; 2, San Francisco; 3, Chicago; 4, New York; 5, Anchorage; 6, London; and 7, Paris.

Table 5 Fares

From \ To	Origin 1	2	Junction 3	4	5	Destination 6	7
Origin 1	0	150	300	350	400	850	775
2	150	0	275	375	350	750	800
Junction 3	300	275	0	275	450	700	750
4	350	375	275	0	375	650	750
5	400	350	450	375	0	450	375
Destination 6	850	750	700	650	450	0	75
7	775	800	750	750	375	75	0

†**20.** NORAD wants to transmit two urgent messages to two of its submarines at sea. Each message will be transmitted from a different site. It does not matter which submarine receives which message. Each message can be sent directly to a submarine or retransmitted via three communications satellites. Each of these satellites can pass the message to either submarine, and each satellite can pass the message for retransmission to another satellite or back to one of the original sending sites. NORAD has associated a number with each transmission route that measures the reliability of the communication channel given the current locations of the satellites and the submarines. We want to formulate the LP transshipment problem that a computer could solve to minimize the sum of the reliability figures over all possible routes from the sending sites to the submarines.

Table 6 displays the reliability figures at the time of transmission. A reliability of $-\infty$ indicates that the message cannot be sent along the associated route. The sources of the messages are numbered 1 and 2; the satellites are numbered 3, 4, and 5; and the submarines are numbered 6 and 7.

Table 6

From \ To	Sources 1	2	Satellites 3	4	5	Submarines 6	7
1	—	∞	27	35	40	120	140
2	∞	—	35	24	25	132	125
3	25	32	—	32	15	54	60
4	35	40	40	—	24	80	45
5	40	20	22	25	—	72	54
6	∞	∞	20	30	25	—	∞
7	∞	∞	22	30	32	∞	—

Use MODELER and SOLVER to solve this problem.

CHECKLIST: CHAPTER 4

Chapter 5

THE REVISED SIMPLEX METHOD

5.1 A FASTER ALGORITHM FOR COMPUTER SOLUTION

You have no doubt noticed that there are a great many zeros in most simplex tableaus. These zeros are not too troublesome in hand calculations because a human recognizes instantly that multiplying a number by zero gives zero and that additions of zeros can be skipped. However, a computing machine cannot recognize such simplifications and takes as much time to multiply a number by zero as by any other quantity. Thus, it is important to reduce the need for arithmetic operations on the zero entries of the simplex tableau. The revised simplex method is not a new method of solving LP problems but a procedure for implementing the ordinary simplex method for machine calculation in a way that avoids most computations involving zeros.

To appreciate the benefits of the revised simplex method, you must understand how computers store numbers and how computer memory can be organized to perform a sequence of calculations efficiently. In Sections 5.2 and 5.3 we explain these computer concepts without assuming that you know anything about computer architecture or programming. Since many of you probably know a great deal about computers and are interested in programming, some of the problems we pose will offer you opportunities to write computer programs; however, such exercises are not necessary to understand the revised simplex method. Those of you who are not interested in programming should be at no disadvantage. In Section 5.3 we compare the number of arithmetic operations (the time complexity) required to solve a problem by the ordinary and the revised simplex methods. We will also compare the amount of computer storage (the space complexity) of the two methods.

In our development of the revised simplex method, we assume that the LP problem to be solved is in the form

$$\text{Maximize:} \quad z = \mathbf{cx}$$

$$\text{Subject to:} \quad A\mathbf{x} = \mathbf{b}$$

$$\mathbf{x} \geq \mathbf{0}$$

where A is an $m \times n$ matrix in canonical form; that is, A contains m unit columns $\mathbf{e}_1$, $\mathbf{e}_2, \ldots, \mathbf{e}_m$ that correspond to an initial basic feasible solution. We also assume that the objective function is initially expressed in terms of nonbasic variables. We use the following notation to refer to tableaus representing various stages of the simplex method:

The initial tableau

$$T = \left[\begin{array}{c|c} A & \mathbf{b} \\ \hline -\mathbf{c} & 0 \end{array} \right]$$

Tableau after the sth pivot

$$T_s = \left[\begin{array}{c|c} A_s & \mathbf{b}_s \\ \hline -\mathbf{c}_s & z_s \end{array} \right]$$

Tableau after an unspecified number of pivots

$$T^* = \left[\begin{array}{c|c} A^* & \mathbf{b}^* \\ \hline -\mathbf{c}^* & z^* \end{array} \right]$$

Notice that this notation is incomplete because it does not define the basic variables corresponding to a tableau. We describe these by associating an injection

$$h(1), h(2), \ldots, h(m)$$

with each tableau.

As we will discover in Section 5.3, implementing the Greedy Entering Basic Variable Rule reduces the efficiency of the revised simplex method. Therefore, we introduce another rule for selecting the entering basic variable, one that has proven efficient in practical applications.

CYCLIC ENTERING BASIC VARIABLE RULE

This rule selects the entering basic variable in the following fashion: Denote the objective row by $-c_1, -c_2, \ldots, -c_n$. In the first pivot of the simplex method, examine the objective-row values in order, starting with $-c_1$, until the first negative value is encountered (or, in the case of an optimal tableau, until the row is exhausted). If the first negative value found is $-c_k$, then halt the search; x_k is the new entering variable. At the next pivot, start the search for the new entering basic variable wherever the previous search ended (at $-c_{k+1}$), and proceed until a negative value is encountered; this corresponds to the new entering basic variable. On reaching $-c_n$, wrap around (cycle) to the beginning of the row ($-c_1$), and continue examining objective-row entries until a negative value is encountered, or until $-c_k$ (the element preceding our starting point $-c_{k+1}$) is encountered. Continue to examine objective-row entries in this cyclic fashion in subsequent steps of the simplex method.

The revised simplex method uses the formulas of Theorem 2 of Section 4.3, which we restate here using the subscript s $(s \geq 0)$ to denote the result of the sth pivot of the simplex method.

$$A_s = B^{-1}A$$

$$\mathbf{b}_s = B^{-1}\mathbf{b}$$

$$-\mathbf{c}_s = \mathbf{c}_B B^{-1}A - \mathbf{c} \tag{1}$$

$$z_s = \mathbf{c}_B B^{-1}\mathbf{b}$$

where, for $s \geq 1$, B^{-1} and $\mathbf{c}_B$ are the matrices associated with the tableau for the $(s-1)$st step of the simplex method; when $s = 0$, B^{-1} is the identity matrix I_m, and $\mathbf{c}_B$ is a $1 \times m$ matrix of zeros; A, $\mathbf{b}$, and $\mathbf{c}$ are the matrix components of the initial tableau. The injection h has entries corresponding to the initial basic variables.

The injection h_{s-1} for the $(s-1)$st tableau determines $\mathbf{c}_B$. The minimum information needed to use formulas (1) to perform the sth pivot consists of the initial tableau T, B^{-1} for the $(s-1)$-tableau, and h_{s-1}. [Thus, as we will discuss further in Section 5.3, implementation of the simplex method based on formulas (1) requires storage for the original tableau T and additional storage for B^{-1}.]

Actually, for the sake of computational efficiency, we are going to store a little more information than is necessary. We explain in Section 5.3 why using this additional storage saves calculation time.

Definition.　The *revised simplex tableau* is

$$R = \left[\begin{array}{c|c} B^{-1} & \mathbf{b}^* \\ \hline \mathbf{w} & z^* \end{array}\right] \tag{2}$$

where

B^{-1} is an $m \times m$ matrix (the inverse of the basis matrix B),

$\mathbf{w} = \mathbf{c}_B B^{-1}$ is a $1 \times m$ matrix of certain objective-row entries (the dual prices introduced in Chapter 6),

$\mathbf{b}^* = B^{-1}\mathbf{b}$ is an $m \times 1$ matrix (the current resource column), and

$z^* = \mathbf{c}_B B^{-1}\mathbf{b} = \mathbf{c}_B \mathbf{b}^*$ is a number (the current value of the objective function).

The basic ideas of the revised simplex method are to avoid storing entire tableaus that result from pivoting and to calculate only those entries needed at a given point, from formulas using the B^{-1} matrix. We apply a similar idea to B^{-1} itself. Rather than store B^{-1}, we store the sequence of pivoting matrices used to update B^{-1} in the course of the revised simplex method; we never actually compute B^{-1}. Since B^{-1} always occurs as a factor in a product, we show that it is faster to multiply together the sequence of pivoting matrices that produce B^{-1} than it is to multiply by B^{-1}.

We next review the way in which the simplex algorithm produces B^{-1}. We use pivoting matrices to describe the pivoting process. Initially, B^{-1} is the identity matrix I_m. The algorithm changes B^{-1} as follows:

PIVOT 1. A pivoting matrix P_1 is formed, and $B^{-1} = I_m$ is replaced by $P_1 B^{-1}$, that is, by $P_1 I_m = P_1$.

PIVOT 2. A pivoting matrix P_2 is formed, and B^{-1} (resulting from pivot 1) is replaced by $P_2 B^{-1}$, that is, by $P_2 P_1$.

$\vdots$

PIVOT s. A pivoting matrix P_s is formed, and B^{-1} (resulting from pivot $s - 1$) is replaced by $P_s B^{-1}$, that is, by $P_s P_{s-1} \cdots P_2 P_1$.

EXAMPLE 1

In Example 3 of Section 4.3 we performed three pivots of the simplex method to obtain Tableau 4 from Tableau 1. We can obtain Tableau 2 by premultiplying Tableau 1 by the pivoting matrix

$$P_1 = \begin{bmatrix} 1 & -\frac{5}{3} & 0 \\ 0 & \frac{1}{3} & 0 \\ 0 & -\frac{2}{3} & 1 \end{bmatrix}$$

In the revised simplex method, we are not concerned with finding the entire new tableau at each stage because we need only the new B^{-1} matrix. As we showed in Example 7 of Section 4.2, we can perform a pivot operation on selected columns of a matrix by premultiplying just those columns by the pivoting matrix. Note that the B^{-1} matrix for Tableau 1 (Example 3, Section 4.3) is the 3×3 identity matrix. Since $P_1 I = P_1$, P_1 is the B^{-1} matrix for Tableau 2.

The pivoting matrix for the second pivot is

$$P_2 = \begin{bmatrix} \frac{3}{4} & 0 & 0 \\ -\frac{1}{4} & 1 & 0 \\ -4 & 0 & 1 \end{bmatrix}$$

The B^{-1} matrix for Tableau 3 is $P_2 P_1$.

The pivoting matrix for the third pivot is

$$P_3 = \begin{bmatrix} 1 & 0 & \frac{5}{24} \\ 0 & 1 & -\frac{1}{8} \\ 0 & 0 & \frac{1}{6} \end{bmatrix}$$

The B^{-1} matrix for Tableau 4 is

$$B^{-1} = P_3 P_2 P_1 = \begin{bmatrix} -\frac{1}{12} & 0 & \frac{5}{24} \\ \frac{1}{4} & 0 & -\frac{1}{8} \\ -\frac{2}{3} & 1 & \frac{1}{6} \end{bmatrix}$$

♦

In general the B^{-1} matrix for a tableau obtained from another by pivoting with a pivoting matrix P is

$$B^{-1}_{\text{new}} = P B^{-1}_{\text{old}}$$

As you will see in the statement of the revised simplex algorithm, we never actually compute the matrix B^{-1} but instead introduce one more pivoting matrix to the sequence $P_s, P_{s-1}, \ldots, P_1$.

Note that if B^{-1} results from the sth pivot as the product of s pivoting matrices, then, for any column A_k of the A matrix, we have

$$B^{-1}A_k = P_s \cdots P_2 P_1 A_k \tag{3}$$

$$\mathbf{c}_B B^{-1} = \mathbf{c}_B P_s \cdots P_2 P_1 \tag{4}$$

In Section 5.2 we find that we can evaluate formulas (3) and (4) with fewer calculations than you might expect.

The Revised Simplex Algorithm

STEP 1. Initialize the problem by storing the problem data in the matrices A, $\mathbf{b}$, $\mathbf{c}$ and the subscripts of the initial basic variables in the injection $h(1)$, $h(2), \ldots, h(m)$. We use s as a counter to keep track of the subscript to be assigned to the pivoting matrix produced in the sth pivot. The initial value of s is 1. Let

$\mathbf{c}_B = [0 \quad 0 \quad \cdots \quad 0]$ (Since we assume the initial objective function is expressed in terms of nonbasic variables)

$\mathbf{b}^* = \mathbf{b}$

$\mathbf{w} = \mathbf{c}_B B^{-1} = [0 \quad 0 \quad \cdots \quad 0]$

$z^* = 0$

STEP 2. Select the entering basic variable x_p by the Cyclic Entering Basic Variable Rule.

Since the cost coefficients $-\mathbf{c}^*$ are not stored at each step, we compute them with the formulas

$$-c_k^* = \mathbf{w}A_k - c_k$$

We already know that $-c_k^* = 0$ when x_k is a basic variable, so we only compute $-c_k^*$ in cyclic order for those k not in the list $h(1), h(2), \ldots, h(m)$. We stop calculating $-c_k^*$ when the first negative value is encountered and set $p = k$ so that x_p is the entering variable. If all the calculated values of $-c_k^*$ are nonnegative, the maximal solution has been found, and the algorithm halts.

STEP 3. Select the departing basic variable x_r.

Compute the pivot column, and store it in an $m \times 1$ matrix $\mathbf{y}$ computed as follows:

If $\quad s = 1 \quad$ then $\quad \mathbf{y} = A_p; \quad$ otherwise $\quad \mathbf{y} = P_{s-1} \cdots P_2 P_1 A_p$

If all the entries of $\mathbf{y}$ are nonpositive, then the objective function is unbounded, and the algorithm halts. Otherwise, choose the departing variable by computing the ratios

$$\frac{b_i^*}{y_i}, \qquad \text{for all} \qquad y_i > 0, \qquad 1 \le i \le m$$

Let r be a value of i corresponding to the smallest of these ratios (so that x_r is the departing basic variable).

STEP 4. Form a new pivoting matrix P_s.

Obtain P_s by replacing the r column of the identity matrix I_m by the column

$$L = \begin{bmatrix} -\dfrac{y_1}{y_r} \\ -\dfrac{y_2}{y_r} \\ \vdots \\ \dfrac{1}{y_r} \\ \vdots \\ -\dfrac{y_m}{y_r} \end{bmatrix} \quad \leftarrow \text{row } r$$

STEP 5. Update the revised simplex tableau by

(a) replacing $\mathbf{b}^*$ by $P_s\mathbf{b}^*$ ($P_s\mathbf{b}^*$ effectively performs the pivot on the column $\mathbf{b}^*$; see Exercise 13),

(b) changing the injection by setting $h(r) = p$,

(c) changing the rth entry of $\mathbf{c}_B$ to c_p,

(d) letting $\mathbf{w} = \mathbf{c}_B P_s \cdots P_2 P_1$, and

(e) letting $z^* = \mathbf{c}_B\mathbf{b}^*$.

Replace s by $s + 1$, and return to step 2.

EXAMPLE 2

We perform the steps of the revised simplex algorithm on the LP problem with the initial tableau

	x_1	x_2	x_3	x_4	x_5	x_6	
x_4	1	2	-1	1	0	0	4
x_5	1	1	1	0	1	0	6
x_6	2	0	1	0	0	1	8
	-2	1	-1	0	0	0	0

STEP 1. The initial values of the matrices are

$$\mathbf{c_B} = [0 \quad 0 \quad 0]$$
$$\mathbf{w} = \mathbf{c_B}B^{-1} = [0 \quad 0 \quad 0]$$
$$\mathbf{b^*} = \mathbf{b} = [4 \quad 6 \quad 8]^\mathrm{T}$$
$$z = 0$$

The current injection is $h(1) = 4; h(2) = 5; h(3) = 6$. By (2), the initial revised simplex tableau is

$$R = \begin{bmatrix} 1 & 0 & 0 & 4 \\ 0 & 1 & 0 & 6 \\ 0 & 0 & 1 & 8 \\ \hline 0 & 0 & 0 & 0 \end{bmatrix}$$

STEP 2.

$$-c_1^* = \mathbf{w}A_1 - c_1 = [0 \quad 0 \quad 0]\begin{bmatrix} 1 \\ 1 \\ 2 \end{bmatrix} - 2 = -2$$

Because we have found a negative entry, we stop calculating entries of $-\mathbf{c}^*$ and select column 1 as the pivot column. Thus, $p = 1$.

STEP 3.

$$\mathbf{y} = A_1 = \begin{bmatrix} 1 \\ 1 \\ 2 \end{bmatrix}$$

$$\text{minimum}\left\{ \frac{b_i^*}{y_i} \,\middle|\, y_i > 0, \quad 1 \le i \le 3 \right\} = \text{minimum}\left\{ \frac{4}{1}, \frac{6}{1}, \frac{8}{2} \right\}$$
$$= 4$$

There is a tie for minimum at $i = 1$ and $i = 3$; we select $r = 1$.

STEP 4.

$$L = \begin{bmatrix} \frac{1}{1} \\ -\frac{1}{1} \\ -\frac{2}{1} \end{bmatrix}; \qquad P_1 = \begin{bmatrix} 1 & 0 & 0 \\ -1 & 1 & 0 \\ -2 & 0 & 1 \end{bmatrix}$$

STEP 5.

(a) The new $\mathbf{b^*}$ is computed from the old $\mathbf{b^*}$ by

$$P_1\mathbf{b^*} = \begin{bmatrix} 1 & 0 & 0 \\ -1 & 1 & 0 \\ -2 & 0 & 1 \end{bmatrix}\begin{bmatrix} 4 \\ 6 \\ 8 \end{bmatrix} = \begin{bmatrix} 4 \\ 2 \\ 0 \end{bmatrix}$$

(b) The new injection is $h(1) = 1, h(2) = 5, h(3) = 6$.

(c) The new $\mathbf{c_B}$ is $[2 \quad 0 \quad 0]$.

(d)

$$\mathbf{w} = \mathbf{c_B}P_1 = \begin{bmatrix} 2 & 0 & 0 \end{bmatrix} \begin{bmatrix} 1 & 0 & 0 \\ -1 & 1 & 0 \\ -2 & 0 & 1 \end{bmatrix} = \begin{bmatrix} 2 & 0 & 0 \end{bmatrix}$$

(e)

$$z^* = \mathbf{c_B}\mathbf{b}^* = \begin{bmatrix} 2 & 0 & 0 \end{bmatrix} \begin{bmatrix} 4 \\ 2 \\ 0 \end{bmatrix} = 8$$

The algorithm now requires us to set $s = 2$ and return to step 2.

STEP 2.

$$-c_2^* = \mathbf{w}A_2 - c_2 = \begin{bmatrix} 2 & 0 & 0 \end{bmatrix} \begin{bmatrix} 2 \\ 1 \\ 0 \end{bmatrix} - (-1) = 5$$

$$-c_3^* = \mathbf{w}A_3 - c_3 = \begin{bmatrix} 2 & 0 & 0 \end{bmatrix} \begin{bmatrix} -1 \\ 1 \\ 1 \end{bmatrix} - 1 = -3$$

Therefore, column 3 is the pivot column and $p = 3$.

STEP 3.

$$\mathbf{y} = P_1 A_3 = \begin{bmatrix} 1 & 0 & 0 \\ -1 & 1 & 0 \\ -2 & 0 & 1 \end{bmatrix} \begin{bmatrix} -1 \\ 1 \\ 1 \end{bmatrix} = \begin{bmatrix} -1 \\ 2 \\ 3 \end{bmatrix}$$

$$\text{minimum} \left\{ \frac{b_i^*}{y_i} \, \middle| \, y_i > 0, \quad 1 \leq i \leq 3 \right\} = \text{minimum} \left\{ \frac{2}{2}, \frac{0}{3} \right\}$$

$$= 0$$

and so $r = 3$.

STEP 4.

$$L = \begin{bmatrix} \frac{1}{3} \\ -\frac{2}{3} \\ \frac{1}{3} \end{bmatrix}; \qquad P_2 = \begin{bmatrix} 1 & 0 & \frac{1}{3} \\ 0 & 1 & -\frac{2}{3} \\ 0 & 0 & \frac{1}{3} \end{bmatrix}$$

STEP 5.

(a) The new $\mathbf{b}^*$ is

$$P_2\mathbf{b}^* = \begin{bmatrix} 1 & 0 & \frac{1}{3} \\ 0 & 1 & -\frac{2}{3} \\ 0 & 0 & \frac{1}{3} \end{bmatrix} \begin{bmatrix} 4 \\ 2 \\ 0 \end{bmatrix} = \begin{bmatrix} 4 \\ 2 \\ 0 \end{bmatrix}$$

(b) The new injection is $h(1) = 1$; $h(2) = 5$; $h(3) = 3$.

(c) The new $\mathbf{c_B}$ is $\begin{bmatrix} 2 & 0 & 1 \end{bmatrix}$.

(d)

$$\mathbf{w} = \mathbf{c_B}P_2P_1 = \begin{bmatrix} 2 & 0 & 1 \end{bmatrix} \begin{bmatrix} \frac{1}{3} & 0 & \frac{1}{3} \\ \frac{1}{3} & 1 & -\frac{2}{3} \\ -\frac{2}{3} & 0 & \frac{1}{3} \end{bmatrix} = \begin{bmatrix} 0 & 0 & 1 \end{bmatrix}$$

(e)

$$z^* = c_B b^* = [2 \quad 0 \quad 1] \begin{bmatrix} 4 \\ 2 \\ 0 \end{bmatrix} = 8$$

The algorithm now requires us to set $s = 3$ and return to step 2.

STEP 2. Because $h(1) = 1$, $h(2) = 5$, and $h(3) = 3$, x_1, x_5, and x_3 are basic variables; so we do not need to calculate $-c_1^*$, $-c_5^*$, and $-c_3^*$. Because the last search for the entering variable ended at $p = 3$, the cyclic rule requires this search to begin at $p = 4$.

$$-c_4^* = w A_4 - c_4 = [0 \quad 0 \quad 1] \begin{bmatrix} 1 \\ 0 \\ 0 \end{bmatrix} - 0 = 0$$

$$-c_6^* = w A_6 - c_6 = [0 \quad 0 \quad 1] \begin{bmatrix} 0 \\ 0 \\ 1 \end{bmatrix} - 0 = 1$$

$$-c_2^* = w A_2 - c_2 = [0 \quad 0 \quad 1] \begin{bmatrix} 2 \\ 1 \\ 0 \end{bmatrix} - (-1) = 1$$

Since all entries in $-c^*$ are nonnegative, we have found the maximum solution, and the algorithm halts. The answer is $x_B = b^*$ and $z^* = 8$; or, explicitly, $x_1 = 4$, $x_5 = 2$, $x_3 = 0$, and the rest of the variables are 0 because they are nonbasic. ◆

We conclude this section by discussing an opportunity to improve the numerical accuracy of the revised simplex method. Because computers can carry only a limited number of decimal places of accuracy, calculations involving real numbers are approximate. Even though such approximations can be very accurate, the errors tend to accumulate, especially when an algorithm involves many subtractions (as the simplex method does). We will not make a study of the propagation of error in executing long computations; this subject is adequately covered in many numerical analysis texts and is not properly part of linear programming. In practice, error propagation does not usually invalidate answers obtained by the simplex method when problems are solved by a computer program that carries at least 16 significant digits in all calculations.

It is, however, useful to consider an interesting phenomenon of the revised simplex method. If, in the course of a calculation, we discover that too much error has accumulated, usually nothing can be done to salvage our work; we have to start the calculation again at the beginning and carry more accuracy. However, if we are executing the revised simplex algorithm and decide that so many calculations have been made that our B^{-1} matrix (stored as $P_1, P_2, \ldots, P_s$) contains unacceptable error, we can recover. The B^{-1} matrix is the inverse of the B matrix, and the B matrix is accurately known because it is a submatrix of the original tableau. We can return to the original tableau and calculate a more accurate version of B^{-1}. The description of the method follows.

Suppose the initial tableau for an LP problem in canonical form is

$$T = \left[\begin{array}{c|c} A & \mathbf{b} \\ \hline -\mathbf{c} & 0 \end{array} \right]$$

and the simplex tableau T^* and revised simplex tableau R^* for a later step are

$$T^* = \left[\begin{array}{c|c} A^* & \mathbf{b}^* \\ \hline -\mathbf{c}^* & z^* \end{array} \right]; \qquad R^* = \left[\begin{array}{c|c} B^{-1} & \mathbf{b}^* \\ \hline \mathbf{w} & z^* \end{array} \right]$$

We do not store T^* during execution of the revised method. However, we can stop the revised simplex method after any pivot and calculate T^*. Then we can *restart* the revised simplex algorithm at step 1, using T^* as the initial tableau (and a new B^{-1} formed from the unit columns of T^* that correspond to its basic variables). There is a problem: We are regarding B^{-1} as being too inaccurate to use, and so we cannot use it to calculate A^*. How can we calculate A^* from B, which is accurately known? Two ways come to mind:

1. Calculate B^{-1} more accurately using some other algorithm. Then we could compute A^* by premultiplying by this better version of B^{-1}.

2. Observe that $A^* = B^{-1}A$ is equivalent to $BA^* = A$. If we regard A^* as unknown and denote its columns by

$$A^* = [X_1 \,|\, X_2 \,|\cdots|\, X_n]$$

then

$$BA^* = [BX_1 \,|\, BX_2 \,|\cdots|\, BX_n]$$
$$= [A_1 \,|\, A_2 \,|\cdots|\, A_n]$$

So, instead of solving $BA^* = A$, we could solve the systems of equations

$$BX_j = A_j, \qquad \text{for } 1 \le j \le n \qquad \text{and} \qquad X_j \text{ nonbasic}$$

(We do not need to solve for the basic columns of A^* because we know exactly what they are.)

As any good text on numerical linear algebra will attest, alternative 2 is the more efficient choice.

The SOLVER program of CALIPSO solves LP problems with the revised simplex method algorithm just described. SOLVER stores its data in the linked list structures that we introduce in the next section.

Exercises

1. For each of Tableaus 1–4 of Example 3 of Section 4.3, give the corresponding pivoting matrix, injection, c_B, and b^*.

2. For each of Tableaus 1–3 of Example 1 of Section 4.4, give the corresponding pivoting matrix, injection, c_B, and b^*.

In Exercises 3–6 you are given the status of a computation using the Revised Simplex Algorithm after two pivots. Perform the next steps of the method to find the next (a) entering variable column, (b) departing variable row, (c) pivoting matrix P_3, (d) b^*, (e) w, and (f) z^*.

3. Suppose the initial tableau for an LP problem is

$$T = \left[\begin{array}{ccccccc|c} 2 & -3 & -1 & 1 & 0 & 0 & 1 \\ -4 & 6 & 4 & 0 & 1 & 0 & 13 \\ 1 & 3 & 1 & 0 & 0 & 1 & 5 \\ \hline -3 & -1 & -2 & 0 & 0 & 0 & 0 \end{array}\right]$$

and the associated injection is $h(1) = 4$, $h(2) = 5$, $h(3) = 6$.

Suppose we have performed two pivots of the revised simplex method to obtain the data

$$P_1 = \begin{bmatrix} \frac{1}{2} & 0 & 0 \\ 2 & 1 & 0 \\ -\frac{1}{2} & 0 & 1 \end{bmatrix}; \qquad P_2 = \begin{bmatrix} 1 & 0 & \frac{1}{3} \\ 0 & 1 & 0 \\ 0 & 0 & \frac{2}{9} \end{bmatrix}$$

$$h(1) = 1, \quad h(2) = 5, \quad h(3) = 2$$

$$b^* = \begin{bmatrix} 2 \\ 15 \\ 1 \end{bmatrix}; \qquad w = \begin{bmatrix} \frac{8}{9} & 0 & \frac{11}{9} \end{bmatrix}; \qquad z^* = 7$$

Column 3 was the last column chosen by the Cyclic Entering Basic Variable Rule.

4. Suppose the initial tableau for an LP problem is

$$T = \left[\begin{array}{ccccccc|c} 6 & 3 & 3 & 1 & 0 & 0 & 12 \\ 5 & -4 & 6 & 0 & 1 & 0 & 6 \\ 6 & 2 & 9 & 0 & 0 & 1 & 9 \\ \hline -12 & -6 & -9 & 0 & 0 & 0 & 0 \end{array}\right]$$

and the associated injection is $h(1) = 4$, $h(2) = 5$, $h(3) = 6$.

Suppose we have performed two pivots of the revised simplex method to obtain the data

$$P_1 = \begin{bmatrix} 1 & -\frac{6}{5} & 0 \\ 0 & \frac{1}{5} & 0 \\ 0 & -\frac{6}{5} & 1 \end{bmatrix}; \qquad P_2 = \begin{bmatrix} 1 & 0 & -\frac{39}{34} \\ 0 & 1 & \frac{4}{34} \\ 0 & 0 & \frac{5}{34} \end{bmatrix}$$

$$h(1) = 4, \quad h(2) = 1, \quad h(3) = 2$$

$$b^* = \begin{bmatrix} \frac{93}{34} \\ \frac{24}{17} \\ \frac{9}{34} \end{bmatrix}; \qquad w = \begin{bmatrix} 0 & -\frac{6}{17} & \frac{39}{17} \end{bmatrix}; \qquad z^* = \frac{315}{17}$$

Column 2 was the last column chosen by the Cyclic Entering Basic Variable Rule.

5. Suppose the initial tableau for an LP problem is

$$T = \begin{bmatrix} 12 & 6 & 5 & 1 & 0 & 0 & 0 & 6 \\ 6 & -3 & 0 & 0 & 1 & 0 & 0 & 9 \\ 9 & 3 & 4 & 0 & 0 & 0 & 1 & 8 \\ 18 & 6 & -4 & 0 & 0 & 1 & 0 & 12 \\ \hline -240 & -200 & -180 & 0 & 0 & 0 & 0 & 0 \end{bmatrix}$$

and the associated injection is $h(1) = 4$, $h(2) = 5$, $h(3) = 7$, $h(4) = 6$. Suppose we have performed two pivots of the revised simplex method to obtain

$$P_1 = \begin{bmatrix} \frac{1}{12} & 0 & 0 & 0 \\ -\frac{1}{2} & 1 & 0 & 0 \\ -\frac{3}{4} & 0 & 1 & 0 \\ -\frac{3}{2} & 0 & 0 & 1 \end{bmatrix}; \qquad P_2 = \begin{bmatrix} 2 & 0 & 0 & 0 \\ 12 & 1 & 0 & 0 \\ 3 & 0 & 1 & 0 \\ 6 & 0 & 0 & 1 \end{bmatrix}$$

$$h(1) = 2, \quad h(2) = 5, \quad h(3) = 7, \quad h(4) = 6$$

$$\mathbf{b}^* = \begin{bmatrix} 1 \\ 12 \\ 5 \\ 6 \end{bmatrix}; \qquad \mathbf{w} = \begin{bmatrix} \frac{100}{3} & 0 & 0 & 0 \end{bmatrix}; \qquad z^* = 200$$

Column 2 was the last column chosen by the Cyclic Entering Basic Variable Rule.

6. Suppose the initial tableau for an LP problem is

$$T = \begin{bmatrix} 0 & 6 & -3 & 2 & 1 & 0 & 0 & 8 \\ 1 & 12 & 6 & 4 & 0 & 0 & 0 & 4 \\ 0 & -8 & -8 & 8 & 0 & 0 & 1 & 12 \\ 0 & 3 & 4 & 6 & 0 & 1 & 0 & 10 \\ \hline 0 & -150 & -80 & -120 & 0 & 0 & 0 & 0 \end{bmatrix}$$

and the associated injection is $h(1) = 5$, $h(2) = 1$, $h(3) = 7$, $h(4) = 6$. Suppose we have performed two pivots of the revised simplex method to obtain the data

$$P_1 = \begin{bmatrix} 1 & -\frac{1}{2} & 0 & 0 \\ 0 & \frac{1}{12} & 0 & 0 \\ 0 & \frac{2}{3} & 1 & 0 \\ 0 & -\frac{1}{4} & 0 & 1 \end{bmatrix}; \qquad P_2 = \begin{bmatrix} 1 & 0 & 12 & 0 \\ 0 & 1 & 2 & 0 \\ 0 & 0 & 8 & 0 \\ 0 & 0 & -5 & 1 \end{bmatrix}$$

$$h(1) = 5, \quad h(2) = 3, \quad h(3) = 7, \quad h(4) = 6$$

$$\mathbf{b}^* = \begin{bmatrix} 10 \\ \frac{2}{3} \\ \frac{52}{3} \\ \frac{22}{3} \end{bmatrix}; \qquad \mathbf{w} = \begin{bmatrix} 0 & \frac{40}{3} & 0 & 0 \end{bmatrix}; \qquad z^* = \frac{160}{3}$$

Column 3 was the last column chosen by the Cyclic Entering Basic Variable Rule.

7. You may have noticed that in Exercises 1 and 2 the values of $\mathbf{w}$ occurred in the objective row of the final tableau in columns corresponding to columns of the B^{-1} matrix. Is this always the case? Explain why or why not.

In Exercises 8, 9, and 10 carry out the steps of the revised simplex method for the given LP problems in a manner similar to that in Example 2 of this section.

8. Example 3 of Section 4.3.   **9.** Example 1 of Section 4.4.

10. Example 1 of Section 3.4.

11. Solve the LP problem of Example 2 of this section by the ordinary simplex method. Compare the entries obtained by the revised simplex method in Example 2 with those obtained by the ordinary simplex method.

12. At each occurrence of step 5 in Example 2, compute the current value of B^{-1}.

13. Explain why it is correct to replace **b*** with *P***b*** in step 5 (b) of the Revised Simplex Algorithm.

†**14.** Write a computer program implementing the Revised Simplex Algorithm.

5.2   *PRELIMINARY RESULTS FOR USE IN COMPLEXITY ANALYSIS*

In this section we lay the background for the complexity analyses of Section 5.3 by discussing some matrix multiplication formulas and computer storage structures that improve the efficiency of the revised simplex method. We begin with a simple theorem about matrix multiplication.

Theorem 1.   If U is an $r \times s$ matrix and V is an $s \times t$ matrix, then calculating the product UV by the formulas defining matrix multiplication requires rst multiplications.

PROOF.   Let $W = UV$, and let w_{ij} be the entry in the i row and j column of W for $1 \le i \le r$ and $1 \le j \le t$. By the definition of multiplication,

$$w_{ij} = \sum_{k=1}^{s} u_{ik}v_{kj} = u_{i1}v_{1j} + u_{i2}v_{2j} + \cdots + u_{is}v_{sj}$$

Thus, it takes s multiplications to compute w_{ij}. Since there are rt of the w_{ij} to compute, we must perform a total of $(rt)s = rst$ multiplications.   ∎

EXAMPLE 1

1. If U is a 2×3 matrix and V is a 3×5 matrix, then calculating UV requires $2 \cdot 3 \cdot 5 = 30$ multiplications.

2. If U is a 1×3 matrix and V is a 3×1 matrix, then calculating UV requires $1 \cdot 3 \cdot 1 = 3$ multiplications.   ◆

The next two theorems and their corollaries enable us to reduce the time spent in multiplying by pivoting matrices.

Theorem 2.   Let P be an $m \times m$ pivoting matrix obtained by replacing column r of an identity matrix by some column of numbers, and let $\mathbf{g}$ be any $m \times 1$ matrix. Then, we can compute the product $P\mathbf{g}$ in m multiplications by the formula

$$P\mathbf{g} = \begin{bmatrix} 1 & 0 & \cdots & p_{1r} & \cdots & 0 \\ 0 & 1 & \cdots & p_{2r} & \cdots & 0 \\ \vdots & & & & & \\ 0 & 0 & \cdots & p_{rr} & \cdots & 0 \\ \vdots & & & & & \\ 0 & 0 & \cdots & p_{mr} & \cdots & 1 \end{bmatrix} \begin{bmatrix} g_1 \\ g_2 \\ \vdots \\ g_r \\ \vdots \\ g_m \end{bmatrix} = \begin{bmatrix} g_1 + p_{1r}g_r \\ g_2 + p_{2r}g_r \\ \vdots \\ p_{rr}g_r \\ \vdots \\ g_m + p_{mr}g_r \end{bmatrix} \tag{1}$$

PROOF.   The proof of this theorem follows from a direct application of the definition of multiplication.   ∎

As a corollary to Theorem 2, we describe how to premultiply an $m \times m$ matrix by a pivoting matrix.

Corollary.   Let P be the pivoting matrix defined in Theorem 2, and let G be any $m \times m$ matrix. Then we can compute PG in m^2 multiplications.

PROOF.   Let $G_1, G_2, \ldots, G_m$ be the columns of G. Then,

$$PG = [PG_1 \,|\, PG_2 \,|\, \cdots \,|\, PG_m]$$

We can compute each of the products PG_j with m multiplications using the formula given in Theorem 2.   ∎

EXAMPLE 2

We use the formula from Theorem 2 to compute

$$\begin{bmatrix} 1 & 5 & 0 & 0 \\ 0 & 6 & 0 & 0 \\ 0 & 7 & 1 & 0 \\ 0 & -2 & 0 & 1 \end{bmatrix} \begin{bmatrix} 2 \\ 3 \\ 4 \\ 5 \end{bmatrix} = \begin{bmatrix} 2 + 5 \cdot 3 \\ 6 \cdot 3 \\ 4 + 7 \cdot 3 \\ 5 - 2 \cdot 3 \end{bmatrix} = \begin{bmatrix} 17 \\ 18 \\ 25 \\ -1 \end{bmatrix}$$ ◆

Theorem 3.   If $\mathbf{h}$ is a $1 \times m$ matrix and P is an $m \times m$ pivoting matrix, then we can compute $\mathbf{h}P$ with m multiplications by the formula

$$[h_1 \quad h_2 \quad \cdots \quad h_m] \begin{bmatrix} 1 & 0 & \cdots & p_{1r} & \cdots & 0 \\ 0 & 1 & \cdots & p_{2r} & \cdots & 0 \\ \vdots & & & & & \\ 0 & 0 & \cdots & p_{rr} & \cdots & 0 \\ \vdots & & & & & \\ 0 & 0 & \cdots & p_{mr} & \cdots & 1 \end{bmatrix} = [h_1 \quad \cdots \quad \underset{\underset{\text{column } r}{\uparrow}}{\sum_{i=1}^{m} h_i p_{ir}} \quad \cdots \quad h_m]$$

PROOF. The proof follows immediately from the definition of matrix multiplication. ∎

Corollary. If $\mathbf{h}$ is a $1 \times m$ matrix, and $P_1, P_2, \ldots, P_s$ is a sequence of pivoting matrices, then we can compute $\mathbf{h}P_s \cdots P_2 P_1$ with sm multiplications.

EXAMPLE 3

We use the formula from Theorem 3 to compute

$$[2 \quad 3 \quad 4]\begin{bmatrix} 1 & 5 & 0 \\ 0 & 6 & 0 \\ 0 & 7 & 1 \end{bmatrix} = [2 \quad 2 \cdot 5 + 3 \cdot 6 + 4 \cdot 7 \quad 4] = [2 \quad 56 \quad 4] \qquad \blacklozenge$$

The next definition introduces a parameter that is useful in discussing matrices containing many zeros.

Definition. The *density* of a matrix is the ratio of the number of nonzero entries in the matrix to the total number of entries in the matrix.

EXAMPLE 4

The matrix

$$\begin{bmatrix} 5 & 1 & 3 & 0 & 1 & 0 \\ 2 & 1 & 4 & 1 & 0 & 0 \\ 1 & 3 & 0 & 0 & 0 & 1 \end{bmatrix}$$

has 11 nonzero entries and 7 zero entries. Its density is $\frac{11}{18}$. $\qquad \blacklozenge$

We now turn to the question of storing matrices that have many zeros in computer memory. As you will see, using well-designed storage structures conserves storage and enhances the speed of matrix multiplications.

Definition. A matrix that has a low density is said to be *sparse*.

Sparse is not an exact term because the meaning of *low density* is subject to interpretation. Nevertheless, it is a useful term when we want to call attention to the fact that a matrix has a lot of zero entries. Commonly occurring LP problems often have very sparse tableaus because of the zeros introduced in adding slack variables and because many of the decision variables do not occur in some constraints. Example 5 of Section 3.2 is a modest example of a sparse matrix; the coefficient matrix in Tableau 1 of that example has density $\frac{2}{5}$. We will see more dramatic examples in the sequel.

Most computer science texts on the subject of "data structures" present a variety of schemes to avoid storing the zero entries of matrices. Which scheme is best for a given application depends on how the entries of the matrices are usually accessed. As we have seen, the calculations of the revised simplex method operate on the initial tableau by columns; therefore, we will introduce a storage structure, called a *linked list*, that makes it easy to access our data by columns.

A *node* is a piece of computer memory that can contain several items of information. For our purpose, we want nodes that can contain three entries: one to hold an entry from the A matrix of a tableau, one to hold a row number, and one to hold a pointer to the next node in a list of nodes. It is not necessary to understand what we mean by a "pointer" in a technical sense. A pointer is just the memory address of the next node in the list; however, you may simply regard it as some mysterious way of getting from one node to the next. We depict a node as follows:

$$\begin{bmatrix} v \\ r \\ \ \end{bmatrix} \begin{matrix} \leftarrow v, \text{ the entry of the array to be stored} \\ \leftarrow r, \text{ a row number} \\ \leftarrow \text{place for the pointer to the next node} \end{matrix}$$

We indicate the end of a list representing a column of a matrix by omitting the arrow leaving the pointer box. These ideas sound complicated, but the next example will show you that it is all very simple.

EXAMPLE 5

Figure 1 gives the linked-list structure for storing the matrix

$$H = \begin{bmatrix} 1 & 2 & 3 & 1 & 0 & 0 \\ 4 & 0 & 6 & 0 & 1 & 0 \\ 7 & 5 & 9 & 0 & 0 & 1 \end{bmatrix}$$

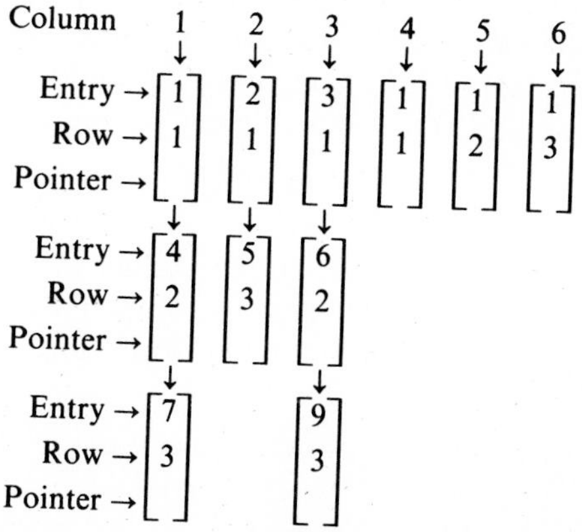

Figure 1

◆

Admittedly, this example does not save any storage. To store 18 numbers, the linked list requires storing 11 nodes, each containing three fields of information. Even so, when we are dealing with large matrices containing many zeros, the savings can be dramatic.

The density of the A matrix of a tableau depends on the problem under consideration. Some of the exercises of this chapter illustrate the fact that many practical problems have very low-density initial tableaus (densities as low as 0.05 are common). It is more difficult to show that B^{-1} matrices for such problems are also likely to have low densities. At the initial step of the simplex method, $B^{-1} = I_m$, and so $d_B = 1/m$ (a small value for large m). Empirical evidence shows that the density of B^{-1} usually remains low during pivoting when the density of A is low; however, it is not true that the density of B^{-1} *must* remain low.

The following theorem illustrates the value of linked-list storage in evaluating an expression such as $c_B B^{-1}$ when B^{-1} is sparse.

Theorem 4.   Suppose g is a $1 \times m$ matrix and H is an $m \times n$ matrix that is stored as a linked list (as in Example 5). If the density of H is d_H, then gH can be evaluated in $d_H mn$ multiplications.

Instead of writing a formal proof, we present an example that fully illustrates the general proof.

___

EXAMPLE 6

Matrix H of Example 5 has density $d_H = \frac{12}{18}$. Let $\mathbf{g} = [g_1 \quad g_2 \quad g_3]$, and let $\mathbf{f} = \mathbf{g}H$. For $1 \le i \le 6$,

$$f_j = g_1 h_{1j} + g_2 h_{2j} + g_3 h_{3j}$$

Consider the calculation of f_2. The second column of H is stored in a linked list containing two nodes. A computer algorithm to evaluate f_2 starts by examining the first node. It discovers that the value 2 is to be regarded as the entry from row 1, and so it multiplies this number by g_1 to obtain $2g_1$. Since the node contains a pointer, the program follows the pointer and examines the second node. There it discovers that the value 5 is to be regarded as being in row 3, and so it multiplies g_3 by 5 to obtain $5g_3$. This value is added to the previously computed value to obtain

$$f_2 = 2g_1 + 5g_3$$

Since the second node is marked as the last on the list, the calculation is complete.

The other five entries in f are calculated in a similar way. Notice that we make only one multiplication for each node (or, equivalently, for each nonzero entry). The density d_H is

$$d_H = \frac{\text{number of nonzero entries}}{mn} = \frac{12}{18}$$

So, the number of multiplications performed is $d_H mn = \left(\frac{12}{18}\right) \cdot 3 \cdot 6 = 12$.   ◆

___

In the next section we streamline the revised simplex method by using the calculation technique illustrated in Example 6 and the formulas given in Theorems 2 and 3.

EXERCISES

1. How many multiplications are required to compute UV when
   (a) U is 4×5 and V is 5×7?
   (b) U is 3×1 and V is 1×4?

2. How many multiplications are required to compute UV when
   (a) U is 1×4 and V is 4×5?
   (b) U is 6×2 and V is 2×1?

3. Suppose U, V, and W have sizes $r \times s$, $s \times t$, and $t \times q$, respectively. Explain why calculating $(UV)W$ requires $rst + rtq$ multiplications. How many multiplications are required to calculate $U(VW)$?

In Exercises 4–9, use Theorems 2 and 3 and their corollaries to compute the matrix products.

4. $\begin{bmatrix} 1 & 0 & 2 & 0 \\ 0 & 1 & 4 & 0 \\ 0 & 0 & 3 & 0 \\ 0 & 0 & 0 & 1 \end{bmatrix} \begin{bmatrix} 1 \\ 2 \\ 3 \\ 4 \end{bmatrix}$

5. $\begin{bmatrix} 1 & 2 & 0 & 0 & 0 & 0 \\ 0 & -1 & 0 & 0 & 0 & 0 \\ 0 & 3 & 1 & 0 & 0 & 0 \\ 0 & 2 & 0 & 1 & 0 & 0 \\ 0 & -2 & 0 & 0 & 1 & 0 \\ 0 & 1 & 0 & 0 & 0 & 1 \end{bmatrix} \begin{bmatrix} 0 \\ 2 \\ 1 \\ 4 \\ 1 \\ 3 \end{bmatrix}$

6. $\begin{bmatrix} 1 & 2 & 3 & 4 \end{bmatrix} \begin{bmatrix} 1 & 0 & 2 & 0 \\ 0 & 1 & 4 & 0 \\ 0 & 0 & 3 & 0 \\ 0 & 0 & 0 & 1 \end{bmatrix}$

7. $\begin{bmatrix} 0 & 2 & 1 & 4 & 1 & 3 \end{bmatrix} \begin{bmatrix} 1 & 2 & 0 & 0 & 0 & 0 \\ 0 & -1 & 0 & 0 & 0 & 0 \\ 0 & 3 & 1 & 0 & 0 & 0 \\ 0 & 2 & 0 & 1 & 0 & 0 \\ 0 & -2 & 0 & 0 & 1 & 0 \\ 0 & 1 & 0 & 0 & 0 & 1 \end{bmatrix}$

8. $\begin{bmatrix} 1 & 0 & 2 \\ 0 & 1 & -1 \\ 0 & 0 & 2 \end{bmatrix} \begin{bmatrix} 1 & 2 & 3 \\ 4 & 5 & 6 \\ 7 & 8 & 9 \end{bmatrix}$

9. $\begin{bmatrix} 1 & 2 & 3 & 1 \\ 4 & 5 & 6 & 2 \\ 3 & 2 & 1 & 1 \\ 2 & 0 & 1 & 2 \end{bmatrix} \begin{bmatrix} 2 & 0 & 0 & 0 \\ 3 & 1 & 0 & 0 \\ 1 & 0 & 1 & 0 \\ 4 & 0 & 0 & 1 \end{bmatrix}$

10. Express the following matrices as linked list, as in Figure 1 of Example 5.

   (a) $\begin{bmatrix} 1 & 0 & 0 & 2 & 3 & 0 \\ 2 & 0 & 1 & 0 & 1 & 1 \\ 0 & 0 & 0 & 0 & 2 & 0 \\ -1 & 1 & 1 & 0 & 4 & 0 \end{bmatrix}$

   (b) $\begin{bmatrix} 1 & 0 & 1 & 2 \\ 0 & 0 & 0 & 0 \\ 0 & 1 & 0 & 0 \\ 0 & 0 & 0 & 1 \\ 0 & 2 & 0 & 0 \\ 0 & 0 & 1 & 0 \end{bmatrix}$

11. Express the following matrices as linked lists, as in Figure 1 of Example 5.

(a) $\begin{bmatrix} 0 & 0 & 0 & 9 & -2 & 0 \\ 0 & 2 & 1 & 0 & 1 & 1 \\ 0 & 0 & 1 & 0 & 0 & 0 \\ -1 & 1 & 1 & 0 & 0 & 0 \end{bmatrix}$ (b) $\begin{bmatrix} 1 & 0 & 1 & 1 \\ 0 & 0 & 0 & 1 \\ 0 & 0 & 0 & 0 \\ 0 & 0 & 1 & 1 \\ 0 & 1 & 0 & 0 \\ 0 & 0 & 1 & 0 \end{bmatrix}$

In Exercises 12–14, construct a pivoting matrix (in the linked-list format given in Example 5) that you can use to obtain the next tableau from the given tableau.

12. Tableau 2 of Example 1 of Section 4.4.

13. Tableau 3 of Example 3 of Section 4.3.

14. Tableau 1 of Example 2 of Section 3.5.

15. What are the densities of the following matrices?
    (a) A and B of Example 4 in Section 4.1.
    (b) The matrix of Exercise 13 of Section 4.2.

16. What are the densities of the following matrices?
    (a) The A matrix corresponding to the transportation problem of Example 1 of Section 1.2.
    (b) The A matrix of Example 2 of Section 3.4.

17. What is the density (in terms of m and n) of the matrix for the assignment problem (3) in Section 1.2?

5.3   SPACE AND TIME COMPLEXITY OF THE SIMPLEX METHOD

Three factors—space, time, and accuracy—limit the capability of computers to solve large problems. The study of these factors is the main focus of the discipline called *analysis of algorithms*. An *algorithm* can be viewed as an explicit description of a sequence of instructions translatable into a computer program to solve a problem in a finite number of steps. The simplex method is a good example of an algorithm. We introduce a few ideas from the analysis of algorithms below.

Space Complexity

Every computer has a component called memory for storing numbers or other information. It is possible to make computers with very large memories, larger than any practical LP problem is likely to require; even so, memory is always a limited resource and a relatively expensive commodity. On a practical level, it is desirable to develop programs to solve LP problems that will run on small personal computers because many enterprises that use linear programming in their operations would rather not buy large, expensive machines. Thus, we set the minimization of memory usage as a primary goal in developing a computationally efficient simplex algorithm.

In the sequel we want to compare the memory requirements of some algorithms. We do not attempt to describe the exact memory requirements of the algorithms because this quantity is dependent on the intricacies of the computer program that implements the algorithm. Instead, we will estimate the major memory requirements by a quantity called space complexity. The *space complexity* of an algorithm is an approximate count of how many numbers must be stored in executing the algorithm. For example, in using the ordinary simplex method to solve an LP problem with n variables and m constraints, we must store the A matrix of coefficients from the constraint equations, the resource column, and the objective row. We will say that such an algorithm has space complexity $(m + 1)(n + 1)$. Certainly it is true that a program executing the simplex method would need to store more than $(m + 1)(n + 1)$ numbers; nevertheless, at least for large values of m and n, the A matrix occupies the bulk of the space required.

Time Complexity

Computers perform operations very quickly; a multiplication requires only a few millionths of a second. Nevertheless, the simplex method requires a stupendous number of operations in solving large problems, and the total computer execution time can be impressive. In fact, execution time is usually the limiting factor in determining how large a problem a particular computer can solve with a particular program. Therefore, we pay most attention to reducing the time required to solve an LP problem.

Making a careful account of the time required to perform each operation for a complex task such as executing the simplex method is difficult and not very rewarding. For purposes of comparing the efficiency of two algorithms that solve the same problem, it is usually enough to count the number of occurrences of a few (often only one or two) operations that occur repeatedly. This number of occurrences is called the *time complexity* of the algorithm. In studying the simplex method, you would probably choose to count only the number of multiplications and divisions as a measure of the time complexity. This is reasonable because approximately the same number of multiplications and divisions are performed as additions and subtractions. The total time required to solve the problem would be approximately some fixed multiple of the time complexity. The exact fixed multiple depends on physical properties of the computer in use and is of little interest in comparing the time requirements of two algorithms for solving the same problem on the same machine. The algorithm with the smaller time complexity is the more efficient one.

Space and Time Complexity of the Ordinary Simplex Method

The only large storage structure is the tableau T. The pivot operation simply changes the values of T in memory. Thus, the space complexity is $(m + 1)(n + 1)$.

For most purposes we could express the time complexity adequately by counting the multiplications and divisions required to perform one pivot; however, the revised simplex method incurs significant time costs in selecting the entering and departing basic variables. So, for the sake of fair comparisons, we include the minor costs of the m divisions used in selecting the departing basic variable.

In performing a pivot, we first divide the pivot row by the pivot element. This takes $n + 1$ divisions (one is for the resource value). Next, we multiply m rows (the objective row and the remaining $m - 1$ rows corresponding to constraints) by the element in their pivot column; this requires $m(n + 1)$ multiplications. Thus, selecting the departing basic variable and performing one pivot requires

$$m + (m + 1)(n + 1) \qquad (1)$$

multiplications or divisions.

Time Complexity of the Revised Simplex Method

We now analyze the time complexity of each step of the Revised Simplex Algorithm given in Section 5.1. Since the complexity increases as more pivots are made, we make the analysis for the sth pivot, for any $s \geq 1$.

STEP 2. We compute $\mathbf{w}A_k$ as the ordinary matrix product of a $1 \times m$ and an $m \times 1$ matrix. The complexity is $(1)(m)(1) = m$.

The fact that we do not know how many times we must calculate $\mathbf{w}A_k$ makes complexity analysis difficult. For purposes of estimating the overall complexity of solving a problem with the revised simplex method, we assume that, on the average, we make $\mu = (n - m)/2$ calculations of $\mathbf{w}A_k$ at each pivot (there are at most $n - m$ values of $\mathbf{w}A_k$ to calculate because we know that the m objective-row entries corresponding to basic variables are 0). Since the actual number of calculations at each pivot ranges from 1 to $n - m$, this estimate is reasonable, at least for a problem requiring many pivots. Thus, we take the complexity of step 2 to be $\mu m = m(n - m)/2$.

STEP 3. We use Theorem 2 of Section 5.2 to premultiply $m \times 1$ matrices by pivoting matrices. Each such premultiplication requires m multiplications. To calculate $\mathbf{y} = P_s \cdots P_2 P_1 A_p$, we start multiplying pairs of $m \times m$ and $m \times 1$ matrices from the right of the expression, as follows:

$$Q_1 = P_1 A_p \quad \text{is an} \quad m \times 1 \quad \text{matrix}$$
$$Q_2 = P_2 Q_1 \quad \text{is an} \quad m \times 1 \quad \text{matrix}$$
$$\vdots$$
$$Q_{s-1} = P_{s-1} Q_{s-2} \quad \text{is an} \quad m \times 1 \quad \text{matrix}$$
$$\mathbf{y} = Q_{s-1}$$

These steps require $s - 1$ premultiplications. We also perform m divisions to determine the departing basic variable. Thus, the complexity of step 3 is $m(s - 1) + m = sm$

STEP 4. We make m divisions to produce L.

STEP 5.

(a) We use Theorem 2 of Section 5.2 to calculate $P_s \mathbf{b}^*$ with m multiplications.

(b) This step uses no multiplications.

(c) This step uses no multiplications.

(d) We use Theorem 3 of Section 5.2 to postmultiply $1 \times m$ matrices by pivoting matrices. Each such step requires m multiplications to compute

$$\mathbf{w} = \mathbf{c}_B P_s \cdots P_2 P_1$$

We start multiplying pairs of $1 \times m$ and $m \times m$ matrices from the left, as follows:

$$Q_1 = \mathbf{c}_B P_s \quad \text{is a} \quad 1 \times m \quad \text{matrix}$$

$$Q_2 = Q_1 P_{s-1} \quad \text{is a} \quad 1 \times m \quad \text{matrix}$$

$$\vdots$$

$$Q_s = Q_{s-1} P_1 \quad \text{is a} \quad 1 \times m \quad \text{matrix}$$

$$\mathbf{w} = Q_s$$

Since these steps require s postmultiplications, the time complexity of this calculation is sm.

(e) We use ordinary matrix multiplication to multiply a $1 \times m$ matrix by an $m \times 1$ matrix. This requires m multiplications.

Adding the complexities of these steps, we conclude that the time complexity of the sth pivot of the revised simplex method is

$$(2s + \mu + 3)m \tag{2}$$

where $\mu = (n - m)/2$.

Since the complexity of a pivot by this method depends on the number of pivots previously performed, we cannot usefully compare the complexities of the ordinary simplex method and this method for a single pivot. We can, however, compare their efficiencies in completely solving an LP problem. Both methods make the same choices of pivots and perform the same number of pivots to solve a problem (when both methods use the cyclic entering basic variable rule). It is easy to compute the complexity of several pivots by the ordinary simplex method because all pivots have the same complexity. The following lemma will help with a similar calculation for the revised simplex method.

Lemma. Let q be any positive integer. Then,

$$\sum_{s=1}^{q} s = 1 + 2 + \cdots + q = \frac{q(q+1)}{2}$$

PROOF. Let

$$S = 1 + 2 + \cdots + (q - 1) + q$$

By reversing the order of addition,

$$S = q + (q - 1) + \cdots + 2 + 1$$

Adding corresponding terms of these two equations gives

$$2S = (q + 1) + (q + 1) + \cdots + (q + 1) + (q + 1) = q(q + 1)$$

The result follows by dividing through by 2.   ∎

The complexity of q pivots of the revised simplex method is

$$\sum_{s=1}^{q} (2s + \mu + 3)m = 2m \sum_{s=1}^{q} s + (\mu + 3) \sum_{s=1}^{q} m = 2m \sum_{s=1}^{q} s + (\mu + 3)qm$$

$$= q(q + 1)m + (\mu + 3)qm = (q^2 + 4q)m + \frac{1}{2}qm(n - m) \quad (3)$$

where $\mu = (n - m)/2$.

We obtain the complexity of q pivots of the ordinary simplex method by multiplying the complexity of one pivot by q. Thus, from (1), this complexity is

$$q[m + (m + 1)(n + 1)] \quad (4)$$

The following example compares (3) and (4) under the assumption that $n = 2m$, an assumption that is often approximately correct for many practical problems because a slack variable is often added for each constraint.

EXAMPLE 1

Suppose we are solving an LP problem with $m = 100$ constraints and $n = 200$ variables. Formula (3) gives the time complexity of q pivots of the revised method as

$$100(q^2 + 4q) + 5000q \quad (5)$$

and (4) gives the time complexity of the ordinary method as

$$[100 + (101)(201)]q = (20401)q \quad (6)$$

Table 1 compares (5) and (6) for several values of q.

Table 1   Comparisons, without considering density

Number of pivots q	Number of multiplications	
	Revised simplex method	Ordinary simplex method
10	64,000	204,010
20	148,000	408,020
100	1,540,000	2,040,100
200	5,080,100	4,080,200

                                                                                        ◆

EXAMPLE 2

We can compare the complexity formulas (3) and (4) in another way by asking how many pivots q we can make before the ordinary method appears more efficient than the revised method. If we assume that $n = 2m$ and that m and q are fairly large numbers, then we can approximate (3) and (4) by

$$(q^2 + 4q)m + \frac{1}{2}qm(n - m) \approx q^2 m + \frac{1}{2}qm^2 \tag{7}$$

$$[m + (m + 1)(2m + 1)]q \approx (2m^2)q \tag{8}$$

because the cubic terms ($m^2 q$, qmn, qm^2, etc.) are very much larger than the others. For example, in the calculation of Example 1, the cubic terms contributed the millions and the other terms the thousands. From (7) and (8), we see that the revised simplex method is more efficient than the ordinary simplex method as long as

$$q^2 m + \frac{1}{2}qm^2 \leq 2qm^2 \tag{9}$$

Combining the qm^2 terms of (9) and dividing both sides by qm gives

$$q \leq \frac{3}{2}m$$

Thus, we conclude that if m is large, and there are about twice as many variables in the problem as there are constraints, then the revised simplex method is more efficient for the first $\frac{3}{2}m$ pivots required to find a solution.   ◆

The advantage displayed by Examples 1 and 2 of the revised over the ordinary simplex method is not that impressive. However, we have yet to consider the density of the matrices. For problems that have low-density matrices, the advantage of the revised method is very great.

Time Complexity for Sparse Problems

We showed in Example 6 of Section 5.2 that we perform products only of nonzero entries when we are multiplying matrices stored as linked lists. Thus, if the A matrix of a tableau has density d, and the pivoting matrices (stored as linked lists) have average density d, then we can multiply formula (3) by d to obtain the time complexity of solution by the revised simplex method. Since the density of most large problems is small (commonly, 0.05 to 0.25), we expect the revised simplex method to perform far better than the ordinary method on most large problems.

EXAMPLE 3

We continue Example 1 with the additional assumption that the LP problem has density $d = 0.1$. We can multiply the calculated time complexity for the revised simplex method by 0.1. The complexity of the ordinary simplex method is unchanged. Table 2 compares the performance of the two methods.

Table 2 **Comparisons, when considering density**

Number of pivots q	Number of multiplications	
	Revised simplex method	Ordinary simplex method
10	6,400	204,010
20	14,800	408,020
100	154,000	2,040,100
200	508,000	4,080,200

It is reasonable to ask if we could make improvements in the performance of the ordinary simplex method for low-density matrices. Unfortunately, the prospects are not good. We are able to improve performance with the revised simplex method primarily because all calculations are made by columns, and so we can reduce the complexity by storing columns as linked lists. The ordinary method accesses the A matrix by both row and column (by row for pivoting operations and by column for departing variable calculations). You would probably need to store the A matrix in a doubly linked list (a list in which each node contains two pointers, one to the next row and one to the next column) to make an appreciable improvement in time complexity. However, the overhead cost of maintaining such a doubly linked list for the many insertions and deletions required in the simplex method would probably consume more time than would be saved by reducing the number of multiplications.

Space Complexity

We must store the initial A matrix (m rows and n columns) as well as the pivoting matrices created at each iteration of the revised simplex method. Usually, the A matrix takes more storage than the pivoting matrices because we must store mn numbers. It is possible to store this matrix on an external medium, such as a disk, when solving a large problem and to bring in one column at a time as needed. Unfortunately, such a technique seriously degrades execution time because input from an external medium is a slow process compared with accessing data in memory.

For a large value of m, it would take a huge amount of storage to contain all of the $m \times m$ pivoting matrices needed to regenerate B^{-1}. Fortunately, a pivoting matrix differs from the identity matrix in just one column. So all we need to store are the contents of one column and the number of the column of the identity it occupies.

Actually, we can do even better. If the initial A matrix is sparse, we expect the pivoting matrices to be sparse also. Therefore, it is appropriate to store them as linked lists.

EXAMPLE 4

Consider the pivoting matrix

$$P = \begin{bmatrix} 1 & 0 & -1 & 0 \\ 0 & 1 & 0 & 0 \\ 0 & 0 & 2 & 0 \\ 0 & 0 & 1 & 1 \end{bmatrix}$$

Define a node structure with three fields:

entry	row number	pointer

In addition to the entries in the third column of P, we need to record the fact that it is the third column that contains these entries. We will store the location of the non-unit column in the first node of our linked list by putting a 3 in the entry field and a blank in the row field. Here is a picture of the linked list for P:

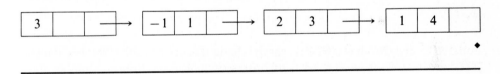

The principal structures that we must store to implement q pivots of the revised simplex method with the product form of B^{-1} are the pivoting matrices $P_1, P_2, \ldots, P_q$ and the initial tableau T. If we assume that the density of A is d_A and the average density of the pivoting matrices is d_B, then the number of nodes required to store these matrices in linked lists is $d_A(m + 1)(n + 1) + d_B qm$.

Summary of Complexity of the Ordinary and Revised Simplex Methods

Suppose that the A matrix has density d and the pivoting matrices have average density d. Table 3 gives the complexities of the simplex method for a problem requiring q pivots.

Table 3

Complexity	Ordinary simplex method	Revised simplex method
Space	$(m + 1)(n + 1)$	$d(m + 1)(n + 1) + dqm$
Time	$q[m + (m + 1)(n + 1)]$	$d[(q^2 + 4q)m + \frac{1}{2}qm(n - m)]$

Exercises

1. Compute $1 + 2 + \cdots + 100$.

2. Compute $\Sigma_{r=1}^{200}\, r$.

In Exercises 3–6, find the time and space complexity required to solve an LP problem in canonical form of the given size (m and n), density d, and number of pivots q by (a) the ordinary simplex method and (b) the revised simplex method with product form of the inverse. In part (b), let $\mu = (n - m)/2$.

3. $m = 50$, $n = 80$, $q = 25$, and $d = 0.15$. **5.** $m = 500$, $n = 1000$, $q = 100$, and $d = 0.10$.

4. $m = 150$, $n = 230$, $q = 50$, and $d = 0.35$. **6.** $m = 500$, $n = 1000$, $q = 250$, and $d = 0.50$.

5.4 LP MODELS

We now consider some practical problems that we can model as LP problems and solve with the revised simplex method. We illustrate the solutions of these problems using the MODELER and SOLVER programs of CALIPSO. The SOLVER program uses the Revised Simplex Algorithm with the product form of the inverse that we developed in Section 5.1.

EXAMPLE 1

Potlatch Timber Products (PTP) owns the Coeur d'Alene and the Nez Perce forests. PTP harvests trees from its forests and processes some of the trees into lumber at its sawmill in Potlatch, processes some into paper at its mill in Lewiston, and exports some to Japan. PTP cuts the trees into 20-foot logs in the forest and grades the logs into three classes—grades 1, 2, and 3—according to the quality of lumber the logs will provide. The grades of lumber are grade A (clear of knots), grade B (a few knots), and grade C (many knots). Some lumber of each grade results from processing each grade of log; however, grade-1 logs yield mostly grade A lumber, grade-2 logs yield mostly grade B lumber, and grade-3 logs yield mostly grade C lumber. Logs of grades 1 and 2 can be exported to Japan, logs of grades 2 and 3 can be made into paper at the Lewiston papermill, and logs of any grade can be sawed into lumber at the Potlatch sawmill. Table 1 gives the average composition of logs from each forest.

Table 1   Composition of logs

Logs	Coeur d'Alene forest (%)	Nez Perce forest (%)
Grade 1	45	54
Grade 2	35	30
Grade 3	20	16

The volume of logs cut from the forests are measured in units of 100 cubic feet (hcf). The Coeur d'Alene forest produces an average of 125 hcf per day, and the Nez Perce forest an average of 200 hcf per day. To model the supply of logs to the three destinations—sawmill, papermill, and export—we designate the Coeur d'Alene forest as forest 1 and the Nez Perce as forest 2, and we introduce the following variables:

$$M_{ij} = \text{hcf of grade-}i \text{ logs sent from forest } j \text{ to the sawmill}$$

$$P_{ij} = \text{hcf of grade-}i \text{ logs sent from forest } j \text{ to the papermill}$$

$$E_{ij} = \text{hcf of grade-}i \text{ logs from forest } j \text{ used for export}$$

By Table 1, the volume of logs of grade 1 produced daily by forest 1 is 45% of 125 or $(0.45) \cdot 125 = 56.25$. The equation

$$M_{11} + P_{11} + E_{11} = 56.25$$

models the disposition of these grade-1 logs. In a similar way, we can obtain the following six equations, which model the disposition of each grade from each forest.

$$
\begin{aligned}
M_{11} + P_{11} + E_{11} &= 56.25 \\
M_{21} + P_{21} + E_{21} &= 43.75 \\
M_{31} + P_{31} + E_{31} &= 25.00 \\
M_{12} + P_{12} + E_{12} &= 108.00 \\
M_{22} + P_{22} + E_{22} &= 60.00 \\
M_{32} + P_{32} + E_{32} &= 32.00
\end{aligned}
\tag{1}
$$

Table 2 gives the cost of transporting 1 hcf of logs from each forest to each mill.

Table 2 Transportation costs

Forest	Potlatch sawmill ($)	Lewiston papermill or export dump ($)
Coeur d'Alene	4	7
Nez Perce	3	4.50

We will maximize an objective function that gives the profit of the enterprise as the difference between money realized from the sale of products and the costs of operations. The transportation costs given in Table 2 will contribute the following amount:

$$
\begin{aligned}
&-4(M_{11} + M_{21} + M_{31}) - 7(P_{11} + P_{21} + P_{31}) - 7(E_{11} + E_{12} + E_{13}) \\
&- 3(M_{12} + M_{22} + M_{32}) - 4.5(P_{12} + P_{22} + P_{32}) - 4.5(E_{12} + E_{22} + E_{32})
\end{aligned}
\tag{2}
$$

Lumber is measured in units of 1000 board feet (kbf). Table 3 gives the average conversion factors (kbf/hcf) for logs at the Potlatch sawmill and the cost of processing 1 hcf of logs.

Table 3   Conversion factors at sawmill

Grade of logs	Lumber (kbf/hcf)	Scraps (hcf/hcf)	Sawdust (hcf/hcf)	Processing time (hr/hcf)
1	0.70	0.28	0.14	0.12
2	0.65	0.30	0.15	0.17
3	0.55	0.41	0.16	0.26

Table 4 gives the average yield of each grade of lumber for each grade of logs from each forest.

Table 4   Lumber yields by grades

Forest	Grade of logs	Grade A lumber (%)	Grade B lumber (%)	Grade C lumber (%)
Coeur d'Alene	1	36	48	16
	2	10	20	70
	3	0	15	85
Nez Perce	1	30	39	31
	2	4	12	84
	3	1	5	94

From Tables 3 and 4 we can compute the amounts of lumber of Grades A, B, and C, the amount of scrap, and the amount of sawdust produced by the sawmill. For instance, 1 hcf of logs cut from forest 1 yields 0.70 kbf of lumber, of which 36% is Grade A, 48% is grade B, and 16% is grade C. Thus, 1 hcf of logs from forest 1 yields $(0.70) \cdot (0.36) = 0.252$ kbf of grade A lumber, $(0.70) \cdot (0.48) = 0.336$ kbf of grade B lumber, $(0.70) \cdot (0.16) = 0.112$ kbf of grade C lumber, 0.28 hcf of scrap, and 0.14 hcf of sawdust. In a similar way you can verify that Table 5 gives the amounts of lumber produced by the sawmill from each grade and each forest.

Table 5   Amount of lumber produced

Forest	Grade of logs	Grade A (kbf)	Grade B (kbf)	Grade C (kbf)	Scrap (hcf)	Sawdust (hcf)
1	1	0.252	0.336	0.116	0.28	0.14
1	2	0.065	0.130	0.455	0.30	0.15
1	3	0	0.0825	0.4675	0.41	0.16
2	1	0.210	0.273	0.217	0.28	0.14
2	2	0.026	0.078	0.546	0.30	0.15
2	3	0.0055	0.0275	0.517	0.41	0.16

If we denote the total amounts of grade A, grade B, grade C, scrap, and sawdust produced at the sawmill by G_1, G_2, G_3, G_4, and G_5, respectively, then

$$G_1 = 0.252M_{11} + 0.065M_{21} + 0.210M_{12} + 0.026M_{22} + 0.0055M_{32}$$
$$G_2 = 0.336M_{11} + 0.130M_{21} + 0.0825M_{31} + 0.273M_{12} + 0.078M_{22} + 0.0275M_{32}$$
$$G_3 = 0.116M_{11} + 0.455M_{21} + 0.4675M_{31} + 0.217M_{12} + 0.546M_{22} + 0.517M_{32} \qquad (3)$$
$$G_4 = 0.280M_{11} + 0.300M_{21} + 0.410M_{31} + 0.280M_{12} + 0.300M_{22} + 0.410M_{32}$$
$$G_5 = 0.140M_{11} + 0.150M_{21} + 0.160M_{31} + 0.140M_{12} + 0.150M_{22} + 0.160M_{32}$$

The sawmill operates 24 hours per day. The times to process each grade from Table 3 give the constraint

$$0.12M_{11} + 0.17M_{21} + 0.26M_{31} + 0.12M_{12} + 0.17M_{22} + 0.26M_{32} \leq 24 \qquad (4)$$

The wholesale prices of lumber at the sawmill are \$200/kbf for grade A, \$150/kbf for grade B, and \$90/kbf for grade C. Scraps from the sawmill are trucked to the papermill at Lewiston at a cost of \$5/hcf. Sawdust is used to generate heat at the sawmill and saves \$10/hcf in heating costs. The contributions to the objective function from these incomes and costs are

$$200G_1 + 150G_2 + 90G_3 - 5G_4 + 10G_5 \qquad (5)$$

The Lewiston papermill produces 75 thousand square feet (tsf) of paper for each 1 hcf of logs or scrap. The papermill has the capacity to produce 12,000 tsf of paper each day, and each tsf of paper sells for \$15. Thus, the amount of paper P produced each day is

$$P = 75(P_{11} + P_{21} + P_{31} + P_{12} + P_{22} + P_{32} + G_4) \qquad (6)$$

The capacity of the papermill is restricted by

$$P \leq 12000 \qquad (7)$$

To meet daily customer demands, PTP must produce at least 35 kbf of grade A lumber, 45 kbf of grade B lumber, 40 kbf of grade C lumber, and 9600 tsf of paper. The constraints corresponding to these minimal demands are

$$G_1 \geq 35$$
$$G_2 \geq 45$$
$$G_3 \geq 40 \qquad (8)$$
$$P \geq 9600$$

This completes the description of the PTP problem. You will notice that many obvious operational costs (such as salaries) were ignored in the formulation. Even so, maximizing the objective function, which is the difference between income and the expenses *that we considered*, will tell us how to allocate the resources *that we considered* in a most profitable way. To obtain a model of tractable size for a complicated enterprise, you will often need to ignore minor aspects of an operation that are unlikely to influence the optimization. Very possibly, we have erred in this example by oversimplifying the operation for the sake of simplifying the model. A more realistic model should include more of the major costs of the sawmill and papermill operations. In Exercise 4 you are asked to introduce more-realistic data.

The objective function for our model is the sum of expressions (2) and (5). The constraints are given in expressions (3), (4), (6), (7), and (8). In entering a problem such as this one with MODELER, you can locate the various pieces of the objective function near the constraints to which they pertain and thereby enhance the readability of your solution, a practice that will help you discover errors. The MODELER program assembles the various pieces of the objective function into one objective function.

The maximal value of the objective function found by SOLVER is $19,499.84. The solution is

$$s_4 = 7.54459 \qquad M_{32} = 32.00000 \qquad P = 9600.00000$$
$$P_{21} = 18.51471 \qquad G_1 = 38.67129 \qquad s_2 = 2400.00000$$
$$P_{31} = 25.00000 \qquad G_2 = 52.54459 \qquad s_5 = 17.98706$$
$$M_{21} = 25.23529 \qquad G_3 = 57.98706 \qquad s_3 = 3.67129$$
$$E_{22} = 42.19529 \qquad G_4 = 66.68059 \qquad M_{12} = 108.00000$$
$$M_{11} = 56.25000 \qquad P_{22} = 17.80471 \qquad G_5 = 31.90029$$

Figure 1 depicts the optimal solution.

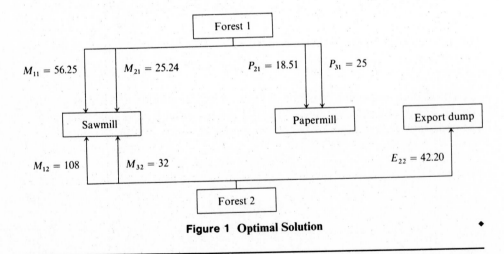

Figure 1 Optimal Solution

The next example is concerned with choosing a set of investments that will provide a series of yearly cash payments.

EXAMPLE 2

In deciding a law suit brought by John Adams, who was injured in an industrial accident, the jury has determined that he should receive a cash payment for each of the next 10 years. The jury awarded increasing yearly payments to compensate for future cost-of-living increases. Table 6 gives the amounts of the payments determined by the jury.

Table 6   Payments determined by jury

Year	1	2	3	4	5	6	7	8	9	10
Cash (in $1000s)	20	21	23	25	26	28	30	31	33	35

The attorneys for each party have agreed to allow John to accept a lump sum of money. This sum is to be paid immediately and is of such a size as to allow him to realize at least the annual amounts in Table 6. John's attorney wants the lump sum to be large, and the industry's attorney wants it to be small. The judge must determine the size of an equitable payment.

John's attorney argues that the payment should be deposited in a government insured savings account because this is the safest form of investment. He assumes that such an account will pay 4.5% interest for each of the next 10 years. To calculate the lump-sum payment, he lets L be the amount of the lump-sum payment and B_1, $B_2, \ldots, B_9$ be the amounts on deposit in the savings account at the beginning of each of the next 9 years. Since it will be most advantageous to his client, he assumes that the man will withdraw his yearly income at the beginning of the year. He also assumes that the bank interest will be added to the account at the time the withdrawal is made. Thus, for $2 \leq j \leq 9$, the account balance in year j is

$$B_j = B_{j-1} + 0.045B_{j-1} - \text{(cash withdrawn)}$$

To find the value of the lump-sum payment L, he solves the system of linear equations

$$B_1 = L - 20000 \qquad \text{(Start of year 1)}$$
$$B_2 = 1.045B_1 - 21000 \qquad \text{(Start of year 2)}$$
$$B_3 = 1.045B_2 - 23000 \qquad \text{(Start of year 3)}$$
$$B_4 = 1.045B_3 - 25000 \qquad \text{(Start of year 4)}$$
$$B_5 = 1.045B_4 - 26000 \qquad \text{(Start of year 5)}$$
$$B_6 = 1.045B_5 - 28000 \qquad \text{(Start of year 6)}$$
$$B_7 = 1.045B_6 - 30000 \qquad \text{(Start of year 7)}$$
$$B_8 = 1.045B_7 - 31000 \qquad \text{(Start of year 8)}$$
$$B_9 = 1.045B_8 - 33000 \qquad \text{(Start of year 9)}$$
$$0 = 1.045B_9 - 35000 \qquad \text{(Start of year 10)}$$

John's lawyer solves this system by solving the equation for the 10th year for B_9, substituting the value of B_9 in the equation for the 9th year to find B_8, and continuing this backwards substitution until he finds L. You can check that $L = \$219,909$.

The industry's lawyer argues that the injured man can invest his money just as safely and more profitably in a mixed portfolio consisting of a security from Blue Chip, Inc., a government bond, and a savings account to hold loose cash. Table 7 describes the investments he selects as an example.

Table 7 Investment plan suggested by industry lawyer

Investment	Initial cost/share ($)	Annual return/share	Years to maturity	Repayment at maturity/share ($)
Blue Chip	950	$60	6	1000
Bond	620	0	8	1000
Savings	—	4.5%	—	—

The lawyer interprets Table 7 as follows. The injured man will purchase C \$1000-shares of Blue Chip securities for $950C$ dollars. He will receive $60C$ dollars each year from Blue Chip and a repayment of $1000C$ dollars after 6 years. The man will purchase E \$1000-government-bonds at a cost of $620E$ dollars and receive $1000E$ dollars after 8 years. He will place in the savings account the amount left after taking his annual payment. The lawyer lets B_i, $1 \le i \le 9$, be the amount deposited in the savings account at the beginning of year i. Thus, the lump-sum payment of L dollars would be used to provide the first year's cash payment of \$20,000, with the rest being divided among the three investments according to the equation

$$L = 950C + 620E + B_1 + 20,000$$

The lawyer calculates that at the beginning of the second year the amount to be paid the injured man and the amount to be left in savings would be determined by the equation

$$\text{income} = 60C + 0.045B_1 + B_1 = B_2 + 21000 = \text{disbursements}$$

Arguing in a similar way for the other years, he formulates the LP problem

Minimize: $z = L$

Subject to:
$$
\begin{aligned}
L &= 950C + 620E + B_1 + 20000 &&\text{(Start of year 1)}\\
60C &+ 1.045B_1 = B_2 + 21000 &&\text{(Start of year 2)}\\
60C &+ 1.045B_2 = B_3 + 23000 &&\text{(Start of year 3)}\\
60C &+ 1.045B_3 = B_4 + 25000 &&\text{(Start of year 4)}\\
60C &+ 1.045B_4 = B_5 + 26000 &&\text{(Start of year 5)}\\
60C &+ 1.045B_5 = B_6 + 28000 &&\text{(Start of year 6)}\\
1000C &+ 1.045B_6 = B_7 + 30000 &&\text{(Start of year 7)}
\end{aligned}
$$

$$1.045B_7 = B_8 + 31000 \qquad \text{(Start of year 8)}$$
$$1000E + 1.045B_8 = B_9 + 33000 \qquad \text{(Start of year 9)}$$
$$1.045B_9 = 35000 \qquad \text{(Start of year 10)}$$

$$B_i \geq 0, \qquad 1 \leq i \leq 9; \qquad L, C, E \geq 0$$

The industry's lawyer enters this LP problem with MODELER and solves it with SOLVER to find that $L = \$168,302$. We do not know what the judge decided. ♦

The LP problems in Example 2 are fairly easy to enter with the MODELER programming by typing all of the expressions. You can, if you like, save some typing time by using the feature that copies lines (the F4 key) to make multiple copies of a line that is repeated many times and then changing some values in the repeated lines. For instance, you could make eight copies of the line for the start of year 2 in the preceding LP problem and then change the subscripts of B and the values of the annual payments to the correct values. The next example concerns a problem that has many more inequalities than the last example. Since many of the inequalities share a common form, you can use the inequality-generation features of MODELER to advantage. This example serves as a model for generating large systems of inequalities with MODELER.

EXAMPLE 3

In Example 2 of Section 1.4 we developed an LP model for managing water flow from two dams to maximize the profits realized from hydroelectric power and irrigation. You can use the following program to instruct MODELER to generate the objective function and constraints we gave in that example.

{&CFj are the cumulative inflows to HMSD.}
assign(&CFj, j = 1 to 12, 7, 9, 12, 16, 19, 21, 23, 24, 26, 29, 32, 34);

{&PHMj are the hydropower profits from HMSD.}
assign(&PHMj, j = 1 to 12, 1.6, 1.7, 1.8, 1.9, 2.0, 2.0, 2.0, 1.9, 1.8, 1.7, 1.6, 1.5);

{&PIRj are the irrigation profits.}
assign(&PIRj, j = 1 to 12, 1.0, 1.2, 1.8, 2.0, 2.2, 2.2, 2.5, 2.2, 1.8, 1.4, 1.1, 1.0);

{&PVj are the hydropower profits from VGD.}
assign(&PVj, j = 1 to 12, 1.4, 1.3, 1.5, 1.9, 2.3, 2.1, 1.8, 1.8, 2.1, 1.8, 1.3, 1.3);

{&MINj are the minimum units needed for irrigation.}
assign(&MINj, j = 1 to 12, 0, 0, 0, 0.5, 1.0, 2.5, 2.8, 1.2, 0.5, 0.5, 0, 0);

{&MAXj are the maximum units needed for irrigation.}
assign(&MAXj, j = 1 to 12, 0.5, 0.2, 0.1, 1.8, 4.0, 5.0, 5.4, 2.5, 0.8, 0.7, 0.5, 0.2);

{The next 12 constraints express the fact that reservoir capacity is 10 units. Note that the add function must come first in the expression.}
for i = 1 to 12 do add(j = 1 to i, − xj) + &CFi <= 10;

{The next constraint requires that 5 units remain in the reservoir at the end of the year.}
add(j = 1 to 12, − xj) + 34 >= 5;

{The next constraint requires that at least 1 unit be allowed to flow downstream.}
for i = 1 to 12 do yi <= xi − 1;

{The next constraints restrict the flow through the turbines.}
for i = 1 to 12 do xi <= 7;
for i = 1 to 12 do yi >= &MINi; {Minimum irrigation requirement}
for i = 1 to 12 do yi <= &MAXi; {Maximum irrigation requirement}
add(i = 1 to 12, &PHMi∗xi); {Hydropower profit from HMSD}
add(i = 1 to 12, &PIRi∗yi); {Profit from irrigation}

{The hydropower profit from VGD is the sum of &PVi(xi − yi). We express this as two sums to avoid generating a statement longer than 160 characters.}
add(i = 1 to 12, &PVi∗xi);
add(i = 1 to 12, − &PVi∗yi);

We gave the solution found by CALIPSO in Example 2 of Section 1.4.   ◆

Exercises

1. A multinational corporation has three divisions: U.S., European, and Asian. In January of every year each division submits project proposals and requests capital investment in these projects from the corporation. This year, the corporation can invest up to 400 million dollars in new projects. Each division has various limitations on the size of venture that it can undertake. Within these limitations, any project proposed can be funded at any level. Money is measured in millions of dollars. The variables and restrictions are as follows:

U.S. Division

Project	1	2	3
Investment dollars	x_1	x_2	x_3
Net profit	$15x_1$	$10x_2$	$6x_3$
Labor restriction	$10x_1 + 8x_2 + 6x_3 \leq 50$		
Facility restriction	$5x_1 + 12x_2 + 8x_3 \leq 35$		

European Division

Project	1	2	3	4
Investment dollars	x_4	x_5	x_6	x_7
Net profit	$7x_4$	$5x_5$	$12x_6$	$8x_7$
Resources restriction	$10x_4 + 8x_5 + 15x_6 + 9x_7 \leq 65$			
Labor restriction	$8x_4 + 9x_5 + 14x_6 + 5x_7 \leq 80$			
Power restriction	$x_4 + 3x_5 + 5x_6 + 12x_7 \leq 45$			

Asian Division

Project	1	2	3	4	5
Investment dollars	x_8	x_9	x_{10}	x_{11}	x_{12}
Net profit	$9x_8$	$6x_9$	$4x_{10}$	$3x_{11}$	$2x_{12}$

Labor restriction $15x_8 + 9x_9 + 6x_{10} + 2x_{11} + 5x_{12} \le 120$

Land restriction $8x_8 + 5x_9 \qquad + 3x_{11} + 6x_{12} \le\ 75$

(a) Formulate an LP problem that maximizes profit.
(b) What is the density of the A matrix?
†(c) Enter the problem with MODELER, and solve it with SOLVER.

2. The construction manager of a road-building project needs to divert five digging machines from their normal uses to move 1000 cubic yards of mud that have slid onto a portion of the completed roadway. Each of the machines must be employed on the main project for part of each day. The machines can be operated for at most 8 hr per day. The mud must be removed in five or fewer working days. Each of the five digging machines is a different type and has a different *capacity*; the capacity is the volume of material the machine can hold in one scoop. Also, each machine has a *cycle time*, which is the time it takes to scoop up a load of its capacity, dump the load in a truck, and return for another load. Table 8 gives the capacity of each digging machine, the cycle time of each machine, the maximum number of hours per day each machine can be diverted for the mud project, and the cost per hour of operating each machine. The supervisor wants to determine what combination of machines she should use to minimize cost. Since it is a big mud pile, she assumes that all five machines could work at the same time without interfering with one another. Formulate and solve an LP problem to solve the supervisor's problem.

Table 8

Digging machine	Capacity (cu. yd.)	Cycle time (min)	Cost ($/hr)	Available time (hr/day)
A	2.2	3.75	17.50	6
B	1.5	2.5	15.00	8
C	3.5	5.0	25.00	4
D	2.5	3.0	20.00	5
E	2.0	2.75	18.00	6

3. In Example 2, John's lawyer did not need to use an LP problem to compute L because he could find L by solving his system of linear equations in a straightforward way. Even so, he could have found L by solving an LP problem.
(a) Formulate an LP problem whose solution gives L.
†(b) Solve the problem with CALIPSO.

4. As we pointed out in Example 1 and in Section 1.3, a mathematical model often excludes some details of a practical problem that are not expected to appreciably affect the decisions under study. We also noted that our model probably erred on the side of oversimplification.
(a) Introduce more data into our formulation of the model that, in your opinion, make the model more realistic; formulate a model that includes your data.
†(b) Solve your problem with CALIPSO.

Solve Exercises 5–8 with CALIPSO.

†5. Solve the following exercises from Section 1.2.
(a) Exercise 3   (b) Exercise 7

†**6.** Solve the following exercises from Section 1.2.
   (a) Exercise 4   (b) Exercise 8

†**7.** Solve the following exercises from Section 1.4.
   (a) Exercise 1   (b) Exercise 5   (c) Exercise 7   (d) Exercise 13

†**8.** Solve the following exercises from Section 1.4.
   (a) Exercise 4   (b) Exercise 6   (c) Exercise 8

CHECKLIST: CHAPTER 5

DEFINITIONS

CONCEPTS AND RESULTS

Chapter 6

DUALITY

6.1 AN ECONOMIC INTERPRETATION OF DUALITY

Duality plays many roles in linear programming and has major theoretical and practical consequences. It provides information about the optimal solutions of LP problems and, in some instances, leads to a more efficient way of solving them. In Section 7.4 we also see how to apply duality theory to sensitivity analysis.

To motivate the notion of duality and to give an economic interpretation of some duality results we consider the following problem.

A small firm, American Coffee, Inc. (ACI), produces four coffee blends, B_1, B_2, B_3, B_4, each of which consists of a mixture of Brazilian, Colombian, and Peruvian coffees. The blends are sold in 10-pound units. Table 1 gives the contents (in pounds) of each blend B_i per 10-pound unit.

Table 1

| | Coffee | | |
Blend	Brazilian (lb)	Colombian (lb)	Peruvian (lb)
B_1	2	4	4
B_2	4	5	1
B_3	3	3	4
B_4	7	2	1

Table 2 gives the dollar profit on the sale of 10-pound units of the various blends.

Table 2

Blend	Profit per 10-lb unit sold
B_1	8
B_2	6
B_3	4
B_4	5

Table 3 lists the number of pounds of each kind of coffee currently held by the firm.

Table 3

Coffee	Pounds available
Brazilian	8000
Colombian	6400
Peruvian	6000

To determine the number of 10-pound units of each blend that the firm should produce to maximize profits (subject to the given constraints), we set up the LP problem

$$\text{Maximize:} \quad z = 8x_1 + 6x_2 + 4x_3 + 5x_4$$

$$\text{Subject to:} \quad \begin{aligned} 2x_1 + 4x_2 + 3x_3 + 7x_4 &\le 8000 \\ 4x_1 + 5x_2 + 3x_3 + 2x_4 &\le 6400 \\ 4x_1 + x_2 + 4x_3 + x_4 &\le 6000 \end{aligned} \tag{1}$$

$$x_i \ge 0, \qquad 1 \le i \le 4$$

where for each i, $1 \le i \le 4$, x_i is the number of 10-pound units of blend B_i sold by the firm. (For simplicity, we assume the firm can sell fractional amounts of each blend.)

You may verify that the following are the initial and final tableaus for this problem:

Initial Tableau

	x_1	x_2	x_3	x_4	s_1	s_2	s_3	
s_1	2	4	3	7	1	0	0	8000
s_2	4	5	3	2	0	1	0	6400
s_3	4	1	4	1	0	0	1	6000
	-8	-6	-4	-5	0	0	0	0

Final Tableau for the Primal Problem

	x_1	x_2	x_3	x_4	s_1	s_2	s_3	
s_3	0	$-\frac{15}{4}$	$\frac{5}{4}$	0	$\frac{1}{6}$	$-\frac{13}{12}$	1	400
x_4	0	$\frac{1}{4}$	$\frac{1}{4}$	1	$\frac{1}{6}$	$-\frac{1}{12}$	0	800
x_1	1	$\frac{9}{8}$	$\frac{5}{8}$	0	$-\frac{1}{12}$	$\frac{7}{24}$	0	1200
	0	$\frac{17}{4}$	$\frac{9}{4}$	0	$\frac{1}{6}$	$\frac{23}{12}$	0	13600

$$\tag{2}$$

We will refer to LP problem (1) as the *primal problem*. To generate the *dual problem* of (1) we suppose that a second firm, Warehouses, Limited (WL), wants to purchase the resources of American Coffee, Inc. In this example the resources are the pounds of Brazilian, Colombian, and Peruvian coffee held by ACI. To determine the prices that WL would be willing to pay for these resources (which, as you will see, can also be interpreted as the *value* of these resources to ACI), we set up the following LP problem.

Let u_1 be the price that WL will pay for a pound of Brazilian coffee, u_2 the price it will pay for a pound of Colombian coffee, and u_3 the price it will pay for a pound of Peruvian coffee. Then the total price that it will pay for all of ACI's resources is

$$y = 8000u_1 + 6400u_2 + 6000u_3 \qquad (3)$$

and it is this quantity that WL would like to minimize.

To determine the dual constraints, we observe that there is no reason for ACI to sell its resources unless it can improve its profit margin. Since the profit on the sale of one 10-pound unit of blend B_1 is 6 dollars, and since the resources needed to produce this unit are 2 pounds of Brazilian coffee, 4 pounds of Colombian coffee, and 4 pounds of Peruvian coffee, it follows that ACI would be willing to sell this resource only if

$$2u_1 + 4u_2 + 4u_3 \geq 6 \qquad (4)$$

Similarly, since ACI's profit from the sale of one 10-pound unit of blend B_2 is 8 dollars, and since the resources needed to produce this unit are 4 pounds of Brazilian coffee, 5 pounds of Colombian coffee, and 1 pound of Peruvian coffee, it follows that ACI would be willing to sell only if

$$4u_1 + 5u_2 + u_3 \geq 8 \qquad (5)$$

Analogous reasoning applied to blends B_3 and B_4 leads to the additional constraints

$$3u_1 + 3u_2 + 4u_3 \geq 4 \qquad (6)$$

and

$$7u_1 + 2u_2 + u_3 \geq 5 \qquad (7)$$

From (3)–(7) we obtain the dual problem to (1),

$$
\begin{aligned}
\text{Minimize:} \quad & y = 8000u_1 + 6400u_2 + 6000u_3 \\
\text{Subject to:} \quad & 2u_1 + 4u_2 + 4u_3 \geq 8 \\
& 4u_1 + 5u_2 + u_3 \geq 6 \\
& 3u_1 + 3u_2 + 4u_3 \geq 4 \\
& 7u_1 + 2u_2 + u_3 \geq 5 \\
& u_i \geq 0, \qquad 1 \leq i \leq 3
\end{aligned}
\qquad (8)
$$

You may verify that the following are the initial and final tableaus for solving LP problem (8) by the two-phase method.

Initial Tableau for the Dual Problem

	u_1	u_2	u_3	t_1	t_2	t_3	t_4	y_1	y_2	y_3	y_4	
y_1	2	4	4	-1	0	0	0	1	0	0	0	8
y_2	4	5	1	0	-1	0	0	0	1	0	0	6
y_3	3	3	4	0	0	-1	0	0	0	1	0	4
y_4	7	2	1	0	0	0	-1	0	0	0	1	5
	8000	6400	6000	0	0	0	0	0	0	0	0	0

Final Tableau for the Dual Problem

	u_1	u_2	u_3	t_1	t_2	t_3	t_4	y_1	y_2	y_3	y_4	
u_1	1	0	$-\frac{1}{6}$	$\frac{1}{12}$	0	0	$-\frac{1}{6}$	$-\frac{1}{12}$	0	0	$\frac{1}{6}$	$\frac{1}{6}$
u_2	0	1	$\frac{13}{12}$	$-\frac{7}{24}$	0	0	$\frac{1}{12}$	$\frac{7}{24}$	0	0	$-\frac{1}{12}$	$\frac{23}{12}$
t_3	0	0	$-\frac{1}{4}$	$-\frac{5}{8}$	0	1	$-\frac{1}{4}$	$\frac{5}{8}$	0	-1	$\frac{1}{4}$	$\frac{2}{4}$
t_2	0	0	$\frac{15}{4}$	$-\frac{9}{8}$	1	0	$-\frac{1}{4}$	$\frac{9}{8}$	-1	0	$\frac{1}{4}$	$\frac{17}{4}$
	0	0	400	1200	0	0	800	-1200	0	0	-800	-13600

(9)

Notice the following similarities between (2), the final tableau for the primal problem, and (9), the optimal tableau for the dual problem:

The objective functions for the two problems have the same optimal value: 13,600. (The $-13{,}600$ appears in (9) because we introduced a minus sign to change the minimization problem to a maximization problem.)

The decision variables of the primal problem have the optimal values

$$(x_1, x_2, x_3, x_4) = (1200, 0, 0, 800) \tag{10}$$

These same numbers appear in the objective row of the optimal tableau (9) for the dual problem in the columns corresponding to the slack variables t_1, t_2, t_3, and t_4.

The decision variables of the dual problem have the optimal values

$$(u_1, u_2, u_3) = \left(\frac{1}{6}, \frac{23}{12}, 0\right) \tag{11}$$

These same numbers appear in the objective row of the final tableau (2) for the primal problem in the columns corresponding to the slack variables s_1, s_2, and s_3.

The values of the slack variables in the primal and dual problems are

$$(s_1, s_2, s_3) = (0, 0, 400) \tag{12}$$

$$(t_1, t_2, t_3, t_4) = \left(0, \frac{17}{4}, \frac{9}{4}, 0\right) \tag{13}$$

Notice that the product of the respective values of the slack variables in (12) and the decision variables in (11) is 0 and that the product of the respective values of the slack variables in (13) and the decision variables in (10) is also 0.

We will explain these connections between the primal and dual problems in later sections of this chapter and discover many interesting and useful applications of the relationship between these two problems.

We can also view the solution of the dual problem (8) as assigning to each resource a "price" or "value" representing the amount ACI would be willing to pay for an additional resource unit. *This amount coincides with the resulting increase in ACI's profit if this additional resource unit were available.* For example, from the final tableau (2) we see that the slack variable s_2 is 0, which means that in producing the four blends, all units of the Colombian coffee resource are used. Suppose we increase the supply of this resource by one unit, that is, from 6400 units to 6401 units; then the solution of (1) is (see Exercise 4):

$$(x_1, x_2, x_3, x_4) = (1200, 0, 0, 800); \qquad z = 13,601.91667 \qquad (14)$$

Note that the increase in the objective function z over its original optimal value of 13,600 is

$$1.91667 = \frac{23}{12}$$

which is precisely the value of u_2 in the optimal solution of the dual problem (8). Therefore, ACI would be willing to pay $u_2 = \frac{23}{12}$ dollars for an additional unit of Colombian coffee because doing so will increase its profit by this amount.

The *marginal price* (or *marginal value*) of a resource is defined to be the increase (or decrease) in profit resulting from a unit increase (or decrease) in the availability of the resource. As you will see in subsequent sections, within certain limitations we can consider dual prices (values of dual variables) as marginal prices.

In the example just discussed, $u_2 = \frac{23}{12}$ is the marginal price of Colombian coffee. You should note that the marginal or dual price does not necessarily reflect the *actual cost* of a resource unit; the actual cost would generally depend on market factors external to the firm's situation. For this reason, the marginal or dual price is often referred to as the *imputed value* of the resource or, more commonly, the *shadow* or *fictitious price* of the resource. We consider all of these terms to be synonymous.

Next, we observe that the dual price u_3 of the Peruvian coffee resource is 0. Note further that in the final tableau (2), the value of the slack variable corresponding to the Peruvian coffee resource is 1200, which indicates that not all of this resource was used in producing the four blends. Thus, ACI would not be willing to buy an additional unit of this resource (i.e., $u_3 = 0$) because this additional unit would not increase profits. The combination $u_3 = 0$, $s_3 \neq 0$ is an example of complementary slackness, which we discuss in Sections 6.2 and 6.4.

One final observation. It is not a coincidence that the optimal value for the objective function z in the LP problem (1), $z = 13,600$, is the same as the optimal objective value for the dual problem (2). In the context of the preceding discussion, this means that the total value of the resources is equal to the optimal profit. This result is known as the Fundamental Principle of Duality and is discussed in detail in Sections 6.2 and 6.4.

From a practical standpoint, the dual (shadow, marginal, etc.) prices in problems such as the one just described are important for determining future managerial action. Solving the original (primal) problem tells us what we should do to maximize profits with the present resources; solving the dual problem for the shadow prices tells us how to increase (or decrease) future profits by increasing (or decreasing) the availability of the resources.

As another example of duality we consider the classic diet problem.

EXAMPLE 1

A hospital meal consists of two foods A and B. Table 4 lists the units of fats, carbohydrates, and proteins found in each ounce of A and B.

Table 4

Constituent	Food A	Food B
Fat	1	4
Carbohydrate	3	2
Protein	5	3

It is necessary that the meal provide at least 12 units of fat, 15 units of carbohydrates, and 22 units of protein. Each ounce of food A costs 30 cents, and each ounce of food B costs 25 cents.

To determine how many ounces of each food should be served to minimize costs (subject to the given constraints), let x_1 denote the number of ounces of A that is served, and let x_2 denote the number of ounces of B that is served. The resulting LP problem is

$$\text{Minimize:} \quad y = 30x_1 + 25x_2$$

$$\text{Subject to:} \quad \begin{aligned} x_1 + 4x_2 &\geq 12 \\ 3x_1 + 2x_2 &\geq 15 \\ 5x_1 + 3x_2 &\geq 22 \end{aligned} \tag{15}$$

$$x_1, x_2 \geq 0$$

You may verify (see Exercise 13) that the optimal tableau of problem (15) is

	x_1	x_2	s_1	s_2	s_3	
x_2	0	1	$-\frac{3}{10}$	$\frac{1}{10}$	0	$\frac{21}{10}$
s_3	0	0	$\frac{1}{10}$	$-\frac{17}{10}$	1	$\frac{23}{10}$
x_1	1	0	$\frac{1}{5}$	$-\frac{2}{5}$	0	$\frac{18}{5}$
	0	0	$\frac{3}{2}$	$\frac{19}{2}$	0	$-\frac{321}{2}$

We refer to (15) as the primal problem. To determine the shadow prices of the resources (fats, carbohydrates, and proteins), we set up the dual problem as follows. Think of yourself as the director of a firm selling artificial foods F, C, and P, in which each unit of F provides one unit of fat, each unit of C one unit of carbohydrates, and each unit of P one unit of protein. The per-unit prices charged for F, C, and P are u_1, u_2, and u_3, respectively. So that your firm can compete with the foods A and B, your products must provide the nutritional values of foods A and B without exceeding the costs of these foods. These conditions lead to the constraints

$$u_1 + 3u_2 + 5u_3 \leq 30 \tag{16}$$

$$4u_1 + 2u_2 + 3u_3 \leq 25 \tag{17}$$

The left-hand side of constraint (16) indicates that F, C, and P provide the same nutritional value as food A (1 unit fat, 3 units carbohydrate, 5 units protein); the inequality in (16) ensures that using artificial foods F, C, and P will be at least as economical as using food A. Similar statements hold for food B for the constraint (17).

Your firm wants to maximize its profits, subject to constraints (16) and (17). Thus, the dual problem of (15) is

$$\begin{aligned} \text{Maximize:} \quad & z = 12u_1 + 15u_2 + 22u_3 \\ \text{Subject to:} \quad & u_1 + 3u_2 + 5u_3 \leq 30 \\ & 4u_1 + 2u_2 + 3u_3 \leq 25 \\ & u_i \geq 0, \quad 1 \leq i \leq 3 \end{aligned} \tag{18}$$

Solving this problem gives us the marginal (dual) values of the primal resources. You may verify (see Exercise 12) that the optimal tableau for LP problem (18) is

	u_1	u_2	u_3	t_1	t_2	
u_2	0	1	$\frac{17}{10}$	$\frac{2}{5}$	$-\frac{1}{10}$	$\frac{19}{2}$
u_1	1	0	$-\frac{1}{10}$	$-\frac{1}{5}$	$\frac{3}{10}$	$\frac{3}{2}$
	0	0	$\frac{23}{10}$	$\frac{18}{5}$	$\frac{21}{10}$	$\frac{321}{2}$

◆

Exercises

Exercises 1–3 concern the following LP problem. A plant manufactures four fertilizers F_1, F_2, F_3, and F_4, whose production involves three steps: manufacture of the ingredients, mixing of the ingredients, and packaging. Table 5 gives the number of hours required to complete each of these steps per gallon produced and the profit per gallon for each fertilizer.

Table 5

Fertilizer	Manufacture (hr)	Mixing (hr)	Packaging (hr)	Profit/unit ($)
F_1	3	4	1	15
F_2	6	8	$\frac{1}{2}$	30
F_3	4	3	1	15
F_4	2	1	1	10

Each week, 150 hours are available for manufacturing ingredients, 120 hours for mixing, and 50 hours for packaging. The plant wants to establish a production schedule that will maximize weekly profit.

1. Formulate and solve an LP problem to solve the plant's problem.

2. Consider the plant's problem from the standpoint of the seller of the resources. The resource seller would like to assign costs u_1, u_2, and u_3 to the resources, where

   u_1 denotes the value per hour for manufacturing ingredients,

   u_2 denotes the value per hour for mixing the ingredients, and

   u_3 denotes the value per hour for packaging.

   (a) Explain why the total value of the resources is

   $$150u_1 + 120u_2 + 50u_3$$

   (b) The profit realized from the sale of one unit of F_1 is \$15. Explain why the constraint

   $$3u_1 + 4u_2 + u_3 \geq 15$$

   should be imposed by the seller.

   (c) Explain why the following constraints should also be imposed by the seller.

   $$6u_1 + 8u_2 + \frac{1}{2}u_3 \geq 30$$

   $$4u_1 + 3u_2 + \quad u_3 \geq 15$$

   $$2u_1 + \quad u_2 + \quad u_3 \geq 10$$

   (d) Formulate and solve the problem from the seller's viewpoint. This is the dual problem. This problem and the one solved in Exercise 1 should have the same optimal value for their objective functions.

   (e) Show that the per-unit profit realized from the sale of F_1 is less than the value of F_1, which is

   $$3u_1^* + 4u_2^* + u_3^*$$

   where u_1^*, u_2^*, and u_3^* are the optimal values found in part (d). Also show that the profit from the sale of F_3 is less than the value of F_3. Conclude that the plant should not manufacture F_1 or F_3.

3. For the ACI problem discussed in this section, show that the per-unit profit from the sale of blend B_3 is less than the value of blend B_3, which is $3u_1^* + 3u_2^* + 4u_3^*$. Also show that the profit from the sale of blend B_4 is not greater than the value of blend B_4. Conclude that ACI should not manufacture blends B_3 and B_4.

†4. Solve LP problem (1) when the supply of Colombian coffee is increased by 1 unit to 6401. Verify that the answer given in (14) is correct.

5. Observe that the slack variable for the manufacturing constraint of Exercise 1 is positive. It follows that an increase in availability of manufacturing time would not increase profits. Explain why this means that u_1^*, found in Exercise 2(d), should be 0.

6. The Moscow Industrial Ghetto (MIG) manufactures large and small iron dragons. The large dragons are used for garden decorations and the small dragons are used for automobile ornaments. Each week up to 1000 pounds of iron can be purchased at \$0.25 per pound. MIG employs five workers, each of whom works 30 hours per week. Their

salary is a fixed cost that does not depend on productivity. The machinery used is available for a maximum of 300 hours per week. Table 6 gives data about dragons.

Table 6

	Small dragon	Large dragon
Iron per dragon (lb)	3	125
Labor per dragon (hr)	2	12
Machine time per dragon (hr)	1.5	9
Selling price per dragon ($)	35	300

Set up the LP problem to determine how many dragons of each size MIG should make to maximize total selling price. Also set up the dual problem to find the marginal cost of material, labor, and machine time.

7. A steel company manufactures three grades of rebar: grade A is mild steel, grade B is high-tensile steel, and grade C is hardened steel. The company can use three processes, each of which produces various quantities of rebar and has a different cost per hour of operation. Table 7 summarizes the operation.

Table 7

	Tons per hour			
	A	B	C	Cost per hour
Process 1	4	5	3	220
Process 2	7	7	8	480
Process 3	7	4	5	375

In the planning period, the company must produce at least 4000 tons of grade A, 2700 tons of grade B, and 3000 tons of grade C. Setup the primal LP problem to minimize manufacturing cost.

To establish a profitable selling price for rebar, the company needs to know the marginal cost of producing each grade of rebar. Set up the dual problem to answer this question.

†8. Solve the primal and dual problems you formed in Exercise 6.

†9. Solve the primal and dual problems you formed in Exercise 7.

10. Form the dual of Exercise 2 of Section 1.1. What are the units of the dual variables?

11. Form the dual of Exercise 1 of Section 1.1. What are the units of the dual variables?

12. Verify the entries in the optimal tableau for LP problem (18).

13. Verify the entries in the optimal tableau for LP problem (15).

6.2   PRIMAL AND DUAL PROBLEMS: A SPECIAL CASE

We now abstract some of the ideas concerning duality that arose in the previous section and examine certain relations between the primal and dual problems. In this section we limit our consideration to LP problems with constraints of the form

$$Ax \leq b$$

We do not assume that the entries of **b** are nonnegative. In subsequent sections we will develop the notion of duality for LP problems with mixed constraints.

In Section 6.1 we saw that the dual of the LP problem

$$\text{Maximize:} \quad z = 8x_1 + 6x_2 + 4x_3 + 5x_4$$

$$\text{Subject to:} \quad 2x_1 + 4x_2 + 3x_3 + 7x_4 \leq 8000$$
$$4x_1 + 5x_2 + 3x_3 + 2x_4 \leq 6400 \qquad (1)$$
$$4x_1 + x_2 + 4x_3 + x_4 \leq 6000$$

$$x_i \geq 0, \qquad 1 \leq i \leq 4$$

is

$$\text{Minimize:} \quad y = 8000u_1 + 6400u_2 + 6000u_3$$

$$\text{Subject to:} \quad 2u_1 + 4u_2 + 4u_3 \geq 8$$
$$4u_1 + 5u_2 + u_3 \geq 6$$
$$3u_1 + 3u_2 + 4u_3 \geq 4 \qquad (2)$$
$$7u_1 + 2u_2 + u_3 \geq 5$$

$$u_i \geq 0, \qquad 1 \leq i \leq 3$$

In this case we referred to (1) as the primal problem and (2) as the dual problem. We make the following observations

1. The resource coefficients of the primal problem are the cost coefficients of the dual problem.

1'. The resource coefficients of the dual problem are the cost coefficients of the primal problem.

2. The constraint coefficients of the ith constraint in the primal problem are the constraint coefficients of the ith variable in the dual problem.

2'. The constraint coefficients of the ith constraint in the dual problem are the constraint coefficients of the ith variable in the primal problem.

3. The constraint matrix of the dual problem is the transpose of the constraint matrix of the primal problem.

4. The inequalities in the primal problem are $\leq$; the inequalities in the dual problem are $\geq$.

5. The primal objective function is maximized; the dual objective function is minimized.

Definition.   The *dual* of the LP problem

$$\text{Maximize:} \quad z = \mathbf{cx} \qquad \mathbf{c}, 1 \times n; \qquad \mathbf{x}, n \times 1$$

$$\text{Subject to:} \quad A\mathbf{x} \leq \mathbf{b} \qquad A, m \times n; \qquad \mathbf{b}, m \times 1 \qquad (3)$$

$$\mathbf{x} \geq 0$$

is

$$\text{Minimize:} \quad y = \mathbf{b}^{\mathsf{T}}\mathbf{u} \qquad \mathbf{b}^{\mathsf{T}}, 1 \times m; \qquad \mathbf{u}, m \times 1$$

$$\text{Subject to:} \quad A^{\mathsf{T}}\mathbf{u} \geq \mathbf{c}^{\mathsf{T}} \qquad A^{\mathsf{T}}, n \times m; \qquad \mathbf{c}^{\mathsf{T}}, n \times 1 \qquad (4)$$

$$\mathbf{u} \geq 0$$

In this context we refer to (3) as the primal problem. Observe that in this definition we do not require the entries of $\mathbf{b}$ to be nonnegative.

EXAMPLE 1

Consider the primal problem

$$\text{Maximize:} \quad z = 4x_1 - 3x_2 + x_3$$

$$\text{Subject to:} \quad 2x_1 - 3x_2 + 4x_3 \leq 5$$
$$-x_1 + x_2 - 3x_3 \leq -3$$
$$x_i \geq 0, \qquad 1 \leq i \leq 3$$

In the notation of (3)

$$A = \begin{bmatrix} 2 & -3 & 4 \\ -1 & 1 & -3 \end{bmatrix} \quad \mathbf{b} = \begin{bmatrix} 5 \\ -3 \end{bmatrix} \quad \mathbf{c} = \begin{bmatrix} 4 & -3 & 1 \end{bmatrix} \quad \mathbf{x} = \begin{bmatrix} x_1 \\ x_2 \\ x_3 \end{bmatrix}$$

The dual problem is

$$\text{Minimize:} \quad y = \mathbf{b}^{\mathsf{T}}\mathbf{u}$$

$$\text{Subject to:} \quad A^{\mathsf{T}}\mathbf{u} \geq \mathbf{c}^{\mathsf{T}}$$

$$\mathbf{u} \geq 0$$

which can also be written

$$\text{Minimize:} \quad y = 5u_1 - 3u_2$$

$$\text{Subject to:} \quad 2u_1 - u_2 \geq 4$$
$$-3u_1 + u_2 \geq -3$$
$$4u_1 - 3u_2 \geq 1$$

$$u_1, u_2 \geq 0$$

Next we examine certain relations between the feasible solutions of the primal and dual problems. This will lead us to the Fundamental Theorem of Duality, which we discuss in more detail in Sections 6.3 and 6.4. We postpone the proofs of the theorems in the remainder of this section until Section 6.4 where we prove them in a more general context.

In the following theorem, known as the Weak Principle of Duality, we use **x** and **u** to represent variable row and column matrices and the corresponding symbols $\hat{\mathbf{x}}$ and $\hat{\mathbf{u}}$ to represent particular solutions. For instance if $\mathbf{x} = [x_1 \quad x_2]^T$, then $\hat{\mathbf{x}}$ might be $[2 \quad 5]^T$.

Theorem 1.   (The Weak Principle of Duality).   Suppose $\hat{\mathbf{x}}$ is a feasible solution of the primal problem

$$\text{Maximize:} \quad z = \mathbf{cx}$$
$$\text{Subject to:} \quad A\mathbf{x} \le \mathbf{b} \tag{5}$$
$$\mathbf{x} \ge \mathbf{0}$$

and suppose $\hat{\mathbf{u}}$ is a feasible solution to the dual problem

$$\text{Minimize:} \quad y = \mathbf{b}^T\mathbf{u}$$
$$\text{Subject to:} \quad A^T\mathbf{u} \ge \mathbf{c}^T \tag{6}$$
$$\mathbf{u} \ge \mathbf{0}$$

Then

(a) $\mathbf{c}\hat{\mathbf{x}} \le \mathbf{b}^T\hat{\mathbf{u}}$

(b) If $\mathbf{c}\hat{\mathbf{x}} = \mathbf{b}^T\hat{\mathbf{u}}$, then $\hat{\mathbf{x}}$ is an optimal solution of (5) and $\hat{\mathbf{u}}$ is an optimal solution of (6).

EXAMPLE 2

Consider the primal problem

$$\text{Maximize:} \quad z = 120x_1 + 100x_2$$
$$\text{Subject to:} \quad \begin{aligned} x_1 + x_2 &\le 4 \\ 5x_1 + 3x_2 &\le 15 \end{aligned} \tag{7}$$
$$x_1, x_2 \ge 0$$

The dual of (7) is

$$\text{Minimize:} \quad y = 4u_1 + 15u_2$$
$$\text{Subject to:} \quad \begin{aligned} u_1 + 5u_2 &\ge 120 \\ u_1 + 3u_2 &\ge 100 \end{aligned} \tag{8}$$
$$u_1, u_2 \ge 0$$

To illustrate conclusion (a) of Theorem 1, take any feasible solution of (7), say $x_1 = \frac{1}{2}$ and $x_2 = 1$, and any feasible solution of (8), say $u_1 = 40$ and $u_2 = 20$. We have

$$z = 120\left(\frac{1}{2}\right) + 100(1) = 60 + 100 = 160$$

and

$$y = 4(40) + 15(20) = 160 + 300 = 460.$$

As Theorem 1 predicts, $z \leq y$.

To illustrate conclusion (b), observe that $x_1 = \frac{3}{2}, x_2 = \frac{5}{2}$ is a feasible solution to (7), and $u_1 = 70, u_2 = 10$ is a feasible solution to (8). Since

$$z = 120\left(\frac{3}{2}\right) + 100\left(\frac{5}{2}\right) = 180 + 250 = 430$$

and

$$y = 4(70) + 15(10) = 280 + 150 = 430$$

we conclude from (b) that $[x_1 \quad x_2]^T = [\frac{3}{2} \quad \frac{5}{2}]^T$ is a maximal solution to the primal problem, $[u_1 \quad u_2]^T = [70 \quad 10]^T$ is a minimal solution to the dual problem, and 430 is the optimal objective value for both problems.   ◆

In Exercise 8 you are asked to prove the following corollary of Theorem 1.

Corollary to Theorem 1. If the objective function of the primal problem is unbounded on the feasible region, then the dual problem has no feasible solutions.

EXAMPLE 3

The LP problem

$$\text{Maximize:} \quad z = 3x_1 + 2x_2$$
$$\text{Subject to:} \quad -x_1 + x_2 \leq 4$$
$$x_1 - 2x_2 \leq 2$$
$$x_1, x_2 \geq 0$$

has unbounded feasible solutions, as you can see from Figure 1(a).

The dual problem

$$\text{Minimize:} \quad y = 4u_1 + 2u_2$$
$$\text{Subject to:} \quad -u_1 + u_2 \geq 3$$
$$u_1 - 2u_2 \geq 2$$
$$u_1, u_2 \geq 0$$

has an empty feasible set, as you can see from Figure 1(b).

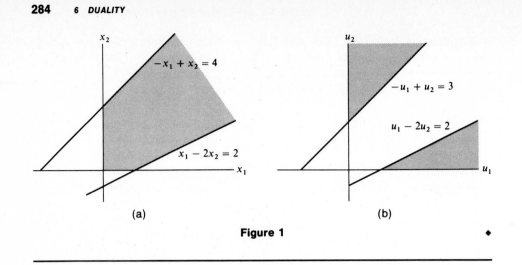

Figure 1

Perhaps the most important result in duality theory is the Fundamental Principle of Duality. This result is not only of considerable theoretical interest but also, as we see in Example 4, has significant practical applications as well.

Theorem 2. (The Fundamental Principle of Duality). If an optimal solution $\mathbf{x}^*$ of the primal problem (5) exists, then an optimal solution $\mathbf{u}^*$ of the dual problem exists and

$$z^* = \mathbf{c}\mathbf{x}^* = \mathbf{b}^T\mathbf{u}^* = y^*$$

Moreover, if an optimal solution $\mathbf{u}^*$ of the dual problem (6) exists, then an optimal solution $\mathbf{x}^*$ of the primal problem (5) exists and

$$y^* = \mathbf{b}^T\mathbf{u}^* = \mathbf{c}\mathbf{x}^* = z^*$$

The next example shows how we can use the Fundamental Principle of Duality to interpret the values of the optimal solution of the dual problem as marginal values of the primal problem.

EXAMPLE 4

We used the following LP problem as an example of marginal values in Section 6.1.

$$\text{Maximize:} \quad z = 8x_1 + 6x_2 + 4x_3 + 5x_4$$

$$\text{Subject to:} \quad 2x_1 + 4x_2 + 3x_3 + 7x_4 \leq 8000$$
$$4x_1 + 5x_2 + 3x_3 + 2x_4 \leq 6400 \qquad (9)$$
$$4x_1 + x_2 + 4x_3 + x_4 \leq 6000$$

$$x_i \geq 0, \qquad 1 \leq i \leq 4$$

The objective function of the dual of (9) is

$$y = 8000u_1 + 6400u_2 + 6000u_3 \qquad (10)$$

You may verify that

$$\mathbf{x}^{*\mathrm{T}} = [x_1^*\ \ x_2^*\ \ x_3^*\ \ x_4^*]^{\mathrm{T}} = [1200\ \ 0\ \ 0\ \ 800]^{\mathrm{T}}$$

is the maximal solution of (9) and

$$\mathbf{u}^{*\mathrm{T}} = [u_1^*\ \ u_2^*\ \ u_3^*]^{\mathrm{T}} = \left[\frac{1}{6}\ \ \frac{23}{12}\ \ 0\right]^{\mathrm{T}}$$

is the minimal solution of the dual problem. By the Fundamental Principle of Duality,

$$z^* = \mathbf{cx}^* = 8000u_1^* + 6400u_2^* + 6000u_3^* \tag{11}$$

We observe that (11) makes exact the discussion of marginal prices given in Section 6.1. For instance, (11) says that a unit increase in the resource value 8000 will result in u_1^* units of increase in the objective function value. Thus, u_1^* gives us the marginal price of the resource. Similarly, a unit increase in the resource value 6400 results in an increase of u_2^* units in the objective function value. Since the third dual price u_3^* equals 0, any change to the resource value 6000 will leave the objective value unchanged. We see then that the optimal values of the dual variables do give us the marginal values of the primal resources.

◆

Next we indicate how we can find the solution of the dual problem from the final primal tableau. We establish and further discuss this result in Section 6.4.

Theorem 3.   Let

$$\text{Maximize:}\quad z = \mathbf{cx}$$
$$\text{Subject to:}\quad A\mathbf{x} \le \mathbf{b} \tag{12}$$
$$\mathbf{x} \ge \mathbf{0}$$

be a primal problem, and let

$$\text{Maximize:}\quad z = \mathbf{cx} + \mathbf{0s}$$
$$\text{Subject to:}\quad A\mathbf{x} + I_m\mathbf{s} = \mathbf{b} \tag{13}$$
$$\mathbf{x} \ge \mathbf{0};\quad \mathbf{s} \ge \mathbf{0}$$

be the corresponding problem with slack variables. Let

$$T = \left[\begin{array}{c|c|c} A & I_m & \mathbf{b} \\ \hline -\mathbf{c} & \mathbf{0} & 0 \end{array}\right]$$

be the initial tableau for (13) and let

$$T^* = \left[\begin{array}{c|c|c} A^* & — & \mathbf{b}^* \\ \hline -\mathbf{c}^* & \mathbf{u}^* & z^* \end{array}\right]$$

be the final tableau for (13). If

$$\begin{bmatrix} \mathbf{x}^* \\ \mathbf{s}^* \end{bmatrix}$$

is an optimal feasible solution for (13), then $\mathbf{x}^*$ is an optimal feasible solution of (12), and $\mathbf{u}^{*T}$ is an optimal feasible solution of the dual of (12).

EXAMPLE 5

Adding slack variables to the LP problem

$$\text{Maximize:}\quad z = 3x_1 + 2x_2 + 3x_3$$

$$\begin{aligned}
\text{Subject to:}\quad 2x_1 + 3x_2 \quad\quad &\le 60 \\
x_1 - x_2 + 4x_3 &\le 10 \\
-2x_1 + 2x_2 + 5x_3 &\le 50
\end{aligned} \tag{14}$$

$$x_i \ge 0, \quad 1 \le i \le 3$$

we obtain

$$\text{Maximize:}\quad z = 3x_1 + 2x_2 + 3x_3 + 0s_1 + 0s_2 + 0s_3$$

$$\begin{aligned}
\text{Subject to:}\quad 2x_1 + 3x_2 \quad\quad + s_1 \quad\quad\quad\quad &= 60 \\
x_1 - x_2 + 4x_3 \quad\quad + s_2 \quad\quad &= 10 \\
-2x_1 + 2x_2 + 5x_3 \quad\quad\quad\quad + s_3 &= 50
\end{aligned} \tag{15}$$

$$x_i \ge 0, \quad s_i \ge 0, \quad 1 \le i \le 3$$

The initial and final tableaus for (15) are

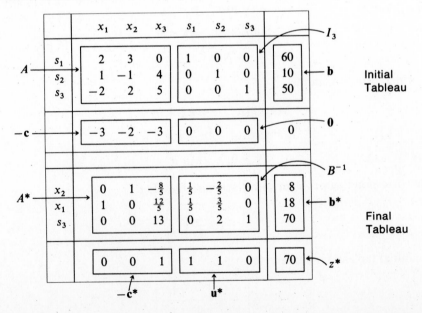

The optimal solution to LP problem (15) is

$$\begin{bmatrix} \mathbf{x}^* \\ \mathbf{s}^* \end{bmatrix} = \begin{bmatrix} 18 \\ 8 \\ 0 \\ 0 \\ 0 \\ 70 \end{bmatrix}$$

The optimal solution to problem (14) is

$$\mathbf{x}^* = \begin{bmatrix} 18 \\ 8 \\ 0 \end{bmatrix}$$

The optimal solution to the dual of problem (14) is

$$\mathbf{u}^* = \begin{bmatrix} 1 \\ 1 \\ 0 \end{bmatrix}$$

◆

The objective function for the final tableau of Example 5 is

$$z = -0x_1 - 0x_2 - x_3 - s_1 - s_2 - 0s_3 \tag{16}$$

Since the dual prices, the negatives of the coefficients of the slack variables in (16), have economic significance, it is natural to ask about the significance of the coefficients of the decision variables of (16). We can interpret the coefficient -1 of x_3 as the amount z will decrease for each unit decrease in x_3 provided all other nonbasic variables remained at value 0. To increase x_3 and hold the other nonbasic variables at 0, we must change the values of the basic variables to compensate for the change in x_3, and, thereby, destroy the optimality of the solution. To clarify this idea we write the constraints given by the final tableau of Example 5 with all nonbasic variables except x_3 omitted (because they are to stay at 0). The constraints are

$$x_2 = \frac{8}{5}x_3 + 8$$

$$x_1 = -\frac{12}{5}x_3 + 18 \tag{17}$$

Equations (17) give the new values of the basic variables if we force x_3 to assume a positive value. Equation (16) gives the amount that the objective value z will change if x_3 is increased. You can see from the equation for x_3 in (17) that x_3 cannot be increased indefinitely without causing x_1 to become infeasible. In fact, x_3 remains nonnegative only when

$$-\frac{12}{5}x_3 + 18 \geq 0$$

which implies that x_3 could only be increased by as much as

$$18\left(\frac{5}{12}\right) = \frac{15}{2}$$

In the next definition we introduce a common terminology for this concept, and in the next example we give an economic interpretation of it.

Definition. If x_i is a decision variable, the *reduced cost* for x_i is the negative of the value in the column corresponding to x_i in the objective row of the optimal tableau for the problem. In other words, the reduced cost of the decision variable x_i is c_i^*.

EXAMPLE 6

Consider again the ACI problem from Section 6.1. In the final tableau (2) of that section, we see that the reduced cost of x_2 is $-\frac{17}{4}$ and the reduced cost of x_3 is $-\frac{9}{4}$. The reduced costs of the basic decision variables x_1 and x_4 are, of course, 0. If we keep all nonbasic variables except x_2 and x_3 at value 0, then the constraints corresponding to tableau (2) are

$$s_3 = \frac{15}{4}x_2 - \frac{5}{4}x_3 + 400$$

$$x_4 = -\frac{1}{4}x_2 - \frac{1}{4}x_3 + 800 \tag{18}$$

$$x_1 = -\frac{9}{8}x_2 - \frac{5}{8}x_3 + 1200$$

and the objective function is

$$z = -\frac{17}{4}x_2 - \frac{9}{4}x_3 + 13{,}600 \tag{19}$$

In this problem, x_2 represents the amount of coffee blend B_2 and x_3 the amount of of blend B_3 that are manufactured. Even though the maximum profit is realized by making neither blend B_2 nor blend B_3, it could be that ACI feels compelled to manufacture some of these blends, perhaps to satisfy a customer of long standing. The reduced costs [see, in particular, (19)] provide a way for ACI to compute the cost of this benevolence. Note, however, that any increase in x_2 or x_3 must be compensated by adjusting the values of x_1, x_4, and s_3 to balance equations (18). For instance, if $x_2 = 1$ and $x_3 = 2$, then the objective function value would be reduced by

$$-\left(\frac{17}{4}\right)1 - \left(\frac{9}{4}\right)2 = -\frac{35}{4}$$

and the new (nonoptimal) solution is

$$(x_1, x_2, x_3, x_4, s_1, s_2, s_3) = (1197.75, 1, 2, 799.25, 0, 0, 401.25) \qquad \blacklozenge$$

To conclude this section we briefly discuss the idea of complementary slackness. If the slack variable of a $\leq$ constraint is not zero in the optimal solution, then the resource value for the constraint is larger than can be profitably used. If the slack variable is positive, economic considerations would place no value on increasing the resource value (because there is already an excess of the resource), and so the dual price would be expected to be zero. The next theorem justifies this economic observation. We will state this theorem in full generality and prove it in Section 6.4.

Theorem 4.   (Complementary Slackness).   Suppose the LP problem

$$\text{Maximize:} \quad z = \mathbf{cx}, \qquad \mathbf{c}, 1 \times n; \qquad \mathbf{x}, n \times 1$$

$$\text{Subject to:} \quad A\mathbf{x} \leq \mathbf{b}, \qquad A, m \times n; \qquad \mathbf{b}, m \times 1 \qquad (20)$$

$$\mathbf{x} \geq \mathbf{0}$$

has the maximal solution

$$\begin{bmatrix} x_1^* & x_2^* & \cdots & x_n^* & x_{n+1}^* & \cdots & x_{n+m}^* \end{bmatrix}^{\mathrm{T}}$$

where $x_{n+1}^*, \ldots, x_{n+m}^*$ are the values of the slack variables. Suppose further that the dual of (20) has the minimal solution

$$\begin{bmatrix} u_1^* & u_2^* & \cdots & u_m^* & u_{m+1}^* & \cdots & u_{m+n}^* \end{bmatrix}^{\mathrm{T}}$$

where $u_1^*, \ldots, u_m^*$ are the dual prices. Then

$$x_{n+i}^* u_i^* = 0, \qquad \text{for} \quad 1 \leq i \leq m \qquad (21)$$

and

$$x_j^* u_{m+j}^* = 0, \qquad \text{for} \quad 1 \leq j \leq n \qquad (22)$$

Note that (21) implies that either the ith slack variable of the primal problem or the ith dual price is 0 and that (22) implies that either the jth slack variable of the dual problem or the jth decision variable of the primal problem is 0.

EXAMPLE 7

By examining the final tableaus (2) and (9) in Section 6.1, which were given for the prototype example of that section, you will see that the optimal value of the third slack variable s_3 is $s_3^* = 400$ and the optimal value of the third dual price u_3 is $u_3^* = 0$. Thus, as expected and as indicated by Theorem 4, since there is an excess of the resource, its marginal price is 0. Also, since in (9) the optimal value of t_2 is $t_2^* = \frac{17}{4}$, it follows that the optimal value of the second primal decision variable x_2 is $x_2^* = 0$; similarly, since in (9) the optimal value of t_3 is $t_3^* = \frac{17}{4}$, we must have $x_3^* = 0$.   ◆

Exercises

Form the duals of the LP problems in Exercises 1–4.

1. Maximize: $z = 24x_1 + 32x_2$

Subject to: $4x_1 + 3x_2 \leq 12$
$2x_1 - x_2 \leq 6$
$5x_1 + 2x_2 \leq 18$
$x_1, x_2 \geq 0$

3. Maximize: $z = -27x_1 + 24x_2 + 15x_3$

Subject to: $3x_1 + 2x_2 + 4x_3 \leq 12$
$2x_1 - 3x_2 + x_3 \leq 21$
$x_2 + 5x_3 \leq 5$
$x_i \geq 0, \quad 1 \leq i \leq 3$

2. Maximize: $z = 21x_1 + 24x_2 + 15x_3$

Subject to: $x_1 + 5x_2 + 3x_3 \leq 20$
$-x_1 + 2x_2 + 4x_3 \leq 15$
$2x_1 + 3x_2 + 5x_3 \leq 18$
$x_i \geq 0, \quad 1 \leq i \leq 3$

4. Maximize: $z = 12x_1 - 15x_2 - 3x_3 + x_4$

Subject to: $2x_1 - 3x_2 + 4x_3 - 2x_4 \leq 12$
$-x_1 + 2x_2 + x_3 + x_4 \leq 10$
$3x_1 + 2x_2 - 4x_3 - 2x_4 \leq 5$
$x_i \geq 0, \quad 1 \leq i \leq 4$

In Exercises 5 and 6, the dual problems of the given primal problems have some negative values on the right-hand side of some of their constraints. Rearrange the constraints so that all have nonnegative right-hand sides, and solve the dual problems with the two-phase method.

5. Maximize: $z = 2x_1 - 3x_2$

Subject to: $x_1 + x_2 \leq 4$
$x_1 - 2x_2 \leq 3$
$x_1, x_2 \geq 0$

6. Maximize: $x = 30x_1 - 20x_2 - 15x_3$

Subject to: $x_1 + x_2 + x_3 \leq 15$
$2x_1 - 3x_2 + x_3 \leq 12$
$3x_1 + 2x_2 - 2x_3 \leq 10$
$x_i \geq 0, \quad 1 \leq i \leq 3$

7. By graphing the feasible sets, find all corner points of LP problems (7) and (8) of Example 2. Show that the value of the objective function at each corner point of problem (7) is less than or equal to the value of the objective function of problem (8) at every one of its corner points.

8. Prove the corollary to Theorem 1. Hint: It is easier to prove the contrapositive statement.

9. The converse of the corollary to Theorem 1 is: if the dual problem has no feasible solutions, then the primal problem has unbounded feasible solutions. Show that this converse is false by giving an example of a primal problem with a bounded feasible solution whose dual has no feasible solutions.

10. The solution of LP problem (1) of Section 3.2 is given in Tableau 3 of that section. Express the optimal value $z^* = 12$ of the objective function of this problem in terms of the dual prices (use Theorem 3 to find the dual prices). By how much will the optimal value $z^* = 12$ change if the cost coefficient of x_1 is changed to 4 from its given value of 3?

11. In Example 5 of Section 3.2, we solved a product mix problem for the Carter Nut Company. The solution is given in Tableau 4 of that example. Express the optimal value $z^* = 213.75$ of the objective function of this problem in terms of the dual prices (use Theorem 3 to find the dual prices).
   (a) By how much will the optimal value $z^* = 213.75$ change if the resource values b_3, b_4, b_5, and b_6 are each increased by 10?
   (b) What are the reduced costs of the decision variables?

12. The solution to the LP problem of Example 3 of Section 4.3 is given in Tableau 4 of that section. Express the optimal solution in terms of the dual prices.

(a) By how much will the optimal solution $z^* = \frac{75}{2}$ change if the resource values are all increased by 1?

(b) What is the reduced cost of the nonbasic variable x_3? By how much will the optimal objective function value z^* be reduced if x_3 is set equal to $\frac{1}{2}$ and all other nonbasic variables stay at 0?

13. We can consider the LP problem of Example 3 of Section 4.4 as having constraints of the form $A\mathbf{x} \leq \mathbf{b}$ by regarding x_2, x_4, and x_6 as slack variables. Find the dual prices for this problem from Tableau 4 of Section 4.4.

(a) How much would the optimal objective value $z^* = 220$ change if the resource value b_3 is changed to 50 from its given value of 60?

(b) What are the reduced costs of the nonbasic decision variables? If we set the nonbasic variable $x_3 = 20$, would there be a feasible solution with all nonbasic variables except x_3 equal to 0? If the answer is yes, what would the corresponding value of the objective function be?

14. Suppose we set $x_2 = 0$ in equation (19) of Example 6. By how much could we increase x_3 without causing any of the basic variables to become infeasible (assume a negative value)?

6.3 DUALITY: THE GENERAL CASE

In this section we extend the concept of duality to LP problems with a mixture of $\leq$, $\geq$, and $=$ constraints. We defer proofs of the results found in this section to Section 6.4.

Definition. The *dual* of the LP problem

$$\begin{aligned} \text{Maximize:} \quad & z = \mathbf{c}\mathbf{x} \qquad \mathbf{c}, 1 \times n; \qquad \mathbf{x}, n \times 1 \\ \text{Subject to:} \quad & A_1\mathbf{x} \leq \mathbf{b}_1 \qquad A_1, m_1 \times n; \qquad \mathbf{b}_1, m_1 \times 1 \\ & A_2\mathbf{x} = \mathbf{b}_2 \qquad A_2, m_2 \times n; \qquad \mathbf{b}_2, m_2 \times 1 \\ & A_3\mathbf{x} \geq \mathbf{b}_3 \qquad A_3, m_3 \times n; \qquad \mathbf{b}_3, m_3 \times 1 \\ & \mathbf{x} \geq \mathbf{0} \end{aligned} \tag{1}$$

is

$$\begin{aligned} \text{Minimize:} \quad & y = \mathbf{b}_1^T\mathbf{u} + \mathbf{b}_2^T\mathbf{v} + \mathbf{b}_3^T\mathbf{w} \qquad \mathbf{u}, m_1 \times 1; \qquad \mathbf{v}, m_2 \times 1; \qquad \mathbf{w}, m_3 \times 1 \\ \text{Subject to:} \quad & A_1^T\mathbf{u} + A_2^T\mathbf{v} + A_3^T\mathbf{w} \geq \mathbf{c}^T \\ & \mathbf{u} \geq \mathbf{0}; \quad \mathbf{v} \text{ unrestricted}; \quad \mathbf{w} \leq \mathbf{0} \end{aligned} \tag{2}$$

Notice that we do not require that $\mathbf{b}_1$, $\mathbf{b}_2$, and $\mathbf{b}_3$ be nonnegative in this definition, even though it is possible to write any LP problem in the form (1) with nonnegative resource values. We do not require nonnegativity in order to more simply state the concept given in Theorem 1 below.

In what follows, whenever a given LP problem does not have constraints of a particular type, we omit all references to the matrices in the definition of the dual that correspond to this type.

EXAMPLE 1

In the notation of (1), the matrices associated with the LP problem

$$\text{Maximize:} \quad z = 2x_1 - x_2 + x_3 + 0x_4 - x_5 + 4x_6$$

$$\begin{aligned}
\text{Subject to:} \quad & 3x_1 + 2x_2 - x_3 && + x_6 \le 2 \\
& x_1 - 3x_2 && + x_4 + x_5 - 2x_6 \le 6 \\
& 4x_1 + x_2 && + x_5 && = 1 \\
& 3x_1 - 2x_2 && + x_4 && + 3x_6 \ge 5 \\
& -2x_1 && + x_3 && - x_6 \ge 0
\end{aligned} \quad (3)$$

$$x_i \ge 0, \qquad 1 \le i \le 6$$

are

$$\mathbf{x} = [x_1 \quad x_2 \quad x_3 \quad x_4 \quad x_5 \quad x_6]^{\mathrm{T}}$$

$$\mathbf{c} = [2 \quad -1 \quad 1 \quad 0 \quad -1 \quad 4]$$

$$A_1 = \begin{bmatrix} 3 & 2 & -1 & 0 & 0 & 1 \\ 1 & -3 & 0 & 1 & 1 & -2 \end{bmatrix}$$

$$A_2 = [4 \quad 1 \quad 0 \quad 0 \quad 1 \quad 0]$$

$$A_3 = \begin{bmatrix} 3 & -2 & 0 & 1 & 0 & 3 \\ -2 & 0 & 1 & 0 & 0 & -1 \end{bmatrix}$$

$$\mathbf{b}_1 = \begin{bmatrix} 2 \\ 6 \end{bmatrix}, \qquad \mathbf{b}_2 = [1], \qquad \mathbf{b}_3 = \begin{bmatrix} 5 \\ 0 \end{bmatrix}$$

The dual of (3) is

$$\text{Minimize:} \quad y = \mathbf{b}_1^{\mathrm{T}}\mathbf{u} + \mathbf{b}_2^{\mathrm{T}}\mathbf{v} + \mathbf{b}_3^{\mathrm{T}}\mathbf{w}$$

$$\text{Subject to:} \quad A_1^{\mathrm{T}}\mathbf{u} + A_2^{\mathrm{T}}\mathbf{v} + A_3^{\mathrm{T}}\mathbf{w} \ge \mathbf{0}$$

$$\mathbf{u} \ge \mathbf{0}; \qquad \mathbf{v} \text{ unrestricted;} \qquad \mathbf{w} \le \mathbf{0}$$

which, when written without matrix notation, becomes

$$\text{Minimize:} \quad y = 2u_1 + 6u_2 + v_1 + 5w_1 + 0w_2$$

$$\begin{aligned}
\text{Subject to:} \quad & 3u_1 + u_2 + 4v_1 + 3w_1 - 2w_2 \ge 2 \\
& 2u_1 - 3u_2 + v_1 - 2w_1 && \ge -1 \\
& -u_1 && + w_2 \ge 1 \\
& u_2 && + w_1 && \ge 0 \\
& u_2 + v_1 && \ge -1 \\
& u_1 - 2u_2 && + 3w_1 - w_2 \ge 4
\end{aligned}$$

$$u_1, u_2 \ge 0; \qquad v_1 \text{ unrestricted;} \qquad w_1, w_2 \le 0 \qquad \blacklozenge$$

We next show that the dual of the dual problem is the primal problem in the following sense. Let (1) be the primal problem.

STEP 1. We restate the dual (2) as a maximization problem.

STEP 2. We convert the resulting problem from step 1 into an equivalent problem in the primal format.

STEP 3. We take the dual of the problem resulting from step 2.

STEP 4. We convert the result of step 3 to an equivalent problem (in primal format) which is the same as the original problem.

Theorem 1.   The dual of the dual of a primal problem is the primal problem (in the sense of the preceding steps 1–4).

PROOF.

STEP 1. The dual of

$$
\begin{aligned}
\text{Maximize:} \quad & z = \mathbf{cx} \\
\text{Subject to:} \quad & A_1\mathbf{x} \le \mathbf{b}_1 \\
& A_2\mathbf{x} = \mathbf{b}_2 \\
& A_3\mathbf{x} \ge \mathbf{b}_2 \\
& \mathbf{x} \ge \mathbf{0}
\end{aligned}
\tag{4}
$$

is

$$
\begin{aligned}
\text{Minimize:} \quad & y = \mathbf{b}_1^\mathsf{T}\mathbf{u} + \mathbf{b}_2^\mathsf{T}\mathbf{v} + \mathbf{b}_3^\mathsf{T}\mathbf{w} \\
\text{Subject to:} \quad & A_1^\mathsf{T}\mathbf{u} + A_2^\mathsf{T}\mathbf{v} + A_3^\mathsf{T}\mathbf{w} \ge \mathbf{c}^\mathsf{T} \\
& \mathbf{u} \ge \mathbf{0}; \quad \mathbf{v} \ \text{unrestricted}; \quad \mathbf{w} \le \mathbf{0}
\end{aligned}
\tag{5}
$$

STEP 2. Note that we can write (5) as

$$
\text{Minimize:} \quad y = [\,\mathbf{b}_1^\mathsf{T} \mid \mathbf{b}_2^\mathsf{T} \mid -\mathbf{b}_3^\mathsf{T}\,]
\begin{bmatrix} \mathbf{u} \\ \mathbf{v} \\ -\mathbf{w} \end{bmatrix}
$$

$$
\text{Subject to:} \quad [\,-A_1^\mathsf{T} \mid -A_2^\mathsf{T} \mid A_3^\mathsf{T}\,]
\begin{bmatrix} \mathbf{u} \\ \mathbf{v} \\ -\mathbf{w} \end{bmatrix} \le -\mathbf{c}^\mathsf{T}
\tag{6}
$$

$$
\mathbf{u} \ge \mathbf{0}; \quad \mathbf{v} \ \text{unrestricted}; \quad \mathbf{w} \le \mathbf{0}
$$

Let $\mathbf{w}' = -\mathbf{w}$, and let $\mathbf{v}^+ \ge \mathbf{0}$, $\mathbf{v}^- \ge \mathbf{0}$ be such that

$$
\mathbf{v}^+ - \mathbf{v}^- = \mathbf{v}
$$

(In Exercises 17 and 18 you are asked to find some matrices $\mathbf{v}^+$ and $\mathbf{v}^-$

having this property.) Then,

$$\mathbf{b}_2^T\mathbf{v} = \mathbf{b}_2^T\mathbf{v}^+ - \mathbf{b}_2^T\mathbf{v}^-, \qquad \text{and} \qquad -A_2^T\mathbf{v} = -A_2^T\mathbf{v}^+ + A_2^T\mathbf{v}^-$$

Thus (6) is equivalent to the LP problem in primal format:

$$\text{Maximize:} \quad y = [-\mathbf{b}_1^T \mid -\mathbf{b}_2^T \mid \mathbf{b}_2^T \mid \mathbf{b}_3^T]\begin{bmatrix} \mathbf{u} \\ \mathbf{v}^+ \\ \mathbf{v}^- \\ \mathbf{w}' \end{bmatrix}$$

$$\text{Subject to:} \quad [-A_1^T \mid -A_2^T \mid A_2^T \mid A_3^T]\begin{bmatrix} \mathbf{u} \\ \mathbf{v}^+ \\ \mathbf{v}^- \\ \mathbf{w}' \end{bmatrix} \le -\mathbf{c}^T \qquad (7)$$

$$\mathbf{u}, \mathbf{v}^+, \mathbf{v}^-, \mathbf{w}' \ge \mathbf{0}$$

STEP 3. From (1) we have that the dual of (7) is

$$\text{Minimize:} \quad z = -\mathbf{cx}$$

$$\text{Subject to:} \quad \begin{bmatrix} -A_1 \\ -A_2 \\ A_2 \\ A_3 \end{bmatrix}\mathbf{x} \ge \begin{bmatrix} -\mathbf{b}_1 \\ -\mathbf{b}_2 \\ \mathbf{b}_2 \\ \mathbf{b}_3 \end{bmatrix} \qquad (8)$$

$$\mathbf{x} \ge \mathbf{0}$$

STEP 4. We can rewrite (8) as

$$\text{Minimize:} \quad z = -\mathbf{cx}$$

$$\begin{aligned} \text{Subject to:} \quad -A_1\mathbf{x} &\ge -\mathbf{b}_1 \\ -A_2\mathbf{x} &\ge -\mathbf{b}_2 \\ A_2\mathbf{x} &\ge \mathbf{b}_2 \\ A_3\mathbf{x} &\ge \mathbf{b}_3 \end{aligned} \qquad (9)$$

$$\mathbf{x} \ge \mathbf{0}$$

But (9) is equivalent to

$$\text{Maximize:} \quad z = \mathbf{cx}$$

$$\begin{aligned} \text{Subject to:} \quad A_1\mathbf{x} &\le \mathbf{b}_1 \\ A_2\mathbf{x} &\le \mathbf{b}_2 \\ A_2\mathbf{x} &\ge \mathbf{b}_2 \\ A_3\mathbf{x} &\ge \mathbf{b}_3 \end{aligned}$$

$$\mathbf{x} \ge \mathbf{0}$$

or

Maximize: $z = \mathbf{cx}$

Subject to: $A_1\mathbf{x} \le \mathbf{b}_1$

$A_2\mathbf{x} = \mathbf{b}_2$

$A_3\mathbf{x} \ge \mathbf{b}_3$

$\mathbf{x} \ge \mathbf{0}$

which is the original LP problem (4).   ∎

We state the Fundamental Principle of Duality in the next theorem and prove it in Section 6.4.

Theorem 2.   (The Fundamental Principle of Duality).   If an optimal solution $\mathbf{x}^*$ of the primal problem (1) exists, then an optimal solution

$$\begin{bmatrix} \mathbf{u}^* \\ \mathbf{v}^* \\ \mathbf{w}^* \end{bmatrix}$$

of the dual problem (2) exists, and

$$z^* = \mathbf{cx}^* = \mathbf{b}_1^T\mathbf{u}^* + \mathbf{b}_2^T\mathbf{v}^* + \mathbf{b}_3^T\mathbf{w}^* = y^*$$

Moreover, if an optimal solution

$$\begin{bmatrix} \mathbf{u}^* \\ \mathbf{v}^* \\ \mathbf{w}^* \end{bmatrix}$$

of the dual problem (2) exists, then an optimal solution $\mathbf{x}^*$ of the primal problem (1) exists, and

$$z^* = \mathbf{cx}^* = \mathbf{b}_1^T\mathbf{u}^* + \mathbf{b}_2^T\mathbf{v}^* + \mathbf{b}_3^T\mathbf{w}^* = y^*$$

The next example illustrates Theorem 2 and also shows how to find the solution of the dual problem from a final tableau of the primal problem. In Section 6.4 we will explain why we can find the solution of the dual problem in this way.

EXAMPLE 2

We consider the primal LP problem

Maximize: $z = 2x_1 - 3x_2 + x_3$

Subject to: $x_1 + x_2 - x_3 \le 10$

$2x_1 + x_2 - x_3 \le 6$

$x_1 \qquad + x_3 = 5$ (10)

$2x_1 - x_2 + x_3 \ge 1$

$x_1 - x_2 - x_3 \ge 1$

$x_i \ge 0, \qquad 1 \le i \le 3$

Introducing slack and artificial variables to (10) we have

Maximize: $z = 2x_1 - 3x_2 + x_3$

Subject to:
$$x_1 + x_2 - x_3 + x_4 \qquad\qquad\qquad\qquad = 10$$
$$2x_1 + x_2 - x_3 \qquad + x_5 \qquad\qquad\qquad = 6$$
$$x_1 \qquad + x_3 \qquad\qquad\qquad\qquad + y_1 = 5 \qquad (11)$$
$$2x_1 - x_2 + x_3 \qquad\qquad\qquad - x_6 + y_2 = 1$$
$$x_1 - x_2 - x_3 \qquad\qquad\qquad - x_7 + y_3 = 1$$

$$x_i \geq 0, \quad 1 \leq i \leq 7; \quad y_i \geq 0, \quad 1 \leq i \leq 3$$

The initial tableau for (11) is

Initial Tableau

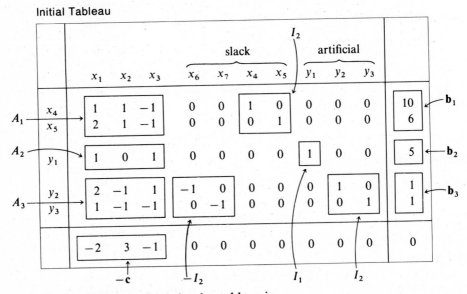

You may verify that the final simplex tableau is

Final Tableau

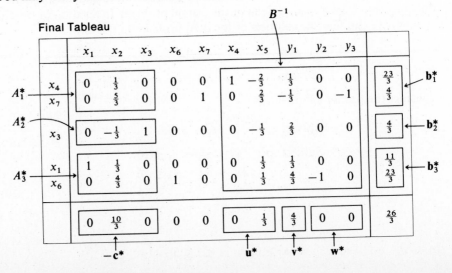

The solution to the dual problem

$$\text{Minimize:} \quad y = 10u_1 + 6u_2 + 5v + w_1 + w_2$$

$$\text{Subject to:} \quad u_1 + 2u_2 + v + 2w_1 + w_2 \geq 2$$
$$u_1 + u_2 \quad - w_1 - w_2 \geq -3$$
$$-u_1 - u_2 + v + w_1 - w_2 \geq 1$$

$$u_1, u_2 \geq 0; \quad v \quad \text{unrestricted;} \quad w_1, w_2 \leq 0$$

is

$$\mathbf{u}^* = \begin{bmatrix} u_1^* \\ u_2^* \end{bmatrix} = \begin{bmatrix} 0 \\ \frac{1}{3} \end{bmatrix}$$

$$\mathbf{v}^* = [v] = \frac{4}{3}$$

$$\mathbf{w}^* = \begin{bmatrix} w_1^* \\ w_2^* \end{bmatrix} = \begin{bmatrix} 0 \\ 0 \end{bmatrix}$$

In the final tableau note that the optimal dual prices are found below the columns of the B^{-1} matrix. In general (as we will see in Section 6.4) the dual price for the ith constraint is found in the objective row of the column corresponding to the ith column of the B^{-1} matrix.

We verify the conclusion of Theorem 2 by the calculation

$$z^* = \frac{26}{3} = \mathbf{b}_1^T \mathbf{u}^* + \mathbf{b}_2^T \mathbf{v}^* + \mathbf{b}_3^T \mathbf{w}^*$$

$$[10 \quad 6]\begin{bmatrix} 0 \\ \frac{1}{3} \end{bmatrix} + [5][\tfrac{4}{3}] + [1 \quad 1]\begin{bmatrix} 0 \\ 0 \end{bmatrix} = 2 + \frac{20}{3} = \frac{26}{3} \qquad \blacklozenge$$

Table 1 summarizes the relations between the primal and dual problems.

Table 1

Primal problem	Dual problem
Maximize objective function	Minimize objective function
Cost coefficients	Resource coefficients
Coefficients of ith constraints	Coefficients of ith variable (in each constraint)
ith constraint $\leq$	ith variable ≥ 0
ith constraint $=$	ith variable unrestricted
ith constraint $\geq$	ith variable ≤ 0
Constraint matrix A	Constraint matrix A^T

In the next example we show how to use Table 1 to form and solve the dual of a problem without resorting to the definition using matrices.

EXAMPLE 3

The dual of the primal LP problem

$$\text{Maximize:} \quad z = 3x_1 - 2x_2$$
$$\text{Subject to:} \quad 3x_1 - 2x_2 + 4x_3 \geq 1$$
$$2x_1 + 3x_2 - x_3 \leq 4$$
$$-x_1 + 2x_2 + 3x_3 = 4$$
$$3x_1 - 2x_2 - x_3 \geq 0$$
$$x_i \geq 0, \quad 1 \leq i \leq 3$$

is

$$\text{Minimize:} \quad y = u_1 + 4u_2 + 4u_3$$
$$\text{Subject to:} \quad 3u_1 + 2u_2 - u_3 + 3u_4 \geq 3$$
$$-2u_1 + 3u_2 + 2u_3 - 2u_4 \geq -2 \tag{12}$$
$$4u_1 - u_2 + 3u_3 - u_4 \geq 0$$
$$u_1 \leq 0, \quad u_2 \geq 0, \quad u_3 \text{ unrestricted}, \quad u_4 \leq 0$$

To solve (12) by the two-phase method, we change it to a maximization problem, make the resource values nonnegative by multiplying inequalities with negative resource values by -1, and then make substitutions for any negative or unrestricted variables to obtain the equivalent problem

$$\text{Maximize:} \quad y' = -u_1 - 4u_2 - 4u_3$$
$$\text{Subject to:} \quad 3u_1 + 2u_2 - u_3 + 3u_4 \geq 3$$
$$2u_1 - 3u_2 - 2u_3 + 2u_4 \leq 2$$
$$4u_1 - u_2 + 3u_3 - u_4 \geq 0$$
$$u_1 \leq 0, \quad u_2 \geq 0, \quad u_3 \text{ unrestricted}, \quad u_4 \leq 0$$

Making the substitutions

$$u_1 = -v_1, \quad u_3 = w_3 - t_3, \quad u_4 = -v_4 \tag{13}$$

we have

$$\text{Maximize:} \quad y' = v_1 - 4u_2 - 4w_3 + 4t_3$$
$$\text{Subject to:} \quad -3v_1 + 2u_2 - w_3 + t_3 - 3v_4 \geq 3$$
$$-2v_1 - 3u_2 - 2w_3 + 2t_3 - 2v_4 \leq 2 \tag{14}$$
$$-4u_1 - u_2 + 3w_3 - 3t_3 + v_4 \geq 0$$
$$u_1, u_2, w_3, t_3, v_4 \geq 0$$

Now we can solve problem (14) by the two-phase method. We can then solve problem (12) by using equations (13).

♦

The next example shows that the duals of two equivalent LP problems may not be the same. Thus, when we speak of *the* dual prices of an LP problem, we must have in mind a particular arrangement of the constraints defining the feasible set.

EXAMPLE 4

The LP problem

$$\text{Maximize:} \quad z = 120x_1 + 100x_2$$
$$\text{Subject to:} \quad 4x_1 + 4x_2 \le 8$$
$$5x_1 + 3x_2 \le 15 \tag{15}$$
$$x_1, x_2 \ge 0$$

and the LP problem

$$\text{Maximize:} \quad z = 120x_1 + 100x_2$$
$$\text{Subject to:} \quad 4x_1 + 4x_2 \le 8$$
$$-5x_1 - 3x_2 \ge -15 \tag{16}$$
$$x_1, x_2 \ge 0$$

are equivalent in the sense that they have the same feasible set and the same optimal solution. The dual of (15) is

$$\text{Minimize:} \quad y = 8u_1 + 15u_2$$
$$\text{Subject to:} \quad 4u_1 + 5u_2 \ge 120$$
$$4u_1 + 3u_2 \ge 100 \tag{17}$$
$$u_1, u_2 \ge 0$$

You may verify that the optimal solution to (16) is $x_1 = \frac{3}{2}$, $x_2 = \frac{5}{2}$ and the dual prices (obtained by using Theorem 3 of Section 6.2) are $u_1 = 35$, $u_2 = 10$.
The dual of (16) is

$$\text{Minimize:} \quad y = 8u_1 - 15u_2$$
$$\text{Subject to:} \quad 4u_1 - 5u_2 \ge 120$$
$$4u_1 - 3u_2 \ge 100 \tag{18}$$
$$u_1 \ge 0, \qquad u_2 \le 0$$

You can verify that the optimal solution of (18) is $u_1 = 35$, $u_2 = -10$.

♦

To conclude this section we consider a class of primal LP problems whose duals are easier to solve by the simplex method than are the primal problems themselves. We can write the dual of any problem of the form

$$\text{Minimize:} \quad z = \mathbf{cx}$$

$$\text{Subject to:} \quad A\mathbf{x} \le \mathbf{b}$$

$$\mathbf{x} \ge \mathbf{0}$$

where $\mathbf{c} \ge \mathbf{0}$ and the entries of $\mathbf{b}$ are positive, negative, or zero as

$$\text{Maximize:} \quad y = \mathbf{b}^{\mathsf{T}}\mathbf{u}$$

$$\text{Subject to:} \quad -A^{\mathsf{T}}\mathbf{u} \le \mathbf{c}^{\mathsf{T}} \tag{19}$$

$$\mathbf{u} \ge \mathbf{0}$$

Problem (19) is easy to solve with the simplex method because the slack variables can be used as an initial basic feasible solution. Example 5 illustrates this idea.

EXAMPLE 5

The LP problem

$$\text{Minimize:} \quad z = 2x_1 + 3x_2 + 4x_3$$

$$\text{Subject to:} \quad \begin{aligned} -x_1 - x_2 - x_3 &\ge -6 \\ 2x_1 - 3x_2 + x_3 &\le -12 \end{aligned} \tag{20}$$

$$x_i \ge 0, \quad 1 \le i \le 3$$

is equivalent to

$$\text{Maximize:} \quad z' = -2x_1 - 3x_2 - 4x_3$$

$$\text{Subject to:} \quad \begin{aligned} x_1 + x_2 + x_3 &\le 6 \\ 2x_1 - 3x_2 + x_3 &\le -12 \end{aligned} \tag{21}$$

$$x_i \ge 0, \quad 1 \le i \le 3$$

The dual of (21) is

$$\text{Minimize:} \quad y = 6u_1 - 12u_2$$

$$\text{Subject to:} \quad \begin{aligned} u_1 + 2u_2 &\ge -2 \\ u_1 - 3u_2 &\ge -3 \\ u_1 + u_2 &\ge -4 \end{aligned} \tag{22}$$

$$u_1, u_2 \ge 0$$

Problem (22) is equivalent to

$$\text{Maximize:} \quad y' = -6u_1 + 12u_2$$

$$\text{Subject to:} \quad -u_1 - 2u_2 \le 2$$
$$-u_1 + 3u_2 \le 3$$
$$-u_1 - u_2 \le 4$$

$$u_1, u_2 \ge 0$$

We also note that if a problem has many more variables than constraints, it is usually more efficient to solve the dual problem than to solve the primal problem. This is evident from our consideration of time complexity in Chapter 5 where we showed that the time complexity of the revised simplex method is highly dependent on m, the number of constraints.

Exercises

In Exercises 1–8, (a) write the LP problem in matrix notation, (b) form the dual in matrix notation, (c) write the matrix expressions of part (b) as a system of linear inequalities and/or equalities, and (d) formulate an LP problem with nonnegative variables that you could use to find the solution of the dual problem.

1. Maximize: $z = 80x_1 + 100x_2$

Subject to: $2x_1 - 3x_2 \le 4$
$3x_1 + 2x_2 \le 3$
$4x_1 + 5x_2 \ge 12$
$2x_1 + 8x_2 \ge 8$

$x_1, x_2 \ge 0$

2. Maximize: $z = 12x_1 - 14x_2$

Subject to: $-2x_1 + 4x_2 \le 5$
$x_1 + 7x_2 \ge 12$
$x_1 - 3x_2 \ge 6$

$x_1, x_2 \ge 0$

3. Maximize: $z = 2x_1 + 3x_2 - 4x_3$

Subject to: $3x_1 + 2x_2 - x_3 \le 12$
$x_1 + x_2 + x_3 = 3$
$4x_1 + 3x_2 + x_3 \ge 8$
$3x_1 + 2x_2 - x_3 \ge 2$

$x_i \ge 0, \quad 1 \le i \le 3$

4. Maximize: $z = 2x_1 - 3x_2 + 5x_3$

Subject to: $2x_1 - x_2 + 3x_3 \le 5$
$3x_1 + 2x_2 - x_3 \le 5$
$4x_1 + 3x_2 + 2x_3 = 10$
$-x_1 + 2x_2 + 3x_3 \ge 1$
$2x_1 + 7x_2 + 13x_3 \ge 27$

$x_i \ge 0, \quad 1 \le i \le 3$

5. Maximize: $z = 2x_1 + 3x_2 + 5x_3 + 7x_4$

Subject to: $x_1 + 3x_2 - 2x_3 + 4x_4 \le 8$
$2x_1 + 3x_2 - 4x_3 + x_4 \le 12$
$x_1 + x_2 + 3x_3 + x_4 = 9$
$2x_1 - 3x_2 + 2x_3 - x_4 = 6$
$x_1 + 3x_2 + 5x_3 + 7x_4 \ge 8$
$2x_1 + 9x_2 + 4x_3 + 8x_4 \ge 15$

$x_i \ge 0, \quad 1 \le i \le 4$

6. Maximize: $z = 2x_1 - 3x_2 + 4x_3 + 5x_4 - 3x_5$

Subject to: $x_1 - 2x_2 + 3x_3 + x_4 \quad\quad \le 8$

$2x_1 + 3x_2 - x_3 \quad\quad + 4x_5 \le 9$

$x_1 - x_2 + x_3 - x_4 + x_5 = 7$

$2x_1 \quad\quad + 2x_3 \quad\quad - 2x_5 = 6$

$x_1 + 3x_2 + 5x_3 \quad\quad\quad \ge 8$

$x_2 + 3x_3 + 4x_4 + x_5 \ge 9$

$x_i \ge 0, \quad 1 \le i \le 5$

7. Minimize: $z = 6x_1 - 3x_2$

Subject to: $3x_1 - 2x_2 \ge 4$

$2x_1 + 3x_2 \ge 5$

$x_1 + x_2 \le 3$

$x_1, x_2 \ge 0$

8. Minimize: $z = -2x_1 + 3x_2 - 4x_3$

Subject to: $2x_1 + 2x_2 - 3x_3 \ge 0$

$x_1 - 2x_2 + x_3 \le 5$

$-x_1 + 2x_2 + x_3 \ge 1$

$x_i \ge 0, \quad 1 \le i \le 3$

9. Use Table 1 to find the dual of the LP problem in
(a) Exercise 1 (b) Exercise 3 (c) Exercise 5 (d) Exercise 7

10. Use Table 1 to find the dual of the LP problem in
(a) Exercise 2 (b) Exercise 4 (c) Exercise 6 (d) Exercise 8

11. Find the dual of

Maximize: $z = 3x_1 - 4x_2 + 9x_3 + x_4$

Subject to: $x_1 - 5x_2 + x_3 \quad\quad \ge 0$

$3x_1 + 8x_2 \quad\quad - x_4 \ge 10$

x_1 unrestricted; $x_2, x_3 \ge 0$; $x_4 \le 0$

12. Find the dual of

Minimize: $z = 2x_1 - x_2 + 4x_3 - x_4$

Subject to: $2x_1 + 3x_2 - x_3 + x_4 \le 12$

$-x_1 - 3x_2 + 2x_3 - x_4 = 9$

$3x_1 + 2x_2 - 9x_3 + 7x_4 \ge 3$

$x_1, x_2 \ge 0$; x_3 unrestricted; $x_4 \le 0$

13. Use Theorem 1 to form the dual of the dual problem (8) of Section 6.1, and show that it is the same as (1) of Section 6.1.

14. Use Theorem 1 to form the dual of the dual problem given in Example 1, and show that it is the same as (3).

†**15.** Solve the primal problem (1) and the dual problem (8) of Section 6.1, and show that the objective functions of both problems have the same optimal value.

†16. Solve the problem

$$\text{Minimize:} \quad z = 30x_1 + 25x_2$$
$$\text{Subject to:} \quad x_1 + 4x_2 \geq 12$$
$$3x_1 + 2x_2 \geq 15$$
$$5x_1 + 3x_2 \geq 22$$
$$x_1, x_2 \geq 0$$

and its dual. Show that both problems have the same optimal value of their objective functions.

17. Write the matrix

$$\mathbf{v} = \begin{bmatrix} 3 \\ -4 \\ -6 \end{bmatrix}$$

as $\mathbf{v} = \mathbf{v}^+ - \mathbf{v}^-$ where $\mathbf{v}^+$ and $\mathbf{v}^-$ have nonnegative entries. (The answer is not unique.)

18. Write the matrix $\mathbf{v} = [-6 \quad 3 \quad 0 \quad 1 \quad -2 \quad -3]^T$ as $\mathbf{v} = \mathbf{v}^+ - \mathbf{v}^-$ where $\mathbf{v}^+$ and $\mathbf{v}^-$ have nonnegative entries.

In Exercises 19–22, you are given an LP problem, its initial tableau, including slack and/or artificial variables, and its optimal tableau, which results from applying the simplex or the two-phase method. Decision and slack variables begin with the letter x, and artificial variables begin with the letter y. For each question, do the following:

(a) Identify the solution to the problem and its dual. Give the number of the row in the initial tableau associated with each dual price.

(b) Evaluate the objective functions for the primal and dual problems, and verify that they have the same value. This serves as a partial check on the correctness of the dual prices you found in (a).

19. Maximize: $z = x_2 + 3x_3$

Subject to: $x_1 + 2x_2 + x_3 \leq 7$
$-3x_1 + 2x_2 + x_3 \leq 3$
$x_i \geq 0, \quad 1 \leq i \leq 3$

	x_1	x_2	x_3	x_4	x_5	
x_4	1	2	1	1	0	7
x_5	-3	2	1	0	1	3
	0	-1	-3	0	0	0
x_1	1	0	0	$\frac{1}{4}$	$-\frac{1}{4}$	1
x_3	0	2	1	$\frac{3}{4}$	$\frac{1}{4}$	6
	0	5	0	$\frac{9}{4}$	$\frac{3}{4}$	18

Initial Tableau / Final Tableau

20. Maximize: $z = 3x_1 + 2x_2 + 3x_3$

Subject to:
$$2x_1 + 3x_2 \quad\quad\; \le 60$$
$$x_1 - x_2 + 4x_3 \le 10$$
$$-2x_1 + 2x_2 + 5x_3 \le 50$$

$$x_i \ge 0, \quad 1 \le i \le 3$$

	x_1	x_2	x_3	x_4	x_5	x_6	
x_4	2	3	0	1	0	0	60
x_5	1	-1	4	0	1	0	10
x_6	-2	2	5	0	0	1	50
	-3	-2	-3	0	0	0	0
x_2	0	1	$-\frac{8}{5}$	$\frac{1}{5}$	$-\frac{2}{5}$	0	8
x_1	1	0	$\frac{12}{5}$	$\frac{1}{5}$	$\frac{3}{5}$	0	18
x_6	0	0	13	0	2	1	70
	0	0	1	1	1	0	70

Initial Tableau / Final Tableau

21. Minimize: $z = 2x_1 - 3x_2 + x_3 + x_4$

Subject to:
$$x_1 - 2x_2 - 3x_3 - 2x_4 = 3$$
$$x_1 - x_2 + 2x_3 + x_4 = 11$$

$$x_i \ge 0, \quad 1 \le i \le 4$$

	x_1	y_1	x_2	x_3	y_2	x_4	
y_1	1	1	-2	-3	0	-2	3
y_2	1	0	-1	2	1	1	11
(z)	2	0	-3	1	0	1	0
(w)	-2	0	3	1	0	1	-14
x_1	1	-1	0	7	2	4	19
x_2	0	-1	1	5	1	3	8
(z)	0	-1	0	2	-1	2	-14

Initial Tableau / Final Tableau

22. Maximize: $z = 4x_1 - 30x_2 - x_3 - 3x_4 + 2x_5 + 11x_6$

Subject to:
$$-2x_1 + 6x_2 \quad\quad - 3x_4 \quad\quad\quad\; + x_7 = 20$$
$$-4x_1 + 7x_2 + x_3 - x_4 \quad\quad + x_6 \quad\;\; = 10$$
$$5x_2 \quad\quad + x_4 + x_5 + 3x_6 \quad = 60$$

$$x_i \ge 0, \quad 1 \le i \le 7$$

Note that we must rewrite the objective function in terms of nonbasic variables before forming the initial tableau. Doing so, we obtain

	x_1	x_2	x_3	x_4	x_5	x_6	x_7		
x_7	-2	6	0	-3	0	0	1	20	Initial
x_3	-4	7	1	-1	0	1	0	10	Tableau
x_5	0	5	0	1	1	3	0	60	
	0	33	0	6	0	-6	0	110	
x_7	0	$\frac{10}{3}$	$-\frac{1}{2}$	$-\frac{7}{3}$	$\frac{1}{6}$	0	1	25	
x_6	0	$\frac{5}{3}$	0	$\frac{1}{3}$	$\frac{1}{3}$	1	0	20	Final
x_1	1	$-\frac{4}{3}$	$-\frac{1}{4}$	$\frac{1}{3}$	$\frac{1}{12}$	0	0	$\frac{5}{2}$	Tableau
	0	43	0	8	2	0	0	230	

In Exercises 23 and 24, show that the two primal problems given are equivalent in the sense that they have the same feasible set and the same optimal solution. Solve the dual of each of the problems, and show that the primal problems have different dual prices.

23. Maximize: $z = 2x_1 + 2x_2$

Subject to: $x_1 + 2x_2 \le 3$

$2x_1 + x_2 \le 2$

$x_1, x_2 \ge 0$

Maximize: $z = 2x_1 + 2x_2$

Subject to: $x_1 + 2x_2 \le 3$

$-2x_1 - x_2 \ge -2$

$x_1, x_2 \ge 0$

24. Maximize: $z = 4x_1 + x_2 + x_3$

Subject to: $3x_1 + 2x_2 + x_3 \le 12$

$2x_1 + x_2 + 2x_3 \le 10$

$-x_1 + x_2 + x_3 = 1$

$x_i \ge 0, \quad 1 \le i \le 3$

Maximize: $z = 4x_1 + x_2 + x_3$

Subject to: $3x_1 + 2x_2 + x_3 \le 12$

$2x_1 + x_2 + 2x_3 \le 10$

$-x_1 + x_2 + x_3 \le 1$

$-x_1 + x_2 + x_3 \ge 1$

$x_i \ge 0, \quad 1 \le i \le 3$

6.4 DUALITY PROOFS

In this section we prove the principal theorems stated in previous sections. We write the primal problem as

$$\text{Maximize:} \quad z = \mathbf{cx}$$

$$\text{Subject to:} \quad A_1\mathbf{x} \le \mathbf{b}_1$$
$$A_2\mathbf{x} = \mathbf{b}_2 \tag{1}$$
$$A_3\mathbf{x} \ge \mathbf{b}_3$$

$$\mathbf{x} \ge \mathbf{0}$$

where $\mathbf{b}_i \geq \mathbf{0}$, $1 \leq i \leq 3$. We write the dual of (1) as

$$\text{Minimize:} \quad y = \mathbf{b}_1^T\mathbf{u} + \mathbf{b}_2^T\mathbf{v} + \mathbf{b}_3^T\mathbf{w}$$

$$\text{Subject to:} \quad A_1^T\mathbf{u} + A_2^T\mathbf{v} + A_3^T\mathbf{w} \geq \mathbf{c}^T \qquad (2)$$

$$\mathbf{u} \geq \mathbf{0}; \quad \mathbf{v} \text{ unrestricted}; \quad \mathbf{w} \leq \mathbf{0}$$

We begin with the Weak Principle of Duality, a special case of which we gave as Theorem 1 in Section 6.2.

Theorem 1. (The Weak Principle of Duality). If $\hat{\mathbf{x}}$ is a feasible solution of the primal problem (1), and

$$\begin{bmatrix} \hat{\mathbf{u}} \\ \hat{\mathbf{v}} \\ \hat{\mathbf{w}} \end{bmatrix}$$

is a feasible solution of the dual problem (2), then

(a) $\mathbf{c}\hat{\mathbf{x}} \leq \mathbf{b}_1^T\hat{\mathbf{u}} + \mathbf{b}_2^T\hat{\mathbf{v}} + \mathbf{b}_3^T\hat{\mathbf{w}}$

(b) If $\mathbf{c}\hat{\mathbf{x}} = \mathbf{b}_1^T\hat{\mathbf{u}} + \mathbf{b}_2^T\hat{\mathbf{v}} + \mathbf{b}_3^T\hat{\mathbf{w}}$, then $\hat{\mathbf{x}}$ is an optimal solution of (1) and

$$\begin{bmatrix} \hat{\mathbf{u}} \\ \hat{\mathbf{v}} \\ \hat{\mathbf{w}} \end{bmatrix}$$

is an optimal solution of (2).

PROOF. (a) Since $\hat{\mathbf{x}}$ is a feasible solution of (1), it follows that

$$A_1\hat{\mathbf{x}} \leq \mathbf{b}_1$$
$$A_2\hat{\mathbf{x}} = \mathbf{b}_2 \qquad (3)$$
$$A_3\hat{\mathbf{x}} \geq \mathbf{b}_3$$

Since the entries of $\hat{\mathbf{u}}$ are nonnegative, the sense of the first inequality in (3) is not changed by premultiplication by $\hat{\mathbf{u}}^T$; therefore,

$$\hat{\mathbf{u}}^TA_1\hat{\mathbf{x}} \leq \hat{\mathbf{u}}^T\mathbf{b}_1 \qquad (4)$$

We can premultiply the equality in (4) by $\hat{\mathbf{v}}^T$ to give

$$\hat{\mathbf{v}}^TA_2\hat{\mathbf{x}} = \hat{\mathbf{v}}^T\mathbf{b}_2 \qquad (5)$$

Since the entries of $\hat{\mathbf{w}}$ are less than or equal to zero, premultiplication by $\hat{\mathbf{w}}^T$ changes the sense of the third inequality of (3) to give

$$\hat{\mathbf{w}}^T A_3 \hat{\mathbf{x}} \le \hat{\mathbf{w}}^T \mathbf{b}_3 \tag{6}$$

Recall (Theorem 3 of Section 4.1) that transposing a product of matrices causes a change of order; for example,

$$(EF)^T = F^T E^T \quad \text{and} \quad (EFS)^T = S^T F^T E^T$$

So, taking transposes in (4), (5), and (6) and adding the results, we have

$$\hat{\mathbf{x}}^T A_1^T \hat{\mathbf{u}} + \hat{\mathbf{x}}^T A_2^T \hat{\mathbf{v}} + \hat{\mathbf{x}}^T A_3^T \hat{\mathbf{w}} \le \mathbf{b}_1^T \hat{\mathbf{u}} + \mathbf{b}_2^T \hat{\mathbf{v}} + \mathbf{b}_3^T \hat{\mathbf{w}} \tag{7}$$

Also, since

$$\begin{bmatrix} \hat{\mathbf{u}} \\ \hat{\mathbf{v}} \\ \hat{\mathbf{w}} \end{bmatrix}$$

is a feasible solution of (2), we have

$$A_1^T \hat{\mathbf{u}} + A_2^T \hat{\mathbf{v}} + A_3^T \hat{\mathbf{w}} \ge \mathbf{c}^T \tag{8}$$

Premultiplying (8) by the nonnegative matrix $\hat{\mathbf{x}}^T$ and using the fact that $\hat{\mathbf{x}}^T \mathbf{c}^T = (\mathbf{c}\hat{\mathbf{x}})^T = \mathbf{c}\hat{\mathbf{x}}$ (since $\mathbf{c}\hat{\mathbf{x}}$ is a 1×1 matrix), we obtain

$$\hat{\mathbf{x}}^T A_1^T \hat{\mathbf{u}} + \hat{\mathbf{x}}^T A_2^T \hat{\mathbf{v}} + \hat{\mathbf{x}}^T A_3^T \hat{\mathbf{w}} \ge \hat{\mathbf{x}}^T \mathbf{c}^T = \mathbf{c}\hat{\mathbf{x}} \tag{9}$$

From (9) and (7) we have

$$\mathbf{c}\hat{\mathbf{x}} \le \mathbf{b}_1^T \hat{\mathbf{u}} + \mathbf{b}_2^T \hat{\mathbf{v}} + \mathbf{b}_3^T \hat{\mathbf{w}}$$

which proves assertion (a).

To prove (b), suppose $\hat{\mathbf{x}}'$ is *any* feasible solution of the primal problem (a). Then from part (a) and the hypothesis of (b), we have

$$\mathbf{c}\hat{\mathbf{x}}' \le \mathbf{b}_1^T \hat{\mathbf{u}} + \mathbf{b}_2^T \hat{\mathbf{v}} + \mathbf{b}_3^T \hat{\mathbf{w}} = \mathbf{c}\hat{\mathbf{x}}$$

Therefore, $\mathbf{c}\hat{\mathbf{x}}$ yields the maximum value of $\mathbf{c}\mathbf{x}$, and so $\hat{\mathbf{x}}$ is the maximal solution of (1).

Similarly, suppose that

$$\begin{bmatrix} \mathbf{u}' \\ \mathbf{v}' \\ \mathbf{w}' \end{bmatrix}$$

is *any* feasible solution of the dual problem (2). Then from part (a) and the hypothesis of (b) we have

$$\mathbf{b}_1^T\hat{\mathbf{u}}' + \mathbf{b}_2^T\hat{\mathbf{v}}' + \mathbf{b}_3^T\hat{\mathbf{w}}' \geq c\hat{\mathbf{x}} = \mathbf{b}_1^T\hat{\mathbf{u}} + \mathbf{b}_2^T\hat{\mathbf{v}} + \mathbf{b}_3^T\hat{\mathbf{w}}$$

and thus,

$$\begin{bmatrix} \hat{\mathbf{u}} \\ \hat{\mathbf{v}} \\ \hat{\mathbf{w}} \end{bmatrix}$$

yields the minimal value for (2). ∎

Our next goal is to prove the Fundamental Principle of Duality, which we stated in Theorem 2 of Section 6.3. First, we introduce the slack variables

$$\mathbf{r} = \begin{bmatrix} r_1 \\ r_2 \\ \vdots \\ r_j \end{bmatrix} \qquad \mathbf{s} = \begin{bmatrix} s_1 \\ s_2 \\ \vdots \\ s_m \end{bmatrix}$$

to LP problem (1) and obtain the problem

$$\begin{aligned}
\text{Maximize:} \quad & z = c\mathbf{x} + 0\mathbf{r} + 0\mathbf{s} \\
\text{Subject to:} \quad & A_1\mathbf{x} \qquad\quad + I_m\mathbf{s} = \mathbf{b}_1 \\
& A_2\mathbf{x} \qquad\qquad\qquad = \mathbf{b}_2 \\
& A_3\mathbf{x} - I_j\mathbf{r} \qquad\quad = \mathbf{b}_3 \\
& \mathbf{x} \geq 0; \quad \mathbf{r} \geq 0; \quad \mathbf{s} \geq 0
\end{aligned} \qquad (10)$$

where the subscript on I indicates the number of rows and the number of columns in the identity matrix I. If we let

$$A = \begin{bmatrix} A_1 & 0 & I_m \\ A_2 & 0 & 0 \\ A_3 & -I_j & 0 \end{bmatrix}$$

$$\mathbf{q} = \begin{bmatrix} \mathbf{x} \\ \mathbf{r} \\ \mathbf{s} \end{bmatrix}, \qquad \mathbf{b} = \begin{bmatrix} \mathbf{b}_1 \\ \mathbf{b}_2 \\ \mathbf{b}_3 \end{bmatrix}$$

then (10) becomes

$$\begin{aligned}
\text{Maximize:} \quad & z = c\mathbf{x} + 0\mathbf{r} + 0\mathbf{s} \\
\text{Subject to:} \quad & A\mathbf{q} = \mathbf{b} \\
& \mathbf{q} \geq 0
\end{aligned} \qquad (11)$$

Introducing artificial variables to the second and third equations of (10) gives us the initial tableau T for phase 1 of the two-phase method (Tableau 1):

Tableau 1

(Slack) (Artificial)

$$T = \begin{bmatrix} A_1 & 0 & I_m & 0 & 0 & b_1 \\ \hline A_2 & 0 & 0 & I_k & 0 & b_2 \\ \hline A_3 & -I_j & 0 & 0 & I_j & b_3 \\ \hline -c & 0 & 0 & 0 & 0 & 0 \end{bmatrix}$$

Next we apply both phase 1 and phase 2 of the two-phase method to Tableau 1 (without eliminating the artificial variables after phase 1). Let

$$\mathbf{q}^* = \begin{bmatrix} \mathbf{x}^* \\ \mathbf{r}^* \\ \mathbf{s}^* \end{bmatrix}$$

be the maximal solution of (11). If B is the final basic matrix, then it follows from Theorem 2 of Section 4.2 that the result of applying the two-phase method to Tableau 1 is Tableau 2:

Tableau 2

(Slack)   (Artificial)

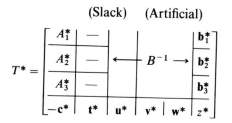

$$T^* = \begin{bmatrix} A_1^* & - & & & & b_1^* \\ \hline A_2^* & - & \leftarrow & B^{-1} & \rightarrow & b_2^* \\ \hline A_3^* & - & & & & b_3^* \\ \hline -c^* & t^* & u^* & v^* & w^* & z^* \end{bmatrix}$$

Theorem 2. Suppose $\mathbf{q}^*$ is a maximal solution of primal problem (11). Then, in the above notation,

(a) $\mathbf{x}^*$ is a feasible solution to the primal problem (1).

(b) $\begin{bmatrix} \mathbf{u}^{*T} \\ \mathbf{v}^{*T} \\ \mathbf{w}^{*T} \end{bmatrix}$ is a feasible solution to the dual problem (2).

(c) $\mathbf{x}^*$ and $\begin{bmatrix} \mathbf{u}^{*T} \\ \mathbf{v}^{*T} \\ \mathbf{w}^{*T} \end{bmatrix}$ are optimal solutions of the primal and dual problems, respectively.

(d) The Fundamental Principle of Duality

$$\mathbf{c}\mathbf{x}^* = \mathbf{b}_1^T\mathbf{u}^{*T} + \mathbf{b}_2^T\mathbf{v}^{*T} + \mathbf{b}_3^T\mathbf{w}^{*T}$$

PROOF.

(a) Since $\mathbf{q}^*$ is a feasible solution of (11)

$$A\mathbf{q}^* = \mathbf{b} \quad \text{and} \quad \mathbf{q}^* \geq 0$$

Therefore,

$$A\mathbf{q}^* = \begin{bmatrix} A_1 & 0 & I_m \\ A_2 & 0 & 0 \\ A_3 & -I_j & 0 \end{bmatrix} \begin{bmatrix} \mathbf{x}^* \\ \mathbf{r}^* \\ \mathbf{s}^* \end{bmatrix} = \begin{bmatrix} A_1\mathbf{x}^* + I_m\mathbf{s}^* \\ A_2\mathbf{x}^* \\ A_3\mathbf{x}^* - I_j\mathbf{r}^* \end{bmatrix} = \begin{bmatrix} \mathbf{b}_1 \\ \mathbf{b}_2 \\ \mathbf{b}_3 \end{bmatrix} \qquad (12)$$

and thus,

$$A_1\mathbf{x}^* \leq \mathbf{b}_1$$
$$A_2\mathbf{x}^* = \mathbf{b}_2 \qquad\qquad (13)$$
$$A_3\mathbf{x}^* \geq \mathbf{b}_3$$

From (13) it follows that $\mathbf{x}^*$ is a feasible solution of (1).

(b) Suppose the final basic matrix B (used to obtain Tableau 2) is a $p \times p$ matrix. Let

$$I_p = \begin{bmatrix} I_m & 0 & 0 \\ 0 & I_k & 0 \\ 0 & 0 & I_j \end{bmatrix}$$

be the identity submatrix of Tableau 1. Then by Theorem 2 in Section 4.3 the objective-row entries $[\mathbf{u}^* \quad \mathbf{v}^* \quad \mathbf{w}^*]$ of Tableau 2 are given by

$$[\mathbf{u}^* \quad \mathbf{v}^* \quad \mathbf{w}^*] = \mathbf{c}_B B^{-1}I_n - [0 \quad 0 \quad \cdots \quad 0] = \mathbf{c}_B B^{-1} \qquad (14)$$

Moreover, since all objective-row entries of Tableau 2 are nonnegative, we have

$$\mathbf{c}_B B^{-1} \begin{bmatrix} A_1 & 0 & I_m \\ A_2 & 0 & 0 \\ A_3 & -I_j & 0 \end{bmatrix} - [\mathbf{c} \quad 0 \quad 0] \geq 0 \qquad (15)$$

From (14) and (15) we have

$$[\mathbf{u}^* \quad \mathbf{v}^* \quad \mathbf{w}^*] \begin{bmatrix} A_1 & 0 & I_m \\ A_2 & 0 & 0 \\ A_3 & -I_j & 0 \end{bmatrix} \geq [\mathbf{c} \quad 0 \quad 0] \qquad (16)$$

Taking transposes of both sides of (16), we obtain

$$\begin{bmatrix} A_1 & 0 & I_m \\ A_2 & 0 & 0 \\ A_3 & -I_j & 0 \end{bmatrix}^{\mathrm{T}} [\mathbf{u}^* \quad \mathbf{v}^* \quad \mathbf{w}^*]^{\mathrm{T}} \geq [\mathbf{c} \quad 0 \quad 0]^{\mathrm{T}}$$

or, equivalently

$$\begin{bmatrix} A_1^{\mathrm{T}} & A_2^{\mathrm{T}} & A_3^{\mathrm{T}} \\ 0 & 0 & -I_j \\ I_m & 0 & 0 \end{bmatrix} \begin{bmatrix} \mathbf{u}^{*\mathrm{T}} \\ \mathbf{v}^{*\mathrm{T}} \\ \mathbf{w}^{*\mathrm{T}} \end{bmatrix} \geq \begin{bmatrix} \mathbf{c}^{\mathrm{T}} \\ 0 \\ 0 \end{bmatrix} \qquad (17)$$

Performing the multiplication in (17) and equating entries of the resulting matrices, we obtain

$$A_1^{\mathrm{T}}\mathbf{u}^{*\mathrm{T}} + A_2^{\mathrm{T}}\mathbf{v}^{*\mathrm{T}} + A_3^{\mathrm{T}}\mathbf{w}^{*\mathrm{T}} \geq \mathbf{c}^{\mathrm{T}}$$

$$-\mathbf{w}^{*\mathrm{T}} \geq 0$$

$$\mathbf{u}^{*\mathrm{T}} \geq 0$$

and hence

$$\begin{bmatrix} \mathbf{u}^{*\mathrm{T}} \\ \mathbf{v}^{*\mathrm{T}} \\ \mathbf{w}^{*\mathrm{T}} \end{bmatrix}$$

is a feasible solution of (2).

To prove parts (c) and (d) it suffices by Theorem 1 to show that

$$\mathbf{cx}^* = \mathbf{b}_1^{\mathrm{T}}\mathbf{u}^{*\mathrm{T}} + \mathbf{b}_2^{\mathrm{T}}\mathbf{v}^{*\mathrm{T}} + \mathbf{b}_3^{\mathrm{T}}\mathbf{w}^{*\mathrm{T}} \qquad (18)$$

We know that

$$\mathbf{cx}^* = z^* = \mathbf{c}_B B^{-1}\mathbf{b}$$

and from (14) that

$$\mathbf{cx}^* = \mathbf{c}_B B^{-1}\mathbf{b} = \begin{bmatrix} \mathbf{u}^* & \mathbf{v}^* & \mathbf{w}^* \end{bmatrix} \begin{bmatrix} \mathbf{b}_1 \\ \mathbf{b}_2 \\ \mathbf{b}_3 \end{bmatrix} \qquad (19)$$

Since $\mathbf{cx}^*$ is a real number, $\mathbf{cx}^*$ also equals the transpose of the right side of (19). Thus, taking transposes in (19), we have

$$\mathbf{cx}^* = \begin{bmatrix} \mathbf{b}_1 \\ \mathbf{b}_2 \\ \mathbf{b}_3 \end{bmatrix}^{\mathrm{T}} \begin{bmatrix} \mathbf{u}^* & \mathbf{v}^* & \mathbf{w}^* \end{bmatrix}^{\mathrm{T}} = \begin{bmatrix} \mathbf{b}_1^{\mathrm{T}} & \mathbf{b}_2^{\mathrm{T}} & \mathbf{b}_3^{\mathrm{T}} \end{bmatrix} \begin{bmatrix} \mathbf{u}^{*\mathrm{T}} \\ \mathbf{v}^{*\mathrm{T}} \\ \mathbf{w}^{*\mathrm{T}} \end{bmatrix}$$

$$= \mathbf{b}_1^{\mathrm{T}}\mathbf{u}^{*\mathrm{T}} + \mathbf{b}_2^{\mathrm{T}}\mathbf{v}^{*\mathrm{T}} + \mathbf{b}_3^{\mathrm{T}}\mathbf{w}^{*\mathrm{T}}$$

and this establishes (d). ∎

We simplified the proof of Theorem 2 by grouping the constraints according to the types $\leq$, $=$, or $\geq$ and by selecting the slack and artificial variables so the tableau contained strategically located identity submatrices. It is not always convenient to rearrange constraints to put a problem in the form (1). In particular, it would be difficult to write a computer program to produce the dual prices by mimicking the proof of Theorem 2. However, it is quite easy to identify the dual prices (the values of the dual variables) corresponding to any row from the injection for the *initial* tableau. Example 1 illustrates the procedure.

EXAMPLE 1

Consider the following LP problem, which is equivalent to the LP problem of Example 2 of Section 6.3 except for the ordering of the constraints.

$$\text{Maximize:} \quad z = 2x_1 - 3x_2 + x_3$$

$$\text{Subject to:} \quad
\begin{aligned}
x_1 + x_2 - x_3 &\le 10 \\
x_1 \qquad\;\; + x_3 &= 5 \\
2x_1 - x_2 + x_3 &\ge 1 \\
2x_1 + x_2 - x_3 &\le 6 \\
x_1 - x_2 - x_3 &\ge 1
\end{aligned}
\tag{20}$$

$$x_i \ge 0, \quad 1 \le i \le 3$$

Introducing slack and artificial variables into (20), we have

$$\text{Maximize:} \quad z = 2x_1 - 3x_2 + x_3$$

$$\text{Subject to:} \quad
\begin{aligned}
x_1 + x_2 - x_3 + x_4 \qquad\qquad\qquad\qquad\qquad\qquad &= 10 \\
x_1 \qquad + x_3 \qquad + x_5 \qquad\qquad\qquad\qquad &= 5 \\
2x_1 - x_2 + x_3 \qquad\qquad - x_6 + x_7 \qquad\qquad &= 1 \\
2x_1 + x_2 - x_3 \qquad\qquad\qquad\qquad + x_8 \qquad &= 6 \\
x_1 - x_2 - x_3 \qquad\qquad\qquad\qquad\qquad - x_9 + x_{10} &= 1
\end{aligned}
\tag{21}$$

$$x_i \ge 0, \quad 1 \le i \le 10$$

The initial tableau for (21) is

Initial Tableau

	x_1	x_2	x_3	x_4	x_5	x_6	x_7	x_8	x_9	x_{10}	
x_4	1	1	-1	1	0	0	0	0	0	0	10
x_5	1	0	1	0	1	0	0	0	0	0	5
x_7	2	-1	1	0	0	-1	1	0	0	0	1
x_8	2	1	-1	0	0	0	0	1	0	0	6
x_{10}	1	-1	-1	0	0	0	0	0	-1	1	1
	-2	3	-1	0	0	0	0	0	0	0	0

We will use the injection for the initial tableau,

$$h(1) = 4, \quad h(2) = 5, \quad h(3) = 7, \quad h(4) = 8, \quad h(5) = 10$$

to identify the dual prices in the final tableau:

Final Tableau

	x_1	x_2	x_3	x_4	x_5	x_6	x_7	x_8	x_9	x_{10}	
x_4	0	$\frac{1}{3}$	0	1	$\frac{1}{3}$	0	0	$-\frac{2}{3}$	0	0	$\frac{23}{3}$
x_3	0	$-\frac{1}{3}$	1	0	$\frac{2}{3}$	0	0	$-\frac{1}{3}$	0	0	$\frac{4}{3}$
x_1	1	$\frac{1}{3}$	0	0	$\frac{1}{3}$	0	0	$\frac{1}{3}$	0	0	$\frac{11}{3}$
x_9	0	$\frac{5}{3}$	0	0	$-\frac{1}{3}$	0	0	$\frac{2}{3}$	1	-1	$\frac{4}{3}$
x_6	0	$\frac{4}{3}$	0	0	$\frac{4}{3}$	1	-1	$\frac{1}{3}$	0	0	$\frac{23}{3}$
	0	$\frac{10}{3}$	0	0	$\frac{4}{3}$	0	0	$\frac{1}{3}$	0	0	$\frac{26}{3}$

We find the dual prices in the objective row in the columns given by the injection for the initial tableau. They are listed in Table 1.

Table 1

Row i	$h(i)$	Dual price
1	4	0
2	5	$\frac{4}{3}$
3	7	0
4	8	$\frac{1}{3}$
5	10	0

You should verify that these are the same dual prices as obtained in Example 2 of Section 6.3.

♦

We conclude this section with a general statement and proof of the principle of complementary slackness introduced in Section 6.2. In the next theorem we use the notation of Theorem 1 and consider LP problem (1) to be written in canonical form (10).

Theorem 3. (Complementary Slackness). Let

$$\begin{bmatrix} \mathbf{x}^* \\ \mathbf{r}^* \\ \mathbf{s}^* \end{bmatrix}$$

be an optimal solution of the primal problem

$$\text{Maximize:} \quad z = \mathbf{cx}$$

$$\begin{aligned}
\text{Subject to:} \quad & A_1\mathbf{x} && + I_m\mathbf{s} = \mathbf{b}_1 \\
& A_2\mathbf{x} && = \mathbf{b}_2 \\
& A_3\mathbf{x} - I_j\mathbf{r} && = \mathbf{b}_3
\end{aligned}$$

$$\mathbf{x} \geq \mathbf{0}; \quad \mathbf{r} \geq \mathbf{0}; \quad \mathbf{s} \geq \mathbf{0}$$

and let

$$\begin{bmatrix} \mathbf{u}^* \\ \mathbf{v}^* \\ \mathbf{w}^* \\ \mathbf{t}^* \end{bmatrix}$$

be an optimal solution of the dual problem

$$\text{Minimize:} \quad y = \mathbf{b}_1^T\mathbf{u} + \mathbf{b}_2^T\mathbf{v} + \mathbf{b}_3^T\mathbf{w}$$

$$\text{Subject to:} \quad A_1^T\mathbf{u} + A_2^T\mathbf{v} + A_3^T\mathbf{w} - I_n\mathbf{t} = \mathbf{c}^T$$

$$\mathbf{u} \geq 0; \quad \mathbf{v} \quad \text{unrestricted}; \quad \mathbf{w} \leq 0; \quad \mathbf{t} \geq 0$$

where $\mathbf{t}$ is an $n \times 1$ matrix of slack variables. Then,

(a) $\mathbf{s}^{*T}\mathbf{u}^* = 0$ and $\mathbf{r}^{*T}\mathbf{w}^* = 0$, or, equivalently,
$s_i^* u_i^* = 0$ for $1 \leq i \leq m$ and $r_i^* w_i^* = 0$ for $1 \leq i \leq j$, and

(b) $\mathbf{t}^{*T}\mathbf{x}^* = 0$, or, equivalently,
$t_i^* x_i^* = 0$ for $1 \leq i \leq n$.

PROOF. We prove part (a) and leave part (b) for you in Exercise 9.
Multiplying both sides of the last equality of (12) by $[\mathbf{u}^{*T} \quad \mathbf{v}^{*T} \quad \mathbf{w}^{*T}]$, we obtain

$$[\mathbf{u}^{*T} \quad \mathbf{v}^{*T} \quad \mathbf{w}^{*T}]\begin{bmatrix} A_1\mathbf{x}^* + I_m\mathbf{s}^* \\ A_2\mathbf{x}^* \\ A_3\mathbf{x}^* - I_j\mathbf{r}^* \end{bmatrix} = [\mathbf{u}^{*T} \quad \mathbf{v}^{*T} \quad \mathbf{w}^{*T}]\begin{bmatrix} \mathbf{b}_1 \\ \mathbf{b}_2 \\ \mathbf{b}_3 \end{bmatrix}$$

Performing the block multiplications gives

$$\mathbf{u}^{*T}A_1\mathbf{x}^* + \mathbf{v}^{*T}A_2\mathbf{x}^* + \mathbf{w}^{*T}A_3\mathbf{x}^* + \mathbf{u}^{*T}\mathbf{s}^* - \mathbf{w}^{*T}\mathbf{r}^* = \mathbf{u}^{*T}\mathbf{b}_1 + \mathbf{v}^{*T}\mathbf{b}_2 + \mathbf{w}^{*T}\mathbf{b}_3$$

Taking transposes and factoring out $\mathbf{x}^{*T}$, we have

$$\mathbf{x}^{*T}(A_1^T\mathbf{u}^* + A_2^T\mathbf{v}^* + A_3^T\mathbf{w}^*) + \mathbf{s}^{*T}\mathbf{u}^* - \mathbf{r}^{*T}\mathbf{w}^* = \mathbf{b}_1^T\mathbf{u}^* + \mathbf{b}_2^T\mathbf{v}^* + \mathbf{b}_3^T\mathbf{w}^* = \mathbf{c}\mathbf{x}^* \quad (22)$$

The last equality in (22) is a consequence of Theorem 2, part (d), the Fundamental
Principle of Duality. Since

is a feasible solution of the dual problem (2), the expression in parentheses on the left
side of equation (22) satisfies

$$A_1^T\mathbf{u}^* + A_2^T\mathbf{v}^* + A_3^T\mathbf{w}^* \geq \mathbf{c}^T$$

Substituting $\mathbf{c}^T$ into (22) gives

$$\mathbf{x}^{*T}\mathbf{c}^T + \mathbf{s}^{*T}\mathbf{u}^* - \mathbf{r}^{*T}\mathbf{w}^* \leq \mathbf{c}\mathbf{x}^* \tag{23}$$

Since $\mathbf{c}\mathbf{x}^*$ is a number (a 1×1 matrix), we have

$$\mathbf{c}\mathbf{x}^* = (\mathbf{c}\mathbf{x}^*)^T = \mathbf{x}^{*T}\mathbf{c}^T$$

Therefore, (23) reduces to

$$\mathbf{s}^{*T}\mathbf{u}^* - \mathbf{r}^{*T}\mathbf{w}^* \leq 0$$

which we can write as

$$s_1^* u_1^* + \cdots + s_m^* u_m^* - r_1^* w_1^* - \cdots - r_j^* w_j^* \leq 0 \tag{24}$$

Since, for $1 \leq i \leq m$ and $1 \leq k \leq j$,

$$s_i^* \geq 0, \qquad u_i^* \geq 0, \qquad r_k^* \geq 0, \qquad \text{and} \qquad w_k^* \leq 0$$

all the terms of (24) are 0; therefore,

$$\mathbf{s}^{*T}\mathbf{u}^* = 0 \qquad \text{and} \qquad \mathbf{r}^{*T}\mathbf{w}^* = 0$$

as was to be shown.   ∎

Exercises

1. Let

$$A_1 = \begin{bmatrix} 1 & 2 & 3 & 4 \\ 5 & 6 & 7 & 8 \end{bmatrix}, \qquad A_2 = \begin{bmatrix} -1 & -2 & -3 & -4 \\ -5 & -6 & -7 & -8 \end{bmatrix}$$

$$A_3 = \begin{bmatrix} 9 & 10 & 11 & 12 \\ 13 & 14 & 15 & 16 \\ 17 & 18 & 19 & 20 \end{bmatrix}$$

Using these values for $A_1, A_2,$ and A_3 and identity matrices of appropriate size, verify the correctness of the following formula which was used to obtain (17).

$$\begin{bmatrix} A_1 & 0 & I_m \\ A_2 & 0 & 0 \\ A_3 & -I_j & 0 \end{bmatrix}^T = \begin{bmatrix} A_1^T & A_2^T & A_3^T \\ 0 & 0 & -I_j \\ I_m & 0 & 0 \end{bmatrix}$$

2. The LP problem

$$\text{Minimize:} \quad z = x_1 + x_2 + x_3$$
$$\text{Subject to:} \quad x_1 - 2x_2 + 2x_3 = 4$$
$$x_1 \qquad\quad + 2x_3 \geq 4$$
$$-x_1 + 2x_2 + \ x_3 \leq 1$$
$$x_i \geq 0, \qquad 1 \leq i \leq 3$$

has the initial and final tableaus

	y_1	y_2	x_1	x_2	x_3	x_4	x_5		
y_1	1	0	1	-2	2	0	0	4	
y_2	0	1	1	0	2	-1	0	4	Initial
x_5	0	0	-1	2	1	0	1	1	Tableau
(z)	0	0	1	1	1	0	0	0	
(w)	0	0	-2	2	-4	1	0	-8	
x_1	$-\frac{2}{3}$	1	1	0	0	-1	$-\frac{2}{3}$	$\frac{2}{3}$	
x_2	$-\frac{1}{2}$	$\frac{1}{2}$	0	1	0	$-\frac{1}{2}$	0	0	Final
x_3	$\frac{1}{3}$	0	0	0	1	0	$\frac{1}{3}$	$\frac{5}{3}$	Tableau
	$\frac{5}{6}$	$-\frac{3}{2}$	0	0	0	$\frac{3}{2}$	$\frac{1}{3}$	$-\frac{7}{3}$	

(a) Find the dual prices.

(b) Verify conclusion (a) of the Complementary Slackness Theorem (Theorem 3).

3. The LP problem

$$\text{Maximize:} \quad z = 4x_1 + 4x_2 - x_3$$

$$\text{Subject to:} \quad 2x_1 + 3x_2 - 4x_3 = 6$$

$$x_1 + 2x_2 + x_3 \le 4$$

$$2x_1 + x_2 + 3x_3 \ge 2$$

$$x_i \ge 0, \quad 1 \le i \le 3$$

has the initial and final tableaus

	x_1	x_2	x_3	s_1	s_2	y_1	y_2		
y_1	2	3	-4	0	0	1	0	6	
s_1	1	2	1	1	0	0	0	4	
y_2	2	1	3	0	-1	0	1	2	Initial
(z)	-4	-4	1	0	0	0	0	0	Tableau
(w)	-4	-4	1	0	1	0	0	-8	
s_2	0	$\frac{19}{6}$	0	$\frac{7}{3}$	1	$-\frac{1}{6}$	-1	$\frac{19}{3}$	
x_3	0	$\frac{1}{6}$	1	$\frac{1}{3}$	0	$-\frac{1}{6}$	0	$\frac{1}{3}$	Final
x_1	1	$\frac{11}{6}$	0	$\frac{2}{3}$	0	$\frac{1}{6}$	0	$\frac{11}{3}$	Tableau
	0	$\frac{19}{6}$	0	$\frac{7}{3}$	0	$\frac{5}{6}$	0	$\frac{43}{3}$	

(a) Find the dual prices.

(b) Verify conclusion (a) of the Complementary Slackness Theorem (Theorem 3).

In Exercises 4–6, solve the given LP problem and identify the solution to the dual problem from the final tableau of the primal problem.

4. Minimize: $z = 3x_1 + x_2$

Subject to: $x_1 \leq x_2 + 3$

$\qquad\quad 2x_1 \leq x_2$

$\qquad\quad x_1 + x_2 \geq 12$

$\qquad\qquad x_1, x_2 \geq 0$

5. Maximize: $z = x_1 + 2x_2 + 3x_3 + 4x_4$

Subject to: $x_1 \qquad + x_3 - 4x_4 = 2$

$\qquad\qquad x_2 - x_3 + 4x_4 = 9$

$\qquad\quad x_1 + x_2 + 2x_3 + 3x_4 = 21$

$\qquad\qquad x_i \geq 0, \qquad 1 \leq i \leq 4$

†6. Maximize: $z = 2x_1 + 3x_2 + x_3 + x_4 + x_5 - 2x_6$

Subject to: $3x_1 + 2x_2 + 4x_3 \qquad\quad - x_5 + x_6 \geq 2$

$\qquad\quad 2x_1 - x_2 + 6x_3 \qquad + x_5 \qquad\quad \geq 3$

$\qquad\quad x_1 + x_2 + x_3 - x_4 + x_5 \qquad\quad = 2$

$\qquad\quad x_1 + x_2 + x_3 + x_4 + x_5 + x_6 \leq 9$

$\qquad\quad 2x_1 \qquad\quad + 3x_3 - x_4 - x_5 \qquad \leq 6$

$\qquad\qquad x_i \geq 0, \qquad 1 \leq i \leq 6$

7. Verify the conclusion of part (a) of Theorem 3 for Example 1. (You can obtain the dual prices from the final tableau given for this problem.)

†8. Solve the dual of the LP problem of Example 1.
   (a) Verify the conclusion of part (b) of Theorem 3.
   (b) Find the values of the slack variables for the dual from the final tableau for the primal problem given in the text.

9. Prove part (b) of Theorem 3. (Hint: Start by substituting the minimal solution of the dual problem into the original constraint system for the dual to obtain

$$A_1^T u^* + A_2^T v^* + A_3^T w^* - t^* = c^T$$

and then multiply both sides by x^{*T}.)

For use in Exercises 10, 11, and 12 (and never again), we define the "twin" of an LP problem as follows. Consider a problem of the form

$$\text{Minimize:} \quad z = cx$$
$$\text{Subject to:} \quad Ax \geq b \tag{25}$$
$$x \geq 0$$

The *twin* of (25) is

$$\text{Maximize:} \quad w = b^T u$$
$$\text{Subject to:} \quad A^T u \leq c^T \tag{26}$$
$$u \geq 0$$

Some authors define the dual as we have defined the twin; however, doing so forces them to develop special rules for finding the correct sign of the dual prices because the dual prices of problem (25) must be ≤ 0 for economic interpretations to be sensible.

10. Prove the analogue of Theorem 1: If x is a feasible solution of (25) and u is a feasible solution of (26), then
   (a) $cx \leq b^T u$
   (b) If $cx = b^T u$, then x is an optimal solution of (25) and u is an optimal solution of (26).

11. Prove the analogue of Theorem 2: If

$$q^* = \begin{bmatrix} x^* \\ s^* \end{bmatrix}$$

is a minimal feasible solution of

Minimize: $z = cx + 0s$

Subject to: $Ax + I_m s = b$ (27)

$x \geq 0;$ $s \geq 0$

then,
(a) x^* is a feasible solution to (25).
(b) u^{*T} is a feasible solution to (26), where u^* is the row matrix of entries from the objective row and B^{-1} columns of the final tableau for problem (27).
(c) x^* and u^{*T} are optimal solutions of the primal and dual problems, respectively.
(d) $cx^* = b^T u^{*T}$

12. Prove the analogue of Theorem 3: If

$$q^* = \begin{bmatrix} x^* \\ s^* \end{bmatrix}$$

is an optimal solution of (25) and

$$\begin{bmatrix} u^* \\ t^* \end{bmatrix}$$

is an optimal solution of its twin (26), then
(a) $s^{*T} u^* = 0$
(b) $t^{*T} x^* = 0$

13. Give an example of a primal problem that has no feasible solution and whose dual also has no feasible solution.

6.5   THE DUAL SIMPLEX MODEL

Thus far, in applying the simplex method we have insisted that each entry in the resource column be nonnegative. However, in integer programming, sensitivity analysis, and other contexts as well, we may encounter a tableau T with the properties:

All objective row entries are nonnegative.

At least one resource column entry is negative (thus indicating that the current basic solution is not a feasible solution).

For problems having an optimal solution, the dual simplex method finds an optimal tableau, that is, a tableau equivalent to T that has nonnegative objective-row and resource-column entries.

The main difference between the simplex method and the dual simplex method lies in determining the pivot elements. In the dual simplex method we select the pivot element to remove from the basis a basic variable with a negative value. To determine this pivot element we first choose the departing basic variable (rather than the entering basic variable as is done in the simplex method), and then we choose the entering basic variable. At each step of this method, the objective row remains nonnegative. If, however, we obtain a tableau with one or more negative entries in the resource column, but for which no pivot element can be found, then there is no feasible solution to the given problem.

The dual simplex method entails the following steps:

1. *Determine the pivot row.* This can be any row with a negative resource value. A common rule is to select a row that contains the most negative entry of the resource column. The corresponding variable is the departing basic variable.

2. *Determine the pivot column.* For each column that has a *negative* entry in its pivot row, form the ratio of the objective-row entry to the pivot-row entry. Choose the *maximum* of these ratios, and select one of the columns that yielded this maximum ratio as the pivot column.

3. Pivot on the entry common to both the pivot row and the pivot column.

4. Continue this process until either

    (a) an optimal feasible solution is obtained, which occurs (as usual) when all objective-row entries and all resource-column entries are nonnegative, or

    (b) it is determined that there is no feasible solution, which occurs whenever, for each negative resource-column entry, there is no pivot element in the corresponding row.

You are asked in Exercise 14 to show that the objective-row entries remain nonnegative at each step of the dual simplex method. In Exercise 16, you are asked to show that for problems for which the dual simplex method is applicable, either case 4(a) or case 4(b) must occur; thus, unbounded feasible solutions cannot result from this method.

EXAMPLE 1

Suppose Tableau 1 arises in connection with an LP problem.

Tableau 1

	x_1	x_2	x_3	x_4	x_5	x_6	
x_4	0	0	3	1	-2	6	-1
x_2	0	1	1	0	$\frac{1}{2}$	-3	10
x_1	1	0	-2	0	1	-4	-2
	0	0	1	0	3	4	10

To apply the dual simplex method to this tableau, first observe that -2 is the most negative entry in the resource column, and therefore by step 1 of this method, the third row becomes the initial pivot row. To determine the pivot column we consider the ratios of the objective-row entries to the negative entries in the pivot row. There are two such ratios:

$$\frac{1}{-2} \quad \text{and} \quad \frac{4}{-4}$$

and the maximum of these ratios is $-\frac{1}{2}$; thus, the pivot column is the third column, and the pivot element is the entry -2 circled above.

Pivoting on this entry, we obtain Tableau 2.

Tableau 2

	x_1	x_2	x_3	x_4	x_5	x_6	
x_4	$\frac{3}{2}$	0	0	1	$\left(-\frac{1}{2}\right)$	0	-4
x_2	$\frac{1}{2}$	1	0	0	1	-5	9
x_3	$-\frac{1}{2}$	0	1	0	$-\frac{1}{2}$	2	1
	$\frac{1}{2}$	0	0	0	$\frac{7}{2}$	2	9

From Tableau 2 we see that the pivot row is the first row. The only ratio we consider is $(\frac{7}{2})/(-\frac{1}{2}) = -7$, and hence the fifth column is the pivot column. The pivot element is the entry circled in Tableau 2.

Pivoting on this entry yields Tableau 3.

Tableau 3

	x_1	x_2	x_3	x_4	x_5	x_6	
x_5	-3	0	0	-2	1	0	8
x_2	$\frac{7}{2}$	1	0	2	0	-5	1
x_3	-2	0	1	-1	0	2	5
	11	0	0	7	0	2	-19

Now we see that all entries in the resource column are nonnegative, and since all entries of the objective row are also nonnegative, it follows that we have obtained an optimal solution $(0, 1, 5, 0, 8, 0)$, and the optimal value of the objective function is $z = -19$. ◆

Example 2 illustrates case 4(b) where there is no feasible solution.

EXAMPLE 2

We apply the dual simplex method to the LP problem

Maximize: $z = -3x_1 - 4x_2$

Subject to: $-x_1 + x_2 \le 2$

$\qquad\qquad\quad 2x_1 - x_2 \le 1$ $\qquad\qquad$ (1)

$\qquad\qquad\; -x_1 - x_2 \le -9$

$\qquad\qquad x_1, x_2 \ge 0$

Introducing (nonnegative) slack variables to (1), we obtain Tableau 1.

Tableau 1

	x_1	x_2	x_3	x_4	x_5	
x_3	-1	1	1	0	0	2
x_4	2	-1	0	1	0	1
x_5	$\boxed{-1}$	-1	0	0	1	-9
	3	4	0	0	0	0

The circled entry in Tableau 1 is the pivot element. Pivoting on this entry, we obtain Tableau 2.

Tableau 2

	x_1	x_2	x_3	x_4	x_5	
x_3	0	2	1	0	-1	11
x_4	0	-3	0	1	2	-17
x_1	1	1	0	0	-1	9
	0	1	0	0	3	-27

Pivoting on the circled entry in Tableau 2, we obtain Tableau 3.

Tableau 3

	x_1	x_2	x_3	x_4	x_5	
x_3	0	0	1	$\frac{2}{3}$	$\frac{1}{3}$	$-\frac{1}{3}$
x_2	0	1	0	$-\frac{1}{3}$	$-\frac{2}{3}$	$\frac{17}{3}$
x_1	1	0	0	$\frac{1}{3}$	$-\frac{1}{3}$	$\frac{10}{3}$
	0	0	0	$\frac{1}{3}$	$\frac{11}{3}$	$-\frac{98}{3}$

Observe that in Tableau 3 the first row is the pivot row, but since there are no negative entries in this row, there is no pivot column (and, hence, no pivot element). It is easy to see that this situation means the original problem is infeasible. The equation corresponding to the first row of Tableau 3 is

$$x_3 + \frac{2}{3}x_4 + \frac{1}{3}x_5 = -\frac{1}{3} \qquad (2)$$

Since the variables are all nonnegative, the left-hand side of (2) is nonnegative and cannot equal the negative number $-\frac{7}{3}$.

$\blacklozenge$

In certain situations we can use the dual simplex method to avoid introducing artificial variables. Example 3 illustrates this idea.

EXAMPLE 3

Consider the LP problem

$$\text{Minimize:} \quad z = 3x_1 + 2x_2$$

$$\text{Subject to:} \quad \begin{aligned} x_1 + x_2 &\geq 1 \\ 4x_1 + x_2 &\geq 2 \\ -x_1 + 2x_2 &\leq 6 \end{aligned} \qquad (3)$$

$$x_1, x_2 \geq 0$$

Rather than introduce artificial variables, we rewrite (3) as

$$\text{Maximize:} \quad z = -3x_1 - 2x_2$$

$$\text{Subject to:} \quad \begin{aligned} -x_1 - x_2 &\leq -1 \\ -4x_1 - x_2 &\leq -2 \\ -x_1 + 2x_2 &\leq 6 \end{aligned} \qquad (4)$$

$$x_1, x_2 \geq 0$$

Introducing (nonnegative) slack variables to (4), we obtain Tableau 1.

Tableau 1

	x_1	x_2	x_3	x_4	x_5	
x_3	-1	-1	1	0	0	-1
x_4	(-4)	-1	0	1	0	-2
x_5	-1	2	0	0	1	6
	3	2	0	0	0	0

Note that if x_1 and x_2 are nonbasic variables in Tableau 1, then this tableau does not represent a feasible solution to (3) because the basic variables x_3 and x_4 take on negative values. However, Tableau 1 is in the correct form for application of the dual simplex method.

Pivoting on the circled entry in Tableau 1, we obtain Tableau 2.

Tableau 2

	x_1	x_2	x_3	x_4	x_5	
x_3	0	$\left(-\frac{3}{4}\right)$	1	$-\frac{1}{4}$	0	$-\frac{1}{2}$
x_1	1	$\frac{1}{4}$	0	$-\frac{1}{4}$	0	$\frac{1}{2}$
x_5	0	$\frac{9}{4}$	0	$-\frac{1}{4}$	1	$\frac{13}{2}$
	0	$\frac{5}{4}$	0	$\frac{3}{4}$	0	$-\frac{3}{2}$

Pivoting on the circled entry in Tableau 2, we obtain the final tableau (Tableau 3).

Tableau 3

	x_1	x_2	x_3	x_4	x_5	
x_2	0	1	$-\frac{4}{3}$	$\frac{1}{3}$	0	$\frac{2}{3}$
x_1	1	0	$\frac{1}{3}$	$-\frac{1}{3}$	0	$\frac{1}{3}$
x_5	0	0	3	-1	1	5
	0	0	$\frac{5}{3}$	$\frac{1}{3}$	0	$-\frac{7}{3}$

In Tableau 3 we have reached a feasible solution; moreover, since all objective-row entries are nonnegative, this basic feasible solution is an optimal solution. Thus, $(\frac{1}{3}, \frac{2}{3}, 0, 0, 5)$ is the optimal solution of (3), and $z = -\frac{7}{3}$ is the maximal value of the objective function; hence, $z = \frac{7}{3}$ is the optimal solution of (3). ◆

Next we show what has happened geometrically in Example 3, which typifies LP problems solved by the dual simplex method. The constraints of (3) give us the feasible region indicated in Figure 1.

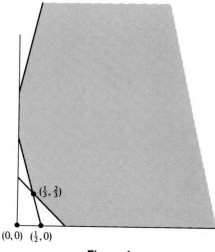

Figure 1

In solving this problem we began at the infeasible constraint intersection point $(0, 0)$, defined by the constraints $x_1 = 0, x_2 = 0$. We see from Tableau 2 of Example 3 that in applying the dual simplex method, we moved to the infeasible constraint intersection point $(\frac{1}{2}, 0)$, defined by the constraints $4x_1 - x_2 = 2$ and $x_1 + x_2 = 1$. The next application of the dual simplex method yielded Tableau 3, from which we see that we moved to the feasible corner point $(\frac{1}{3}, \frac{2}{3})$, and this corner point corresponded to the optimal solution.

In essence then, the dual simplex method, in attempting to find a feasible solution, moves from one infeasible constraint intersection point to another until it reaches a feasible point. Using arguments similar to those given in Theorem 1 in Section 3.3, you can show that for maximization problems each dual pivot operation leads to a decrease (or no change) in the objective function (see Exercise 15). Since the various constraints result in only a finite number of constraint intersection points (i.e., only a finite number of basic though not necessarily feasible solutions), it follows as in Theorem 1 of Section 3.3 that, unless cycling occurs, the dual simplex method must eventually arrive at a corner point of the feasible region.

To start the dual simplex method we need to obtain a tableau with nonnegative entries in its objective row and at least one negative entry in its resource column. Although there are procedures for putting a general LP column in this form, we do not consider them because they do not have generally useful applications. We will use the dual simplex method in Chapter 8 to solve integer programming problems by adding a new constraint to the final tableau of an LP problem. Example 4 illustrates this technique.

EXAMPLE 4

You may verify that applying the simplex method to the LP problem

$$\text{Maximize:} \quad z = 2x_1 - x_2 + x_3$$

$$\text{Subject to:} \quad \begin{aligned} x_2 + x_3 + x_4 \qquad\qquad &= 5 \\ x_1 + 2x_2 + x_3 \qquad + x_5 \qquad &= 4 \\ x_1 \qquad - x_3 \qquad\qquad + x_6 &= 2 \end{aligned} \tag{5}$$

$$x_i \geq 0, \qquad 1 \leq i \leq 6$$

results in the final tableau (Tableau 1).

Tableau 1

	x_1	x_2	x_3	x_4	x_5	x_6	
x_4	0	0	0	1	$-\frac{1}{2}$	$\frac{1}{2}$	4
x_3	0	1	1	0	$\frac{1}{2}$	$-\frac{1}{2}$	1
x_1	1	1	0	0	$\frac{1}{2}$	$\frac{1}{2}$	3
	0	4	0	0	$\frac{3}{2}$	$\frac{1}{2}$	7

We see from this tableau that the optimal solution of (5) is

$$(3, 0, 1, 4, 0, 0) \tag{6}$$

and the maximal objective value is $z = 7$.

Suppose now we introduce the constraint

$$x_1 - x_2 + x_3 \leq 2 \tag{7}$$

to (5). Observe that the optimal solution (6) does not satisfy this constraint and hence is not a feasible solution of the LP problem that has (7) as one of its constraints.

To obtain an optimal feasible solution to the LP problem (5) with the added constraint (7), we proceed as follows. Introducing a slack variable x_7 to (7) and augmenting the final tableau (Tableau 1) of the original problem with a row corresponding to this new constraint, we obtain Tableau 2.

Tableau 2

	x_1	x_2	x_3	x_4	x_5	x_6	x_7	
(x_4)	0	0	0	1	$-\frac{1}{2}$	$\frac{1}{2}$	0	4
(x_3)	0	1	①	0	$\frac{1}{2}$	$-\frac{1}{2}$	0	1
(x_1)	①	1	0	0	$\frac{1}{2}$	$\frac{1}{2}$	0	3
(x_7)	1	-1	1	0	0	0	1	2
	0	4	0	0	$\frac{3}{2}$	$\frac{1}{2}$	0	7

In Tableau 2 the basic variables are x_1, x_3, and x_4. It is reasonable that these variables, and the variable x_7, be the basic variables for Tableau 2. However, if the variables x_1, x_3, and x_4 are to be basic, then we must clear their columns in Tableau 2 of nonzero entries. We do this by pivoting on the circled entries in Tableau 2 to obtain Tableau 3.

Tableau 3

	x_1	x_2	x_3	x_4	x_5	x_6	x_7	
x_4	0	0	0	1	$-\frac{1}{2}$	$\frac{1}{2}$	0	4
x_3	0	1	1	0	$\frac{1}{2}$	$-\frac{1}{2}$	0	1
x_1	1	1	0	0	$\frac{1}{2}$	$\frac{1}{2}$	0	3
x_7	0	⊖-3	0	0	-1	0	1	-2
	0	4	0	0	$\frac{3}{2}$	$\frac{1}{2}$	0	7

Note now that because of the negative resource entry we do not have a feasible solution. However, applying the dual simplex method to this tableau will restore feasibility. The circled entry in Tableau 3 is the initial pivot element. Pivoting on this entry, we obtain Tableau 4.

Tableau 4

	x_1	x_2	x_3	x_4	x_5	x_6	x_7	
x_4	0	0	0	1	$-\frac{1}{2}$	$\frac{1}{2}$	0	4
x_3	0	0	1	0	$\frac{1}{6}$	$-\frac{1}{2}$	$\frac{1}{3}$	$\frac{1}{3}$
x_1	1	0	0	0	$\frac{1}{6}$	$\frac{1}{2}$	$\frac{1}{3}$	$\frac{7}{3}$
x_2	0	1	0	0	$\frac{1}{3}$	0	$-\frac{1}{3}$	$\frac{2}{3}$
	0	0	0	0	$\frac{1}{6}$	$\frac{1}{2}$	$\frac{4}{3}$	$\frac{13}{3}$

We now see that there are no negative entries in the resource column; hence, all of the basic variables are positive. This signals the end to this procedure, and the optimal solution to (5) with the added constraint (7) is

$$x_1 = \frac{7}{3}, \qquad x_2 = \frac{2}{3}, \qquad x_3 = \frac{1}{3}; \qquad z = \frac{13}{3}$$

●

To conclude this section we illustrate the origin of the name "dual simplex method" by an example that makes a connection between the solution of a primal problem with the simplex method and its dual with the dual simplex method.

EXAMPLE 5

We consider the pair of primal and dual problems

	Primal		*Dual*
Maximize:	$z = 120x_1 + 100x_2$	Minimize:	$y = 8u_1 + 15u_2$
Subject to:	$2x_1 + 2x_2 \le 8$	Subject to:	$2u_1 + 2u_2 \ge 120$
	$5x_1 + 3x_2 \le 15$		$5u_1 + 3u_2 \ge 100$
	$x_1, x_2 \ge 0$		$u_1, u_2 \ge 0$

We solve the primal problem with the simplex method and the dual problem with the dual simplex method to obtain the two sequences of tableaus

Simplex Method

	x_1	x_2	s_1	s_2	
s_1	2	2	1	0	8
s_2	5	3	0	1	15
	-120	-100	0	0	0
s_1	0	$\frac{4}{5}$	1	$-\frac{2}{5}$	2
x_1	1	$\frac{3}{5}$	0	$\frac{1}{5}$	3
	0	-28	0	24	360
x_2	0	1	$\frac{5}{4}$	$-\frac{1}{2}$	$\frac{5}{2}$
x_1	1	0	$-\frac{3}{4}$	$\frac{1}{2}$	$\frac{3}{2}$
	0	0	35	10	430

Dual Simplex Method

	u_1	u_2	t_1	t_2	
t_1	-2	-5	1	0	-120
t_2	-2	-3	0	1	-100
	8	15	0	0	0
u_2	$\frac{2}{5}$	1	$-\frac{1}{5}$	0	24
t_2	$-\frac{4}{5}$	0	$-\frac{3}{5}$	1	-28
	2	0	3	0	-360
u_2	0	1	$-\frac{1}{2}$	$\frac{1}{2}$	10
u_1	1	0	$\frac{3}{4}$	$-\frac{5}{4}$	35
	0	0	$\frac{3}{2}$	$\frac{5}{2}$	-430

Figure 2 shows that the simplex method seeks the optimal solution by moving through the sequence of feasible corner points P_1, P_2, P_3 and that the dual simplex method seeks the optimal solution by moving through the sequence of infeasible corner points Q_1, Q_2 to the feasible corner point Q_3. By comparing the preceding

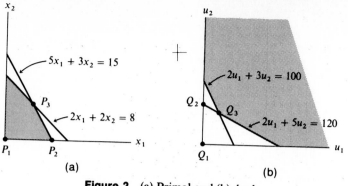

Figure 2 (a) Primal and (b) dual

pairs of tableaus, you will see that the solutions of each tableau for the primal problem correspond to the dual prices in the corresponding tableau for the dual problem, and conversely, that the solutions given by the dual tableaus correspond to the dual prices given by the primal tableaus.

◆

Exercises

In Exercises 1–6, use the dual simplex method to restore feasibility to the tableaus or to conclude that there are no feasible solutions.

1.

	x_1	x_2	x_3	x_4	x_5	x_6	
x_4	0	0	-3	1	-2	6	-1
x_2	0	1	1	0	2	-3	3
x_1	1	0	-2	0	1	-6	-3
	0	0	1	0	4	6	21

2.

	x_1	x_2	x_3	x_4	x_5	x_6	
x_2	0	1	0	1	-2	6	-1
x_3	0	0	1	0	2	3	-3
x_1	1	0	0	0	-1	-6	-3
	0	0	0	0	6	5	12

3.

	x_1	x_2	x_3	x_4	x_5	x_6	
x_6	0	0	-4	-2	-2	1	-6
x_2	0	1	1	0	2	0	4
x_1	1	0	-2	2	1	0	-4
	0	0	3	1	2	0	16

4.

	x_1	x_2	x_3	x_4	x_5	x_6	
x_4	0	0	-3	1	-2	6	1
x_2	0	1	-1	0	-2	-4	-4
x_1	1	0	-2	0	1	-6	-3
	0	0	2	0	4	4	12

5.

	x_1	x_2	x_3	x_4	x_5	x_6	
x_4	0	0	-3	1	-2	6	1
x_2	0	1	1	0	2	4	-4
x_1	1	0	-2	0	1	-6	-3
	0	0	2	0	4	4	12

6.

	x_1	x_2	x_3	x_4	x_5	x_6	
x_4	0	-2	1	1	0	-2	-6
x_5	0	4	4	0	1	-3	-2
x_1	1	-4	2	0	0	-4	-8
	0	8	3	0	0	6	45

Solve Exercises 7–12 by the dual simplex method.

7. Minimize: $z = 4x_1 + 15x_2$

Subject to: $x_1 + 5x_2 \geq 120$
$x_1 + 3x_2 \geq 100$

$x_1, x_2 \geq 0$

8. Minimize: $z = 150x_1 + 120x_2 + 50x_3$

Subject to: $3x_1 + 4x_2 + x_3 \geq 15$
$12x_1 + 16x_2 + x_3 \geq 30$
$4x_1 + 3x_2 + x_3 \geq 15$
$2x_1 + x_2 + x_3 \geq 10$

$x_i \geq 0, \quad 1 \leq i \leq 3$

9. Minimize: $z = 10x_1 + 6x_2 + x_3 + x_4$

Subject to: $x_1 + 3x_2 - 2x_3 - x_4 \geq 2$
$-x_1 - x_2 - x_3 - x_4 \leq 3$
$-x_1 - x_2 - x_3 + x_4 \geq 1$

$x_i \geq 0, \quad 1 \leq i \leq 4$

10. Maximize: $z = -x_1 - 3x_2 - x_3 - x_4$

Subject to: $x_1 + x_2 - 3x_3 + 2x_4 \geq 6$
$-2x_1 + 3x_2 + x_3 - 2x_4 \leq 9$
$- x_2 + x_3 \leq 0$
$x_1 - 2x_2 - 3x_3 + x_4 \geq 4$

$x_i \geq 0, \quad 1 \leq i \leq 4$

11. Minimize: $z = x_1 + x_2 + x_3$

Subject to: $-x_1 + 2x_2 + x_3 \leq 1$
$x_1 - 2x_3 \leq -4$
$x_1 - x_2 + 2x_3 \leq 3$

$x_i \geq 0, \quad 1 \leq i \leq 3$

12. Minimize: $z = x_1 + x_2 + x_3$

Subject to: $-x_1 + 2x_2 + x_3 \leq 1$
$x_1 - 2x_3 \leq -4$
$x_1 - x_2 + 2x_3 \geq 4$

$x_i \geq 0, \quad 1 \leq i \leq 3$

13. Show that the dual simplex method can be used to solve any LP problem in which the objective function has nonnegative coefficients and is to be minimized.

14. With your help, we now show that the dual simplex method never causes a negative entry to appear in the objective row. For simplicity we consider the initial tableau for a problem with three decision variables; however, it is clear that the idea of the proof we give applies to any stage of any problem. Consider the simplex tableau

	x_1	x_2	x_3	x_4	x_5	x_6	
x_4	a_{11}	a_{12}	a_{13}	1	0	0	b_1
x_5	a_{21}	a_{22}	a_{23}	0	1	0	b_2
x_6	a_{31}	a_{32}	a_{33}	0	0	1	b_3
	d_1	d_2	d_3	0	0	0	0

where $d_1, d_2, d_3 \geq 0$ and $b_1 < 0$. Suppose we choose the first row for the pivot row.

(a) Suppose the dual simplex method chooses a_{11} as the pivot element. Then $a_{11} < 0$ and d_1/a_{11} is the largest negative number in the set $\{d_i/a_{1i} \mid a_{1i} < 0, i = 1, 2, 3\}$. In performing the pivot, we begin by dividing the first row by a_{11} to obtain the row

$$\left[1 \quad \frac{a_{21}}{a_{11}} \quad \frac{a_{31}}{a_{11}} \quad \frac{1}{a_{11}} \quad 0 \quad 0 \quad \frac{b_1}{a_{11}} \right] \tag{8}$$

To perform the pivot operation on the objective row, we multiply (8) by d_1 and subtract it from the objective row to obtain the new objective row

$$\left[0 \quad d_2 - (d_1)\frac{a_{21}}{a_{11}} \quad d_3 - (d_1)\frac{a_{31}}{a_{11}} \quad \frac{-d_1}{a_{11}} \quad 0 \quad 0 \right]$$

Obviously, $(-d_1)/a_{11} \geq 0$ since $d_1 \geq 0$ and $a_{11} < 0$. Consider $d_2 - (d_1)a_{21}/a_{11}$. If $a_{21} \geq 0$, then $a_{21}/a_{11} < 0$, and so $d_2 - (d_1)a_{21}/a_{11} \geq 0$. If $a_{21} < 0$, write

$$d_2 - (d_1)\frac{a_{21}}{a_{11}} = (-a_{21})\left(\frac{d_1}{a_{11}} - \frac{d_2}{a_{21}} \right)$$

Since d_1/a_{11} was the largest of the negative ratios,

$$\frac{d_1}{a_{11}} - \frac{d_2}{a_{21}} \geq 0$$

Since $-a_{21} > 0$,

$$d_2 - (d_1)\frac{a_{21}}{a_{11}} \geq 0$$

Use a similar argument to show that

$$d_3 - (d_1)\frac{a_{31}}{a_{11}} \geq 0$$

(b) Use similar calculations to show that pivoting on a_{12} or a_{13} cannot introduce negative numbers in the objective row.

15. Using the tableau in Exercise 14, show that a pivot on any negative a_{1j} yields the new objective value $(-d_j)b_1/a_{1j}$. Show that this is a nonpositive number, and explain why this observation illustrates the reason the objective function cannot increase when we perform a pivot of the dual simplex algorithm.

16. Continue the argument of Exercise 14 to show that unbounded feasible solutions cannot result from application of the dual simplex method.

17. Suppose a tableau contains a row with all nonnegative entries except for a negative entry in the resource column. Explain why the tableau corresponds to a problem with an empty feasible set.

18. Use the final simplex tableau

	x_1	x_2	x_3	x_4	x_5	x_6	
x_2	0	1	$\frac{4}{5}$	$\frac{3}{5}$	$-\frac{1}{5}$	0	2
x_1	1	0	$-\frac{3}{5}$	$-\frac{1}{5}$	$\frac{2}{5}$	0	0
x_6	0	0	$-\frac{2}{5}$	$\frac{1}{5}$	$-\frac{7}{5}$	1	6
	0	4	$\frac{3}{5}$	$\frac{1}{5}$	$\frac{3}{5}$	0	2

for the LP problem

$$\text{Maximize:} \quad z = 2x_1 + x_2 - x_3$$
$$\text{Subject to:} \quad x_1 + 2x_2 + x_3 \leq 4$$
$$3x_1 + x_2 - x_3 \leq 2$$
$$4x_1 + x_2 - 2x_3 \leq 8$$
$$x_i \geq 0, \quad 1 \leq i \leq 3$$

to solve this problem given the additional constraint
(a) $-x_1 + 3x_2 + 2x_3 \leq 1$
(b) $2x_1 + 2x_2 + x_3 \leq 5$
(c) $2x_1 + x_2 - x_3 \geq 4$
(d) $2x_1 + 3x_2 - x_3 = 4$

19. Use the final simplex tableau

	x_1	x_2	x_3	x_4	
x_3	0	$-\frac{5}{2}$	1	$-\frac{1}{2}$	1
x_1	1	$\frac{3}{2}$	0	$\frac{1}{2}$	4
	0	4	0	2	16

for the LP problem

$$\text{Maximize:} \quad z = 4x_1 + 2x_2$$
$$\text{Subject to:} \quad x_1 - x_2 \leq 5$$
$$2x_1 + 3x_2 \leq 8$$
$$x_1, x_2 \geq 0$$

to solve this problem given the additional constraints

(a) $3x_1 + x_2 \le 6$
(b) $2x_1 - 3x_2 \le 4$
(c) $x_1 + 2x_2 \ge 5$
(d) $x_1 - 2x_2 = 2$

6.6 LP MODELS

We now discuss in some detail three examples that illustrate how to use duality theory in analyzing and, in some cases, altering certain optimal solutions. In the next chapter we will apply duality theory to sensitivity analysis.

EXAMPLE 1

In this example we use the optimal dual prices to make changes in the original primal problem so that a nonbasic variable appearing in the optimal primal solution becomes basic. To illustrate this procedure we return to the problem discussed in the first two exercises of Section 6.1.

A plant manufactures four liquid fertilizers F_1, F_2, F_3, and F_4. The production of each of these products involves three steps: manufacture of the ingredients, mixing of the ingredients, and packaging. Table 1 gives the number of hours required to complete each of these steps per unit produced and the unit profit for each product.

Table 1

Product	Manufacture (hr/unit)	Mixing (hr/unit)	Packaging (hr/unit)	Profit ($/unit)
F_1	3	4	1	15
F_2	6	8	$\frac{1}{2}$	30
F_3	4	3	1	15
F_4	2	1	1	10

Each week, 150 hours are available for manufacture of the ingredients, 120 hours for mixing, and 50 hours for packaging. The plant wants to establish a production schedule that will maximize weekly profit.

We let x_1, x_2, x_3, x_4 denote the number of gallons of F_1, F_2, F_3, F_4, respectively, that are manufactured in a week. As you may verify, the initial and final tableaus for this problem are as follows.

Initial Tableau

	x_1	x_2	x_3	x_4	s_1	s_2	s_3	
s_1	3	6	4	2	1	0	0	150
s_2	4	8	3	1	0	1	0	120
s_3	1	$\frac{1}{2}$	1	1	0	0	1	50
	-15	-30	-15	-10	0	0	0	0

Final Tableau

	x_1	x_2	x_3	x_4	s_1	s_2	s_3	
s_1	-1	0	$\frac{2}{3}$	0	1	$-\frac{2}{3}$	$-\frac{4}{3}$	$\frac{10}{3}$
x_2	$\frac{2}{5}$	1	$\frac{4}{15}$	0	0	$\frac{2}{15}$	$-\frac{2}{15}$	$\frac{28}{3}$
x_4	$\frac{4}{5}$	0	$\frac{13}{15}$	1	0	$-\frac{1}{15}$	$\frac{16}{15}$	$\frac{136}{3}$
	5	0	$\frac{5}{3}$	0	0	$\frac{10}{3}$	$\frac{20}{3}$	$\frac{2200}{3}$

From the final tableau we see that in the optimal solution, $x_3 = 0$; that is, to maximize profits the plant should not produce any gallons of product F_3. We now use the shadow (dual) prices to determine what changes we can make in the original problem so that the new optimal solution will call for production of F_3.

The shadow value (which, in this instance, we call the *shadow cost*) of producing F_3 is defined by the third column of the initial tableau as

$$4u_1 + 3u_2 + u_3 \tag{1}$$

From the final tableau we see that optimal values of the dual variables are

$$u_1 = 0, \qquad u_2 = \frac{10}{3}, \qquad u_3 = \frac{20}{3} \tag{2}$$

Substituting these values in (1), we find that the shadow cost of F_3 is

$$4(0) + 3\left(\frac{10}{3}\right) + 1\left(\frac{20}{3}\right) = \frac{50}{3}$$

This "cost" is larger than the \$15 profit resulting from the sale of one gallon of product F_3; hence, as we expected, the optimal solution of the primal solution calls for no gallons of F_3 to be produced. If, however, the unit shadow cost of F_3 were less than the profit per gallon (\$15), then, to maximize its profits, the plant would produce some gallons of F_3.

We now investigate what changes in the original unit times for manufacturing, mixing, or packaging will result in the production of F_3. Since $u_1 = 0$ in (2), it will do no good to make changes in the manufacturing time. Note too that, since the optimal values of u_2 and u_3 are $\frac{10}{3}$ and $\frac{20}{3}$, respectively, a unit change in the packaging time will lead to a greater change in the shadow cost of F_3 than will a unit change in the time for mixing. If we let k represent this change in the original packaging time, it follows from (1) and the preceding discussion that, for F_3 to be produced, we must have

$$4u_1 + 3u_2 + (1 - k)u_3 < 15 \tag{3}$$

Substituting the optimal values (2) in (3), we obtain

$$4(0) + 3\left(\frac{10}{3}\right) + (1 - k)\left(\frac{20}{3}\right) < 15$$

from which it follows that if $k > \frac{1}{4}$ (or, equivalently, if the packaging time is reduced by more than 0.25 hr), then the optimal solution will call for the production of F_3. The number of gallons of F_3 that are produced will depend on the particular value assigned to k. ◆

In Exercise 1 you are asked to determine what change in the original mixing cost will lead to the production of F_3. In Exercise 2 you are to make a similar analysis for the product F_1.

EXAMPLE 2

Wily Turkey has hired us to determine the most economical way of blending four spirits to make its distinctively flavored bourbon whisky. Since Wily Turkey does not want competitors to discover its recipe, it has disguised the nature of the three flavor ingredients by describing them to us with the names mellow, spark, and spice. Table 2 gives the percentage of each of these ingredients in each spirit. The last row of Table 2 gives the cost per barrel of each spirit.

Table 2

Ingredient	Spirit 1 (%)	Spirit 2 (%)	Spirit 3 (%)	Spirit 4 (%)
Mellow	24	12	8	4
Spark	12	8	20	24
Spice	32	12	8	4
Cost/barrel ($)	120	100	80	60

Wily Turkey also tells us that a batch of whisky should contain at least 14% mellow, no more than 12% spark, and exactly 16% spice. The rest of the batch is spring water, which costs essentially nothing. Wily Turkey plans to sell the blend for $200 per barrel and wants to mix the spirits in the way that will maximize profit, which is the difference between the selling price and the cost of the spirits.

The easiest way to deal with a mix described as percentages is to think of making a batch of some fixed size, say 100 barrels; the fact that 14% of the mixture is mellow is equivalent to supposing that 14 barrels of the batch are mellow. So, we let x_i be the number of barrels of spirit i used in the blend and require that

$$x_1 + x_2 + x_3 + x_4 = 100$$

The profit is

$$z = (200 - 120)x_1 + (200 - 100)x_2 + (200 - 80)x_3 + (200 - 60)x_4$$
$$= 80x_1 + 100x_2 + 120x_3 + 140x_4$$

and the extreme amounts of mellow, spark, and spice in a 100-barrel batch are 14, 12, and 16 barrels, respectively. The LP problem is

$$\text{Maximize:} \quad z = 80x_1 + 100x_2 + 120x_3 + 140x_4$$

$$
\begin{aligned}
\text{Subject to:} \quad 0.24x_1 + 0.12x_2 + 0.08x_3 + 0.04x_4 &\geq 14 \\
0.12x_1 + 0.08x_2 + 0.20x_3 + 0.24x_4 &\leq 12 \\
0.32x_1 + 0.12x_2 + 0.08x_3 + 0.04x_4 &= 16 \\
x_1 + x_2 + x_3 + x_4 &= 100
\end{aligned}
$$

$$x_i \geq 0, \quad 1 \leq i \leq 4$$

The solution found by CALIPSO is

Row	Basic variable	Value	Dual price
1	x_2	50	-1000
2	x_4	0	0
3	x_1	25	500
4	x_2	25	160

and the maximum profit is $z^* = \$10,000$.

We can use the Fundamental Principle of Duality to examine the effect on profit of varying the resource values. If we let $b_1 = 14$, $b_2 = 12$, $b_3 = 16$, and $b_4 = 100$, we have

$$z^* = 10{,}000 = -1000b_1 + 0b_2 + 500b_3 + 160b_4$$

as you can easily check. If we add a change Δb_i to each b_i, we obtain a new profit $\hat{z}$ given by

$$\hat{z} = -1000(b_1 + \Delta b_1) + 0(b_2 + \Delta b_2) + 500(b_3 + \Delta b_3) + 160(b_4 + \Delta b_4)$$

$$= 10{,}000 - 1000\,\Delta b_1 + 0\,\Delta b_2 + 500\,\Delta b_3 + 160\,\Delta b_4$$

For instance, increasing b_1 by 0.2 barrels ($\Delta b_1 = 0.2$), but not changing the other resources, costs Wily Turkey \$200. Decreasing b_3 by 0.2, but leaving the other resources the same, costs Wily Turkey \$100. Increasing b_2 has no effect on profit.

Notice what we have not said. The Fundamental Principle of Duality can be used to examine the effect on z^* of changing resource values; it does not say that the solution remains optimal or even feasible after making changes to the resource values. In fact, if we increase b_1 by more than 0.4 barrels, the problem has no feasible solution. We chose the changes we made to the resource values in the last paragraph carefully to ensure that the set of basic variables in the optimal solution was unchanged. In the next chapter we use sensitivity analysis to see that we can change each resource value within certain bounds without changing the set of basic variables. If we make changes outside these bounds, the solution may no longer be optimal. ◆

EXAMPLE 3

A new competitor, Alternative Telephone and Telegraph (AT&T), is trying to capture part of the telephone market now dominated by DASH. Since DASH owns many telephone lines joining cities all over the world, it can usually route a call in many ways. For instance, Figure 1 gives the alternative routes for transmitting a call from New York to Moscow. DASH's pricing policy is to charge the customer the sum of the charges between cities along the route. Thus, a call from New York to Moscow through Tokyo and Paris would cost $3 + 5 + 4 = 12$ dollars.

To formulate an LP problem that finds the cheapest route for sending a call from New York to Moscow, we introduce a variable x_{ij} that is 1 if the call is sent from City i to City j and 0 if it is not. We introduce the following constraints to model the network given in Figure 1.

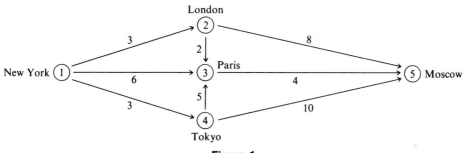

Figure 1

The call originates in New York:	$x_{12} + x_{13} + x_{14} = 1$
The call is not lost in London:	$x_{12} = x_{23} + x_{25}$
The call is not lost in Paris:	$x_{13} + x_{23} + x_{43} = x_{35}$
The call is not lost in Tokyo:	$x_{14} = x_{43} + x_{45}$
The call gets to Moscow:	$x_{25} + x_{35} + x_{45} = 1$

Notice that these constraints are not independent. In fact, the last constraint is a consequence of the other four constraints, and so we omit this constraint. The LP problem is

$$\text{Maximize:} \quad z = -3x_{12} - 6x_{13} - 3x_{14} - 2x_{23} \\ - 5x_{43} - 8x_{25} - 4x_{35} - 10x_{45}$$

$$\text{Subject to:} \quad \begin{aligned} x_{12} + x_{13} + x_{14} \quad\quad\quad\quad\quad\quad\quad\quad &= 1 \\ -x_{12} \quad\quad\quad + x_{23} \quad\quad + x_{25} \quad\quad &= 0 \\ - x_{13} \quad\quad - x_{23} - x_{43} \quad\quad + x_{35} \quad &= 0 \\ - x_{14} \quad\quad + x_{43} \quad\quad\quad\quad + x_{45} &= 0 \end{aligned} \quad (4)$$

$$x_{ij} \geq 0, \quad 1 \leq i \leq 4, \quad 1 \leq j \leq 5$$

This problem, and any problem that originates in a similar way from a network such as the one in Figure 1, is a *natural binary* problem; that is, the variables assume only the values 0 or 1 in the optimal solution. We do not prove this fact because doing so requires a greater knowledge of linear algebra than we presume. The solution found by CALIPSO is

Row	Basic variable	Value	Dual price
1	x_{12}	1	-9
2	x_{23}	1	-6
3	x_{35}	1	-4
4	x_{43}	0	-9

To interpret the meaning of the dual prices, we formulate the dual of problem (4).

$$\text{Minimize:} \quad y = u_1$$

$$
\begin{aligned}
\text{Subject to:} \quad u_1 - u_2 & & & \geq -3 \\
u_1 & \quad - u_3 & & \geq -6 \\
u_1 & & - u_4 & \geq -3 \\
& u_2 - u_3 & & \geq -2 \\
& & - u_3 + u_4 & \geq -5 \\
& u_2 & & \geq -8 \\
& u_3 & & \geq -4 \\
& & u_4 & \geq -10
\end{aligned}
\tag{5}
$$

$$u_i \quad \text{unrestricted,} \quad 1 \leq i \leq 4$$

We can interpret the variables u_i as the price AT&T would have to charge (in negatives of dollars) for a call from City i to Moscow in order to compete with DASH. We can rewrite the first constraint of (5) as

$$-u_1 \leq -u_2 + 3$$

and can interpret it as saying the price of calling Moscow from City 1 should not exceed the cost of calling from City 2 by more than $3. This is reasonable because a customer can call London from New York for $3. You can make similar interpretations of the second and third constraints of (5). The fourth constraint of (1) is

$$-u_2 \leq -u_3 + 2$$

and says that the cost of calling Moscow from City 2 should not exceed the cost of calling from City 3 by more than $2.

We can interpret the dual price $u_1^* = -9$ as saying that decreasing the first resource value of problem (4) by 1 will increase z^* by 9. This is reasonable because DASH loses its whole profit of $9 if the call is never made. We can interpet the dual price $u_2^* = -6$ as saying that increasing the second resource value of problem (4) by 1 will decrease z^* by -6. This makes sense to DASH because they lost the $2 to send the call on to Paris and the $4 to send it from there to Moscow; it probably makes less sense to the customer calling Moscow. You can interpret the other dual prices in a similar manner.

In general, the Fundamental Principle of Duality shows that the objective value corresponding to a change Δb in the resource column $\mathbf{b}$ is given by

$$\hat{z} = z^* - 9\,\Delta b_1 - 6\,\Delta b_2 - 4\,\Delta b_3 - 9\,\Delta b_4$$

As we observed in Example 2, the new value of $\hat{z}$ may not correspond to an optimal solution. In the present example, if the change in Δb_i does not result in a resource value that is 0 or 1, the very nature of the LP problem as a model of message transmission may be altered.

We can also draw economic conclusions by studying the reduced costs of the optimal solution. Tableau 1 gives the optimal tableau for problem (4).

Tableau 1

	x_{12}	x_{13}	x_{14}	x_{23}	x_{43}	x_{25}	x_{35}	x_{45}	
x_{12}	1	1	1	0	0	0	0	0	1
x_{23}	0	1	1	1	0	1	0	0	1
x_{35}	0	0	0	0	0	1	1	1	1
x_{43}	0	0	−1	0	1	0	0	1	0
	0	1	3	0	0	2	0	1	−9

The reduced cost of x_{13} is -1. We interpret this as meaning that forcing x_{13} to be 1 will cost AT&T \$1 in revenue. This is sensible because forcing the call to be routed from New York to Paris costs \$6, which is \$1 more than sending it from New York to London and then to Paris.

The reduced cost of x_{14} is \$3. We interpret this as meaning that forcing x_{14} to be 1 will cost AT&T \$3 in revenue. This is sensible because if the call goes first to Tokyo, then to Paris, and then to Moscow, the cost is \$12, which is \$3 more than the minimum cost. You are asked to interpret the reduced costs of x_{25} and x_{45} in Exercise 10.

◆

Exercises

1. In Example 1, what change in the original mixing cost will lead to the production of F_3?

2. In Example 1, what changes in the original manufacture, mixing, and production costs will lead to the production of F_1?

3. What is the cost to the plant described in Example 1 of producing 1 unit of product F_1 and continuing to use the same set of basic variables found in the optimal solution to the LP problem? How many gallons of F_1 could they produce without making one of the basic variables negative? (Hint: See Example 6 of Section 6.2.)

4. Answer the question posed in Exercise 3 for product F_3.

†5. Show that increasing b_1 by 1 barrel in Example 2 results in an infeasible problem.

6. Assuming Wily Turkey uses the same set of basic variables (continues to use all spirits except Spirit 3) that we found in the optimal solution for Example 2, by how much will its profit change if b_1 is increased by 0.3, b_2 is increased by 4, b_3 is decreased by 2, and b_4 is unchanged?

7. Use the technique of Example 2 to write an expression for the profit $\hat{z}$ that results from changing b_i to $b_i + \Delta b_i$, $1 \leq i \leq 4$, for the problem of Example 1.

8. Use the technique of Example 1 to determine the amount of change in the cost of Spirit 3 that will lead to its inclusion in the blend.

9. Use Tableau 1 to show that setting either $x_{13} = 1$ or $x_{14} = 1$ causes the basic variables x_{12} and x_{23} to become 0. Interpret this result using Figure 1.

10. Describe the effect of forcing x_{25} or x_{45} to be 1 in the optimal solution to the problem in Example 3.

11. Figure 2 shows the options for flying from Moscow to Pocatello on Serendipity Airways (affectionately known as Dipity). Dipity requires travelers to change planes at each city on a route and charges the fare shown in Figure 2 for each trip between two cities.

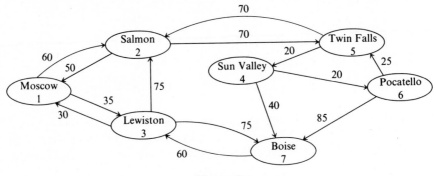

Figure 2

(a) Denote each city in Figure 2 by the number given by the name of the city in Figure 2. Let $x_{ij} = 1$ if a trip is taken between cities i and j, and let $x_{ij} = 0$ if a trip is not taken between cities i and j. Formulate an LP problem to find the minimum cost route to travel from Moscow to Pocatello.

†(b) Solve the LP problem.

†(c) The reduced costs of x_{52}, x_{73}, and x_{32} are positive in the optimal tableau for this problem. Explain the economic significance of each of these reduced costs.

12. Answer questions similar to those in Exercise 11 for the minimum cost of a return trip from Pocatello to Moscow.

CHECKLIST: CHAPTER 6

DEFINITIONS

Primal problem   273
Dual problem   273, 280, 291
Marginal price (value)   275
Shadow (fictitious) price   275
Reduced costs   288

CONCEPTS AND RESULTS

Chapter 7

SENSITIVITY ANALYSIS

7.1 WHAT IS SENSITIVITY ANALYSIS?

In broad terms, *sensitivity analysis* is the study of the "stability" of LP problems; it analyzes how changes to an already formulated and solved LP problem affect the optimal solution of the original problem. These changes can include additions of new variables or constraints to the original problem and alterations in the various cost, constraint, or resource coefficients of the problem.

The following simple product-mix problem illustrates some of the basic questions arising in sensitivity analysis.

EXAMPLE 1

A small firm wants to package various mixtures of cashews, peanuts, and pecans. The firm, which has 125 lb (2000 oz) of cashews, 180 lb (2880 oz) of pecans, and 250 lb (4000 oz) of peanuts available each week, has decided to market 1-lb packages of three different mixtures.

> Mixture A: 3 oz of cashews, 5 oz of pecans, 8 oz of peanuts
>
> Mixture B: 4 oz of cashews, 6 oz of pecans, 6 oz of peanuts
>
> Mixture C: 6 oz of cashews, 5 oz of pecans, 5 oz of peanuts

The firm estimates that its profits on each package produced are

> Mixture A: $0.50
>
> Mixture B: $0.75
>
> Mixture C: $1.00

and it would like to maximize its weekly profits. If we let x_1, x_2, and x_3 be the number of packages of mixtures A, B, and C, respectively, produced each week, then we have the LP problem

$$\text{Maximize:} \quad z = 0.50x_1 + 0.75x_2 + 1.00x_3$$

$$\text{Subject to:} \quad 3x_1 + 4x_2 + 6x_3 \leq 2000$$
$$5x_1 + 6x_2 + 5x_3 \leq 2880 \qquad (1)$$
$$8x_1 + 6x_2 + 5x_3 \leq 4000$$

$$x_i \geq 0, \qquad 1 \leq i \leq 3$$

Using the simplex method, you may verify that the optimal solution is

$$(x_1, x_2, x_3, s_1, s_2, s_3) = (0, 455, 30, 0, 0, 1120) \qquad (2)$$

where s_1, s_2, and s_3 are slack variables for the constraints in (1), and $\{x_2, x_3, s_3\}$ is the set of basic variables in the solution (2). The maximal objective value is

$$z = 371.25$$

Because of imprecise or new data, or the desire for more information, it may become necessary to make changes in (1). These changes can take on the following forms and lead to the following questions.

1. *Changes in the resource column.* Almost any change in the entries in the resource column changes the *values* of the basic variables in the optimal solution. Nevertheless, it is useful to ask to what extent we can vary the entries in the resource column of (1) without altering the *set* of basic variables that appear in the optimal solution (2). In Section 7.2 we show that if we replace $b_1 = 2000$ with any value

$$1920 \leq b_1 \leq 3456 \qquad (3)$$

then the set of basic variables $\{x_2, x_3, s_3\}$ given by solution (2) remains the optimal set of basic variables. Thus, in our example, we can reduce the pounds of cashews available per week to 1920 or increase the number to 3456 without affecting the mixtures of nuts we use for the optimal solution, although we will vary the amounts of each mixture used and the maximal profit. In the case of the resource entry b_2, we find that as long as

$$1666.6667 \leq b_2 \leq 3000 \qquad (4)$$

the same basic variables $\{x_2, x_3, s_3\}$ will yield the optimal solution, and as long as b_3 is in the range

$$2880 \leq b_3 < \infty$$

the set of basic variables will not change. (By $b_3 < \infty$ we mean that we can replace b_3 with as large a number as we want.)

However, you will see that if we want to change both b_1 and b_2 simultaneously, say to

$$b_1 = 3000; \qquad b_2 = 1700$$

then, although these changes are within the limits defined by (3) and (4), they will result in a new set of basic variables.

2. *Changes in the cost coefficients.* To what extent can we change one or more of the cost coefficients

$$c_1 = 0.5; \qquad c_2 = 0.75; \qquad c_3 = 1$$

without changing the optimal solution (2)? In Section 7.3 we develop a formula that shows, for instance, that if

$$0.6667 \le c_2 \le 1.2 \tag{5}$$

then the solution (2) is still optimal. If, however, we replace $c_2 = 0.75$ with $c_2 = 1.4$, then (2) is no longer an optimal solution. We also find that if

$$0.625 \le c_3 \le 1.125 \tag{6}$$

or if

$$-\infty < c_1 \le 0.57813 \tag{7}$$

then (2) still provides an optimal solution. (By $-\infty < c_1$, we mean that we can replace c_1 by as small a number as we want.)

As was the case with resource values, we must be careful in changing more than one cost coefficient at a time. For instance, although changes to

$$c_2 = 1.1; \qquad c_3 = 0.7$$

are within the limits given by (5) and (6), the solution defined by (2) is no longer optimal for these values of c_2 and c_3. We discuss problems of this nature in Section 7.3.

3. *Introduction of a new variable.* Suppose the firm is considering introducing a new mixture to its line of products:

     Mixture D: 8 oz of cashews, 6 oz of pecans, 2 oz of peanuts

Suppose further that each unit of this mixture that is produced results in a profit of $c_4 = \$1.40$. Two questions arise (which we deal with in Section 7.5):

(a) Is it worthwhile for the firm to introduce this new mixture? That is, if x_4 denotes the number of units of mixture D produced each week, will the optimal solution of the LP problem

$$\text{Maximize:} \quad z = 0.5x_1 + 0.75x_2 + 1.00x_3 + 1.40x_4$$

$$\text{Subject to:} \quad 3x_1 + 4x_2 + 6x_3 + 8x_4 \le 2000$$
$$5x_1 + 6x_2 + 5x_3 + 6x_4 \le 2880 \tag{8}$$
$$8x_1 + 6x_2 + 5x_3 + 2x_4 \le 4000$$

$$x_i \ge 0, \qquad 1 \le i \le 4$$

indicate a higher profit than $z = 371.25$, found in (2), or does the optimal solution defined by (2) (and with $x_4 = 0$) remain optimal?

(b) If $x_4 \neq 0$ in the final tableau associated with LP problem (8), is there a way of using the information obtained in solving the *original* LP problem (1) to find the optimal solution of (8)?

4. *Changes in the constraint coefficients.* To what extent can we change constraint coefficients without affecting the optimal solution? Suppose, for instance, that in Example 1 we want to vary one or more of the mixtures. As you can verify, changing mixture A to a new mixture A′, where

Mixture A′: 2 oz of cashews, 5 oz of pecans, 9 oz of peanuts

results in a change in the optimal solution; but changing mixture A to Mixture A″, where

Mixture A″: 4 oz of cashews, 5 oz of pecans, 7 oz of peanuts

does not lead to any change in the optimal solution.

Alterations in constraint coefficients are difficult to analyze, and the theoretical results that have been obtained are of limited practical value; so it is usually more effective simply to solve the new problem using the altered coefficients. We do not discuss this subject.

From this discussion it should be apparent why sensitivity analysis is often referred to as *postoptimality analysis*. It involves analyzing the results of changes occurring after a given problem has been solved. As you will see, we can take advantage of the calculations performed in solving the original problem to study the effects of changes in the various parameters of the problem. It should be clear why sensitivity or postoptimality analysis is of considerable practical importance.

In Example 6 of Section 2.2 we gave a geometric interpretation of a sensitivity analysis of cost coefficients. In the next example we perform a similar analysis for changes in the resource column.

EXAMPLE 2

We consider the problem

$$\text{Maximize:} \quad z = 12x_1 + 5x_2$$
$$\text{Subject to:} \quad 4x_1 + 3x_2 \leq 12$$
$$3x_1 + 6x_2 \leq 18$$
$$8x_1 + 2x_2 \leq 16$$
$$x_1, x_2 \geq 0$$

Introducing slack variables gives us the problem in canonical form:

$$\text{Maximize:} \quad z = 12x_1 + 5x_2$$

$$\text{Subject to:} \quad \begin{aligned} 4x_1 + 3x_2 + s_1 \qquad\qquad &= 12 \\ 3x_1 + 6x_2 \qquad + s_2 \qquad &= 18 \\ 8x_1 + 2x_2 \qquad\qquad + s_3 &= 16 \end{aligned} \tag{9}$$

$$x_1, x_2 \geq 0; \qquad s_i \geq 0, \qquad 1 \leq i \leq 3$$

Figure 1 shows the feasible region for this problem.

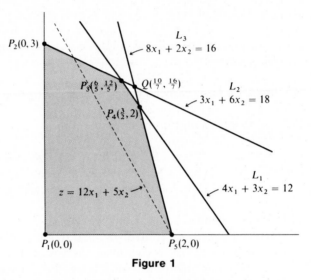

Figure 1

Tableau 1 gives the final tableau for LP problem (9).

Tableau 1   Final Tableau

	x_1	x_2	s_1	s_2	s_3	
x_1	1	0	$-\frac{1}{8}$	0	$\frac{3}{16}$	$\frac{3}{2}$
x_2	0	1	$\frac{1}{2}$	0	$-\frac{1}{4}$	2
s_2	0	0	$-\frac{21}{8}$	1	$\frac{15}{16}$	$\frac{3}{2}$
	0	0	1	0	1	28

Tableau 1 shows that the maximal solution is at corner point P_4. This point is the intersection of lines L_1 and L_3 which correspond to the first and third constraints of (9). Suppose we are interested in varying the resource value of the first constraint. That is, we want to find the range over which b_1 can vary in the constraint

$$4x_1 + 3x_2 \leq b_1 \tag{10}$$

without changing the basic variables $\{x_1, x_2, s_2\}$ (these are the basic variables for the optimal tableau, Tableau 1). As b_1 varies, the boundary line

$$4x_1 + 3x_2 = b_1 \tag{11}$$

corresponding to (10) moves parallel to L_1.

As you can see from Figure 1, such parallel motion does not change the fact that the maximal solution occurs at the intersection of L_1 and L_3 until the line (11) moves to the left of P_5 or to the right of Q (the intersection of L_2 and L_3). Parallel motion between these limits does not change the fact that Tableau 1 is optimal; that is, $\{x_1, x_2, s_2\}$ continue to be the basic variables of the optimal tableau. Of course, the values of these basic variables and of the objective function do change as the point of intersection of L_1 and L_3 moves.

In Figure 1, we see that the optimal solution switches from the intersection of L_1 and L_3 to the intersection of L_1 and the line $x_2 = 0$ when line (11) passes through P_5. We can determine the value of b_1 when L_1 passes through P_5 by substituting the coordinates of P_5 in (11). We obtain

$$b_1 = 4(2) + 3(0) = 8$$

To find the maximum that b_1 can increase, we need to determine the value of b_1 for which line (11) passes through the intersection $Q(\frac{10}{7}, \frac{16}{7})$ of L_2 and L_3. Thus, the maximum increase allowed b_1 without a change of basis is

$$4\left(\frac{10}{7}\right) + 3\left(\frac{16}{7}\right) = \frac{88}{7}$$

Therefore, b_1 can vary over the interval

$$8 \le b_1 \le \frac{88}{7}$$

without a change in the set of basic variables.

You are asked to discuss the variation of the other resource values in Exercises 13 and 14.

Exercises

You can solve these problems by hand; however, you may prefer to use TUTOR or SOLVER.

1. Solve LP problem (1) when b_1 is increased to 3000. [Note that this change is within the allowable range given by (3).] Verify that the set of basic variables in the optimal solution is unchanged. What is the new solution and the new maximal value of z?

2. Solve LP problem (1) when b_2 is decreased to 1600. [Note that this change is not within the allowable range given by (4).] Verify that the set of basic variables in the optimal solution is changed. What is the new solution and the new maximal value of z?

3. Solve LP problem (1) when b_1 is changed to 3000 and b_2 is changed to 1700. [Note that each of these changes is within the allowable ranges given by (3) and (4) for changes to only one of b_1 or b_2.] Verify that the basic solution changes, and give the new solution and the new maximal value of z.

4. Solve LP problem (1) when c_2 is changed to 0.7. [Note that this change is within the allowable range given in (5).] Verify that the optimal solution is the same as (2). What is the new maximal value of z?

5. Solve LP problem (1) when c_3 is changed to 0.8. [Note that this change is within the allowable range given in (6).] Verify that the optimal solution is the same as (2). What is the new maximal value of z?

6. Solve LP problem (1) when c_3 is changed to 0.5. [Note that this change is not within the allowable range given in (6).] Find the new optimal solution and the new maximal value of z.

7. Solve LP problem (1) when c_3 is changed to 1.5. [Note that this change is not within the allowable range given in (5).] Find the new optimal solution and the new maximal value of z.

8. Solve LP problem (1) when c_2 is increased to 1.1 and c_3 is decreased to 0.7. Verify that the basic solution changes.

9. Solve LP problem (1) when c_2 is increased to 0.9 and c_3 is decreased to 0.95. Verify that the basic solution is unchanged.

10. Determine the coordinates of the points P_1, P_2, P_3, P_4, and P_5 of Example 2 by performing pivots on Tableau 1.

11. Solve LP problem (8). Is the addition of Mixture D profitable?

12. Use the method presented in Example 6 of Section 2.2 to perform geometrically the following cost coefficient analyses on the LP problem given in Example 2 (see Figure 1).
    (a) Show that the solution is unchanged as long as

$$-\frac{8}{2} \le -\frac{c_1}{c_2} \le -\frac{4}{3}$$

    (b) Show that the objective function can be changed to $z = 8x_1 + 3x_2$ without changing the optimal solution. What is the new optimal value of z?
    (c) Show that the objective function can be changed to $z = 2x_1 + x_2$ without changing the optimal solution. What is the new optimal value of z?

13. In Example 2, suppose the resource values of the first and third constraints are kept at 12 and 16, respectively, and that b_2, the resource value for the second constraint, is allowed to vary from its value of 18. Over what range can b_2 vary without a change in the set of basic variables?

14. In Example 2, suppose the resource values of the first and second constraints are kept at 12 and 18, respectively, and that b_3, the resource value for the third constraint, is allowed to vary from its value of 16. Over what range can b_3 vary without a change in the set of basic variables?

15. Find the final tableau for LP problem (1). Use it to explain why b_3 can be increased indefinitely without changing the set of basic variables in the optimal solution.

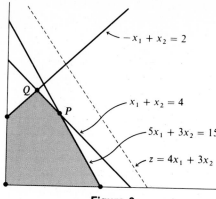

Figure 2

16. Figure 2 shows the feasible region for the LP problem

$$\text{Maximize:} \quad z = 4x_1 + 3x_2$$

$$\text{Subject to:} \quad 5x_1 + 3x_2 \leq 15$$
$$-x_1 + x_2 \leq 2$$
$$x_1 + x_2 \leq 4$$

$$x_1, x_2 \geq 0$$

(a) Let $z = c_1 x_1 + c_2 x_2$. Over what range can the slope $m = -c_1/c_2$ of z vary without changing the fact that the maximal solution is at P?

(b) For what range of values of $m = -c_1/c_2$ does the maximal solution occur at Q?

(c) Let $5x_1 + 3x_2 = b_1$. Over what range can b_1 vary without changing the fact that the maximal solution occurs at point P, which is the intersection of the lines $5x_1 + 3x_2 = b_1$ and $x_1 + x_2 = 4$?

(d) Let $-x_1 + x_2 = b_2$. Over what range can b_2 vary without changing the fact that the maximal solution occurs at point P, which is the intersection of the lines $5x_1 + 3x_2 = 15$ and $x_1 + x_2 = 4$?

(e) Let $x_1 + x_2 = b_3$. Over what range can b_3 vary without changing the fact that the maximal solution occurs at point P, which is the intersection of the lines $5x_1 + 3x_2 = 15$ and $x_1 + x_2 = b_3$?

17. Figure 3 shows the feasible region for the LP problem

$$\text{Maximize:} \quad z = 9x_1 + 3x_2$$

$$\text{Subject to:} \quad 2x_1 + x_2 \leq 10$$
$$4x_1 - x_2 \leq 8$$
$$-x_1 + x_2 \leq 4$$

$$x_1, x_2 \geq 0$$

(a) Let $z = c_1 x_1 + c_2 x_2$. Over what range can the slope $m = -c_1/c_2$ of z vary without changing the fact that the maximal solution is at P? (Be careful! What is the direction of increasing values of z when a level line of $z = c_1 x_1 + c_2 x_2$ passes through P and has a positive slope?)

(b) For what range of values of $m = -c_1/c_2$ does the maximal solution occur at Q?

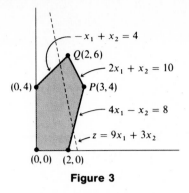

Figure 3

(c) Let $2x_1 + x_2 = b_1$. Over what range can b_1 vary without changing the fact that the maximal solution occurs at point P, which is the intersection of the lines $2x_1 + x_2 = b_1$ and $4x_1 - x_2 = 8$?

(d) Let $4x_1 - x_2 = b_2$. Over what range can b_2 vary without changing the fact that the maximal solution occurs at point P, which is the intersection of the lines $4x_1 - x_2 = b_2$ and $2x_1 + x_2 = 10$?

(e) Let $-x_1 + x_2 = b_3$. Over what range can b_3 vary without changing the fact that the maximal solution occurs at point P, which is the intersection of the lines $4x_1 - x_2 = 8$ and $2x_1 + x_2 = 10$?

7.2   CHANGES IN THE RESOURCE COLUMN

Suppose the LP problem

$$\text{Maximize:}\quad z = c_1x_1 + c_2x_2 + \cdots + c_nx_n$$

$$\text{Subject to:}\quad a_{11}x_1 + a_{12}x_2 + \cdots + a_{1n}x_n = b_1$$
$$a_{21}x_1 + a_{22}x_2 + \cdots + a_{2n}x_n = b_2 \qquad (1)$$
$$\vdots$$
$$a_{m1}x_1 + a_{m2}x_2 + \cdots + a_{mn}x_n = b_m$$

$$x_i \geq 0, \qquad 1 \leq i \leq n$$

is in canonical form and that we have obtained an optimal solution to it by the simplex method.

In this section we investigate to what extent we can change the kth resource entry b_k without altering the *set* of basic variables corresponding to the optimal solution of (1). As you will see, changing resource values in (1) causes changes in the resource column of the optimal tableau; consequently, we cannot expect the *values* of the basic variables in the optimal solution to be unchanged.

In the remainder of this chapter, we make extensive use of the matrix relationships among tableaus generated by the simplex method introduced in Theorem 2 of Section 4.3. For convenience, we reproduce this theorem here.

Theorem 1. Suppose that

$$T = \left[\begin{array}{c|c} A & \mathbf{b} \\ \hline -\mathbf{c} & z_0 \end{array} \right]$$

is the tableau for an LP problem in canonical form, where A contains the unit columns $\mathbf{e}_1, \mathbf{e}_2, \ldots, \mathbf{e}_m$ in some order. Let T^* be any tableau derived from T during application of the simplex method, and let B be the basis matrix for T^*. Denote the submatrices of T^* by

$$T^* = \left[\begin{array}{c|c} A^* & \mathbf{b}^* \\ \hline -\mathbf{c}^* & z^* \end{array} \right]$$

Then,

$$A^* = B^{-1}A$$

$$\mathbf{b}^* = B^{-1}\mathbf{b}$$

$$-\mathbf{c}^* = \mathbf{c}_B B^{-1}A - \mathbf{c}$$

$$z^* = \mathbf{c}_B B^{-1}\mathbf{b} + z_0$$

Definition. In what follows, whenever we make a change in a resource value (or, later in this chapter, in a cost coefficient), we call the resulting problem a *perturbation* of the original problem. Similarly, we refer to the tableaus for the perturbed problem as *perturbations of the tableaus* for the original problem. We often identify a perturbed value by putting a caret ($\hat{}$) above it, as in $\hat{b}$; we indicate the amount (plus or minus) of the change by putting a Δ before the variable, as in $\hat{b} = b + \Delta b$.

First note that a change in the resource column does not affect the entries in the objective row of the final simplex tableau corresponding to (1). This follows, because by Theorem 1 the entries in the objective row are determined by

$$\mathbf{c}_B B^{-1}A - \mathbf{c} \tag{2}$$

and (2) does not involve entries from the resource column. In particular then, objective-row entries will remain nonnegative despite changes in the resource column. What may occur, however, is that if we change an entry in the resource column, the basic variables may take on negative values and, hence, no longer give a feasible solution.

Suppose we change b_k to $b_k + \Delta b_k$. If we let $\mathbf{e}_k$ denote the unit column vector with a 1 in the k position and zeros elsewhere, then the new resource column is

$$\hat{\mathbf{b}} = \mathbf{b} + \Delta b_k \, \mathbf{e}_k = \begin{bmatrix} b_1 \\ \vdots \\ b_k + \Delta b_k \\ \vdots \\ b_m \end{bmatrix}$$

The basic solution corresponding to the inverse basis matrix in the final tableau of the original problem and the perturbed resource column $\hat{\mathbf{b}}$ is defined by

$$\mathbf{x}_B = B^{-1}\hat{\mathbf{b}} = B^{-1}(\mathbf{b} + \Delta b_k \, \mathbf{e}_k) = B^{-1}\mathbf{b} + \Delta b_k \, B^{-1}\mathbf{e}_k \qquad (3)$$

It follows from (3) that the perturbed final tableau for the solution of (1) remains optimal as long as the entries in the column matrix

$$B^{-1}\mathbf{b} + \Delta b_k \, B^{-1}\mathbf{e}_k$$

remain nonnegative. Note that $B^{-1}\mathbf{e}_k$ is the kth column of B^{-1} and $\mathbf{x}_B = B^{-1}\mathbf{b}$ is the original optimal basic solution.

EXAMPLE 1

We consider again the LP problem

$$\text{Maximize:} \quad z = \frac{1}{2}x_1 + \frac{3}{4}x_2 + x_3$$

$$\text{Subject to:} \quad \begin{aligned} 3x_1 + 4x_2 + 6x_3 &\leq 2000 \\ 5x_1 + 6x_2 + 5x_3 &\leq 2880 \\ 8x_1 + 6x_2 + 5x_3 &\leq 4000 \end{aligned} \qquad (4)$$

$$x_i \geq 0, \qquad 1 \leq i \leq 3$$

that we used as an example in Section 7.1. Its final tableau is

	x_1	x_2	x_3	x_4	x_5	x_6	
x_3	$-\frac{1}{8}$	0	1	$\frac{3}{8}$	$-\frac{1}{4}$	0	30
x_2	$\frac{15}{16}$	1	0	$-\frac{5}{16}$	$\frac{3}{8}$	0	455
x_6	3	0	0	0	-1	1	1120
	$\frac{5}{64}$	0	0	$\frac{9}{64}$	$\frac{1}{32}$	0	$\frac{1485}{4}$

(5)

If B is the basis matrix corresponding to the final tableau (5), then from this tableau we see that

$$B^{-1} = \begin{bmatrix} \frac{3}{8} & -\frac{1}{4} & 0 \\ -\frac{5}{16} & \frac{3}{8} & 0 \\ 0 & -1 & 1 \end{bmatrix} \qquad (6)$$

$$\mathbf{x}_B = \begin{bmatrix} x_3 \\ x_2 \\ x_6 \end{bmatrix} = \begin{bmatrix} 30 \\ 455 \\ 1120 \end{bmatrix} = \mathbf{b}^* = B^{-1}\mathbf{b}$$

We now find the allowable increments Δb_1 corresponding to the resource value $b_1 = 2000$, which will not cause a change in the set of optimal basic variables

x_B (although we expect the values $b*$ to change). Since in the notation used above,

$$\hat{b}* = B^{-1}b + \Delta b_1\, B^{-1}e_1 = \begin{bmatrix} 30 \\ 455 \\ 1120 \end{bmatrix} + \Delta b_1 \begin{bmatrix} \frac{3}{8} \\ -\frac{5}{16} \\ 0 \end{bmatrix} \tag{7}$$

it follows that for x_B to be feasible we must have

$$\begin{bmatrix} 30 \\ 455 \\ 1120 \end{bmatrix} + \Delta b_1 \begin{bmatrix} \frac{3}{8} \\ -\frac{5}{16} \\ 0 \end{bmatrix} \geq \begin{bmatrix} 0 \\ 0 \\ 0 \end{bmatrix}$$

That is, it must be the case that

$$30 + (\Delta b_1)\left(\frac{3}{8}\right) \geq 0$$

$$455 + (\Delta b_1)\left(-\frac{5}{16}\right) \geq 0$$

$$1120 + (\Delta b_1)(0) \geq 0$$

or, equivalently,

$$\Delta b_1 \geq -80$$
$$\Delta b_1 \leq 1456 \tag{8}$$
$$\Delta b_1 \quad \text{(no restriction)}$$

Thus, from (8) we obtain the restriction

$$-80 \leq \Delta b_1 \leq 1456$$

on the increments Δb_1.

To find the limits of Δb_2 we consider

$$\hat{b}* = B^{-1}b + \Delta b_2\, B^{-1}e_2 = \begin{bmatrix} 30 \\ 455 \\ 1120 \end{bmatrix} + \Delta b_2 \begin{bmatrix} -\frac{1}{4} \\ \frac{3}{8} \\ -1 \end{bmatrix} \tag{9}$$

It follows from (9) that for x_B to be feasible we must have

$$30 + (\Delta b_2)\left(-\frac{1}{4}\right) \geq 0$$

$$455 + (\Delta b_2)\left(\frac{3}{8}\right) \geq 0$$

$$1120 + (\Delta b_2)(-1) \geq 0$$

or, equivalently,

$$\Delta b_2 \leq 120$$

$$\Delta b_2 \geq -\frac{3640}{3} \qquad (10)$$

$$\Delta b_2 \leq 1120$$

From (10) we obtain the restriction

$$-\frac{3640}{3} \leq \Delta b_2 \leq 120$$

Finally, for the case of Δb_3 we have

$$\hat{\mathbf{b}}^* = B^{-1}\mathbf{b} + \Delta b_3\, B^{-1}\mathbf{e}_3 = \begin{bmatrix} 30 \\ 455 \\ 1120 \end{bmatrix} + \Delta b_3 \begin{bmatrix} 0 \\ 0 \\ 1 \end{bmatrix} \qquad (11)$$

It follows from (11) that for $\mathbf{x}_B$ to be feasible we must have

$$30 + (\Delta b_3)(0) \geq 0$$

$$455 + (\Delta b_3)(0) \geq 0 \qquad (12)$$

$$1120 + (\Delta b_3)(1) \geq 0$$

Thus from (12) we obtain the restriction

$$-1120 \leq \Delta b_3 < \infty$$

on Δb_3.

In summary, the restrictions on changes to *one* of the resource values $b_1, b_2,$ or b_3 while holding the other two fixed are

$$-80 \leq \Delta b_1 \leq 1456$$

$$-\frac{3640}{3} \leq \Delta b_2 \leq 120 \qquad (13)$$

$$-1120 \leq \Delta b_3 < \infty \qquad \blacklozenge$$

The next theorem gives a formula for computing bounds on perturbations of a single resource value. In Exercise 24 you are asked to prove this theorem (a proof can be modeled on the calculations in Example 1).

Theorem 2.   Suppose that LP problem (1) has an optimal tableau obtained by the simplex method. Suppose the rth resource value of the original problem is changed by an amount Δb_r, and the other resource values are unchanged. Denote the resource values in the optimal tableau by $\mathbf{b}^*$, and let $B^{-1} = [t_{ij}]_{m \times m}$. Then the set of basic variables in the optimal solution is unchanged if and only if

$$\max_i \left\{ \frac{-b_i^*}{t_{ir}} \,\middle|\, t_{ir} > 0 \right\} \le \Delta b_r \le \min_i \left\{ \frac{-b_i^*}{t_{ir}} \,\middle|\, t_{ir} < 0 \right\} \tag{14}$$

If the rth column of B^{-1} has no negative entries, then the right-hand side of inequality (14) is ∞ (i.e., there is no upper bound on Δb_i). If the rth column of B^{-1} has no positive entries, then the left-hand side of inequality (14) is $-\infty$ (i.e., there is no lower bound on Δb_i).

We can obtain the ranges (13) from inequalities (14) and tableau (5) by the calculation

$$-80 = \max \left\{ \frac{-30}{\frac{3}{8}} \right\} \le \Delta b_1 \le \min \left\{ \frac{-455}{\frac{5}{16}} \right\} = 1456$$

$$-\frac{3640}{3} = \max \left\{ \frac{-455}{\frac{3}{8}} \right\} \le \Delta b_2 \le \min \left\{ \frac{-30}{-1}, \frac{-1120}{-1} \right\} = 120$$

$$-1120 = \max \left\{ \frac{-1120}{1} \right\} \le \Delta b_3 < \infty$$

You must take care to use the correct B^{-1} matrix when you solve a problem with the two-phase method. You should use the B^{-1} matrix that corresponds to the basis matrix for the original tableau that was augmented with artificial variables, and not the B^{-1} matrix that corresponds to the initial tableau for phase 2. The next example illustrates this point.

EXAMPLE 2

The initial tableau for phase 1 of Example 1 of Section 3.4 is Tableau 1; the final tableau for phase 1 is Tableau 2.

Tableau 1   Initial Tableau

	x_1	x_2	x_3	x_4	x_5	y_1	y_2	
x_4	2	3	1	1	0	0	0	12
y_1	5	6	3	0	-1	1	0	15
y_2	2	-3	1	0	0	0	1	8
(z)	-80	-60	-42	0	0	0	0	0
(w)	-7	-3	-4	0	1	0	0	-23

Tableau 2   Final Tableau for Phase 1

	x_1	x_2	x_3	x_4	x_5	y_1	y_2	
x_4	0	6	0	1	0	0	-1	4
x_1	1	$-\frac{3}{2}$	$\frac{1}{2}$	0	0	0	$\frac{1}{2}$	4
x_5	0	$-\frac{27}{2}$	$-\frac{1}{2}$	0	1	-1	$\frac{5}{2}$	5
(z)	0	-180	-2	0	0	0	40	320
(w)	0	0	0	0	0	1	1	0

If we intend to perform a sensitivity analysis on the resource values, we must retain the artificial variable columns during phase 2. The row corresponding to the artificial objective function w can be deleted. The final tableau (Tableau 3) results from applying the phase 2 procedure to Tableau 2.

Tableau 3   Final Tableau for Phase 2

	x_1	x_2	x_3	x_4	x_5	y_1	y_2	
x_2	0	1	0	$\frac{1}{6}$	0	0	$-\frac{1}{6}$	$\frac{2}{3}$
x_3	2	0	1	$\frac{1}{2}$	0	0	$\frac{1}{2}$	10
x_5	1	0	0	$\frac{5}{2}$	1	-1	$\frac{1}{2}$	19
(z)	4	0	0	31	0	0	11	460

The B^{-1} matrix is found in the columns of Tableau 3 labeled x_4, y_1, and y_2.

We obtain the following sensitivity ranges for changes to *only one* of Δb_1, Δb_2, Δb_3 from (14).

$$-4 = \max\left\{ \frac{-\frac{2}{3}}{\frac{1}{6}}, \frac{-10}{\frac{1}{2}}, \frac{-19}{\frac{5}{2}} \right\} \le \Delta b_1 < \infty$$

$$-\infty < \Delta b_2 \le \min\left\{ \frac{-19}{-1} \right\} = 19$$

$$-20 = \max\left\{ \frac{-10}{\frac{1}{2}}, \frac{-19}{\frac{1}{2}} \right\} \le \Delta b_3 \le \min\left\{ \frac{-\frac{2}{3}}{-\frac{1}{6}} \right\} = 4$$

Note that the B^{-1} matrix in Tableau 3 that corresponds to the initial tableau (Tableau 2) of *phase 2* consists of the columns x_4, x_1, x_5 of Tableau 3. You would not obtain the correct answer by using this matrix in formula (14). You would obtain sensitivity ranges for the resource values of Tableau 2.   ◆

The next example shows how to restore feasibility when a change to a resource value introduces negative values in the resource column.

EXAMPLE 3

Suppose we change b_2 by $\Delta b_2 = 140$ in the problem of Example 1. Notice that this change is not within the range given in (13). We can use relation (9) to compute the new resource column as

$$
\hat{\mathbf{b}}^* = \begin{bmatrix} 30 \\ 455 \\ 1120 \end{bmatrix} + 140 \begin{bmatrix} -\frac{1}{4} \\ \frac{3}{8} \\ -1 \end{bmatrix} = \begin{bmatrix} -5 \\ \frac{1015}{2} \\ 980 \end{bmatrix}
$$

and the new objective function value as

$$
\hat{z}^* = \mathbf{c}_B \hat{\mathbf{b}}^* - 0 = \begin{bmatrix} 1 & \frac{3}{4} & 0 \end{bmatrix} \hat{\mathbf{b}}^* = \frac{3005}{8}
$$

The new tableau obtained from the final tableau (5) of the original problem is

	x_1	x_2	x_3	x_4	x_5	x_6	
x_3	$-\frac{1}{8}$	0	1	$\frac{3}{8}$	$-\frac{1}{4}$	0	-5
x_2	$\frac{15}{16}$	1	0	$-\frac{5}{16}$	$\frac{3}{8}$	0	$\frac{1015}{2}$
x_6	3	0	0	0	-1	1	980
	$\frac{5}{64}$	0	0	$\frac{9}{64}$	$\frac{1}{32}$	0	$\frac{3005}{8}$

(15)

We use the dual simplex method to restore feasibility by pivoting on the circled entry of tableau (15) to obtain the new optimal tableau

	x_1	x_2	x_3	x_4	x_5	x_6	
x_5	$\frac{1}{2}$	0	-4	$-\frac{3}{2}$	1	0	20
x_2	$\frac{3}{4}$	1	$\frac{3}{2}$	$\frac{1}{4}$	0	0	500
x_6	$\frac{7}{2}$	0	-4	$-\frac{3}{2}$	0	1	1000
	$\frac{1}{16}$	0	$\frac{1}{8}$	$\frac{3}{16}$	0	0	375

Observe that the set of basic variables in this tableau differ from the set of basic variables in the final tableau (5). ◆

As was noted in Section 7.1, we must exercise care if we change more than one entry in the resource column at a time. For instance, if in Example 1 we set

$$
\Delta b_1 = 1000
$$

$$
\Delta b_2 = -1000
$$

$$
\Delta b_3 = -1000
$$

then, even though each of these increments falls within the individual limits given in (13), it is easy to verify that the optimal solution of the original problem is no longer a feasible solution.

We now allow simultaneous changes in some or all of the resource values and perform an analysis that is similar to the previous one.

Suppose that $\mathbf{x}_B = \mathbf{b}^*$ is an optimal basic solution of the LP problem (1). Let

$$\Delta\mathbf{b} = \begin{bmatrix} \Delta b_1 \\ \Delta b_2 \\ \vdots \\ \Delta b_m \end{bmatrix}$$

The basic solution $\mathbf{x}_B = \mathbf{b}^*$ corresponding to the inverse basis matrix of the original problem and the perturbed resource column $\hat{\mathbf{b}}$ is defined by

$$\hat{\mathbf{x}}_B = \hat{\mathbf{b}}^* = B^{-1}\hat{\mathbf{b}} = B^{-1}(\mathbf{b} + \Delta\mathbf{b}) = B^{-1}\mathbf{b} + B^{-1}\Delta\mathbf{b} = \mathbf{b}^* + B^{-1}\Delta\mathbf{b}$$

The perturbed final tableau remains optimal (and hence $\hat{\mathbf{x}}_B = \mathbf{x}_B$) provided all entries of $\mathbf{b}^* + B^{-1}\Delta\mathbf{b}$ are nonnegative. Thus, if

$$B^{-1} = \begin{bmatrix} b_{11} & b_{12} & \cdots & b_{1m} \\ b_{12} & b_{22} & \cdots & b_{2m} \\ \vdots & & & \\ b_{m1} & b_{m2} & \cdots & b_{mm} \end{bmatrix}$$

then $\Delta\mathbf{b}$ must satisfy the system of equations

$$b_1^* + b_{11}\,\Delta b_1 + b_{12}\,\Delta b_2 + \cdots + b_{1m}\,\Delta b_m \geq 0$$
$$b_2^* + b_{21}\,\Delta b_1 + b_{22}\,\Delta b_2 + \cdots + b_{2m}\,\Delta b_m \geq 0$$
$$\vdots \qquad\qquad\qquad\qquad\qquad\qquad (16)$$
$$b_m^* + b_{m1}\,\Delta b_1 + b_{m2}\,\Delta b_2 + \cdots + b_{mm}\,\Delta b_m \geq 0$$

EXAMPLE 4

The system of equations (16) for the LP problem of Example 1 is

$$30 + \left(\frac{3}{8}\right)\Delta b_1 + \left(-\frac{1}{4}\right)\Delta b_2 + (0)\,\Delta b_3 \geq 0$$

$$455 + \left(-\frac{5}{16}\right)\Delta b_1 + \left(\frac{3}{8}\right)\Delta b_2 + (0)\,\Delta b_3 \geq 0$$

$$1120 + (0)\,\Delta b_1 + (-1)\,\Delta b_2 + (1)\,\Delta b_3 \geq 0$$

We can simplify the system to obtain

$$-3\,\Delta b_1 + 2\,\Delta b_2 \qquad\leq\ 240$$
$$5\,\Delta b_1 - 6\,\Delta b_2 \qquad\leq 7280 \qquad (17)$$
$$\Delta b_2 - \Delta b_3 \leq 1120$$

It can be very difficult to solve a system of inequalities like (17), especially when the number of constraints is large. Nevertheless, we can effectively use such systems of inequalities as (17) to determine if a change in resource values $\Delta \mathbf{b}$ requires us to find a new set of basic variables. In fact, in many practical applications, changes in resource values are imposed by uncontrollable circumstances, such as an unexpected shortage of raw materials. A system like (17) can be used to determine if the optimal solution is changed. For instance, you can verify that

$$\Delta b_1 = -100, \qquad \Delta b_2 = -100, \qquad \Delta b_3 = -900$$

satisfies (17). ◆

The sensitivity report produced by CALIPSO gives the B^{-1} matrix so that you can form the system of inequalities (16); however, most computer packages for solving LP problems do not produce B^{-1}, and so it can be difficult to study the effects of simultaneous changes in resource values. The next theorem gives a test that can be performed on some simultaneous changes of resource values to ensure that the basis does not change. Unfortunately, the test is only sufficient: It is possible for simultaneous changes that would not cause a change in the set of basic variables to fail this test (see Exercise 14 for an example).

Theorem 3. Suppose that $\mathbf{x_B} = \mathbf{b}^*$ is an optimal basic solution to LP problem (1) and that $\mathbf{q}_1, \mathbf{q}_2, \ldots, \mathbf{q}_k$ are $m \times 1$ column matrices such that for each i, $1 \leq i \leq k$

$$B^{-1}(\mathbf{b} + \mathbf{q}_i) \geq \mathbf{0}$$

Let $\mathbf{q} = \lambda_1 \mathbf{q}_1 + \lambda_2 \mathbf{q}_2 + \cdots + \lambda_k \mathbf{q}_k$, where $\Sigma_{i=1}^k \lambda_i \leq 1$, and $\lambda_i \geq 0$ for each i, $1 \leq i \leq k$. Then,

$$B^{-1}(\mathbf{b} + \mathbf{q}) \geq \mathbf{0}$$

[and hence, $\mathbf{x_B}$ remains as the basis of the perturbed LP problem obtained from (1) by replacing $\mathbf{b}$ with $\mathbf{b} + \mathbf{q}$].

PROOF. Let $\lambda_0 = 1 - \Sigma_{i=1}^k \lambda_i \geq 0$, so that $\Sigma_{i=0}^k \lambda_i = 1$. Then we have

$$B^{-1}(\mathbf{b} + \mathbf{q}) = B^{-1}\mathbf{b} + B^{-1}\mathbf{q}$$
$$= B^{-1}\mathbf{b} + B^{-1}(\lambda_1 \mathbf{q}_1 + \cdots + \lambda_k \mathbf{q}_k)$$
$$= (\lambda_0 + \lambda_1 + \lambda_2 + \cdots + \lambda_k)B^{-1}\mathbf{b}$$
$$\quad + \lambda_1 B^{-1}\mathbf{q}_1 + \lambda_2 B^{-1}\mathbf{q}_2 + \cdots + \lambda_k B^{-1}\mathbf{q}_k$$
$$= \lambda_0 B^{-1}\mathbf{b} + \lambda_1(B^{-1}\mathbf{b} + B^{-1}\mathbf{q}_1) + \lambda_2(B^{-1}\mathbf{b} + B^{-1}\mathbf{q}_2) + \cdots$$
$$\quad + \lambda_k(B^{-1}\mathbf{b} + B^{-1}\mathbf{q}_k)$$

Since for each i, $\lambda_i \geq 0$, and since $\lambda_0 B^{-1}\mathbf{b} \geq \mathbf{0}$ and (by hypothesis) $B^{-1}\mathbf{b} + B^{-1}\mathbf{q}_i \geq \mathbf{0}$, it follows that $B^{-1}(\mathbf{b} + \mathbf{q}) \geq \mathbf{0}$. ∎

EXAMPLE 5

For the problem of Example 1, let

$$\mathbf{q}_1 = \begin{bmatrix} 100 \\ 200 \\ 40 \end{bmatrix}, \qquad \mathbf{q}_2 = \begin{bmatrix} -50 \\ 40 \\ -100 \end{bmatrix}$$

You can verify that the entries in $\mathbf{q}_1$ and $\mathbf{q}_2$ satisfy inequalities (16) and, hence, that $B^{-1}(\mathbf{b} + \mathbf{q}_1) \geq 0$ and $B^{-1}(\mathbf{b} + \mathbf{q}_2) \geq 0$, where B^{-1} is defined by (6). Let $\lambda_1 = \frac{2}{5}$ and $\lambda_2 = \frac{3}{5}$. It follows from Theorem 3 that the set of optimal basic variables $\mathbf{x}_B$ is still the set of basic variables if we replace

$$\mathbf{b} = \begin{bmatrix} 2000 \\ 2880 \\ 4000 \end{bmatrix}$$

in (4) with $\mathbf{b} + \mathbf{q}$, where

$$\mathbf{q} = \frac{2}{5}\mathbf{q}_1 + \frac{3}{5}\mathbf{q}_2 = \begin{bmatrix} 10 \\ 104 \\ -44 \end{bmatrix} \qquad \blacklozenge$$

The following corollary to Theorem 3 is often called the *100% rule*.

Corollary.   Let $\mathbf{x}_B = \mathbf{b}^*$ be an optimal basic solution to LP problem (1). Suppose that, for $1 \leq i \leq m$,

$U_i \geq 0$ is the maximal allowable increase for b_i, and

$L_i \geq 0$ is the maximal allowable decrease for b_i

for which the current set of basic variables remains unchanged. Let

$$\Delta\mathbf{b} = \begin{bmatrix} \Delta b_1 \\ \Delta b_2 \\ \vdots \\ \Delta b_m \end{bmatrix}$$

be a perturbation of $\mathbf{b}$, and let

$$\lambda_i = \begin{cases} \dfrac{\Delta b_i}{U_i} & \text{if} \quad \Delta b_i \geq 0, \quad U_i \neq \infty \\[2mm] -\dfrac{\Delta b_i}{L_i} & \text{if} \quad \Delta b_i \leq 0, \quad L_i \neq \infty \\[2mm] 0 & \text{otherwise} \end{cases} \qquad (18)$$

If $\Sigma_{i=1}^{m} \lambda_i \leq 1$, then the set of optimal basic variables $\mathbf{x_B}$ remains unchanged when $\mathbf{b}$ is replaced by

$$\hat{\mathbf{b}} = \mathbf{b} + \begin{bmatrix} \lambda_1 \, \Delta b_1 \\ \lambda_2 \, \Delta b_2 \\ \vdots \\ \lambda_m \, \Delta b_m \end{bmatrix}$$

PROOF.   For $1 \leq i \leq m$, let

$$\mathbf{q}_i = \begin{bmatrix} 0 \\ \vdots \\ \Delta b_i \\ \vdots \\ 0 \end{bmatrix} \quad \leftarrow \text{row } i$$

be the $m \times 1$ matrix whose ith row is Δb_i and whose other entries are 0. Since $0 \leq \lambda_i \leq 1$, it follows from (18) that for $1 \leq i \leq m$

$$-L_i \leq \Delta b_i \leq U_i$$

and from the definitions of L_i and U_i that

$$B^{-1}(\mathbf{b} + \mathbf{q}_i) \geq \mathbf{0}$$

Thus, the corollary follows from Theorem 3.   ∎

EXAMPLE 6

For the problem of Example 1, the values $\Delta b_1 = 240$, $\Delta b_2 = 100$, and $\Delta b_3 = 300$ are within the ranges given in (13). Let

$$\mathbf{q}_1 = \begin{bmatrix} 240 \\ 0 \\ 0 \end{bmatrix}, \qquad \mathbf{q}_2 = \begin{bmatrix} 0 \\ 100 \\ 0 \end{bmatrix}, \qquad \mathbf{q}_3 = \begin{bmatrix} 0 \\ 0 \\ 300 \end{bmatrix}$$

Then, for $i = 1, 2, 3$, $B^{-1}(\mathbf{b} + \mathbf{q}_i) \geq \mathbf{0}$, where B^{-1} is defined by (6). Let $\lambda_1 = 240/1456$, $\lambda_2 = 100/120$, $\lambda_3 = 1/1000$. Then $\lambda_1 + \lambda_2 + \lambda_3 = 0.999 \leq 1$. It follows from the corollary to Theorem 3 that the optimal set of basic variables $\mathbf{x_B}$ is unchanged if we replace

$$\mathbf{b} = \begin{bmatrix} 2000 \\ 2880 \\ 4000 \end{bmatrix}$$

in (4) with $\mathbf{b} + \mathbf{q}$, where

$$\mathbf{q} = \frac{240}{1456}\mathbf{q}_1 + \frac{100}{120}\mathbf{q}_2 + \frac{1}{1000}\mathbf{q}_3 = \begin{bmatrix} 39.56 \\ 83.33 \\ 0.3 \end{bmatrix}$$

◆

Exercises

Exercises 1–14 concern the problem solved in Examples 1 and 2.

1. Give the ranges over which the resource values $\hat{b}_1, \hat{b}_2, \hat{b}_3$ can vary when $\Delta b_1, \Delta b_2, \Delta b_3$ vary over the ranges given in (13).

2. (a) Show that the change $\Delta b_1 = 1000$, $\Delta b_2 = -1000$, $\Delta b_3 = -1000$ causes the solution given by tableau (5) to become infeasible.
   (b) Find the new optimal solution to the perturbed problem using the dual simplex method.

3. Find the solution to the problem of Example 1 when $b_1 = 2000$ is changed to $\hat{b}_1 = 3500$. Do not solve the problem again from the beginning but start with final tableau (5).

4. Follow the instructions of Exercise 3, except change b_2 to 140.

5. Obtain the restrictions on Δb_1, Δb_2, and Δb_3 given in (13) from the system of inequalities (17).

In Exercises 6, 7, and 8, use the system of inequalities (17) to find the ranges over which resource values of the LP problem in Example 1 can vary without a change in the set of basic variables.

6. Suppose $\Delta b_2 = 80$. Find the ranges over which Δb_1 and Δb_2 can vary.

7. Suppose $\Delta b_3 = 200$ and $\Delta b_2 = -100$. Find the range for Δb_1.

8. Suppose $\Delta b_1 = 100$. Find the ranges for Δb_2 and Δb_3.

9. For the B^{-1} matrix given by tableau (5), verify that $\mathbf{q}_1, \mathbf{q}_2, \mathbf{q}_3$ satisfy $B^{-1}(\mathbf{b} + \mathbf{q}_i) \geq 0$ for

$$\mathbf{q}_1 = \begin{bmatrix} 240 \\ 120 \\ 140 \end{bmatrix}, \qquad \mathbf{q}_2 = \begin{bmatrix} -30 \\ 0 \\ 210 \end{bmatrix}, \qquad \mathbf{q}_3 = \begin{bmatrix} 180 \\ -900 \\ 300 \end{bmatrix}$$

Then use Theorem 3 to show that the perturbation by

$$\mathbf{q} = \frac{1}{3}\mathbf{q}_1 + \frac{1}{3}\mathbf{q}_2 + \frac{1}{6}\mathbf{q}_3$$

does not change the set of basic variables for the optimal solution to the LP problem in Example 1.

10. For the B^{-1} matrix given by tableau (5), verify that $\mathbf{q}_1, \mathbf{q}_2, \mathbf{q}_3$ satisfy $B^{-1}(\mathbf{b} + \mathbf{q}_i) \geq 0$ for

$$\mathbf{q}_1 = \begin{bmatrix} 40 \\ 150 \\ -300 \end{bmatrix}, \qquad \mathbf{q}_2 = \begin{bmatrix} -100 \\ -200 \\ -400 \end{bmatrix}, \qquad \mathbf{q}_3 = \begin{bmatrix} 0 \\ 20 \\ 50 \end{bmatrix}$$

Then use Theorem 3 to show that the perturbation by

$$\mathbf{q} = \frac{1}{2}\mathbf{q}_1 + \frac{1}{4}\mathbf{q}_2 + \frac{1}{4}\mathbf{q}_3$$

does not change the set of basic variables in the optimal solution to the LP problem in Example 1.

11. Use the 100% rule (the corollary to Theorem 3) to show that the simultaneous changes $\Delta b_1 = 700$ and $\Delta b_2 = -600$ do not change the set of optimal basic variables of the LP problem of Example 1.

12. Use the 100% rule (the corollary to Theorem 3) to show that the simultaneous changes $\Delta b_1 = -40$, $\Delta b_2 = -300$, and $\Delta b_3 = -250$ do not change the set of optimal basic variables of the LP problem of Example 1.

13. The perturbations $\Delta b_1 = 1400$, $\Delta b_2 = 100$, and $\Delta b_3 = -1000$ are within the ranges for changes to a *single* resource value without changing the set of basic variables of the LP problem in Example 1. Find appropriate values for λ_1, λ_2, and λ_3, and apply Theorem 3 to show that the *simultaneous* changes $\Delta b_1 = 560$, $\Delta b_2 = 50$, and $\Delta b_3 = -100$ do not change the set of basic variables.

14. Show that the simultaneous changes $\Delta b_1 = 1500$, $\Delta b_2 = 40$, and $\Delta b_3 = 0$ do not cause a change of the set of basic variables for the LP problem in Example 1 by using (17). Explain why it is not possible to draw this conclusion from the individual ranges given in (13) and Theorem 3.

In Exercises 15–21, calculate the ranges over which each of the resource values can vary individually (with others held constant) without a change of the set of basic variables. In Exercises 18–21, you must be careful in determining the appropriate B^{-1} matrix because the two-phase method is used.

15. Example 2 of Section 7.1.

16. LP problem (1) of Section 3.2.

17. Example 1 of Section 3.3.

18. LP problem (4) of Section 3.4. Most of the required information is in Tableaus 1 and 7; however, you need to do some work to find B^{-1}.

19. Exercise 13 of Section 3.4.

20. Example 1 of Section 3.5.

21. Example 1 of Section 3.6.

22. Find the system of inequalities that must be satisfied by simultaneous changes to the resource values if the set of basic variables is to be preserved for
(a) Exercise 16   (b) Exercise 18   (c) Exercise 20

23. Follow the instructions of Exercise 22 for
(a) Exercise 15   (b) Exercise 17   (c) Exercise 19   (d) Exercise 21

24. Prove Theorem 2.

25. Solve

$$\text{Maximize:} \quad z = 80x_1 + 50x_2$$

$$\text{Subject to:} \quad 4x_1 + 3x_2 \le 12$$

$$6x_1 + 3x_2 \le 15$$

$$x_1, x_2 \ge 0$$

by the simplex method. Find the system of inequalities that must be satisfied by any simultaneous perturbation of $b_1 = 12$ and $b_2 = 15$ that leaves the set of basic variables unchanged.

26. Suppose the second constraint in Exercise 25 is changed to $6x_1 + 3x_2 \leq 3$. From the final tableau of Exercise 25, find the new optimal solution by the dual simplex method.

27. Add the constraint $x_1 + 2x_2 \leq 1$ to the final tableau for the problem you solved in Exercise 25, and solve with the dual simplex method. Find the system of inequalities that must be solved by any simultaneous perturbation of the resource values that leaves the set of basic variables unchanged.

Solve Exercises 28–31 with SOLVER. Choose the sensitivity analysis from the menu of reports (we do not discuss the analysis of cost coefficients that is included in this report until Section 7.3).

†28. Maximize: $z = 4x_1 + 8x_2 + 3x_3$

Subject to: $2x_1 + x_2 + x_3 \leq 20$
$$x_1 - 3x_2 + 2x_3 \leq 36$$
$$3x_1 - 4x_2 + x_3 \leq 60$$
$$x_1 \geq 0, \qquad 1 \leq i \leq 3$$

†29. Maximize: $z = 3x_1 + 2x_2 - x_3$

Subject to: $2x_1 + 3x_2 + 3x_3 \leq 10$
$$x_1 - x_2 - x_3 \leq 3$$
$$x_1 + x_2 + x_3 \leq 4$$
$$x_i \geq 0, \qquad 1 \leq i \leq 3$$

†30. Solve Exercise 3 of Section 1.1 with SOLVER. Examine the final tableau and the sensitivity analysis, and answer the following questions.
   (a) The report says that a $2000 resource value can be reduced by as much as $4000 without changing the set of basic variables. Explain how this is possible in view of the fact that such a change introduces a negative value in the resource column of the initial tableau. How does a $4000 change alter Adam Smith's theory?
   (b) Without changing the basis, can we reduce Adam's $14,000 constraint to $13,000 and his $18,000 constraint to $16,000?
   (c) Does the basis change if we reduce the $14,000 constraint by $4,000 and increase the $8,000 constraint by $4,000?
   (d) Does the basis change if we decrease the $18,000 constraint by $1,000 and decrease the 11% rate of return to 10%?

†31. The Moscow Bicycle Makers (MBM) make four kinds of bicycles. Type A is a touring bike, type B a racing bike, type C a mountain bike, and type D a mountain racing bike. The profits on a single bike are 90, 22, 75, and 120 dollars for types A, B, C, and D, respectively. For next month, MBM predicts that it will sell no more than 24, 16, 16, and 6 bikes of types A, B, C, and D, respectively. MBM has 175 hr of labor available for the next month and $10,000 in capital to invest on materials to make bikes. Table 1 gives the materials cost and the hours of labor required to make each type of bike.
   (a) Find the solution that maximizes profit and produce a sensitivity analysis. Then use this solution and the sensitivity analysis to answer parts (b)–(e); do not solve the problems by changing the initial tableau.

Table 1

Type	Materials cost (\$)	Labor (hr)
A	90	2
B	140	3
C	150	4
D	180	5

(b) How much can the available capital be decreased without requiring a change of the types of bicycles made? Will any decrease in capital change the number of bikes of each type made?

(c) How much can available labor be decreased without requiring a change of the types of bicycles made?

(d) Should MBM change the production schedule if the predicted demand for bikes of type B drops from 16 to 10?

(e) Should MBM change the production schedule if the predicted demand for bikes of type A drops to 17?

†32. The LP problems

Problem A	*Problem B*

Maximize: $z = x_1 + 2x_2 + 5x_3$ Maximize: $z = x_1 + 2x_2 + 5x_3$

Subject to:
$$x_1 + x_2 + x_3 \le 12$$
$$2x_1 - 3x_2 + 2x_3 = 3$$

$$-x_1 + 2x_2 + 5x_3 \ge 15$$
$$x_i \ge 0, \quad 1 \le i \le 3$$

Subject to:
$$x_1 + x_2 + x_3 \le 12$$
$$2x_1 - 3x_2 + 2x_3 \le 3$$
$$2x_1 - 3x_2 + 2x_3 \ge 3$$
$$-x_1 + 2x_2 + 5x_3 \ge 15$$
$$x_i \ge 0, \quad 1 \le i \le 3$$

are equivalent.

(a) Show that the sensitivity ranges for the resource values are

Problem A

$$-8.526 \le \Delta b_1 < \infty$$
$$-39 \le \Delta b_2 \le 21$$
$$-\infty < \Delta b_3 \le 32.4$$

Problem B

$$-8.526 \le \Delta b_1 < \infty$$
$$0 \le \Delta b_2 \le 21$$
$$-\infty < \Delta b_3 \le 0$$
$$-\infty < \Delta b_4 \le 32.4$$

(b) Replace b_2 by 2 in both problems, and solve each resulting problem. Note that this change is within the sensitivity range for a solution to Problem A, but not for Problem B.

(c) Explain why the sensitivity ranges given in part (a) are both correct.

7.3   CHANGES IN THE COST COEFFICIENTS

In this and subsequent sections we use the matrix notation and matrix relationships between simplex tableaus developed in Sections 4.1–4.3.

Suppose the LP problem

$$\text{Maximize:} \quad z = \mathbf{cx}$$
$$\text{Subject to:} \quad A\mathbf{x} = \mathbf{b} \tag{1}$$
$$\mathbf{x} \geq \mathbf{0}$$

is in canonical form, where A is the $m \times n$ constraint matrix, $\mathbf{b}$ the $m \times 1$ resource column, $\mathbf{c}$ the $1 \times n$ cost matrix, and $\mathbf{x}$ the $n \times 1$ variable matrix. We want to determine to what extent we can vary one entry, c_s, in the cost matrix

$$\mathbf{c} = [\, c_1 \quad c_2 \quad \cdots \quad c_s \quad \cdots \quad c_n \,]$$

without affecting the optimal solution obtained in solving (1). Our principal result is the following.

Theorem 1.   Suppose in (1) we replace

$$\mathbf{c} = [\, c_1 \quad c_2 \quad \cdots \quad c_s \quad \cdots \quad c_n \,]$$

with

$$\hat{\mathbf{c}} = [\, c_1 \quad c_2 \quad \cdots \quad c_s + \Delta c_s \quad \cdots \quad c_n \,]$$

Let

$$T^* = \left[\begin{array}{c|c} A^* & \mathbf{b}^* \\ \hline -\mathbf{c}^* & z^* \end{array} \right] \tag{2}$$

be the final simplex tableau derived in solving (1), and let $h(1), h(2), \ldots, h(m)$ be the associated injection that defines the final basis. If in (1), $\mathbf{c}$ is replaced by $\hat{\mathbf{c}}$, there is no change in the optimal solution given by T^* provided that

(i) if x_s is a nonbasic variable in T^*, then $\Delta c_s \leq -c_s^*$; or

(ii) if x_s is a basic variable in T^* and $h(k) = s$, then

$$\max_{j} \left\{ -\frac{c_j^*}{a_{kj}^*} \,\middle|\, a_{kj}^* > 0 \right\} \leq \Delta c_s \leq \min_{j} \left\{ -\frac{c_j^*}{a_{kj}^*} \,\middle|\, a_{kj}^* < 0 \right\} \tag{3}$$

In (3), k is the row corresponding to the basic variable x_s $[h(k) = s]$, a_{kj}^* is the k, j entry in this row of T^*, c_j^* is the jth entry in $\mathbf{c}^*$, and j ranges over subscripts of the *nonbasic* variables. If the set on the left side of (3) is empty, then we interpret the lower bound to be $-\infty$; if the set on the right side of (3) is empty, then we interpret the upper bound to be ∞.

Before proving Theorem 1, we give some examples. The first example shows how to apply part (i) of Theorem 1 to analyze the range over which cost coefficients of nonbasic variables can vary.

EXAMPLE 1

The LP problem

$$\text{Maximize:}\quad z = 9x_1 + 6x_2 + x_3 + 9x_4$$

$$\text{Subject to:}\quad 2x_1 + \frac{1}{3}x_2 + \frac{1}{3}x_3 + \ x_4 \le \ 8$$

$$2x_1 + \ x_2 \qquad\qquad + 4x_4 \le 12$$

$$23x_1 + \frac{7}{3}x_2 + \frac{4}{3}x_3 - 2x_4 \le 78$$

$$x_i \ge 0, \qquad 1 \le i \le 4$$

has the initial tableau

	x_1	x_2	x_3	x_4	x_5	x_6	x_7	
x_5	2	$\frac{1}{3}$	$\frac{1}{3}$	1	1	0	0	8
x_6	2	1	0	4	0	1	0	12
x_7	23	$\frac{7}{3}$	$\frac{4}{3}$	-2	0	0	1	78
	-9	-6	-1	-9	0	0	0	0

(4)

and the final tableau

	x_1	x_2	x_3	x_4	x_5	x_6	x_7	
x_3	4	0	1	-1	3	-1	0	12
x_2	2	1	0	4	0	1	0	12
x_7	13	0	0	-10	-4	-1	1	34
	7	0	0	14	3	5	0	84

(5)

Since x_1 is a nonbasic variable in (5), conclusion (i) of Theorem 1 says we can increase c_1 by as much as 7 or decrease it by any amount (i.e., $-\infty < \Delta c_1 \le 7$) without changing the optimal basic solution

$$[x_3 \quad x_2 \quad x_7]^T = [12 \quad 12 \quad 34]^T$$

Since the initial value of c_1 is 9, we could also say that the solution is unchanged so long as

$$-\infty < c_1 \le 16 = 9 + 7$$

Since the value of the nonbasic variable x_1 is 0 in the optimal solution, the objective value $z^* = 84$ is also unchanged as c_1 varies in this range.

Similarly, without changing the solution, we can decrease the cost coefficients of the other nonbasic variables x_4, x_5, and x_6 indefinitely or increase them by as much as $\Delta c_4, \Delta c_5$, and Δc_6, respectively, where $\Delta c_4 \le 14, \Delta c_5 \le 3, \Delta c_6 \le 5$.   ◆

The next example shows how to apply part (ii) of Theorem 1 to analyze the range over which cost coefficients of basic variables can vary.

EXAMPLE 2

Let Δc_3 be the amount that we can vary the cost coefficient c_3 of the LP problem in Example 1 without changing the solution. Since x_3 is the basic variable for the first row, the value of k in (3) is 1 $[h(1) = 3]$. To find the left side of inequality (3), we look in the first row of tableau (5) for positive entries in columns corresponding to *nonbasic* variables. Thus, the positive values to consider from tableau (5) are 4 and 3. To find the right side of inequality (3), we look for negative values in the first row of (5). They are -1 and -1. Thus, inequality (3) is

$$\max\left\{-\frac{7}{4}, -\frac{3}{3}\right\} \le \Delta c_3 \le \min\left\{\frac{14}{1}, \frac{5}{1}\right\}$$

and so

$$-1 \le \Delta c_3 \le 5$$

Alternatively, we can give the range in terms of c_3 by setting $\hat{c}_3 = 1 + \Delta c_3$ (1 is the initial value of c_3); thus,

$$1 - 1 = 0 \le \hat{c}_3 \le 6 = 1 + 5$$

Note that the *value* of the objective function does change when we change the cost coefficient of a basic variable. It is the solution (the values of the basic variables) that does not change.

Let Δc_2 denote the amount that we can vary c_2 without changing the solution. Since x_2 is the basic variable for the second row, the k of (3) is 2. Notice that there are no negative entries in the second row of tableau (5). In this event, the right-hand side of inequality (3) is interpreted as ∞. (If a row had no positive entries in columns corresponding to basic variables, the left-hand side of (3) would be interpreted as $-\infty$.) Thus, inequality (3) gives the range

$$\max\left\{-\frac{7}{2}, -\frac{14}{4}, -\frac{5}{1}\right\} \le \Delta c_2 < \infty$$

or

$$-\frac{7}{2} \le \Delta c_2 < \infty$$

Let Δc_7 be the amount that c_7 can vary. Then

$$\max\left\{-\frac{7}{13}\right\} \le \Delta c_7 \le \min\left\{\frac{14}{10}, \frac{3}{4}, \frac{5}{1}\right\}$$

or

$$-\frac{7}{13} \le \Delta c_7 \le \frac{3}{4}$$

In summary, the ranges over which the cost coefficients can vary individually without changing the solution are

Basic variables	Nonbasic variables
$-\frac{7}{2} \leq \Delta c_2 < \infty$	$-\infty < \Delta c_1 \leq 7$
$-1 \leq \Delta c_3 \leq 5$	$-\infty < \Delta c_4 \leq 14$
$-\frac{7}{13} \leq \Delta c_7 \leq \frac{3}{4}$	$-\infty < \Delta c_5 \leq 3$
	$-\infty < \Delta c_6 \leq 5$

Note that these ranges apply only when one cost coefficient changes. If more than one (basic or nonbasic) cost coefficient changes, Theorem 1 does not apply. (See Exercises 5–8.)

♦

The proof of Theorem 1 is not difficult, but it is lengthy. It can be omitted without loss of continuity.

PROOF OF THEOREM 1. We consider two cases: c_s is either the coefficient of a nonbasic variable or the coefficient of a basic variable.

CASE I.

Suppose x_s is a nonbasic variable in the final simplex tableau (2) and $\hat{c}_s = c_s + \Delta c_s$. We examine the objective row of the perturbed tableau resulting from changing c_s to $\hat{c}_s$ and see to what extent we can change Δc_s without allowing any entries of the objective row to become negative. Since x_s is nonbasic, c_s is not an entry in $\mathbf{c_B}$; hence, changing c_s does not affect $\mathbf{c_B} B^{-1} A_j = \mathbf{c_B} A_j^*$.

Subcase Ia. Suppose $j \neq s$. Then,

$$-\hat{c}_j^* = \mathbf{c_B} A_j^* - \hat{c}_j = -c_j^* \tag{6}$$

Since the optimality of tableau (2) implies that all entries $-c_j^*$ in its objective row are nonnegative, it follows from (6) that $-\hat{c}_j^*$ is nonnegative.

Subcase Ib. Suppose $j = s$. Then,

$$-\hat{c}_j^* = \mathbf{c_B} A_s^* - \hat{c}_s = \mathbf{c_B} A_s^* - c_s - \Delta c_s = -c_s^* - \Delta c_s$$

We must have $-c_s^* - \Delta c_s \geq 0$ for the perturbed final tableau to remain optimal.

From Subcase Ia and Ib, we see that if x_s is a nonbasic variable, then the only restriction on Δc_s is

$$\Delta c_s \leq -c_s^*$$

You should also note that since x_s is nonbasic, it has value 0 in the basic solution; therefore, changes in Δc_s do not change the optimal value of the objective function z.

CASE II.

Suppose x_s is a basic variable in the final tableau (2) and $\hat{c}_s = c_s + \Delta c_s$. Since x_s is a basic variable, $s = h(k)$ for some k, $1 \le k \le m$, and $c_s = c_{h(k)}$. In

$$\mathbf{c_B} = [\, c_{h(1)} \quad \cdots \quad c_{h(k)} \quad \cdots \quad c_{h(m)} \,]$$

replace $c_{h(k)}$ with $c_{h(k)} + \Delta c_s$ to obtain

$$\hat{\mathbf{c}}_\mathbf{B} = [\, c_{h(1)} \quad \cdots \quad c_{h(k)} + \Delta c_s \quad \cdots \quad c_{h(m)} \,] \tag{7}$$

In

$$\mathbf{c} = [\, c_1 \quad \cdots \quad c_s \quad \cdots \quad c_n \,]$$

replace c_s with $c_s + \Delta c_s$ to obtain

$$\hat{\mathbf{c}} = [\, c_1 \quad \cdots \quad c_s + \Delta c_s \quad \cdots \quad c_n \,] \tag{8}$$

We can use the unit row matrices

$$\mathbf{r}_j = [\, 0 \quad 0 \quad \cdots \quad 1 \quad \cdots \quad 0 \,]$$
$$\underset{\text{position } j}{\uparrow}$$

to write (7) and (8) as

$$\hat{\mathbf{c}}_\mathbf{B} = \mathbf{c_B} + \Delta c_s \mathbf{r}_k \tag{9}$$

and

$$\hat{\mathbf{c}} = \mathbf{c} + \Delta c_s \mathbf{r}_s \tag{10}$$

We separate the analysis of cost coefficient c_j (for any j, $1 \le j \le n$) into three subcases: (a) x_j is basic but $j \ne s$, (b) x_j is basic and $j = s$, and (c) x_j is nonbasic. Subcase (c) turns out to be the most interesting because it is the case that produces restrictions on Δc_s.

Subcase IIa. Suppose x_j is basic and $j \ne s$. Thus, $s = h(k)$ and $j = h(i)$ for some $i \ne k$.

The jth column A_j^* of the final tableau (2) is the unit column $\mathbf{e}_i$; in particular, the cost coefficient c_j^* is 0. Thus,

$$0 = -c_j^* = \mathbf{c_B} B^{-1} A_j - c_j = \mathbf{c_B} A_j^* - c_j \tag{11}$$

in the final tableau before we change c_s to $c_s + \Delta c_s$. After we change c_s, the jth entry of the objective row of the perturbation of final tableau (2) becomes

$$
\begin{aligned}
-\hat{c}_j^* &= \hat{\mathbf{c}}_\mathbf{B} A_j^* - c_j \\
&= (\mathbf{c_B} + \Delta c_s \mathbf{r}_k) A_j^* - c_j \qquad \text{[Using (9)]} \\
&= \mathbf{c_B} A_j^* - c_j + \Delta c_s \mathbf{r}_k A_j^* \\
&= 0 + \Delta c_s \mathbf{r}_k A_j^* \qquad\qquad \text{[Using (11)]}
\end{aligned}
\tag{12}
$$

Since $\mathbf{r}_k$ has a 1 in the k position and zeros elsewhere, and A_j^* has a 1 in the i position and zeros elsewhere, and since $k \neq i$, it follows that the product of the $1 \times m$ matrix $\mathbf{r}_k$ and the $m \times 1$ matrix A_j^* is the number 0. Thus, expression (12) is 0; and consequently, the entry in the jth column of the objective row is unchanged.

Subcase IIb. Suppose x_j is basic and $j = s = h(k)$. Then the entry in the jth column of the final tableau (2) is $-c_j^* = 0$. We need to calculate the new value $-\hat{c}_j^*$ of the jth entry after replacing c_j by $\hat{c}_j = c_j + \Delta c_s$. It is

$$-\hat{c}_j^* = \hat{\mathbf{c}}_{\mathbf{B}} A_j^* - \hat{c}_j$$

$$= (\mathbf{c}_{\mathbf{B}} + \Delta c_s \mathbf{r}_k) A_j^* - (c_j + \Delta c_s) \qquad \text{[Using (9)]}$$

$$= (\mathbf{c}_{\mathbf{B}} A_j^* - c_j) + (\Delta c_s \mathbf{r}_k A_j^* - \Delta c_s) \qquad (13)$$

The first parenthesized term in (13) is 0 because it equals $-c_j^*$. The second term is also 0 because $\mathbf{r}_k A_j^* = 1$. This is true because $\mathbf{r}_k$ is a $1 \times m$ unit row with its 1 in the k position, and A_j^* is an $m \times 1$ unit column with its 1 in the $k = h(j)$ position. Therefore, $-\hat{c}_j^* = 0$; and again there is no change in the jth column of the objective row.

Subcase IIc. Suppose x_j is nonbasic. Then,

$$\hat{\mathbf{c}}_{\mathbf{B}} = [\, c_{h(1)} \quad \cdots \quad c_{h(k)} + \Delta c_s \quad \cdots \quad c_{h(m)} \,] = \mathbf{c}_{\mathbf{B}} + \Delta c_s \mathbf{r}_k$$

and $\hat{c}_j = c_j$. Thus, the jth entry of the perturbation of tableau (2) is

$$-\hat{c}_j^* = \hat{\mathbf{c}}_{\mathbf{B}} A_j^* - \hat{c}_j = (\mathbf{c}_{\mathbf{B}} + \Delta c_s \mathbf{r}_k) A_j^* - c_j$$

$$= (\mathbf{c}_{\mathbf{B}} A_j^* - c_j) + \Delta c_s \mathbf{r}_k A_j^*$$

$$= -c_j^* + \Delta c_s \mathbf{r}_k A_j^* \qquad (14)$$

We can write

$$A_j^* = \begin{bmatrix} a_{1j}^* \\ \vdots \\ a_{kj}^* \\ \vdots \\ a_{mj}^* \end{bmatrix}$$

and

$$\mathbf{r}_k A_j^* = a_{kj}^*$$

Thus, we can write the condition that (14) is nonnegative as

$$-\hat{c}_j^* = -c_j^* + \Delta c_s a_{kj}^* \geq 0$$

or as

$$\Delta c_s a_{kj}^* \geq c_j^* \qquad (15)$$

where $h(k) = s$.

To solve (15) for Δc_s, we must consider three cases:

If $a_{kj}^* > 0$, then $\Delta c_s \geq c_j^*/a_{kj}^*$.

If $a_{kj}^* < 0$, then $\Delta c_s \leq c_j^*/a_{kj}^*$.

If $a_{kj}^* = 0$, then (15) places no restriction on Δc_s.

Since these inequalities must hold for all j, $1 \leq j \leq n$, Δc_s must satisfy

$$\max_j \left\{ -\frac{c_j^*}{a_{kj}^*} \,\middle|\, a_{kj}^* > 0 \right\} \leq \Delta c_s \leq \min_j \left\{ -\frac{c_j^*}{a_{kj}^*} \,\middle|\, a_{kj}^* < 0 \right\} \qquad (16)$$

where j runs over all subscripts of *nonbasic* variables x_j. If the set of j such that $a_{kj}^* > 0$ is empty, then the maximum is taken to be $-\infty$ (that is to say, there is no lower bound on Δc_s). Similarly, if the set of j such that $a_{kj}^* < 0$ is empty, then the minimum is taken to be ∞ (there is no upper bound on Δc_s).

This concludes the proof of Theorem 1. ∎

You should note that Theorem 1 applies only when just one cost coefficient is changed. The next theorem gives a test that you can perform on some simultaneous changes of cost coefficients to ensure that the solution does not change. Unfortunately, the test is only sufficient; it is possible that simultaneous changes that would not change the solution will fail this test (see Exercise 15). In the next section, we consider a test that is more generally applicable though it is more difficult to apply.

Theorem 2.   Suppose that $\mathbf{x_B} = \mathbf{b}^*$ is an optimal solution to LP problem (1) and that $\mathbf{q}_1, \mathbf{q}_2, \ldots, \mathbf{q}_k$ are $m \times 1$ column matrices such that for each i, $1 \leq i \leq k$, the optimal solution is not changed by replacing $\mathbf{c}$ with $\mathbf{c} + \mathbf{q}_i$. Let

$$\lambda_1 + \lambda_2 + \cdots + \lambda_k \leq 1; \qquad \lambda_i \geq 0, \qquad 1 \leq i \leq k$$

Then the optimal solution is unchanged by replacing $\mathbf{c}$ with $\mathbf{c} + \Sigma_{i=1}^k \lambda_i \mathbf{q}_i$.

PROOF.   The proof is outlined in Exercise 16. ∎

Exercises

Exercises 1–4 give initial and final tableaus. Find the ranges over which you can vary each cost coefficient individually without changing the solution.

1. Initial Tableau

	x_1	x_2	x_3	x_4	x_5	x_6	x_7	x_8	
x_6	$-\frac{3}{10}$	0	$-\frac{1}{2}$	$\frac{3}{10}$	0	1	$\frac{1}{10}$	$-\frac{1}{10}$	$\frac{161}{10}$
x_5	$\frac{7}{5}$	0	1	$-\frac{2}{5}$	1	0	$\frac{1}{5}$	$\frac{9}{5}$	$\frac{86}{5}$
x_2	$\frac{14}{5}$	1	0	$-\frac{9}{5}$	0	0	$\frac{2}{5}$	$\frac{48}{5}$	$\frac{157}{5}$
	0	0	-1	1	0	0	-1	-4	111

Final Tableau

	x_1	x_2	x_3	x_4	x_5	x_6	x_7	x_8	
x_2	-2	1	-4	0	-3	2	0	4	12
x_4	-2	0	-2	1	-1	2	0	-2	15
x_7	3	0	1	0	3	4	1	5	116
	5	0	2	0	4	2	0	3	212

2. Initial Tableau

	x_1	x_2	x_3	x_4	x_5	x_6	x_7	x_8	
x_7	$\frac{2}{3}$	0	$-\frac{1}{3}$	-1	0	$\frac{1}{3}$	1	$-\frac{4}{3}$	9
x_2	0	1	-3	-6	0	-1	0	1	38
x_5	$\frac{17}{3}$	0	$\frac{5}{3}$	5	1	$-\frac{2}{3}$	0	$\frac{35}{3}$	36
	$\frac{1}{3}$	0	$\frac{40}{3}$	9	0	$-\frac{4}{3}$	0	$\frac{28}{3}$	75

Final Tableau

	x_1	x_2	x_3	x_4	x_5	x_6	x_7	x_8	
x_6	2	0	-1	-3	0	1	3	-4	27
x_2	2	1	-4	-9	0	0		-3	65
x_5	7	0	1	3	1	0	2	9	54
	3	0	12	5	0	0	4	4	111

3. Initial Tableau

	x_1	x_2	x_3	x_4	x_5	x_6	x_7	x_8	
x_8	4	0	8	$-\frac{8}{3}$	$\frac{4}{3}$	$-\frac{8}{3}$	0	1	$\frac{20}{3}$
x_2	-6	1	-11	$\frac{10}{3}$	$-\frac{4}{3}$	$\frac{23}{3}$	0	0	$\frac{16}{3}$
x_7	$\frac{7}{3}$	0	$\frac{38}{3}$	$-\frac{41}{9}$	$\frac{16}{9}$	$-\frac{5}{9}$	1	0	$\frac{134}{9}$
	-46	0	-64	41	-16	47	0	0	-5

Final Tableau

	x_1	x_2	x_3	x_4	x_5	x_6	x_7	x_8	
x_5	3	0	6	-2	1	-2	0	$\frac{3}{4}$	5
x_2	-2	1	-3	$\frac{2}{3}$	0	5	0	1	12
x_7	-3	0	2	-1	0	3	1	$-\frac{4}{3}$	6
	2	0	32	9	0	15	0	12	75

4. Initial Tableau

	x_1	x_2	s_1	s_2	s_3	
s_1	2	-3	1	0	0	1
s_2	-3	4	0	1	0	13
s_3	1	1	0	0	1	5
	-3	-2	0	0	0	0

Final Tableau

	x_1	x_2	s_1	s_2	s_3	
x_1	1	0	$\frac{1}{5}$	0	$\frac{3}{5}$	$\frac{16}{5}$
s_2	0	0	$\frac{7}{5}$	1	$\frac{1}{5}$	$\frac{77}{5}$
x_2	0	1	$-\frac{1}{5}$	0	$\frac{2}{5}$	$\frac{9}{5}$
	0	0	$\frac{1}{5}$	0	$\frac{13}{5}$	$\frac{66}{5}$

Exercises 5–8 refer to Examples 1 and 2. In each problem you are given changes to two cost coefficients, both of which are within the allowable ranges for changing one cost coefficient at a time. Solve the perturbed problems to show that the simultaneous changes do change the solution.

5. $\Delta c_2 = -3$ and $\Delta c_3 = 4$ (both correspond to basic variables).

6. $\Delta c_3 = 5$ and $\Delta c_7 = \frac{3}{4}$ (both correspond to basic variables).

7. $\Delta c_4 = 12$ and $\Delta c_7 = \frac{1}{2}$ (one corresponds to a basic and one to a nonbasic variable).

8. $\Delta c_3 = 4$ and $\Delta c_6 = 5$ (one corresponds to a basic and one to a nonbasic variable).

In Exercises 9–13 you are given changes to a cost coefficient of a basic variable in the solution to the LP problem given in Examples 1 and 2. Each change is within the ranges given in Example 2 for changes that do not alter the solution. Find the objective function value after each change is made.

9. $\Delta c_3 = 4$

10. $\Delta c_3 = -\frac{1}{2}$

11. $\Delta c_7 = \frac{1}{2}$

12. $\Delta c_2 = -3$

13. Show that the optimal solution to the LP problem in Example 1 is unchanged by the simultaneous changes $\Delta c_1 = -2$, $\Delta c_2 = -4$, and $\Delta c_4 = -2$, even though $\Delta c_2 = -4$ is not within the allowable range for changing c_2.

14. By considering the proof of case (I) of Theorem 1, show that the ranges obtained for changes to individual nonbasic variables still apply when several nonbasic variables (and no basic variables) are changed.

15. This problem shows that you cannot depend on Theorem 2 to find all possible simultaneous changes of several cost coefficients by combining individual changes that lie within the sensitivity ranges given in Theorem 1. Let

$$\mathbf{q} = [-2 \quad -4 \quad 0 \quad -2 \quad 0 \quad 0 \quad 0]$$

be a perturbation of the cost coefficient matrix of the LP problem in Example 1.

(a) Use Exercise 13 to show that changing $\mathbf{c}$ to $\mathbf{c} + \mathbf{q}$ does not change the optimal solution of Example 1.

(b) Show that $\mathbf{q}$ cannot be written as

$$\mathbf{q} = \sum_{i=1}^{s} \lambda_i \mathbf{q}_i$$

where $\sum_{i=1}^{s} \lambda_i \leq 1$; $\lambda_i \geq 0$ for $1 \leq i \leq s$; and for $1 \leq i \leq s$, $\mathbf{q}_i = [0 \quad 0 \quad \cdots \quad \Delta c_i \quad 0 \quad \cdots \quad 0]$, where Δc_i lies within the sensitivity ranges given in Examples 1 and 2.

16. Prove Theorem 2 by answering the following sequence of questions.

(a) Let $\lambda_0 = 1 - \sum_{k=1}^{s} \lambda_k$. Explain why $\lambda_0 \geq 0$ and $\sum_{k=0}^{s} \lambda_k = 1$.

(b) Explain why $c_j = \sum_{k=0}^{s} \lambda_k c_j$ and why

$$c_j + \sum_{k=0}^{s} \lambda_i q_{jk} = \sum_{k=0}^{s} \lambda_k (c_j + q_{jk})$$

where q_{kj} is the jth entry in the $1 \times n$ row matrix $\mathbf{q}_k$.

(c) Let $h(1), h(2), \ldots, h(m)$ be the injection corresponding to the optimal tableau. Explain why

$$[c_{h(1)} + q_{k,h(1)} \quad c_{h(2)} + q_{k,h(2)} \quad \cdots \quad c_{h(m)} + q_{k,h(m)}] A_j^* - (c_j + q_{kj}) \geq 0$$
$$\text{for} \quad 1 \leq j \leq n$$

implies that replacing $\mathbf{c}$ by $\mathbf{c} + \mathbf{q}_k$ does not change the optimal basis.

(d) Show that the inequality in part (c) can be written as

$$\sum_{i=1}^{m} (c_{h(i)} + q_{k,h(i)}) a_{ij}^* - (c_j + q_{kj}) \geq 0$$

(e) Show that Theorem 1 has been proved if we show that for each j, $1 \leq j \leq n$,

$$\sum_{i=1}^{m} \left(c_{h(i)} + \sum_{k=1}^{s} \lambda_k q_{k,h(i)} \right) a_{ij}^* - \left(c_j + \sum_{k=1}^{s} \lambda_k q_{kj} \right) \geq 0$$

(f) Show that the inequality in part (e) can be written as

$$\sum_{i=1}^{m} \left[\sum_{k=0}^{s} \lambda_k (c_{h(i)} + q_{k,h(i)}) \right] a_{ij}^* - \sum_{k=0}^{s} \lambda_k (c_j + q_{kj}) \geq 0$$

where $q_{0j} = 0$, $1 \leq j \leq n$.

(g) Show that the inequality in part (f) can be written as

$$\sum_{k=0}^{s} \lambda_k \left[\sum_{i=1}^{m} (c_{h(i)} + q_{k,h(i)}) a_{ij}^* - (c_j + q_{kj}) \right] \geq 0$$

(h) Use part (c) to show that the expression in part (g) is greater than or equal to 0 for each j, $1 \leq j \leq n$ and conclude that we have proved the theorem.

†**17.** The Moscow Industrial Ghetto (MIG) manufactures large and small iron dragons. The large dragons are used for garden decorations and the small dragons for automobile ornaments. Each week up to 1000 pounds of iron can be purchased at $0.25 per pound. MIG employs five workers, each of whom works 30 hours per week. Their salary is a fixed cost that does not depend on productivity. The machinery used is available for a maximum of 300 hours per week. Table 1 gives data about dragons.

(a) Assume that the measure of profit is the selling price minus the cost of iron. Solve the problem of maximizing profit with CALIPSO. Do not require an integer solution. Produce a sensitivity analysis.

Table 1

	Small dragon	Large dragon
Iron/dragon (lb)	3	125
Labor/dragon (hr)	2	12
Machine time/dragon (hr)	1.5	9
Selling price ($)	35	300

(b) How much would the selling price of small dragons need to be increased for it to become profitable to make more small dragons and fewer large dragons? (Observe that the cost coefficient of the variable giving the number of small dragons produced is the difference between the selling price and the unit cost of materials.)

(c) How much would the selling price of large dragons need to be decreased before it become profitable to make fewer of them?

†**18.** Use SOLVER to obtain a sensitivity analysis of Example 1. Compare the sensitivity ranges given by the program to those obtained in the text.

†**19.** The LP problems A and B are equivalent:

Problem A	*Problem B*
Maximize: $z = x_1 + 2x_2 + 5x_3$	Maximize: $z = x_1 + 2x_2 + 5x_3$

Subject to:
$$x_1 + x_2 + x_3 \le 12$$
$$2x_1 - 3x_2 + 2x_3 = 3$$

$$-x_1 + 2x_2 + 5x_3 \ge 15$$
$$x_i \ge 0, \quad 1 \le i \le 3$$

Subject to:
$$x_1 + x_2 + x_3 \le 12$$
$$2x_1 - 3x_2 + 2x_3 \le 3$$
$$2x_1 - 3x_2 + 2x_3 \ge 3$$
$$-x_1 + 2x_2 + 5x_3 \ge 15$$
$$x_i \ge 0, \quad 1 \le i \le 3$$

(a) Show that the sensitivity ranges for the cost coefficients are

Problem A	Problem B
$-\infty < \Delta c_1 \le 4$	$-\infty < \Delta c_1 \le 4$
$-9.5 \le \Delta c_2 < \infty$	$-9.5 \le \Delta c_2 \le 3$
$-4 \le \Delta c_3 < \infty$	$-3 \le \Delta c_3 < \infty$

(b) Replace c_2 by 6 in both problems, and solve each resulting problem. Note that this change is within the sensitivity range for Problem A but not for Problem B.

(c) Explain how the sensitivity ranges given in part (a) can both be correct.

7.4   *USING DUALITY IN SENSITIVITY ANALYSIS*

In Sections 6.3 and 6.4, we considered the solution of the LP problem

$$\text{Maximize:} \quad z = \mathbf{cx}$$

$$\text{Subject to:} \quad A_1\mathbf{x} \le \mathbf{b}_1$$
$$A_2\mathbf{x} = \mathbf{b}_2 \tag{1}$$
$$A_3\mathbf{x} \ge \mathbf{b}_3$$

$$\mathbf{x} \ge \mathbf{0}; \quad \mathbf{b}_i \ge \mathbf{0}, \quad 1 \le i \le 3$$

and its dual

$$\text{Minimize:} \quad y = \mathbf{b}_1^T\mathbf{u} + \mathbf{b}_2^T\mathbf{v} + \mathbf{b}_3^T\mathbf{w}$$

$$\text{Subject to:} \quad A_1^T\mathbf{u} + A_2^T\mathbf{v} + A_3^T\mathbf{w} \geq \mathbf{c}^T \tag{2}$$

$$\mathbf{u} \geq \mathbf{0}; \qquad \mathbf{v} \quad \text{unrestricted}; \qquad \mathbf{w} \leq \mathbf{0}$$

The sizes of the matrices of (1) and (2) are the same as given in Sections 6.3 and 6.4.

We solved the primal problem (1) by the two-phase method. We added slack variables to (1) to obtain the equivalent problem

$$\text{Maximize:} \quad z = \mathbf{cx} + \mathbf{0r} + \mathbf{0s}$$

$$\text{Subject to:} \quad \begin{aligned} A_1\mathbf{x} \quad\;\; + I_m\mathbf{s} &= \mathbf{b}_1 \\ A_2\mathbf{x} \quad\quad\quad\;\; &= \mathbf{b}_2 \\ A_3\mathbf{x} - I_j\mathbf{r} \quad\;\; &= \mathbf{b}_3 \end{aligned} \tag{3}$$

$$\mathbf{x} \geq \mathbf{0}; \qquad \mathbf{r} \geq \mathbf{0}; \qquad \mathbf{s} \geq \mathbf{0}$$

We then added artificial variables to obtain the initial tableau (Tableau 1):

Tableau 1

$$
T = \left[
\begin{array}{c|c|c|c|c|c}
 & & \text{(Slack)} & \multicolumn{2}{c}{\text{(Artificial)}} & \\
\hline
A_1 & 0 & I_m & 0 & 0 & \mathbf{b}_1 \\
\hline
A_2 & 0 & 0 & I_k & 0 & \mathbf{b}_2 \\
\hline
A_3 & -I_j & 0 & 0 & I_j & \mathbf{b}_3 \\
\hline
-\mathbf{c} & 0 & 0 & 0 & 0 & 0
\end{array}
\right]
$$

(The row corresponding to the artificial objective function has been omitted from Tableau 1 because it is irrelevant to this discussion.)

Suppose that Tableau 2,

Tableau 2

$$
T^* = \left[
\begin{array}{c|ccc|cc|c}
 & & \text{(Slack)} & & \text{(Artificial)} & & \\
\hline
A_1^* & \cdots & & & & & \mathbf{b}_1^* \\
A_2^* & \cdots & & \multicolumn{3}{c}{\longleftarrow \; B^{-1} \longrightarrow} & \mathbf{b}_2^* \\
A_3^* & \cdots & & & & & \mathbf{b}_3^* \\
\hline
-\mathbf{c}^* & \mathbf{t}^* & \mathbf{u}^* & \mathbf{v}^* & \mathbf{w}^* & & z^*
\end{array}
\right]
$$

results from performing any number of pivots of the two-phase method on Tableau 1. Theorem 1 of Section 6.4 shows that if the solution

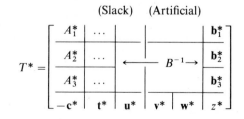

corresponding to Tableau 2, is an optimal solution to the primal problem (1), then

$$\begin{bmatrix} \mathbf{u} \\ \mathbf{v} \\ \mathbf{w} \end{bmatrix} = \begin{bmatrix} \mathbf{u}^* \\ \mathbf{v}^* \\ \mathbf{w}^* \end{bmatrix}$$

is a feasible solution to the dual problem (2).

The next example illustrates several ways of performing sensitivity analyses on an LP problem by studying its dual.

EXAMPLE 1

Consider the LP problem

$$\text{Maximize:} \quad z = 2x_1 - 3x_2 + x_3$$

$$\begin{aligned}
\text{Subject to:} \quad & x_1 + x_2 - x_3 \le 8 \\
& 2x_1 + x_2 - x_3 \le 6 \\
& x_1 \quad\quad + x_3 \le 5 \\
& 2x_1 - x_2 + x_3 \ge 5 \\
& -3x_1 - x_2 + 2x_3 \ge 3
\end{aligned} \tag{4}$$

$$x_i \ge 0, \quad 1 \le i \le 3$$

The dual of (4) is

$$\text{Minimize:} \quad y = 8u_1 + 6u_2 + 5u_3 + 5w_1 + 3w_2$$

$$\begin{aligned}
\text{Subject to:} \quad & u_1 + 2u_2 + u_3 + 2w_1 - 3w_2 \ge 2 \\
& u_1 + u_2 \quad\quad - w_1 - w_2 \ge -3 \\
& -u_1 - u_2 + u_3 + w_1 + 2w_2 \ge 1
\end{aligned} \tag{5}$$

$$u_1, u_2, u_3 \ge 0; \quad w_1, w_2 \le 0$$

Introducing slack and artificial variables to (4) we have

$$\text{Maximize:} \quad z = 2x_1 - 3x_2 + x_3$$

$$\begin{aligned}
\text{Subject to:} \quad & x_1 + x_2 - x_3 + s_1 \quad\quad\quad\quad\quad\quad\quad\quad = 8 \\
& 2x_1 + x_2 - x_3 \quad + s_2 \quad\quad\quad\quad\quad\quad = 6 \\
& x_1 \quad\quad + x_3 \quad\quad\quad + s_3 \quad\quad\quad\quad = 5 \\
& 2x_1 - x_2 + x_3 \quad\quad\quad\quad\quad - r_1 + y_1 \quad\quad = 5 \\
& -3x_1 - x_2 + 2x_3 \quad\quad\quad\quad\quad\quad\quad\quad - r_2 + y_2 = 3
\end{aligned} \tag{6}$$

$$x_i, s_i \ge 0, \quad 1 \le i \le 3; \quad r_i, y_i \ge 0, \quad 1 \le i \le 2$$

The initial tableau for (6) is

Initial Tableau

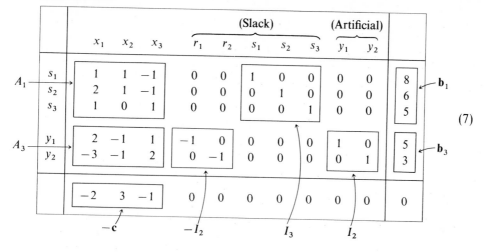

$$(7)$$

You may verify that the final simplex tableau is

Final Tableau

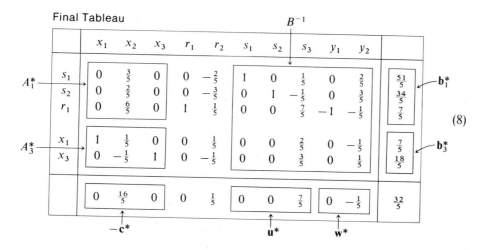

$$(8)$$

By Theorem 2 of Section 6.4, the dual prices are

$$[u_1 \quad u_2 \quad u_3 \quad w_1 \quad w_2] = [0 \quad 0 \quad \tfrac{7}{5} \quad 0 \quad -\tfrac{1}{5}] \qquad (9)$$

If we write the objective function corresponding to LP problem (4) as

$$z = c_1 x_1 + c_2 x_2 + c_3 x_3 + c_4 s_1 + c_5 s_2 + c_6 s_3 + c_7 r_1 + c_8 r_2$$

we can use the formulas given in Section 7.3 (or the CALIPSO program) to obtain the following ranges over which a single cost coefficient of a decision variable can vary (with the others held constant) without changing the solution.

Ranges for Δc_i	Ranges for c_i	
$-1 \leq \Delta c_1 < \infty$	$1 \leq c_1 < \infty$	
$-\infty < \Delta c_2 \leq \frac{16}{5}$	$-\infty < c_2 \leq \frac{1}{5}$	(10)
$-\frac{7}{3} \leq \Delta c_3 \leq 1$	$-\frac{4}{3} \leq c_3 \leq 2$	◆

The next example shows how to use the solution to the dual problem to find the range over which the cost coefficient of a nonbasic variable can vary. While illuminating, this technique is not especially useful because the same result can be obtained at a glance from the final tableau for the primal problem.

EXAMPLE 2

Since in (8) x_2 is a nonbasic variable in the optimal solution to the primal problem, changing c_2 does not change the values of any entries in the objective row of the final tableau (8) except the value $\frac{16}{5}$ in the x_2 column (see Case I of the proof of Theorem 1 in Section 7.3). Therefore, only the second constraint of the dual problem (5) is changed. If the change in c_2 is small enough so that the solution to the primal problem does not change, then the dual prices given in (9) must be a feasible solution to the dual problem (by Theorem 2 of Section 6.4). For the second constraint, this means that the dual prices satisfy

$$u_1 + u_2 - w_1 - w_2 \geq \hat{c}_2$$

Substituting these prices, we obtain

$$0 + 0 - 0 - \left(-\frac{1}{5}\right) = \frac{1}{5} \geq \hat{c}_2$$

This is the same range for c_2 obtained from tableau (8) by observing that $\Delta c_2 \leq \frac{16}{5}$, since $\hat{c}_2 = -3 + \Delta c_2$.   ◆

We cannot easily perform an analysis similar to the one in Example 2 for the cost coefficient of a basic variable because such a change alters the values of all cost coefficients in the optimal tableau (see Case II of the proof of Theorem 1 in Section 7.3). In fact, the analysis is more difficult than that given in Section 7.3 and provides no additional information; therefore, we will not present it. We can, however, use duality to gather information about simultaneous changes to several coefficients of a nonbasic variable. Example 3 illustrates the idea.

EXAMPLE 3

We continue to consider the nonbasic variable x_2 of the LP problem of Example 1. The coefficients of x_2 from the initial tableau (7) are $a_{12} = 1$, $a_{22} = 1$, $a_{32} = 0$, $a_{42} = -1$, and $a_{52} = -1$ in the constraint equations and $c_2 = -3$ in the objective function. We want to consider simultaneous changes to some or all of these coefficients and to find a condition that guarantees that such changes do not alter the optimal solution.

Suppose the new values of the coefficients of x_2 are $\hat{a}_{12}, \hat{a}_{22}, \hat{a}_{32}, \hat{a}_{42}, \hat{a}_{52}$, and $\hat{c}_2$. These changes in the initial problem result in changes only in the second column of the optimal tableau (8). We already discussed the fact that changing c_2 changes only the second entry in the objective row. The new constraint coefficients in the jth column of the optimal tableau are

$$B^{-1}A_j^* \tag{11}$$

Since x_2 is nonbasic, B^{-1} is not changed by perturbing the coefficients of x_2, and so (11) shows that only the second column ($j = 2$) is changed. Thus, if c_2 is not changed by so much that the entry in the final tableau becomes negative, tableau (8) remains optimal for the perturbed problem. Therefore, the dual prices given in (9) are still a feasible solution to the dual problem. The second constraint of the perturbed dual problem is

$$\hat{a}_{12}u_1 + \hat{a}_{22}u_2 + \hat{a}_{32}v + \hat{a}_{42}w_1 + \hat{a}_{52}w_2 \geq \hat{c}_2 \tag{12}$$

Substituting the dual prices (9) into (12) gives

$$\frac{7}{5}\hat{a}_{32} - \frac{1}{5}\hat{a}_{52} \geq \hat{c}_2 \tag{13}$$

as the condition that must be satisfied by any set of changes to the coefficients of x_2 to leave the optimal solution unchanged.   ◆

We now consider the problem of finding a condition that guarantees that the optimal solution is unchanged under a set of simultaneous perturbations to the cost coefficients of a primal problem. We will use a technique that involves the inverse basis matrix for the dual problem. Unfortunately, we cannot obtain this matrix from the optimal tableau for the primal problem. However, the expense of solving the dual problem to obtain the inverse basis matrix may be justified in an application that involves frequently fluctuating cost coefficients. In the next example we find B^{-1} corresponding to the optimal tableau for the dual problem (5) and then use this matrix to perform a sensitivity analysis.

EXAMPLE 4

We begin by forming an equivalent problem to (5) that is in the form to which the simplex method can be applied. First we multiply the second constraint by -1 to obtain a constraint with a nonnegative resource value, and then we let

$$w'_1 = -w_1$$
$$w'_2 = -w_2 \tag{14}$$

After making these substitutions in the dual problem (5), we obtain the equivalent problem

$$\text{Maximize:} \quad z' = -8u_1 - 6u_2 - 5u_3 + 5w'_1 + 3w'_2$$

$$\text{Subject to:} \quad u_1 + 2u_2 + u_3 - 2w'_1 + 3w'_2 \geq 2$$
$$-u_1 - u_2 \qquad\quad - w'_1 - w'_2 \leq 3$$
$$-u_1 - u_2 + u_3 - w'_1 - 2w'_2 \geq 1$$
$$u_1, u_2, u_3, w'_1, w'_2 \geq 0$$

We introduce slack variables s_1, s_2, s_3 and artificial variables y_1, y_3 to obtain the initial tableau:

Initial Tableau for the Dual

	u_1	u_2	u_3	w'_1	w'_2	s_1	s_2	s_3	y_1	y_3	
y_1	1	2	1	-2	3	-1	0	0	1	0	2
s_2	-1	-1	0	-1	-1	0	1	0	0	0	3
y_3	-1	-1	1	-1	-2	0	0	-1	0	1	1
(z')	8	6	5	-5	-3	0	0	0	0	0	0
(w)	0	-1	-2	3	-1	1	0	1	0	0	-3

You can verify that the final tableau, with the artificial objective row (w) suppressed, that results from applying the two-phase method is

Final Tableau for the Dual

	u_1	u_2	u_3	w'_1	w'_2	s_1	s_2	s_3	y_1	y_3	
w'_2	$\frac{2}{5}$	$\frac{3}{5}$	0	$-\frac{1}{5}$	1	$-\frac{1}{5}$	0	$\frac{1}{5}$	$\frac{1}{5}$	$-\frac{1}{5}$	$\frac{1}{5}$
s_2	$-\frac{3}{5}$	$-\frac{2}{5}$	0	$-\frac{6}{5}$	0	$-\frac{1}{5}$	1	$\frac{1}{5}$	$\frac{1}{5}$	$-\frac{1}{5}$	$\frac{16}{5}$
u_3	$-\frac{1}{5}$	$\frac{1}{5}$	1	$-\frac{7}{5}$	0	$-\frac{2}{5}$	0	$-\frac{3}{5}$	$\frac{2}{5}$	$\frac{3}{5}$	$\frac{7}{5}$
(z')	$\frac{51}{5}$	$\frac{34}{5}$	0	$\frac{7}{5}$	0	$\frac{7}{5}$	0	$\frac{18}{5}$	$-\frac{7}{5}$	$-\frac{18}{5}$	$-\frac{32}{5}$

The B^{-1} matrix for the dual is

$$B^{-1} = \begin{bmatrix} \frac{1}{5} & 0 & -\frac{1}{5} \\ \frac{1}{5} & 1 & -\frac{1}{5} \\ \frac{2}{5} & 0 & \frac{3}{5} \end{bmatrix} \tag{15}$$

Suppose we perturb the cost coefficients of the primal problem (4) by letting $\hat{c}_1 = 2 + \Delta c_1$, $\hat{c}_2 = -3 + \Delta c_2$, and $\hat{c}_3 = 1 + \Delta c_3$. Then the perturbed resource values of the dual problem are $b_1 = \hat{c}_1$, $b_2 = -\hat{c}_2$, and $b_3 = \hat{c}_3$ ($b_2 = -\hat{c}_2$ because we multiplied the second constraint by -1). We can write the condition that the basis of the dual problem is unchanged as

$$b_1^* + b_{11}\,\Delta c_1 + b_{12}(-\Delta c_2) + b_{13}\,\Delta c_3 \geq 0$$
$$b_2^* + b_{21}\,\Delta c_1 + b_{22}(-\Delta c_2) + b_{23}\,\Delta c_3 \geq 0 \tag{16}$$
$$b_3^* + b_{31}\,\Delta c_1 + b_{32}(-\Delta c_2) + b_{33}\,\Delta c_3 \geq 0$$

where b_1^*, b_2^*, b_3^* are the values of the variables in the final tableau, and the b_{ij} are the entries in the B^{-1} matrix corresponding to this final tableau. Substituting these values in (16) we obtain

$$\frac{1}{5} + \left(\frac{1}{5}\right)\Delta c_1 - (0)\,\Delta c_2 + \left(-\frac{1}{5}\right)\Delta c_3 \geq 0$$

$$\frac{16}{5} + \left(\frac{1}{5}\right)\Delta c_1 - (1)\,\Delta c_2 + \left(-\frac{1}{5}\right)\Delta c_3 \geq 0 \tag{17}$$

$$\frac{7}{5} + \left(\frac{2}{5}\right)\Delta c_1 - (0)\,\Delta c_2 + \left(\frac{3}{5}\right)\Delta c_3 \geq 0$$

Simplifying (17) gives

$$-\Delta c_1 \qquad\quad + \ \ \Delta c_3 \leq \ \ 1$$
$$-\Delta c_1 + 5\,\Delta c_2 + \ \ \Delta c_3 \leq 16 \tag{18}$$
$$-2\,\Delta c_1 \qquad\quad - 3\,\Delta c_3 \leq \ \ 7$$

You can verify by setting pairs of Δc_i equal to 0 that (18) gives the same ranges for changes to one cost coefficient at a time as we found in (10). Note also that you cannot obtain the sensitivity ranges for slack variables (which are sometimes of practical interest) by this method.

You should avoid using this method of analyzing cost coefficients for primal problems involving equality constraints because it is difficult to keep track of the signs in determining the correct B^{-1} matrix. Instead you should replace $=$ constraints with an equivalent pair of $\leq$ and $\geq$ constraints. (See Exercise 3.)

As a final application of duality, we give a simple test for determining if the basis is changed when a new variable is added to a solved problem.

EXAMPLE 5

Consider the following LP problem, which we obtained from primal problem (4) by adding a new variable x_4 with the indicated cost and constraint coefficients.

$$\text{Maximize:}\quad z = 2x_1 - 3x_2 + x_3 + 2x_4$$

$$\text{Subject to:}\quad \begin{aligned} x_1 + x_2 - x_3 + 3x_4 &\leq 8 \\ 2x_1 + x_2 - x_3 - 2x_4 &\leq 6 \\ x_1 \quad\quad + x_3 + 2x_4 &\leq 5 \\ 2x_1 - x_2 + x_3 - 3x_4 &\geq 5 \\ -3x_1 - x_2 + 2x_3 + x_4 &\geq 3 \end{aligned}$$

$$x_i \geq 0, \quad 1 \leq i \leq 4$$

The addition of x_4 adds the constraint

$$3u_1 - 2u_2 + 2u_3 - 3w_1 + w_2 \geq 2$$

to the dual problem (5). If this constraint is satisfied by the dual prices (9), then the basis obtained in tableau (8) remains optimal. Substituting the dual prices (9) gives

$$3 \cdot 0 - 2 \cdot 0 + 2 \cdot \frac{7}{5} - 3 \cdot 0 - \frac{1}{5} = \frac{13}{5} \geq 2$$

Therefore, adding this variable does not change the solution.   ◆

Exercises

1. The dual of the American Coffee, Inc., problem of Section 6.1 [see (1) of that section] is

$$\text{Minimize:}\quad y = 8000u_1 + 6400u_2 + 6000u_3$$

$$\text{Subject to:}\quad \begin{aligned} 2u_1 + 4u_2 + 4u_3 &\geq 8 \\ 4u_1 + 5u_2 + u_3 &\geq 6 \\ 3u_1 + 3u_2 + 4u_3 &\geq 4 \\ 7u_1 + 2u_2 + u_3 &\geq 5 \end{aligned}$$

$$u_i \geq 0, \quad 1 \leq i \leq 3$$

The initial tableau for the dual problem is

	u_1	u_2	u_3	s_1	s_2	s_3	s_4	y_1	y_2	y_3	y_4	
y_1	2	4	4	-1	0	0	0	1	0	0	0	8
y_2	4	5	1	0	-1	0	0	0	1	0	0	6
y_3	3	3	4	0	0	-1	0	0	0	1	0	4
y_4	7	2	1	0	0	0	-1	0	0	0	1	5
(z)	8000	6400	6000	0	0	0	0	0	0	0	0	0
(w)	-16	-14	-10	1	1	1	1	0	0	0	0	-23

and the optimal tableau resulting from the two-phase method is

	u_1	u_2	u_3	s_1	s_2	s_3	s_4	y_1	y_2	y_3	y_4	
s_3	0	0	$-\frac{5}{4}$	$-\frac{5}{8}$	0	1	$-\frac{1}{4}$	$\frac{5}{8}$	0	-1	$\frac{1}{4}$	$\frac{9}{4}$
u_2	0	1	$\frac{13}{12}$	$-\frac{7}{24}$	0	0	$\frac{1}{12}$	$\frac{7}{24}$	0	0	$-\frac{1}{12}$	$\frac{23}{12}$
s_2	0	0	$\frac{15}{4}$	$-\frac{9}{8}$	1	0	$-\frac{1}{4}$	$\frac{9}{8}$	-1	0	$\frac{1}{4}$	$\frac{17}{4}$
u_1	1	0	$-\frac{1}{6}$	$\frac{1}{12}$	0	0	$-\frac{1}{6}$	$-\frac{1}{12}$	0	0	$\frac{1}{6}$	$\frac{1}{6}$
(z)	0	0	400	1200	0	0	800	-1200	0	0	-800	$-13{,}600$

(a) Find the ranges for individual changes to the cost coefficients of the decision variables of the primal problem from tableau (2) of Section 6.1. Express these ranges in terms of c_i rather than Δc_i.

(b) Find the ranges for individual changes to the cost coefficients of the nonbasic *decision* variables of the primal problem using the technique given in Example 2. Compare your answers to the ones you obtained in part (a).

(c) Show that the optimal basis does not change when the coefficients of x_2 are changed to $\hat{a}_{12} = 3$, $\hat{a}_{22} = 6$, $\hat{a}_{32} = 4$, and $\hat{c}_2 = 8$.

(d) Show that the optimal basis does change when the coefficients of x_3 are changed to $\hat{a}_{13} = 2$, $\hat{a}_{23} = 2$, $\hat{a}_{33} = 4$, and $\hat{c}_3 = 5$.

(e) Using the method of Example 4, find a system of inequalities that must be satisfied by a set of perturbations of the cost coefficients of the primal problem so that the optimal solution is unchanged.

(f) Use the result of part (e) to calculate the ranges over which each cost coefficient can vary when the other cost coefficients are fixed. Compare these ranges to those you obtained in part (a).

(g) Suppose a new variable x_5 with cost coefficient $c_5 = 3$ and constraint coefficients $a_{15} = 2$, $a_{25} = -2$, and $a_{35} = 4$ is added to the primal problem. Does the optimal solution change?

2. Solving the LP problem

$$\text{Maximize:} \quad z = 2x_1 - 3x_2 + x_3$$
$$\text{Subject to:} \quad x_1 + 2x_2 + x_3 \le 15$$
$$x_1 + 8x_2 + x_3 \ge 20 \tag{19}$$
$$2x_1 + 4x_2 + 5x_3 \ge 5$$
$$x_i \ge 0, \quad 1 \le i \le 3$$

by the two-phase method (after adding slack variables s_1, s_2, s_3 and artificial variables y_1, y_2), we obtain the optimal tableau

	x_1	x_2	x_3	s_1	s_2	s_3	y_1	y_2	
x_1	1	0	1	$\frac{4}{3}$	$\frac{1}{3}$	0	$-\frac{1}{3}$	0	$\frac{40}{3}$
s_3	0	0	-3	2	0	1	0	-1	25
x_2	0	1	0	$-\frac{1}{6}$	$-\frac{1}{6}$	0	$\frac{1}{6}$	0	$\frac{5}{6}$
	0	0	1	$\frac{19}{6}$	$\frac{7}{6}$	0	$-\frac{7}{6}$	0	$\frac{145}{6}$

(a) Use the technique of Example 3 to find the ranges over which the coefficients of the nonbasic variable x_3 can vary without changing the solution.

(b) Show that we can write the dual of (19) as

Minimize: $\quad y = 15u_1 - 20u_2 - 5u_3$

Subject to: $\quad u_1 - u_2 - 2u_3 \geq 2$

$$-2u_1 + 8u_2 + 4u_3 \leq 3 \qquad (20)$$

$$u_1 - u_2 - 5u_3 \geq 1$$

$\quad u_i \geq 0, \qquad 1 \leq i \leq 3$

After adding slack variables t_1, t_2, t_3 and artificial variables r_1, r_2, we solve (20) by the two-phase method to obtain the optimal tableau

	u_1	u_2	u_3	t_1	t_2	t_3	r_1	r_2	
t_3	0	0	3	-1	0	1	1	-1	1
u_2	0	1	0	$-\frac{1}{3}$	$\frac{1}{6}$	0	$\frac{1}{3}$	0	$\frac{7}{6}$
u_1	1	0	-2	$-\frac{4}{3}$	$\frac{1}{6}$	0	$\frac{4}{3}$	0	$\frac{19}{6}$
	0	0	25	$\frac{40}{3}$	$\frac{5}{6}$	0	$-\frac{40}{3}$	0	$-\frac{145}{6}$

(c) Using the method of Example 4, find a system of inequalities that must be satisfied by a set of perturbations of the cost coefficients of the primal problem so that the optimal solution is unchanged.

(d) Use the result of part (c) to calculate the ranges over which each cost coefficient can vary when the other cost coefficients are fixed.

3. Consider the LP problem

Maximize: $\quad z = x_1 + 2x_2 + 5x_3$

Subject to: $\quad x_1 + x_2 + x_3 \leq 12$

$$2x_1 - 3x_2 + 2x_3 = 3 \qquad (21)$$

$$-x_1 + 2x_2 + 5x_3 \geq 15$$

$\quad x_i \geq 0, \qquad 1 \leq i \leq 3$

We replace the equality constraint in (21) by a pair of $\leq$ and $\geq$ constraints to make it easier to find the correct B^{-1} matrix in the analysis that follows. The problem

$$\text{Maximize:} \quad z = x_1 + 2x_2 + 5x_3$$

$$\text{Subject to:} \quad \begin{aligned} x_1 + x_2 + x_3 &\leq 12 \\ 2x_1 - 3x_2 + 2x_3 &\leq 3 \\ 2x_1 - 3x_2 + 2x_3 &\geq 3 \\ -x_1 + 2x_2 + 5x_3 &\geq 15 \end{aligned} \tag{22}$$

$$x_i \geq 0, \quad 1 \leq i \leq 3$$

is equivalent to (21). We solve (22) by the two-phase method (after adding slack variables s_1, s_2, s_3, s_4 and artificial variables y_1, y_2) to obtain the optimal tableau

	x_1	x_2	x_3	s_1	s_2	s_3	s_4	y_1	y_2	
s_4	6	0	0	$\frac{19}{5}$	$\frac{3}{5}$	0	1	0	-1	$\frac{162}{5}$
s_3	0	0	0	0	1	1	0	-1	0	0
x_3	1	0	1	$\frac{3}{5}$	$\frac{1}{5}$	0	0	0	0	$\frac{39}{5}$
x_2	0	1	0	$\frac{2}{5}$	$-\frac{1}{5}$	0	0	0	0	$\frac{21}{5}$
	4	0	0	$\frac{19}{5}$	$\frac{3}{5}$	0	0	0	0	$\frac{237}{5}$

(a) Use the technique of Example 3 to find the ranges over which the coefficients of the nonbasic variable x_1 can vary without changing the solution.

(b) Show that we can write an equivalent LP problem for the dual of (22) as

$$\text{Minimize:} \quad y = 12u_1 + 3u_2 - 3u_3 - 15u_4$$

$$\text{Subject to:} \quad \begin{aligned} u_1 + 2u_2 - 2u_3 + u_4 &\geq 1 \\ u_1 - 3u_2 + 3u_3 - 2u_4 &\geq 2 \\ u_1 + 2u_2 - 2u_3 - 5u_4 &\geq 5 \end{aligned} \tag{23}$$

$$u_i \geq 0, \quad 1 \leq i \leq 4$$

After adding slack variables t_1, t_2, t_3 and artificial variables r_1, r_2, r_3, we solve (23) by the two-phase method to obtain the optimal tableau

	u_1	u_2	u_3	u_4	t_1	t_2	t_3	r_1	r_2	r_3	
u_1	1	0	0	$-\frac{19}{5}$	0	$-\frac{2}{5}$	$-\frac{3}{5}$	0	$\frac{2}{5}$	$\frac{3}{5}$	$\frac{19}{5}$
t_1	0	0	0	-6	1	0	-1	-1	0	1	4
u_2	0	1	-1	$-\frac{3}{5}$	0	$\frac{1}{5}$	$-\frac{1}{5}$	0	$-\frac{1}{5}$	$\frac{1}{5}$	$\frac{3}{5}$
	0	0	0	$\frac{162}{5}$	0	$\frac{21}{5}$	$\frac{39}{5}$	0	$-\frac{21}{5}$	$-\frac{39}{5}$	$-\frac{237}{5}$

(c) Using the method of Example 4, find a system of inequalities that must be satisfied by a set of perturbations of the cost coefficients of the primal problem so that the optimal solution is unchanged.

(d) Use the result of part (c) to calculate the ranges over which each cost coefficient can vary when the other cost coefficients are fixed.

†**4.** Solve the dual of the LP problem in Example 2 of Section 7.1. Use the technique from Example 4 to write a system of inequalities that must be satisfied by any set of perturbations of the cost coefficients of the primal problem if the optimal solution is to be unchanged.

†**5.** Follow the instructions given in Exercise 4 for the LP problem in Example 1 of Section 7.3.

6. Suppose we add a new variable t to the LP problem in Example 1 of Section 3.4. Let the cost coefficient of t be 40 and the constraint coefficients in the first, second, and third constraints be 3, 2, and -1, respectively. Use the technique of Example 5 to determine whether or not the basis remains optimal.

7. Follow the instructions given in Exercise 6 for a new variable t that is added to the LP problem in Example 1 of Section 3.4. Suppose the cost coefficient of t is 3 and the constraint coefficients are the same as those given in Exercise 6.

7.5 INTRODUCTION OF A NEW VARIABLE

We now describe how to bring a new variable into a solved LP problem. We assume that the final tableau for the optimal solution is available or, at least, that we know the B^{-1} matrix from the final tableau. We will use the final tableau (or B^{-1}) to solve the new problem instead of solving the entire problem again.

We point out, however, that in practice neither the final tableau nor the B^{-1} matrix is usually available at a time when a new variable is likely to be added. Commercial linear programming programs do not usually store final tableaus in machine readable form; in fact, most of them do not even allow you to print the final tableau. You could have the CALIPSO program print the final tableau and retain it in case you need to add a variable later; even so, this would not be particularly useful because you would have to enter the final tableau with a keyboard. You would also lose considerable accuracy in working from printed output because CALIPSO does not print the same number of significant digits as it uses in calculations. So, when solving problems with a computer, you usually have to add a new variable to the original LP problem and solve the new problem.

Suppose we have solved the LP problem

$$\text{Maximize:} \quad z = c_1 x_1 + c_2 x_2 + \cdots + c_n x_n$$

$$\text{Subject to:} \quad \begin{aligned} a_{11}x_1 + a_{12}x_2 + \cdots + a_{1n}x_n &= b_1 \\ a_{21}x_1 + a_{22}x_2 + \cdots + a_{2n}x_n &= b_2 \\ &\vdots \\ a_{m1}x_1 + a_{m2}x_2 + \cdots + a_{mn}x_n &= b_m \\ x_i &\geq 0, \quad 1 \leq i \leq n \end{aligned} \quad (1)$$

and want to introduce another variable x_{n+1}. For instance, if $x_1, x_2, \ldots, x_n$ represent units of products manufactured, then the new variable x_{n+1} could represent the units of a new product being considered for manufacture.

Corresponding to the variable x_{n+1}, we introduce a cost coefficient c_{n+1} and constosnt coefficients

$$A_{n+1} = \begin{bmatrix} a_{1,\,n+1} \\ a_{2,\,n+1} \\ \vdots \\ a_{m,\,n+1} \end{bmatrix}$$

Augmenting (1) with the new variable x_{n+1}, the new cost coefficient c_{n+1}, and the new constraint column A_{n+1}, we obtain the *augmented* problem

Maximize: $z = c_1 x_1 + c_2 x_2 + \cdots + c_n x_n + c_{n+1} x_{n+1}$

Subject to: $a_{11} x_1 + a_{12} x_2 + \cdots + a_{1n} x_n + a_{1,\,n+1} x_{n+1} = b_1$

$a_{21} x_1 + a_{22} x_2 + \cdots + a_{2n} x_n + a_{2,\,n+1} x_{n+1} = b_2$

$\vdots$

$a_{m1} x_1 + a_{m2} x_2 + \cdots + a_{mn} x_n + a_{m,\,n+1} x_{n+1} = b_m$ (2)

$x_i \geq 0, \qquad 1 \leq i \leq n+1$

We could, of course, solve (2) by the simplex method; however, we are going to assume that the solution to (1) was costly to obtain and that we want to find the solution of (2) from the final tableau for (1) obtained previously in some manner.

Consider how introducing this variable affects the optimal tableau for problem (1). Suppose that B is the basic matrix corresponding to this tableau and that $\mathbf{c}_B$ is the corresponding basic cost matrix. If, as usual, we denote the entries of the optimal tableau by placing asterisks on the entries of the original tableau, the formulas

$$A_j^* = \mathbf{c}_B B^{-1} A_j \tag{3}$$

$$-c_j^* = \mathbf{c}_B B^{-1} A_j^* - c_j \tag{4}$$

show that the introduction of a new $(n + 1)$st column does not affect the entries in the old optimal tableau, provided $1 \leq j \leq n$. When $j = n + 1$, formulas (3) and (4) provide the values for the $(n + 1)$st column. If the value of $-c_{n+1}^*$ is nonnegative, then we are still at an optimal solution (with $x_{n+1} = 0$). If the value of $-c_{n+1}^*$ is negative, we must apply the simplex method until all entries in the objective row of the augmented tableau become nonnegative (or, possibly, until we discover that the new problem has unbounded solutions). Examples 1 and 2 illustrate the technique.

EXAMPLE 1

You can verify that the following tableaus,

Initial Tableau

	x_1	x_2	x_3	x_4	x_5	x_6	
x_4	1	-1	2	1	0	0	4
x_5	3	-2	1	0	1	0	6
x_6	1	0	1	0	0	1	9
	-3	3	-4	0	0	0	0

(5)

Final Tableau

	x_1	x_2	x_3	x_4	x_5	x_6	
x_3	0	$-\frac{1}{5}$	1	$\frac{3}{5}$	$-\frac{1}{5}$	0	$\frac{6}{5}$
x_1	1	$-\frac{3}{5}$	0	$-\frac{1}{5}$	$\frac{2}{5}$	0	$\frac{8}{5}$
x_6	0	$\frac{4}{5}$	0	$-\frac{2}{5}$	$-\frac{1}{5}$	1	$\frac{31}{5}$
	0	$\frac{2}{5}$	0	$\frac{9}{5}$	$\frac{2}{5}$	0	$\frac{48}{5}$

(6)

are the initial and final tableaus obtained in applying the simplex method to the LP problem

$$\text{Maximize:} \quad z = 3x_1 - 3x_2 + 4x_3$$

$$\text{Subject to:} \quad \begin{aligned} x_1 - x_2 + 2x_3 + x_4 \qquad\qquad &= 4 \\ 3x_1 - 2x_2 + x_3 \qquad + x_5 \qquad &= 6 \\ x_1 \qquad + x_3 \qquad\qquad + x_6 &= 9 \end{aligned}$$

(7)

$$x_i \geq 0, \quad 1 \leq i \leq 6$$

From these tableaus we see that the final inverse basis matrix is

$$B^{-1} = \begin{bmatrix} \frac{3}{5} & -\frac{1}{5} & 0 \\ -\frac{1}{5} & \frac{2}{5} & 0 \\ -\frac{2}{5} & -\frac{1}{5} & 1 \end{bmatrix}$$

and the final basic cost matrix is

$$\mathbf{c_B} = [c_3 \quad c_1 \quad c_6] = [4 \quad 3 \quad 0]$$

and the optimal basic solution is

$$\mathbf{x_B} = [x_3 \quad x_1 \quad x_6]^T = [\tfrac{6}{5} \quad \tfrac{8}{5} \quad \tfrac{31}{5}]^T$$

Suppose we now introduce to (4) the variable x_7 with cost coefficient

$$c_7 = -3$$

and constraint coefficient

$$A_7 = \begin{bmatrix} -1 \\ 2 \\ 4 \end{bmatrix}$$

The resulting new entry in the objective row of the final tableau is

$$-c_7^* = \mathbf{c_B}B^{-1}A_7 - c_7 = [4 \quad 3 \quad 0] \begin{bmatrix} \frac{3}{5} & -\frac{1}{5} & 0 \\ -\frac{1}{5} & \frac{2}{5} & 0 \\ -\frac{2}{5} & -\frac{1}{5} & 1 \end{bmatrix} \begin{bmatrix} -1 \\ 2 \\ 4 \end{bmatrix} - (-3) = 2$$

Since $-c_7^* \geq 0$, the final tableau remains optimal, and the solution is unchanged by the addition of the new variable.   ◆

EXAMPLE 2

Suppose in Example 1, the coefficients of the new variable x_7 are

$$c_7 = 4 \quad \text{and} \quad A_7 = \begin{bmatrix} 2 \\ -1 \\ 2 \end{bmatrix}$$

then the new entry in the objective row is

$$-c_7^* = \mathbf{c}_B B^{-1} A_7 - c_7 = \begin{bmatrix} 4 & 3 & 0 \end{bmatrix} \begin{bmatrix} \frac{3}{5} & -\frac{1}{5} & 0 \\ -\frac{1}{5} & \frac{2}{5} & 0 \\ -\frac{2}{5} & -\frac{2}{5} & 1 \end{bmatrix} \begin{bmatrix} 2 \\ -1 \\ 2 \end{bmatrix} - (4) = -\frac{4}{5} \quad (8)$$

and hence the augmented tableau is not optimal. Since we must continue the simplex method on the augmented tableau, we calculate the column of coefficients of x_7 by

$$B^{-1} A_7 = \begin{bmatrix} \frac{3}{5} & -\frac{1}{5} & 0 \\ -\frac{1}{5} & \frac{2}{5} & 0 \\ -\frac{2}{5} & -\frac{1}{5} & 1 \end{bmatrix} \begin{bmatrix} 2 \\ -1 \\ 2 \end{bmatrix} = \begin{bmatrix} \frac{7}{5} \\ -\frac{4}{5} \\ \frac{7}{5} \end{bmatrix} \quad (9)$$

From (5), (6), and (8) we obtain the tableau

	x_1	x_2	x_3	x_4	x_5	x_6	x_7	
x_3	0	$-\frac{1}{5}$	1	$\frac{3}{5}$	$-\frac{1}{5}$	0	$\left(\frac{7}{5}\right)$	$\frac{6}{5}$
x_1	1	$-\frac{3}{5}$	0	$-\frac{1}{5}$	$\frac{2}{5}$	0	$-\frac{4}{5}$	$\frac{8}{5}$
x_6	0	$\frac{4}{5}$	0	$-\frac{2}{5}$	$-\frac{1}{5}$	1	$\frac{7}{5}$	$\frac{31}{5}$
	0	$\frac{2}{5}$	0	$\frac{9}{5}$	$\frac{2}{5}$	0	$-\frac{4}{5}$	$\frac{48}{5}$

Pivoting on the circled entry, we obtain

	x_1	x_2	x_3	x_4	x_5	x_6	x_7	
x_7	0	$-\frac{1}{7}$	$\frac{5}{7}$	$\frac{3}{7}$	$-\frac{1}{7}$	0	1	$\frac{6}{7}$
x_1	1	$-\frac{5}{7}$	$\frac{4}{7}$	$\frac{1}{7}$	$\frac{2}{7}$	0	0	$\frac{16}{7}$
x_6	0	1	-1	-1	0	1	0	5
	0	$\frac{2}{7}$	$\frac{4}{7}$	$\frac{15}{7}$	$\frac{2}{7}$	0	0	$\frac{72}{7}$

(10)

From this tableau we see that introducing a new variable x_7 has resulted in the optimal basic solution

$$\mathbf{x_B} = \begin{bmatrix} x_7 & x_1 & x_6 \end{bmatrix}^T = \begin{bmatrix} \frac{6}{7} & \frac{16}{7} & 5 \end{bmatrix}^T$$

and the new maximal objective value

$$z = \frac{72}{7}$$

corresponding to the new objective function

$$z = 3x_1 - 3x_2 + 4x_3 - 3x_7 \qquad \blacklozenge$$

Exercises

1. From the final tableau (6) of Example 1, find the solution to the problem when a new variable x_7 is added to (6) with cost coefficient $c_7 = 5$ and constraint column $A_7 = [1 \quad 1 \quad -1]^{\mathrm{T}}$.

2. From the final tableau (6) of Example 1, find the solution to the problem when a new variable x_7 is added to (6) with cost coefficient $c_7 = 4$ and constraint column $A_7 = [0 \quad 1 \quad 0]^{\mathrm{T}}$.

3. From the final tableau (10) of Example 2, find the solution to the problem solved in Example 2 when a new variable x_8 is added with cost coefficient $c_8 = 6$ and constraint column $A_8 = [2 \quad -1 \quad 1]^{\mathrm{T}}$.

4. From the final tableau (10) of Example 2, find the solution to the problem when a new variable x_8 is added with cost coefficient $c_8 = 6$ and constraint column $A_8 = [-2 \quad -1 \quad -1]^{\mathrm{T}}$.

5. Add a new variable x_8 to the problem of Example 1 of Section 7.3. Find the new maximal solution beginning with the final tableau (6). First use the coefficients in (a) and then the coefficients in (b).
    (a) $c_8 = \frac{3}{2}, a_{18} = 4, a_{28} = 4, a_{38} = 5$
    (b) $c_8 = \frac{1}{2}, a_{18} = -2, a_{28} = 1, a_{38} = 1$

6. Suppose that a new variable y is added to a previously solved LP problem and that the cost coefficient of y is positive and all of the constraint coefficients of y are negative. Show that the resulting problem has unbounded feasible solutions.

7. Use the technique of Example 5 of Section 7.4 to show that the changes given in Exercise 5, part (a), do not change the optimal solution and that the changes given in part (b) do change the optimal solution.

7.6 LP MODELS

In this section we apply our knowledge of duality theory and sensitivity analysis to a case study. Moscow Operations Management, Inc. (MOM) offers production management services to businesses. A typical client presents MOM with data describing its manufacturing processes and the various supplies, demands, and costs that affect the production process. MOM uses LP methods to find an optimal

Table 1

Process	Resources				Output of slurry	
	Labor (worker-hr)	Power (kWh)	White clay (tons)	Gray clay (tons)	Grade A (tons)	Grade B (tons)
1	24	900	3.0	15.0	40	50
2	32	1200	3.2	10.0	45	40
3	12	800	5.0	20.0	75	0
4	28	750	7.0	8.5	0	100

solution to the client's problem and describes to the client the best way to set up the manufacturing process to optimize the objective function of interest. After the initial consultation, the client can call MOM for advice about the best response to changing conditions. We will describe one of MOM's cases and discuss the responses MOM gives to several customer queries.

Potlatch Forest Products (PFP) operates a clay mine and processing plant in Bovill to supply clay slurry for use in its paper production plant in Lewiston. PFP uses two kinds of slurry: grade A to make magazine stock and grade B to make newsprint. The Bovill mine produces a gray clay at an extraction cost of $30 per ton from one mining site and a white clay from a second site at an extraction cost of $70 per ton. The slurry plant can be set up to use any one of four processes to manufacture grade A and grade B slurry. Table 1 summarizes the amounts of labor (worker-hours), electrical power (kilowatt-hours, kwh), and clay (tons) of each color used in each process, and the tons of slurry of each type produced by each process. The mines can produce up to 200 tons of white clay and 600 tons of gray clay each week. The Bovill plant employs 30 workers who work 8-hour days for a total labor supply of 1200 worker-hours per week. Bovill's union contract requires that workers be paid for a full week even if they cannot all be used to make slurry. Electricity costs $0.04 per kWh, and the supply is virtually unlimited.

PFP wants to determine the most economical way of operating the Bovill plant to produce 2000 tons of grade A and 1800 tons of grade B slurry each week.

MOM's Solution

Let $x_1, x_2, x_3,$ and x_4 denote the number of hours per week that PFP uses processes 1, 2, 3, and 4, respectively. Since PFP gave no data on wages and appears to have no intention of varying the size of its workforce, we regard labor as a fixed cost that need not be included in the objective function. To find the cost of extracting white clay, first note that the amount (in tons) of white clay extracted is

$$3x_1 + 3.2x_2 + 5x_3 + 7x_4 \tag{1}$$

Since the extraction cost of a ton of white clay is $70, the cost of extracting white clay is

$$70(3x_1 + 3.2x_2 + 5x_3 + 7x_4) = 210x_1 + 224x_2 + 350x_3 + 490x_4 \tag{2}$$

The amount (in tons) of gray clay extracted is

$$15x_1 + 10x_2 + 20x_3 + 8.5x_4 \tag{3}$$

Since the cost of extracting a ton of gray clay is $30, the cost of extracting gray clay is

$$30(15x_1 + 10x_2 + 20x_3 + 8.5x_4) = 450x_1 + 300x_2 + 600x_3 + 255x_4 \tag{4}$$

The cost of electricity is

$$0.04(900x_1 + 1200x_2 + 800x_3 + 750x_4) = 36x_1 + 48x_2 + 32x_3 + 30x_4 \tag{5}$$

The objective function is the sum of the expressions in (2), (4), and (5). The LP problem is

Minimize: $z = 696x_1 + 572x_2 + 982x_3 + 775x_4$

Subject to:
$$
\begin{aligned}
24x_1 + 32x_2 + 12x_3 + 28x_4 &\le 1200 && \text{(Labor constraint)} \\
3x_1 + 3.2x_2 + 5x_3 + 7x_4 &\le 200 && \text{(White clay supply)} \\
15x_1 + 10x_2 + 20x_3 + 8.5x_4 &\le 600 && \text{(Gray clay supply)} \\
40x_1 + 45x_2 + 75x_3 &\ge 2000 && \text{(Grade A demand)} \\
50x_1 + 40x_2 + 100x_4 &\ge 1800 && \text{(Grade B demand)}
\end{aligned} \tag{6}
$$

$$x_i \ge 0, \qquad 1 \le i \le 4$$

MOM expects that PFP will be calling throughout the year with questions about the effect of various changes in the labor force, the supplies of clay, the demand for slurry, and the costs of operation. To respond quickly to queries from PFP and to do so with minimum effort, MOM plans to gather all of the data from the solution to the LP problem that might prove useful in responding. Obviously, the sensitivity ranges on resource values will be useful; however, the sensitivity ranges on the cost coefficients will not be very useful because any change in the cost of extraction or electricity will affect all of the cost coefficients. For instance, changing the extraction cost of white clay from $70 to $75 would affect all of the cost coefficients in (2). Therefore, MOM plans also to solve the dual of (6) to obtain a system of equations that can be used to test whether a certain change in the cost of one aspect of the operation will change the solution. MOM uses CALIPSO to find the following solution (Table 2) to the primal problem (6).

Table 2

The maximum of the objective function is: $-28,825.27$

Row	Basic Variable	Value	Dual Price	
1	x_2	22.99	14.46	
2	s_2	56.20	0	
3	s_3	36.32	0	(7)
4	x_3	3.48	-15.41	
5	x_1	17.62	-8.54	

Table 2 (*continued*)

SENSITIVITY ANALYSIS

Ranges in which the basis is unchanged

COST COEFFICIENT RANGES

Basic Variables

	Current Coefficient	Allowable Decrease	Allowable Increase	
x_2	−572.00	155.01	485.90	(8)
s_2	0	51.90	Infinite	
s_3	0	28.36	85.03	
x_3	−982.00	1715.60	343.44	(9)
x_1	−696.00	128.62	184.47	(10)

Nonbasic Variables

x_4	−775.00	Always infinite	326.35	(11)
s_1	0	Always infinite	14.46	
s_4	0	Always infinite	15.41	
s_5	0	Always infinite	8.54	

RESOURCE VALUE RANGES

Row Number	Current Value	Allowable Decrease	Allowable Increase	
1	1200.00	71.22	215.38	(12)
2	200.00	56.20	Infinite	(13)
3	600.00	36.32	Infinite	(14)
4	2000.00	218.75	104.29	(15)
5	1800.00	380.65	136.45	(16)

The B^{-1} matrix and the resource column $\mathbf{b}^*$ from the final tableau are

B_1	B_2	B_3	B_4	B_5	$\mathbf{b}^*$	
0.0933	0.0000	0.0000	−0.0149	−0.0328	22.9851	
0.0062	1.0000	0.0000	−0.0677	−0.0089	56.1990	
0.5100	0.0000	1.0000	−0.3483	−0.2662	36.3184	(17)
−0.0162	0.0000	0.0000	0.0159	−0.0050	3.4826	
−0.0746	0.0000	0.0000	0.0119	0.0463	17.6119	

Next, MOM forms the dual of (6) as

Minimize: $y = 1200u_1 + 200u_2 + 600u_3 + 2000u_4 + 1800u_5$

Subject to: $24u_1 + 3u_2 + 15u_3 + 40u_4 + 50u_5 \geq -696$

$32u_1 + 3.2u_2 + 10u_3 + 45u_4 + 40u_5 \geq -572$

$12u_1 + 5u_2 + 20u_3 + 75u_4 \geq -982$ $\qquad$ (18)

$28u_1 + 7u_2 + 8.5u_3 + 100u_5 \geq -775$

$u_i \geq 0, \qquad 1 \leq i \leq 3; \qquad u_4, u_5 \leq 0$

To solve (18) by the simplex method, multiply each constraint by -1, substitute $v_4 = -u_4$, $v_5 = -u_5$ (because u_4, $u_5 \leq 0$), and multiply the objective function by -1 to obtain the equivalent problem

Maximize: $y = -1200u_1 - 200u_2 - 600u_3 + 2000v_4 + 1800v_5$

Subject to: $-24u_1 - 3u_2 - 15u_3 + 40v_4 + 50v_5 \leq 696$

$-32u_1 - 3.2u_2 - 10u_3 + 45v_4 + 40v_5 \leq 572$

$-12u_1 - 5u_2 - 20u_3 + 75v_4 \leq 982$ $\qquad$ (19)

$-28u_1 - 7u_2 - 8.5u_3 + 100v_5 \leq 775$

$u_i \geq 0, \qquad 1 \leq i \leq 3; \qquad v_4, v_5 \geq 0$

MOM uses CALIPSO to find the solution of (19) and to print the B^{-1} matrix and final resource column. The solution of (19) is

The maximum of the objective function is 28,825.27

Row	Basic variable	Value	Dual price
1	u_1	14.46	17.61
2	v_4	15.41	22.99
3	s_4	326.35	3.48
4	v_5	8.54	0

The B^{-1} matrix and **b*** column is

B_1	B_2	B_3	B_4	b*
0.0746	-0.0933	0.0162	0.0000	14.4602
0.0119	-0.0149	0.0159	0.0000	15.4070
-2.5373	0.6176	0.9502	1.0000	326.3532
0.0463	-0.0328	-0.0050	0.0000	8.5353

(20)

From (20), MOM can derive the system of inequalities that must be satisfied by perturbations of the cost coefficients of problem (6). Since the constraints of (18) were multiplied by -1 to obtain problem (19), the resource values b_1, b_2, b_3, b_4 of (19) are the negatives $-c_1 = 696$, $-c_2 = 572$, $-c_3 = 982$, $-c_4 = 775$ of the cost coefficients of (6) (see Example 5 of Section 7.4). From (20) we obtain the system of inequalities (rounded to two decimal places)

$$0.07(-\Delta c_1) - 0.09(-\Delta c_2) + 0.02(-\Delta c_3) + 0(-\Delta c_4) \geq -14.46$$
$$0.01(-\Delta c_1) - 0.01(-\Delta c_2) + 0.02(-\Delta c_3) + 0(-\Delta c_4) \geq -15.41$$
$$-2.54(-\Delta c_1) - 0.62(-\Delta c_2) + 0.95(-\Delta c_3) + 1(-\Delta c_4) \geq -326.35 \tag{21}$$
$$0.05(-\Delta c_1) - 0.03(-\Delta c_2) - 0.01(-\Delta c_3) + 0(-\Delta c_4) \geq -8.54$$

that must be satisfied by the changes to the cost coefficients of (6) for the solution to be unchanged.

REPORT TO PFP

MOM reports the solution (6) to PFP as

Process	Hours per week
1	17.62
2	22.99
3	3.48
4	0

PFP will need to mine 143.8 tons of white clay and 563.68 tons of gray clay per week at a total cost of $28,825.27.

PFP sets up its operation according to MOM's advice and, at various times, asks MOM the following questions.

Question 1. One of our employees is leaving for a two-week vacation. We cannot replace him because it takes about two weeks to train a replacement. Should we alter the mix of processes while he is gone? How much will this vacation cost us?

MOM sees from (12) that a decrease in labor time of 40 hours per week will not change the set of basic variables and concludes that no change in the mix of processes is called for. By the Fundamental Principle of Duality, the new cost can be expressed in terms of the dual prices from (7) as

$$\hat{z}^* = 14.46(b_1 + \Delta b_1) + 0b_2 + 0b_3 - 15.41b_4 - 8.54b_5$$
$$= z^* + 14.46\,\Delta b_1 = 28,825.27 + 14.46 \ (-40)$$

Therefore, the cost of the two-week vacation is

$$2(14.46)40 = \$1156.80$$

To calculate the new mix of processes, MOM uses the B^{-1} matrix (17) and calculates

$$\mathbf{x_B} = \begin{bmatrix} x_2 \\ s_2 \\ s_3 \\ x_3 \\ x_1 \end{bmatrix} = B^{-1} \begin{bmatrix} 1160 \\ 200 \\ 600 \\ 2000 \\ 1800 \end{bmatrix} = \begin{bmatrix} 19.25 \\ 55.95 \\ 15.92 \\ 4.13 \\ 20.60 \end{bmatrix}$$

Question 2. A second employee has asked to take vacation at the same time as the other employee. Can we meet our production demands with two people off? How much will it cost us?

MOM sees from (12) that a reduction in labor force of 80 hours per week will change the optimal mix of processes. MOM did not save the final tableau for the primal problem because the CALIPSO program cannot store a final tableau in machine readable form. MOM could have printed approximations to the entries in the final tableau and used these values to start the simplex method; however, it was judged that such a procedure would introduce unacceptable error. Using the B^{-1} matrix, which *was* printed out, also would lead to calculation errors. Therefore, MOM elects to solve the primal problem again and finds it possible to meet the production demands. The solution is

$$(x_1, x_2, x_3, x_4, s_1, s_2, s_3, s_4, s_5) = (20.62, 16.31, 5.88, 1.17, 0, 48.37, 0, 0, 0)$$

$$z = \$30,362.80$$

Question 3. We have a spare front-loader that we could use to mine gray clay for one week. It will increase our supply of gray clay by 75 tons. How much can we save by using it?

MOM sees from (7) that $s_3 > 0$ and, hence, that there is no use for more gray clay. Therefore, MOM reports that the front-loader should not be used because the production cost will not change.

Question 4. We must produce an extra 100 tons of grade A slurry next week. Can we meet this demand? What mix of processes should we use, and what will the production cost be?

MOM sees from (15) that an increase of 100 tons of grade A slurry will not require a change in the mix of processes. By using the B^{-1} matrix (17), MOM calculates the new solution to be

$$\mathbf{x_B} = \begin{bmatrix} x_2 \\ s_2 \\ s_3 \\ x_3 \\ x_1 \end{bmatrix} = B^{-1} \begin{bmatrix} 1200 \\ 200 \\ 600 \\ 2100 \\ 1800 \end{bmatrix} = \begin{bmatrix} 21.49 \\ 49.43 \\ 1.49 \\ 5.07 \\ 18.81 \end{bmatrix}$$

From (7), the dual price of grade A slurry is -15.41 dollars per ton. Therefore, the increase in cost will be

$$100(15.41) = \$1541$$

PFP will need to mine

$$3x_1 + 3.2x_2 + 5x_3 + 7x_4 = 3(18.81) + 3.2(21.49) + 5(5.07) + 7(0)$$

$$= 150.55$$

tons of white clay, and

$$15x_1 + 10x_2 + 20x_3 + 8.5x_4 = 598.45$$

tons of gray clay.

Question 5.   Our electricity rate has increased from 4 cents to 5 cents per kwh. Should we change the mix of processes? How much will this increase cost us?
    MOM uses the expression in (5) to calculate the increase in electricity cost to be

$$0.01(900x_1 + 1200x_2 + 800x_3 + 750x_4) = 9x_1 + 12x_2 + 8x_3 + 7.5x_4$$

Therefore, the changes in the cost coefficients of (6) are

$$\Delta c_1 = 9, \qquad \Delta c_2 = 12, \qquad \Delta c_3 = 8, \qquad \text{and} \qquad \Delta c_4 = 7.5$$

MOM substitutes these changes in (21) and finds that all of the inequalities are satisfied. Therefore, this change does not affect the solution. Using the solution (7), MOM concludes that the increase in cost is

$$9(17.62) + 12(22.99) + 8(3.48) + 7.5(0) = \$462.30$$

Exercises

Answer the following questions that PFP poses to MOM. When it is practical, use the sensitivity analyses we gave in preference to solving new LP problems.

1. One of our employees is taking vacation for all of next week, and another employee is taking three days of sick leave. Should we alter the mix of processes while they are gone? How much will this cost us?

2. Our supply of white clay will be reduced to 175 tons next week. How should we alter the mix of processes? How much white and how much gray clay should we mine? How much will our costs increase?

3. We will have one person on sick leave next week, and we must produce an additional 100 tons of grade A slurry. Can we meet this demand? If so, how should we alter the mix of processes, how much of each type of clay should we mine, and how much will our costs be?

4. We must produce an extra 100 tons each of Grade A and B slurry next week. How should we adjust our processing and mining to meet this demand. What will our costs be?

5. The cost of extracting gray clay has changed to \$35 per ton, and the cost of extracting white clay has changed to \$75 per ton. How should we adjust our mining and processing plan? How much will our weekly costs be?

6. We are considering giving each of our workers a \$1 per hour raise. If we do so, will we need to change our mix of processes? How much will the raise cost us each week?

7. We are considering giving each of our workers a \$1.50 per hour raise. If we do so, will we need to change our mix of processes? How much will the raise cost us each week?

8. We are considering replacing process 4 with a new process that requires 25 worker-hours of labor and uses 800 kwh of power, 8 tons of white clay, 9 tons of gray clay, and produces 20 tons of grade A slurry and 90 tons of grade B slurry. Will this new process reduce our costs?

9. A potter has offered to buy 50 tons of white clay next week at $85 per ton. Can we make this sale without changing our processing schedule? How much will we make from the sale?

10. Suppose we offer to sell the potter (see Exercise 9) 30 tons of gray clay at $40 per ton in addition to the 50 tons of white clay. Can we make this sale without changing our processing schedule? How much would we make?

CHECKLIST: CHAPTER 7

DEFINITION

CONCEPTS AND RESULTS

Chapter 8

AN INTRODUCTION TO INTEGER PROGRAMMING

8.1 A SAMPLING OF IP PROBLEMS

From time to time we have encountered LP problems that require integer (rather than noninteger or fractional) solutions. For instance, optimal solutions such as $17\frac{1}{3}$ or 1.71 do not make sense for problems dealing with number of tractors sold, pieces of furniture produced, and so forth; and, as you will see, rounding off such solutions does not necessarily yield optimal (or even feasible) integer solutions.

A less-apparent use for LP problems requiring integer solutions is in solving problems that involve yes/no decisions. Such problems are solved by using binary variables, which take on only the values 0 and 1. For example, assignment problems have this property, as we saw in Chapter 1. In this chapter, we will investigate other situations in which variables can only assume the values 0 or 1.

This section introduces the basic ideas of integer programming and presents examples that illustrate some widely applicable techniques for formulating integer problems. Sections 8.2–8.5 develop three methods for solving integer problems, and Section 8.6 introduces some more-specialized techniques.

Integer programming (IP) refers to LP problems for which some or all of the variables take on integer values. If all of the variables are integer, then we say that the problem is a *pure integer problem*; otherwise, the problem is a *mixed integer problem*. If all of the variables of an IP problem are restricted to the values 0 or 1, we often refer to the problem as a *pure binary problem* (sometimes called a *zero–one problem*).

Before taking up a rather diverse group of IP problems, we show that rounding off a fractional optimal solution does not always yield the desired IP optimum.

EXAMPLE 1

We consider the LP problem

$$\text{Maximize:} \quad z = 5x_1 + 27x_2$$
$$\text{Subject to:} \quad 2x_1 + 11x_2 \leq 59$$
$$x_1 - \quad x_2 \leq 7 \tag{1}$$
$$x_1, x_2 \geq 0 \quad \text{and} \quad \text{integer}$$

If we ignore the integer constraints on x_1 and x_2, the optimal solution of (1) is

$$x_1 = 10\tfrac{6}{13}; \qquad x_2 = 3\tfrac{6}{13}; \qquad z = 145\tfrac{10}{13}$$

Suppose we round off the values of x_1 and x_2 in an attempt to obtain the optimal integer solution. Table 1 lists the various possibilities

Table 1

x_1	x_2	z	Feasible?
10	3	131	Yes
10	4	158	No
11	3	136	No
11	4	163	No

It is clear from Figure 1, which gives the feasible region for (1), that the maximal integer solution occurs at the point $x_1 = 2$, $x_2 = 5$ where $z = 145$. Note that this point and the value for z differ considerably from the only feasible round-off solution found in Table 1. Thus, rounding off optimal fractional solutions is generally not an appropriate procedure for solving IP problems.

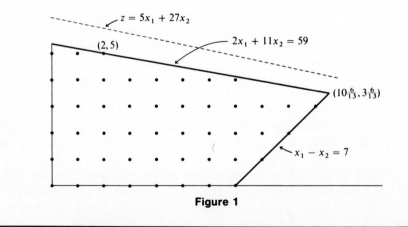

Figure 1

We now consider a variety of problems that require integer or mixed integer solutions. We introduce more problems of this sort in the exercises.

In the next example we see how to convert a problem, which at first glance does not even appear to be an LP problem, into an IP problem.

___

EXAMPLE 2

(Fixed Cost Problem). Suppose a plant has five machines, each of which can produce a product P. For each machine M_i ($1 \leq i \leq 5$), there is a fixed maintenance cost f_i that applies whenever the machine is started during a working day, regardless of the number of units of P produced. In addition, for each machine M_i, there is a unit cost c_i for each unit of P produced by that machine. The daily demand for product P is d, and each machine M_i can produce up to B_i units of P in a working day. The plant manager wants to minimize costs while meeting the demand requirement.

To formulate the problem mathematically, we let x_i denote the number of units produced by machine M_i, and for $1 \leq i \leq 5$ we define

$$h(x_i) = \begin{cases} c_i x_i + f_i & \text{if} \quad x_i > 0 \\ 0 & \text{if} \quad x_i = 0 \end{cases}$$

Then our problem becomes

$$\text{Minimize:} \quad z = h(x_1) + h(x_2) + h(x_3) + h(x_4) + h(x_5)$$
$$\text{Subject to:} \quad x_1 + x_2 + x_3 + x_4 + x_5 \geq d \tag{2}$$
$$x_i \geq 0 \quad \text{and} \quad \text{integer;} \quad x_i \leq B_i, \quad 1 \leq i \leq 5$$

What differentiates this problem from those previously studied is that the objective function is not a linear function (i.e., it is not of the form $d_1 x_1 + d_2 x_2 + \cdots + d_5 x_5$), and you cannot directly apply the techniques learned so far.

To convert this problem into an LP (actually an IP) problem, we introduce the binary variable w_i ($1 \leq i \leq 5$) by

$$w_i = \begin{cases} 1 & \text{if} \quad x_i > 0 \\ 0 & \text{if} \quad x_i = 0 \end{cases} \tag{3}$$

We can ensure that w_i satisfies condition (3) by introducing constraints

$$x_i \leq B_i w_i \tag{4}$$
$$w_i \leq x_i \tag{5}$$

Since w_i can be only 0 or 1, inequality (4) ensures that $w_i = 1$ whenever $x_i > 0$. If $w_i = 1$, inequality (4) becomes $x_i \leq B_i$, which is the desired upper bound on x_i. Inequality (5) ensures that $w_i = 0$ whenever $x_i = 0$. Since x_i is an integer and w_i is 0 or 1, inequality (5) does not impose any unwanted constraint on x_i or w_i.

From the preceding remarks, we see that the following IP problem is equivalent to (2).

$$\text{Minimize:} \quad z = c_1 x_1 + c_2 x_2 + c_3 x_3 + c_4 x_4 + c_5 x_5$$
$$+ f_1 w_1 + f_2 w_2 + f_3 w_3 + f_4 w_4 + f_5 w_5$$

$$\text{Subject to:} \quad x_1 + x_2 + x_3 + x_4 + x_5 \geq d$$
$$x_i \leq B_i w_i, \quad 1 \leq i \leq 5$$
$$w_i \leq x_i, \quad 1 \leq i \leq 5$$
$$x_i \geq 0 \quad \text{and} \quad \text{integer;} \quad w_i \quad \text{binary,} \quad 1 \leq i \leq 5 \qquad \blacklozenge$$

To develop an integer program we often need to find bounds such as $x_i \leq B_i$ on variables. Sometimes such bounds are given in the problem statement (as was the case in Example 2); at other times the bounds must be inferred from the constraints. Exercises 5–9 show how this can be done.

The next example shows how to use binary variables to select alternative sets of constraints.

EXAMPLE 3

(Alternative Sets of Constraints). Occasionally a problem requires that solutions satisfy alternative sets of constraints. For instance, a problem might require that either the constraint

$$2x_1 + 3x_2 \leq 12 \qquad (6)$$

is satisfied or that the two constraints

$$3x_1 + 4x_2 \leq 24$$
$$-x_1 + x_2 \geq 1 \qquad (7)$$

are satisfied. (The word *or* is used in the inclusive sense; i.e., either (6) or (7) or both could be true.)

The first step in formulating an IP problem to model this situation (as well as many other applications of binary variables) requires that we find "bounds" for expressions that occur in certain constraints. For instance, if there are positive numbers M_1, M_2, and M_3 such that

$$2x_1 + 3x_3 - 12 \leq M_1$$
$$3x_1 + 4x_2 - 24 \leq M_2 \qquad (8)$$
$$-x_1 + x_2 - 1 \geq -M_3$$

for all feasible solutions to our problem, then the constraints

$$2x_1 + 3x_3 - 12 \leq M_1 y \tag{9}$$

$$3x_1 + 4x_2 - 24 \leq M_2(1 - y) \tag{10}$$

$$-x_1 + x_2 - 1 \geq -M_3(1 - y) \tag{11}$$

$$y \quad \text{binary}$$

are equivalent to the either/or conditions of (6) and (7). To see this, note that if $y = 1$, then constraint (9) has no effect since M_1 is a bound for $2x_1 + 3x_3 - 12$ [see (8)]; however, (10) and (11) reduce to constraints (7). By similar reasoning, if $y = 0$, then constraint (6) must be satisfied whereas constraints (7) may or may not be satisfied.

♦

The next example illustrates a practical application of alternative sets of constraints.

EXAMPLE 4

(Scheduling Problem). A company uses five machines M_i, $(1 \leq i \leq 5)$ to manufacture three products: A_1, A_2, and A_3. Figure 2 gives the machine order and the time required on each machine to produce each of the products A_i.

The production of A_1, A_2, and A_3 is subject to the restrictions.

1. Two machines cannot work simultaneously on a product.

2. Each machine must finish processing one product before beginning another.

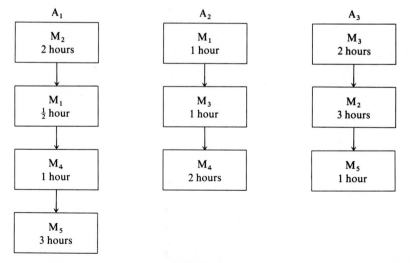

Figure 2

We want to determine the production schedule that minimizes the time needed to process all of the products. To this end, for $j = 1, 2, 4, 5$ we let x_{1j} be the starting time for processing A_1 on machine j (time can be measured in hours from any convenient reference point, say the beginning of the workday). Similarly, for $j = 1, 3, 4$ we let x_{2j} be the starting time for processing A_2 on machine j, and for $j = 2, 3, 5$ we let x_{3j} be the starting time for processing A_3 on machine j.

We can express the condition that M_1 cannot begin on A_1 until M_2 finishes working on A_1 with the inequality

$$x_{12} + 2 \le x_{11}$$

Note that we used $\le$ rather than $=$ in this constraint because we cannot be sure that M_1 can begin on A_1 as soon as M_2 is finished since M_1 might be working on another product.

Similarly, we can express the condition that M_4 cannot start working on A_1 until M_1 is finished with the inequality

$$x_{11} + \frac{1}{2} \le x_{14}$$

Continuing in this manner, we obtain the following constraints from Figure 2.

$$\text{Product } A_1 \begin{cases} x_{12} + 2 \le x_{11} & (12) \\ x_{11} + \dfrac{1}{2} \le x_{14} & (13) \\ x_{14} + 1 \le x_{15} & (14) \end{cases}$$

$$\text{Product } A_2 \begin{cases} x_{21} + 1 \le x_{23} & (15) \\ x_{23} + 1 \le x_{24} & (16) \end{cases}$$

$$\text{Product } A_3 \begin{cases} x_{33} + 2 \le x_{32} & (17) \\ x_{32} + 3 \le x_{35} & (18) \end{cases}$$

Next we deal with the restriction that each machine must finish processing one product before beginning another. This means, for example, that machine M_1 cannot work on products A_1 and A_2 simultaneously; that is, it either works on A_1 before it works on A_2 or vice versa. Thus, we have either the constraint

$$x_{11} + \frac{1}{2} \le x_{21}$$

or the constraint

$$x_{21} + 1 \le x_{11}$$

As in Example 3 this leads to the constraints

$$x_{11} + \frac{1}{2} \le x_{21} + By_1 \tag{19}$$

$$x_{21} + 1 \le x_{11} + B(1 - y_1) \tag{20}$$

where B is suitably large (the hour at which the production shift ends would do), and y_i is a binary variable. In a similar way, we obtain four more pairs of constraints

$$x_{12} + 2 \leq x_{32} + By_2 \tag{21}$$

$$x_{32} + 3 \leq x_{12} + B(1 - y_2) \tag{22}$$

$$x_{23} + 1 \leq x_{33} + By_3 \tag{23}$$

$$x_{33} + 2 \leq x_{23} + B(1 - y_3) \tag{24}$$

$$x_{14} + 1 \leq x_{24} + By_4 \tag{25}$$

$$x_{24} + 2 \leq x_{14} + B(1 - y_4) \tag{26}$$

$$x_{15} + 3 \leq x_{35} + By_5 \tag{27}$$

$$x_{35} + 1 \leq x_{15} + B(1 - y_5) \tag{28}$$

where for $1 \leq i \leq 5$, y_i is a binary variable.

Let t be the time needed to process all of the products. Note that from Figure 2 we have

$$t \geq x_{15} + 3 \tag{29}$$

$$t \geq x_{24} + 2 \tag{30}$$

$$t \geq x_{35} + 1 \tag{31}$$

Thus, our problem is to minimize t, subject to the constraints (12)–(31).

In Exercise 25 of Section 8.2 you are asked to show that Table 2 gives the solution to this problem.

Table 2

Machine 1	Machine 2	Machine 3	Machine 4	Machine 5
$x_{11} = 2$	$x_{12} = 0$	$x_{23} = 2.5$	$x_{14} = 2.5$	$x_{15} = 3.5$
$x_{21} = 1$	$x_{32} = 2$	$x_{33} = 0$	$x_{24} = 3.5$	$x_{35} = 6.5$

Minimum time required to complete the work is $t = 7.5$.

Table 3 displays the schedule in time sequence.

Table 3

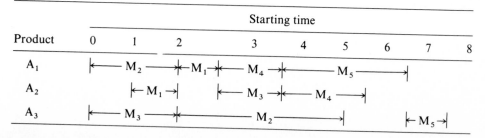

Other solutions to this problem are possible. The one displayed in Table 3 is the one found by CALIPSO for the order in which the constraints were entered. Of course, all optimal solutions have the same finishing time. ♦

The next example introduces another useful "trick" involving binary variables.

EXAMPLE 5

(If ..., then...). Some problems require that a constraint be satisfied when some variable assumes a positive value. For instance, an automobile manufacturer might insist that if any cars are to be painted red on a given day, then at least 50 cars must be painted red.

In general, suppose that a problem situation requires that whenever

$$a_1 x_1 + a_2 x_2 + \cdots + a_n x_n > b_1 \tag{32}$$

then the constraint

$$d_1 x_1 + d_2 x_2 + \cdots + d_n x_n \le b_2 \tag{33}$$

must be satisfied. If (32) is not satisfied by a solution, then (33) is not to be a constraint (i.e., it may or may not be satisfied by the solution).

To simplify the notation, let

$$f(x_1, x_2, \ldots, x_n) = a_1 x_1 + a_2 x_2 + \cdots + a_n x_n - b_1$$

$$g(x_1, x_2, \ldots, x_n) = d_1 x_1 + d_2 x_2 + \cdots + d_n x_n - b_2$$

Then the problem posed in (32) and (33) becomes:

$$\text{If} \quad f(x_1, x_2, \ldots, x_n) > 0, \quad \text{then} \quad g(x_1, x_2, \ldots, x_n) \le 0 \tag{34}$$

To express (34) in a more manageable form, we assume bounds M_1 and M_2 such that

$$f(x_1, x_2, \ldots, x_n) \le M_1 \tag{35}$$

and

$$g(x_1, x_2, \ldots, x_n) \le M_2$$

[Bounds such as (35) are usually found from other constraints of the problem. See Exercises 5–9.]

The constraints

$$f(x_1, x_2, \ldots, x_n) \le M_1 y$$
$$g(x_1, x_2, \ldots, x_n) \le M_2(1 - y) \tag{36}$$

$$y \quad \text{binary}$$

are equivalent to the "If, then" condition (34). To see this, note that the constraints (36) ensure that y is 1 whenever $f(x_1, x_2, \ldots, x_n) > 0$. If $y = 1$, then constraint (33) is imposed. If $y = 0$, then constraint (33) has no effect. ◆

Enterprises often find it undesirable, for one reason or another, to make fewer than a fixed number of units of a product. Such problems are usually called *minimum batch size* problems. The next example applies the idea of Example 5 to such a problem.

EXAMPLE 6

The Lambhorghetty Motorcar Company makes three styles of cars: Dove, Hawk, and Eagle. Doves and Hawks differ only in minor ways, and their manufacture involves only minor changes in the assembly procedure, whereas Eagles require major adjustments in the assembly procedure. Therefore, Lambhorghetty makes Doves or Hawks only when it must make at least 100 of these models during a production week. For this week's production, Lambhorghetty has 8000 tons of steel and 80,000 worker-hours of labor available. Table 4 gives the resources and profits for each type of car. Formulate an IP problem to maximize Lambhorghetty's profit.

Table 4

	Dove	Hawk	Eagle
Steel required (tons)	1.5	1.6	5
Labor (worker-hours)	35	38	45
Profit ($)	2000	2300	4200

Let x_1 be the number of Doves produced, x_2 the number of Hawks produced, and x_3 the number of Eagles produced. The profit in thousands of dollars is $z = 2x_1 + 2.3x_2 + 4.2x_3$.

The constraints are

$$\text{If} \quad x_1 + x_2 > 0, \quad \text{then} \quad x_1 + x_2 \geq 100 \tag{37}$$

$$1.5x_1 + 1.6x_2 + 5x_3 \leq 8{,}000$$
$$35x_1 + 38x_2 + 45x_3 \leq 80{,}000 \tag{38}$$

$$x_1, x_2, x_3 \geq 0$$

Using the notation of Example 5, let

$$f(x_1, x_2) = x_1 + x_2$$

and

$$g(x_1, x_2) = -x_1 - x_2 + 100$$

By setting two variables at a time equal to 0 in (38), we obtain the bounds

$$x_1 \leq 5334, \qquad x_2 \leq 5000$$

Hence,

$$f(x_1, x_2) = x_1 + x_2 \leq 5334 + 5000 = 10{,}334$$

$$g(x_1, x_2) \leq -x_1 - x_2 + 100 \leq 0 + 0 + 100 = 100$$

From (36) we see that constraint (37) is expressed by

$$f(x_1, x_2) = x_1 + x_2 \leq 10{,}334y$$

$$g(x_1, x_2) = -x_1 - x_2 + 100 \leq 100(1 - y) \qquad (39)$$

$$y \quad \text{binary}$$

After simplifying (39), we find the corresponding IP problem to be

Maximize: $z = 2x_1 + 2.3x_2 + 4.2x_3$

Subject to: $1.5x_1 + 1.6x_2 + 5x_3 \leq 8{,}000$

$35x_1 + 38x_2 + 45x_3 \leq 80{,}000$

$x_1 + x_2 \qquad\quad \leq 10{,}334y$

$x_1 + x_2 \qquad\quad \geq 100y$

$x_1, x_2, x_3 \geq 0$ and   integer; y binary

In Exercise 22 of Section 8.2 you are asked to show that the solution is $x_1 = 5$, $x_2 = 335$, $x_3 = 1491$, and $z = 7042.7$.   ◆

Exercises

1. The LP problem

Maximize: $z = 6x_1 + 9x_2$

Subject to: $4x_1 + 18x_2 \leq 25$

$3x_1 + 4x_2 \leq 12$

$x_1, x_2 \geq 0$

has the optimal solution $x_1 = \frac{58}{19}$, $x_2 = \frac{27}{38}$; $z = \frac{939}{38}$. Find the optimal integer solution graphically.

2. The LP problem

Maximize: $z = 9x_1 + 2x_2$

Subject to: $5x_1 + x_2 \leq 24$

$x_1 + 2x_2 \leq 20$

$x_1, x_2 \geq 0$ and   integer

has the optimal solution $x_1 = 3\frac{1}{9}$, $x_2 = 8\frac{4}{9}$; $z = 44\frac{8}{9}$. Find the optimal integer solution graphically.

3. (Stock Cutting Problem). A certain pipe comes in 25-ft lengths. Suppose a contractor needs at least 200 sections of 6-ft pipe; 250 sections of 9-ft pipe; and 300 sections of 12-ft pipe. The contractor can cut each 25-ft pipe as indicated in Table 5.

Table 5

Cutting pattern	Number of 6-ft sections	Number of 9-ft sections	Number of 12-ft sections	Feet of scrap
1	4	0	0	1
2	2	1	0	4
3	2	0	1	1
4	1	2	0	1
5	0	1	1	4
6	0	0	2	1

The problem is to determine to what extent the contractor should use each cutting pattern so that the demands are met and the amount of scrap is minimized.

Let x_i, $1 \le i \le 6$, be the number of times that cutting pattern i is used. Formulate an IP problem to minimize the amount of scrap.

4. Let

$$h(x_1) = \begin{cases} 120x_1 - 150 & \text{if} \quad x_1 > 0 \\ 0 & \text{if} \quad x_1 = 0 \end{cases}$$

$$h(x_2) = \begin{cases} 100x_2 - 220 & \text{if} \quad x_2 > 0 \\ 0 & \text{if} \quad x_2 = 0 \end{cases}$$

Write the following problem as an IP problem.

$$\text{Maximize:} \quad z = h(x_1) + h(x_2)$$
$$\text{Subject to:} \quad x_1 + x_2 \le 4$$
$$5x_1 + 3x_2 \le 15$$
$$x_1, x_2 \ge 0$$

5. Suppose an LP problem contains the constraints

$$2x_1 + 3x_2 + 4x_3 \le 24$$
$$x_1 + 2x_2 - x_3 \le 15$$
$$x_1, x_2, x_3 \ge 0$$

Explain why $x_1 \le 12$, $x_2 \le 8$, and $x_3 \le 6$.

6. Suppose an LP problem contains the constraints

$$4x_1 + 5x_2 + 3x_3 \le 12$$
$$-3x_1 + 2x_2 - x_3 \le 20$$

Explain why $x_1 \le 3$, $x_2 \le \frac{12}{5}$, and $x_3 \le 4$.

7. Suppose an LP problem contains the constraints

$$x_1 + 5x_2 + 8x_3 \leq 40$$

$$x_i \geq 0, \quad 1 \leq i \leq 3$$

and either

$$2x_1 + 3x_2 + 4x_3 \leq 12 \tag{40}$$

or

$$x_1 + x_2 + x_3 \geq 3 \tag{41}$$

(a) Show that $-x_1 - x_2 - x_3 + 3 \leq 3$.
(b) Show that $2x_1 + 3x_2 + 4x_3 - 12 \leq 112$.
(c) Use a binary variable to formulate a pair of constraints that require that either (40) or (41) be effective (provided the first two constraints are valid).

8. Suppose $0 \leq x_1 \leq 3, 0 \leq x_2 \leq 4$, and $0 \leq x_3 \leq 2$. For each of the following expressions, find a value for M that will serve as the indicated bound. Make M no larger or smaller than it needs to be.
(a) $2x_1 + x_2 + 5x_3 - 14 \leq M$
(b) $x_1 + 2x_2 + 3x_3 - 54 \leq M$
(c) $-x_1 - x_2 - x_3 + 10 \geq M$
(d) $2x_1 - 3x_2 + 4x_3 + 6 \leq M$

9. Suppose $0 \leq x_1 \leq 5, 0 \leq x_2 \leq 6$, and $0 \leq x_3 \leq 8$. For each of the following expressions, find a value for M that will serve as the indicated bound. Make M no larger or smaller than it needs to be.
(a) $3x_1 + 2x_2 + 6x_3 - 24 \leq M$
(b) $2x_1 + 3x_2 + 5x_3 - 4 \leq M$
(c) $3x_1 - 2x_2 - 3x_3 - 10 \leq M$
(d) $6x_1 - 3x_2 - 4x_3 + 5 \leq M$

10. Formulate the following problem as an IP problem.

$$\text{Maximize:} \quad z = 2x_1 + 3x_2 - 5x_3$$

$$\text{Subject to:} \quad x_1 + 3x_2 + 4x_3 \leq 12$$

$$x_1, x_2, x_3 \geq 0$$

$$\text{and either} \quad 2x_1 + 4x_2 - x_3 \leq 10$$

$$\text{or} \quad 3x_1 - 2x_2 + 5x_3 \leq 5$$

Do not introduce any bounds that are larger than they need to be.

11. Formulate the following problem as an IP problem

$$\text{Maximize:} \quad z = 2x_1 + 3x_2 + 5x_3$$

$$\text{Subject to:} \quad x_1 + 5x_2 + 4x_3 \leq 15$$

$$x_1, x_2, x_3 \geq 0$$

and use at least one of the following three sets of constraints

$$\{x_1 + 2x_2 + 3x_3 \leq 11\}$$

$$\{3x_1 + 2x_2 + 4x_3 \leq 15\}$$

$$\{4x_1 + 6x_2 + x_3 \leq 18 \quad \text{and} \quad x_1 + 5x_2 + 2x_3 \leq 13\}$$

(Suggestion: Use three binary variables, and place a constraint on their sum.)

12. Solve Exercise 11 when at least two of the three sets of constraints must be satisfied.

13. We want either the constraint

$$2x_1 + 3x_2 + 4x_3 \leq 12 \tag{42}$$

or the constraint

$$x_1 + x_2 + x_3 \leq 4 \tag{43}$$

to be satisfied for $x_1, x_2, x_3 \geq 0$.
(a) Show that $x_1 \leq 6$, $x_2 \leq 4$, and $x_3 \leq 4$, regardless of which of (42) or (43) is satisfied.
(b) Use a binary variable to formulate a pair of constraints that require that either (42) or (43) be effective.

14. Suppose $a_i > 0$ and $b_i > 0$ for $1 \leq i \leq 3$. Find an IP problem that is equivalent to the problem

$$\text{Maximize:} \quad z = x_1 + 2x_2 + 3x_3$$
$$\text{Subject to:} \quad x_1, x_2, x_3 \geq 0$$

and either

$$a_1x_1 + a_2x_2 + a_3x_3 \leq 10$$

or

$$b_1x_1 + b_2x_2 + b_3x_3 \leq 12$$

15. Solve Exercise 14 when the last inequality is replaced by

$$b_1x_1 + b_2x_2 + b_3x_3 \geq 12$$

16. Write an IP problem that is equivalent to

$$\text{Maximize:} \quad z = x_1 + 2x_2 + 9x_3$$
$$\text{Subject to:} \quad x_1 + x_2 + 4x_3 \leq 85$$
$$-x_1 + 2x_2 + x_3 \geq 27$$
$$\text{If} \quad x_2 > 0, \quad \text{then} \quad x_2 \geq 24$$
$$x_1, x_2, x_3 \geq 0$$

17. Write an IP problem that is equivalent to

$$\text{Maximize:} \quad z = 3x_1 + 4x_2 - 3x_3$$
$$\text{Subject to:} \quad x_1 + x_2 + 4x_3 \leq 60$$
$$-x_1 + 2x_2 + x_3 \geq 12$$
$$\text{If} \quad x_2 + x_3 > 0, \quad \text{then} \quad x_1 + x_3 \leq 54$$
$$x_1, x_2, x_3 \geq 0$$

18. Explain why the constraints $w_i \leq x_i$, $1 \leq i \leq 3$ can be omitted from the LP problem in Example 2.

19. Suppose u and v are binary variables. Write inequalities or equalities that ensure that
(a) either u or v is 1
(b) exactly one of u and v is 1
(c) at most one of u and v is 1
(d) if $u = 1$, then $v = 1$
(e) if $v = 0$, then $u = 0$

8.2   A BRANCH AND BOUND METHOD

Most of the methods available for solving integer (and mixed integer) programming problems fall into two broad categories: *Branch and Bound* methods and *Cutting Plane* methods. In this section we consider one common branch and bound method that can be used to solve any pure or mixed integer problem. Sections 8.3 and 8.4 describe another branch and bound method for solving pure binary problems, and Section 8.5 describes a standard cutting plane procedure.

The branch and bound method discussed in this section is often called the *best bound* method. We now illustrate the procedure by applying it to the problem

$$\text{Maximize:} \quad z = 9x_1 + 4x_2$$

$$\text{Subject to:} \quad x_1 + 5x_2 \leq 20$$
$$5x_1 + 2x_2 \leq 17 \tag{1}$$

$$x_1, x_2 \geq 0 \quad \text{and} \quad \text{integer}$$

The first step is to solve (1) without considering the integer constraints. Introducing slack variables we have the initial tableau

	x_1	x_2	x_3	x_4	
x_3	1	5	1	0	20
x_4	5	2	0	1	17
	-9	-4	0	0	0

(2)

Applying the simplex method to (2) we eventually obtain the final tableau

	x_1	x_2	x_3	x_4	
x_2	0	1	$\frac{5}{23}$	$-\frac{1}{23}$	$\frac{83}{23}$
x_1	1	0	$-\frac{2}{23}$	$\frac{5}{23}$	$\frac{45}{23}$
	0	0	$\frac{2}{23}$	$\frac{41}{23}$	$32\frac{1}{23}$

(3)

from which we see that the optimal solution of (1) (without the integer constraints) is

$$x_1 = \frac{45}{23} = 1.9565, \qquad x_2 = \frac{83}{23} = 3.6087$$

and

$$z = 32\frac{1}{23} = 32.0435$$

If at this stage x_1 and x_2 had integer values, we would be done. Since this is not the case, we begin the branching process. We choose to branch at x_1 (though we could branch at x_2) and consider two new LP problems (the branches). Since $x_1 = 1.9565$, one of these problems is formed by adding the constraint $x_1 \geq 2$ to (1), and the other is formed by adding the constraint $x_1 \leq 1$ to (1). For these two new problems we again ignore the integer constraints and thus consider the problems

$$\text{Maximize:} \quad z = 9x_1 + 4x_2$$
$$\text{Subject to:} \quad x_1 + 5x_2 \leq 20$$
$$5x_1 + 2x_2 \leq 17 \qquad (4)$$
$$x_1 \qquad \geq 2$$
$$x_1, x_2 \geq 0$$

and

$$\text{Maximize:} \quad z = 9x_1 + 4x_2$$
$$\text{Subject to:} \quad x_1 + 5x_2 \leq 20$$
$$5x_1 + 2x_2 \leq 17 \qquad (5)$$
$$x_1 \qquad \leq 1$$
$$x_1, x_2 \geq 0$$

Although we could solve problem (4) with the two-phase method, it is usually easier to apply the dual simplex method developed in Section 6.5 (especially see Example 4).

We incorporate the constraint $x_1 \geq 2$ into the final tableau (3) by adding a slack variable x_5 to the inequality $x_1 \geq 2$ to obtain

$$x_1 - x_5 = 2$$

and then adding a row for this equation to obtain tableau (6).

	x_1	x_2	x_3	x_4	x_5	
(x_2)	0	1	$\frac{5}{23}$	$-\frac{1}{23}$	0	$\frac{83}{23}$
(x_1)	①	0	$-\frac{2}{23}$	$\frac{5}{23}$	0	$\frac{45}{23}$
(x_5)	1	0	0	0	⊝1	2
	0	0	$\frac{2}{23}$	$\frac{41}{23}$	0	$32\frac{1}{23}$

(6)

Next we restore the basis by pivoting on the circled entries to obtain

	x_1	x_2	x_3	x_4	x_5	
x_2	0	1	$\frac{5}{23}$	$-\frac{1}{23}$	0	$\frac{83}{23}$
x_1	1	0	$-\frac{2}{23}$	$\frac{5}{23}$	0	$\frac{45}{23}$
x_5	0	0	$\left(-\frac{2}{23}\right)$	$\frac{5}{23}$	1	$-\frac{1}{23}$
	0	0	$\frac{2}{23}$	$\frac{41}{23}$	0	$32\frac{1}{23}$

(7)

Applying the dual simplex method to (7), we obtain

	x_1	x_2	x_3	x_4	x_5	
x_2	0	1	0	$\frac{1}{2}$	$\frac{5}{2}$	$\frac{7}{2}$
x_1	1	0	0	0	-1	2
x_3	0	0	1	$-\frac{5}{2}$	$-\frac{23}{2}$	$\frac{1}{2}$
	0	0	0	2	1	32

$$(8)$$

and we see that

$$x_1 = 2, \qquad x_2 = 3.5; \qquad z = 32$$

In a similar manner we obtain the optimal solution of (5)

$$x_1 = 1, \qquad x_2 = 3.8; \qquad z = 24.2000$$

It is convenient to use a diagram to keep track of the branching process. Figure 1 displays the convention we will follow in making these diagrams. We call the boxes on the diagram *nodes*. Usually the nodes contain information about the solution, such as the value z of the objective function and the values of the integer variables that affect the branching process. We call such a diagram a *binary tree* (binary because there can be only two branches from a node). Two nodes obtained by branching are called the *children* of a node. We number the nodes as follows:

the node at the top is node 1,

the left child of node k is node $2k$, and

the right child of node k is node $2k + 1$

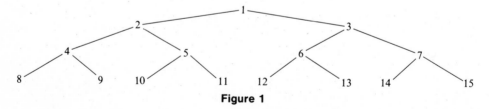

Figure 1

Figure 2 displays a diagram of the results obtained so far.

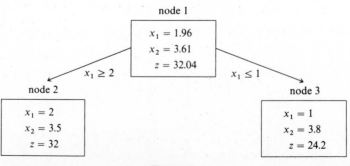

Figure 2

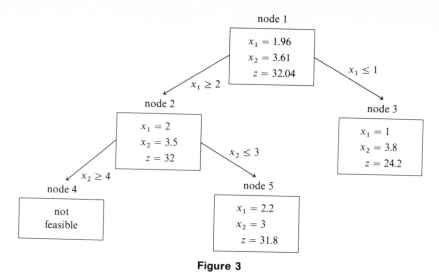

node 1

$x_1 = 1.96$
$x_2 = 3.61$
$z = 32.04$

$x_1 \geq 2$

$x_1 \leq 1$

node 2

$x_1 = 2$
$x_2 = 3.5$
$z = 32$

$x_2 \leq 3$

node 3

$x_1 = 1$
$x_2 = 3.8$
$z = 24.2$

$x_2 \geq 4$

node 4

not
feasible

node 5

$x_1 = 2.2$
$x_2 = 3$
$z = 31.8$

Figure 3

Since we have not found an integer solution at node 2 or node 3, we continue branching. We will select one node to work on first; and, if necessary, we will examine the other node later. We call nodes that must be examined further *dangling nodes*. As you can imagine, the number of dangling nodes can grow quite large if many branches are required to solve a problem. As a general rule, we select the dangling node with the largest z value in the hope that the maximum value of z will be found in this branch; however, there is no guarantee that this will occur. In Figure 2, node 2 has the largest z value. Since at node 2, x_1 is an integer, we concentrate on the variable x_2 and form two new problems by adding the new constraints $x_2 \geq 4$ and $x_2 \leq 3$ to the final tableau corresponding to node 2 in Figure 2. We obtain the solutions to these two problems by the dual simplex method. Figure 3 gives the solutions at nodes 4 and 5.

Since node 4 yields an infeasible problem, it is eliminated from further consideration (any branches from it would also be infeasible). Of the two dangling nodes, 3 and 5, we select node 5 because it has the larger z value (though the answer may not be along this branch, in which case we will have to return to examine node 3). Figure 4 shows the result of branching to node 10 with $x_1 \geq 3$ and to node 11 with $x_1 \leq 2$.

Node 10 yields an integer solution with $x_1 = 3$, $x_3 = 1$, and $z = 31$. The *bound* aspect of the method begins with the discovery of an integer solution. Usually we cannot conclude that the first integer solution discovered is the maximal solution; however, we can use it to eliminate dangling nodes that offer no hope for improvement. Thus, we can eliminate node 3 from our list of dangling nodes because its objective value of $z = 24.2$ is less than our known feasible solution $z = 31$. The objective values of nodes reached by branching from node 3 could not have a larger objective value than 24.2 because adding more constraints *reduces* the size of the feasible set and *cannot increase* the maximal objective value. We often say that a node has been *fathomed* when we can eliminate it (and all its descendants) from the search.

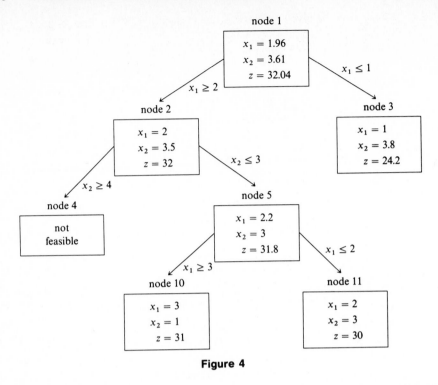

node 1
$x_1 = 1.96$
$x_2 = 3.61$
$z = 32.04$

$x_1 \geq 2$

$x_1 \leq 1$

node 2
$x_1 = 2$
$x_2 = 3.5$
$z = 32$

node 3
$x_1 = 1$
$x_2 = 3.8$
$z = 24.2$

$x_2 \geq 4$

$x_2 \leq 3$

node 4
not feasible

node 5
$x_1 = 2.2$
$x_2 = 3$
$z = 31.8$

$x_1 \geq 3$

$x_1 \leq 2$

node 10
$x_1 = 3$
$x_2 = 1$
$z = 31$

node 11
$x_1 = 2$
$x_2 = 3$
$z = 30$

Figure 4

Notice that we can also fathom node 11 because its objective value ($z = 30$) is less than 31. Thus, no dangling nodes remain, and node 10 has the maximal solution.

A Branch and Bound Algorithm

If an IP problem has a bounded feasible set, its optimal solution can be found using the best bound form of the branch and bound method. We describe the method in five steps.

STEP 1. *Initialization.* Ignore the integer constraints, and solve the problem by any appropriate method. We have so far developed three methods: the simplex method, the two-phase method, and the dual simplex method. (The revised simplex method could be used to implement either the simplex or the two-phase method.)

If the solution you find happens to satisfy the integer constraints, proceed no further; you have the answer. Otherwise, set $z^* = \infty$. (By ∞ we mean a value larger than anything we might compare it to.) We use z^* to keep track of the current best bound: the largest objective value we have discovered for a solution meeting all the integer constraints.

STEP 2. *Pick a dangling node K.* We maintain a list of dangling nodes that may lead to the maximal solution. In the beginning this list contains only node 1 (the solution to the given LP problem without considering integer constraints). We select the dangling node with largest objective value to examine next; however, you could select any dangling node.

STEP 3. *Branch at node K.* Select a variable x_i that is constrained to have an integer value but does not currently have one. Suppose the current value of x_i is w. Let j be the integer such that $j < w < j + 1$. Add the constraint $x_i \le j$ to the optimal tableau for node K, find the optimal solution to this problem by using the dual simplex method, and store the answer at node $2K$. Then add the constraint $x_i \ge j + 1$ to the optimal tableau for node K, find the optimal solution by the dual simplex method, and store the answer at node $2K + 1$.

STEP 4. *Examine nodes $2K$ and $2K + 1$.* For each of the two nodes, do the following:

(a) If all variables that are constrained to be integers have integer values, then compare the objective value to the current bound z^*. (The current bound is the largest objective value of all feasible integer solutions discovered so far.) If this is the first integer solution found, or if the value of this integer solution exceeds the bound, then change the bound to the new value.

Eliminate all dangling nodes whose objective value is less than the new bound.

Note that step (a) must be performed at both nodes $2K$ and $2K + 1$. In general, you cannot draw any valid conclusion without examining the values at both nodes.

(b) If some of the variables that are constrained to be integers are not integers, then compare the objective value for this node to z^*. If the value is greater than z^*, add this node to the list of dangling nodes. If the value is less than or equal to z^*, then fathom the node. Return to step 2.

STEP 5. *Termination.* The algorithm halts when the list of dangling nodes is empty. If integer solutions have been found, then the integer solution corresponding to the current bound is the maximal solution. If no integer solution has been found, the problem is infeasible.

It should be apparent that even for small problems, the calculations needed to find an optimal integer solution can be formidable; nevertheless, branch and bound is probably the best (certainly the most commonly used) method for solving pure and mixed integer problems. Branch and bound techniques can sometimes be developed to solve problems more efficiently by using information about the form of the constraints or the probable ranges of the integer variables. Section 8.3 presents a variation for solving pure binary problems that does not use the simplex method.

EXAMPLE 1

We apply the branch and bound method to the mixed integer problem

$$\text{Maximize:} \quad z = 2x_1 - x_2 + x_3$$

$$\text{Subject to:} \quad 3x_1 - x_2 + x_3 \leq 10$$

$$2x_1 - 3x_2 + 4x_3 \leq 9$$

$$x_i \geq 0, \quad 1 \leq i \leq 3; \quad x_1 \text{ and } x_3 \text{ integer}$$

(9)

Ignoring the integer constraints we obtain the first node

node 1

$x_1 = 3.10$
$x_2 = 0$
$x_3 = 0.70$
$z = 6.90$

Branching on x_3 we obtain nodes 2 and 3 in Figure 5.

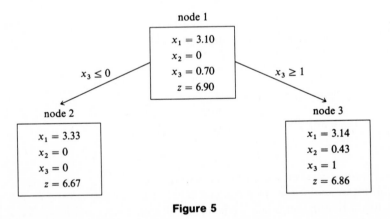

Figure 5

Since the value of the objective function is larger at node 3, we branch at this node. Since x_2 is not required to be integer, we branch on x_1 to obtain nodes 6 and 7 in Figure 6.

Note that at both nodes 6 and 7 we have obtained feasible solutions to (9) that satisfy the integer constraints on x_1 and x_3. Moreover, the objective value at node 6 is larger than the objective value at all other dangling nodes (nodes 2 and 7). Therefore, nodes 2 and 7 are fathomed and node 6 gives the maximal solution to (9).

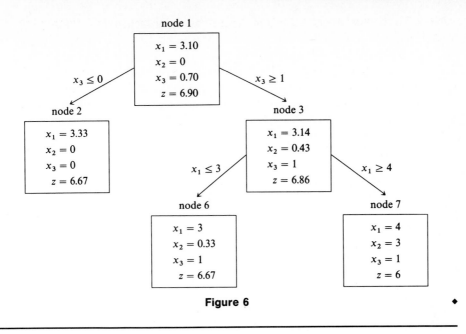

Figure 6   ◆

EXAMPLE 2

We apply the branch and bound method to the IP problem of Example 1 of Section 8.1. Branching on node 1 gives nodes 2 and 3 shown in Figure 7. Since node 3 is infeasible, we branch on node 2 to obtain nodes 4 and 5. Node 4 yields an integer solution; so we set the bound $z^* = 131$.

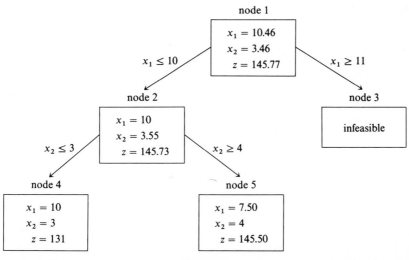

Figure 7

Node 5 is the only dangling node in Figure 7; so we branch on it to obtain nodes 10 and 11 as shown in Figure 8. Node 11 is fathomed because it is infeasible. Node 10 becomes a dangling node because its objective value of 145.5 exceeds $z^* = 131$. Therefore, we branch on node 10 to obtain nodes 20 and 21 as shown in Figure 8.

Node 20 gives an integer solution with objective value 143. This exceeds $z^* = 131$; so we set $z^* = 143$ and fathom node 4. Node 21 gives an integer solution with objective value 145. Since this exceeds $z^* = 143$, we set $z^* = 145$ and fathom node 20. Because no dangling nodes remain, the maximal solution occurs at node 21.

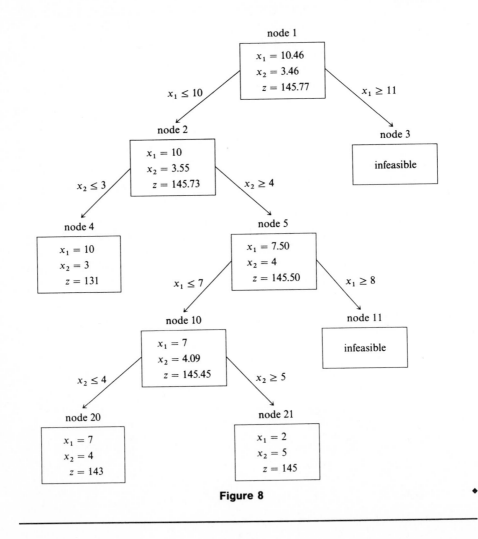

Figure 8

We conclude this section by noting that you must exercise caution when seeking an integer solution to an LP problem if some of the constraint coefficients are not integers. Consider, for instance, the constraint

$$2.3x_1 + 3.5x_2 \leq 6 \tag{10}$$

If you add a slack variable s_1 to (10), you obtain

$$2.3x_1 + 3.5x_2 + s_1 = 6 \tag{11}$$

If you were to require that x_1, x_2, and s_1 assume only integer values, then (11) would not be feasible for all integer assignments to x_1 and x_2; for instance, $(x_1, x_2, s_1) = (1, 1, 0.2)$ is a solution to (11), and constraint (11) could lead to an erroneous answer. Therefore, if you introduce a slack variable to a constraint in which some of the coefficients or the constant on the right-hand side are not integers, then you must not require that the slack variable be an integer.

As an alternative to introducing a noninteger slack variable, we can clear an inequality of fractional coefficients by multiplying through by a constant. Thus, we could multiply (10) by 10 to obtain

$$23x_1 + 35x_2 \leq 60$$

and then add a slack variable s_2 to obtain

$$23x_1 + 35x_2 + s_2 = 60 \tag{12}$$

There is no harm in requiring s_2 to be an integer in (12) since s_2 must be an integer in any solution of (12) having integer values of x_1 and x_2.

Exercises

In Exercises 1–4, describe the next step that you would take in performing the branch and bound method. Each of the problems has two decision variables.

1.

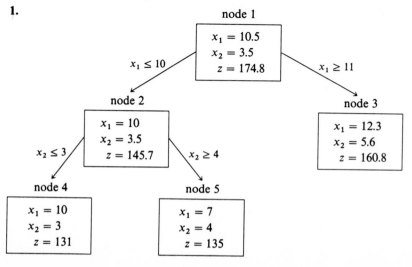

2.

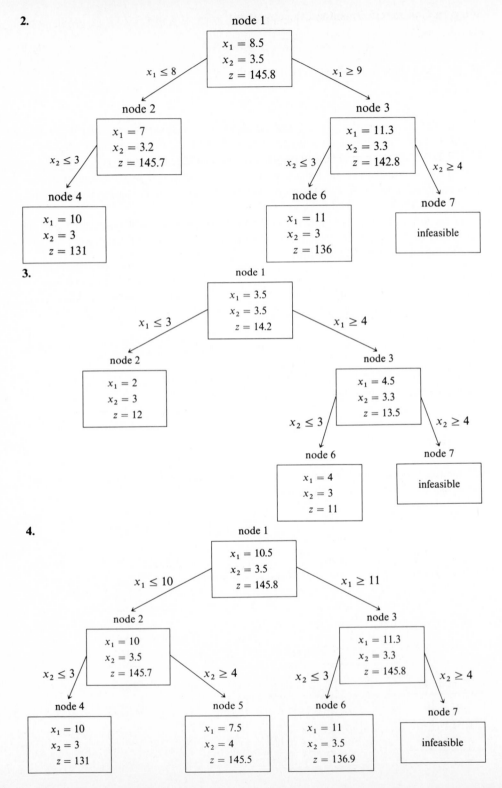

3.

4.

Solve Exercises 5–9 by the branch and bound method. You can solve these problems by making one branch.

5. Maximize: $z = 3x_1 + 2x_2$

Subject to: $x_1 + 2x_2 \le 4$
$2x_1 + x_2 \le 6$

$x_1, x_2 \ge 0$ and integer

6. Maximize: $z = 2x_1 + 3x_2$

Subject to: $x_1 - 3x_2 \ge 2$
$2x_1 + x_2 \le 6$

$x_1, x_2 \ge 0$ and integer

7. Maximize: $z = 2x_1 + 6x_2$

Subject to: $4x_1 - x_2 \ge 2$
$2x_1 + x_2 \le 6$

$x_1, x_2 \ge 0$ and integer

8. Maximize: $z = 5x_1 + 3x_2 + 3x_3$

Subject to: $3x_1 + x_2 + 2x_3 \le 12$
$x_1 + 2x_2 + x_3 \le 10$

$x_1, x_2, x_3 \ge 0$ and integer

9. A small (but successful) manufacturer produces two mechanical products A and B. Assembly of the nonmechanical components of A and B requires 1 and 6 hours, respectively, and assembly of the mechanical components requires 3 and 8 hours, respectively. The manufacturer has 36 hours of labor per day available for the nonmechanical assembly and 60 hours of labor per day available for the mechanical assembly. If the profit on each unit of A produced is $100 and for each unit of B is $400, how many units of each product should be manufactured on a daily basis to maximize profit?

Solve Exercises 10–12 by branch and bound. Since each problem requires two branches, you may want to use the branch and bound tutorial of TUTOR. (See Appendix A to learn how to use the branch and bound tutorial.)

10. Maximize: $z = 7x_1 + 6x_2$

Subject to: $x_1 + 3x_2 \le 6$
$2x_1 + x_2 \le 9$

$x_1, x_2 \ge 0$ and integer

11. Maximize: $z = 3x_1 + x_2 + 3x_3$

Subject to: $x_1 - x_2 + x_3 \le 4$
$2x_1 + 2x_2 + x_3 \le 12$

$x_1, x_2, x_3 \ge 0$ and integer

12. Maximize: $z = 2x_1 + 3x_2 - x_3$

Subject to: $2x_1 - x_2 - 2x_3 \le 12$
$x_1 + 3x_2 + x_3 \le 8$

$x_1, x_2, x_3 \ge 0$ and integer

Solve the mixed integer problems in Exercises 13–15 by branch and bound. They are not too long to be done by hand, but you may prefer to use the branch and bound tutorial or the automatic branch and bound method of SOLVER. (See Appendix A to learn how to use CALIPSO to solve these problems.)

13. Maximize: $z = 2x_1 + 8x_2 + 3x_3$

Subject to: $x_1 + x_2 - x_3 \le 5$
$-x_1 + 2x_2 + 3x_3 \le 8$

$x_1, x_2, x_3 \ge 0$; x_2 integer

14. Maximize: $z = 3x_1 + 2x_2 + 2x_3$

Subject to: $2x_1 + x_2 + 2x_3 \le 4$
$x_1 + 3x_2 - x_3 \le 6$

$x_1, x_2, x_3 \ge 0$; x_1 and x_3 integer

†15. Maximize: $z = 3x_1 - x_2 + 3x_3$

Subject to: $2x_1 + x_2 + 2x_3 \le 10$
$x_1 - x_2 + 2x_3 \le 4$

$x_1, x_2, x_3 \ge 0$; x_2 and x_3 integer

You can solve Exercises 16–20 with the branch and bound tutorial or the automatic solution programs of SOLVER. (To learn how to use these programs see Appendix A.)

†16. Solve Example 3 from Section 1.1 if x_1 and x_2 are assumed to be integer valued.

†17. Solve Exercise 8 from Section 1.1. Note that the slack variables cannot be expected to have integer values because some of the coefficients in the constraints are not integers.

†18. Suppose a plant has four machines, each of which can produce a product P. For each machine M_1, M_2, M_3, M_4, there is a fixed maintenance cost of \$40, \$35, \$37, and \$42, respectively, that applies whenever the machine is started on a working day. The machines M_1, M_2, M_3, M_4 can produce up to 100, 90, 92, 105 units per day of P, respectively. There is a cost of \$4 for materials for each unit produced by any machine. Which machines should the plant use to produce 275 units of product P in one day at minimum cost?

†19. Solve

$$\text{Maximize:} \quad z = 2x_1 + 3x_2 - 5x_3$$
$$\text{Subject to:} \quad x_1 + 3x_2 + 4x_3 \leq 12$$
$$x_1, x_2, x_3 \geq 0$$

and either

$$2x_1 + 4x_2 - \quad x_3 \leq 10$$

or

$$3x_1 - 2x_2 + 5x_3 \leq \quad 5$$

†20. Solve the following problems from the exercises of Section 8.1.
    (a) Exercise 4   (b) Exercise 10   (c) Exercise 12   (d) Exercise 16

You should solve Exercises 21–24 with the branch and bound program of SOLVER. They are too large to solve with the branch and bound tutorial.

†21. Solve the following problems from the exercises of Section 8.1.
    (a) Exercise 3   (b) Exercise 11   (c) Exercise 17

†22. Solve Example 6 from Section 8.1. (Note that this is a mixed integer problem since the slack variables in the first constraint need not be integer-valued.)

†23. An oil refinery has three cracking plants that can be used to meet demands for gasoline. An operating day is divided into two periods. All gasoline that is produced on a given day must be purchased by customers and hauled away in tankers. None can be stored for future sale. For today's operation, 2950 barrels are required in the first period and 4000 barrels in the second period. Table 1 gives the characteristics of the cracking plants.

Table 1

Plant number	Fixed startup cost (\$)	Fixed cost per period (\$)	Cost per barrel (\$)	Maximum production per period (barrels)
1	3000	700	5	2100
2	2400	800	4	1900
3	1800	900	6	3000

A cracking plant started in the first period can be used in the second period without incurring an additional start-up cost. Find a production schedule that minimizes cost. (This problem can be solved by the branch and bound facility of SOLVER.)

†**24.** Suppose the refinery described in Exercise 23 operates its cracking plants for three periods each day. The demands for today are 1500 barrels in period 1, 1800 barrels in period 2, and 3000 barrels in period 3. Find a minimum cost production schedule for the plant characteristics given in Table 2.

Table 2

Plant number	Fixed startup cost ($)	Fixed cost per period ($)	Cost per barrel ($)	Maximum production per period (barrels)
1	2500	700	14	2100
2	3000	800	6	1800
3	2000	1000	8	1500

†**25.** Solve Example 4 from Section 8.1.

*__26.__ Prove that the branch and bound method converges to a solution of an IP problem unless the original LP problem fails to have an optimal solution or the simplex method cycles at some node. Assume the feasible region is bounded.

8.3 IMPLICIT ENUMERATION: PART I

The branch and bound method described in Section 8.2 is a general method that we can apply to any LP problem for which some of the variables are constrained to have integer values. Essentially, it is an orderly way of enumerating (listing) the finite, but potentially large, number of values that can be assigned to certain selected variables (the integer variables). We are able to avoid considering all of the possible values the integer variables can assume by establishing a bound whenever a feasible solution to the problem is discovered. But the generality of the branch and bound method becomes a weakness when it is applied to some pure integer problems, especially those in which each integer variable is constrained to assume only a few values. As you will see in this and the next section, it is not necessary to perform the simplex method at every step of the branch and bound method when *all* of the variables are constrained to have integer values.

We begin this section with *binary* (also called 0–1) problems in which the decision variables are allowed to assume only the values 0 or 1. In Section 8.1 we considered several important problems that involve binary variables. Section 8.4 will show how to formulate any pure integer problem as a binary problem.

To begin our discussion of implicit enumeration we note that for a binary problem in n variables, there are 2^n solutions. This follows from the observation that since there are two possibilities (0 and 1) for each of the n variables, there are

$$2 \cdot 2 \cdot \cdots \cdot 2 = 2^n$$

n-tuples consisting of 0's and 1's, each of which corresponds to a possible solution of the problem. In theory, we could examine each of these 2^n n-tuples to find the optimal feasible solution. However, because for even modest values of n, 2^n is such a large number that checking all possibilities might be a life's work, even for a computer (for instance, $2^{30} = 1,073,741,824$).

The basic idea behind implicit enumeration is to systematically eliminate blocks of the 2^n possible solutions from further consideration. We do this (without resorting to the simplex method) by testing candidate solutions for feasibility and establishing bounds to eliminate useless branching.

To describe our implicit enumeration method we will need the following concepts.

Definition.   Suppose that $0 \leq c_1 \leq c_2 \leq \cdots \leq c_n$. The problem

$$\text{Minimize:}\quad z = c_1 x_1 + c_2 x_2 + \cdots + c_n x_n$$

$$\text{Subject to:}\quad
\begin{aligned}
a_{11}x_1 + a_{12}x_2 + \cdots + a_{1n}x_n &\geq b_1 \\
a_{12}x_1 + a_{22}x_2 + \cdots + a_{2n}x_n &\geq b_2 \\
&\vdots \qquad\qquad\vdots \\
a_{m1}x_1 + a_{m2}x_2 + \cdots + a_{mn}x_n &\geq b_m
\end{aligned}
\tag{1}$$

$$x_i \quad \text{binary,} \quad 1 \leq i \leq n$$

is a *binary problem in standard form.* (Note that the b_i are *not* required to be nonnegative, and the constants are not required to be integers.)

Example 1 introduces a binary problem in standard form for use throughout this section.

EXAMPLE 1

$$\text{Minimize:}\quad z = 2x_1 + 3x_2 + 5x_3 + 5x_4 + 8x_5 + 10x_6$$

$$\text{Subject to:}\quad
\begin{aligned}
-2x_1 - 5x_2 - 3x_3 + 7x_4 + 3x_5 - x_6 &\geq 2 \\
x_1 + x_2 - 5x_3 + 3x_4 - 5x_5 - 2x_6 &\geq -3 \\
3x_1 - 7x_2 - x_3 + 2x_4 + x_5 + x_6 &\geq 1
\end{aligned}
\tag{2}$$

$$x_i \quad \text{binary,} \quad 1 \leq i \leq 6 \qquad\qquad \blacklozenge$$

In a standard-form problem we require that the cost coefficients be in ascending order because the method we develop proceeds by trying to set variables equal to 1 in the order $x_1, x_2, \ldots, x_n$. Thus, we expect to have a better chance of finding the minimum value of the objective function by trying to set variables with smaller cost coefficients to 1 before trying to set variables with larger cost coefficients to 1. This is called a "greedy strategy" because it repeatedly selects from alternative choices the one that gives the best immediate progress toward the goal. Unfortunately, this greedy strategy does not always work better than some other strategy; though a greedy approach usually works better than a random guess.

Definition.   For an integer k $(1 \leq k \leq n)$, a *partial solution of* (1) *with k elements* is a k-tuple $P = [p_1, p_2, \ldots, p_k]$ where each p_i is 0 or 1.

If P is a partial solution with k elements and if

$$\mathbf{x} = [x_1, x_2, \ldots, x_n] = [p_1, p_2, \ldots, p_k, 0, 0, \ldots, 0]$$

is a feasible solution of (1), then P is called a *feasible partial solution of* (1) *with k elements*. We also sometimes describe this situation by saying that P can be *trivially extended* by adding zeros to become a feasible solution.

If

$$\mathbf{x} = [x_1, x_2, \ldots, x_n] = [p_1, p_2, \ldots, p_k, p_{k+1}, \ldots, p_n]$$

is not a feasible solution for any assignment of binary values to $p_{k+1}, \ldots, p_n$, then P is called an *infeasible partial solution of* (1) *with k elements*. We also describe this situation by saying P cannot be *extended* to a feasible solution.

If

$$P = [p_1, p_2, \ldots, p_k]$$

is a partial solution of (1) with k elements, then an extension

$$P' = [p_1, p_2, \ldots, p_k, p_{k+1}, \ldots, p_n]$$

of P that is a feasible solution of (1) is called a *feasible extension of P*.

EXAMPLE 2

We illustrate the various kinds of partial solutions for the problem given in Example 1.

The 3-tuple $[1, 0, 0]$ is a partial solution of (2) with 3 elements. It is neither a feasible nor an infeasible partial solution. It is not a feasible partial solution because its extension by zeros to $[1, 0, 0, 0, 0, 0]$ is not feasible. It is not an infeasible partial solution because some of its extensions (for instance, $[1, 0, 0, 1, 0, 0]$) are feasible.

The 4-tuple $[1, 0, 1, 1]$ is a feasible partial solution of (2) because its extension by zeros yields the feasible solution $[1, 0, 1, 1, 0, 0]$.

The 3-tuple $[1, 1, 1]$ is an infeasible partial solution of (2) because there is no way to extend it (with 0's or 1's) to a feasible solution. ◆

Definition. For a given partial solution $P = [p_1, p_2, \ldots, p_k]$, the *value of the objective function at P* is

$$z_P = c_1 p_1 + c_2 p_2 + \cdots + c_k p_k$$

EXAMPLE 3

For the problem of Example 1 and the partial solution $P = [1, 0, 1, 1]$, $z_P = 2 \cdot 1 + 3 \cdot 0 + 5 \cdot 1 + 5 \cdot 1 = 12$. ◆

Definition.   A partial solution $P = [p_1, p_2, \ldots, p_k]$ is *fathomed* if it is not possible to find the minimal solution to (1) by extending P to a feasible solution

$$\mathbf{x} = [p_1, p_2, \ldots, p_k, p_{k+1}, \ldots, p_n]$$

with at least one of $p_{k+1}, \ldots, p_n$ having value 1. We identify the following three *types* of fathoming:

EOV   Excessive objective value. $z_P \geq z^*$ where z^* is known to be a feasible solution to (1). (In this case, there is nothing to be gained by adding more elements to P or even by checking to see if there are feasible completions of P.)

FPS   Feasible partial solution. By adding 0's we can extend P to a feasible solution of (1). (In this case, no nonzero extension P' of P could yield a smaller objective function value than P because the cost coefficients are all nonnegative.)

IPS   Infeasible partial solution. P is not a feasible partial solution, and it is not possible to extend P to a feasible solution of (1).

We describe our algorithm for implicit enumeration in terms of its moving from one partial solution to another. When the problem is not too large, it is often helpful to follow the progress of the algorithm by drawing a graph connecting the partial solutions. Figure 1 is a graph of all of the partial solutions for a problem with 3 variables.

An implicit enumeration algorithm provides a way of moving between partial solutions on a graph such as the one in Figure 1. The objective is to reach a minimal solution by examining as few partial solutions as possible. The following definition gives the two ways we will move between partial solutions, depending on the results of fathoming tests.

Definition.   Let P be a partial solution of (1) having k entries, and suppose P has been fathomed by one of the fathoming tests: EOV, FPS, or IPS. The *sequel* P' to P is

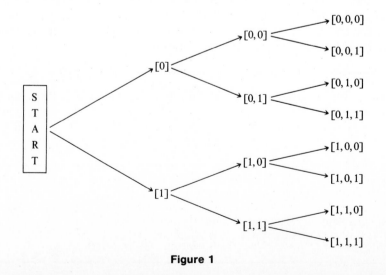

Figure 1

defined as follows.

1. If P contains any entries that are 1, then

$$P' = [p_1, p_2, \ldots, 1 - p_t]$$

where t is the largest subscript i in P for which $p_i = 1$.

2. If all the entries p_i of P are 0, and if $k < n$, then P' is the partial solution that sets the first $k + 1$ variables equal to 0.

3. If all the entries p_i of P are 0, and $k = n$, then there is no sequel to P (we will have reached a termination point of the algorithm).

In what follows, if a partial solution P has not been fathomed and if there are feasible extensions of P that must be examined, then we form the *adjunct P''* of P by adding a 1 to the end of P.

EXAMPLE 4

We give examples of partial solutions P and their sequels P' and their adjuncts P'' for the case $n = 7$.

P	P' (sequel to P)	P'' (adjunct to P)
$[1,1,0,0,1]$	$[1,1,0,0,0]$	$[1,1,0,0,1,1]$
$[1,1,0,0,0]$	$[1,0]$	$[1,1,0,0,0,1]$
$[1,1,1,1,1]$	$[1,1,1,1,0]$	$[1,1,1,1,1,1]$
$[0,0,0,0,0]$	$[0,0,0,0,0,0]$	$[0,0,0,0,0,1]$
$[0]$	$[0,0]$	$[0,1]$
$[1]$	$[0]$	$[1,1]$
$[0,0,0,0,0,0,0]$	none	none (since $n = 7$)

You should satisfy yourself that you can traverse (follow a path through all the nodes of) the graph in Figure 1 by performing a sequence of adjuncts and sequels. Here is the way to begin: From $[1]$, form the adjunct to get to $[1,1]$, and then form another adjunct to get to $[1,1,1]$. Next, form the sequel of $[1,1,1]$ to get to $[1,1,0]$. Continue in this manner until you have reached every partial solution in Figure 1.

To illustrate the essential idea behind implicit enumeration methods, we now present a very simple algorithm. There are a number of variations of this algorithm, which are intended to improve the average efficiency of solution (see Section 8.4 for some possibilities).

An Implicit Enumeration Algorithm for Solving (1)

Figure 2 shows this algorithm in the form of a flowchart. Details concerning the numbered boxes are described in the following numbered paragraphs. The variable k contains the length of the current partial solution.

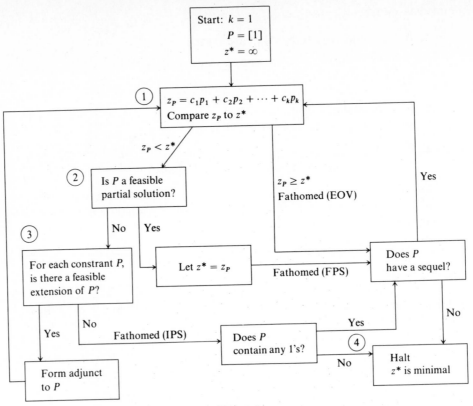

Figure 2

1. We calculate z_P from the current partial solution P every time this box is entered from one of the three incoming arrows; z^* is the current bound (i.e., the smallest value of z_P for all feasible partial solutions previously discovered). If z_P is smaller than z^*, we must examine P to see if it is feasible or has feasible extensions that might lead to a smaller value of z^*. If $z_P \geq z^*$, there is nothing to be gained by testing P for feasibility or attempting to extend it; so we fathom P and form its sequel.

2. If we reach box 2 and P is feasible, then we have found a better solution than z^*. Therefore, we change the value of z^* to z_P and form the sequel to P to continue the enumeration.

    If you are interested in writing a computer program to perform this algorithm, it is important to consider the most efficient way of testing a partial solution for feasible extensions because this sort of calculation is the most costly part of this algorithm. We discuss this calculation in detail in Exercise 9.

3. If we were unable to fathom P, we must decide whether or not to search for feasible extensions of P that might yield an improvement in z^*. This is a difficult point in the algorithm, and many strategies have been developed to make this decision.

    It is not practical to try to determine if there are extensions of P that *simultaneously* satisfy all of the constraints. For instance, one way to do that would be to produce *all* of the extensions of P that satisfy the first constraint and then eliminate those extensions that do not satisfy at least one of the remaining constraints. A little consideration should convince you that this could be a very time-consuming task when the problem has many variables. It is also not a very rewarding way to spend solution time because there is no guarantee that the feasible solutions you find will result in an improvement to z^*.

We adopt a cruder but much faster approach. We will test each of the constraints to see if there is any possible extension of P that could satisfy the constraint. If for *every* constraint there is *some* extension of P satisfying the constraint, we will arbitrarily decide to extend the current partial solution P with a 1 in the next place and go back to step 1 of the flowchart. This procedure cannot miss any feasible solutions, though it might cause you to test more partial solutions than would other strategies.

4. The algorithm always terminates with a partial solution that has been fathomed and has *only* zero entries. If this solution was fathomed by IPS, then it is pointless to extend P by adding a 1 or 0 because the resulting partial solutions are already known to be infeasible. If a partial solution that sets all decision variables to zero was fathomed by FPS or EOV, then the algorithm should terminate since all variables have been examined.

EXAMPLE 5

We perform the algorithm for the problem from Example 1.

P formed as	P	z_P	z^*	Fathomed?	Reason
Initial	[1]	2	∞	no	
Adjunct	[1,1]	5	∞	yes	IPS (constraint 3)
Sequel	[1,0]	2	∞	no	
Adjunct	[1,0,1]	7	∞	no	
Adjunct	[1,0,1,1]	12	12	yes	FPS
Sequel	[1,0,1,0]	7	12	yes	IPS (constraint 1)
Sequel	[1,0,0]	2	12	no	
Adjunct	[1,0,0,1]	7	7	yes	FPS
Sequel	[1,0,0,0]	2	7	yes	IPS (constraint 1)
Sequel	[0]	0	7	no	
Adjunct	[0,1]	3	7	yes	IPS (constraint 3)
Sequel	[0,0]	0	7	no	
Adjunct	[0,0,1]	5	7	no	
Adjunct	[0,0,1,1]	10	7	yes	EOV
Sequel	[0,0,1,0]	0	7	yes	IPS (constraint 1)
Sequel	[0,0,0]	0	7	no	
Adjunct	[0,0,0,1]	5	5	yes	FPS
Sequel	[0,0,0,0]	0	5	no	
Adjunct	[0,0,0,0,1]	8	5	yes	EOV
Sequel	[0,0,0,0,0]	0	5	yes	IPS (constraint 1)

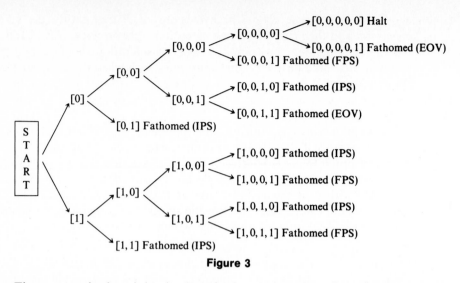

Figure 3

Thus, we attain the minimal value of 5 for z by setting $x_4 = 1$ and the rest of the variables to 0. Figure 3 illustrates the progress of the algorithm with a tree diagram.

♦

We cannot claim that implicit enumeration is always a more efficient method for solving binary problems than the branch and bound method given in Section 8.2. Some problems can be solved more quickly by implicit enumeration and others by branch and bound. There is no way to tell in advance which method is faster for a given problem. You should take care in comparing the time complexity of a problem solved by branch and bound and the same problem solved by implicit enumeration. Each branch of the branch and bound method requires the solution of two problems by the simplex method, whereas a branch of the implicit enumeration problem requires only the verification of feasibility and the evaluation of the objective function. Moreover, a distinct advantage of implicit enumeration is that it virtually eliminates round-off errors that arise in applying the simplex method to large problems.

In Section 8.4 we see how to convert general integer problems to binary problems in standard form; we also discuss briefly some more-sophisticated fathoming tests.

Exercises

Solve Exercises 1–4 by the implicit enumeration algorithm developed in this section.

1. Minimize: $z = x_1 + 3x_2 + 4x_3$

   Subject to: $3x_1 + x_2 - x_3 \geq 4$

   $2x_1 - x_2 - 4x_3 \geq -2$

   x_1, x_2 binary

2. Minimize: $\quad z = x_1 + x_2 + x_3 + 2x_4 + 3x_5$

Subject to: $\quad -2x_1 - x_2 + 2x_3 - 3x_4 + x_5 \geq 2$
$\qquad\qquad -x_1 - 3x_2 \qquad\qquad + 4x_5 \geq -2$

$\qquad\quad x_i \quad$ binary, $\quad 1 \leq i \leq 5$

3. Minimize: $\quad z = x_1 + x_2 + 2x_3 + 3x_4$

Subject to: $\quad 4x_1 - 5x_2 - 3x_3 + 2x_4 \geq 4$
$\qquad\qquad -2x_1 - x_2 + 3x_3 + 2x_4 \geq 0$
$\qquad\qquad 2x_1 + 4x_2 - 4x_3 + 3x_4 \geq 1$

$\qquad\quad x_i \quad$ binary, $\quad 1 \leq i \leq 4$

4. Minimize: $\quad z = x_1 + x_2 + 3x_3 + 6x_4$

Subject to: $\quad 3x_1 - x_2 + 2x_3 - 4x_4 \geq 0$
$\qquad\qquad 3x_1 - 2x_2 - 4x_3 - 2x_4 \geq -1$
$\qquad\qquad 6x_1 - 4x_2 - 7x_3 - 5x_4 \geq 5$

$\qquad\quad x_i \quad$ binary, $\quad 1 \leq i \leq 4$

You can solve the following problems with the implicit enumeration procedure of TUTOR. See Appendix A for instructions on using this tutorial. Note that you can enter problems for solution by the implicit enumeration program with the tableau editor of TUTOR or with the MODELER program. If you enter a problem with the editor of TUTOR, do not include any slack variables; the implicit enumeration tutorial assumes that all inequalities are of the $\geq$ type. If you use the MODELER program, be sure to choose the option for making a file for the implicit enumeration program. Do not insert constraints making the binary variables ≤ 1; the implicit enumeration algorithm assigns only 0 or 1 values to variables. Also note that the implicit enumeration tutorial gives you the option of selecting the partial solution yourself or of having the program select the partial solutions automatically according to the algorithm we gave in Figure 2.

Exercises 5–8 are of a class called *set covering* problems because they require dividing a set into disjoint subsets (subsets whose intersection is empty) to optimize some property. They can be formulated as binary problems and solved using the implicit enumeration algorithm we developed in this section.

†**5.** Shoshone County is planning to build fire stations in some of its cities to provide emergency service to all of its six cities. The county wants to build the stations as cheaply

Table 1

From	To (in minutes)					
	City 1	City 2	City 3	City 4	City 5	City 6
City 1	0	21	40	59	61	42
City 2	20	0	48	65	41	20
City 3	41	48	0	28	58	39
City 4	55	72	30	0	27	48
City 5	58	39	61	30	0	28
City 6	41	22	40	50	25	0
Cost (thousand \$)	36	42	45	48	54	60

as possible to ensure that a fire in any city can be reached from a fire station in no more than 30 minutes. The county commissioners have measured the travel time between all six cities and estimated the cost (in thousands of dollars) of building a station in each city. Table 1 gives their data. The last row gives the estimated cost of building each station. Determine the locations of the stations they should build.

†6. The National Business Machine Company (NBM) is planning to locate no more than six service centers to provide repair service to eight cities. They want to locate these service centers so that at least two centers will be within 50 miles of each city they serve. Table 2 gives the mileage between the cities. The last row of Table 2 gives the estimated cost (in thousands of dollars) of locating a service center in each of cities $C_1, C_2, C_3, C_4, C_5,$ and C_6. They have determined that sites in cities C_7 and C_8 are impractical. Find the locations that minimize their cost of building service centers.

Table 2

From	To (in miles)							
	C_1	C_2	C_3	C_4	C_5	C_6	C_7	C_8
C_1	0	22	45	54	59	21	63	28
C_2	22	0	27	49	65	43	27	42
C_3	45	27	0	65	28	35	27	81
C_4	54	49	65	0	43	42	57	72
C_5	59	65	28	28	0	54	99	36
C_6	21	43	35	35	54	0	12	87
Cost (thousand $)	27	35	42	45	47	48	51	54

†7. Irate citizens have demanded that the Moscow City Council undertake eight road repair projects $(P_1, P_2, \ldots, P_8)$ this year. The Council has privately agreed that it is politically expedient to do four projects. They have also determined that project P_1 can only be done if P_4 is also done, that P_2 cannot be done if both P_4 and P_5 are done, and that P_3 cannot be done if either P_4 or P_8 is done. Table 3 gives the costs of the projects (in thousands of dollars). Which projects should the council schedule in order to complete four of them at the least cost?

Table 3

Project	P_1	P_2	P_3	P_4	P_5	P_6	P_7	P_8
Cost (thousand $)	12	14	15	20	21	22	25	27

†8. The Friends of the Mallard–Larkins Wilderness (FMLW) wants to mail a mass solicitation of funds for its project. FMLW has identified six occupational groups that it would like to reach and has found eight companies that are willing to sell it mailing lists. Each of these eight mailing lists contains names from some of the groups it wants to reach. Table 4 contains a *Y* in the row for a mailing list when it contains names from the group it wants to reach and an *N* when it does not. The cost of each mailing list (in dollars) is given in the right-most column. Since FMLW considers it too expensive to attempt to purge duplicate names from the mailing lists used, it will suffer the expense and other consequences of mailing more than one solicitation to an individual. Determine

which mailing lists FMLW should buy to minimize its cost and reach all of the targeted occupational groups.

Table 4

List	Teachers	Doctors	Lawyers	Students	Hunters	Anglers	Cost ($)
L_1	N	Y	Y	N	N	N	450
L_2	N	N	Y	N	Y	Y	495
L_3	N	N	N	N	Y	Y	510
L_4	N	N	Y	N	N	Y	525
L_5	N	N	N	Y	N	N	540
L_6	N	Y	Y	Y	Y	N	550
L_7	Y	N	Y	N	Y	N	600
L_8	Y	Y	N	Y	Y	N	650

Since solving this problem requires an impressively large number of branches, you may want to perform a few branches yourself and then invoke the automatic solution procedure of the tutorial to have the problem completed automatically.

9. If you are interested in writing a computer program to perform our implicit enumeration algorithm, you should consider in some detail the arithmetic required to test for a feasible extension to one constraint since this calculation is the most expensive (in computer time) part of the algorithm. Because addition takes considerably less computer time than multiplication, you should avoid unnecessary multiplications.

Suppose that the current partial solution P has k entries. The ith constraint of problem (1) is

$$a_{i1}x_1 + a_{i2}x_2 + \cdots + a_{ik}x_k + a_{i, k+1}x_{k+1} + \cdots + a_{in}x_n \geq b_i$$

(a) Describe a way to determine if P is feasible for this constraint without using multiplication.

(b) Describe a way to tell (without using multiplication) if P has any extensions that satisfy this constraint.

†10. Solve Example 5 with the implicit enumeration tutorial.

8.4 IMPLICIT ENUMERATION: PART II

We continue the discussion of implicit enumeration by showing how to convert a binary problem to standard form and how to convert any integer problem to a binary problem. We also briefly consider the use of "surrogate constraints," one of the more promising improvements to the basic implicit enumeration algorithm that we developed.

We can solve any binary IP by transforming it into a standard form problem using the following five steps. We can derive the solution to the original problem from the solution to the standard problem by reversing these steps, as we will demonstrate in Example 1.

1. If the problem is a maximization problem, change the sign of all coefficients in the objective function and treat it as a minimization problem.

2. For each cost coefficient c_i that is negative, replace its corresponding variable x_i with $1 - u_i$.

3. Renumber (if necessary) the cost coefficients c_i and their corresponding variables x_i (or u_i) so that

$$0 \le c_1 \le c_2 \cdots \le c_n$$

4. Replace any $\le$ inequalities with $\ge$ inequalities by multiplying both sides of the $\le$ inequality by -1 (it does not matter if negative numbers occur on the right-hand sides).

5. Replace any constraint equalities with two equivalent $\ge$ inequalities.

EXAMPLE 1

Suppose we are given the binary problem

$$\text{Maximize:} \quad z = 3x_1 - 4x_2 + x_3 - 2x_4$$

$$\text{Subject to:} \quad \begin{aligned} x_1 + 2x_2 + 3x_3 - \quad x_4 &\le 2 \\ 2x_1 + 3x_2 - \quad x_3 + 2x_4 &\ge 7 \\ 5x_1 - 2x_2 + \quad x_3 - \quad x_4 &= 2 \end{aligned} \tag{1}$$

$$x_i \quad \text{binary}, \quad 1 \le i \le 4$$

The objective function of the corresponding minimization problem is

$$z' = -3x_1 + 4x_2 - x_3 + 2x_4 \tag{2}$$

To eliminate the negative cost coefficients in (2) we make the changes in variables

$$x_1 = 1 - u_1 \quad \text{and} \quad x_3 = 1 - u_3 \tag{3}$$

These changes give us the problem

$$\text{Minimize:} \quad \begin{aligned} z' &= -3(1 - u_1) + 4x_2 - 1(1 - u_3) + 2x_4 \\ &= 3u_1 + 4x_2 + u_3 + 2x_4 - 4 \end{aligned}$$

$$\text{Subject to:} \quad \begin{aligned} (1 - u_1) + 2x_2 + 3(1 - u_3) - \quad x_4 &\le 2 \\ 2(1 - u_1) + 3x_2 - \quad (1 - u_3) + 2x_4 &\ge 7 \\ 5(1 - u_1) - 2x_2 + \quad (1 - u_3) - \quad x_4 &= 2 \end{aligned} \tag{4}$$

$$u_1, x_2, u_3, x_4 \quad \text{binary}$$

We now put the cost coefficients of (4) in ascending order and rename the variables as

$$y_1 = u_3; \quad y_2 = x_4; \quad y_3 = u_1; \quad \text{and} \quad y_4 = x_2 \tag{5}$$

Thus, problem (1) becomes

$$\text{Minimize:} \quad z' = y_1 + 2y_2 + 3y_3 + 4y_4 - 4$$

$$\text{Subject to:} \quad (1 - y_3) + 2y_4 + 3(1 - y_1) - \quad y_2 \leq 2$$
$$2(1 - y_3) + 3y_4 - \quad (1 - y_1) + 2y_2 \geq 7$$
$$5(1 - y_3) - 2y_4 + \quad (1 - y_1) - \quad y_2 = 2$$
$$y_i \quad \text{binary,} \quad 1 \leq i \leq 4$$

By rearranging the order of variables in the constraints and moving constants to the right-hand sides, we obtain

$$\text{Minimize:} \quad z' = y_1 + 2y_2 + 3y_3 + 4y_4 - 4$$

$$\text{Subject to:} \quad -3y_1 - \quad y_2 - \quad y_3 + 2y_4 \leq -2$$
$$y_1 + 2y_2 - 2y_3 + 3y_4 \geq \quad 6 \qquad (6)$$
$$-y_1 - \quad y_2 - 5y_3 - 2y_4 = -4$$
$$y_i \quad \text{binary,} \quad 1 \leq i \leq 4$$

Finally, we multiply the first inequality of (6) by -1 and replace the fourth inequality by two $\geq$ inequalities to obtain the standard form binary problem

$$\text{Minimize:} \quad z' = y_1 + 2y_2 + 3y_3 + 4y_4 - 4$$

$$\text{Subject to:} \quad 3y_1 + \quad y_2 + \quad y_3 - 2y_4 \geq \quad 2$$
$$y_1 + 2y_2 - 2y_3 + 3y_4 \geq \quad 6$$
$$-y_1 - \quad y_2 - 5y_3 - 2y_4 \geq -4 \qquad (7)$$
$$y_1 + \quad y_2 + 5y_3 + 2y_4 \geq \quad 4$$
$$y_i \quad \text{binary,} \quad 1 \leq i \leq 4$$

Table 1 gives the solution to problem (7) found by using the implicit enumeration algorithm of Section 8.3.

Table 1

P formed as	P	z_P	z^*	Fathomed?	Reason
Initial	[1]	-3	∞	no	
Adjunct	[1, 1]	-1	∞	no	
Adjunct	[1, 1, 1]	2	∞	yes	IPS (Constraint 2)
Sequel	[1, 1, 0]	-1	∞	no	
Adjunct	[1, 1, 0, 1]	3	3	yes	FPS
Sequel	[1, 1, 0, 0]	-1	3	yes	IPS (Constraint 2)
Sequel	[1, 0]	-3	3	yes	IPS (Constraint 2)
Sequel	[0]	-4	3	yes	IPS (Constraint 2)
Sequel	[0, 0]	-4	3	yes	IPS (Constraint 2)

We see that the answer is $z^* = 3$ at $y_1 = 1, y_2 = 1, y_3 = 0, y_4 = 1$. By using (3) and (5) we find the solution of (1) to be

$$x_1 = 1, \qquad x_2 = 1, \qquad x_3 = 0, \qquad x_4 = 1, \qquad \text{and} \qquad z^* = -3 \qquad \blacklozenge$$

We can transform any pure integer problem that has a bounded feasible set into a binary problem; however, the process may introduce a great many variables. For instance, suppose x is an integer variable and we know that $x < 2^{n+1}$, for some integer $n \geq 0$. Then we can write x (uniquely, in fact) as

$$x = u_0 + 2u_1 + 2^2 u_2 + \cdots + 2^n u_n,$$

where u_i is 0 or 1, $1 \leq i \leq n$.

EXAMPLE 2

Consider the problem

$$\text{Minimize:} \quad z = -3x_1 + 2x_2$$
$$\text{Subject to:} \quad 2x_1 - x_2 \geq -3$$
$$3x_1 + x_2 \geq \quad 4 \tag{8}$$
$$x_1, x_2 \quad \text{integer} \quad \text{and} \quad 0 \leq x_1 \leq 2, \qquad 0 \leq x_2 \leq 5$$

We can express x_1 as a pair of binary digits (u_0 and u_1) and x_2 as a triple of binary digits ($t_0, t_1,$ and t_2) by

$$x_1 = u_0 + 2u_1$$
$$x_2 = t_0 + 2t_1 + 4t_2 \tag{9}$$
$$u_0, u_1, t_0, t_1, t_2 \quad \text{binary}$$

To solve problem (8), we solve the binary problem

$$\text{Minimize:} \quad z = -3u_0 - 6u_1 + 2t_0 + 4t_1 + 8t_2$$
$$\text{Subject to:} \quad 2u_0 + 4u_1 - t_0 - 2t_1 - 4t_2 \geq -3$$
$$3u_0 + 6u_1 + t_0 + 2t_1 + 4t_2 \geq \quad 4$$
$$u_0, u_1, t_0, t_1, t_2 \quad \text{binary}$$

and then find the solution to (8) by using equations (9). $\qquad \blacklozenge$

There are many ways of improving the performance of implicit enumeration algorithms on some problems or classes of problems. The general idea is to seek more-sophisticated fathoming tests that shorten the search. We will briefly discuss the use of "surrogate constraints," which seem to offer the best chance of reducing the average solution time for a broad class of binary programming problems.

A *surrogate constraint* is a constraint that is added to the original set of constraints and that does not exclude any of the original feasible solutions. A simple way of forming a surrogate constraint is to add nonnegative multiples of other constraints. Example 3 illustrates this technique.

EXAMPLE 3

The solution of the LP problem

$$\text{Minimize:} \quad z = x_1 + x_2 + x_3 + x_4 + x_5$$

$$\begin{aligned}
\text{Subject to:} \quad & 3x_1 + 2x_2 + x_3 + 2x_4 + x_5 \geq 4 \\
& -2x_1 + 6x_2 + 5x_3 - 2x_4 - 4x_5 \geq 2 \\
& 4x_1 - 4x_2 - 14x_3 + 2x_4 + 3x_5 \geq 4
\end{aligned} \tag{10}$$

$$x_i \quad \text{binary,} \quad 1 \leq i \leq 5$$

requires many iterations of the implicit enumeration method because each constraint can be satisfied in several ways. If we add the three constraints of this problem, we obtain the constraint

$$5x_1 + 4x_2 - 8x_3 + 2x_4 \geq 10 \tag{11}$$

Obviously, the only solution to (11) is $x_1 = 1$, $x_2 = 1$, $x_3 = 0$, and $x_4 = 1$. Thus, constraint (11) is an unusually strong surrogate constraint because it rejects every solution except the optimal one. If we place constraint (11) before the three constraints of (10), the implicit enumeration method only needs to evaluate (11) to reject every partial solution except the optimal one. ◆

Devising surrogate constraints that fathom partial solutions at early stages of the enumeration algorithm is an interesting and challenging research area. Unfortunately, further discussion of these ideas is beyond the scope of this book. If you are interested in reading some of the clever methods that have been developed, you could start by reading [13].

The next two examples show how a problem with a simple statement can lead to an intractably large binary problem. It also introduces a class of problems of significant practical application for which researchers have been unable to discover a method of solution that does not require exhorbitant computation time.

EXAMPLE 4

(Traveling Salesman Problem). In this classic problem a salesman plans to visit six cities. He will start at City 1, visit each of the other cities exactly once, and then return to City 1. He wants a tour that minimizes the miles traveled. Table 2 gives the mileages between the various cities.

Table 2

City	City					
	1	2	3	4	5	6
1	—	40	70	25	60	45
2	40	—	30	15	50	25
3	70	30	—	80	20	30
4	25	15	80	—	10	42
5	60	50	20	10	—	25
6	45	25	30	42	25	—

As a first attempt to formulate this problem as a binary IP problem we define, for $1 \le i \le 6$, $1 \le j \le 6$,

$$x_{ij} = \begin{cases} 1 & \text{if the salesman travels from City } i \text{ to City } j \\ 0 & \text{if the salesman does not travel from City } i \text{ to City } j \end{cases}$$

From Table 2 we see that we want to minimize the objective function

$$
\begin{aligned}
z = \quad & 40x_{12} + 70x_{13} + 25x_{14} + 60x_{15} + 45x_{16} \\
& + 40x_{21} \qquad\quad + 30x_{23} + 15x_{24} + 50x_{25} + 25x_{26} \\
& + 70x_{31} + 30x_{32} \qquad\quad + 80x_{34} + 20x_{35} + 30x_{36} \\
& + 25x_{41} + 15x_{42} + 80x_{43} \qquad\quad + 10x_{45} + 42x_{46} \\
& + 60x_{51} + 50x_{52} + 20x_{53} + 10x_{54} \qquad\quad + 25x_{56} \\
& + 45x_{61} + 25x_{62} + 30x_{63} + 42x_{64} + 25x_{65}
\end{aligned}
\tag{12}
$$

To determine the constraints we note that the salesman must enter and leave each city exactly once.

The constraints

$$
\begin{aligned}
x_{12} + x_{13} + x_{14} + x_{15} + x_{16} &= 1 \\
x_{21} \qquad\quad + x_{23} + x_{24} + x_{25} + x_{26} &= 1 \\
x_{31} + x_{32} \qquad\quad + x_{34} + x_{35} + x_{36} &= 1 \\
x_{41} + x_{42} + x_{43} \qquad\quad + x_{45} + x_{46} &= 1 \\
x_{51} + x_{52} + x_{53} + x_{54} \qquad\quad + x_{56} &= 1 \\
x_{61} + x_{62} + x_{63} + x_{64} + x_{65} \qquad\quad &= 1
\end{aligned}
\tag{13}
$$

ensure that he leaves each city exactly once because one and only one of the variables in each equation of (13) is 1. The constraints

$$x_{21} + x_{31} + x_{41} + x_{51} + x_{61} = 1$$
$$x_{12} \quad\quad + x_{32} + x_{42} + x_{52} + x_{62} = 1$$
$$x_{13} + x_{23} \quad\quad + x_{43} + x_{53} + x_{63} = 1$$
$$x_{14} + x_{24} + x_{34} \quad\quad + x_{54} + x_{64} = 1 \tag{14}$$
$$x_{15} + x_{25} + x_{35} + x_{45} \quad\quad + x_{65} = 1$$
$$x_{16} + x_{26} + x_{36} + x_{46} + x_{56} \quad\quad = 1$$

ensure that he enters each city exactly once because one and only one of the variables in each equation of (14) is 1.

We can write the constraints in (13) more compactly using summation (Σ) notation

$$\sum_{\substack{j=1 \\ j \neq i}}^{6} x_{ij} = 1 \quad\quad (1 \leq i \leq 6) \tag{13'}$$

and we can write the constraints in (14) as

$$\sum_{\substack{i=1 \\ i \neq j}}^{6} x_{ij} = 1 \quad\quad (1 \leq j \leq 6) \tag{14'}$$

♦

Unfortunately, the constraints (13') and (14') fail to model the Traveling Salesman Problem correctly because they do not eliminate disconnected tours, such as the one in Figure 1.

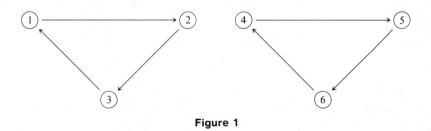

Figure 1

In the next example, we give a formulation of the Traveling Salesman Problem that eliminates the possibility of disconnected tours.

EXAMPLE 5

To eliminate disconnected tours we present an IP formulation of the Traveling Salesman Problem that involves additional variables to keep track of the order in which cities are visited.

Suppose there are n cities to be toured and that the mileage between City i and City j is c_{ij}. Let

$$x_{ijk} = \begin{cases} 1 & \text{if the salesman travels from City } i \text{ to City } j \\ & \text{as the } k\text{th trip of his tour} \\ 0 & \text{if the salesman does not travel from City } i \\ & \text{to City } j \text{ as the } k\text{th trip of his tour} \end{cases}$$

Figure 2 gives one of the possible tours for the Traveling Salesman Problem given in Example 4. We have labeled the arrows of Figure 2 with the corresponding values of x_{ijk}.

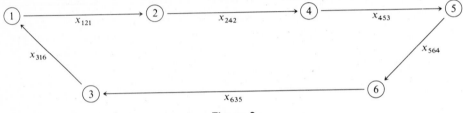

Figure 2

The IP problem

$$\text{Minimize:} \quad z = \sum_{i=1}^{n} \sum_{\substack{j=1 \\ j \neq i}}^{n} \sum_{k=1}^{n} c_{ij} x_{ijk}$$

$$\text{Subject to:} \quad \sum_{k=1}^{n} \sum_{\substack{j=1 \\ j \neq i}}^{n} x_{ijk} = 1 \quad \text{for each} \quad i, \quad 1 \leq i \leq n \qquad (15)$$

$$\sum_{k=1}^{n} \sum_{\substack{i=1 \\ i \neq j}}^{n} x_{ijk} = 1 \quad \text{for each} \quad j, \quad 1 \leq j \leq n \qquad (16)$$

$$\sum_{i=1}^{n} \sum_{\substack{j=1 \\ j \neq i}}^{n} x_{ijk} = 1 \quad \text{for each} \quad k, \quad 1 \leq k \leq n \qquad (17)$$

$$x_{ijk} \geq 0 \text{ and binary}, \quad 1 \leq i, j, k \leq n$$

correctly models the Traveling Salesman Problem. The constraints (15) ensure that the salesman leaves each city, the constraints (16) ensure that the salesman enters each city, and the constraints (17) ensure that the salesman leaves each city as a different trip of his tour.

EXAMPLE 6

In the IP problem of Example 4, the first constraint of (13) becomes

$$x_{121} + x_{122} + x_{123} + x_{124} + x_{125} + x_{126}$$
$$+ x_{131} + x_{132} + x_{133} + x_{134} + x_{135} + x_{136}$$
$$+ x_{141} + x_{142} + x_{143} + x_{144} + x_{145} + x_{146} \qquad (18)$$
$$+ x_{151} + x_{152} + x_{153} + x_{154} + x_{155} + x_{156}$$
$$+ x_{161} + x_{162} + x_{163} + x_{164} + x_{165} + x_{166} = 1$$

Exactly one of the variables in (18) can be 1. If that variable is x_{1jk}, then the salesman traveled from City 1 to City j as the kth trip of his tour.

There are five more constraints similar to (18) for cities 2 through 6, six more constraints of type (16), and six more constraints of type (17). The problem has $6^3 = 216$ decision variables. ◆

Note that to use the implicit enumeration method to solve a problem with equality constraints, such as the one in Example 4, you must replace each equality constraint by a pair of $\geq$ constraints, as we described in step 5 at the beginning of this section.

For a tour of n cities, the IP problem in Example 4 has n^3 decision variables. For a large value of n, computer solution of this problem is often impractical. A number of techniques exist for approximating solutions to the Traveling Salesman Problem, but at present there is no efficient or practical way of solving the problem completely. We present a model of the Traveling Salesman Problem that uses fewer (though still a great many) variables in Exercise 12.

Exercises

Solve Exercises 1–7 by formulating an appropriate standard form problem, solving the standard form problem by implicit enumeration, and deducing the solution to the original problem.

1. Maximize: $z = 2x_1 + 4x_2 - 3x_3$

Subject to: $x_1 - 2x_2 + x_3 \leq 5$
$2x_1 + x_2 - x_3 \geq 2$
$x_1 + x_2 + 2x_3 \leq 4$

x_i binary, $1 \leq i \leq 3$

2. Minimize: $z = 3x_1 + 2x_2 + x_3 - 4x_4$

Subject to: $2x_1 + 2x_2 - x_3 + 3x_4 \geq 6$
$x_1 + x_2 + 2x_3 + 4x_4 \leq 7$
$3x_1 - 2x_2 + 4x_3 - 5x_4 \geq 2$

x_i binary, $1 \leq i \leq 4$

3. Minimize: $z = 2x_1 + 3x_2 - x_3$

   Subject to: $x_1 + x_2 + x_3 \geq 1$

   $2x_1 - x_2 + x_3 \geq 2$

   $3x_1 - x_2 + 2x_3 \geq \dfrac{3}{2}$

   x_i binary, $1 \leq i \leq 3$

4. Maximize: $z = 2x_1 + 3x_2 + x_3$

   Subject to: $x_1 + 2x_2 + 3x_3 \leq 5$

   $2x_1 - x_2 + 4x_3 \leq 2$

   $-x_1 + x_2 + 2x_3 \geq 2$

   x_i binary, $1 \leq i \leq 3$

5. Maximize: $z = 120x + 100y$

   Subject to: $x + y \leq 4$

   $5x + 3y \leq 15$

   $x, y \geq 0$ and   integer

6. Minimize: $z = x_1 + 2x_2 + 3x_3$

   Subject to: $x_1 + 2x_2 + x_3 \geq 2$

   $x_1 + 2x_2 + 3x_3 \leq 8$

   $x_i \geq 0$ and   integer, $1 \leq i \leq 3$

7. Maximize: $z = 40x + 50y$

   Subject to: $2x + 3y \leq 6$

   $4x + y \leq 4$

   $x, y \geq 0$ and   integer

Formulate Exercises 8–10 in the standard form of a binary problem for solution by implicit enumeration. Try to introduce as few binary variables as possible. You do not need to solve the problems.

8. Minimize: $z = 59x_1 + 7x_2$

   Subject to: $2x_1 + 2x_2 \geq 7$

   $11x_1 - x_2 \geq 27$

   $x_1, x_2 \geq 0$ and   integer

(Hint: Any feasible solution of the two constraints provides an upper bound z^*, which in turn can be used to find upper bounds on x_1 and x_2.)

9. Minimize: $z = 12x_1 + 14x_2 + 6x_3$

   Subject to: $2x_1 + 3x_2 + 4x_3 \geq 12$

   $4x_1 + 3x_2 + 2x_3 \geq 12$

   $x_1 + x_2 + x_3 \geq 5$

   $x_1, x_2, x_3 \geq 0$ and   integer

10. Maximize: $z = 10x_1 - 3x_2 + 5x_3$

   Subject to: $x_2 + 2x_3 \geq 5$

   $x_1 + 2x_2 + 3x_3 \leq 12$

   $x_1, x_2, x_3 \geq 0$ and   integer

11. Suppose the constraints

$$x_1 + x_2 - 2x_3 \geq 2$$

$$x_1 - 2x_2 + x_3 \geq 1$$

occur in a binary problem. Find a surrogate constraint that shows immediately that the problem has no feasible solutions.

12. With your help, we will develop an alternative model of the Traveling Salesman Problem that requires fewer variables than the one we gave earlier. Although it has fewer variables, this model is not necessarily easier to solve than the previous one because it is not a pure binary IP problem and must be solved by a method such as branch and bound.
    (a) Suppose the distances between the n cities are given by c_{ij} $(1 \le i, j \le n)$ and that the variable x_{ij} is 1 if a trip is made from City i to City j and 0 if such a trip is not made. Write the objective function, a set of constraints guaranteeing that each city is entered exactly once, and a set of constraints guaranteeing that each city is left exactly once.
    (b) For $i = 2, 3, \ldots, n$, let t_i be the position in the tour at which City i is visited. For instance, for the tour described in Figure 2, $t_2 = 1, t_3 = 5, t_4 = 2, t_5 = 3$, and $t_6 = 4$. For each i and j with $i \ne j$ and $i, j = 2, 3, \ldots, n$, introduce a constraint

    $$t_i - t_j + nx_{ij} \le n - 1 \tag{19}$$

    (c) Show that a tour that starts at City 1, visits each of the n cities, ends at City 1, and has no disconnected subtours satisfies constraint (19). You can do this by considering two cases:

    (i) If $x_{ij} = 1$, then City j is visited immediately after City i. Thus, $t_j = t_i + 1$. Use this to show that (19) holds.

    (ii) If $x_{ij} = 0$, show that $t_i - t_j \le n - 2$ and deduce from this fact that (19) holds.

    (d) Form the three constraints of type (19) for the subtour from Figure 1 that does not contain City 1. Add the three constraints and draw a conclusion.
    (e) Show that a tour consisting of two or more disconnected subtours cannot occur. To this end, suppose there are two or more disconnected subtours. Choose one of the subtours that does not contain City 1. Suppose this subtour starts and ends at City p and includes visits to r cities (where, of course, $r < n$). By adding the r constraints of type (19) corresponding to cities on this subtour, conclude that $nr \le (n - 1)r$. Since that is impossible, there could not be such a subtour.
    (f) In terms of n, how many variables and how many constraints appear in this model?

13. Use the method of Exercise 12 to formulate the binary problem in Example 4.

14. Suppose the traveling salesman of Example 3 is relieved of the job of visiting City 6.
    (a) Explain why all disconnected tours have the form of a loop through three cities and a loop between two cities.
    (b) Add appropriate constraints to those given in (13) and (14) to correctly model the Traveling Salesman Problem for a tour of five cities.
    (c) Is it possible to extend your model to more than five cities?

15. An industrial plant has a robot to drill holes in steel plates. (A robot is a machine that is controlled by a computer.) The robot is programmed by sending it the locations of the

Table 3

Hole	Hole		
	1	2	3
1	—	42.5	68.7
2	39.6	—	31.2
3	69.9	48.7	—

holes to be drilled. The computer computes the distances between each pair of holes and produces a table of these distances. Table 3 is an example of such a table for a pattern of 3 holes. The computer then solves an IP problem to determine the order in which the holes are to be drilled in a way that minimizes the distance traveled by the drill. Solve the IP with the implicit enumeration tutorial of TUTOR using the data in Table 3.

†**16.** Use the branch and bound program of SOLVER to solve Exercise 1. Is it necessary to restrict the variables to ≤ 1, or does the objective function impose this limitation for you?

†**17.** Bill Voxman is planning a three-day trip into the Idaho primitive area and wants to take no more than 40 pounds of equipment in his backpack. He would like to take the items listed in Table 4, but he cannot take them all because their weight exceeds 40 pounds. To select the items that he wants most, he assigned a number to each item that reflects its relative satisfaction for him. Use the implicit enumeration tutorial of TUTOR to help him maximize his satisfaction.

Table 4

Item	Weight (lb)	Satisfaction
Sleeping bag	6	50
Air mattress	5	15
Tent	8	30
Book	4	25
Fishing gear	7	75
Rubber raft	8	40
Camera	5	65
Change of clothes	6	10
Flashlight	1	20
Beer	4	55
Food	3	50
More food	3	20

8.5 *GOMORY CUTTING PLANE METHOD*

We now turn to an entirely different approach to solving IP problems, an approach based on repeatedly reducing the feasible region of the problem until an integer (or mixed integer) solution is obtained. This procedure, due to Gomory [6], is known as the Gomory cutting plane method.

To carry out this method we will introduce a succession of constraints to the original problem. Each new constraint

decreases (cuts) the current feasible region (and, in particular, eliminates the current noninteger solution from the feasible region) and

does not eliminate any feasible integer solution to the original problem.

We call these new constraints *Gomory cuts.* Geometrically, these cuts usually pass through feasible integer points (points whose coordinates are integer valued).

To illustrate the cutting plane procedure we consider the LP problem

$$\text{Maximize:} \quad z = -x_1 + 4x_2$$

$$\text{Subject to:} \quad \begin{aligned} 2x_1 + x_2 &\leq 10 \\ -2x_1 + 3x_2 &\leq 8 \end{aligned} \tag{1}$$

$$x_1, x_2 \geq 0 \quad \text{and} \quad \text{integer}$$

If we (temporarily) ignore the integer constraints in (1), then application of the simplex method to (1) results in the initial and final tableaus

Initial Tableau

	x_1	x_2	x_3	x_4	
x_3	2	1	1	0	10
x_4	−2	3	0	1	8
	1	−4	0	0	0

$$(2)$$

Final Tableau

	x_1	x_2	x_3	x_4	
x_1	1	0	$\frac{3}{8}$	$-\frac{1}{8}$	$\frac{11}{4}$
x_2	0	1	$\frac{1}{4}$	$\frac{1}{4}$	$\frac{9}{2}$
	0	0	$\frac{5}{8}$	$\frac{9}{8}$	$\frac{61}{4}$

$$(3)$$

If at this stage x_1 and x_2 were integer valued, we would be done, since we would have attained an optimal integer solution. However, we see from (3) that $x_1 = \frac{11}{4}$ and $x_2 = \frac{9}{2}$.

The next step is to select the row in (3) that corresponds to the basic variable whose current value has the *largest* fractional part; since the fractional part of $x_1 = \frac{11}{4} = 2\frac{3}{4}$ is $\frac{3}{4}$, and the fractional part of $x_2 = \frac{9}{2} = 4\frac{1}{2}$ is $\frac{1}{2}$, we select the row in the final tableau that is labeled by x_1. This row corresponds to the equation

$$x_1 + \frac{3}{8}x_3 - \frac{1}{8}x_4 = 2\frac{3}{4} \tag{4}$$

We rewrite (4) so that all fractional terms are positive and less than 1. Thus, we write (4) in the form

$$x_1 + \frac{3}{8}x_3 + \left(-1 + \frac{7}{8}\right)x_4 = 2 + \frac{3}{4} \tag{5}$$

Next, we bring all fractional terms of (5) to the left-hand side of the equation and all integer terms to the right-hand side. This gives us

$$\frac{3}{8}x_3 + \frac{7}{8}x_4 - \frac{3}{4} = -x_1 + x_4 + 2 \tag{6}$$

Now, if we insist that all variables (including slack variables) are integer valued, then the right-hand side of (6) is integer valued and therefore the left-hand side is also integer valued. Since x_3 and x_4 are nonnegative,

$$\frac{3}{8}x_3 + \frac{7}{8}x_4 - \frac{3}{4} \geq -\frac{3}{4} \tag{7}$$

Moreover, since the left-hand side of (7) is an integer, this integer must be greater than or equal to 0 (the first integer larger than $-\frac{3}{4}$). This gives us the first Gomory constraint

$$\frac{3}{8}x_3 + \frac{7}{8}x_4 - \frac{3}{4} \geq 0$$

or, equivalently,

$$-\frac{3}{8}x_3 - \frac{7}{8}x_4 \leq -\frac{3}{4} \tag{8}$$

which as we will see corresponds geometrically to a "cutting plane" that reduces the size of the original feasible region. It is easy to see that constraint (8) "cuts off" the optimal noninteger solution given by tableau (3) since

$$(x_1, x_2, x_3, x_4) = \left(\frac{11}{4}, \frac{9}{2}, 0, 0\right)$$

does not satisfy this constraint.

If we add the new constraint (8) to the final tableau (3), we have the first cutting plane tableau

First Cutting Plane Tableau

	x_1	x_2	x_3	x_4	x_5	
x_1	1	0	$\frac{3}{8}$	$-\frac{1}{8}$	0	$\frac{11}{4}$
x_2	0	1	$\frac{1}{4}$	$\frac{1}{4}$	0	$\frac{9}{2}$
x_5	0	0	$-\frac{3}{8}$	$-\frac{7}{8}$	1	$-\frac{3}{4}$
	0	0	$\frac{5}{8}$	$\frac{9}{8}$	0	$\frac{61}{4}$

(9)

Applying the dual simplex method to (9), we obtain

Final Cutting Plane Tableau

	x_1	x_2	x_3	x_4	x_5	
x_1	1	0	$\frac{3}{7}$	0	$-\frac{1}{7}$	$\frac{20}{7}$
x_2	0	1	$\frac{1}{7}$	0	$\frac{2}{7}$	$\frac{30}{7}$
x_4	0	0	$\frac{3}{7}$	1	$-\frac{8}{7}$	$\frac{6}{7}$
	0	0	$\frac{1}{7}$	0	$\frac{9}{7}$	$\frac{100}{7}$

(10)

Observe that x_1, x_2, and x_4 are not integer valued, and so we continue by adding another cutting plane. To obtain the next constraint we again select the row corresponding to the basic variable with the largest fractional part. Since the fractional part of x_2 is $\frac{2}{7}$ and the fractional part of x_1 and of x_4 is $\frac{6}{7}$, we arbitrarily choose the row in tableau (10) corresponding to x_1 (rather than x_4) to rewrite. This row corresponds to the equation

$$x_1 + \frac{3}{7}x_3 - \frac{1}{7}x_5 = 2\frac{6}{7} \tag{11}$$

Rewriting (11) so that all fractional coefficients are nonnegative, we have

$$x_1 + \frac{3}{7}x_3 + \left(-1 + \frac{6}{7}\right)x_5 = 2\frac{6}{7} \tag{12}$$

Bringing all fractional terms of (12) to the left side and moving all integer terms to the right side gives us

$$\frac{3}{7}x_3 + \frac{6}{7}x_5 - \frac{6}{7} = -x_1 + x_5 + 2 \tag{13}$$

Since the right-hand side of (13) is an integer, the left-hand side is also an integer. Since x_3 and x_5 are nonnegative,

$$\frac{3}{7}x_3 + \frac{6}{7}x_5 - \frac{6}{7} \geq -\frac{6}{7} \tag{14}$$

Moreover, since the left-hand side of (14) is an integer, this integer must be greater than or equal to 0 (the first integer larger than $-\frac{6}{7}$). This gives us the second Gomory constraint

$$\frac{3}{7}x_3 + \frac{6}{7}x_5 - \frac{6}{7} \geq 0$$

or, equivalently,

$$-\frac{3}{7}x_3 - \frac{6}{7}x_5 \leq -\frac{6}{7} \tag{15}$$

Notice that (15) cuts off the optimal solution given in tableau (10) since the nonbasic variables x_3 and x_5 of that solution are 0 and do not satisfy constraint (15).

If we add the new constraint (15) to tableau (10) (the previous final cutting plane tableau), we obtain the next cutting plane tableau

Second Cutting Plane Tableau

	x_1	x_2	x_3	x_4	x_5	x_6	
x_1	1	0	$\frac{3}{7}$	0	$-\frac{1}{7}$	0	$\frac{20}{7}$
x_2	0	1	$\frac{1}{7}$	0	$\frac{2}{7}$	0	$\frac{30}{7}$
x_4	0	0	$\frac{3}{7}$	1	$-\frac{8}{7}$	0	$\frac{6}{7}$
x_6	0	0	$-\frac{3}{7}$	0	$-\frac{6}{7}$	1	$-\frac{6}{7}$
	0	0	$\frac{1}{7}$	0	$\frac{9}{7}$	0	$\frac{100}{7}$

(16)

Applying the dual simplex method to (16), we obtain

	x_1	x_2	x_3	x_4	x_5	x_6	
x_1	1	0	0	0	-1	1	2
x_2	0	1	0	0	0	$\frac{1}{3}$	4
x_4	0	0	0	1	-2	1	0
x_3	0	0	1	0	2	$-\frac{7}{3}$	2
	0	0	0	0	1	$\frac{1}{3}$	14

$$(17)$$

From (17) we see that we have obtained a maximal integer solution

$$x_1 = 2, \qquad x_2 = 4, \qquad z = 14$$

and thus solved the problem.

To abstract and summarize the steps of the Gomory cutting plane method, we introduce the following notation. If a is any real number, then $[a]$ denotes the greatest integer less than or equal to a. Thus, for example,

$$\left[\frac{21}{5}\right] = 4, \qquad [3] = 3, \qquad \left[-\frac{2}{3}\right] = -1, \qquad \text{and} \qquad \left[-\frac{16}{5}\right] = -4$$

For any real number a, we define the *fractional part* of a as the difference: $a - [a]$. For instance, if $a = \frac{13}{3}$, then

$$a - [a] = \frac{13}{3} - \left[\frac{13}{3}\right] = \frac{13}{3} - 4 = \frac{1}{3}$$

and if $a = -\frac{11}{4}$, then

$$a - [a] = -\frac{11}{4} - \left[-\frac{11}{4}\right] = -\frac{11}{4} - (-3) = \frac{1}{4}$$

With this notation we can state a general method for solving pure integer problems.

The Gomory Cutting Plane Method

Let a_{ij} $(1 \leq i \leq m, 1 \leq j \leq n)$ be integers and b_i $(1 \leq i \leq m)$ be nonnegative integers. The problem

$$\text{Maximize:} \quad z = c_1 x_1 + c_2 x_2 + \cdots + c_n x_n$$

$$\text{Subject to:} \quad a_{11} x_1 + a_{22} x_2 + \cdots + a_{1n} x_n \leq b_1$$
$$a_{21} x_1 + a_{22} x_2 + \cdots + a_{2n} x_n \leq b_2$$
$$\vdots$$
$$a_{m1} x_1 + a_{m2} x_2 + \cdots + a_{mn} x_n \leq b_m$$

$$x_i \geq 0 \quad \text{and} \quad \text{integer}, \qquad 1 \leq i \leq n$$

$$(18)$$

can be solved by performing the following steps.

STEP 1. Solve the original LP problem (without the integer constraints). If the optimal solution is integer, then you are done. If one or more of the variables has a noninteger value, proceed to Step 2.

STEP 2. Determine the Gomory cutting plane constraint. To do this, select the row of the current final tableau corresponding to the current resource value (the value of the basic variable) having the largest fractional part. Suppose this row corresponds to the equality

$$d_1 x_1 + d_2 x_2 + \cdots + d_n x_n = b$$

or, equivalently,

$$([d_1] + f_1)x_1 + ([d_2] + f_2)x_2 + \cdots + ([d_n] + f_n)x_n = [b] + f \quad (19)$$

where f and the f_i are the fractional parts of b and the d_i, respectively. Move the fractional parts of (19) to the left-hand side and the integer parts to the right-hand side to obtain

$$f_1 x_1 + f_2 x_2 + \cdots + f_n x_n - f = -[d_1]x_1 - [d_2]x_2 - \cdots - [d_n]x_n + [b] \quad (20)$$

Since the right-hand side of (20) is an integer for any integer values of $x_1, x_2, \ldots, x_n$, so is the left-hand side. Since $x_i \geq 0$ and $0 \leq f_i < 1$ for $1 \leq i \leq n$, the constraint

$$f_1 x_1 + f_2 x_2 + \cdots + f_n x_n - f \geq -f$$

where $0 \leq f < 1$, is satisfied by any feasible integer solution to the original problem (18). Since the smallest integer greater than or equal to $-f$ is 0, the stronger constraint

$$f_1 x_1 + f_2 x_2 + \cdots + f_n x_n - f \geq 0$$

must be satisfied by any feasible integer solution to (18). Thus, the Gomory cutting plane constraint is

$$-f_1 x_1 - f_2 x_2 - \cdots - f_n x_n \leq -f \quad (21)$$

Note that the optimal solution $(x_1^*, x_2^*, \ldots, x_n^*)$ to the last LP problem solved does not satisfy (21). This is a consequence of the fact that

$$f_1 x_1^* + f_2 x_2^* + \cdots + f_n x_n^* = 0 \quad (22)$$

To see that (22) holds, consider the basic and nonbasic variables separately. If x_i is a basic variable, its coefficient in a constraint from an optimal tableau is either 0 or 1. In either event, the fractional part of its coefficient f_i is 0, so that $f_i x_i^* = 0$. If x_i is a nonbasic variable, then $x_i^* = 0$, so that $f_i x_i^* = 0$. Thus, all of the terms in (22) are 0.

STEP 3. Add the constraint (21) to the previously obtained final tableau, and use the dual simplex method to solve the new problem. If it is not possible to restore feasibility of the tableau by the dual simplex method, the original problem does not have any integer solutions.

STEP 4. If step 3 yields an integer solution you are done; otherwise, repeat steps 2 and 3 until you obtain an optimal integer solution or conclude that the problem is infeasible.

Next, we see what happened geometrically in solving the LP problem (1). We will see how the Gomory constraints correspond to cutting planes (lines in the case of two variables) which successively reduce the feasible region.

Figure 1 gives the feasible region of the original problem (without the integer constraints).

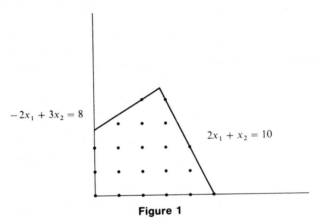

Figure 1

In (8) we obtained the first Gomory constraint

$$-\frac{3}{8}x_3 - \frac{7}{8}x_4 \leq -\frac{3}{4} \tag{23}$$

From (2) we have

$$x_3 = 10 - 2x_1 - x_2$$
$$x_4 = 8 + 2x_1 - 3x_2$$

Substituting these values into (23) and simplifying, we obtain the equivalent constraint

$$-x_1 + 3x_2 \leq 10 \tag{24}$$

Constraint (24) gives the cutting plane indicated in Figure 2.
Note:

1. This new constraint has reduced the original feasible region and has excluded the noninteger optimal solution $(\frac{11}{4}, \frac{9}{2})$ of the original problem.

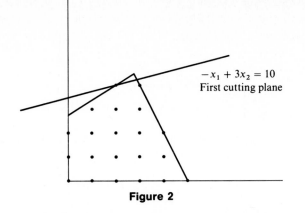

$-x_1 + 3x_2 = 10$
First cutting plane

Figure 2

2. The new constraint does not exclude any of the feasible "integer points" in the original feasible region.

The second Gomory constraint we added was

$$-\frac{3}{7}x_3 - \frac{6}{7}x_5 \le -\frac{6}{7} \tag{25}$$

From tableau (2) we have

$$x_3 = 10 - 2x_1 - x_2$$

From tableaus (2) and (9) we have

$$x_5 = \frac{3}{8}x_3 + \frac{7}{8}x_4 - \frac{3}{4}$$

$$= \frac{3}{8}(10 - 2x_1 - x_2) + \frac{7}{8}(8 + 2x_1 - 3x_2) - \frac{3}{4}$$

$$= 10 + x_1 - 3x_2$$

Substituting these values into (25) and simplifying, we obtain the equivalent constraint

$$x_2 \le 4 \tag{26}$$

This constraint gives the cutting plane indicated in Figure 3.
Note:

1. This new constraint (26) has reduced the feasible region given in Figure 2 and has excluded the optimal solution $(\frac{20}{7}, \frac{30}{7})$ corresponding to that feasible region (where integer constraints are ignored).

2. The new constraint (26) does not exclude any of the feasible "integer points" in the original feasible region.

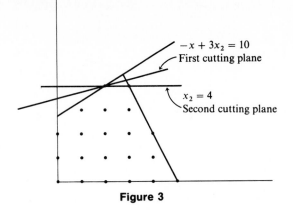

Figure 3

As should be apparent from the work we did (and the work we omitted showing) in solving (1), the calculations involved in carrying out the Gomory cutting plane procedure can be formidable. Generally, branch and bound methods are favored over cutting plane methods, although combinations of the two methods are often employed. It can be shown that if an optimal integer solution exists, then use of the Gomory cutting plane method (theoretically) produces this solution. In practice, however, all is not well:

1. The number of iterations required to obtain the optimal solution may be large. Results of the first few iterations will often tend fairly rapidly towards the optimal solution, but subsequent iterations may make slow progress.

2. Numerical errors that result from the inability of computing machines to carry enough significant digits can deter a program from converging to the solution.

Finally, we mention that cutting plane formulas for mixed-integer problems differ from those for pure-integer problems. You can find a discussion of cutting planes for mixed integer problems in [2].

Exercises

1. Find the integer and fractional parts of the numbers
   (a) $\frac{3}{2}$ (b) 2   (c) -2 (d) $-\frac{3}{2}$ (e) 1.7
   (f) -1.7 (g) 0   (h) 0.2   (i) -0.2 (j) -13.7

2. Find the integer and fractional parts of the numbers
   (a) $\frac{7}{3}$ (b) 5   (c) -5 (d) $-\frac{7}{3}$ (e) -0.2
   (f) -3 (g) 5.3   (h) -5.3 (i) -0.05 (j) -20.3

In Exercises 3 and 4, suppose that each of the given constraints occurs in the optimal solution to an LP problem solved by applying the Gomory cutting plane method to a problem for which all variables are to have integer values. In each constraint, suppose that x_3

is the basic variable. Find the Gomory cutting plane constraint that cuts off the noninteger value of x_3.

3. (a) $(\frac{3}{4})x_1 + (\frac{3}{2})x_2 + x_3 + (\frac{5}{4})x_4 = \frac{5}{4}$

  (b) $(\frac{3}{4})x_1 - (\frac{3}{2})x_2 + x_3 - (\frac{5}{4})x_4 = \frac{5}{4}$

  (c) $-1.2x_1 + 3.5x_2 + x_3 - 2.3x_4 = 23.7$

  (d) $2.5x_1 - 3.6x_2 + x_3 - 10.8x_4 + 3x_5 = 17.6$

4. (a) $(\frac{2}{7})x_1 + 3x_2 + x_3 - (\frac{17}{4})x_4 = \frac{23}{2}$

  (b) $-(\frac{2}{7})x_1 - 3x_2 + x_3 - (\frac{210}{25})x_4 = \frac{17}{4}$

  (c) $-1.07x_1 + 2.2x_2 + x_3 - 1.27x_4 = 21.6$

  (d) $3.17x_1 + 3.12x_2 + x_3 - 13.15x_4 - 4x_5 = 69.69$

In Exercises 5–10, we give an IP problem and the optimal tableau that results from applying the simplex method ignoring the requirement that the variables have integer values. Complete the solution of the IP problem by the Gomory cutting plane method.

5. Maximize: $z = x_1 - x_2 + x_3$

Subject to: $3x_1 + 2x_2 - x_3 \le 4$

$\qquad\qquad x_1 + 3x_2 + 2x_3 \le 12$

$x_1, x_2, x_3 \ge 0$ and   integer

The optimal tableau ignoring the integer requirement is

	x_1	x_2	x_3	s_1	s_2	
x_1	1	1	0	$\frac{2}{7}$	$\frac{1}{7}$	$\frac{20}{7}$
x_3	0	1	1	$-\frac{1}{7}$	$\frac{3}{7}$	$\frac{32}{7}$
	0	3	0	$\frac{1}{7}$	$\frac{4}{7}$	$\frac{52}{7}$

6. Maximize: $z = 2x_1 + 3x_2$

Subject to: $x_1 - 3x_2 \ge 2$

$\qquad\qquad 2x_1 + x_2 \le 6$

$x_1, x_2 \ge 0$ and   integer

The optimal tableau ignoring the integer requirement is

	x_1	x_2	s_1	s_2	
x_1	1	0	$-\frac{1}{7}$	$\frac{3}{7}$	$\frac{20}{7}$
x_2	0	1	$\frac{2}{7}$	$\frac{1}{7}$	$\frac{2}{7}$
	0	0	$\frac{4}{7}$	$\frac{9}{7}$	$\frac{46}{7}$

7. Maximize: $z = x_1 - x_2 + 4x_3$

Subject to: $-3x_1 + 2x_2 + 2x_3 \le 5$

$\qquad\qquad x_1 + 3x_2 + x_3 \le 13$

$x_1, x_2, x_3 \ge 0$ and   integer

The optimal tableau ignoring the integer requirement is

	x_1	x_2	x_3	s_1	s_2	
x_3	0	$\frac{11}{5}$	1	$\frac{1}{5}$	$\frac{3}{5}$	$\frac{44}{5}$
x_1	1	$\frac{4}{5}$	0	$-\frac{1}{5}$	$\frac{2}{5}$	$\frac{21}{5}$
	0	$\frac{53}{5}$	0	$\frac{3}{5}$	$\frac{14}{5}$	$\frac{197}{5}$

8. Maximize: $z = 4x_1 + x_2$

Subject to: $2x_1 - x_2 \leq 4$

$x_1 + 6x_2 \leq 6$

$x_1, x_2 \geq 0$ and   integer

The optimal tableau ignoring the integer requirement is

	x_1	x_2	s_1	s_2	
x_1	1	0	$\frac{6}{13}$	$\frac{1}{13}$	$\frac{30}{13}$
x_2	0	1	$-\frac{1}{13}$	$\frac{2}{13}$	$\frac{8}{13}$
	0	0	$\frac{23}{13}$	$\frac{6}{13}$	$\frac{128}{13}$

9. Maximize: $z = 7x_1 + 6x_2$

Subject to: $x_1 + 3x_2 \leq 6$

$2x_1 + x_2 \leq 9$

$x_1, x_2 \geq 0$ and   integer

The optimal tableau ignoring the integer requirement is

	x_1	x_2	s_1	s_2	
x_2	0	1	$\frac{2}{5}$	$-\frac{1}{5}$	$\frac{3}{5}$
x_1	1	0	$-\frac{1}{5}$	$\frac{3}{5}$	$\frac{21}{5}$
	0	0	1	3	33

10. Maximize: $z = 5x_1 + 3x_2 + 3x_3$

Subject to: $3x_1 + x_2 + 2x_3 \leq 12$

$x_1 + 2x_2 + x_3 \leq 10$

$x_1, x_2, x_3 \geq 0$ and   integer

The optimal tableau ignoring the integer requirement is

	x_1	x_2	x_3	s_1	s_2	
x_1	1	0	$\frac{3}{5}$	$\frac{2}{5}$	$-\frac{1}{5}$	$\frac{14}{5}$
x_2	0	1	$\frac{1}{5}$	$-\frac{1}{5}$	$\frac{3}{5}$	$\frac{18}{5}$
	0	0	$\frac{3}{5}$	$\frac{7}{5}$	$\frac{4}{5}$	$\frac{124}{5}$

Solve Exercises 11–13 with TUTOR. After you solve the LP problem (ignoring the integer requirement) with TUTOR, you can add a cutting plane constraint to the tableau by increasing the number of rows (m) by 1 and the number of columns (n) by 1.

†11. Maximize: $z = 2x_1 + 6x_2$

Subject to: $4x_1 - x_2 \geq 2$

$2x_1 + x_2 \leq 6$

$x_1, x_2 \geq 0$ and integer

†12. Maximize: $z = x_1 + x_2 + x_3$

Subject to: $3x_1 + 2x_2 + x_3 \leq 12$

$x_1 + 2x_2 + 3x_3 \leq 6$

$x_1, x_2, x_3 \geq 0$ and integer

†13. Maximize: $z = 2x_1 + 4x_2 + x_3$

Subject to: $x_1 + 2x_2 - x_3 \leq 6$

$2x_1 + x_2 + 3x_3 \leq 12$

$x_1, x_2, x_3 \geq 0$ and integer

8.6 LP MODELS

In this section we discuss three ways of modeling aspects of practical problems with integer programming techniques. The first model involves representing a class of nonlinear objective functions by linear objective functions. The second model shows how to deal with problems that depend on the time sequence in which events occur. The third model illustrates a technique that can sometimes be used to solve a problem in which the number of possible variables is too large or too difficult to determine in advance.

Model 1 Piecewise Linear Objective Functions

A vendor often offers reductions in cost per unit to a buyer who purchases large quantities. For instance, a vendor of microchips might charge $5 a piece for the first 50 chips ordered, $4.50 a piece for the next 50 chips ordered, and $4 a piece for the next 100 ordered. If we let x be the number of chips purchased, then we can write the cost $f(x)$ as

$$f(x) = \begin{cases} 5x, & 0 \leq x \leq 50 \\ 4.5(x - 50) + 250 = 4.5x + 25, & 50 \leq x \leq 100 \\ 4(x - 100)x + 475 = 4x + 75, & 100 \leq x \leq 200 \end{cases} \quad (1)$$

Figure 1 gives the graph of $f(x)$.

Since $f(x)$ is nonlinear, we cannot optimize it with the LP techniques we have developed. However, by using binary variables we will be able to formulate an IP problem whose optimal value is the same as the optimal value of such a cost function. To this end, first notice that we can write each point Q on the line segment joining P_1 and P_2 in Figure 1 as

$$Q = \lambda P_1 + (1 - \lambda)P_2 \quad (2)$$

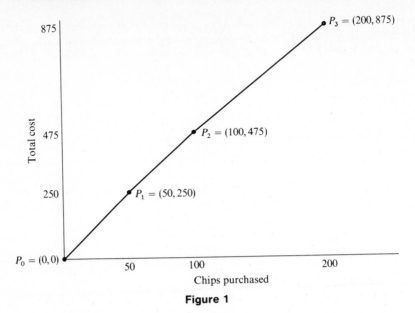

Figure 1

where $0 \leq \lambda \leq 1$ (see the definition of convex combinations in Section 2.1). Writing (2) in terms of the coordinates of $Q = [x, f(x)]$, P_1, and P_2, we obtain

$$[x, f(x)] = \lambda(50, 250) + (1 - \lambda)(100, 475) \tag{3}$$

Separating the components in (3), we obtain

$$\begin{aligned} x &= 50\lambda + 100(1 - \lambda) \\ f(x) &= 250\lambda + 475(1 - \lambda) \\ &\ \ 0 \leq \lambda \leq 1 \end{aligned} \tag{4}$$

You can write formulas similar to (4) for each pair of consecutive corner points on the graph in Figure 1. This discussion motivates us to make the following definition, in which we make use of the fact that if $c \leq x \leq d$, then there is a unique number λ, $0 \leq \lambda \leq 1$, such that $x = c\lambda + d(1 - \lambda)$.

Definition.   A function $f(x)$ with domain $a \leq x \leq b$ is *piecewise linear* if there is a partition

$$a = a_1 < a_2 < \cdots < a_n = b$$

of the interval $[a, b]$ such that we can represent $f(x)$ as

$$f(x) = f(a_i)\lambda + f(a_{i+1})(1 - \lambda)$$

whenever

$$a_i \leq x \leq a_{i+1} \quad \text{and} \quad x = a_i\lambda + a_{i+1}(1 - \lambda), \quad 0 \leq \lambda \leq 1$$

EXAMPLE 1

If we assume that no more than 200 chips will be purchased, the cost function (1) is piecewise linear with partition points

$$a_1 = 0, \qquad a_2 = 50, \qquad a_3 = 100, \qquad a_4 = 200$$

and corresponding function values

$$f(a_1) = 0, \qquad f(a_2) = 250, \qquad f(a_3) = 475, \qquad f(a_4) = 875$$

To see this, observe that

If $0 \le x \le 50$ and $x = 0\lambda + 50(1 - \lambda)$, then

$$f(x) = 5x = 0\lambda + 250(1 - \lambda) = f(0)\lambda + f(5)(1 - \lambda), \qquad 0 \le \lambda \le 1$$

If $50 \le x \le 100$ and $x = 50\lambda + 100(1 - \lambda)$, then

$$f(x) = 4.5x + 25 = 250\lambda + 475(1 - \lambda)$$
$$= f(50)\lambda + f(100)(1 - \lambda), \qquad 0 \le \lambda \le 1$$

If $100 \le x \le 200$ and $x = 100\lambda + 200(1 - \lambda)$, then

$$f(x) = 4x + 75 = 475\lambda + 875(1 - \lambda)$$
$$= f(100)\lambda + f(200)(1 - \lambda), \qquad 0 \le \lambda \le 1$$

The next two theorems show how to use binary variables to represent any piecewise linear function $f(x)$ in a form suitable for inclusion in an LP program.

Theorem 1. If $z_1, z_2, \ldots, z_n$ and $y_2, y_3, \ldots, y_n$ satisfy the constraints

$$z_1 + z_2 + \cdots + z_n = 1 \tag{5}$$
$$y_2 + \cdots + y_n = 1 \tag{6}$$
$$z_1 \le y_2 \tag{7}$$
$$z_2 \le y_2 + y_3 \tag{8}$$
$$z_3 \le y_3 + y_4$$
$$\vdots$$
$$z_{n-1} \le y_{n-1} + y_n$$
$$z_n \le y_n \tag{9}$$

$$0 \le z_i \le 1, \qquad 1 \le i \le n; \qquad y_i \text{ binary}, \qquad 2 \le i \le n$$

then for some i, $1 \le i \le n - 1$

$$z_i + z_{i+1} = 1 \tag{10}$$

and

$$z_j = 0 \quad \text{for all} \quad j \ne i, \quad j \ne i + 1$$

PROOF.   We first note that (6) implies that there is exactly one binary variable y_i such that $y_i = 1$. If $i = 2$ ($y_2 = 1$), then (7) and (8) allow z_1 and z_2 to assume nonzero values, and the following constraints imply that $z_j = 0$, $j > 2$. Condition (5) implies that $z_1 + z_2 = 1$. For $i > 2$ ($y_i = 1$), the situation for the pair z_i, z_{i+1} is similar.   ∎

Theorem 2.   If $f(x)$ with domain $[a, b]$ is piecewise linear with partition

$$a = a_1 < a_2 < \cdots < a_n = b$$

then there are variables $z_1, z_2, \ldots, z_n$ and $y_2, y_3, \ldots, y_n$ satisfying the constraints in Theorem 1 such that

$$f(x) = f(a_1)z_1 + f(a_2)z_2 + \cdots + f(a_n)z_n$$

PROOF.   Suppose that $a_i \le x \le a_{i+1}$ and $x = a_i \lambda + a_{i+1}(1 - \lambda)$. Let $z_i = \lambda$, $z_{i+1} = 1 - \lambda$, and set $z_j = 0$ for $j \ne i$, $j \ne i + 1$. Let $y_i = 1$, and set $y_j = 0$ for $j \ne i$. Then

$$f(x) = f(a_1)z_1 + \cdots + f(a_i)z_i + f(a_{i+1}z_{i+1}) + \cdots + f(a_n)z_n \qquad ∎$$

The next two examples illustrate the use of Theorems 1 and 2.

EXAMPLE 2

By Theorem 2 we can model the piecewise linear cost function $z = f(x)$ given in (1) by

$$z = 0z_1 + 250z_2 + 475z_3 + 875z_4$$

where

$$z_1 + z_2 + z_3 + z_4 = 1$$

$$y_2 + y_3 + y_4 = 1$$

$$z_1 \le y_2$$

$$z_2 \le y_2 + y_3$$

$$z_3 \le y_3 + y_4$$

$$z_4 \le y_4$$

$$z_i \ge 0, \quad 1 \le i \le 4; \quad y_i \text{ binary}, \quad 2 \le i \le 4 \qquad ◆$$

In Exercise 5 you are asked to use binary variables to achieve an IP representation of "step functions," another class of functions that are linear on subintervals of their domain.

The next example shows how to incorporate a nonlinear objective function into a simple product mix problem.

EXAMPLE 3

Burger Bar Supply (BBS) sells two kinds of cooking oil: oil 1 to cook french fries and oil 2 to cook onion rings. Each of these oils is made by mixing two kinds of grease: grease 1 and grease 2. Each gallon of oil 1 contains at least 50% of grease 1, and the rest is grease 2. Each gallon of oil 2 contains at least 60% of grease 1, and the rest is grease 2. Oil 1 sells for $0.85 per gallon, and oil 2 sells for $1.00 per gallon. BBS's grease supplier offers to sell BBS up to 2000 gallons of grease 1 at the per gallon prices: $0.45 for the first 500 gallons, $0.42 for the next 500 gallons, and $0.38 for the next 1000 gallons. The supplier offers up to 3000 gallons of grease 2 at the per gallon prices: $0.36 for the first 500 gallons, $0.33 for the next 1000 gallons, $0.30 for the next 500 gallons, and $0.27 for the next 1000 gallons. BBS wants to make a purchase that will maximize its profit (revenues from oil sales − grease costs). Let x_{ij} = gallons of grease i used in making the amount sold of oil j, $1 \le i, j \le 2$. The cost of buying $g_1 = x_{11} + x_{12}$ gallons of grease 1 is

$$f(g_1) = \begin{cases} 0.45g_1, & 0 \le g_1 \le 500 \\ 0.42(g_1 - 500) + 225, & 500 \le g_1 \le 1000 \\ 0.38(g_1 - 1000) + 435, & 1000 \le g_1 \le 2000 \end{cases}$$

We can model $f(g_1)$ by

$$f(g_1) = 0z_1 + 225z_2 + 435z_3 + 815z_4$$

$$z_1 + z_2 + z_3 + z_4 = 1$$

$$y_2 + y_3 + y_4 = 1$$

$$z_1 \le y_2$$

$$z_2 \le y_2 + y_3$$

$$z_3 \le y_3 + y_4$$

$$z_4 \le y_4$$

$$z_i \ge 0, \quad 1 \le i \le 4; \quad y_i \text{ binary}, \quad 2 \le i \le 4$$

In Exercise 4 you are asked to write a similar expression to model the cost $f(g_2)$ of buying $g_2 = x_{21} + x_{22}$ gallons of grease 2.

The objective function representing BBS's profit is

$$z = 0.85x_{11} + 1.00x_{12} + 0.85x_{21} + 1.00x_{22} - f(g_1) - f(g_2)$$

The constraints relating the purchase of grease 1 and $f(g_1)$ are

$$500 - x_{11} - x_{12} \leq 2000 y_2 \qquad\qquad 1000 - x_{11} - x_{12} \leq 2000(y_2 + y_3)$$

$$2000 - x_{11} - x_{12} \leq 2000(y_2 + y_3 + y_4) \qquad\qquad x_{11} + x_{12} \leq 2000$$

The constraints controlling the mixture of grease in each oil are

$$\frac{x_{11}}{x_{11} + x_{21}} \geq 0.5, \qquad \text{or} \qquad 0.5x_{11} - 0.5x_{21} \geq 0$$

$$\frac{x_{12}}{x_{12} + x_{22}} \geq 0.6, \qquad \text{or} \qquad 0.4x_{12} - 0.6x_{22} \geq 0 \qquad \blacklozenge$$

Model 2   Scheduling

In Example 4 of Section 8.1 we used binary variables to solve a problem of scheduling time on five machines for work on three products. The solution was made simple (perhaps, deceptively so) by having each machine work on only two products. In fact, general scheduling problems are very hard to solve with IP techniques because even a small problem can require an impressive number of constraints and binary variables. The next example illustrates the difficulty.

EXAMPLE 4

In Figure 2 we have modified the problem given in Example 4 of Section 8.1 to require each of machines M_1, M_2, and M_3 to work on all three products.

For each product A_1, A_2, A_3 we must have a set of constraints guaranteeing that each machine completes its work on one product before starting work on another product. The constraints accomplishing this for the scenario described in Figure 2 are similar to constraints (12)–(18) of Section 8.1; so we do not write them. Consider constraints (19) and (20) of Section 8.1. These constraints ensure that machine M_1 completes its work on product A_1 before beginning work on product A_2. Since in the earlier example, M_1 works on only two products, there are only two possible orders in which it can do its job. However, there are now six possible orders in which M_1 can work on the three products. If we denote the order in which M_1 works on the three products by a permutation (an ordered triple) of the numbers 1, 2, 3, then the possible orders are

$$(123) \quad (132) \quad (213) \quad (231) \quad (312) \quad (321)$$

where, for instance, (213) indicates that M_1 works first on product A_2, next on product A_1, and finally on product A_3. We need to introduce six binary variables in order to select any one of these six alternatives for M_1. We let the subscripts on the binary variable y correspond to the six possible permutations, and we let B be a constant that is larger than the time the entire job could take. As before, we let x_{ij}

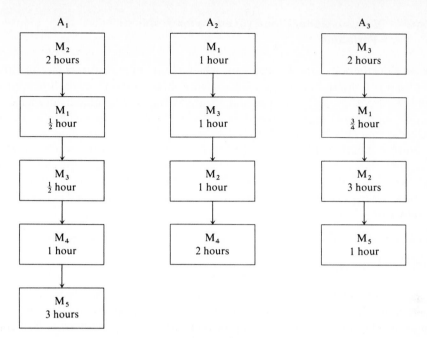

Figure 2

be the starting time for process A_i on machine M_j. The constraints

Permutation	Constraints

(123) $\quad x_{11} + \dfrac{1}{2} \le x_{21} + By_{123}; \qquad x_{21} + 1 \le x_{31} + By_{123}$

(132) $\quad x_{11} + \dfrac{1}{2} \le x_{31} + By_{132}; \qquad x_{31} + \dfrac{3}{4} \le x_{21} + By_{132}$

(213) $\quad x_{21} + 1 \le x_{11} + By_{213}; \qquad x_{11} + \dfrac{1}{2} \le x_{31} + By_{213}$

(231) $\quad x_{21} + 1 \le x_{31} + By_{231}; \qquad x_{31} + \dfrac{3}{4} \le x_{11} + By_{231}$

(312) $\quad x_{31} + \dfrac{3}{4} \le x_{11} + By_{312}; \qquad x_{11} + \dfrac{1}{2} \le x_{21} + By_{312}$

(321) $\quad x_{31} + \dfrac{3}{4} \le x_{21} + By_{321}; \qquad x_{21} + 1 \le x_{11} + By_{321}$

$$y_{123} + y_{132} + y_{213} + y_{231} + y_{312} + y_{321} \ge 5$$

allow the integer solution method to set one binary variable y_{ijk} equal to 0 and the other binary variables equal to 1. Since setting a binary variable equal to 1 makes the constraint in which it occurs ineffectual, the integer solution method can try each permutation in seeking to minimize total processing time. In Exercises 8 and 9 you are asked to write a similar set of constraints for machines M_2 and M_3. $\quad \bullet$

Model 3 Column Generation Procedure for Problems with Many Variables

We consider "stock cutting" problems, which many manufacturers face in cutting quantities of material of various sizes from a large piece of stock. Stock cutting problems are often difficult to model because they may involve listing all of the possible patterns for cutting the sizes you need from the stock. The resulting IP problems can also be difficult to solve because the problems have no uniform structure and may require a general IP solution method such as branch and bound or cutting planes. We will develop a general approach for solving these problems that, to some extent, reduces both the difficulty of formulation and the complexity of solution. We begin by reviewing the stock cutting problem described in Exercise 3 of Section 8.1.

EXAMPLE 5

A certain pipe comes in 25-ft lengths. Suppose a contractor needs at least 200 sections of 6-ft pipe; 250 sections of 9-ft pipe; and 300 sections of 12-ft pipe. The contractor decides to cut each 25-ft pipe by using the patterns given in Table 1.

Table 1

Cutting pattern	Number of 6-ft sections	Number of 9-ft sections	Number of 12-ft sections	Feet of scrap
1	4	0	0	1
2	2	1	0	4
3	2	0	1	1
4	1	2	0	1
5	0	1	1	4
6	0	0	2	1

The problem is to determine how frequently each cutting pattern should be used so that the demands are met and the amount of scrap is minimized. If we let x_i, $1 \le i \le 6$, be the number of times that cutting pattern i is used, then the objective function to be minimized is

$$w = x_1 + 4x_2 + x_3 + x_4 + 4x_5 + x_6$$

Note that we can also write this objective function as the difference between the total length of 25-ft pipe sections used and the total length used. Thus,

$$w = 25(x_1 + x_2 + x_3 + x_4 + x_5 + x_6) - 200 \cdot 6 - 250 \cdot 9 - 300 \cdot 12$$
$$= 25(x_1 + x_2 + x_3 + x_4 + x_5 + x_6) - 7050$$

Clearly, the solution that minimizes

$$z = x_1 + x_2 + x_3 + x_4 + x_5 + x_6 \qquad (11)$$

also minimizes w. As you will soon see, it is crucial to the method we will develop to solve cutting-stock problems that the objective function be expressible in a form in which the cost coefficients do not depend on the particular pattern [in (11) all of the cost coefficients are 1].

We can model the contractor's problem with the IP problem

Minimize: $z = x_1 + x_2 + x_3 + x_4 + x_5 + x_6$

Subject to: $4x_1 + 2x_2 + 2x_3 + x_4 \qquad\qquad\quad \geq 200 \qquad$ (6-ft sections)

$\qquad\qquad\quad x_2 \qquad\quad + 2x_4 + x_5 \qquad\quad \geq 250 \qquad$ (9-ft sections)

$\qquad\qquad\qquad\quad x_3 \qquad\quad + x_5 + 2x_6 \geq 300 \qquad$ (12-ft sections)

$\qquad\qquad x_i \geq 0 \quad$ and $\quad$ integer, $\quad 1 \leq i \leq 6$

If you solve this problem with the branch and bound program of SOLVER, you will discover (after a great many iterations) the solution

$$(x_1, x_2, x_3, x_4, x_5, x_6) = (15, 1, 7, 124, 1, 146); \qquad z = 294$$

(Note that this problem has many alternative optimal solutions.)

We now take a different approach to solving the problem in Example 5. To find the solution in this example, we first listed possible patterns for cutting the stock. Note, however, that Table 1 does not contain all possible patterns. For instance, we could cut one 6-ft section from a 25-ft length of pipe and throw the rest away. Of course, it is reasonable to eliminate such a foolish pattern from consideration because the optimal solution would never select it. Nevertheless, though it is not too difficult to see that Table 1 does list all possible patterns that might appear in the optimal solution, it is clear that for larger and more complex cutting problems, enumerating all reasonable patterns could be an immense and error-prone activity.

A technique involving *column generation*, which is based on ideas underlying our development of the revised simplex method in Chapter 5, enables us to avoid enumerating all possible patterns in advance. We will use column generation again in Section 10.2 to study another class of difficult problems.

The column generation method we use will solve LP problems, rather than IP problems. For instance, when this method is applied to the LP problem given in Example 5, it produces the "solution"

$$(x_1, x_2, x_3, x_4, x_5, x_6) = \left(\frac{75}{4}, 0, 0, 125, 0, 150\right) \qquad (12)$$

$$z = 293.75$$

Solution (12) is not a practical solution to the problem because we cannot use the first cutting pattern $\frac{75}{4}$ times. Even so, such a solution may be useful because we can round fractional values to the next integer and obtain a solution that is probably close to the optimal one. The contractor of Example 5 may prefer to accept a small amount of waste rather than go to the trouble and expense of solving the IP problem exactly. In this case, rounding $\frac{75}{4}$ to 19 happens to give the exact solution

$$(x_1, x_2, x_3, x_4, x_5, x_6) = (19, 2, 0, 125, 0, 150)$$

$$z = 294$$

This solution is an alternative optimal solution to the one found by branch and bound in Example 5. As we have seen before, rounding does not necessarily yield an optimal solution; nevertheless, rounding does yield an heuristic solution that may be close enough for practical purposes. (See Exercises 13 and 15 for a discussion of a situation in which rounding always yields an optimal solution.)

We will use Example 5 to illustrate the method of column generation. The LP problem in Example 5 does not have a readily identifiable initial basic solution. Rather than introducing artificial variables to manufacture an initial basic solution, we will add three new patterns, each of which produces only one useful length of pipe. Suppose pattern 7 produces one 6-ft section and 19 ft of scrap, pattern 8 produces one 9-ft section and 16 ft of scrap, and pattern 9 produces one 12-ft section and 13 ft of scrap. We let x_7, x_8, x_9 be the numbers of times that the patterns 7, 8, 9 are used, respectively. Since we will (as usual) solve a maximization problem, the objective function is

$$z = -x_1 - x_2 - x_3 - x_4 - x_5 - x_6 - x_7 - x_8 - x_9$$

Tableau 1 gives the A matrix for the LP problem in Example 1, with the additional patterns added and slack variables included.

Tableau 1

	x_1	x_2	x_3	x_4	x_5	x_6	x_7	x_8	x_9	s_1	s_2	s_3	
x_7	4	2	2	1	0	0	1	0	0	-1	0	0	200
x_8	0	1	0	2	1	0	0	1	0	0	-1	0	250
x_9	0	0	1	0	1	2	0	0	1	0	0	-1	300

We are going to imagine that we do not know all of the entries in Tableau 1, and we will use the revised simplex method to find the optimal solution to our problem. We will not even assume we know how many variables (how many patterns) the tableau contains. In what follows, note that

$$\mathbf{c}_B = \begin{bmatrix} -1 & -1 & -1 \end{bmatrix}$$

for each basic matrix B, since the objective function is

$$z = -x_1 - x_2 - \cdots - x_n$$

regardless of how many patterns n there may be.

ITERATION 0.

We know an initial basis matrix for Tableau 1 because we added the patterns x_7, x_8, and x_9 to produce it. It is

$$B = \begin{bmatrix} 1 & 0 & 0 \\ 0 & 1 & 0 \\ 0 & 0 & 1 \end{bmatrix} \qquad B^{-1} = \begin{bmatrix} 1 & 0 & 0 \\ 0 & 1 & 0 \\ 0 & 0 & 1 \end{bmatrix}$$

If

$$\begin{bmatrix} a_1 \\ a_2 \\ a_3 \end{bmatrix}$$

is any column of Tableau 1, then the entry in the objective row of this column is

$$\mathbf{c}_B B^{-1} \begin{bmatrix} a_1 \\ a_2 \\ a_3 \end{bmatrix} - (-1) = -a_1 - a_2 - a_3 + 1 \tag{13}$$

If

$$\begin{bmatrix} a_1 \\ a_2 \\ a_3 \end{bmatrix}$$

is a column of Tableau 1 that corresponds to a decision variable, then it also represents a pattern; therefore, it satisfies

$$6a_1 + 9a_2 + 12a_3 \leq 25$$

The greedy rule for selecting the next entering basic variable chooses the smallest negative entry in the objective row. By (13), the smallest such entry is the solution to the IP problem

$$\begin{aligned} \text{Minimize:} \quad & w = -a_1 - a_2 - a_3 + 1 \\ \text{Subject to:} \quad & 6a_1 + 9a_2 + 12a_3 \leq 25 \\ & a_i \geq 0 \quad \text{and} \quad \text{integer,} \quad 1 \leq i \leq 3 \end{aligned} \tag{14}$$

You can check (by using the branch and bound method, or otherwise) that the solution to (14) is

$$\begin{bmatrix} a_1 \\ a_2 \\ a_3 \end{bmatrix} = \begin{bmatrix} 4 \\ 0 \\ 0 \end{bmatrix}; \qquad w = -3 \tag{15}$$

(By expressing the objective function in terms of the nonbasic variables of Tableau 1, you can see that we have found an appropriate entering basic variable and discovered one of the patterns. However, remember: We are supposing that we have not enumerated the patterns, and so we are not really using Tableau 1.)

To find the departing basic variable, we must form ratios of the current resource values to the column of the current tableau corresponding to (15). In this first step of the algorithm this is easy because we know the initial resource values; in later steps we will need to calculate them. Since

$$\min\left\{\frac{200}{4}, -, -\right\} = 50$$

the departing basic variable is in row 1.

Next, we obtain the pivoting matrix for row 1 from (15). It is

$$P_1 = \begin{bmatrix} \frac{1}{4} & 0 & 0 \\ 0 & 1 & 0 \\ 0 & 0 & 1 \end{bmatrix}$$

ITERATION 1.

The inverse of the new basis matrix B_1 and the resource column $\mathbf{b}_1$ for the next tableau are (see Example 1 of Section 5.1)

$$B_1^{-1} = P_1 B^{-1} = \begin{bmatrix} \frac{1}{4} & 0 & 0 \\ 0 & 1 & 0 \\ 0 & 0 & 1 \end{bmatrix}; \quad \mathbf{b}_1 = P_1 \mathbf{b} = \begin{bmatrix} 50 \\ 250 \\ 300 \end{bmatrix} \tag{16}$$

To find the next pattern to enter the basis, we calculate

$$\mathbf{c}_B B_1^{-1} \begin{bmatrix} a_1 \\ a_2 \\ a_3 \end{bmatrix} - (-1) = -\frac{1}{4}a_1 - a_2 - a_3 + 1$$

and solve the IP problem

$$\text{Minimize:} \quad w = -\frac{1}{4}a_1 - a_2 - a_3 + 1$$

$$\text{Subject to:} \quad 6a_1 + 9a_2 + 12a_3 \le 25 \tag{17}$$

$$a_i \ge 0 \quad \text{and} \quad \text{integer}, \quad 1 \le i \le 3$$

The solution to (17) is

$$\begin{bmatrix} a_1 \\ a_2 \\ a_3 \end{bmatrix} = \begin{bmatrix} 1 \\ 2 \\ 0 \end{bmatrix}; \quad w = -\frac{5}{4} \tag{18}$$

To find the departing basic variable we first calculate the column of the current tableau corresponding to pattern (18). By premultiplying the column in (18) by B_1^{-1}, we obtain the new column

$$
\begin{bmatrix} \frac{1}{4} & 0 & 1 \\ 0 & 1 & 0 \\ 0 & 0 & 1 \end{bmatrix} \begin{bmatrix} 1 \\ 2 \\ 0 \end{bmatrix} = \begin{bmatrix} \frac{1}{4} \\ 2 \\ 0 \end{bmatrix}
\tag{19}
$$

Next we form ratios using $\mathbf{b}_1$ from (16) and the new column (19). Since

$$
\min \left\{ \frac{50}{(\frac{1}{4})}, \frac{250}{2}, - \right\} = 125
$$

the departing basic variable is in row 2. The pivoting matrix obtained from the column (19) is

$$
P_2 = \begin{bmatrix} 1 & -\frac{1}{8} & 0 \\ 0 & \frac{1}{2} & 0 \\ 0 & 0 & 1 \end{bmatrix}
$$

ITERATION 2.

The inverse of the new basis matrix B_2 and the resource column $\mathbf{b}_2$ for the next tableau are

$$
B_2^{-1} = P_2 B_1^{-1} = \begin{bmatrix} \frac{1}{4} & -\frac{1}{8} & 0 \\ 0 & \frac{1}{2} & 0 \\ 0 & 0 & 1 \end{bmatrix}; \qquad \mathbf{b}_2 = P_2 \mathbf{b}_1 = \begin{bmatrix} \frac{75}{4} \\ 125 \\ 300 \end{bmatrix}
\tag{20}
$$

To find the next pattern to enter the basis, we calculate

$$
\mathbf{c}_B B_2^{-1} \begin{bmatrix} a_1 \\ a_2 \\ a_3 \end{bmatrix} - (-1) = -\frac{1}{4} a_1 - \frac{3}{8} a_2 - a_3 + 1
$$

and solve the IP problem

$$
\text{Minimize:} \quad w = -\frac{1}{4} a_1 - \frac{3}{8} a_2 - a_3 + 1
$$

$$
\text{Subject to:} \quad 6a_1 + 9a_2 + 12a_3 \le 25
\tag{21}
$$

$$
a_i \ge 0 \quad \text{and} \quad \text{integer}, \qquad 1 \le i \le 3
$$

The solution to (21) is

$$
\begin{bmatrix} a_1 \\ a_2 \\ a_3 \end{bmatrix} = \begin{bmatrix} 0 \\ 0 \\ 2 \end{bmatrix}; \qquad w = -1
\tag{22}
$$

To find the departing basic variable we first calculate the column of the current tableau corresponding to pattern (22). By premultiplying the column in (22) by B_2^{-1}

we obtain

$$\begin{bmatrix} \frac{1}{4} & -\frac{1}{8} & 0 \\ 0 & \frac{1}{2} & 0 \\ 0 & 0 & 1 \end{bmatrix} \begin{bmatrix} 0 \\ 0 \\ 2 \end{bmatrix} = \begin{bmatrix} 0 \\ 0 \\ 2 \end{bmatrix} \tag{23}$$

Next, we form ratios using $\mathbf{b}_2$ from (20) and the new column from (23). Since

$$\min\left\{-, -, \frac{300}{2}\right\} = 150$$

the departing variable is in row 3. The pivoting matrix obtained from the column in (23) is

$$P_2 = \begin{bmatrix} 1 & 0 & 0 \\ 0 & 1 & 0 \\ 0 & 0 & \frac{1}{2} \end{bmatrix}$$

ITERATION 3.

The inverse of the new basis matrix B_3 and the resource column $\mathbf{b}_3$ for the next tableau are

$$B_3^{-1} = P_3 B_2^{-1} = \begin{bmatrix} \frac{1}{4} & -\frac{1}{8} & 0 \\ 0 & \frac{1}{2} & 0 \\ 0 & 0 & \frac{1}{2} \end{bmatrix} \qquad \mathbf{b}_3 = P_3 \mathbf{b}_2 = \begin{bmatrix} \frac{75}{4} \\ 125 \\ 150 \end{bmatrix} \tag{24}$$

To find the next pattern to enter the basis, we calculate

$$\mathbf{c}_B B_3^{-1} \begin{bmatrix} a_1 \\ a_2 \\ a_3 \end{bmatrix} - (-1) = -\frac{1}{4}a_1 - \frac{3}{8}a_2 - \frac{1}{2}a_3 + 1$$

and solve the IP problem

$$\text{Minimize:} \quad w = -\frac{1}{4}a_1 - \frac{3}{8}a_2 - \frac{1}{2}a_3 + 1$$

$$\text{Subject to:} \quad 6a_1 + 9a_2 + 12a_3 \leq 25 \tag{25}$$

$$a_i \geq 0 \quad \text{and} \quad \text{integer,} \quad 1 \leq i \leq 3$$

The solution of (25) is

$$\begin{bmatrix} a_1 \\ a_2 \\ a_3 \end{bmatrix} = \begin{bmatrix} 0 \\ 0 \\ 2 \end{bmatrix}; \quad w = 0$$

Since $w = 0$, there are no other columns corresponding to decision variables (patterns) to enter the basis. It is possible that we have not found the optimal solution because some of the slack variables have negative objective values. To check them we calculate

$$\mathbf{c_B} B_3^{-1} \begin{bmatrix} -1 \\ 0 \\ 0 \end{bmatrix} - 0 = \frac{1}{4}$$

$$\mathbf{c_B} B_3^{-1} \begin{bmatrix} 0 \\ -1 \\ 0 \end{bmatrix} - 0 = \frac{3}{8}$$

$$\mathbf{c_B} B_3^{-1} \begin{bmatrix} -1 \\ 0 \\ 0 \end{bmatrix} - 0 = \frac{1}{2}$$

Since these values are nonnegative, we have found the optimal solution. From the final resource column given in (24), we see that the solution is to use the pattern given in (15) $\frac{75}{4}$ times, the pattern given in (18) 125 times, and the pattern given in (22) 150 times. The minimum cost is

$$\frac{75}{4} + 125 + 150 = \frac{1175}{4} = 293.75$$

Exercises

In Exercises 1–3 you are given a piecewise linear cost function. Represent each function in a form suitable to use in formulating an IP problem (see Example 2).

1. $$f(x) = \begin{cases} 10x, & 0 \le x \le 10 \\ 15x - 50, & 10 \le x \le 20 \\ 20x - 150, & 20 \le x \le 30 \end{cases}$$

2. $$f(x) = \begin{cases} 100x, & 0 \le x \le 20 \\ 90x + 200, & 20 \le x \le 50 \\ 80x + 700, & 50 \le x \le 75 \\ 70x + 1450, & 75 \le x \le 100 \\ 60x + 2450, & 100 \le x \le 200 \end{cases}$$

3. $$f(x) = \begin{cases} 10x, & 5 \le x \le 10 \\ 15x - 50, & 10 \le x \le 20 \\ 20x - 150, & 20 \le x \le 30 \end{cases}$$

4. Write an objective function and appropriate constraints to model the cost function $f(g_2)$ in Example 3.

5. A *step function* assumes a constant value on a finite set of consecutive intervals. More exactly, suppose $f(x)$ is defined for $a \le x \le b$ and $a = a_1 < a_2 < \cdots < a_n = b$. If $f(x) = b_i$ when $a_i < x \le a_{i+1}$, then $f(x)$ is a step function. Show that, if $b_i \le b_{i+1}$, we can

model an objective function to *minimize* $z = f(x)$ for use in an LP problem as follows. Let

$$z = b_1 + (b_2 - b_1)z_1 + (b_3 - b_2)z_2 + \cdots + (b_n - b_{n-1})z_{n-1}$$

$$x - a_1 \quad \leq a_n - a_1$$

$$x - a_2 \quad \leq (a_n - a_2)z_1$$

$$\vdots$$

$$x - a_{n-1} \leq (a_n - a_{n-1})z_{n-1}$$

$$x \qquad \geq a_1$$

$$x \geq 0, \qquad z_i \quad \text{binary}, \qquad 1 \leq i \leq n$$

6. Suppose it costs 22 cents to mail a letter weighing 1 ounce or less, 44 cents to mail a letter weighing more than one ounce but not more than 2 ounces, and 66 cents to mail a letter weighing more than two ounces but not more than 3 ounces. Let a variable x in an LP problem give the weight of a letter. Use the method given in Exercise 5 to write an objective function and appropriate inequalities to minimize the cost of mailing a letter.

7. Table 2 gives the freight charges for shipping x pounds of goods from Pocatello to Seattle.

Table 2

Weight (lb)	Freight charge ($)
$0 < x \leq 10$	2.00
$10 < x \leq 30$	5.00
$30 < x \leq 50$	7.50

Use the method given in Exercise 5 to model an objective function that could be used to minimize freight charges.

8. For the scheduling problem given in Example 4, write the constraints guaranteeing that machine M_2 completes its work on each product before starting on another.

9. For the scheduling problem given in Example 4, write the constraints guaranteeing that machine M_3 completes its work on each product before starting on another.

10. In our solution of the problem in Example 5 by column generation, each of the IP problems solved to find the entering basic variable had only one constraint. How would you formulate these IP problems if the stock sizes of pipe came in two lengths: 16 feet and 25 feet?

11. Suppose a builder has a large stock of 20-ft boards from which he wants to cut 25 3-ft boards, 50 6-ft boards, and 40 9-ft boards.
    (a) Write an objective function to minimize waste in a form suitable for solving the problem by the column generation method.
    (b) Perform the first pivot of the column generation method.
    (c) Perform the second pivot.
    (d) Complete the solution by the column generation method.
    (e) Enumerate all of the practical patterns, and formulate an LP problem to solve the builder's problem. As we did in Tableau 1, introduce appropriate patterns to obtain a starting basis of unit columns in your tableau.

†(f) Use SOLVER or the branch and bound tutorial of TUTOR to find the LP and integer solutions to the problem in part (e). (Express the objective function in terms of

(e) the *nonbasic* variables before using SOLVER or TUTOR.)

12. Suppose rolls of newsprint come in 51-inch widths. A printer wants to saw these rolls into rolls of widths 9, 15, and 27 inches. She needs at least 25 9-inch rolls, 20 15-inch rolls, and 15 27-inch rolls.

(a) Enumerate all of the practical cutting patterns, formulate an LP problem to minimize waste, and solve the LP problem.

†(b) Solve the problem from part (a) as an IP problem using the branch and bound procedure of SOLVER or the branch and bound tutorial of TUTOR. (Express the objective function in terms of the *nonbasic* variables before using SOLVER or TUTOR.)

(c) Formulate this problem for solution by the column generation method. Perform two iterations of the method.

13. Prove the following theorem: Suppose the LP problem

$$\text{Minimize:} \quad z = x_1 + x_2 + \cdots + x_n$$

$$\text{Subject to:} \quad A\mathbf{x} \geq \mathbf{b}$$

$$\mathbf{x} \geq \mathbf{0}$$

has a minimal feasible solution $\mathbf{x}^*$ and corresponding objective function value z^*. Let $\hat{\mathbf{x}}^*$ be obtained from $\mathbf{x}^*$ by rounding all the entries in $\mathbf{x}^*$ to the next higher integer; that is,

$$\mathbf{x}^* = \begin{bmatrix} a_1 \\ a_2 \\ \vdots \\ a_n \end{bmatrix}, \qquad \hat{\mathbf{x}}^* = \begin{bmatrix} a_1 + d_1 \\ a_2 + d_2 \\ \vdots \\ a_n + d_n \end{bmatrix}, \qquad 0 \leq d_i < 1$$

If $\mathbf{x}^*$ and $\hat{\mathbf{x}}^*$ are feasible solutions and $d_1 + d_2 + \cdots + d_n < 1$, then $\hat{\mathbf{x}}^*$ is the minimal feasible solution.

14. The Moscow Industrial Ghetto employs one person to paint its line of metal dragons. While the painter was on vacation, a backlog of 20 small red dragons, 15 small green dragons, 25 large red dragons, and 30 large green dragons accumulated. It takes the painter 1.5 hours to paint a small red dragon, 2.2 hours to paint a small green dragon, 2.7 hours to paint a large red dragon, and 4.1 hours to paint a large green dragon. Once the painter starts on a dragon, she must complete painting it before the paint dries; therefore, she cannot let an unfinished dragon stand overnight. She works 8-hour days. Her problem is to plan a work schedule that enables her to paint all of the dragons in the fewest number of work days.

(a) Formulate the problem for solution by the column generation technique.

(b) Perform one iteration of the column generation technique.

(c) Perform another iteration.

*(d) Complete the solution of the problem.

(e) Enumerate all of the "reasonable" daily schedules, and formulate an LP problem to solve the painter's problem.

†(f) Use SOLVER or the branch and bound tutorial of TUTOR to find the LP and the integer solution to the problem from part (e). (Express the objective function in terms of the *nonbasic* variables before using SOLVER or TUTOR.)

15. Explain why rounding x_1 in (12) necessarily yields an optimal solution to the integer problem.

CHECKLIST: CHAPTER 8

DEFINITIONS
Integer programming   399
Pure integer problem   399
Mixed integer problem   399
Pure binary problem (zero-one problem)   399
Node   414
Binary tree   414
Dangling node   415
Fathomed node   415
Binary problem in standard form   426
Partial solution with k elements   426
Feasible partial solution with k elements   427
Infeasible partial solution with k elements   427
Feasible extension   427
Fathomed partial solution   428
Sequel   428
Adjunct   429
Surrogate constraint   439
Gomory cut   446
Piecewise linear function   458
Step function   471

CONCEPTS AND RESULTS
Alternative sets of constraints   402
If ... then constraints   406
Branch and Bound Algorithm   416
Implicit Enumeration Algorithm   429
Traveling Salesman Problem   440
Gomory Cutting Plane Method   450
Piecewise linear objective functions   (Theorems 1 and 2, Section 8.6)   459
Column generating procedure   464
Stock-cutting problem   409, 464

Chapter 9

THE TRANSPORTATION PROBLEM

9.1 CHOOSING THE ENTERING BASIC VARIABLE

In this and the next two sections we discuss in some detail the transportation problem that we introduced in Section 1.2 and solved with the simplex method in Section 4.5. Historically, the transportation problem provided one of the first significant applications of linear programming. This kind of problem arises frequently not only in the context of resource allocation but also in such diverse areas as scheduling, physics, and finance.

In Section 1.2 we described the general transportation problem as follows. Suppose there are m supply points, $S_1, S_2, \ldots, S_m$, for a product P and n demand points, $D_1, D_2, \ldots, D_n$, for this product. For $1 \leq i \leq m$, $1 \leq j \leq n$, define

c_{ij}:   the per-unit cost of shipping product P from supply point S_i to demand point D_j;

s_i: the number of units of product P available for shipment at supply point S_i;

s: the total supply ($s = s_1 + s_2 + \cdots + s_m$);

d_j: the number of units of product P required by demand point D_j;

d: the total demand ($d = d_1 + d_2 + \cdots + d_n$); and

x_{ij}:   the number of units of product P shipped from supply point S_i to demand point D_j.

The existence of feasible solutions to transportation problems depends on the relative size of the total supply s and the total demand d. If $s < d$, then some of the demand cannot be met; and if $s > d$, then some of the supply will not be used. We call problems with $s \neq d$ *unbalanced* transportation problems. We can expand any unbalanced problem to a balanced one by adding either a "dummy" supply or demand point, and we can obtain a solution to the original problem from the expanded problem by deleting quantities shipped from a dummy supply point or to a dummy destination point. Examples 1 and 2 illustrate the conversion of unbalanced problems to balanced ones.

EXAMPLE 1

In Section 1.2 we learned how to represent transportation problems by cost tables, such as Table 1,

Table 1

Supply points	Demand points				Supply
	D_1	D_2	D_3	D_4	
S_1	1	3	12	15	100
S_2	5	4	9	7	120
S_3	7	6	20	10	90
Demand	60	80	70	60	

Since $s = 100 + 120 + 90 = 310$ and $d = 60 + 80 + 70 + 60 = 270$, we have an unbalanced cost table. To balance it, we introduce a dummy destination point D^*. We assign a cost of 0 to ship from any supply point to D^* and establish a demand at D^* of $s - d = 40$. Thus, all excess supplies will be shipped to D^*. The augmented cost table (Table 2) is

Table 2

Supply points	Demand points					Supply
	D_1	D_2	D_3	D_4	D^*	
S_1	1	3	12	15	0	100
S_2	5	4	9	7	0	120
S_3	7	6	20	10	0	90
Demand	60	80	70	60	40	

We obtain the solution to the original problem from the solution to the augmented problem by leaving whatever goods were shipped to D^* at the supply points. ◆

EXAMPLE 2

Consider the cost table (Table 3)

Table 3

Supply points	Demand points					Supply
	D_1	D_2	D_3	D_4	D_5	
S_1	1	3	12	15	10	100
S_2	5	4	9	7	8	120
Demand	60	80	70	60	40	

Since $d = 60 + 80 + 70 + 60 + 40 = 310$ and $s = 100 + 120 = 220$, we have an unbalanced problem. This time we introduce a dummy supply point S^*, from which we ship goods to any destination at ∞ cost (actually, in solving the problem, we replace ∞ by a higher cost than any other supply route). The supply at S^* is $s^* = 310 - 220 = 90$. The augmented cost table (Table 4) is

Table 4

Supply points	Demand points					Supply
	D_1	D_2	D_3	D_4	D_5	
S_1	1	3	12	15	10	100
S_2	5	4	9	7	8	120
S^*	∞	∞	∞	∞	∞	90
Demand	60	80	70	60	40	

After solving the augmented problem, we obtain the solution to the original problem by deleting all supplies sent from S^*. In this way, the most expensive demands to supply will go unfilled. ◆

Since we can convert any unbalanced problem to a balanced one using the method of either Example 1 or Example 2, it suffices to develop a method for solving the balanced problem. In standard LP notation, the balanced problem is

$$\text{Minimize:} \quad z = \sum_{i=1}^{m} \sum_{j=1}^{n} c_{ij} x_{ij}$$

$$\text{Subject to:} \quad \sum_{j=1}^{n} x_{ij} = s_i; \quad 1 \le i \le m \quad \text{(Supply constraints)}$$

$$\sum_{i=1}^{m} x_{ij} = d_j; \quad 1 \le j \le n \quad \text{(Demand constraints)} \tag{1}$$

$$x_{ij} \ge 0 \quad \text{and} \quad \text{integer,} \quad 1 \le i \le m, \quad 1 \le j \le n$$

We could solve (1) using the simplex method. However, by taking advantage of the special structure of the constraints of (1), we can derive a more efficient way for solving transportation problems. For small problems this is of no great consequence; but many transportation problems involve hundreds or even thousands of constraints, and for such problems the simplex method is often impractical.

The simplex method always produces an integer solution to a transportation problem. This fact is a consequence of the special structure of the constraint equations. We will not prove this fact since doing so requires a greater knowledge of linear algebra than we presume.

We will use the following example to motivate (and clarify) the ideas underlying the method we introduce for solving transportation problems.

EXAMPLE 3

A manufacturer of a single product A has two plants P_1 and P_2 and three warehouses W_1, W_2, and W_3. Table 5 is the cost table giving the unit costs for shipping a unit of product A from plant P_i to warehouse W_j. The weekly number of units of product A produced by the plants is found to the right of the table, and the weekly number of units demanded by each warehouse is found below the table.

Table 5

Plant	Warehouse			Supply
	W_1	W_2	W_3	
P_1	4	2	3	30
P_2	2	4	3	20
Demand	15	10	25	

From Table 5 and (1) we see that the LP formulation of this transportation problem is

$$\text{Minimize:} \quad z = 4x_{11} + 2x_{12} + 3x_{13} + 2x_{21} + 4x_{22} + 3x_{23}$$

$$
\begin{aligned}
\text{Subject to:} \quad & x_{11} + x_{12} + x_{13} && && = 30 \\
& && x_{21} + x_{22} + x_{23} && = 20 \\
& x_{11} && + x_{21} && = 15 \\
& \quad x_{12} && + x_{22} && = 10 \\
& \quad\quad x_{13} && + x_{23} && = 25
\end{aligned}
\tag{2}
$$

$$x_{ij} \geq 0 \quad \text{and} \quad \text{integer}, \quad 1 \leq i \leq 2, \quad 1 \leq j \leq 3$$

In Section 9.3 we will discuss two methods for finding an initial basic feasible solution to (2). However, at this stage it is instructive to apply the two-phase method.

Introducing artificial variables to (2), we obtain the initial tableau

	x_{11}	x_{12}	x_{13}	x_{21}	x_{22}	x_{23}	y_1	y_2	y_3	y_4	y_5		
y_1	1	1	1	0	0	0	1	0	0	0	0	30	
y_2	0	0	0	1	1	1	0	1	0	0	0	20	
y_3	1	0	0	1	0	0	0	0	1	0	0	15	(3)
y_4	0	1	0	0	1	0	0	0	0	1	0	10	
y_5	0	0	1	0	0	1	0	0	0	0	1	25	
(z)	4	2	3	2	4	3	0	0	0	0	0	0	
(w)	-2	-2	-2	-2	-2	-2	0	0	0	0	0	-100	

Application of phase 1 of the two-phase method to (3) yields the tableau

	x_{11}	x_{12}	x_{13}	x_{21}	x_{22}	x_{23}	y_1	y_2	y_3	y_4	y_5		
x_{13}	0	0	1	0	0	1	1	1	-1	-1	0	25	
x_{22}	-1	0	0	0	1	1	0	1	-1	0	0	5	
x_{21}	1	0	0	1	0	0	0	0	1	0	0	15	(4)
x_{12}	1	1	0	0	0	-1	0	-1	1	1	0	5	
y_5	0	0	0	0	0	0	-1	-1	1	1	1	0	
(z)	4	0	0	0	0	-2	-3	-5	3	1	0	-135	
(w)	0	0	0	0	0	0	2	2	0	0	0	0	

Note that in tableau (4) an artificial variable remains as a basic variable. This is always the case because, as we show next, at least one constraint in any transportation problem is redundant. Thus, a transportation problem with $m + n$ constraints has bases containing no more than $m + n - 1$ *nonartificial* variables. ◆

Theorem 1. At least one of the constraints in (1) is redundant.

PROOF. We outline the proof and in Exercise 8 ask you to fill in the details.

(i) Add all of the supply constraints in (1).

(ii) Eliminate the nth demand constraint in (1), and add the remaining demand constraints together.

(iii) Show that the difference of the sums found in (i) and (ii) is the demand constraint

$$x_{1n} + x_{2n} + \cdots + x_{mn} = d_n \qquad (5)$$

(iv) Conclude that the demand constraint (5) is a combination of the other supply and demand constraints and hence is redundant.   ∎

Suppose we have obtained a basic feasible solution to (1). We use what is commonly called the *uv-method* to find the new entering basic variable. To understand why and how this method works, we first convert (1) to the equivalent maximization problem and use matrix notation to rewrite (1) as

$$\text{Maximize:} \quad z = -\mathbf{cx}$$

$$\text{Subject to:} \quad A\mathbf{x} = \mathbf{b} \tag{6}$$

$$\mathbf{x} \geq \mathbf{0} \quad \text{and} \quad \text{integer}$$

where **c** is the cost matrix of (1), A is the $(m + n) \times n$ constraint matrix, and **b** is the resource column. Introducing $m + n$ artificial variables to (6), we obtain the initial simplex tableau

$$\left[\begin{array}{c|c|c} A & I & \mathbf{b} \\ \hline \mathbf{c} & 0 & 0 \end{array} \right] \tag{7}$$

Observe that I is the $m + n$ identity matrix and that **c** has the correct sign since the cost matrix in (6) is $-\mathbf{c}$. Suppose B is the basis matrix corresponding to the current basic feasible solution. By our previous comments, one column of B corresponds to an artificial variable. As we showed in Theorem 2 of Section 4.4, to obtain the next tableau it suffices to premultiply (7) by

$$\left[\begin{array}{c|c} B^{-1} & 0 \\ \hline -\mathbf{c}_B B^{-1} & 1 \end{array} \right] \tag{8}$$

Doing so, we obtain the tableau

$$\left[\begin{array}{c|ccccccccc|c} B^{-1}A & & & & & B^{-1} & & & & & \mathbf{b}^* \\ \hline \mathbf{c}^* & -u_1 & -u_2 & \cdots & -u_m & -v_1 & -v_2 & \cdots & -v_n & & z^* \end{array} \right] \tag{9}$$

For notational convenience we have used negative signs to write the uv entries, which result from premultiplying the middle column of (7) by the bottom row of (8) as

$$-\mathbf{c}_B B^{-1}I + [0 \quad 0 \quad \cdots \quad 0] = -\mathbf{c}_B B^{-1}$$

$$= [-u_1 \quad -u_2 \quad \cdots \quad -u_m \quad -v_1 \quad -v_2 \quad \cdots \quad -v_n] \tag{10}$$

The entries $-u_i$ and $-v_i$ in (10) are the dual prices. Now we use (10) to express the objective-row entries comprising $\mathbf{c}^*$ in terms of these dual prices. The entry c_{ij}^* in the column of (9) corresponding to x_{ij} is

$$c_{ij}^* = -\mathbf{c}_B B^{-1}A_{ij} + c_{ij} \tag{11}$$

where A_{ij} is the constraint matrix column corresponding to the variable x_{ij}. It is at this point that we take advantage of the special nature of the transportation constraint matrix. Note that for each i,j the constraint column A_{ij} in (7) has a 1 in its ith row, a 1 in its $(m + j)$th row, and 0's elsewhere. Thus, from (10)

$$-\mathbf{c}_B B^{-1} A_{ij} = \begin{bmatrix} -u_1 & -u_2 & \cdots & -u_m & -v_1 & -v_2 & \cdots & -v_n \end{bmatrix} \begin{bmatrix} 0 \\ 0 \\ \vdots \\ 1 \\ \vdots \\ 1 \\ \vdots \\ 0 \end{bmatrix} = -u_i - v_j \tag{12}$$

and from (11) and (12) we have

$$c_{ij}^* = -\mathbf{c}_B B^{-1} A_{ij} + c_{ij} = -u_i - v_j + c_{ij} \tag{13}$$

Now observe that if x_{ij} is a basic variable, then the corresponding objective-row entry c_{ij}^* is 0; therefore, if x_{ij} is a basic variable, then

$$c_{ij} - u_i - v_j = 0 \tag{14}$$

By Theorem 1 there are fewer equations of the form (14) than there are unknowns, where the unknowns are the u_i and the v_j. This enables us to set one of the unknowns equal to 0 and solve for the remaining ones.

From (13) it also follows that the new entering basic variable corresponds to the most negative of the values

$$c_{ij}^* = c_{ij} - u_i - v_j \qquad (x_{ij} \text{ is a nonbasic variable}) \tag{15}$$

As usual, if all of the values defined by (15) are nonnegative, then we have obtained an optimal solution; if one or more of these values is negative, we select the new entering basic variable to correspond to the most negative of these values (actually, we could choose any negative value).

We now apply these ideas to Example 3. From tableau (4) we see that the current basic variables are $x_{12}, x_{13}, x_{21}, x_{22}$. (We ignore the artificial basic variables.) From (14) and Table 5 in Example 3 we have

$$u_1 + v_2 = 2$$
$$u_1 + v_3 = 3$$
$$u_2 + v_1 = 2 \tag{16}$$
$$u_2 + v_2 = 4$$

Observe that we have four equations in five unknowns (which is consistent with our previous discussion). This allows us to set one of the unknowns, say u_1, equal to 0 and solve for the others. If we set $u_1 = 0$, then from (16) we have

$$
\begin{aligned}
u_1 &= 0 & v_1 &= 0 \\
u_2 &= 2 & v_2 &= 2 \\
& & v_3 &= 3
\end{aligned}
\tag{17}
$$

To find the entering basic variable we now use (15) and (17) to calculate the objective-row values corresponding to the current nonbasic variables (Table 6).

Table 6

Nonbasic variable x_{ij}	Objective-row entry corresponding to x_{ij}
x_{11}	$c_{11} - u_1 - v_1 = 4 - 0 - 0 = \quad 4$
x_{23}	$c_{23} - u_2 - v_3 = 3 - 2 - 3 = -2$

From Table 6 we see that x_{23} corresponds to the most (in this case, the only) negative objective-row entry and that, therefore, this variable is the entering basic variable.

In the next section we will see how to find departing basic variables (and additional entering basic variables) for this problem.

EXAMPLE 4

In Section 9.3 we will obtain the feasible solution

$$
x_{11} = 70,\ x_{12} = 10,\ x_{22} = 80,\ x_{23} = 60,\ x_{33} = 40,\ x_{34} = 50 \tag{18}
$$

to the transportation problem

Minimize: $\quad z = 4x_{11} + 8x_{12} + 6x_{13} + 2x_{14} + 5x_{21} + x_{22} + 9x_{23}$
$\qquad\qquad + 6x_{24} + 2x_{31} + 9x_{32} + 7x_{33} + 8x_{34}$

Subject to:

$$
\begin{aligned}
x_{11} + x_{12} + x_{13} + x_{14} & & & & & = 80 \\
x_{21} + x_{22} + x_{23} + x_{24} & & & & & = 140 \\
x_{31} + x_{32} + x_{33} + x_{34} & & & & & = 90 \\
x_{11} \qquad\qquad + x_{21} \qquad\qquad + x_{31} & & & & & = 70 \\
x_{12} \qquad\qquad + x_{22} \qquad\qquad + x_{32} & & & & & = 90 \\
x_{13} \qquad\qquad + x_{23} \qquad\qquad + x_{33} & & & & & = 100 \\
x_{14} \qquad\qquad + x_{24} \qquad\qquad + x_{34} & & & & & = 50
\end{aligned}
\tag{19}
$$

$$
x_{ij} \ge 0 \quad \text{and} \quad \text{integer,} \qquad 1 \le i \le 3, \qquad 1 \le j \le 4
$$

To find the next entering basic variable we proceed as before. The current basic variables are $x_{11}, x_{12}, x_{22}, x_{23}, x_{33},$ and x_{34}. Thus, by (14) we have

$$
u_1 + v_1 = 4
$$
$$
u_1 + v_2 = 8
$$

$$u_2 + v_2 = 1$$

$$u_2 + v_3 = 9$$

$$u_3 + v_3 = 7$$

$$u_3 + v_4 = 8$$

Note that we have six equations in seven unknowns, as we would expect. Setting $u_1 = 0$, we obtain

$$\begin{aligned} u_1 &= 0 & v_1 &= 4 \\ u_2 &= -7 & v_2 &= 8 \\ u_3 &= -9 & v_3 &= 16 \\ & & v_4 &= 17 \end{aligned} \tag{20}$$

To find the entering basic variable, we use (15) and (20) to calculate the objective row entries corresponding to the current nonbasic variables (Table 7).

Table 7

Nonbasic variable x_{ij}	Objective row entry corresponding to x_{ij}
x_{13}	$c_{13} - u_1 - v_3 = 6 - 0 - 16 = -10$
x_{14}	$c_{14} - u_1 - v_4 = 2 - 0 - 17 = -15$
x_{21}	$c_{21} - u_2 - v_1 = 5 + 7 - 4 = 8$
x_{24}	$c_{24} - u_2 - v_4 = 6 + 7 - 17 = -4$
x_{31}	$c_{31} - u_3 - v_1 = 2 + 9 - 4 = 7$
x_{32}	$c_{32} - u_3 - v_2 = 9 + 9 - 8 = 10$

From Table 7 we see that x_{14} is the entering basic variable. In the next section we will determine the departing basic variables (and additional entering basic variables) for this problem.

♦

Exercises

The cost tables given in Exercises 1–4 describe unbalanced transportation problems. Augment the tables with a dummy supply or demand point to form a balanced problem.

1.

Supply points	Demand points					Supply
	D_1	D_2	D_3	D_4	D_5	
S_1	4	6	10	5	15	220
S_2	15	7	10	8	9	110
S_3	17	6	22	12	12	190
Demand	70	80	90	40	50	

2.

Supply points	Demand points					Supply
	D_1	D_2	D_3	D_4	D_5	
S_1	6	3	12	15	5	150
S_2	15	8	14	4	12	190
S_3	17	8	24	12	10	120
Demand	70	80	90	40	50	

3.

Supply points	Demand points					Supply
	D_1	D_2	D_3	D_4	D_5	
S_1	6	12	9	15	5	50
S_2	25	18	14	4	12	90
S_3	13	8	24	12	10	30
S_4	21	7	18	9	12	60
Demand	70	80	90	40	50	

4.

Supply points	Demand points					Supply
	D_1	D_2	D_3	D_4	D_5	
S_1	12	6	9	12	7	85
S_2	25	16	12	4	12	55
S_3	13	8	18	12	10	95
S_4	18	6	15	9	12	40
Demand	65	85	98	70	60	

5. In Example 3, let $s = 50$ be the total supply, $s_1 = 30$, $s_2 = 20$, $d_1 = 15$, $d_2 = 10$, $d_3 = 25$. Show that

$$x_{11} = \frac{s_1 d_1}{s} = 9 \qquad x_{12} = \frac{s_1 d_2}{s} = 6 \qquad x_{13} = \frac{s_1 d_3}{s} = 15$$

$$x_{21} = \frac{s_2 d_1}{s} = 6 \qquad x_{22} = \frac{s_2 d_2}{s} = 4 \qquad x_{23} = \frac{s_2 d_3}{s} = 10$$

is a solution to the constraints of (2). Also observe that this is not a basic feasible solution and hence cannot be used to start the *uv*-method.

6. For any balanced transportation problem, show that $x_{ij} = (s_i d_j)/s$ is a solution of the constraints of (1). (Notice the last sentence of Exercise 5.)

7. Find another basic feasible solution to the transportation problem (2) by performing a pivot on tableau (4).

8. Complete the proof of Theorem 1.

In Exercises 9–16 you are given a balanced transportation problem and a basic feasible solution. Use the uv-method to find the next entering basic variable.

9. Minimize: $z = \quad 6x_{11} + 7x_{12} + 3x_{13}$
$$+ 5x_{21} + 4x_{22} + 9x_{23}$$
$$+ 4x_{31} + 6x_{32} + 3x_{33}$$

Subject to: $x_{11} + x_{12} + x_{13} = 45$
$$x_{21} + x_{22} + x_{23} = 65$$
$$x_{31} + x_{32} + x_{33} = 75$$
$$x_{11} + x_{21} + x_{31} = 35$$
$$x_{12} + x_{22} + x_{32} = 85$$
$$x_{13} + x_{23} + x_{33} = 65$$

$x_{ij} \geq 0$ and integer, $1 \leq i, j \leq 3$

Basic feasible solution:

$$x_{32} = 40, \qquad x_{22} = 45, \qquad x_{23} = 20, \qquad x_{13} = 45, \qquad x_{31} = 35$$

10. Minimize: $z = \quad 5x_{11} + 3x_{12} + 4x_{13}$
$$+ 3x_{21} + 7x_{22} + 8x_{23}$$
$$+ 5x_{31} + 8x_{32} + 9x_{33}$$

Subject to: $x_{11} + x_{12} + x_{13} = 50$
$$x_{21} + x_{22} + x_{23} = 40$$
$$x_{31} + x_{32} + x_{33} = 75$$
$$x_{11} + x_{21} + x_{31} = 35$$
$$x_{12} + x_{22} + x_{32} = 95$$
$$x_{13} + x_{23} + x_{33} = 35$$

$x_{ij} \geq 0$ and integer, $1 \leq i, j \leq 3$

Basic feasible solution:

$$x_{12} = 15, \qquad x_{22} = 40, \qquad x_{13} = 35, \qquad x_{31} = 35, \qquad x_{32} = 40$$

11. Minimize: $z = \quad 6x_{11} + 7x_{12} + 3x_{13} + 5x_{14}$
$$+ 5x_{21} + 4x_{22} + 9x_{23} + 6x_{24}$$
$$+ 4x_{31} + 3x_{32} + 3x_{33} + 5x_{34}$$
$$+ 8x_{41} + 9x_{42} + 7x_{43} + 9x_{44}$$

Subject to: $x_{11} + x_{12} + x_{13} + x_{14} = 40$
$$x_{21} + x_{22} + x_{23} + x_{24} = 40$$
$$x_{31} + x_{32} + x_{33} + x_{34} = 45$$
$$x_{41} + x_{42} + x_{43} + x_{44} = 95$$
$$x_{11} + x_{21} + x_{31} + x_{41} = 35$$
$$x_{12} + x_{22} + x_{32} + x_{42} = 75$$
$$x_{13} + x_{23} + x_{33} + x_{43} = 65$$
$$x_{14} + x_{24} + x_{34} + x_{44} = 45$$

$x_{ij} \geq 0$ and integer, $1 \leq i, j \leq 4$

Basic feasible solution:

$$x_{41} = 35, \qquad x_{22} = 15, \qquad x_{32} = 45, \qquad x_{42} = 15,$$
$$x_{13} = 40, \qquad x_{23} = 25, \qquad x_{44} = 45$$

12. Minimize: $z = \quad 5x_{11} + 2x_{12} + 13x_{13} + 3x_{14}$
$$+ \;\; 8x_{21} + 6x_{22} + \;\; 3x_{23} + 4x_{24}$$
$$+ \;14x_{31} + 6x_{32} + \;\; 5x_{33} + 6x_{34}$$
$$+ \;\; 7x_{41} + 8x_{42} + \;\; 3x_{43} + 7x_{44}$$

Subject to: $x_{11} + x_{12} + x_{13} + x_{14} = 25$
$$x_{21} + x_{22} + x_{23} + x_{24} = 65$$
$$x_{31} + x_{32} + x_{33} + x_{34} = 45$$
$$x_{41} + x_{42} + x_{43} + x_{44} = 85$$
$$x_{11} + x_{21} + x_{31} + x_{41} = 35$$
$$x_{12} + x_{22} + x_{32} + x_{42} = 35$$
$$x_{13} + x_{23} + x_{33} + x_{43} = 85$$
$$x_{14} + x_{24} + x_{34} + x_{44} = 65$$

$$x_{ij} \geq 0 \quad \text{and} \quad \text{integer}, \qquad 1 \leq i, j \leq 4$$

Basic feasible solution:

$$x_{31} = 15, \qquad x_{13} = 25, \qquad x_{32} = 30, \qquad x_{41} = 20,$$
$$x_{23} = 60, \qquad x_{22} = 5, \qquad x_{44} = 65$$

13.

Supply points	Demand points			Supply
	D_1	D_2	D_3	
S_1	7	9	4	100
S_2	3	2	5	120
S_3	6	1	4	210
S_4	2	5	6	150
Demand	220	160	200	

Basic feasible solution:

$x_{32} = 140, \qquad x_{41} = 150, \qquad x_{13} = 100, \qquad x_{23} = 100, \qquad x_{31} = 70, \qquad x_{22} = 20$

14.

Supply points	Demand points			Supply
	D_1	D_2	D_3	
S_1	6	5	8	150
S_2	9	1	3	100
S_3	9	5	6	200
S_4	9	6	2	150
Demand	250	240	110	

Basic feasible solution:

$$x_{12} = 40, \qquad x_{13} = 110, \qquad x_{22} = 100,$$
$$x_{32} = 100, \qquad x_{31} = 100, \qquad x_{41} = 150$$

15.

Supply points	Demand points					Supply
	D_1	D_2	D_3	D_4	D_5	
S_1	3	4	2	9	7	250
S_2	6	2	4	3	3	200
S_3	7	2	4	5	6	140
S_4	7	3	6	2	7	260
Demand	250	240	110	90	160	

Basic feasible solution:

$$x_{42} = 10, \qquad x_{32} = 140, \qquad x_{22} = 90, \qquad x_{24} = 0,$$
$$x_{15} = 160, \qquad x_{23} = 110, \qquad x_{41} = 250, \qquad x_{14} = 90$$

16.

Supply points	Demand points					Supply
	D_1	D_2	D_3	D_4	D_5	
S_1	2	1	6	3	9	125
S_2	9	3	7	3	5	320
S_3	4	5	7	2	4	230
S_4	6	2	1	4	9	420
Demand	150	340	210	190	205	

Basic feasible solution:

$$x_{32} = 70, \qquad x_{15} = 125, \qquad x_{23} = 50, \qquad x_{25} = 80,$$
$$x_{33} = 160, \qquad x_{24} = 190, \qquad x_{41} = 150, \qquad x_{42} = 270$$

9.2 CHOOSING THE DEPARTING BASIC VARIABLE

We now turn to the procedure for selecting departing basic variables in transportation problems. This procedure involves the use of appropriate tables and loops. To illustrate this procedure we return to the transportation problem in Example 4 of Section 9.1. With this problem we will associate transportation cost tables of the following form.

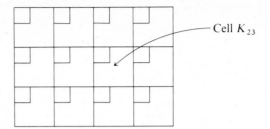

Cell K_{23}

We refer to the larger rectangles in this table as *cells*, letting K_{ij} denote the cell lying in the ith row and jth column of the table. Cell K_{ij} corresponds to the variable x_{ij}. In the rectangle in the upper left corner of each cell K_{ij}, we will give the cost coefficient c_{ij}. In addition, to the right of the cells we will list the supply constraints, and below the cells we will list the demand constraints of the problem.

In the appropriate cells in Table 1 we also include the values of the initial basic variables (as given in Example 4 of Section 9.1). The blank cells correspond to nonbasic variables with value 0.

Table 1

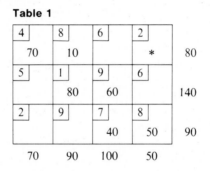

The asterisk in cell K_{14} indicates that x_{14} is the entering basic variable (as we saw in Example 4, Section 9.1). To determine the departing basic variable we form a *loop* that includes cell K_{14} together with some (but not all) of the cells corresponding to the current basic variables. To find this loop we first cross out (see shaded portion of Table 1a) a row or a column in Table 1 that has a single positive entry. (We count the asterisk * as a positive entry.) In the resulting table we again cross out a row or column containing a single positive entry and continue this process until there is no row or column with a single positive entry. The cells that have positive entries and are not crossed out, together with the cell corresponding to the entering basic variable (the cell marked by *), form the desired loop. As we will see, the following properties characterize a loop:

1. Two consecutive cells in the loop lie in the same row or column.

2. No three consecutive cells in the loop lie in the same row or column.

3. The first and last cells of the loop lie in the same row or column.

4. No cell occurs twice in the loop.

In our example we find the loop after just one step.

Table 1a

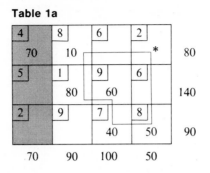

Next, we find the smallest value in the loop that is in a cell adjacent (in the loop) to the cell containing ∗. From Table 1a we see that this value is 10. We augment the cell with the asterisk by this value and then alternately subtract and add this amount as we go around the loop. Thus, we obtain Table 2.

Table 2

4	8	6	2	
70	0		10	80
5	1	9	6	
	90	50		140
2	9	7	8	
		50	40	90
70	90	100	50	

From Table 2 we see that the variable x_{12} has taken on the value 0 and, hence, is the departing basic variable. The current basic feasible solution is

$$x_{11} = 70, \qquad x_{14} = 10, \qquad x_{22} = 90, \qquad x_{23} = 50, \qquad x_{33} = 50, \qquad x_{34} = 40$$

It is interesting to note that the result of this procedure is the same as the result obtained by applying the simplex method to the problem in Example 4 of Section 9.1. In fact, this will always be the case, although a proof of this statement is beyond the scope of this text. For future reference, we now list the tableaus (Tableaus 1–4) obtained by applying the simplex method to the LP problem in Example 4 of Section 9.1, where the initial basic solution is that given in Example 4 of that section. We obtain the initial tableau by first forming a tableau for the LP problem (19) of Section 9.1 with the last constraint omitted (recall that it is redundant). We then perform pivots (without using artificial variables) to bring the variables in (18) of Section 9.1 into the basis.

Tableau 1 Corresponds to the Initial Basic Feasible Solution

	x_{11}	x_{12}	x_{13}	x_{14}	x_{21}	x_{22}	x_{23}	x_{24}	x_{31}	x_{32}	x_{33}	x_{34}	
x_{11}	1	0	0	0	1	0	0	0	1	0	0	0	70
x_{22}	0	0	-1	-1	1	1	0	0	1	1	0	0	80
x_{33}	0	0	0	-1	0	0	0	-1	1	1	1	0	40
x_{12}	0	1	1	1	-1	0	0	0	-1	0	0	0	10
x_{23}	0	0	1	1	0	0	1	1	-1	-1	0	0	60
x_{34}	0	0	0	1	0	0	0	1	0	0	0	1	50
	0	0	-10	-15	8	0	0	-4	7	10	0	0	-1660

Tableau 2

	x_{11}	x_{12}	x_{13}	x_{14}	x_{21}	x_{22}	x_{23}	x_{24}	x_{31}	x_{32}	x_{33}	x_{34}	
x_{11}	1	0	0	0	1	0	0	0	1	0	0	0	70
x_{22}	0	1	0	0	0	1	0	0	0	1	0	0	90
x_{33}	0	1	1	0	-1	0	0	-1	0	1	1	0	50
x_{14}	0	1	1	1	-1	0	0	0	-1	0	0	0	10
x_{23}	0	-1	0	0	1	0	1	1	0	-1	0	0	50
x_{34}	0	-1	-1	0	1	0	0	1	1	0	0	1	40
	0	15	5	0	-7	0	0	-4	-8	10	0	0	-1510

Tableau 3

	x_{11}	x_{12}	x_{13}	x_{14}	x_{21}	x_{22}	x_{23}	x_{24}	x_{31}	x_{32}	x_{33}	x_{34}	
x_{11}	1	1	1	0	0	0	0	-1	0	0	0	-1	30
x_{22}	0	1	0	0	0	1	0	0	0	1	0	0	90
x_{33}	0	1	1	0	-1	0	0	-1	0	1	1	0	50
x_{14}	0	0	0	1	0	0	0	1	0	0	0	1	50
x_{23}	0	-1	0	0	1	0	1	1	0	-1	0	0	50
x_{31}	0	-1	-1	0	1	0	0	1	1	0	0	1	40
	0	7	-3	0	1	0	0	4	0	10	0	8	-1190

Tableau 4 Final

	x_{11}	x_{12}	x_{13}	x_{14}	x_{21}	x_{22}	x_{23}	x_{24}	x_{31}	x_{32}	x_{33}	x_{34}	
x_{13}	1	1	1	0	0	0	0	-1	0	0	0	-1	30
x_{22}	0	1	0	0	0	1	0	0	0	1	0	0	90
x_{33}	-1	0	0	0	-1	0	0	0	0	1	1	1	20
x_{14}	0	0	0	1	0	0	0	1	0	0	0	1	50
x_{23}	0	-1	0	0	1	0	1	1	0	-1	0	0	50
x_{31}	1	0	0	0	1	0	0	0	1	0	0	0	70
	3	10	0	0	1	0	0	1	0	10	0	5	-1100

Observe that Table 2 corresponds to Tableau 2. From Tableau 2 we see that the current transportation cost is $1510 (which appears as -1510 in the tableau because we converted the original minimization problem to a maximization problem). We can also obtain this value from Table 2 since

$$4 \cdot 70 + 2 \cdot 10 + 1 \cdot 90 + 9 \cdot 50 + 7 \cdot 50 + 8 \cdot 40 = 1510$$

The current basic variables are $x_{11}, x_{14}, x_{22}, x_{23}, x_{33}$, and x_{34}. To find the next entering basic variable we proceed as in Section 9.1 and use (14) of that section to obtain the equations

$$
\begin{array}{ll}
u_1 + v_1 = 4 & u_2 + v_3 = 9 \\
u_1 + v_4 = 2 & u_3 + v_3 = 7 \\
u_2 + v_2 = 1 & u_3 + v_4 = 8
\end{array}
\tag{1}
$$

If we set $u_1 = 0$ in (1), we have

$$
\begin{array}{ll}
u_1 = 0 & v_1 = 4 \\
u_2 = 8 & v_2 = -7 \\
u_3 = 6 & v_3 = 1 \\
 & v_4 = 2
\end{array}
\tag{2}
$$

From (2) we obtain the objective-row entries corresponding to the current nonbasic variables.

Nonbasic variable	Objective-row entry
x_{12}	$c_{12} - u_1 - v_2 = 8 + 7 = 15$
x_{13}	$c_{13} - u_1 - v_3 = 6 - 1 = 5$
x_{21}	$c_{21} - u_2 - v_1 = 5 - 8 - 4 = -7$
x_{24}	$c_{24} - u_2 - v_4 = 6 - 8 - 2 = -4$
x_{31}	$c_{31} - u_3 - v_1 = 2 - 6 - 4 = -8$
x_{32}	$c_{32} - u_3 - v_2 = 9 - 6 + 7 = 10$

Since, in this table, -8 is the most negative value assigned to a nonbasic variable, we let x_{31} be the entering basic variable. Applying to Table 2 the procedure described above for determining the departing basic variable we obtain Tables 2a, 2b, and 2c.

Table 2a

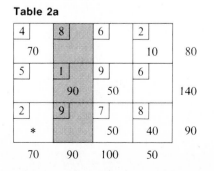

Table 2b

Table 2c

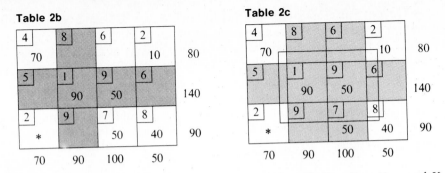

From Table 2c we find that our loop consists of the cells K_{11}, K_{14}, K_{34}, and K_{31}. Since 40 is the smallest value assigned to a cell adjacent (in the loop) to $*$, we augment cell K_{31} by this amount and then alternately decrease and increase the remaining cells in the loop to obtain Table 3.

Table 3

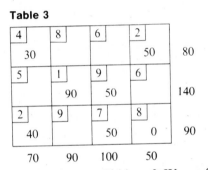

Observe that this table corresponds to Tableau 3. We see from Table 3 that x_{41} is the departing basic variable and that the current basic variables have values

$$x_{11} = 30, \qquad x_{14} = 50, \qquad x_{22} = 90, \qquad x_{23} = 50, \qquad x_{31} = 40, \qquad x_{33} = 50$$

The current transportation cost is

$$4 \cdot 30 + 2 \cdot 50 + 1 \cdot 90 + 9 \cdot 50 + 2 \cdot 40 + 7 \cdot 50 = 1190$$

We now determine the next entering basic variable. Since the current basic variables are $x_{11}, x_{14}, x_{22}, x_{23}, x_{31}, x_{33}$, we proceed as before to obtain the equations

$$
\begin{aligned}
u_1 + v_1 &= 4 & u_2 + v_3 &= 9 \\
u_1 + v_4 &= 2 & u_3 + v_1 &= 2 \\
u_2 + v_2 &= 1 & u_3 + v_3 &= 7
\end{aligned}
\tag{3}
$$

Setting $u_1 = 0$ in (3), we have

$$
\begin{aligned}
u_1 &= 0 & v_1 &= 4 \\
u_2 &= 0 & v_2 &= 1 \\
u_3 &= -2 & v_3 &= 9 \\
 & & v_4 &= 2
\end{aligned}
\tag{4}
$$

From (4) we obtain the following values.

Nonbasic variable	Objective-row entry
x_{12}	$c_{12} - u_1 - v_2 = 8 - 1 = 7$
x_{13}	$c_{13} - u_1 - v_3 = 6 - 9 = -3$
x_{21}	$c_{21} - u_2 - v_1 = 5 - 4 = 1$
x_{24}	$c_{24} - u_2 - v_4 = 6 - 2 = 4$
x_{32}	$c_{32} - u_3 - v_2 = 9 + 2 - 1 = 10$
x_{34}	$c_{34} - u_3 - v_4 = 8 + 2 - 2 = 8$

From the preceding values we find that x_{13} is the new entering basic variable. Applying our procedure for finding the departing basic variable, we obtain Tables 3a and 3b.

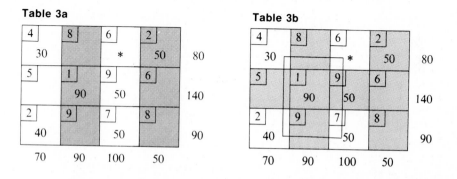

Table 3a

Table 3b

Table 4

From Table 3b we see that the loop consists of the cells $K_{11}, K_{13}, K_{33}, K_{31}$. Since 30 is the smallest value assigned to a cell adjacent to *, we augment cell K_{13} by 30 and then alternately subtract and add 30 to the remaining cells in the loop to obtain Table 4.

Note that $x_{11} = 0$; therefore, x_{11} is the departing basic variable. The new transportation cost is

$$6 \cdot 30 + 2 \cdot 50 + 1 \cdot 90 + 9 \cdot 50 + 2 \cdot 70 + 7 \cdot 20 = 1100 \tag{5}$$

The current basic variables are $x_{13}, x_{14}, x_{22}, x_{23}, x_{31}, x_{33}$. To find the new entering basic variable we use the equations

$$u_1 + v_3 = 6 \qquad u_2 + v_3 = 9$$
$$u_1 + v_4 = 2 \qquad u_3 + v_1 = 2 \qquad (6)$$
$$u_2 + v_2 = 1 \qquad u_3 + v_3 = 7$$

Setting $u_1 = 0$ in (6), we obtain

$$\begin{aligned} u_1 &= 0 & v_1 &= 1 \\ u_2 &= 3 & v_2 &= -2 \\ u_3 &= 1 & v_3 &= 6 \\ & & v_4 &= 2 \end{aligned} \qquad (7)$$

From (7) we obtain the following objective-row entries.

Nonbasic variables	Objective-row entries
x_{11}	$c_{11} - u_1 - v_1 = 4 - 1 = 3$
x_{12}	$c_{12} - u_1 - v_2 = 8 + 2 = 10$
x_{21}	$c_{21} - u_2 - v_1 = 5 - 3 - 1 = 1$
x_{24}	$c_{24} - u_2 - v_4 = 6 - 3 - 2 = 1$
x_{32}	$c_{32} - u_3 - v_2 = 9 - 1 + 2 = 10$
x_{34}	$c_{34} - u_3 - v_4 = 8 - 1 - 2 = 5$

Since all objective-row entries are now nonnegative we have arrived at an optimal solution with basic variables

$$x_{13} = 30, \qquad x_{14} = 50, \qquad x_{22} = 90, \qquad x_{23} = 50, \qquad x_{41} = 70, \qquad x_{43} = 20$$

Thus (5) gives the minimal shipping cost. Observe that Table 4 corresponds to Tableau 4.

We conclude this section with a review of the steps of the transportation algorithm from this section and Section 9.1.

STEP 1. Find an initial basic feasible solution and form the transportation cost table. (We will show you how to do this in Section 9.3.)

STEP 2. Find the entering basic variable.

(a) Solve the equations $u_i + v_j = c_{ij}$ for each pair i, j occurring as subscripts of a basic variable x_{ij}.

(b) Calculate $c_{ij}^* = c_{ij} - u_i - v_j$ for each pair i, j occurring as subscripts of a nonbasic variable x_{ij}.

(c) Select as entering basic variable the variable x_{ij} corresponding to the most negative c_{ij}^* from (b). If all the c_{ij}^* are nonnegative, stop; you have found the optimal solution.

STEP 3. Find the departing basic variable by forming a loop beginning and ending in the cell corresponding to the entering basic variable.

When you find a loop, consider the two cells adjacent to the entering basic variable in the loop, and select the one with the smallest value. The variable corresponding to this cell is the departing basic variable. Place this smallest value in the entering variable cell, and pass around the loop, alternately subtracting and adding this value to the entry in each cell of the loop. Continue performing steps 1 and 2 until all of the cost coefficients calculated in step 2 (c) are nonnegative.

Note: We have suggested that you begin your search for the loop by striking off rows and columns containing only one cell with a positive entry because such cells could not be used to form a loop; however, this step is not necessary. Any loop that you discover will work. Just as it is possible to have ties in choosing the departing basic variable, it is possible to have alternative choices of loops.

Exercises

In Exercises 1–11 you are given a transportation cost table with the entering basic variable marked by *. Find the departing basic variable and the new cost table after the departing basic variable is replaced in the basis with the entering variable.

1.

[7]	[9]	[4]	[3]	supply
		*	210	210
[2]	[5]	[6] 270	[1] 35	305
[4]	[2] 140	[5] 280	[6]	420
[3] 200	[2] 110	[6]	[2]	310
200	250	550	245	

2.

[3]	[3]	[5]	[9]	[7]	supply
				225	225
[7]	[3]	[3] 100	[2] 155	[5] 45	300
[4]	[5] 25	[7] 100	[3] *	[8]	125
[2] 150	[2] 75	[7]	[7]	[7]	225
150	100	200	155	270	

3.

[7]	[9]	[4]	[3]	supply
		210		210
[2]	[5]	[6] 60	[1] 245	305
*				
[4]	[2] 140	[5] 280	[6]	420
[3] 200	[2] 110	[6]	[2]	310
200	250	550	245	

4.

[3]	[3]	[5]	[9]	[7]	supply
	*			225	225
[7]	[3]	[3] 200	[2] 55	[5] 45	300
[4]	[5] 25	[7]	[3] 100	[8]	125
[2] 150	[2] 75	[7]	[7]	[7]	225
150	100	200	155	270	

5.

[7]	[9]	[4]	[3]	supply
		130	80	210
[2]	[5] 140	[6] *	[1] 165	305
[4]	[2]	[5] 420	[6]	420
[3] 60	[2] 250	[6]	[2]	310
60	390	550	245	

6.

[2]	[8]	[4]	[5]	supply
*		295	155	450
[8]	[3] 170	[2] 155	[6]	325
[5] 40	[1] 180	[9]	[4]	220
[2] 110	[2]	[5]	[6]	110
150	350	450	155	

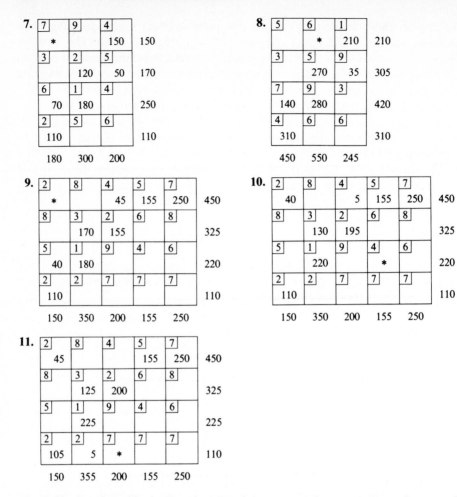

12. Verify that the tables in Exercises 2 and 4 correspond to a sequence of stages of the transportation algorithm. The table obtained by applying the transportation method to the table in Exercise 4 is optimal. What is the optimal solution?

13. Verify that the tables in Exercises 1 and 3 correspond to a sequence of stages of the transportation algorithm. The table obtained by applying the transportation method to the table in Exercise 3 is optimal. What is the optimal solution?

In Exercises 14–17 solve the transportation problems starting at the given basic feasible solution.

14. This problem requires three steps of the transportation method.

$$\text{Minimize:} \quad z = \quad 12x_{11} + 3x_{12} + 6x_{13}$$
$$+ \quad 3x_{21} + 9x_{22} + 4x_{23}$$
$$+ \quad 4x_{31} + 5x_{32} + 2x_{33}$$

Subject to:
$$x_{11} + x_{12} + x_{13} = 600$$
$$x_{21} + x_{22} + x_{23} = 400$$
$$x_{31} + x_{32} + x_{33} = 500$$
$$x_{11} + x_{21} + x_{31} = 100$$
$$x_{12} + x_{22} + x_{32} = 800$$
$$x_{13} + x_{23} + x_{33} = 600$$
$$x_{ij} \geq 0, \qquad 1 \leq i, j \leq 3$$

Initial basic solution:

$$x_{13} = 600, \qquad x_{22} = 400, \qquad x_{23} = 0, \qquad x_{31} = 100, \qquad x_{32} = 400$$

15. Minimize: $z = 3x_{11} + 5x_{12} + 7x_{13} + 6x_{21} + 5x_{22} + 9x_{23}$

Subject to:
$$x_{11} + x_{12} + x_{13} \qquad\qquad = 80$$
$$x_{21} + x_{22} + x_{23} = 40$$
$$x_{11} \qquad\quad + x_{21} \qquad\qquad = 45$$
$$x_{12} \qquad\quad + x_{22} \qquad\quad = 55$$
$$x_{13} \qquad\quad + x_{23} = 20$$
$$x_{ij} \geq 0, \qquad 1 \leq i \leq 2, \qquad 1 \leq j \leq 3$$

Initial basic solution:

$$x_{11} = 5, \qquad x_{12} = 55, \qquad x_{13} = 20, \qquad x_{21} = 40$$

16. This problem requires four steps of the transportation method.

Minimize: $z = \quad 3x_{11} + 6x_{12} + 9x_{13} + 6x_{21} + 5x_{22} + 4x_{23}$
$$\qquad\qquad + 7x_{31} + 8x_{32} + 7x_{33} + 8x_{41} + 2x_{42} + 7x_{43}$$

Subject to:
$$x_{11} + x_{12} + x_{13} = 210$$
$$x_{21} + x_{22} + x_{23} = 205$$
$$x_{31} + x_{32} + x_{33} = 320$$
$$x_{41} + x_{42} + x_{43} = 210$$
$$x_{11} + x_{21} + x_{31} = 350$$
$$x_{12} + x_{22} + x_{32} = 450$$
$$x_{13} + x_{23} + x_{33} = 145$$
$$x_{ij} \geq 0, \qquad 1 \leq i \leq 4, \qquad 1 \leq j \leq 3$$

Initial basic solution:

$$x_{12} = 65, \qquad x_{13} = 145, \qquad x_{22} = 205, \qquad x_{31} = 140, \qquad x_{32} = 180, \qquad x_{41} = 210$$

17. Minimize: $z = 10x_{11} + 7x_{12} + 3x_{13} + 4x_{21} + 5x_{22} + 8x_{23} + 4x_{31} + 3x_{32} + 2x_{33}$

Subject to:
$$x_{11} + x_{12} + x_{13} = 150$$
$$x_{21} + x_{22} + x_{23} = 250$$
$$x_{31} + x_{32} + x_{33} = 300$$
$$x_{11} + x_{21} + x_{31} = 350$$
$$x_{12} + x_{22} + x_{32} = 200$$
$$x_{13} + x_{23} + x_{33} = 150$$
$$x_{ij} \geq 0, \qquad 1 \leq i, j \leq 3$$

Initial basic solution:

$$x_{11} = 150, \qquad x_{21} = 200, \qquad x_{22} = 50, \qquad x_{32} = 150, \qquad x_{33} = 150$$

*18. This problem assumes that you have studied linear algebra, understand the concept of linear dependence, and know what a basis of a vector space is.

(a) Verify that the transportation problem given in Example 4 of Section 9.1 can be written as $Ax = b$, where

$$A = \begin{bmatrix} 1 & 1 & 1 & 1 & 0 & 0 & 0 & 0 & 0 & 0 & 0 & 0 \\ 0 & 0 & 0 & 0 & 1 & 1 & 1 & 1 & 0 & 0 & 0 & 0 \\ 0 & 0 & 0 & 0 & 0 & 0 & 0 & 0 & 1 & 1 & 1 & 1 \\ 1 & 0 & 0 & 0 & 1 & 0 & 0 & 0 & 1 & 0 & 0 & 0 \\ 0 & 1 & 0 & 0 & 0 & 1 & 0 & 0 & 0 & 1 & 0 & 0 \\ 0 & 0 & 1 & 0 & 0 & 0 & 1 & 0 & 0 & 0 & 1 & 0 \\ 0 & 0 & 0 & 1 & 0 & 0 & 0 & 1 & 0 & 0 & 0 & 1 \end{bmatrix}$$

$$\mathbf{x} = \begin{bmatrix} x_{11} \\ x_{12} \\ x_{13} \\ x_{21} \\ x_{22} \\ x_{23} \end{bmatrix} \qquad \mathbf{b} = \begin{bmatrix} 80 \\ 140 \\ 90 \\ 70 \\ 100 \\ 50 \end{bmatrix}$$

(b) The cells in the loop of Table 1a, starting with the entering basic variable x_{14} and proceeding around the loop in a clockwise direction, correspond to the basic variables $x_{14}, x_{34}, x_{33}, x_{23}, x_{22}$, and x_{12}. Verify that the corresponding columns from A are

x_{14}	x_{34}	x_{33}	x_{23}	x_{22}	x_{12}
1	0	0	0	0	1
0	0	0	1	1	0
0	1	1	0	0	0
0	0	0	0	0	0
0	0	0	0	1	1
0	0	0	1	0	0
1	1	0	0	0	0

(c) Verify that you obtain the zero vector when you alternately subtract and add the six vectors given in part (b).

(d) Conclude from part (c) that the six vectors in part (b) are linearly dependent and that, therefore, the entering basic variable x_{14} can be written as a linear combination of the other five vectors.

Although we have avoided saying so earlier in the text, because we want to avoid using much linear algebra, the pivoting process of the simplex method is essentially a way of expressing the column of the A matrix corresponding to the entering basic variable as a linear combination of the columns corresponding to the current basic variables. The loops formed in the transportation method are a means of finding this dependency relation using the simple structure of the A matrix. (For a more complete discussion see [10].)

†19. Solve Exercise 14 with SOLVER.

†20. Solve Exercise 16 with SOLVER.

9.3 FINDING AN INITIAL BASIC FEASIBLE SOLUTION

In Section 9.1 we used phase 1 of the two-phase method to find a basic feasible solution for the transportation problem

$$\text{Minimize:} \quad z = 4x_{11} + 2x_{12} + 3x_{13} + 2x_{21} + 4x_{22} + 3x_{23}$$

$$
\begin{aligned}
\text{Subject to:} \quad x_{11} + x_{12} + x_{13} &\phantom{{}+{}} &= 30 \\
x_{21} + x_{22} + x_{23} &= 20 \\
x_{11} \phantom{+ x_{12} + x_{13}} + x_{21} \phantom{+ x_{22} + x_{23}} &= 15 \\
x_{12} \phantom{+ x_{13}} + x_{22} \phantom{+ x_{23}} &= 10 \\
x_{13} \phantom{+ x_{23}} + x_{23} &= 25
\end{aligned}
$$

$$x_{ij} \geq 0 \quad \text{and} \quad \text{integer,} \quad 1 \leq i \leq 2, \quad 1 \leq j \leq 3$$

We now examine two other methods for finding such a solution, both of which avoid the introduction of artificial variables (and, in fact, avoid the simplex procedure altogether). The first of these methods, the Northwest Corner Method, is perhaps the simplest procedure for finding an initial basic feasible solution to the transportation problem. However, although this method is easy to implement, it usually does not yield a solution that is "close" to the optimal solution of the problem. The second method, Vogel's Approximation Method, which we discuss later in this section, requires more calculations but tends to result in a basic feasible solution that better approximates the optimal solution. We do not explain why either of these methods works because an explanation requires a greater knowledge of linear algebra than we expect most students to have.

Northwest Corner Method

Let K be an $m \times n$ transportation cost table with cells K_{ij}, $1 \leq i \leq m, 1 \leq j \leq n$. The Northwest Corner Method, applied to K, consists of the following steps.

STEP 1. Assign the largest possible value to the cell K_{11} (the cell in the northwest corner of K), taking into account the supply and demand constraints. We call K_{11} a *basic cell.*

STEP 2. Let cell K_{ij} be the most recently determined basic cell. If $j < n$, move to cell $K_{i,j+1}$, and assign to this cell the largest possible positive value, taking into account the supply and demand constraints. Designate $K_{i,j+1}$ as a basic cell.

  If, because $j = n$ or because of the supply and demand constraints, it is impossible to assign a positive value to the cell $K_{i,j+1}$, move instead to cell $K_{i+1,j}$ and assign this cell the largest possible positive value, taking into account the supply and demand constraints. Designate the cell $K_{i+1,j}$ as a basic cell.

If it is impossible to assign a positive value to either of the cells $K_{i,j+1}$ or $K_{i+1,j}$, then assign 0 to one of these cells and designate it as a basic cell.

STEP 3. Repeat step 2 until $m + n - 1$ cells have been designated as basic cells.

EXAMPLE 1

We apply the Northwest Corner Method to Table 5 of Example 3 in Section 9.1 (given here as Table 1).

Table 1

Plant	Warehouse			Supply
	W_1	W_2	W_3	
P_1				30
P_2				20
Demand	15	10	25	

We have not included the cost coefficients in this table because they do not play a role in the Northwest Corner Method.

We first assign the largest possible value (15) to the northwest corner of Table 1, thus obtaining Table 2.

Table 2

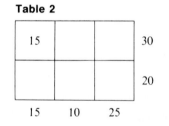

Moving one cell to the right of the cell K_{11} (to cell K_{12}) we see that we can assign the value 10 to the cell K_{12} without violating the supply and demand constraints (Table 3).

Table 3

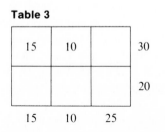

Moving one cell to the right of cell K_{12} (to cell K_{13}) we see that we can assign the value 5 to cell K_{13} (Table 4).

Table 4

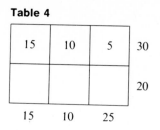

Finally, we move one cell down from cell K_{13} (to cell K_{23}) and assign the value 20 to this cell (Table 5).

Table 5

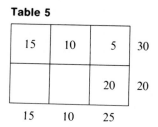

We now have $3 + 2 - 1 = 4$ basic cells, and this concludes the procedure. The basic cells are K_{11}, K_{12}, K_{13}, and K_{23}. These cells and the values assigned to them correspond to the basic feasible solution

$$x_{11} = 15, \qquad x_{12} = 10, \qquad x_{13} = 5, \qquad x_{23} = 20 \qquad \blacklozenge$$

EXAMPLE 2

You can verify that applying the Northwest Corner Method to the LP problem in Example 4 of Section 9.2 yields the basic cells and values in Table 6.

Table 6

70	10			80
	80	60		140
		40	50	90
70	90	100	50	

The basic feasible solution corresponding to Table 6 is

$$x_{11} = 70, \qquad x_{12} = 10, \qquad x_{22} = 80, \qquad x_{23} = 60, \qquad x_{33} = 40, \qquad x_{34} = 50 \quad \blacklozenge$$

EXAMPLE 3

Table 7 is a slight modification of Table 6.

Table 7

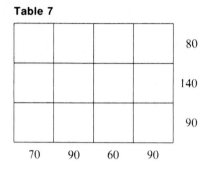

Table 8 gives the first four entries that result from applying the Northwest Corner Method to Table 7.

Table 8

70	10			80
	80	60		140
				90
70	90	60	90	

Note that at this point we cannot assign a positive value to either cell K_{24} or cell K_{33}. Therefore, we assign cell K_{24} the value 0 and designate this cell as a basic cell. Continuing the Northwest Corner Method, we obtain

Table 9

70	10			80
	80	60	0	140
			90	90
70	90	60	90	

This table corresponds to the (degenerate) basic feasible solution

$$x_{11} = 70, \qquad x_{12} = 10, \qquad x_{22} = 80, \qquad x_{23} = 60, \qquad x_{24} = 0, \qquad x_{34} = 90 \qquad \blacklozenge$$

As we have seen, the cost coefficients play no role in the Northwest Corner Method. Thus, there is no reason to expect that this method will produce a basic feasible solution near the optimal solution.

Vogel's Approximation Method

Vogel's Approximation Method (VAM) is somewhat more difficult to implement than the Northwest Corner Method, but it usually results in a basic feasible solution that is reasonably close to the optimal solution. Thus, VAM usually requires fewer iterations to arrive at an optimal solution.

We describe VAM by applying it to Table 1 in Section 9.2 (given here as Table 10).

Table 10

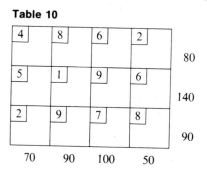

STEP 1. For each row and each column of the cost table, calculate the difference between the second smallest cost coefficient and the smallest cost coefficient.

Table 11 indicates these row differences (RD) and column differences (CD) for Table 10.

Table 11

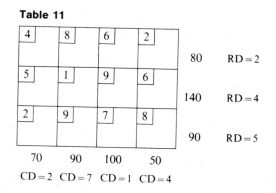

STEP 2. Select the row or column with the largest row or column difference obtained in step 1 (break ties arbitrarily).

In our example we see that column 2 has the largest cost difference (7).

STEP 3. Find the cell with the smallest cost coefficient in the row or column selected in step 2, and assign the largest possible value to this cell, taking into account the supply and demand constraints.

In our example, K_{22} is the cell with the smallest cost coefficient (1) in column 2; we are able to assign 90 to this cell.

STEP 4. Cross out any row or column for which it is now impossible to assign additional positive values.

In our example, steps 2, 3, 4 give us Table 12.

Table 12

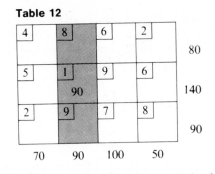

STEP 5. Return to step 1, and apply steps 1–4 to the rows and columns that have not been crossed out. If only one cell remains in a row or column, use its cost coefficient as the difference in step 1.

In our example you can verify that we obtain Tables 13–22.

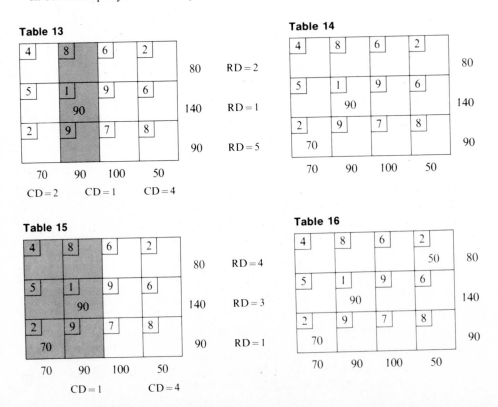

Table 13

Table 14

Table 15

Table 16

Table 17

4	8	6	2		
			50	80	RD = 6
5	1	9	6		
	90			140	RD = 9
2	9	7	8		
70				90	RD = 7
70	90	100	50		

CD = 1

Table 18

4	8	6	2	
			50	80
5	1	9	6	
	90	50		140
2	9	7	8	
70				90
70	90	100	50	

Table 19

4	8	6	2		
			50	80	RD = 6
5	1	9	6		
	90	50		140	
2	9	7	8		
70				90	RD = 7
70	90	100	50		

CD = 1

Table 20

4	8	6	2	
			50	80
5	1	9	6	
	90	50		140
2	9	7	8	
70		20		90
70	90	100	50	

Table 21

4	8	6	2		
			50	80	RD = 6
5	1	9	6		
	90	50		140	
2	9	7	8		
70		20		90	
70	90	100	50		

CD = 6

Table 22

4	8	6	2	
		30	50	80
5	1	9	6	
	90	50		140
2	9	7	8	
70		20		90
70	90	100	50	

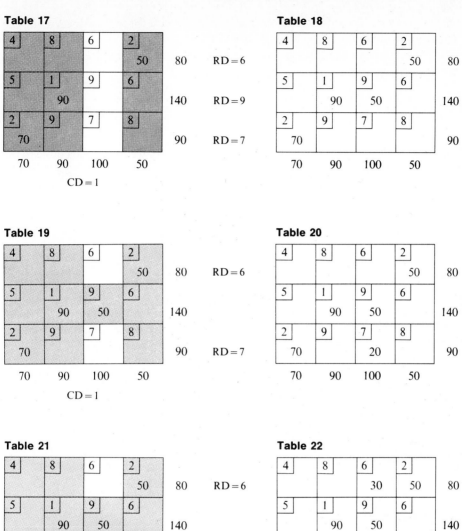
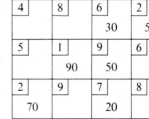
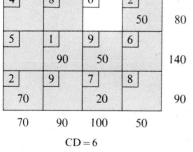

Thus, the resulting basic feasible solution is

$$x_{13} = 30, \qquad x_{14} = 50, \qquad x_{22} = 90, \qquad x_{23} = 50, \qquad x_{31} = 70, \qquad x_{33} = 20$$

Note that the corresponding transportation cost is

$$2 \cdot 70 + 1 \cdot 90 + 6 \cdot 30 + 9 \cdot 50 + 7 \cdot 20 + 2 \cdot 50 = 1100$$

which is less than the value (1270) obtained in Example 2 for this problem using the Northwest Corner Method.

In the next example, we suggest a notation that you may find useful in applying Vogel's method.

EXAMPLE 4

Instead of making a new cost table for each step of VAM, we construct a table to the right of the cost table, in which we record the row differences, and a table below, in which we record the column differences. Even though it is somewhat difficult to reconstruct the steps of VAM from the final tables of row and column differences, it is fairly easy to work through the method constructing the table as you proceed. Table 23 gives the final table of this sort, corresponding to Tables 11, 13, 15, 17, 19, and 21 of Example 3.

Table 23

[4]	[8]	[6] 30	[2] 50	80
[5]	[1] 90	[9] 50	[6]	140
[2] 70	[9]	[7] 20	[8]	90
70	90	100	50	

Step

1	2	3	4	5	6	
2	2	4	6	6	6	Row Differences
4	1	3	9	—	—	
5	5	1	7	7	—	

Step

1	2	7	1	4
2	2	—	1	4
3	—	—	1	4
4	—	—	1	—
5	—	—	1	—
6	—	—	6	—

Column Differences

◆

Exercises

In Exercises 1–6, find an initial basic solution using (a) The Northwest Corner Method and (b) Vogel's Approximation Method.

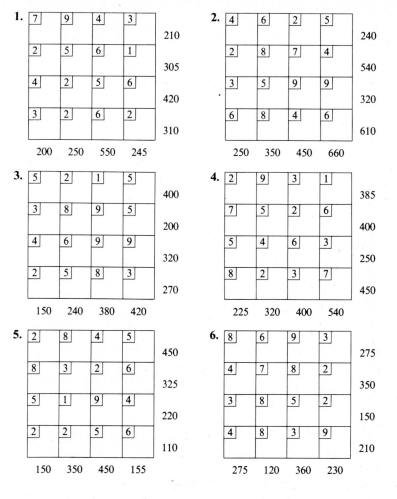

1.

7	9	4	3	210
2	5	6	1	305
4	2	5	6	420
3	2	6	2	310
200	250	550	245	

2.

4	6	2	5	240
2	8	7	4	540
3	5	9	9	320
6	8	4	6	610
250	350	450	660	

3.

5	2	1	5	400
3	8	9	5	200
4	6	9	9	320
2	5	8	3	270
150	240	380	420	

4.

2	9	3	1	385
7	5	2	6	400
5	4	6	3	250
8	2	3	7	450
225	320	400	540	

5.

2	8	4	5	450
8	3	2	6	325
5	1	9	4	220
2	2	5	6	110
150	350	450	155	

6.

8	6	9	3	275
4	7	8	2	350
3	8	5	2	150
4	8	3	9	210
275	120	360	230	

CHECKLIST: CHAPTER 9

CONCEPTS AND RESULTS

Chapter 10

LARGE-SCALE PROBLEMS

10.1 LP PROBLEMS WITH BOUNDED CONSTRAINTS

In this chapter we consider some methods for solving problems that are too large for practical solution by the methods we have developed so far. In Sections 10.1–10.3 we consider two methods that take advantage of the specialized structure of two kinds of problems. In Section 10.4 we discuss briefly a newly developed procedure that uses nonlinear programming techniques to solve linear programming problems.

In practice, a given model may indicate that there are natural bounds on some or all of the variables of an LP problem. For instance, previous experience may show that sales of a particular product will not exceed a certain number of units. Thus, LP problems such as

$$\begin{aligned}
\text{Maximize:} \quad & z = 4x_1 + 7x_2 + 3x_3 \\
\text{Subject to:} \quad & x_1 + x_2 + 3x_3 \leq 18 \\
& 3x_1 + 2x_2 - x_3 \leq 15 \\
& x_1 \qquad\qquad \leq 4 \\
& \qquad x_2 \qquad \leq 3 \\
& \qquad\qquad x_3 \leq 5 \\
& x_i \geq 0, \qquad 1 \leq i \leq 3
\end{aligned} \tag{1}$$

are not uncommon. We could, of course, solve (1) directly using the simplex method; however, as you will see, the procedure we now introduce is, for problems with many bounded variables, considerably more efficient than is a direct application of the simplex method. Unfortunately, this technique does not maintain the inverse basis matrix B^{-1} at each step. Consequently, we cannot use the revised simplex method or perform a sensitivity analysis on the solution. There is a method, known as the *generalized upper bounding (GUB)* method, that does not suffer this defect, but it requires a greater knowledge of linear algebra than we presume (for a discussion of GUB, see Murty [9]).

We will develop a procedure called the *bounded variable method* to solve problems in which some of the variables x_i have bounds of the form

$$0 \le x_i \le B_i$$

The essential idea behind the bounded variable method is to remove the bounded variable constraints, such as the last three constraints in (1), from the LP formulation of the problem and to treat them in a special way. This will mean that at each step it is necessary to do more than just a pivot operation. However, the extra work involved is more than compensated for by the reduction in the size of the overall problem.

The usual rules of the simplex method apply, but we will also set up other conditions for determining the basic and nonbasic variables. In particular, when a bounded variable x_i takes on its maximal value B_i we replace x_i with a new variable x_i' defined by

$$x_i' = B_i - x_i$$

Since at this stage $x_i = B_i$, it follows that $x_i' = 0$, and hence x_i' can be considered as a nonbasic variable. Note that x_i' also satisfies $0 \le x_i' \le B_i$. We use LP problem (1) to illustrate the underlying ideas for this procedure. We first express (1) as

$$\text{Maximize:} \quad z = 4x_1 + 7x_2 + 3x_3$$

$$\text{Subject to:} \quad x_1 + x_2 + 3x_3 \le 18$$
$$3x_1 + 2x_2 - x_3 \le 15 \tag{2}$$

$$0 \le x_1 \le 4, \qquad 0 \le x_2 \le 3, \qquad 0 \le x_3 \le 5$$

STEP 1.

The initial tableau corresponding to (2) is Tableau 1,

Tableau 1

	x_1	x_2	x_3	s_1	s_2	
s_1	1	1	3	1	0	18
s_2	3	2	-1	0	1	15
	-4	-7	-3	0	0	0

which indicates that x_2 is to be considered as the entering basic variable. If x_2 is to enter, then we are interested in to what extent we can increase x_2 without leaving the feasible region. Because of the bounds on the variables, there are three possible constraints on x_2.

1. $x_2 \le \min\{\frac{18}{1}, \frac{15}{2}\} = \frac{15}{2}$. This constraint is the result of the departing basic variable rule.

2. $x_2 \leq 3$. This limit is imposed by the original constraint on x_2.

3. If any of the bounded variables are current basic variables, then we must ensure that x_2 does not become so large that one of these variables would exceed its bound. Since, at the present step, none of the bounded variables is basic, this consideration does not produce any additional constraint on x_2.

From constraints 1 and 2 we see that $x_2 \leq 3$. Since in this case, constraint 2 is the operative constraint, we define a new (nonbasic) variable x_2' to replace x_2, where $x_2' = 3 - x_2$ or, equivalently, $x_2 = 3 - x_2'$. Note that if $x_2 = 3$, then $x_2' = 0$, and thus x_2' can be regarded as a nonbasic variable at this stage. Replacing x_2 with $3 - x_2'$ in Tableau 1 gives us Tableau 2.

Tableau 2

	x_1	x_2'	x_3	s_1	s_2	
s_1	1	-1	3	1	0	15
s_2	3	-2	-1	0	1	9
	-4	7	-3	0	0	21

To see how Tableau 2 arose from Tableau 1, consider the first row of Tableau 1. This row corresponds to the equation

$$x_1 + x_2 + 3x_3 + s_1 = 18 \tag{3}$$

Substituting $3 - x_2'$ for x_2 in (3) gives us

$$x_1 + 3 - x_2' + 3x_3 + s_1 = 18$$

or

$$x_1 - x_2' + 3x_3 + s_1 = 15$$

and this equation is represented by the first row of Tableau 2.
  Similarly, the second row of Tableau 1 corresponds to the equation

$$3x_1 + 2x_2 - x_3 + s_2 = 15 \tag{4}$$

Substituting $3 - x_2'$ for x_2 in (4) gives us

$$3x_1 + 2(3 - x_2') - x_3 + s_2 = 15$$

or, equivalently,

$$3x_1 - 2x_2' - x_3 + s_2 = 9$$

and this equation is represented by the second row of Tableau 2.
  Finally, the objective row of Tableau 1 corresponds to the equation

$$z - 4x_1 - 7x_2 - 3x_3 = 0 \tag{5}$$

Substituting $3 - x_2'$ for x_2 in (5) gives us

$$z - 4x_1 - 7(3 - x_2') - 3x_3 = 0$$

or

$$z - 4x_1 + 7x_2' - 3x_3 = 21$$

which corresponds to the objective row of Tableau 2.

STEP 2.

We now apply the same procedure to Tableau 2, which indicates that x_1 is to be considered as the entering basic variable. If x_1 is to enter, then we are interested in to what extend we can increase x_1 without leaving the feasible region. The constraints on x_1 are given by

1. $x_1 \le \min\{\frac{15}{1}, \frac{9}{3}\} = 3$

2. $x_1 \le 4$

3. Since at this stage there are still no bounded basic variables, such variables do not impose any restriction on x_1.

From constraints 1 and 2 we see that the operative constraint on x_1 is given by constraint 1, and thus we perform the usual pivot operation on entry 3 in Tableau 2. This gives us Tableau 3.

Tableau 3

	x_1	x_2'	x_3	s_1	s_2	
s_1	0	$-\frac{1}{3}$	$\frac{10}{3}$	1	$-\frac{1}{3}$	12
x_1	1	$-\frac{2}{3}$	$-\frac{1}{3}$	0	$\frac{1}{3}$	3
	0	$\frac{13}{3}$	$-\frac{13}{3}$	0	$\frac{4}{3}$	33

STEP 3.

Tableau 3 indicates that x_3 is to be considered the entering basic variable. The constraints on x_3 are

1. $x_3 \le \frac{36}{10}$

2. $x_3 \le 5$

3. Note that x_1 is a bounded basic variable. To determine what effect this has on x_3, we first observe that since x_2' and s_2 are nonbasic variables, the second row of Tableau 3 (with the other nonbasic variables set to 0) corresponds to the equation

$$x_1 - \frac{1}{3}x_3 = 3 \tag{6}$$

Since $x_1 \le 4$, it follows from (6) that $3 + (\frac{1}{3})x_3 \le 4$ or $x_3 \le 3$.

(If there were other bounded basic variables, we would make the same kind of analysis for each of them, and then bound x_3 by the most restrictive of the resulting values.) From constraints 1, 2, and 3 we see that constraint 3 provides the operative constraint on x_3. If $x_3 = 3$, then from (6) we see that x_1 attains its bound, and thus we introduce the new (nonbasic) variable

$$x_1' = 4 - x_1$$

Using the same arguments as in Step 1, you can verify that Tableau 4, which results from Tableau 3 when you substitute $4 - x_1'$ for x_1 (and multiply the second row of the tableau through by -1), is

Tableau 4

	x_1'	x_2'	x_3	s_1	s_2	
s_1	0	$-\frac{1}{3}$	$\frac{10}{3}$	1	$-\frac{1}{3}$	12
x_1'	1	$\frac{2}{3}$	$\frac{1}{3}$	0	$-\frac{1}{3}$	1
	0	$\frac{13}{3}$	$-\frac{13}{3}$	0	$\frac{4}{3}$	33

STEP 4.

Tableau 4 indicates that x_3 is still to be considered the entering basic variable. The current constraints on x_3 are

1. $x_3 \leq 3$

2. $x_3 \leq 5$

3. Note that x_1 is a bounded basic variable. Since from the second row of Tableau 4 we have $x_1' + (\frac{1}{3})x_3 = 1$, and since $x_1' \leq 4$, we have

$$1 - \frac{1}{3}x_3 \leq 4 \tag{7}$$

Observe that (7) imposes no restriction on x_3.

It follows from constraints 1, 2, and 3 that 1 is the operative constraint, and hence we pivot on the entry $\frac{1}{3}$ in Tableau 4 to obtain Tableau 5,

Tableau 5

	x_1'	x_2'	x_3	s_1	s_2	
s_1	-10	-7	0	1	3	2
x_3	3	2	1	0	-1	3
	13	13	0	0	-3	46

STEP 5.

Tableau 5 indicates that s_2 is to be considered the entering basic variable. The constraints on s_2 are

1. $s_2 \leq \frac{2}{3}$

2. s_2 is not a bounded variable; thus, there are no constraints for this reason.

3. Note that x_3 is a bounded basic variable. The second row of Tableau 5 corresponds to the equation

$$x_3 - s_2 = 3 \tag{8}$$

Since $x_3 \leq 5$, it follows from (8) that $s_2 \leq 2$

From constraints 1, 2, and 3 we see that the operative constraint arises from 1, and hence we pivot on entry 3 in the s_2 column of Tableau 5 to obtain Tableau 6,

Tableau 6

	x_1'	x_2'	x_3	s_1	s_2	
s_2	$-\frac{10}{3}$	$-\frac{7}{3}$	0	$\frac{1}{3}$	1	$\frac{2}{3}$
x_3	$-\frac{1}{3}$	$-\frac{1}{3}$	1	$\frac{1}{3}$	0	$\frac{11}{3}$
	13	6	0	1	0	48

Tableau 6 shows that we have reached an optimal solution

$$x_1' = 0, \qquad x_2' = 0, \qquad x_3 = \frac{11}{3}, \qquad s_1 = 0, \qquad s_2 = \frac{2}{3}; \qquad z = 48$$

which gives us the optimal solution

$$x_1 = 4 - x_1' = 4, \qquad x_2 = 3 - x_2' = 3, \qquad x_3 = \frac{11}{3}$$

$$z = 4 \cdot 4 + 7 \cdot 3 + 3 \cdot \frac{11}{3} = 48$$

to the original problem.

To formulate the general rules for the bounded variable method, suppose that at a given stage we have the tableau

	w_1	w_2	w_3		w_n	
$w_{h(1)}$	a_{11}^*	a_{12}^*	a_{13}^*	$\cdots$	a_{1n}^*	b_1^*
$w_{h(2)}$	a_{21}^*	a_{22}^*	a_{23}^*	$\cdots$	a_{2n}^*	b_2^*
$\vdots$						
$w_{h(m)}$	a_{m1}^*	a_{m2}^*	a_{m3}^*	$\cdots$	a_{mn}^*	b_m^*
	$-c_1^*$	$-c_2^*$	$-c_3^*$	$\cdots$	$-c_m^*$	z^*

(9)

where w_i represents x_i if x_i is not a bounded variable and w_i is either x_i or x_i' if x_i is a bounded variable; if x_i is a bounded variable, we suppose $0 \le w_i \le B_i$. The symbols $w_{h(1)}, w_{h(2)}, \ldots, w_{h(m)}$ represent the current basic variables.

Suppose $-c_i^*$ is a negative entry in the objective row of (9) and that w_i is the entering basic variable. The constraints on w_i are given by

1. $m_1 = \min\{b_j^*/a_{ji}^* \,|\, a_{ji}^* > 0, \quad 1 \le j \le m\}$ (the Departing Basic Variable Rule).

2. $m_2 = \begin{cases} B_i & \text{if } w_i \text{ is a bounded variable and } w_i \le B_i. \\ \infty & \text{if } w_i \text{ is not a bounded variable.} \end{cases}$

3. $m_3 = \min\{(B_{h(j)} - b_j^*)/(-a_{ji}^*) \,|\, a_{ji}^* < 0$ and $w_{h(j)}$ is a bounded basic variable, $1 \le j \le m\}$. If at this stage there are no bounded basic variables, then $m_3 = \infty$.

Once the values m_1, m_2, and m_3 are calculated, then

1. if $m_1 = \min\{m_1, m_2, m_3\}$, pivot on the entry determined by the Departing Basic Variable Rule;

2. if $m_2 = \min\{m_1, m_2, m_3\}$, let $w_i' = B_i - w_i$ and substitute $B_i - w_i'$ for w_i in Tableau (9) to obtain the next tableau;

3. if $m_3 = \min\{m_1, m_2, m_3\}$, let p be the value such that

$$m_3 = \frac{B_{h(p)} - b_p^*}{-a_{pi}^*}$$

Let $w_{h(p)}' = B_{h(p)} - w_{h(p)}$, and substitute $B_{h(p)} - w_{h(p)}'$ for $w_{h(p)}$ in Tableau (9); then multiply the new row by -1.

EXAMPLE 1

Suppose that Tableau 1 arises in applying the bounded variable method to an LP problem that has the bounded variables

$$0 \le x_1 \le B_1$$

$$0 \le x_3 \le B_3$$

$$0 \le x_5 \le B_5$$

$$0 \le x_7 \le B_7$$

Tableau 1

	x_1'	x_2	x_3	x_4	x_5	x_6	x_7'	
x_3	0	0	1	-3	2	4	0	b_1^*
x_1'	1	0	0	2	4	-3	0	b_2^*
x_2	0	1	0	-1	-5	-2	0	b_3^*
x_7'	0	0	0	-4	0	1	1	b_4^*
	0	0	0	$-c_4^*$	$-c_5^*$	$-c_6^*$	0	

Suppose that x_4 is the entering basic variable.

1. The departing basic variable rule gives

$$m_1 = \min\left\{ -, \frac{b_2^*}{2}, -, - \right\}$$

2. Since x_4 is not a bounded variable, $m_2 = \infty$.

3. Since there are bounded basic variables, we write the equations corresponding to the rows of Tableau 1 with all nonbasic variables except x_4 set equal to 0.

$$x_3 - 3x_4 = b_1^*$$
$$x_1' + 2x_4 = b_2^*$$
$$x_2 - x_4 = b_3^*$$
$$x_7' - 4x_4 = b_4^*$$

The bounds on x_1, x_3, and x_7 imply that

$$x_3 = 3x_4 + b_1^* \le B_3 \tag{10}$$
$$x_1' = -2x_4 + b_2^* \le B_1 \tag{11}$$
$$x_2 = x_4 + b_3^* \tag{12}$$
$$x_7' = 4x_4 + b_4^* \le B_7 \tag{13}$$

Since $b_2^* \le B_1$ and the coefficient of x_4 is negative, constraint (11) imposes no restriction on x_4. Since x_2 is not a bounded variable, constraint (12) imposes no restriction on x_4. From (10) and (13) we obtain

$$x_4 \le \frac{B_3 - b_1^*}{3}$$

$$x_4 \le \frac{B_7 - b_4^*}{4}$$

Therefore,

$$m_3 = \min\left\{ \frac{B_3 - b_1^*}{3}, -, \frac{B_7 - b_4^*}{4} \right\} \tag{14}$$

If $m_1 < m_3$, then we use the ordinary Departing Basic Variable Rule to choose the departing variable. If $m_1 = m_3$, then we have a choice between this rule or the one described in the next paragraph.

If $m_3 < m_1$, then we choose the basic variable corresponding to the smallest entry in (14) as the departing basic variable. If this smallest entry corresponds to B_3, we replace x_3 in Tableau 1 by $x'_3 = B_3 - x_3$. If this smallest entry corresponds to B_7, we replace x'_7 by $x_7 = B_7 - x'_7$. After making one of these replacements, we multiply the new constraint through by -1.

You are asked to consider the cases when either x_5 or x_6 is the entering basic variable in Exercises 1 and 2.

◆

To conclude this section we discuss briefly why this procedure is preferred over applying the simplex method to the original problem with the bounded variables listed among the constraints. Consider a rather extreme case.

$$\text{Maximize:} \quad z = c_1 x_1 + c_2 x_2 + \cdots + c_{100} x_{100}$$

$$\text{Subject to:} \quad a_1 x_1 + a_2 x_2 + \cdots + a_{100} x_{100} \le b \tag{15}$$

$$x_i \ge 0, \quad 1 \le i \le 100$$

Suppose that for each i, $1 \le i \le 100$

$$x_i \le B_i \tag{16}$$

To apply the simplex method to this problem we would have to introduce 100 additional constraints to (15) of the form (16) and then work with tableaus having 102 rows and 202 columns. In cases such as this it is far better to use the procedure discussed in this section. Although in solving (15), at each choice of departing basic variable, there are potentially two additional calculations to be made; the work is always done with tableaus of just two rows and 102 columns. This generally represents a more efficient way to proceed than to apply the simplex method.

As we observed earlier, some of the computational advantage of this method is lost since we cannot use the revised simplex method to carry out the steps of the bounded variable method (because the B^{-1} matrix is not readily available after we make a substitution of one of the forms $x'_j = B_j - x_j$ or $x_j = B_j - x'_j$). Also, we cannot perform a sensitivity analysis based on using the B^{-1} matrix.

Exercises

1. Discuss the selection of the departing basic variable for Tableau 1 of Example 1 when x_5 is the entering basic variable.

2. Discuss the selection of the departing basic variable for Tableau 1 of Example 1 when x_6 is the entering basic variable.

Suppose Tableau 2 arises in the course of performing the bounded variable method. In Exercises 3–8 you are given bounds on some variables and a choice for the entering basic variable. Select the departing basic variable, and form the next tableau resulting from applying the bounded variable method.

Tableau 2

	x_1'	x_2	x_3	x_4	x_5	x_6	x_7'	
x_3	0	0	1	2	2	0	0	2
x_1'	1	0	0	-2	-4	-2	0	6
x_2	0	1	0	-1	-4	-3	0	3
x_7'	0	0	0	-3	0	1	1	4
	0	0	0	-3	-4	-5	0	27

3. Suppose $x_1 \le 7$, $x_2 \le 4$, $x_4 \le 4$, and $x_5 \le 3$. Let the entering basic variable be x_4.

4. Suppose $x_1 \le 8$, $x_2 \le 3.5$, $x_4 \le 4$, and $x_5 \le 3$. Let the entering basic variable be x_4.

5. Suppose $x_1 \le 20$, $x_2 \le 18$, and $x_5 \le 3.5$. Let the entering basic variable be x_5.

6. Suppose $x_1 \le 12$, $x_2 \le 10$, and $x_5 \le 1$. Let the entering basic variable be x_5.

7. Suppose $x_1 \le 12$, $x_2 \le 20$, $x_5 \le 1$, and $x_7 \le 5$. Let the entering basic variable be x_6.

8. Suppose $x_1 \le 15$, $x_2 \le 24$, $x_5 \le 8$, and $x_7 \le 9$. Let the entering basic variable be x_6.

9. Solve the LP problem

$$\text{Maximize:} \quad z = 5x_1 + 3x_2 + 4x_3$$
$$\text{Subject to:} \quad x_1 + 2x_2 + 2x_3 \le 9$$
$$-x_1 - x_2 + 4x_3 \le 6$$
$$x_1 + 2x_2 \qquad \le 5$$
$$0 \le x_1 \le 2, \quad 0 \le x_2 \le 3, \quad 0 \le x_3 \le 4$$

by the bounded variable method.

10. Solve the LP problem

$$\text{Maximize:} \quad z = 3x_1 + 2x_2$$
$$\text{Subject to:} \quad 2x_1 - x_2 \le 1$$
$$-3x_1 + 4x_2 \le 13$$
$$x_1 + x_2 \le 5$$
$$0 \le x_1 \le 1, \quad 0 \le x_2 \le 2$$

by the bounded variable method.

10.2 A DECOMPOSITION PROCEDURE: THE MASTER PROBLEM

In practice it is not uncommon that an LP problem arises of such magnitude that direct application of the simplex or revised simplex method is impractical. However, it is sometimes possible to take advantage of the structure of such a problem to "decompose" it into a number of subproblems, each of which is solvable in a reasonably efficient manner. An optimal solution to the original problem is then obtained from the subproblems.

In this section we consider certain potentially large-scale LP problems that can be broken down into smaller components. We begin with a problem of the form (in matrix notation)

$$\text{Maximize:}\quad z = \mathbf{c}_1\mathbf{v}_1 + \mathbf{c}_2\mathbf{v}_2 + \cdots + \mathbf{c}_k\mathbf{v}_k$$

$$\begin{aligned}
\text{Subject to:}\quad F_1\mathbf{v}_1 + F_2\mathbf{v}_2 + \cdots + F_k\mathbf{v}_k &\le \mathbf{b}_0 \\
D_1\mathbf{v}_1 \qquad\qquad\qquad &\le \mathbf{b}_1 \\
D_2\mathbf{v}_2 \qquad\qquad &\le \mathbf{b}_2 \\
&\vdots \\
D_k\mathbf{v}_k &\le \mathbf{b}_k
\end{aligned} \tag{1}$$

$$\mathbf{v}_i \ge \mathbf{0}, \qquad 1 \le i \le k$$

In what follows we will refer to the constraints

$$F_1\mathbf{v}_1 + F_2\mathbf{v}_2 + \cdots + F_k\mathbf{v}_k \le \mathbf{b}_0$$

as the *common constraints*. The constraints

$$D_i\mathbf{v}_i \le \mathbf{b}_i, \qquad 1 \le i \le k$$

are called the *independent constraints*.

We stated problem (1) with only $\le$ constraints; however, the method we described will also work for a problem with a mixture of $\le$, $\ge$, and $=$ constraints. In Example 3 we will show how to use the two-phase method to solve a mixed constraint problem.

EXAMPLE 1

We apply the notation of (1) to the LP problem

$$\text{Maximize:}\quad z = 3x_1 + x_2 + 2x_3 + x_4$$

$$\begin{aligned}
\text{Subject to:}\quad 2x_1 + 3x_2 + x_3 + 2x_4 &\le 10 \\
x_1 \phantom{{}+2x_2} - 2x_3 + 3x_4 &\le 8 \\
-x_1 + x_2 \phantom{{}+3x_4} &\le 2 \\
2x_1 + x_2 \phantom{{}+3x_4} &\le 8 \\
x_3 + 3x_4 &\le 12 \\
3x_3 - 2x_4 &\le 3
\end{aligned}$$

$$x_i \ge 0, \qquad 1 \le i \le 4$$

The matrices corresponding to (1) are

$$\mathbf{c}_1 = [3 \quad 1] \qquad \mathbf{c}_2 = [2 \quad 1]$$

$$\mathbf{v}_1 = \begin{bmatrix} x_1 \\ x_2 \end{bmatrix} \qquad \mathbf{v}_2 = \begin{bmatrix} x_3 \\ x_4 \end{bmatrix}$$

$$F_1 = \begin{bmatrix} 2 & 3 \\ 1 & 0 \end{bmatrix} \qquad F_2 = \begin{bmatrix} 1 & 2 \\ -2 & 3 \end{bmatrix} \qquad \mathbf{b}_0 = \begin{bmatrix} 10 \\ 8 \end{bmatrix}$$

$$D_1 = \begin{bmatrix} -1 & 1 \\ 2 & 1 \end{bmatrix} \qquad \mathbf{b}_1 = \begin{bmatrix} 2 \\ 8 \end{bmatrix}$$

$$D_2 = \begin{bmatrix} 1 & 3 \\ 3 & -2 \end{bmatrix} \qquad \mathbf{b}_2 = \begin{bmatrix} 12 \\ 3 \end{bmatrix} \qquad \qquad \blacklozenge$$

For simplicity, we will now work with problems having one block of common constraints and just two or three blocks of independent constraints. As will be obvious, the discussion is equally valid for problems having any number of blocks of independent constraints.

Observe that the constraints in each block of independent constraints define a feasible region. We make the assumption that each of these feasible regions is bounded; if this were not the case, then we could bound them by introducing additional constraints with appropriately large numbers that bound the variables (see Exercises 5 and 6).

Suppose then we have an LP problem of the form

$$\text{Maximize:} \quad z = \mathbf{c}_1 \mathbf{v}_1 + \mathbf{c}_2 \mathbf{v}_2 + \mathbf{c}_3 \mathbf{v}_3$$

$$\text{Subject to:} \quad F_1 \mathbf{v}_1 + F_2 \mathbf{v}_2 + F_3 \mathbf{v}_3 \leq \mathbf{b}_0$$

$$D_1 \mathbf{v}_1 \qquad\qquad\qquad \leq \mathbf{b}_1$$

$$D_2 \mathbf{v}_2 \qquad\qquad \leq \mathbf{b}_2 \qquad\qquad (2)$$

$$D_3 \mathbf{v}_3 \leq \mathbf{b}_3$$

$$\mathbf{v}_i \geq \mathbf{0}, \qquad 1 \leq i \leq 3$$

We will break down (decompose) this problem by working with the three feasible regions determined by the three blocks of independent constraints. Let

$$P_1, P_2, \ldots, P_r; \qquad Q_1, Q_2, \ldots, Q_s; \qquad R_1, R_2, \ldots, R_t$$

be the corner points of the feasible regions determined by

$$D_1 \mathbf{v}_1 \leq \mathbf{b}_1, \quad \mathbf{v}_1 \geq \mathbf{0}; \qquad D_2 \mathbf{v}_2 \leq \mathbf{b}_2, \quad \mathbf{v}_2 \geq \mathbf{0}; \qquad D_3 \mathbf{v}_3 \leq \mathbf{b}_3, \quad \mathbf{v}_3 \geq \mathbf{0}$$

respectively. Then as we saw in Chapter 2, any point $\mathbf{v}_1$ lying in the feasible region determined by $D_1 \mathbf{v}_1 \leq \mathbf{b}_1$, $\mathbf{v}_1 \geq \mathbf{0}$ can be written as a convex combination of the corner points of this region; that is,

$$\mathbf{v}_1 = \alpha_1 P_1 + \alpha_2 P_2 + \cdots + \alpha_r P_r \tag{3}$$

where each $\alpha_i \geq 0$ and

$$\alpha_1 + \alpha_2 + \cdots + \alpha_r = 1 \tag{4}$$

Similarly, points $\mathbf{v}_2$ and $\mathbf{v}_3$ lying in the feasible regions determined by the constraints $D_2\mathbf{v}_2 \leq \mathbf{b}_2$, $\mathbf{v}_2 \geq \mathbf{0}$ and $D_3\mathbf{v}_3 \leq \mathbf{b}_3$, $\mathbf{v}_3 \geq \mathbf{0}$ respectively, can be written as

$$\mathbf{v}_2 = \beta_1 Q_1 + \beta_2 Q_2 + \cdots + \beta_s Q_s \tag{5}$$

where each $\beta_i \geq 0$ and

$$\beta_1 + \beta_2 + \cdots + \beta_s = 1 \tag{6}$$

and

$$\mathbf{v}_3 = \gamma_1 R_1 + \gamma_2 R_2 + \cdots + \gamma_t R_t \tag{7}$$

where each $\gamma_i \geq 0$ and

$$\gamma_1 + \gamma_2 + \cdots + \gamma_t = 1 \tag{8}$$

Our next task is to reformulate (2) as an LP problem with variables

$$\alpha_1, \alpha_2, \ldots, \alpha_r, \qquad \beta_1, \beta_2, \ldots, \beta_s, \qquad \gamma_1, \gamma_2, \ldots, \gamma_t$$

Substituting (3), (5), and (7) into the objective function of (2) we obtain

$$\begin{aligned}
z &= \mathbf{c}_1(\alpha_1 P_1 + \alpha_2 P_2 + \cdots + \alpha_r P_r) \\
&\quad + \mathbf{c}_2(\beta_1 Q_1 + \beta_2 Q_2 + \cdots + \beta_s Q_s) \\
&\quad + \mathbf{c}_3(\gamma_1 R_1 + \gamma_2 R_2 + \cdots + \gamma_t R_t) \\
&= (\mathbf{c}_1 P_1)\alpha_1 + (\mathbf{c}_1 P_2)\alpha_2 + \cdots + (\mathbf{c}_1 P_r)\alpha_r \\
&\quad + (\mathbf{c}_2 Q_1)\beta_1 + (\mathbf{c}_2 Q_2)\beta_2 + \cdots + (\mathbf{c}_2 Q_s)\beta_s \\
&\quad + (\mathbf{c}_3 R_1)\gamma_1 + (\mathbf{c}_3 R_2)\gamma_2 + \cdots + (\mathbf{c}_3 R_t)\gamma_t
\end{aligned}$$

Substituting (3), (5), and (7) into the left-hand side of the common constraints of (2), we have

$$\begin{aligned}
F_1\mathbf{v}_1 + F_2\mathbf{v}_2 + F_3\mathbf{v}_3 &= F_1(\alpha_1 P_1 + \alpha_2 P_2 + \cdots + \alpha_r P_r) \\
&\quad + F_2(\beta_1 Q_1 + \beta_2 Q_2 + \cdots + \beta_s Q_s) \\
&\quad + F_3(\gamma_1 R_1 + \gamma_2 R_2 + \cdots + \gamma_t R_t) \\
&= (F_1 P_1)\alpha_1 + (F_1 P_2)\alpha_2 + \cdots + (F_1 P_r)\alpha_r \\
&\quad + (F_2 Q_1)\beta_1 + (F_2 Q_2)\beta_2 + \cdots + (F_2 Q_s)\beta_s \\
&\quad + (F_3 R_1)\gamma_1 + (F_3 R_2)\gamma_2 + \cdots + (F_3 R_t)\gamma_t
\end{aligned}$$

Next observe that equations (3)–(8) define every point corresponding to the constraints $D_1 v_1 \le b_1$, $D_2 v_2 \le b_2$, and $D_3 v_3 \le b_3$ in (2). Thus we can express these constraints in terms of α_i, β_i, and γ_i. From the preceding remarks we see that we can reformulate the LP problem (1) as

$$
\begin{aligned}
\text{Maximize:} \quad z = \ & (c_1 P_1)\alpha_1 + (c_1 P_2)\alpha_2 + \cdots + (c_1 P_r)\alpha_r \\
& + (c_2 Q_1)\beta_1 + (c_2 Q_2)\beta_2 + \cdots + (c_2 Q_s)\beta_s \\
& + (c_3 R_1)\gamma_1 + (c_3 R_2)\gamma_2 + \cdots + (c_3 R_t)\gamma_t
\end{aligned}
$$

$$
\begin{aligned}
\text{Subject to:} \quad & (F_1 P_1)\alpha_1 + (F_1 P_2)\alpha_2 + \cdots + (F_1 P_r)\alpha_r \\
& + (F_2 Q_1)\beta_1 + (F_2 Q_2)\beta_2 + \cdots + (F_2 Q_s)\beta_s \\
& + (F_3 R_1)\gamma_1 + (F_3 R_2)\gamma_2 + \cdots + (F_3 R_t)\gamma_t + S = b_0 \qquad (9)
\end{aligned}
$$

$$
\begin{aligned}
\alpha_1 + \quad & \alpha_2 + \cdots + & \alpha_r & = 1 \\
\beta_1 + \quad & \beta_2 + \cdots + & \beta_s & = 1 \\
\gamma_1 + \quad & \gamma_2 + \cdots + & \gamma_t & = 1
\end{aligned}
$$

$$
\alpha_i \ge 0, \quad 1 \le i \le r; \quad \beta_i \ge 0, \quad 1 \le i \le s;
$$

$$
\gamma_i \ge 0, \quad 1 \le i \le t; \quad S \ge 0
$$

where S is a column matrix of nonnegative slack variables. This problem is called the *master problem*, and it is this problem that we will solve.

At first glance it would appear that the master problem (9) is no easier to solve than the original problem. However, we will be able to take advantage of its special structure to set up a number of solvable subproblems. In particular, we will see that we do not need to explicitly find all of the corner points $P_1, P_2, \ldots, P_r; Q_1, Q_2, \ldots, Q_s; R_1, R_2, \ldots, R_t$. In fact, at each iteration we will find only one corner point for each of as few blocks as possible. We will then use a column-generating procedure (as we did in Chapter 5 when developing the revised simplex method and again in Example 5 of Section 8.6) to determine the entering basic variable. The departing basic variables are chosen in the usual manner.

EXAMPLE 2

The master problem for the LP problem given in Example 1 is

$$
\begin{aligned}
\text{Maximize:} \quad z = \ & (c_1 P_1)\alpha_1 + (c_1 P_2)\alpha_2 + \cdots + (c_1 P_r)\alpha_r \\
& + (c_2 Q_1)\beta_1 + (c_2 Q_2)\beta_2 + \cdots + (c_2 Q_s)\beta_s
\end{aligned}
$$

$$
\begin{aligned}
\text{Subject to:} \quad & (F_1 P_1)\alpha_1 + (F_1 P_2)\alpha_2 + \cdots + (F_1 P_r)\alpha_r \\
& + (F_2 Q_1)\beta_1 + (F_2 Q_2)\beta_2 + \cdots + (F_2 Q_s)\beta_s + \begin{bmatrix} s_1 \\ s_2 \end{bmatrix} = b_0 \qquad (10)
\end{aligned}
$$

$$
\begin{aligned}
\alpha_1 + \quad & \alpha_2 + \cdots + & \alpha_r & = 1 \\
\beta_1 + \quad & \beta_2 + \cdots + & \beta_s & = 1
\end{aligned}
$$

$$
\alpha_i \ge 0, \quad 1 \le i \le r; \quad \beta_i \ge 0, \quad 1 \le i \le s; \quad s_1, s_2 \ge 0
$$

where the matrices $c_1, c_2, F_1,$ and F_2 are given in Example 1. As we will see in the next section, it is not necessary to determine the values of r and s or the corner points P_i and Q_i before solving the master problem.

As we will also show in the next section, the master problem (10) is relatively easy to solve because the corner points $P_1 = \mathbf{0}$ and $Q_1 = \mathbf{0}$ determine an initial solution (the origin) of the master problem. The corresponding basis matrix is

$$B = \begin{array}{c} s_1 \\ s_2 \\ \alpha_1 \\ \beta_1 \end{array} \begin{bmatrix} 1 & 0 & 0 & 0 \\ 0 & 1 & 0 & 0 \\ 0 & 0 & 1 & 0 \\ 0 & 0 & 0 & 1 \end{bmatrix}$$

$\blacklozenge$

Next we see how to deal with problems that have mixed constraints.

EXAMPLE 3

Repeating the arguments of this section for the LP problem

$$\text{Maximize:} \quad z = 3x_1 + x_2 + 2x_3 + x_4$$

$$
\begin{aligned}
\text{Subject to:} \quad & 2x_1 + 3x_2 + x_3 + 2x_4 \geq 10 \\
& x_1 \qquad\quad - 2x_3 + 3x_4 \leq 8 \\
& -x_1 + x_2 \qquad\qquad\qquad \geq 2 \\
& 2x_1 + x_2 \qquad\qquad\qquad \leq 8 \\
& \qquad\qquad\quad x_3 + 3x_4 \leq 12 \\
& \qquad\qquad\quad 3x_3 - 2x_4 \geq 3
\end{aligned}
$$
(11)

$$x_i \geq 0, \qquad 1 \leq i \leq 4$$

you can show that the corresponding master problem is

$$
\begin{aligned}
\text{Maximize:} \quad z = & (c_1 P_1)\alpha_1 + (c_1 P_2)\alpha_2 + \cdots + (c_1 P_r)\alpha_r \\
& + (c_2 Q_1)\beta_1 + (c_2 Q_2)\beta_2 + \cdots + (c_2 Q_s)\beta_s
\end{aligned}
$$

$$
\begin{aligned}
\text{Subject to:} \quad & (F_1 P_1)\alpha_1 + (F_1 P_2)\alpha_2 + \cdots + (F_1 P_r)\alpha_r \\
& + (F_2 Q_1)\beta_1 + (F_2 Q_2)\beta_2 + \cdots + (F_2 Q_s)\beta_s + \begin{bmatrix} -s_1 \\ s_2 \end{bmatrix} = \begin{bmatrix} 10 \\ 8 \end{bmatrix}
\end{aligned}
$$
(12)

$$
\begin{aligned}
\alpha_1 + \quad & \alpha_2 + \cdots + \quad \alpha_r \quad = 1 \\
\beta_1 + \quad & \beta_2 + \cdots + \quad \beta_s \quad = 1
\end{aligned}
$$

$$0 \leq \alpha_i \leq 1, \qquad 1 \leq i \leq r; \qquad 0 \leq \beta_i \leq 1, \qquad 1 \leq i \leq s; \qquad s_1, s_2 \geq 0$$

In (12), s_1 and s_2 are slack variables, $P_1, P_2, \ldots, P_r$ are the corner points of the feasible region determined by the constraints

$$-x_1 + x_2 \geq 2$$

$$2x_1 + x_2 \leq 8$$

$$x_1, x_2 \geq 0$$

and $Q_1, Q_2, \ldots, Q_s$ are the corner points of the feasible region determined by the constraints

$$x_3 + 3x_4 \leq 12$$

$$3x_3 - 2x_4 \geq 3$$

$$x_3, x_4 \geq 0$$

The matrices are defined by

$$\mathbf{c}_1 = \begin{bmatrix} 3 & 1 \end{bmatrix} \qquad \mathbf{c}_2 = \begin{bmatrix} 2 & 1 \end{bmatrix}$$

$$\mathbf{v}_1 = \begin{bmatrix} x_1 \\ x_2 \end{bmatrix} \qquad \mathbf{v}_2 = \begin{bmatrix} x_3 \\ x_4 \end{bmatrix}$$

$$F_1 = \begin{bmatrix} 2 & 3 \\ 1 & 0 \end{bmatrix} \qquad F_2 = \begin{bmatrix} 1 & 2 \\ -2 & 3 \end{bmatrix} \qquad \mathbf{b}_0 = \begin{bmatrix} 10 \\ 8 \end{bmatrix}$$

$$D_1 = \begin{bmatrix} -1 & 1 \\ 2 & 1 \end{bmatrix} \qquad \mathbf{b}_1 = \begin{bmatrix} 2 \\ 8 \end{bmatrix}$$

$$D_2 = \begin{bmatrix} 1 & 3 \\ 3 & -2 \end{bmatrix} \qquad \mathbf{b}_2 = \begin{bmatrix} 12 \\ 3 \end{bmatrix}$$

The master problem (12) does not have an immediately identifiable basis since the origin is not a corner point of the feasible regions defined by the two sets of independent constraints. We will use the two-phase method to determine an initial basis. Introducing three artificial variables y_1, y_2, and y_3 to (12), we form the *auxiliary master problem*

Maximize: $w = -y_1 - y_2 - y_3$

Subject to: $(F_1 P_1)\alpha_1 + (F_1 P_2)\alpha_2 + \cdots + (F_1 P_r)\alpha_r$

$$+ (F_2 Q_1)\beta_1 + (F_2 Q_2)\beta_2 + \cdots + (F_2 Q_s)\beta_s + \begin{bmatrix} -s_1 + y_1 \\ s_2 \end{bmatrix} = \begin{bmatrix} 10 \\ 8 \end{bmatrix} \quad (13)$$

$$
\begin{array}{ccccc}
\alpha_1 + & \alpha_2 + \cdots + & \alpha_r + & y_2 & = 1 \\
\beta_1 + & \beta_2 + \cdots + & \beta_s + & y_3 & = 1
\end{array}
$$

$$0 \leq \alpha_i \leq 1, \quad 1 \leq i \leq r; \quad 0 \leq \beta_i \leq 1, \quad 1 \leq i \leq s;$$

$$s_1, s_2 \geq 0; \quad y_i \geq 0, \quad 1 \leq i \leq 3$$

Before starting phase 1 of the two-phase procedure, we must express $w = -y_1 - y_2 - y_3$ in terms of nonbasic variables. We do this by solving the constraint equations for $-y_1$, $-y_2$, and $-y_3$. We easily obtain

$$-y_2 = \alpha_1 + \alpha_2 + \cdots + \alpha_r - 1$$
$$-y_3 = \beta_1 + \beta_2 + \cdots + \beta_s - 1 \tag{14}$$

It is a bit more difficult to solve for $-y_1$ because it occurs in the first row of the matrix expression for the common constraint. Each coefficient of α_i in the common constraint is a 2×2 matrix $F_1 P_i$. We obtain the first row of $F_1 P_i$ by premultiplying it by

$$\mathbf{e}_1 = [1 \quad 0]$$

For instance, if $F_1 P_1$ were equal to

$$\begin{bmatrix} 1 & 2 \\ 3 & 4 \end{bmatrix}$$

then

$$\mathbf{e}_1 \begin{bmatrix} 1 & 2 \\ 3 & 4 \end{bmatrix} = [1 \quad 2]$$

Thus we can write $-y_1$ as

$$-y_1 = \quad (\mathbf{e}_1 F_1 P_1)\alpha_1 + (\mathbf{e}_1 F_1 P_2)\alpha_2 + \cdots + (\mathbf{e}_1 F_1 P_r)\alpha_r$$
$$+ (\mathbf{e}_1 F_2 Q_1)\beta_1 + (\mathbf{e}_1 F_2 Q_2)\beta_2 + \cdots + (\mathbf{e}_1 F_2 Q_s)\beta_s - s_1 - 10 \tag{15}$$

From (14) and (15) we obtain

$$w = \quad (\mathbf{e}_1 F_1 P_1 + 1)\alpha_1 + (\mathbf{e}_1 F_1 P_2 + 1)\alpha_2 + \cdots + (\mathbf{e}_1 F_1 P_r + 1)\alpha_r$$
$$+ (\mathbf{e}_1 F_2 Q_1 + 1)\beta_1 + (\mathbf{e}_1 F_2 Q_2 + 1)\beta_2 + \cdots + (\mathbf{e}_1 F_2 Q_s + 1)\beta_s - s_1 - 12$$

In this way we have written problem (13) in canonical form with initial basis matrix

$$B = \begin{array}{c} y_1 \\ s_2 \\ y_2 \\ y_3 \end{array} \begin{bmatrix} 1 & 0 & 0 & 0 \\ 0 & 1 & 0 & 0 \\ 0 & 0 & 1 & 0 \\ 0 & 0 & 0 & 1 \end{bmatrix} \quad \blacklozenge$$

We will solve the master problems we formulated in Examples 2 and 3 in the next section, and we will use their solutions to find the solutions of the LP problems to which they correspond.

Exercises

In Exercises 1 and 2, formulate the decomposable problem in the matrix form (2). Define all matrices that you use. Formulate the master problem as we did in (9). Do not find the corner points of the feasible regions corresponding to the independent constraints because these will not be needed in the solution method we give in Section 10.3.

1. Maximize: $z = x_1 + 2x_2 + 3x_3 - x_4 + x_5 + x_6$

Subject to:
$$
\begin{array}{rl}
x_1 - x_2 + 2x_3 + x_4 - x_5 & \leq 12 \\
2x_1 + x_2 - x_3 + 3x_4 + x_5 + 2x_6 & \leq 10 \\
x_1 + x_2 & \leq 4 \\
5x_1 + 3x_2 & \leq 15 \\
x_3 - 3x_4 & \leq 4 \\
x_3 + 2x_4 & \leq 6 \\
4x_5 + 3x_6 & \leq 12 \\
3x_5 + 4x_6 & \leq 12 \\
5x_5 + 2x_6 & \leq 10
\end{array}
$$

$$x_i \geq 0, \qquad 1 \leq i \leq 6$$

2. Maximize: $z = 2x_1 - x_2 + 3x_3 + x_4$

Subject to:
$$
\begin{array}{rl}
-2x_1 + x_2 + 2x_3 + x_4 & \leq 10 \\
x_1 - x_2 + x_3 + 2x_4 & \leq 8 \\
3x_1 + 4x_2 + 2x_3 + 3x_4 & \leq 24 \\
x_1 - 3x_2 & \leq 4 \\
x_1 + 2x_2 & \leq 6 \\
x_3 - 4x_4 & \leq 4 \\
2x_3 + 4x_4 & \leq 10
\end{array}
$$

$$x_i \geq 0, \qquad 1 \leq i \leq 4$$

In Exercises 3 and 4, formulate the decomposable problem in the matrix form (2). Define all matrices that you use. Formulate the auxiliary master problem as we did in Example 3. Do not find the corner points of the feasible regions corresponding to the independent constraints since these will not be needed in the solution method we give in Section 10.3.

3. Maximize: $z = x_1 + 2x_2 + x_3 - x_4 + 2x_5 + x_6 + 3x_7$

Subject to:
$$
\begin{array}{rl}
2x_1 + 3x_2 + x_3 - x_4 + x_5 + 2x_6 + x_7 & \geq 6 \\
x_1 + x_2 + 2x_3 + x_4 - x_5 + x_6 - x_7 & \leq 12 \\
x_1 - x_3 + x_5 - x_7 & = 4 \\
x_1 + x_2 & \leq 4 \\
5x_1 + 3x_2 & \leq 15 \\
x_3 + x_4 & \leq 5 \\
x_3 + x_4 & \geq 2 \\
x_5 - x_6 + 2x_7 & = 3 \\
2x_5 + 3x_6 + x_7 & \leq 12
\end{array}
$$

$$x_i \geq 0, \qquad 1 \leq i \leq 7$$

4. Maximize: $z = x_1 + x_2 + x_3 + x_4 + x_5 + x_6 + x_7$

Subject to:

$$-2x_1 + 3x_2 + \qquad 2x_4 - x_5 + x_6 - x_7 = 6$$
$$x_1 - 3x_2 + 2x_3 + 6x_4 + x_5 - x_6 + 3x_7 \geq 8$$
$$2x_1 - x_2 + 3x_3 \qquad\qquad\qquad \leq 12$$
$$x_1 + x_2 - x_3 \qquad\qquad\qquad = 6$$
$$3x_4 + 4x_5 \qquad\qquad \leq 12$$
$$x_4 - x_5 \qquad\qquad \geq 2$$
$$x_6 - x_7 \geq 1$$
$$4x_6 + 5x_7 \leq 20$$

$$x_i \geq 0, \qquad 1 \leq i \leq 7$$

The feasible sets corresponding to the independent constraints in Exercises 5 and 6 are unbounded. Use the common constraints to determine bounds on the variables. Use these bounds to add constraints to the sets of independent constraints in such a way that the feasible set of the entire problem is unchanged, but each set of independent constraints determines a bounded feasible set. Formulate the auxiliary master problem.

5. Maximize: $z = 2x_1 + 3x_2 + 4x_3 + 2x_4$

Subject to:

$$2x_1 + 4x_2 + 3x_3 + 2x_4 \leq 12$$
$$x_1 - x_2 + 2x_3 - x_4 \leq 3$$
$$x_1 - x_2 \qquad\qquad \leq 1$$
$$-x_1 + x_2 \qquad\qquad \leq 2$$
$$x_3 + x_4 \geq 5$$
$$2x_3 + 5x_4 \geq 12$$

$$x_i \geq 0, \qquad 1 \leq i \leq 4$$

6. Maximize: $z = x_1 - 3x_2 + 4x_3 - x_4$

Subject to:

$$x_1 - x_2 + x_3 - 2x_4 \leq 12$$
$$x_1 + 3x_2 + 2x_3 + 6x_4 \leq 12$$
$$-3x_1 + 4x_2 \qquad\qquad \leq 6$$
$$3x_1 + 5x_2 \qquad\qquad \geq 15$$
$$-x_3 + x_4 \geq 5$$
$$-2x_3 + 5x_4 \geq 12$$

$$x_i \geq 0, \qquad 1 \leq i \leq 4$$

10.3   SOLVING THE MASTER PROBLEM

In this section we will use the revised simplex method (see Chapter 5) to solve the master problem

$$
\begin{aligned}
\text{Maximize:}\quad z = \ & (\mathbf{c}_1 P_1)\alpha_1 + (\mathbf{c}_1 P_2)\alpha_2 + \cdots + (\mathbf{c}_1 P_r)\alpha_r \\
& + (\mathbf{c}_2 Q_1)\beta_1 + (\mathbf{c}_2 Q_2)\beta_2 + \cdots + (\mathbf{c}_2 Q_s)\beta_s \\
& + (\mathbf{c}_3 R_1)\gamma_1 + (\mathbf{c}_3 R_2)\gamma_2 + \cdots + (\mathbf{c}_3 R_t)\gamma_t
\end{aligned}
$$

Subject to:
$$(F_1P_1)\alpha_1 + (F_1P_2)\alpha_2 + \cdots + (F_1P_r)\alpha_r$$
$$+ (F_2Q_1)\beta_1 + (F_2Q_2)\beta_2 + \cdots + (F_2Q_s)\beta_s$$
$$+ (F_3R_1)\gamma_1 + (F_3R_2)\gamma_2 + \cdots + (F_3R_t)\gamma_t + S = \mathbf{b}_0 \qquad (1)$$
$$\alpha_1 + \quad \alpha_2 + \cdots + \quad \alpha_r \quad = 1$$
$$\beta_1 + \quad \beta_2 + \cdots + \quad \beta_s \quad = 1$$
$$\gamma_1 + \quad \gamma_2 + \cdots + \quad \gamma_t \quad = 1$$

$$\alpha_i \geq 0, \quad 1 \leq i \leq r; \quad \beta_i \geq 0, \quad 1 \leq i \leq s;$$
$$\gamma_i \geq 0, \quad 1 \leq i \leq t; \quad S \geq \mathbf{0}$$

where S is a column matrix containing whatever slack or surplus variables are needed. Specifically, if there are m common constraints we let S_i denote a column of the initial tableau corresponding to a slack or surplus variable, where

$$S_i = \begin{bmatrix} 0 \\ \vdots \\ \pm 1 \\ \vdots \\ 0 \\ 0 \\ 0 \end{bmatrix} \begin{matrix} \\ \\ \leftarrow \text{row } i \\ \\ \leftarrow \text{row } m+1 \\ \leftarrow \text{row } m+2 \\ \leftarrow \text{row } m+3 \end{matrix} \qquad (+ \text{ for a slack or } - \text{ for a surplus variable})$$

We let $A_1, A_2, \ldots, A_r; B_1, B_2, \ldots, B_s;$ and $C_1, C_2, \ldots, C_t$ denote the columns of the constraint matrix of the initial tableau of the master problem (1). That is,

$$A_i = \begin{bmatrix} F_1P_i \\ 1 \\ 0 \\ 0 \end{bmatrix} \qquad B_i = \begin{bmatrix} F_2Q_i \\ 0 \\ 1 \\ 0 \end{bmatrix} \qquad C_i = \begin{bmatrix} F_3R_i \\ 0 \\ 0 \\ 1 \end{bmatrix} \qquad (2)$$

To determine the entering basic variable for a tableau arising in solving the master problem, we proceed as follows. Let B be the basic matrix for the current tableau. Then, as we saw in Chapter 4, a typical objective-row entry of the current tableau is determined by one of four expressions:

$$\mathbf{c}_B B^{-1}A_i - \mathbf{c}_1 P_i; \qquad \mathbf{c}_B B^{-1}B_i - \mathbf{c}_2 Q_i; \qquad \mathbf{c}_B B^{-1}C_i - \mathbf{c}_3 R_i; \qquad \mathbf{c}_B B^{-1}S_i \qquad (3)$$

We will determine the entering basic variable by selecting a negative objective value from one of the four expressions in (3). We use the Cyclic Entering Basic Variable Rule introduced in Section 5.1.

The matrices B and B^{-1} are $(m+3) \times (m+3)$ matrices. We decompose B^{-1} into blocks

$$B^{-1} = [J \mid U \mid V \mid W] \qquad (4)$$

where J is an $(m+3) \times (m)$ matrix and U, V, and W are column matrices of size $(m+3) \times (1)$. From (2), (3), and (4) we have that the objective-row entry corresponding to column A_i is

$$c_B B^{-1} A_i - c_1 P_i = c_B J F_1 P_i + c_B U - c_1 P_i$$
$$= (c_B J F_1 - c_1) P_i + c_B U$$

We would like to find the minimum of these values over the feasible region given by the constraints

$$D_1 v_1 \leq b_1$$

This gives us the first subproblem that we must solve,

Minimize: $\{z_1 = (c_B J F_1 - c_1) P_i + c_B U \mid 1 \leq i \leq r\}$

Subject to: P_i is a corner point of $\{v_1 \mid D_1 v_1 \leq b_1, v_1 \geq 0\}$

Since the optimal solution of an LP problem with a bounded feasible set must occur at a corner point of the feasible set, it follows that the subproblem is equivalent to the LP problem

Minimize: $z_1 = (c_B J F_1 - c_1) v_1 + c_B U$

Subject to: $D_1 v_1 \leq b_1$ (5)

$$v_1 \geq 0$$

In (5) note that $c_B U$ is a constant. Let z_1^* be the minimal value obtained by solving (5) with the simplex (or any other) method.

For the second block of independent constraints we obtain in a similar way the LP subproblem

Minimize: $z_2 = (c_B J F_2 - c_2) v_2 + c_B V$

Subject to: $D_2 v_2 \leq b_2$ (6)

$$v_2 \geq 0$$

Let z_2^* be the optimal solution of problem (6).

For the third block of independent constraints, we obtain the LP subproblem

Minimize: $z_3 = (c_B J F_3 - c_3) v_3 + c_B W$

Subject to: $D_3 v_3 \leq b_3$ (7)

$$v_3 \geq 0$$

Let z_3^* be the optimal solution of problem (7).

Since we will be using the Cyclic Entering Basic Variable Rule, we will solve just enough of problems (5), (6), and (7) to find a negative value of z_1^*, z_2^*, or z_3^*. If $z_1^* < 0$, then we select a variable α_i that we have not used before as the entering basic variable; if $z_1^* \geq 0$ and if $z_2^* < 0$, then we select a new variable β_i; and if $z_1^* \geq 0$, $z_2^* \geq 0$, and $z_3^* < 0$, then we select a new variable γ_i. As indicated previously, the customary Departing Basic Variable Rule is used to determine the departing basic variable.

Finally we may have to examine some of the objective values corresponding to the slack variable columns S_i. These values are calculated with the formula

$$-\mathbf{c}_i^* = \mathbf{c_B}B^{-1}S_i \tag{8}$$

This completes one iteration. To carry out the next iteration, we update the matrix B^{-1} as was done in Chapter 5 and repeat the above steps until all objective-row entries are nonnegative.

What has been gained from this relatively complicated procedure? For large-scale problems with many blocks of independent variables, direct use of the revised simplex method would require working with extremely large B^{-1} matrices, often too large to be computationally feasible. The decomposition procedure breaks the original problem down into smaller chunks, each of which may be relatively easy to solve. As a result, an impossible-to-solve (directly) large problem becomes solvable through the device of solving a number of smaller problems.

Before applying this procedure to an example, we list the steps of the decomposition algorithm for problems that have only $\leq$ constraints. In Example 2 we see how to deal with mixed constraints.

STEP 1. Formulate the master problem by rewriting the original problem in terms of the new variables α_i, β_i, and γ_i. Use α_1, β_1, γ_1, and the slack variables of the independent constraints as initial basic variables.

STEP 2. Use the Cyclic Entering Basic Variable Rule to find the entering basic variable by solving as many problems of type (5)–(8) as necessary. If all objective values are nonnegative, an optimal solution has been found. If a negative minimal value arises in problem (5), (6), or (7), then assign α_i, β_i, or γ_i, respectively, to be the entering basic variable. The choice α_i, β_i, or γ_i is made arbitrarily, except that you must be sure never to label two variables as corresponding to the same corner point. If a negative value arises in a calculation of type (8), assign the corresponding value of s_i to be the entering basic variable.

STEP 3. Determine the departing basic variable in the usual way.

STEP 4. Find the new matrix B^{-1}, and repeat step 2.

EXAMPLE 1

In Examples 1 and 2 of Section 10.2 we considered the LP problem

$$\text{Maximize:} \quad z = 3x_1 + x_2 + 2x_3 + x_4$$

$$\text{Subject to:} \quad 2x_1 + 3x_2 + x_3 + 2x_4 \leq 10$$
$$x_1 \qquad\quad - 2x_3 + 3x_4 \leq 8$$
$$-x_1 + x_2 \qquad\qquad\quad \leq 2$$
$$2x_1 + x_2 \qquad\qquad\quad \leq 8 \tag{9}$$
$$x_3 + 3x_4 \leq 12$$
$$3x_3 - 2x_4 \leq 3$$

$$x_i \geq 0, \qquad 1 \leq i \leq 4$$

We found the master problem to be

Maximize: $(\mathbf{c}_1 P_1)\alpha_1 + (\mathbf{c}_1 P_2)\alpha_2 + \cdots + (\mathbf{c}_1 P_r)\alpha_r$
$+ (\mathbf{c}_2 Q_1)\beta_1 + (\mathbf{c}_2 Q_2)\beta_2 + \cdots + (\mathbf{c}_2 Q_s)\beta_s$

Subject to: $(F_1 P_1)\alpha_1 + (F_1 P_2)\alpha_2 + \cdots + (F_1 P_r)\alpha_r$

$$+ (F_2 Q_1)\beta_1 + (F_2 Q_2)\beta_2 + \cdots + (F_2 Q_s)\beta_s + \begin{bmatrix} s_1 \\ s_2 \end{bmatrix} = \begin{bmatrix} 10 \\ 8 \end{bmatrix}$$

$$\begin{array}{ccccc} \alpha_1 + & \alpha_2 + \cdots + & \alpha_r & = 1 \\ \beta_1 + & \beta_2 + \cdots + & \beta_s & = 1 \end{array}$$

$$0 \le \alpha_i \le 1, \, 1 \le i \le r; \qquad 0 \le \beta_i \le 1, \, 1 \le i \le s; \qquad s_1, s_2 \ge 0$$

ITERATION 0.

If we let $P_1 = \mathbf{0}$ and $Q_1 = \mathbf{0}$, then

$$\mathbf{x}_B = [s_1 \quad s_2 \quad \alpha_1 \quad \beta_1]^T$$

is a basic feasible solution. We also have

$$\mathbf{c}_1 = [3 \quad 1] \qquad \mathbf{c}_2 = [2 \quad 1]$$

$$\mathbf{c}_B = [0 \quad 0 \quad 0 \quad 0]$$

$$B = B^{-1} = \begin{array}{c} s_1 \\ s_2 \\ \alpha_1 \\ \beta_1 \end{array} \begin{bmatrix} 1 & 0 & 0 & 0 \\ 0 & 1 & 0 & 0 \\ 0 & 0 & 1 & 0 \\ 0 & 0 & 0 & 1 \end{bmatrix}$$

We partition B^{-1} as $B^{-1} = [J \mid U \mid V]$, where

$$J = \begin{bmatrix} 1 & 0 \\ 0 & 1 \\ 0 & 0 \\ 0 & 0 \end{bmatrix} \qquad U = \begin{bmatrix} 0 \\ 0 \\ 1 \\ 0 \end{bmatrix} \qquad V = \begin{bmatrix} 0 \\ 0 \\ 0 \\ 1 \end{bmatrix}$$

The first subproblem is

Minimize: $z_1 = (\mathbf{c}_B J F_1 - \mathbf{c}_1)\mathbf{v}_1 + \mathbf{c}_B U = -3x_1 - x_2$

Subject to: $-x_1 + x_2 \le 2$
$2x_1 + x_2 \le 8$

$$x_1, x_2 \ge 0$$

The optimal solution to this problem is

$$\mathbf{v}_1^* = \begin{bmatrix} 4 \\ 0 \end{bmatrix} \qquad z_1^* = -12$$

Since z_1^* is negative, we label the corner point $\mathbf{v}_1^*$ as P_2. Note that the decision to label this corner point P_2 was arbitrary. By this choice, we have associated the variable α_2 with P_2. Since

$$F_1 P_2 = \begin{bmatrix} 2 & 3 \\ 1 & 0 \end{bmatrix} \begin{bmatrix} 4 \\ 0 \end{bmatrix} = \begin{bmatrix} 8 \\ 4 \end{bmatrix}$$

it follows that the α_2 column in the master problem is

$$\begin{bmatrix} 8 \\ 4 \\ 1 \\ 0 \end{bmatrix} \qquad (10)$$

To determine the departing basic variable we use the Departing Basic Variable Rule and see that since

$$\min \left\{ \frac{10}{8}, \frac{8}{4}, 1, - \right\} = 1$$

it follows that α_1 is the departing basic variable.

ITERATION 1.

From iteration 0 we have

$$\mathbf{x}_B = [s_1 \quad s_2 \quad \alpha_2 \quad \beta_1]^T$$

and since

$$\mathbf{c}_1 P_2 = [3 \quad 1] \begin{bmatrix} 4 \\ 0 \end{bmatrix} = 12$$

it follows that

$$\mathbf{c}_B = [0 \quad 0 \quad 12 \quad 0]$$

To calculate the new inverse basic matrix B^{-1}, we premultiply the old B^{-1} matrix by the pivoting matrix formed from the column (10); the pivot is on the 1 since α_2 is the basic variable for the third row. (We discussed pivoting matrices in Chapter 4.) Thus we obtain the new B^{-1} by calculating

$$B^{-1} = \begin{bmatrix} 1 & 0 & -8 & 0 \\ 0 & 1 & -4 & 0 \\ 0 & 0 & 1 & 0 \\ 0 & 0 & 0 & 1 \end{bmatrix} \begin{bmatrix} 1 & 0 & 0 & 0 \\ 0 & 1 & 0 & 0 \\ 0 & 0 & 1 & 0 \\ 0 & 0 & 0 & 1 \end{bmatrix} = \begin{bmatrix} 1 & 0 & -8 & 0 \\ 0 & 1 & -4 & 0 \\ 0 & 0 & 1 & 0 \\ 0 & 0 & 0 & 1 \end{bmatrix}$$

We partition B^{-1} into $B^{-1} = [J \mid U \mid V]$, where

$$J = \begin{bmatrix} 1 & 0 \\ 0 & 1 \\ 0 & 0 \\ 0 & 0 \end{bmatrix} \qquad U = \begin{bmatrix} -8 \\ -4 \\ 1 \\ 0 \end{bmatrix} \qquad V = \begin{bmatrix} 0 \\ 0 \\ 0 \\ 1 \end{bmatrix}$$

By the Cyclic Entering Basic Variable Rule, the first subproblem for this iteration is

$$\text{Minimize:} \quad z_2 = (\mathbf{c}_\mathrm{B} J F_2 - \mathbf{c}_2)\mathbf{v}_2 + \mathbf{c}_\mathrm{B} V = -2x_3 - x_4$$

$$\text{Subject to:} \quad x_3 + 3x_4 \le 12$$
$$3x_3 - 2x_4 \le 3$$

$$x_3, x_4 \ge 0$$

The optimal solution to this problem is

$$\mathbf{v}_2^* = \begin{bmatrix} 3 \\ 3 \end{bmatrix} \qquad z_2^* = -9$$

Since we found a negative value, we label the corner point $\mathbf{v}_2^*$ as Q_2 (an arbitrary choice from the set $\{Q_2, \ldots, Q_s\}$). We designate the entering basic variable as β_2. Since

$$F_2 Q_2 = \begin{bmatrix} 1 & 2 \\ -2 & 3 \end{bmatrix}\begin{bmatrix} 3 \\ 3 \end{bmatrix} = \begin{bmatrix} 9 \\ 3 \end{bmatrix}$$

the column corresponding to β_2 in the original tableau is

To find the departing basic variable, we first calculate the β_2 column in the current tableau; this column is

$$B^{-1}\begin{bmatrix} 9 \\ 3 \\ 0 \\ 1 \end{bmatrix} = \begin{bmatrix} 1 & 0 & -8 & 0 \\ 0 & 1 & -4 & 0 \\ 0 & 0 & 1 & 0 \\ 0 & 0 & 0 & 1 \end{bmatrix}\begin{bmatrix} 9 \\ 3 \\ 0 \\ 1 \end{bmatrix} = \begin{bmatrix} 9 \\ 3 \\ 0 \\ 1 \end{bmatrix} \qquad (11)$$

The resource column in the current tableau is

$$B^{-1}\begin{bmatrix} 10 \\ 8 \\ 1 \\ 1 \end{bmatrix} = \begin{bmatrix} 1 & 0 & -8 & 0 \\ 0 & 1 & -4 & 0 \\ 0 & 0 & 1 & 0 \\ 0 & 0 & 0 & 1 \end{bmatrix}\begin{bmatrix} 10 \\ 8 \\ 1 \\ 1 \end{bmatrix} = \begin{bmatrix} 2 \\ 4 \\ 1 \\ 1 \end{bmatrix}$$

Since

$$\min\left\{\frac{2}{9}, \frac{4}{3}, -, 1\right\} = \frac{2}{9}$$

it follows that s_1 is the departing basic variable.

ITERATION 2.

From Iteration 1 we have

$$\mathbf{x_B} = [\beta_2 \quad s_2 \quad \alpha_2 \quad \beta_1]^T$$

and since

$$\mathbf{c}_2 Q_2 = [2 \quad 1]\begin{bmatrix} 3 \\ 3 \end{bmatrix} = 9$$

it follows that

$$\mathbf{c_B} = [9 \quad 0 \quad 12 \quad 0]$$

To calculate the new inverse basis matrix, we premultiply the old B^{-1} matrix by the pivoting matrix corresponding to the first row (the s_1 row) of the β_2 column (11) to obtain

$$B^{-1} = \begin{bmatrix} \frac{1}{9} & 0 & 0 & 0 \\ -\frac{1}{3} & 1 & 0 & 0 \\ 0 & 0 & 1 & 0 \\ -\frac{1}{9} & 0 & 0 & 1 \end{bmatrix}\begin{bmatrix} 1 & 0 & -8 & 0 \\ 0 & 1 & -4 & 0 \\ 0 & 0 & 1 & 0 \\ 0 & 0 & 0 & 1 \end{bmatrix} = \begin{bmatrix} \frac{1}{9} & 0 & -\frac{8}{9} & 0 \\ -\frac{1}{3} & 1 & -\frac{4}{3} & 0 \\ 0 & 0 & 1 & 0 \\ -\frac{1}{9} & 0 & \frac{8}{9} & 1 \end{bmatrix}$$

We partition B^{-1} into $B^{-1} = [J \mid U \mid V]$ to obtain

$$J = \begin{bmatrix} \frac{1}{9} & 0 \\ -\frac{1}{3} & 1 \\ 0 & 0 \\ -\frac{1}{9} & 0 \end{bmatrix} \qquad U = \begin{bmatrix} -\frac{8}{9} \\ -\frac{4}{3} \\ 1 \\ \frac{8}{9} \end{bmatrix} \qquad V = \begin{bmatrix} 0 \\ 0 \\ 0 \\ 1 \end{bmatrix}$$

The Cyclic Entering Basic Variable Rule calls for us to examine the slack variable column S_1 next. Using (8) we obtain

$$-c_1^* = \mathbf{c_B} B^{-1} S_1 = [9 \quad 0 \quad 12 \quad 0]\begin{bmatrix} \frac{1}{9} & 0 & -\frac{8}{9} & 0 \\ -\frac{1}{3} & 1 & -\frac{4}{3} & 0 \\ 0 & 0 & 1 & 0 \\ -\frac{1}{9} & 0 & \frac{8}{9} & 1 \end{bmatrix}\begin{bmatrix} 1 \\ 0 \\ 0 \\ 0 \end{bmatrix} = 1$$

Since $-c_1^*$ is not negative, we next consider the S_2 column. However, since s_2 is a basic variable, we know the corresponding objective-row entry is 0 without making a calculation. Therefore, we cycle back to the first block of independent constraints and consider the subproblem

Minimize: $z_1 = (\mathbf{c_B} J F_1 - \mathbf{c}_1)\mathbf{v}_1 + \mathbf{c_B} U = -x_1 + x_2 + 4$

Subject to: $-x_1 + x_2 \le 2$

$2x_1 + x_2 \le 8$

$x_1, x_2 \ge 0$

The optimal solution of this problem is

$$\mathbf{v}_1^* = \begin{bmatrix} 4 \\ 0 \end{bmatrix} \qquad z_1^* = 0$$

Since z_1^* is not negative, we cycle to the next block of independent constraints and consider the subproblem

$$\text{Minimize:} \quad z_2 = (\mathbf{c_B}JF_2 - \mathbf{c_2})\mathbf{v}_2 + \mathbf{c_B}V = -x_3 + x_4$$

$$\text{Subject to:} \quad x_3 + 3x_4 \leq 12$$
$$3x_3 - 2x_4 \leq 3$$
$$x_3, x_4 \geq 0$$

The optimal solution to this problem is

$$\mathbf{v}_2^* = \begin{bmatrix} 1 \\ 0 \end{bmatrix} \qquad z_2^* = -1$$

Since z_2^* is negative, we designate $Q_3 = \mathbf{v}_2^*$ as a new corner point. Thus, β_3 is the entering basic variable. Since

$$F_2Q_3 = \begin{bmatrix} 1 & 2 \\ -2 & 3 \end{bmatrix} \begin{bmatrix} 1 \\ 0 \end{bmatrix} = \begin{bmatrix} 1 \\ -2 \end{bmatrix}$$

the column corresponding to β_3 in the original tableau is

$$\begin{bmatrix} 1 \\ -2 \\ 0 \\ 1 \end{bmatrix}$$

The current β_3 column is

$$B^{-1} \begin{bmatrix} 1 \\ -2 \\ 0 \\ 1 \end{bmatrix} = \begin{bmatrix} \frac{1}{9} & 0 & -\frac{8}{9} & 0 \\ -\frac{1}{3} & 1 & -\frac{4}{3} & 0 \\ 0 & 0 & 1 & 0 \\ -\frac{1}{9} & 0 & \frac{8}{9} & 1 \end{bmatrix} \begin{bmatrix} 1 \\ -2 \\ 0 \\ 1 \end{bmatrix} = \begin{bmatrix} \frac{1}{9} \\ -\frac{7}{3} \\ 0 \\ \frac{8}{9} \end{bmatrix} \qquad (12)$$

The current resource column is

$$B^{-1} \begin{bmatrix} 10 \\ 8 \\ 1 \\ 1 \end{bmatrix} = \begin{bmatrix} \frac{1}{9} & 0 & -\frac{8}{9} & 0 \\ -\frac{1}{3} & 1 & -\frac{4}{3} & 0 \\ 0 & 0 & 1 & 0 \\ -\frac{1}{9} & 0 & \frac{8}{9} & 1 \end{bmatrix} \begin{bmatrix} 10 \\ 8 \\ 1 \\ 1 \end{bmatrix} = \begin{bmatrix} \frac{2}{9} \\ \frac{10}{3} \\ 1 \\ \frac{7}{9} \end{bmatrix}$$

Since

$$\min\left\{2, -, -, \frac{7}{8}\right\} = \frac{7}{8}$$

it follows that β_1 is the departing basic variable.

ITERATION 3.

From Iteration 2 we have

$$\mathbf{x}_B = [\beta_2 \ \ s_2 \ \ \alpha_2 \ \ \beta_3]^\mathrm{T}$$

and since

$$\mathbf{c}_2 Q_3 = [2 \ \ 1]\begin{bmatrix} 1 \\ 0 \end{bmatrix} = 2$$

it follows that

$$\mathbf{c}_B = [9 \ \ 0 \ \ 12 \ \ 2]$$

To calculate the new inverse basis matrix, we premultiply the old B^{-1} matrix by the pivoting matrix corresponding to the fourth row (the β_1 row) of the β_3 column (12) to obtain

$$B^{-1} = \begin{bmatrix} 1 & 0 & 0 & -\frac{1}{8} \\ 0 & 1 & 0 & \frac{21}{8} \\ 0 & 0 & 1 & 0 \\ 0 & 0 & 0 & \frac{9}{8} \end{bmatrix}\begin{bmatrix} \frac{1}{9} & 0 & -\frac{8}{9} & 0 \\ -\frac{1}{3} & 1 & -\frac{4}{3} & 0 \\ 0 & 0 & 1 & 0 \\ -\frac{1}{9} & 0 & \frac{8}{9} & 1 \end{bmatrix} = \begin{bmatrix} \frac{1}{8} & 0 & -1 & -\frac{1}{8} \\ -\frac{5}{8} & 1 & 1 & \frac{21}{8} \\ 0 & 0 & 1 & 0 \\ -\frac{1}{8} & 0 & 1 & \frac{9}{8} \end{bmatrix}$$

We partition B^{-1} into $B^{-1} = [J \mid U \mid V]$, where

$$J = \begin{bmatrix} \frac{1}{8} & 0 \\ -\frac{5}{8} & 1 \\ 0 & 0 \\ -\frac{1}{8} & 0 \end{bmatrix} \qquad U = \begin{bmatrix} -1 \\ 1 \\ 1 \\ 1 \end{bmatrix} \qquad V = \begin{bmatrix} -\frac{1}{8} \\ \frac{21}{8} \\ 0 \\ \frac{9}{8} \end{bmatrix}$$

We next examine the objective entry in the S_1 column by calculating

$$-c_1^* = \mathbf{c}_B B^{-1} S_1 = \frac{7}{8}$$

Since $-c_1^* > 0$, we move on to the S_2 column; however, we know that $-c_2^* = 0$ because s_2 currently is a basic variable. Next we move to the first block of independent constraints and consider the subproblem

Minimize: $\quad z_1 = (\mathbf{c}_B J F_1 - \mathbf{c}_1)\mathbf{v}_1 + \mathbf{c}_B U = -\frac{5}{4}x_1 + \frac{13}{8}x_2 + 5$

Subject to: $-x_1 + x_2 \leq 2$

$2x_1 + x_2 \leq 8$

$x_1, x_2 \geq 0$

The solution to this problem is

$$\mathbf{v}_1^* = \begin{bmatrix} 4 \\ 0 \end{bmatrix} \qquad z_1^* = 0$$

Since $z_1^* = 0$, we move on to the next block and consider the problem

Minimize: $z_2 = (\mathbf{c}_B J F_2 - \mathbf{c}_2)\mathbf{v}_2 + \mathbf{c}_B V = -\dfrac{9}{8}x_3 + \dfrac{3}{4}x_4 + \dfrac{9}{8}$

Subject to: $x_3 + 3x_4 \leq 12$

$3x_3 - 2x_4 \leq 3$

$x_3, x_4 \geq 0$

The solution to this problem is

$$\mathbf{v}_2^* = \begin{bmatrix} 1 \\ 0 \end{bmatrix} \qquad z_2^* = 0$$

Since we have determined that all values in the objective row are nonnegative, we have found the optimal solution. To find this solution we compute the new resource column as

$$\mathbf{b}^* = B^{-1} \begin{bmatrix} 10 \\ 8 \\ 1 \\ 0 \end{bmatrix} = \begin{bmatrix} \frac{1}{8} & 0 & -1 & -\frac{1}{8} \\ -\frac{5}{8} & 1 & 1 & \frac{21}{8} \\ 0 & 0 & 1 & 0 \\ -\frac{1}{8} & 0 & 1 & \frac{9}{8} \end{bmatrix} \begin{bmatrix} 10 \\ 8 \\ 1 \\ 0 \end{bmatrix} = \begin{bmatrix} \frac{1}{8} \\ \frac{43}{8} \\ 1 \\ \frac{7}{8} \end{bmatrix}$$

Therefore, the solution to the master problem is

$$\mathbf{x}_B = \begin{bmatrix} \beta_2 \\ s_2 \\ \alpha_2 \\ \beta_3 \end{bmatrix} = \begin{bmatrix} \frac{1}{8} \\ \frac{43}{8} \\ 1 \\ \frac{7}{8} \end{bmatrix}$$

and the optimal value is $z^* = \mathbf{c}_B \mathbf{b}^* = \frac{119}{8}$.
    The solution to the original problem (9) is

$$\begin{bmatrix} x_1 \\ x_2 \end{bmatrix} = \alpha_2 \begin{bmatrix} 4 \\ 0 \end{bmatrix} = \begin{bmatrix} 4 \\ 0 \end{bmatrix}$$

$$\begin{bmatrix} x_3 \\ x_4 \end{bmatrix} = \beta_2 \begin{bmatrix} 3 \\ 3 \end{bmatrix} + \beta_3 \begin{bmatrix} 1 \\ 0 \end{bmatrix} = \begin{bmatrix} \frac{5}{4} \\ \frac{3}{8} \end{bmatrix}$$

In the preceding example we were able to start the decomposition procedure because we knew that the origin provided us with an initial corner point (and an initial basic matrix, whose inverse was the identity matrix). If such an initial basic corner point or matrix is not obvious, then we can use the two-phase method. The next example illustrates how this is done.

EXAMPLE 2

In Example 3 of Section 10.2, we considered the LP problem

$$\text{Maximize:} \quad z = 3x_1 + x_2 + 2x_3 + x_4$$

$$\text{Subject to:} \quad 2x_1 + 3x_2 + x_3 + 2x_4 \geq 10$$
$$x_1 \quad\quad - 2x_3 + 3x_4 \leq 8$$
$$-x_1 + x_2 \quad\quad\quad \geq 2$$
$$2x_1 + x_2 \quad\quad\quad \leq 8$$
$$x_3 + 3x_4 \leq 12$$
$$3x_3 - 2x_4 \geq 3$$

$$x_i \geq 0, \quad 1 \leq i \leq 4$$

In that example we saw that the associated auxiliary master problem is

$$w = (\mathbf{e}_1 F_1 P_1 + 1)\alpha_1 + (\mathbf{e}_1 F_1 P_2 + 1)\alpha_2 + \cdots + (\mathbf{e}_1 F_1 P_r + 1)\alpha_r$$
$$+ (\mathbf{e}_1 F_2 Q_1 + 1)\beta_1 + (\mathbf{e}_1 F_2 Q_2 + 1)\beta_2 + \cdots + (\mathbf{e}_1 F_2 Q_s + 1)\beta_s$$
$$- s_1 - 12$$

$$\text{Subject to:} \quad (F_1 P_1)\alpha_1 + (F_1 P_2)\alpha_2 + \cdots + (F_1 P_r)\alpha_r$$
$$+ (F_2 Q_1)\beta_1 + (F_2 Q_2)\beta_2 + \cdots + (F_2 Q_s)\beta_s + \begin{bmatrix} -s_1 + y_1 \\ s_2 \end{bmatrix} = \begin{bmatrix} 10 \\ 8 \end{bmatrix} \quad (13)$$

$$\alpha_1 + \alpha_2 + \cdots + \alpha_r + y_2 = 1$$
$$\beta_1 + \beta_2 + \cdots + \beta_s + y_3 = 1$$
$$0 \leq \alpha_i \leq 1, \quad 1 \leq i \leq r; \quad 0 \leq \beta_i \leq 1, \quad 1 \leq i \leq s;$$
$$s_1, s_2 \geq 0; \quad y_i \geq 0, \quad 1 \leq i \leq 3$$

Taking an approach similar to that for LP problem (1), we let A_i, B_i, S_i, and Y_i be typical columns of the initial tableau for (13), where

$$A_i = \begin{bmatrix} F_1 P_i \\ 1 \\ 0 \end{bmatrix} \quad B_i = \begin{bmatrix} F_2 Q_i \\ 0 \\ 1 \end{bmatrix}$$

$$S_1 = \begin{bmatrix} -1 \\ 0 \\ 0 \\ 0 \end{bmatrix} \quad S_2 = \begin{bmatrix} 0 \\ 1 \\ 0 \\ 0 \end{bmatrix}$$

$$Y_1 = \begin{bmatrix} 1 \\ 0 \\ 0 \\ 0 \end{bmatrix} \qquad Y_2 = \begin{bmatrix} 0 \\ 0 \\ 1 \\ 0 \end{bmatrix} \qquad Y_3 = \begin{bmatrix} 0 \\ 0 \\ 0 \\ 1 \end{bmatrix}$$

As before, we partition B^{-1} into

$$B^{-1} = [J \mid U \mid V]$$

where J is a 4×2 matrix and U and V are 4×1 matrices. A typical objective-row entry of the current tableau is determined by one of the expressions

$$c_B B^{-1} A_i - (e_1 F_1 P_i + 1) = c_B J F_1 P_i + c_B U - e_1 F_1 P_i - 1 \qquad (14)$$

$$c_B B^{-1} B_i - (e_1 F_1 Q_i + 1) = c_B J F_2 Q_i + c_B V - e_1 F_2 Q_i - 1 \qquad (15)$$

$$c_B B^{-1} S_1 + 1 \qquad (16)$$

$$c_B B^{-1} S_2 - 1 \qquad (17)$$

$$c_B B^{-1} Y_i \qquad (18)$$

We perform phase 1 on the auxiliary master problem (13) just as we solved the problem in Example 1, except that we use expressions (14) or (15) to determine the objective functions of the subproblems. On completion of phase 1, we start phase 2 with the objective functions of the subproblems determined by the expressions we obtained earlier for the objective-row entries (2).

ITERATION 0.

We observe that the basic matrix associated with (17) is

$$B = \begin{matrix} y_1 \\ s_2 \\ y_2 \\ y_3 \end{matrix} \begin{bmatrix} 1 & 0 & 0 & 0 \\ 0 & 1 & 0 & 0 \\ 0 & 0 & 1 & 0 \\ 0 & 0 & 0 & 1 \end{bmatrix} = B^{-1}$$

It follows that

$$x_B = [y_1 \quad s_2 \quad y_2 \quad y_3]^T$$

is a basic feasible solution. We have

$$c_B = [0 \quad 0 \quad 0 \quad 0]$$

The Cyclic Entering Basic Variable Rule ordinarily begins with the first block of independent constraints; however, we will begin with the second block of constraints because in this example the sequence of iterations happens to illustrate all of the possibilities better. Our first subproblem is

$$\text{Minimize:} \quad z = \mathbf{c_B} J F_2 \mathbf{v}_2 + \mathbf{c_B} V - \mathbf{e}_1 F_2 \mathbf{v}_2 - 1$$

$$= 0 + 0 - \begin{bmatrix} 1 & 0 \end{bmatrix} \begin{bmatrix} 1 & 3 \\ 3 & -2 \end{bmatrix} \begin{bmatrix} x_3 \\ x_4 \end{bmatrix} - 1$$

$$= -x_3 - 3x_4 - 1$$

$$\text{Subject to:} \quad x_3 + 3x_4 \leq 12$$

$$3x_3 - 2x_4 \geq 3$$

$$x_3, x_4 \geq 0$$

The optimal solution is $x_3 = 12$, $x_4 = 0$; $z^* = -13$. We label this corner point as

$$Q_1 = \begin{bmatrix} 12 \\ 0 \end{bmatrix}$$

and choose β_1 as the entering variable. Since

$$F_2 Q_1 = \begin{bmatrix} 1 & 2 \\ -2 & 3 \end{bmatrix} \begin{bmatrix} 12 \\ 0 \end{bmatrix} = \begin{bmatrix} 12 \\ -24 \end{bmatrix}$$

it follows that the β_1 column in the auxiliary master problem is

$$\begin{bmatrix} 12 \\ -24 \\ 0 \\ 1 \end{bmatrix} \tag{19}$$

To determine the departing basic variable, we use the Departing Basic Variable Rule and see that since

$$\min \left\{ \frac{10}{12}, -, -, \frac{1}{1} \right\} = \frac{10}{12}$$

it follows that y_1 is the departing basic variable.

ITERATON 1.

From Iteration 0 we have

$$\mathbf{x_B} = \begin{bmatrix} \beta_1 & s_2 & y_2 & y_3 \end{bmatrix}^\mathsf{T}$$

We obtain the cost coefficient of β_1 in the original auxiliary master problem (13) from

$$\mathbf{e}_1 F_2 Q_1 + 1 = \begin{bmatrix} 1 & 0 \end{bmatrix} \begin{bmatrix} 1 & 2 \\ -2 & 3 \end{bmatrix} \begin{bmatrix} 12 \\ 0 \end{bmatrix} + 1 = 13$$

It follows that the new c_B is

$$c_B = [13 \quad 0 \quad 0 \quad 0]$$

We calculate the new inverse basis matrix B^{-1} by premultiplying the old B^{-1} matrix by the pivoting matrix formed from column (19).

$$B^{-1} = \begin{bmatrix} \frac{1}{12} & 0 & 0 & 0 \\ 2 & 1 & 0 & 0 \\ 0 & 0 & 1 & 0 \\ -\frac{1}{12} & 0 & 0 & 1 \end{bmatrix} \begin{bmatrix} 1 & 0 & 0 & 0 \\ 0 & 1 & 0 & 0 \\ 0 & 0 & 1 & 0 \\ 0 & 0 & 0 & 1 \end{bmatrix} = \begin{bmatrix} \frac{1}{12} & 0 & 0 & 0 \\ 2 & 1 & 0 & 0 \\ 0 & 0 & 1 & 0 \\ -\frac{1}{12} & 0 & 0 & 1 \end{bmatrix}$$

We partition B^{-1} into $B^{-1} = [J \mid U \mid V]$ in the usual way.

The Cyclic Entering Basic Variable Rule requires us to examine the objective-row entry in column S_1 next. By expression (16), it is

$$c_B B^{-1} S_1 + 1 = [13 \quad 0 \quad 0 \quad 0] \begin{bmatrix} \frac{1}{12} & 0 & 0 & 0 \\ 2 & 1 & 0 & 0 \\ 0 & 0 & 1 & 0 \\ -\frac{1}{12} & 0 & 0 & 1 \end{bmatrix} \begin{bmatrix} -1 \\ 0 \\ 0 \\ 0 \end{bmatrix} + 1 = -\frac{1}{12}$$

We conclude that s_1 is the entering basic variable. To determine the departing basic variable, we first compute the new S_1 column. It is

$$B^{-1} S_1 = \begin{bmatrix} \frac{1}{12} & 0 & 0 & 0 \\ 2 & 1 & 0 & 0 \\ 0 & 0 & 1 & 0 \\ -\frac{1}{12} & 0 & 0 & 1 \end{bmatrix} \begin{bmatrix} -1 \\ 0 \\ 0 \\ 0 \end{bmatrix} = \begin{bmatrix} -\frac{1}{12} \\ -2 \\ 0 \\ \frac{1}{12} \end{bmatrix} \qquad (20)$$

Next, we compute the new resource column

$$B^{-1} \begin{bmatrix} 10 \\ 8 \\ 1 \\ 1 \end{bmatrix} = \begin{bmatrix} \frac{1}{12} & 0 & 0 & 0 \\ 2 & 1 & 0 & 0 \\ 0 & 0 & 1 & 0 \\ -\frac{1}{12} & 0 & 0 & 1 \end{bmatrix} \begin{bmatrix} 10 \\ 8 \\ 1 \\ 1 \end{bmatrix} = \begin{bmatrix} \frac{5}{6} \\ 28 \\ 1 \\ \frac{1}{6} \end{bmatrix}$$

Since $\min\{-, -, -, (\frac{1}{6})/(\frac{1}{12})\} = 2$, the departing basic variable is y_3.

ITERATION 2.

From Iteration 1 we have

$$x_B = [\beta_1 \quad s_2 \quad y_2 \quad s_1]^T$$

Since the cost coefficient of s_1 in the original auxiliary master problem (17) is -1, it follows that the new c_B is

$$c_B = [13 \quad 0 \quad 0 \quad -1]$$

We calculate the new inverse basis matrix B^{-1} by premultiplying the old B^{-1} matrix by the pivoting matrix formed from the new column S_1 given in (20).

$$B^{-1} = \begin{bmatrix} 1 & 0 & 0 & 1 \\ 0 & 1 & 0 & 24 \\ 0 & 0 & 1 & 0 \\ 0 & 0 & 0 & 12 \end{bmatrix} \begin{bmatrix} \frac{1}{12} & 0 & 0 & 0 \\ 2 & 1 & 0 & 0 \\ 0 & 0 & 1 & 0 \\ -\frac{1}{12} & 0 & 0 & 1 \end{bmatrix} = \begin{bmatrix} 0 & 0 & 0 & 1 \\ 0 & 1 & 0 & 24 \\ 0 & 0 & 1 & 0 \\ -1 & 0 & 0 & 12 \end{bmatrix}$$

We partition B^{-1} into $B^{-1} = [J \mid U \mid V]$ in the usual way.

The Cyclic Entering Basic Variable Rule tells us to begin with column S_2; however, we skip it because s_2 is basic. The next nonbasic variable is y_1. We use (18) to calculate the objective-row entry in column Y_1 as

$$c_B B^{-1} Y_1 = \begin{bmatrix} 13 & 0 & 0 & -1 \end{bmatrix} \begin{bmatrix} 0 & 0 & 0 & 1 \\ 0 & 1 & 0 & 24 \\ 0 & 0 & 1 & 0 \\ -1 & 0 & 0 & 12 \end{bmatrix} \begin{bmatrix} 1 \\ 0 \\ 0 \\ 0 \end{bmatrix} = 1$$

Since this value is positive, we try column Y_2 next. By (18),

$$c_B B^{-1} Y_2 = \begin{bmatrix} 13 & 0 & 0 & -1 \end{bmatrix} \begin{bmatrix} 0 & 0 & 0 & 1 \\ 0 & 1 & 0 & 24 \\ 0 & 0 & 1 & 0 \\ -1 & 0 & 0 & 12 \end{bmatrix} \begin{bmatrix} 0 \\ 0 \\ 1 \\ 0 \end{bmatrix} = 0$$

Since this value is 0, we try Y_3 next. By (18),

$$c_B B^{-1} Y_3 = \begin{bmatrix} 13 & 0 & 0 & -1 \end{bmatrix} \begin{bmatrix} 0 & 0 & 0 & 1 \\ 0 & 1 & 0 & 24 \\ 0 & 0 & 1 & 0 \\ -1 & 0 & 0 & 12 \end{bmatrix} \begin{bmatrix} 0 \\ 0 \\ 0 \\ 1 \end{bmatrix} = 1$$

Since this value is positive, we try the first block of independent constraints by solving the subproblem

Minimize:

$z = c_B J F_1 v_1 + c_B U - e_1 F_1 v_1 - 1$

$$= \begin{bmatrix} 13 & 0 & 0 & -1 \end{bmatrix} \begin{bmatrix} 0 & 0 \\ 0 & 1 \\ 0 & 0 \\ -1 & 0 \end{bmatrix} \begin{bmatrix} 2 & 3 \\ 1 & 0 \end{bmatrix} \begin{bmatrix} x_1 \\ x_2 \end{bmatrix}$$

$$+ \begin{bmatrix} 13 & 0 & 0 & -1 \end{bmatrix} \begin{bmatrix} 0 \\ 0 \\ 1 \\ 0 \end{bmatrix} - \begin{bmatrix} 1 & 0 \end{bmatrix} \begin{bmatrix} 2 & 3 \\ 1 & 0 \end{bmatrix} \begin{bmatrix} x_1 \\ x_2 \end{bmatrix} - 1 = -1 \qquad (21)$$

Subject to: $-x_1 + x_2 \geq 2$

$2x_1 + x_2 \leq 8$

$x_1, x_2 \geq 0$

Notice that the auxiliary objective function has a constant value of -1. This means that any point in the feasible region gives an optimal solution. This is a fairly common occurrence in this procedure; it means that all the auxiliary objective-row entries in the tableau have the same value. It is not a major difficulty because we would usually solve a problem like (21) by the two-phase method, and the tableau at the end of phase 1 gives a corner point of the feasible region of (21). We can use this corner point to proceed. You can verify that

$$P_1 = \begin{bmatrix} 2 \\ 4 \end{bmatrix}$$

is a corner point for (21) that can be found by phase 1. (We made the choice to label this point P_1 arbitrarily.) The entering basic variable is α_1. Since

$$F_1 P_1 = \begin{bmatrix} 2 & 3 \\ 1 & 0 \end{bmatrix} \begin{bmatrix} 2 \\ 4 \end{bmatrix} = \begin{bmatrix} 16 \\ 2 \end{bmatrix}$$

the column corresponding to α_1 in the original auxiliary master problem is

$$\begin{bmatrix} 16 \\ 2 \\ 1 \\ 0 \end{bmatrix}$$

The current α_1 column is

$$B^{-1} \begin{bmatrix} 16 \\ 2 \\ 1 \\ 0 \end{bmatrix} = \begin{bmatrix} 0 \\ 2 \\ 1 \\ -16 \end{bmatrix} \tag{22}$$

The current resource column is

$$B^{-1} \begin{bmatrix} 10 \\ 8 \\ 1 \\ 1 \end{bmatrix} = \begin{bmatrix} 1 \\ 32 \\ 1 \\ 2 \end{bmatrix}$$

Since $\min\{-, \frac{32}{2}, \frac{1}{1}, -\} = 1$, it follows that y_2 is the departing basic variable.

ITERATION 3.

From Iteration 2 we have

$$\mathbf{x_B} = [\beta_1 \quad s_2 \quad \alpha_1 \quad s_1]^T$$

We obtain the cost coefficient of α_1 in the original auxiliary master problem (13) from

$$\mathbf{e}_1 F_1 P_2 + 1 = \begin{bmatrix} 1 & 0 \end{bmatrix} \begin{bmatrix} 2 & 3 \\ 1 & 0 \end{bmatrix} \begin{bmatrix} 2 \\ 4 \end{bmatrix} + 1 = 17$$

It follows that the new $\mathbf{c}_B$ is

$$\mathbf{c}_B = \begin{bmatrix} 13 & 0 & 17 & -1 \end{bmatrix}$$

We calculate the new inverse basis matrix by premultiplying the old B^{-1} matrix by the pivoting matrix formed from the column in (22).

$$B^{-1} = \begin{bmatrix} 1 & 0 & 0 & 0 \\ 0 & 1 & -\frac{1}{2} & 0 \\ 0 & 0 & 1 & 0 \\ 0 & 0 & 16 & 1 \end{bmatrix} \begin{bmatrix} 0 & 0 & 0 & 1 \\ 0 & 1 & 0 & 24 \\ 0 & 0 & 1 & 0 \\ -1 & 0 & 0 & 12 \end{bmatrix}$$

$$= \begin{bmatrix} 0 & 0 & 0 & 1 \\ 0 & 1 & -2 & 24 \\ 0 & 0 & 1 & 0 \\ -1 & 0 & 16 & 12 \end{bmatrix}$$

Since all of the artificial variables have left the basis, we check that the auxiliary objective function has value 0 by calculating

$$\mathbf{c}_B B^{-1} \begin{bmatrix} 10 \\ 8 \\ 1 \\ 1 \end{bmatrix} - 12 = 0$$

Thus, phase 1 has successfully terminated, and we have found an initial basic solution. We start phase 2 by switching to the objective function given in (12) of Section 10.2.

ITERATION 4.

We begin phase 2 by calculating the current value of $\mathbf{c}_B$ for the master problem. The corner points corresponding to the current basis

$$\mathbf{x}_B = \begin{bmatrix} \beta_1 & s_2 & \alpha_1 & s_1 \end{bmatrix}^T$$

are

$$P_1 = \begin{bmatrix} 2 \\ 4 \end{bmatrix} \qquad Q_1 = \begin{bmatrix} 12 \\ 0 \end{bmatrix}$$

The corresponding cost coefficients are

$$\mathbf{c}_1 P_1 = \begin{bmatrix} 3 & 1 \end{bmatrix} P_1 = 10 \qquad \mathbf{c}_2 Q_1 = \begin{bmatrix} 2 & 1 \end{bmatrix} Q_1 = 24$$

The cost coefficients of the slack variables are 0, thus c_B is

$$c_B = [24 \quad 0 \quad 10 \quad 0]$$

The first subproblem is

Minimize: $z = (c_B J F_1 - c_1)v_1 + c_B U$

$$= ([24 \quad 0 \quad 10 \quad 0] \begin{bmatrix} 0 & 0 \\ 0 & 1 \\ 0 & 0 \\ -1 & 0 \end{bmatrix} \begin{bmatrix} 2 & 3 \\ 1 & 0 \end{bmatrix} - [3 \quad 1]) \begin{bmatrix} x_1 \\ x_2 \end{bmatrix}$$

$$+ [24 \quad 0 \quad 10 \quad 0] \begin{bmatrix} 0 \\ -2 \\ 1 \\ 16 \end{bmatrix} = -3x_1 - x_2 + 10$$

Subject to: $-x_1 + x_2 \geq 2$
$\qquad\qquad\quad 2x_1 + x_2 \leq 8$

$$x_1, x_2 \geq 0$$

The optimal solution is $x_1 = 2$, $x_2 = 4$; $z^* = 0$. We conclude that all of the current entries in the objective row for the first block of independent constraints are nonnegative.

In Exercise 8, you are asked to show that the objective-row entries for the second block of independent constraints and for the slack variable columns are also nonnegative. Therefore, we have found the optimal solution of the master problem to be

$$x_B = \begin{bmatrix} \beta_1 \\ s_2 \\ \alpha_1 \\ s_1 \end{bmatrix} = b^* = B^{-1} \begin{bmatrix} 10 \\ 8 \\ 1 \\ 1 \end{bmatrix} = \begin{bmatrix} 1 \\ 30 \\ 1 \\ 18 \end{bmatrix}$$

$$z^* = cb^* = [24 \quad 0 \quad 10 \quad 0]b^* = 34$$

The solution to the original problem is

$$\begin{bmatrix} x_1 \\ x_2 \end{bmatrix} = \alpha_1 \begin{bmatrix} 2 \\ 4 \end{bmatrix} = \begin{bmatrix} 2 \\ 4 \end{bmatrix}$$

$$\begin{bmatrix} x_3 \\ x_4 \end{bmatrix} = \beta_2 \begin{bmatrix} 12 \\ 0 \end{bmatrix} = \begin{bmatrix} 12 \\ 0 \end{bmatrix}$$

$\qquad\qquad\qquad\qquad\qquad\qquad\qquad\qquad\qquad\qquad\qquad\qquad\qquad\qquad$ ◆

Exercises

1. Find the corner points of the two blocks of independent constraints for the LP problem in Example 1. Write the master problem in terms of these corner points, and form the tableau for this problem.

†2. Solve the problem you formulated in Exercise 1 with TUTOR. Perform the pivots in the same way as we did in Example 1, and observe that all calculations we made are correct.

3. Find the corner points of the two blocks of independent constraints for the LP problem in Example 2. Write the auxiliary master problem in terms of these corner points, and form the tableau for this problem.

†4. Solve the problem you formulated in Exercise 3 with TUTOR. Perform the pivots in the same way as we did in Example 2, and observe that all calculations we made are correct.

5. Solve the LP problem in Exercise 1 of Section 10.2 using the method given in Example 1.

6. Solve the LP problem in Exercise 2 of Section 10.2 using the method given in Example 1.

7. Solve the LP problem in Exercise 3 of Section 10.2 using the method given in Example 2.

8. Complete the solution of the problem in Example 2 from where we left off by showing that the objective-row entries for the second block of independent constraints and for the slack variable columns are nonnegative.

10.4 FINAL WORDS

Karmarkar's Method

As we have discovered, the simplex method provides a powerful procedure for solving both small- and large-scale problems. Remarkably, this method usually generates a solution by checking only a relatively small number of the feasible corner points. However, *usually* does not necessarily mean *always*. For example, if we use the greedy method to select the entering basic variables, then the LP problem in Example 1 requires $2^n - 1$ pivots of the simplex method to find a solution.

EXAMPLE 1 (Klee [8])

$$\text{Maximize:} \quad z = \sum_{j=1}^{n} 10^{n-j} x_j$$

$$\text{Subject to:} \quad x_1 \leq 1$$

$$2 \sum_{j=1}^{i-1} 10^{i-j} x_j + x_i \leq 100^{i-1} \quad (2 \leq i \leq n) \tag{1}$$

$$x_i \geq 0, \quad 1 \leq i \leq n$$

For instance, if $n = 3$, then LP problem (1) is

$$\text{Maximize:} \quad z = 100x_1 + 10x_2 + x_3$$

$$\text{Subject to:} \quad \begin{array}{rcl} x_1 & \leq & 1 \\ 20x_1 + x_2 & \leq & 100 \\ 200x_1 + 20x_2 + x_3 & \leq & 10{,}000 \end{array}$$

$$x_i \geq 0, \quad 1 \leq i \leq 3$$

You can verify that the simplex method requires $2^3 - 1 = 7$ pivots to solve this problem.   ◆

From Example 1 it can be deduced that the simplex method is of exponential complexity; that is, if N is the number of variables in an LP problem, then in the "worst case," the amount of time needed to solve the problem is proportional to 2^N. For instance, at the rate of 100 iterations per second, problem (1) with $n = 50$ would take more than 300,000 years to solve! To solve a single problem with 200 variables (a relatively small problem in many contexts), the simplex method might conceivably require billions of centuries of computer time to arrive at a solution.

As a result of these considerations, much research has gone into finding an algorithm of polynomial complexity for solving LP problems. For such an algorithm, the time needed to solve a problem encoded with N variables would be proportional to N^r for some number r. This would, at least in theory and quite possibly in practice, represent a substantial improvement over the simplex method. For instance, if $N = 1000$, then $2^N \approx 10^{301}$, whereas $N^3 = 10^9$.

In 1979, the Russian mathematician L. Khatchian described a method for solving LP problems (often referred to as the *ellipsoid method*) that has polynomial complexity. However, Khatchian's procedure, though of considerable theoretical interest, has not proven to be of great practical significance. In fact, the ellipsoid method usually performs less satisfactorily than the simplex method, in part because it requires an unusual degree of numerical precision.

In 1983, a young mathematician at Bell Labs, Nerenda Karmarkar, made a startling announcement; he had found a method for solving LP problems that not only was of polynomial complexity but could also solve large-scale LP problems (problems, for instance, with more than 5000 variables) 50 times more rapidly than the currently used simplex procedures. Although at present there is some question as to the validity of this claim, many think that with additional refinements Karmarkar's procedure will prove able to solve a variety of large LP problems much more quickly than is currently possible. A thorough description of Karmarkar's method properly belongs to the area of nonlinear programming. We limit ourselves to mentioning the fundamental ideas behind his procedure.

Karmarkar's method rests on two basic ideas, which we briefly describe as follows. As we know, to find the optimal solution, the simplex method starts at some corner point of the feasible region and then moves from corner point to adjacent corner point until the optimal solution is reached. Karmarkar's method begins at an interior point of the feasible region and then moves in the direction of "steepest ascent" towards an optimal solution; that is, it moves from an initial feasible point in the direction that gives the most favorable improvement of the objective function. Moving in this direction, it stops just short of the boundary of the feasible region. This is the first basic idea underlying the Karmarkar procedure.

The second fundamental idea involves making a change of variables that "moves" the stopping point to a point near the center of the transformed feasible region. Then a new direction of steepest ascent is found, and again there is movement along this direction to a point just short of the boundary. This is done in such a way that the original problem remains essentially the same, but progress is continually made toward approximating the optimal solution. Karmarkar has shown that this progress can be amazingly good (in comparison with the simplex method) for many large-scale problems, and further refinements of his procedure may provide a practical method for solving problems that are impractical to solve with the simplex method.

A Sampling of Applications We Have Not Considered

As we have seen, the simplex method has a remarkably wide range of applications to a multitude of seemingly disparate problems. In addition to the problems we have seen, there are many other areas in which linear programming has important theoretical and practical significance. For instance, many problems involving network flows (of which the transportation and traveling salesman problems are examples) can be formulated as LP problems. Game theory is another rich area involving linear programming that yields results of theoretical and practical significance in military and business applications. It should be noted, however, that a number of special algorithms are often more efficient than the simplex method in solving LP problems that have special structures. Such special algorithms include the Critical Path Method (used for many scheduling problems), the Hungarian Method (used for the assignment problem), and a variety of methods for solving many graph theory problems.

Another collection of problems we have not considered are classified under the names Goal Programming and Multiattribute Decision Theory. Such problems seek solutions that optimize some activity that is subject to a number of essentially unrelated constraints. For instance, a person planning to purchase a car might consider four attributes: size, economy of operation, purchase price, and style. These four attributes may not be quantitatively related; even so, the purchaser wants to make a decision that minimizes his dissatisfaction with each attribute.

Finally, many practical problems involve nondeterministic optimization, that is, decisions that must be made to optimize satisfaction and minimize potential losses under constraints that cannot be exactly predicted in advance. Problems related to long-range agricultural planning, military ventures, robotics, insurance, pensions, and macroeconomics often fall into this category. Some of the LP techniques we have developed can be adapted to such probabilistic models, and other techniques we have not discussed, such as dynamic programming, are often useful.

The importance and wide-ranging applicability of linear programming can hardly be overestimated; a 1979 survey indicates that about 79% of large corporations and banks use LP methods to make decisions. Yet, as a subject of mathematical investigation, the field is in its infancy. We will surely see dramatic new techniques resulting from mathematical research on optimization problems.

CHECKLIST: CHAPTER 10

DEFINITIONS

CONCEPTS AND RESULTS

INSTRUCTIONS FOR CALIPSO

A.1 INTRODUCTION

CALIPSO is a system of computer programs designed to help you learn to formulate and solve practical linear and integer programming problems. Its acronym was derived from the words CAlculate, Linear, Integer, Programming, and SOlution. Although CALIPSO can solve large problems, problem solving is not its principal purpose. Throughout the programs, you will be expected to understand the algorithms of linear and integer programming and will be called upon to make informed choices. We have divided the programs into three groups— TUTOR, MODELER, and SOLVER—and have summarized their main features below. Occasionally in this appendix, we indicate in brackets the chapter in which a topic is introduced, such as [Chapter 3]. We advise you not to attempt to use a program until you have reached the appropriate point in the text.

TUTOR contains

1. Tutorials to learn:

   a. The Simplex Algorithm. This program enables you to enter a tableau and perform pivots on it [Chapter 3].

   b. The branch and bound technique for integer problems. This program displays a binary tree and enables you to add constraints to solve an integer problem with the branch and bound method [Chapter 8].

   c. The implicit enumeration method for binary integer programming problems [Chapter 8].

2. An editor that enables you to:

   a. Enter tableaus with as many as 39 constraints and 63 variables on a full screen display.

b. Change entries in the tableaus to study the effects of changing values of the constraint and objective function coefficients.

c. Enter additional variables or constraints in a tableau.

MODELER contains an inequality editor, with which you can

1. Enter inequalities with a full screen editor and have tableau files automatically created for use by the TUTOR or SOLVER programs [Chapter 3].

2. Generate large systems of inequalities with less typing [Chapter 4].

SOLVER is an automatic solution program that enables you to

1. Solve problems having as many as 250 constraints and 500 variables by the simplex method [Chapter 3].

2. Solve an integer problem by the branch and bound method [Chapter 8].

Getting Started

If you receive your copy of CALIPSO from your instructor, there is probably little or nothing you must do to install the program on a computer. If you receive the program on a distribution disk accompanying the text, you will need to perform a few simple steps before running the program. We describe the installation steps in Section A.8.

The programs TUTOR, MODELER, and SOLVER are ordinarily called from a program named CALIPSO. You initiate the program by starting the computer with the disk operating system (DOS) and then typing CALIPSO. A menu appears on the screen that enables you to execute TUTOR, MODELER, or SOLVER. (Actually, you can call these three programs directly by typing their names when the DOS prompt appears, but we do not recommend this method. If you do use it, be sure that the SETUP file is correct for your computer.)

The SETUP File

The SETUP file contains information about the computer you are using to run CALIPSO and is described in detail in Section A.8. If you encounter any trouble in seeing displays on your monitor or in executing some of the programs, the SETUP file probably contains wrong information about your computer. Since it is common for students working in a college environment to use a variety of computers, we have made it easy to change the SETUP file. The first panel that appears on your monitor after starting CALIPSO offers you the opportunity to change the SETUP file. You do so by typing the letter S and responding to several questions. *If anything goes wrong, check the SETUP file first.*

A.2 INTRODUCTION TO TUTOR

It is easy to learn to use CALIPSO programs, but you cannot progress far just by reading this appendix. You should perform the exercises on a computer while reading these instructions. We assume you have set up your program disk as described in Section A.8 and have created an appropriate SETUP file for your computer.

How to Get Out of Trouble

If you make a mistake or ever want to change the SETUP file again, you can do so by restarting the program. Unless you have done something very unusual, you can restart the program by following the instructions appearing on every screen that tell you what key to press to exit the screen. You may have to exit several screens to get back to the DOS prompt (the DOS prompt is usually a > symbol). If really bad things have happened, you can usually restart the program by holding down the control key (it is labeled Ctrl) and pressing the letter C. If you have somehow achieved such a disastrous situation that Ctrl-C won't stop the program, you can probably recover by rebooting the computer. To do this, hold down keys Ctrl and Alt and press Del. Although the need should never occur, the last resort for making the computer behave is to turn off its power. There is no harm in doing this. The worst that can happen is to lose the file you have been working on before it was saved to a disk. Actually, there is little likelihood that you will do anything so wrong that the programs cannot recover. Usually, they give you some sort of error message to help you.

Using the Simplex Tutorial

STEP 1
Start up the computer and get to the DOS prompt for the disk drive containing the CALIPSO programs. If you don't know how to do this, read your computer manual or get someone to help you.

STEP 2
Type CALIPSO. A display appears that contains a little advertising and asks you whether or not you want to change the SETUP file. If you need to do so, respond by typing S.

STEP 3

After you create the SETUP file (if it was necessary to do so), a screen appears offering you the option of running TUTOR, MODELER, SOLVER, QUIT, LIST FILES, or ERASE FILES. Notice that the disk drives you chose in the SETUP file are displayed. This is done to remind you to keep the disks in the correct drive. There are six sets of square brackets [ ], before the words TUTOR, MODELER, SOLVER, QUIT, LIST FILES, and ERASE FILES. The one by TUTOR contains a field in reverse video. We call this field the *cursor*. You can move the cursor between sets of square brackets with the up and down arrows on the keyboard (try it, but do not press any other key yet). To select a program or QUIT, place the cursor in the desired place and press any key. (The phrase *press any key* occurs many times in this appendix and in instructions displayed by the programs. We should probably say *almost any key* because there are a few that do not work. The Shift, Ctrl, and Alt keys are used to change the character produced by some keys and do not send a signal to the computer. The arrow keys are used to move the cursor.)

Place the cursor by TUTOR and press any key. One of the disk drives will whir a bit, and then the menu for TUTOR will appear. We are going to use the simplex tutorial, so move the cursor down to the tutorial and press any key.

STEP 4

We stored a small tableau named POTATO on the original diskette for use in this exercise. The instructions in Section A.8 asked you to copy this file to the disk drive designated to hold files. If you did so and have constructed the SETUP to look on this same drive for files, everything that follows will work. If you neglected doing any of these things, you will have to stop and fix them now. (To stop, select QUIT from successive menus until the DOS prompt appears.)

You should now be looking at a screen with lots of blank space in the middle and some instructions at the top and bottom. Read the menu that is at the bottom of the screen. You will be instructed to make various choices from this menu in exercises below. You make a choice by pressing a function key. These are the keys labeled F1, F2,..., F10. Now press F8 to read a file. When you do so, a box illuminated in reverse video or some color appears at the bottom of the screen. When it does, you should type POTATO and then press ENTER. You can type POTATO in upper or lower case, or a mixture of the two if you like. When you type a word in response to a query, usually nothing happens until you press ENTER to signal that you have finished typing the word.

A tableau appears on the screen. The format is similar to those in the text. Across the top are column labels. These labels are letters or numbers of no more than three characters. Since subscripts cannot be displayed, a label such as S1 represents an S subscripted with a 1. You will notice that some of the column and row labels use capital letters and others use small letters. You are advised not to use capital letters in tableaus because you will probably forget to capitalize when selecting pivots by specifying column and row labels. We used capitals in this example to impress you with the fact that the program cannot find a label such as S1 if you refer to it as s1. Notice that the resource column at the right is headed RHS (standing for Right-Hand Side). In this exercise, the label RHS is not really necessary; but later, when we feed this tableau to the SOLVER program, the RHS will be necessary. Although the resource values are customarily placed at the right of the tableau, you can place them in any column headed with RHS. Similarly, the bottom row is labeled OBJ for OBJective. You can place the objective function in any row labeled OBJ. Later we will show you how to create your own tableaus. Right now, we will demonstrate the features of this tutorial using the POTATO example.

STEP 5

Notice that the menu advertises that F5 prints a tableau. Never press F5 unless you have a printer attached to your computer and have turned it on. If you have a printer, press F5. You are then offered an opportunity to type a line to be printed above the tableau. This can be useful in identifying the printed output later. Press ENTER if you don't want to type anything, or type something and press ENTER.

STEP 6

We will now perform some pivots to implement the simplex algorithm. Press F6. The menu at the bottom of the screen now has changed, and you are asked to choose the column in which you want to pivot. A small box is outlined in reverse video, in which you can type up to three characters. The best pivoting choice is the first column (which is labeled xf). So, type xf and press ENTER. A new column appears. It is labeled RAT (for ratios) and is highlighted in reverse video. This column contains the ratios of the entries from the resource column to the entries from the entering basic variable column; you can select the departing basic variable by studying the entries in this column. For our example, these ratios are $\frac{8}{2} = 4$ and $\frac{15}{5} = 3$. You should choose the smaller, which is 3. Therefore, you want to pivot on the row labeled S2. Type S2 in the entry box. (Beware! This is a capital S introduced to confuse you.) Now press ENTER and observe that the tableau was pivoted.

The entry box has now moved back to the column position. According to the simplex method, you now want to select column F (capital F). So, type F. From studying the ratios of the RHS entries to the pivot column entries, you will conclude that you should pivot on row S1. Enter S1 and press ENTER. The tableau is pivoted, and you will discover that it indicates a maximum has been found. The maximum is 430 and appears in the lower right-hand corner. If you wish, you can print a copy of the final tableau by pressing F5.

If you make a mistake in entering a row or column name, you can correct it by pressing the Backspace key to erase letters you typed. The Backspace key is labeled Backspace on some keyboards or with a bold left arrow or others.

Notice the other instructions at the bottom of the screen. If you hold down the Shift key and press PrtSc, the entire screen is printed by the printer. You probably won't choose to do this often because F5 gives you a better-looking output; however, PrtSc is worth knowing about. You can usually print any screen that you are looking at by pressing it. (On some keyboards, it isn't necessary to hold Shift down.)

You are also told at the bottom of the screen that Esc is the key to press when you want to return to the main menu to do something else. Don't press it now!

STEP 7

Now we will create our own tableau and store it on disk. Press F10 to get a fresh screen. We are going to enter the tableau from Example 2 of Section 3.3. Here it is:

	x1	x2	x3	x4	x5	x6	x7	RHS
x5	1/2	$-11/2$	$-5/2$	9	1	0	0	0
x6	1/2	$-3/2$	$-1/2$	1	0	1	0	0
x7	1	0	0	0	0	0	1	1
OBJ	-10	57	9	24	0	0	0	0

There are four rows of data (the objective row is included). The value of M referred to at the top of the screen is 4. There are eight columns including the resource value column (the row labels are not included). The value of N referred to at the top of the screen is 8. You can enter these values by pressing key F1. Do so now. Type 4 for the value of M. Press the right arrow key and type 8 for the value of N. You now have a choice of pressing F2 to enter column labels or pressing ENTER to return to the menu. Press ENTER this time.

You see that a box has been drawn on the screen to contain the tableau. The box is open on the right. There are enough spaces across the screen to display only seven columns, and we have eight. The other column is on a screen to the right. To see it, hold down the Ctrl key and press the right arrow. Look at the upper right-hand corner of the screen. This screen is numbered (1, 2), meaning it is the *first* horizontal and *second* vertical screen. A maximum of three horizontal and nine vertical screens are available, providing for 39 rows with 13 on each screen and 63 columns with 7 on each screen. You move to a screen to the right with Ctrl-right arrow, to one to the left with Ctrl-left arrow, to one below with PgDn, and to one above with PgUp. You won't be able to look at all these screens right now. If you try, you will hear a beep. Try it if you want. The program only lets you view screens on which data could appear. Since you specified eight columns, you get to observe only two vertical screens and one horizontal. Press Ctrl-left arrow or whatever it takes to get to screen (1, 1) again.

Now we will enter the column labels. Press F2, which the menu advertises for entry of column labels. The little entry box is at the top and can be moved with left and right arrows. Enter the column labels x1, x2, . . . , x7. We also want to enter RHS as a label for the resource column; however, that column doesn't appear on this screen. To reach it, hold down Ctrl and press right arrow. When you have entered RHS, press F3 and enter the row labels x5, x6, x7, and OBJ. (Once again, observe that RHS and OBJ serve no purpose in this tutorial program; nevertheless, enter them so that you will develop the habit of doing so. They are essential when you process a tableau by the SOLVER program.)

Now press F4 to enter the tableau values. First try moving the entry cell. You can also use the up and down arrows now. Notice that the entry cell is wider than before; it is ten characters wide. In these fields you can enter only the digits 0–9, /, −, or a decimal point. Entering other characters will produce a beep. The simplex tutorial uses fractional arithmetic. Thus, you can enter numbers such as 27, −69, 234/123, and $-1234/4321$. You cannot enter decimals such as 12.34 in a tableau to be processed by the simplex tutorial. Other editing screens, which you will encounter later, allow you to enter decimals. The simplex tutorial does only fractional arithmetic. We may as well point out now that the tutorials do not solve all problems. The simplex tutorial can display only numbers with a maximum of four digits in the numerator or denominator. Sometimes while pivoting, you may be told that the display capacity has been exceeded and the problem cannot be solved. This is not an important limitation. There are other programs in the package to solve practical problems. These tutorials are to help you learn how the algorithms work.

By now you should be able to figure out how to enter the rest of the values in the tableau. If you make a mistake, erase it with Backspace and then type the correct value. If you see that you made a mistake in some cell after leaving it, simply return and type over the old value. Here is a useful feature: When a coefficient is zero, you can leave the cell blank. The first time you pivot, the empty cells will be filled with 0's. When you have entered the tableau, press ENTER.

Now we are going to save this file on disk. This is always a good thing to do because you might make a mistake while pivoting and want to start over. You might also want to return to this problem in a later session. Don't be cheap about saving files. The disk can hold a huge number of them. We will describe a way of erasing files from your disk in Section A.4.

To save a file, press F7. You are asked to type a name. The name can be any word of eight or fewer letters or numbers. You should never include a file extension. A file extension is a suffix beginning with a period and containing up to three letters; if you don't understand this, don't worry: You're not supposed to use them anyway. You might use the name EXAMPLE2 for this tableau. This time you don't need to worry about distinguishing upper and lower case; the program recognizes file names in either case, even with a mixture. When you have typed a name, press ENTER. Nothing much will appear to happen except for a whir from a disk drive; however, the file is stored and you can get it back anytime.

Now press F6 to do some pivots. If you are not yet adept with the simplex procedure, you may turn to Example 2 of Section 3.3 and perform the pivots done there. Keep at it until you're comfortable with the technique. When you have had enough, press Esc to get back to the menu.

STEP 8

Notice the offer of help from key F9. It works only if you copied the file HELP. LP to the disk as you were told to do in the installation procedure. If you did not do that, you are about to experience your first crash. Press F9 to see what happens. There are other help texts available, in fact, one for each function on the screen. For example, to get help with size entry, press F1 and then F9. Try it.

STEP 9

You have seen all the features of the simplex tutorial. There also are tutorials for integer programming, which we will discuss in Section A.7. Press Esc to return to the main selection menu. You can then either continue through the lessons or quit for now. Beginners sometimes worry about ending the session. Actually, it's easy. Just take out your disks and leave. Turn off the machine, if that is appropriate.

You can turn it off almost anytime, even while it is running a program. The only way you could do damage is by turning the machine off when a disk drive is running. This could result in destroying the data on a disk.

A.3 USING TUTOR TO CREATE FILES

There are two ways to enter LP problems for processing by the programs of CALIPSO: as tableaus with the simplex tutorial (discussed in Section A.2) or as systems of inequalities with the MODELER program. In this section, we discuss the first way. We introduce the MODELER program in Section A.5.

In Step 7 of Section A.2, you learned how to enter a tableau with the simplex tutorial and how to save such a file to disk using the F7 key. We saved the file created in that step under the name EXAMPLE2. Actually, the TUTOR program added the extension .TAB to this file name and stored it on disk as EXAMPLE2.TAB. The extensions are added to help you identify the files on your disks according to the program that created them. You will learn in the next section that the MODELER program creates files with one of four extensions: .TAB, .RSM, .INQ, or .GEN. TUTOR can process files with extension .TAB; SOLVER can process those with .TAB or .RSM. In Section A.2 we discussed reading such a file into the simplex tutorial of TUTOR; we will discuss using such files with the integer programming options of TUTOR in Section A.7.

Using Tutor to Create a File for Use by SOLVER

When you entered tableaus in Section A.2, you labeled every row with the variable for which the row was basic, in the sense that the row and column labeled by this variable contained a 1 and the other entries in its column were 0. If a problem contains some $\geq$ or $=$ constraints, you cannot always use slack variables for an initial basis. Usually, you would use artificial variables and the two phase or the Big-M method to proceed; the SOLVER program uses the two-phase method. Here is the main point: To signal a program that a basic variable has not been identified for a row, *leave the row label blank*. The next example shows how to enter a problem as a tableau with the simplex tutorial of TUTOR and solve it by the two-phase method with SOLVER. You should enter this example on a computer.

EXAMPLE 1

Here is how you store a tableau with TUTOR.

1. Choose TUTOR from the main CALIPSO menu.
2. Enter the tableau with the simplex tutorial (see step 7 of Section A.2).
3. Press the F7 key to store the tableau in a disk file.

4. Press Esc to leave the simplex tutorial.

5. Choose QUIT to return to the main CALIPSO menu.

Follow steps 1–5 to enter and store the following problem under the name NOBASIS. (Note that we use $<=$ for $\leq$ and $>=$ for $\geq$ because that is the way inequality signs are entered on a computer.)

$$\text{Maximize:}\quad z = -x1 - x2 - x3$$
$$\text{Subject to:}\quad -x1 + 2.5x2 + x3\ <= 1$$
$$-x1\qquad\qquad + 2x3 >= 2$$
$$x1 - x2\quad + 2x3 =\quad 4$$

Enter this problem as the following tableau.

	x1	x2	x3	x4	x5	RHS
x1	−1	2.5	1	1	0	1
	−1	0	2	0	−1	2
	1	−1	2	0	0	4
OBJ	1	1	1	0	0	0

Notice that we included slack variables in this tableau but not artificial variables. The SOLVER program will add artificial variables to those rows that we did not label with basic variables and solve the resulting problem with the two-phase method.

You should now be back at the CALIPSO menu. Choose SOLVER. Move the cursor down to the file name entry box and type NOBASIS. Then move the cursor down to the box indicating that you want to solve an LP problem and press ENTER. The solution to the problem will appear on the screen. You can ignore the column headed DUAL PRICES until you reach Chapter 6. When you have finished reading the solution, press any key. The next menu that appears offers you options for displaying or printing various reports. Be sure your printer is ready before choosing a print option. One of the available reports is a sensitivity analysis. You will learn about sensitivity in Chapter 7.   ◆

___

A.4 LISTING AND ERASING FILES

The CALIPSO menu offers the option of listing or erasing files. These functions can also be performed with DOS; in fact, the CALIPSO program calls the directory listing or erase features of DOS to perform the tasks. You should experiment with only the listing option until you are familiar with the ideas. If you make a mistake when listing files, no harm will be done; however, if you make a mistake when erasing files, you can lose information.

For use in examples, suppose we have the following files stored, some on the A disk and some on the B disk.

A disk	B disk
PEACH1.TAB	APPLE.TAB
PEACH1.INQ	APPLES.TAB
PEAR.TAB	APPLES.RSM
PEACH2.TAB	APRICOT.INQ

When we select LIST FILES (or ERASE FILES), we are asked for a file specifier. File specifiers are used by many DOS functions, and a more complete discussion than we provide can be found in a DOS manual; however, what we say here should be adequate for our purposes. A file specifier is formed with the symbols *, ?, and portions of a file name. The * and ? are called *wild-card* characters. If * occurs to the left of the period in a file name, it stands for any file name. If * occurs to the right of the period, it stands for any extension. The ? stands for any character that could occupy the spot in which it appears. In the following examples we suppose that the B drive is the default drive specified in the SETUP file and that the diskettes in the drives contain the files listed previously.

Specifier	Files listed (or erased)
.	APPLE.TAB, APPLES.TAB, APPLES.RSM, APRICOT.TAB
A:*.TAB	PEACH1.TAB, PEAR.TAB, PEACH2.TAB
APPLE?.*	APPLE.TAB, APPLES.TAB, APPLES.RSM
A:PEAR.TAB	PEAR.TAB
APPLES.*	APPLES.TAB, APPLES.RSM

A.5 MODELER: AN INEQUALITY EDITOR

The MODELER program enables you to enter problems as systems of inequalities and/or equalities. It also can generate large files of inequalities by interpreting a set of instructions.

Since MODELER uses many different file types, we will begin with a discussion of the file structure used by CALIPSO. All files used by the programs of CALIPSO have a file extension of the form .XXX added to the names you are accustomed to using in response to requests for file names by the programs. Actually, you do not need to know much about file types because the programs never ask you to know one. The only time you need to know about them is when using the LIST or ERASE features. Nevertheless, a full discussion will help give you a perspective on what is going on.

TAB

Both the TUTOR and SOLVER programs can process a file of type TAB. TAB files are stored in character format to accommodate fractional numbers and allow for display by the editing programs of TUTOR. A major disadvantage of this sort of storage is that files are larger than they need to be. This is not of much concern because TUTOR handles only fairly small files (39 rows and 63 columns); so there is usually plenty of room for storage on disks. However, you should avoid creating TAB files for large problems that will be processed only by the SOLVER program.

RSM

Files of type RSM can be created only by MODELER and used only by SOLVER. The data are stored in coded form as real numbers. Tableaus are stored as linked lists. (You will learn about linked lists in Chapter 5.) SOLVER files are usually much smaller than TAB files, primarily because they don't store zeros. In large tableaus, most of the entries usually are zeros.

INQ

The MODELER program uses files of type INQ for storing inequalities. You can enter these inequalities directly into the editor of MODELER. The GEN facility of MODELER also creates files of type INQ.

GEN

When you enter inequality generation functions with the editor of MODELER, the instructions are stored for later recall in a file of type GEN.

Since the rules for constructing files with MODELER are more restrictive for TAB files than for RSM files, we will discuss TAB files first. After that we will take up RSM files and finally GEN files.

The data for all three types of files are physically entered in the same way. Immediately after you choose MODELER from the CALIPSO menu, a screen appears requesting a file name. The file name you give must not have an extension. MODELER first searches for a file with the extension GEN. If no file with this extension exists, MODELER searches for a file with extension INQ. If such a file does not exist, MODELER assumes a new file is to be created. After you have entered your data, MODELER inspects it for instructions to generate inequalities. If there are such instructions, MODELER stores your data as a GEN file, processes the instructions to form a set of inequalities, and stores the inequalities as an INQ file. After it has stored your data, MODELER asks you whether you want to create a TAB file, an RSM file, or both types.

Producing TAB Files

You should read these instructions while working at the computer so you can try out the features we describe. Select MODELER from the CALIPSO menu and enter a file name that you have not used previously. MODELER's editing screen has instructions at the top and bottom. Look at the top first. You are offered help with the most common difficulties you may encounter. Look at some of the help messages. Start with General Help, which you get by holding down Ctrl and pressing F1. You can look at these help functions anytime. All are activated by holding down Ctrl and pressing one of the function keys. Don't be concerned if your work appears to be written over. It will all be restored.

Now look at the instructions at the bottom of the screen. You are already familiar with the arrow keys, which move the cursor. The other instructions are described below.

PGUP/PGDN

You use PgUp/PgDn to scroll through the six screens that are available for typing. You can type inequalities anywhere on these pages. You can even type several inequalities on a line. The current page number is reported at the upper right of the screen. Try it!

BACKSPACE

You use Backspace to erase a character at the immediate left of the cursor. Type something on the screen, and erase it with Backspace if you like.

ESC

Esc is an abnormal quit. It can be used if you change your mind about creating or changing a file.

INS

You can insert characters in a previously typed line by pressing Ins (on some keyboards it is labeled Insert). Ins is a toggle key: You press it once to turn the feature on and again to turn it off. Turning it on causes characters following the typing position to be moved to the right along the line. If you try to insert so many characters that some would have to move off the line, you will hear a beep. Try it. Type the following sentence: THIS IS A LINE. Now use the left arrow to move the cursor back to the L in LINE. Press the Ins key. Type SHORT. The line moves to the right as the word is inserted. Press Ins again to turn the insert mode off.

DEL

You use Del to delete characters from a line. Characters that follow on the line are moved to the left. Try using Del to erase the line you just typed. Move the cursor to the start of the line. Press Del repeatedly, or hold it down, to erase the line. You can also erase a line by typing spaces over it or by just holding down the space bar. Experiment with typing characters on the screen and using Backspace, Del, and Ins to change things.

F1, NORMAL RETURN

This is the normal way to end a session. When F1 is pressed, the inequalities you have typed are stored on disk.

F2, INSERT A NEW LINE

You can use this key to insert a blank line between two lines you have typed previously. The new line is inserted below the line on which the cursor is located.

F3, DELETE A LINE

You can delete a whole line by placing the cursor anywhere in the line and pressing F6. Try it!

F4, COPY A LINE

The line containing the cursor is copied to the line below. If there is a line of data below the line, it will not be lost, only moved down. This feature is useful in producing lists of inequalities that are similar. After making the desired number of copies of a line, you can change some letters in the copies.

F5, PRINT A FILE

Use F5 to print the data you have entered. Any information on any of the six screens will be printed.

EXAMPLE 2

Here is an example of an inequality file. As an experiment, you can type it and process the resulting TAB file with the TUTOR program. The information in curly brackets { } are comments. You can place comments anywhere to document your problem. The processing program ignores them.

$$120x1 + 100x2; \qquad \{\text{Objective function}\}$$
$$2x1 + 2x2 <= 8; \qquad \{\text{First constraint}\}$$
$$5x1 + 3x2 <= 15; \qquad \{\text{Second constraint}\}$$
$$-x1 + x2 >= 1; \qquad \{\text{Third constraint}\}$$

Press F1 to return. Choose to make a file of type TAB. EXIT the MODELER and go to TUTOR. From TUTOR, you can view the tableau you created with the simplex tutorial (select the simplex tutorial, and use F8 to retrieve the file). Notice that slack variables are added to all constraints. The slack variables for the first two constraints are used as initial basis variables (that is, the names s1 and s2 are used as row labels). Since the slack variable s3 for the third constraint was subtracted, it could not be used as an initial basis variable, and so the row label is left blank. The SOLVER program automatically adds an artificial variable for the third constraint.

◆

GENERAL RULES

1. You must always mark the end of an inequality or objective function with a semicolon. Omission of a semicolon is the most likely source of error. The program that creates the tableau will usually detect a missing semicolon. You may get a message saying that an inequality contains too many comparisons ($<, >, =$) or a message saying that a constraint is too long.

2. The programs detect the objective function by sensing the absence of $<$, $>$, or $=$. You can type the objective function anywhere.

3. You can insert comments anywhere. A comment is a phrase enclosed between curly braces, such as

$$\{\text{This is a comment}\}$$

4. The words *max* or *min* can precede the objective function. The word *max* never does anything; but *min* causes a sign change for all coefficients in the objective function, thus changing a minimization problem into a maximization problem. (Note that the processing programs only solve maximization problems.) Since *max* and *min* have this special meaning, they should never be used for variable names. For example, here are three ways to code the same objective function.

$$5x1 - 2x2 + 3x3;$$
$$\min -5x1 + 2x2 - 3x3;$$
$$\max \quad 5x1 - 2x2 + 3x3;$$

5. A variable name is any string of five or fewer characters that begins with a letter. For example, valid variable names are x1, trial, and x1234. *However*, you cannot use more than three characters in a variable name used by the TUTOR program (i.e., for a TAB file). We impose this limitation to allow the display of more columns by the editing programs of TUTOR.

6. To multiply a variable by a constant, type the constant before the variable. If you prefer, you can put an $*$ between them to indicate multiplication. Here is an example of a constraint.

$$2x1 + 13*x2 + 1/2two <= 23/67;$$

7. When making a TAB file, all variables must appear on the left of the $<=$, $>=$, or $=$ sign, and a single constant must appear on the right. As you will see later, these restrictions do not apply to RSM files.

8. Blanks (spaces) can occur anywhere in expressions. The first thing the processing program does is to compress the file by removing all blanks and comments.

9. You can enter constants in decimal or exponential notation for input to the SOLVER program, but you cannot read such files with the simplex tutorial program of TUTOR. You can use such files with the integer programming programs of TUTOR. You cannot use the ordinary E notation that is common in programming for exponential notation. So, $1.23E-2$ would not be an acceptable coding of .0123. Instead you must use a $\#$ in place of the E. Thus, $1.23\#-2$ is acceptable. We did this because the letter E is a valid variable name. The inequality

$$0.5x1 + 1.23\#-12x2 + 3/5x3 <= 831.5$$

is acceptable in a TAB file for any program except the simplex tutorial of TUTOR.

10. Parentheses cannot occur in any calculation; for instance, expressions like $2 * (x1 + x2)$ are not allowed.

Producing RSM Files for Use by SOLVER

You can use all of the features given for TAB files in creating RSM files. The following rules are in addition to those given above.

1. You can put variables on the right-hand side of inequalities and constants on the left. For example,

$$2x1 + 3x4 - 5 <= x2 - 3x3;$$

is an acceptable way to enter

$$2x1 - x2 + 3x3 + 3x4 <= 5;$$

A variable or constant can occur in more than one place. For example,

$$2apple + 5.3peach - 3 >= apple - 7/10 * peach + 2;$$

is an acceptable way to enter

$$apple + 6peach >= 5;$$

2. The objective function can occur in more than one place, even in places located far apart. For example, including the two objective functions

$$3x1 + 2x2 + x1;$$
$$5x1 + x3;$$

is the same as writing

$$9x1 + 2x2 + x3;$$

3. Variable names for RSM files can have as many as five characters.

Limitation

No single expression can exceed 160 characters (excluding blanks). In this context, an expression is any string terminated by a semicolon. If you need to include an objective function with more than 160 characters, you must split it into several objective functions, each having 160 or fewer characters. (You cannot do this for TAB files; however, there is no need to make a TAB file with such a long objective function because such a large problem cannot be practically solved with the simplex

tutorial.) If a constraint exceeds 160 characters, you must introduce more variables to reduce its size. For instance, the inequality

$$x1 + 2x2 + 3x3 + 4x4 + 5x5 + 6x6 + 7x7 + 8x8 + 9x9 +$$
$$10x10 + 11x11 + 12x12 + 13x13 + 14x14 + 15x15 + 16x16 +$$
$$17x17 + 18x18 + 19x19 + 20x20 + 21x21 + 22x22 + 23x23 +$$
$$24x24 + 25x25 + 26x26 + 27x27 + 28x28 + 29x29 <= 9;$$

can be entered as

$$x1 + 2x2 + 3x3 + 4x4 + 5x5 + 6x6 + 7x7 + 8x8 + 9x9 +$$
$$10x10 + 11x11 + 12x12 + 13x13 + 14x14 + 15x15 + 16x16 = y;$$

$$17x17 + 18x18 + 19x19 + 20x20 + 21x21 + 22x22 + 23x23 +$$
$$24x24 + 25x25 + 26x26 + 27x27 + 28x28 + 29x29 + y <= 9;$$

GENERATING SETS OF INEQUALITIES

LP problems may require typing many inequalities having a common form. For instance, there may be many variables that must be $<= 1$. Also, many LP problems have the same form but differ in the number of constraints, the number of variables, or the values of coefficients used. Transportation and assignment problems are good examples of problems with such a uniform structure.

The generation features of MODELER offer a way of producing files of inequalities with less typing and in which values can be changed easily. We will begin with a discussion of how to create a generation program for which you can easily change values of constants.

SYMBOLIC VARIABLES

Symbolic variables (or *symbols* for short) are variables that consist of the character & followed by a string of no more than nine alphabetic or numeric characters. Here are some examples of symbols:

$$\&s, \&S, \&tree, \&r12, \&BigFellow$$

You can assign a constant value to a symbol for use as a number in expressions that follow it. You can also assign to a symbol a string that involves variables such as $x1 + 2x2 - x3$. After you assign a value to a symbol, it becomes a replacement character in expressions that follow. That is, wherever the symbol occurs in an inequality or objective function, the symbol is replaced by the current value assigned to it.

Assignment of a value to a symbol can occur in three ways: by an assignment statement, by a Concat (concatenation) statement, or by an Assign statement. An assignment is made to a symbol with the operation $:=$. You can assign values to the symbols &t and &tree by

$$\&t := 27.9;$$

$$\&tree := x1 + 2x2 - x3;$$

The semicolons at the ends of assignments are not part of the value assigned (they delimit the expression). If you have made the assignments above, and the program encounters the expression

$$\&tree = \&t;$$

it generates the equality

$$x1 + 2x2 - x3 = 27.9;$$

and places it in the file of type INQ being created.

An inequality file can contain as many as 150 symbols. The total length of strings assigned to all symbols cannot exceed 3000 characters. That is more capacity than you should ever need.

There is an important distinction between the use of := and = in a generation file. The statement

$$\&s := 9;$$

assigns the value 9 to &s, whereas the statement

$$\&s = 9;$$

inserts an equation into the output file with &s replaced by its current value.

The Assign statement is useful when you need to assign values to many subscripted symbolic variables. For instance, the expression

$$assign(\&ci, i = 1 \text{ to } 5, 1, 2, 1/2, 3.9, -4);$$

is equivalent to the assignment statements

$$\&c1 := 1; \&c2 := 2; \&c3 := 1/2; \&c4 := 3.9; \&c5 := -4;$$

and the expression

$$assign(\&aij, i = 1 \text{ to } 2, j = 1 \text{ to } 3, 1, 2, 3, 1/2, 2.5, -3.4);$$

is equivalent to the assignment statements

$$\&a11 := 1; \quad \&a12 := 2; \quad \&a13 := 3;$$
$$\&a21 := 1/2; \&a22 := 2.5; \&a23 := -3.4;$$

The Assign statement does not accept a symbolic variable with more than two subscripts.

You can change the value of a symbol at any time by another assignment. For instance, if you had assigned to &t the value given above at one point in the file, you could change its value to 54 by inserting &t := 54 at a later point. There is another option that is useful in the concatenation operation discussed below. Writing &t := null erases the current value of &t and, in fact, removes it from the list of symbol names.

You must exercise care in using a symbol as the coefficient of a variable. Suppose that &t now has value 54 and that you are interested in generating the term 54x1 using &t. If you were to code

&tx1

the program would consider &tx1 to be a symbol name; you should code &t*x1. The program discovers a symbol name by searching for one of the special symbols *, +, −, >, <, =, or ; to delimit the symbol name. A blank does not work because the program removes all blanks from expressions.

Here is an example of a generation file that uses symbolic variables. It isn't a very good example of the power of generation statements because it is longer to type than the file of inequalities it generates; however, it may help you see how symbols can be used. To enter the example, execute the MODELER program from the CALIPSO menu.

{A GEN file for the POTATO problem}

&c1:= 120; &c2:= 100;	{Cost coefficients}
&b1:= 8; &b2:= 15;	{Resource values}
&a11 := 2; &a12:= 2;	{Constraint coefficients}
&a21 := 5; &a22:= 3;	
&c1*xf + &c2*F;	{Objective function}
&a11*xf + &a12*F <= &b1;	{Constraints}
&a21*xf + &a22*F <= &b2;	

You probably don't want to save this file under the name POTATO because doing so will cause the INQ that is created to overlay the old POTATO.INQ. Maybe you could call it GENPOT. Anyway, save it and then return by pressing F1. You will be told that an INQ file was created and offered an opportunity to view the INQ file to make sure it is correct. It will look like this:

$$120*xf + 100*F; 2*xf + 2*F <= 8; 5*xf + 3*F <= 15;$$

This is a little hard to read, especially when you are looking at a long file; however, the computer programs will be pleased with it. The storage format is compressed because files created by MODELER can be very large. Usually, you won't need to look at them at all, except possibly to satisfy yourself that they turned out as planned. Changes to the file would more naturally be made by altering the GEN file and generating another INQ file than by attempting to change this compressed file directly.

CONCATENATION

Concatenation is a powerful operation that you can use to generate complicated forms; however, it is a bit difficult to master. You will probably prefer to use the Add operation described below when it can do the job. You can assign a value to a symbol by an operation such as

concat(&s,string)

where *string* is any expression. The string can itself involve symbols. Thus, the sequence

$$\&s := x1;$$
$$concat(\&s, + x2);$$
$$concat(\&s, + 2x3);$$
$$\&s <= 12;$$

results in the inequality

$$x1 + x2 + 2x3 <= 12;$$

The utility of Concat will become apparent when we introduce For loops. Notice that the program remembers the contents of a symbol until you change that symbol by an assignment. If you were to code

$$concat(\&s, - 3x4); \&s = 2;$$

later in the program, the result would be

$$x1 + x2 + 2x3 - 3x4 = 2;$$

If the first occurrence of a symbol is in a Concat statement, the symbol is initially assigned a null value. Thus, if &Peach has never occurred before, then

$$concat(\&Peach, - 2x13); concat(\&Peach, + 5x42); \&Peach <= 1;$$

will generate

$$-2x13 + 5x42 <= 1;$$

Also note that &Peach and &peach are different symbols; that is, the program distinguishes between upper and lower case letters

FOR LOOPS

We use For loops to generate lists of inequalities. We describe the use of For loops in Examples 3–6.

EXAMPLE 3

$$for\ i = 1\ to\ 5\ do\ xi <= 1;$$

generates

$$x1 <= 1; x2 <= 1; x3 <= 1; x4 <= 1; x5 <= 1;$$

The letter i becomes a replacement symbol in the expression that follows the word do. Each occurrence of i is replaced successively by 1, 2, 3, 4, or 5, and a copy of whatever expression follows "do" is made for each value. ◆

EXAMPLE 4

Symbols can be subscripted and can occur in For loops. The expressions

$$\&a1:=1; \&a2:=3; \&a3:=5;$$
$$\&b1:=27.2; \&b2:=18.3; \&b3:=3/2;$$
$$\text{for } j = 1 \text{ to } 3 \text{ do } \&aj*xj <= \&bj;$$

result in

$$1*x1 <= 27.2; 3*x2 <= 18.3; 5*x3 <= 3/2; \qquad \blacklozenge$$

You can use any letter as a replacement letter in a For loop. Example 3 used i, and Example 4 used j.

EXAMPLE 5

You can nest For loops.

$$\text{for } i = 1 \text{ to } 2 \text{ do for } j = 2 \text{ to } 4 \text{ do for } k = 12 \text{ to } 14 \text{ do } xijk <= 1;$$

generates

$$x1212 <= 1; x1213 <= 1; x1214 <= 1;$$
$$x1312 <= 1; x1313 <= 1; x1314 <= 1;$$
$$x1412 <= 1; x1413 <= 1; x1414 <= 1;$$
$$x2212 <= 1; x2213 <= 1; x2214 <= 1;$$
$$x2312 <= 1; x2313 <= 1; x2314 <= 1;$$
$$x2412 <= 1; x2413 <= 1; x2414 <= 1; \qquad \blacklozenge$$

The sequence of events in Example 5 is (1) i is set to 1, j is set to 2, and k varies from 12 to 14; (2) with i held at 1, j is advanced to 3, and k varies from 12 to 14; (3) with i still at 1, j is advanced to 4, and k varies from 12 to 14; (4) i is advanced to 2, and the process repeats for all the k and j values. Computers are better at this than humans.

EXAMPLE 6

Here is an example of the way to create a generic program that can be easily modified to change the values of constants and the number of variables. You may recognize it as a transportation problem.

assign(&cij, i = 1 to 2, j = 1 to 3, 12, 9, 16, 3, 5, 7);
&a1:= 30; &a2:= 50;
&b1:= 20; &b2:= 25; &b3:= 35;

&m:= 2; &n:= 3;

for i = 1 to &m do for j = 1 to &n do concat(&r, + &cij*xij);

min &r;

for i = 1 to &m do for j = 1 to &n do concat(&si, + xij);
for i = 1 to &m do &si <= &ai;

for j = 1 to &n do for i = 1 to &m do concat(&uj, + xij);
for j = 1 to &n do &uj = &bj;

The generated inequalities are

$$\text{min} + 12*x11 + 9*x12 + 16*x13 + 3*x21 + 5*x22 + 27*x32;$$

$$+ x11 + x12 + x13 <= 30;$$
$$+ x21 + x22 + x23 <= 50;$$
$$+ x11 + x21 = 20;$$
$$+ x12 + x22 = 25;$$
$$+ x13 + x23 = 35;$$

In this example, the GEN file is longer than the inequality file; however, you can see that increasing the values of &n and &m would alter that. ◆

CAUTION!

The replacement feature of For loops replaces all occurrences of its parameter within its range. Consider what will happen if you enter

$$\text{for c} = 1 \text{ to 2 concat(&s, + xc);}$$

The substitution for c will also be made in Concat, causing an error.

ADD

The most common expression you would want to generate is a sum of constants multiplied by variables for use in an objective function or a constraint. You can do this more easily with Add than with Concat. This simple example conveys the idea.

$$\text{add(i} = 1 \text{ to 5,xi)} <= 27;$$

generates

$$x1 + x2 + x3 + x4 + x5 <= 27;$$

(If you forget how to code the Add function, or any other function, you can get a quick review from the on-screen Help functions of MODELER.)

The i used in the Add function above is a substitution character that behaves exactly like those used in For loops. The substitution occurs only in the expression within the parentheses and following the comma. *You cannot expect substitution to occur on the right side of the inequality.* Here is another example that uses symbolic variables. The expressions

$$\&a1:= 5; \&a2:= 1/2; \&a3:= -0.69; \&b:= 9;$$

$$add(i = 1 \text{ to } 3, \&ai*yi) >= \&b;$$

generates

$$5*y1 + 1/2*y2 + -0.69*y3 >= 9;$$

(The $+ -$ is interpreted as $-$ by the processing program.) Note that an Add function cannot have more than one substitution variable; thus

$$add(i = 1 \text{ to } 2, j = 1 \text{ to } 3, \&aij*yij) >= 2;$$

is not meaningful.

You can use the Add function within For loops. In the example below, we assume the symbols were assigned the indicated values earlier in the problem.

$$for i = 1 \text{ to } 2 \text{ do } add(j = 1 \text{ to } 3, \&aij*xij) <= \&bi;$$

generates

$$5*x11 + 7*x12 + 9*x13 <= 5;$$
$$-4*x21 + 0*x22 + 2/3*x23 <= 12;$$

The Add function can appear only on the left side of a comparison. Expressions such as

$$add(i = 1 \text{ to } 3,xi) <= add(j = 1 \text{ to } 5,yj)$$

do not work.

A.6   SOLVER: THE AUTOMATIC SOLUTION PROGRAM

SOLVER uses the revised simplex method introduced in Chapter 5. Since you will not notice any difference between the performance of SOLVER and that of the ordinary simplex method used in the text before Chapter 5, you can use it to solve problems before you reach Chapter 5.

Before starting SOLVER, you must have created a file of type RSM or TAB to process. If you don't have one on your disk, use MODELER to create one now so you can use it to try SOLVER.

Choose SOLVER from the CALIPSO menu, and examine the menu that appears. The menus are self-explanatory with the possible exception of the option of displaying the tableau after each pivot, which is available from the first menu. This option can produce a lot of output. You should use it only if you need to study the pivoting process. After the problem is solved, you are given a chance to display or print the initial and final tableaus. This is all you would ordinarily need.

SOLVER uses the revised simplex method with product form of the inverse basis matrix (see Chapter 5). The algorithm is implemented in an efficient manner. The columns of the tableau and of the vectors holding the inverse basis matrix are stored in linked lists. We describe this form of storage in the text. Linked lists conserve storage and avoid the storage of and multiplication by zeros.

A.7 INTEGER PROBLEMS

TUTOR contains a branch and bound tutorial that allows you to control execution of the branch and bound method for problems that do not require too many branches. You must make a file of type TAB with the simplex tutorial or the MODELER program before executing it. Its operation is self-explanatory. (Note that TAB files used by the integer programs of TUTOR can contain decimal or exponential constants.)

TUTOR also contains an implicit enumeration program to execute the algorithm given in Sections 8.3 and 8.4. You must make a file of a special form with the simplex tutorial or choose the option of the MODELER program to make a special sort of TAB file for this program. The execution of this program will be obvious after you read Chapter 8.

SOLVER also enables you to solve integer problems with the branch and bound method. You must make a file of type RSM or TAB before execution. After choosing to solve an IP problem from the SOLVER menu, you are offered a chance to specify which variables are to have integer values. One of the ways to specify integer variables for SOLVER is by the first letter of the variable. If you use letters from the word *integer* as the first character of a variable, it will be easier to recognize them as integer variables.

A.8 CONFIGURING YOUR SYSTEM

The programs should run on any computer using an MS-DOS operating system. They will certainly run on any IBM or IBM-compatible machine with DOS version 2.0 or higher. The programs are distributed on a single diskette containing the following files:

CALIPSO.EXE
TUTOR.EXE
MODELER.EXE
SOLVER.EXE
HELP.LP
SETUP
POTATO.TAB
POTATO.INQ
CONFIG.SYS
AUTOEXEC.BAT
README
LOGO.HBJ

The file README may not appear on your diskette. If it does, it contains additional instructions on using CALIPSO. Although these files use most of the space on the diskette, there is enough space left to hold about 20 problem data files. Eventually, however, unless you erase old files, you will run out of space. If you want to store more than a few problem files, it is easy to store them on a separate diskette. Several ways of configuring diskettes are described below.

You will need to know a little about the DOS used by your computer. Since there are many versions, we will not attempt to instruct you on DOS. The DOS manual that came with your operating system will tell you far more than you need to know to execute these programs.

Backup

The CALIPSO diskette is not copy protected. If you received your copy in a distribution to a class, you probably don't need to worry about making a backup because you can easily get another copy. If you are working alone, you should keep a backup in case some programs are destroyed inadvertently. Probably the best way to make a backup is to create a program diskette as described in the next paragraph and store the distribution diskette away.

Making a Program Diskette

If your computer has a hard disk, you can copy CALIPSO to it as described below in the paragraph headed Configuration 3. If you do not have a hard disk, you will probably want to format your Program Diskette as a system disk so you can use it to start your computer. In the instructions we give for making a system disk, we assume that the diskette you are copying from is in drive A and the diskette you are copying to is in drive B. If you have only one drive, you will need to change the drive letters in the instructions to A and switch diskettes when told to do so by messages on the screen.

1. Put your DOS diskette in drive A and start the computer.

2. Put a blank diskette in drive B.

3. Enter the DOS commands, one at a time.

> FORMAT B: /S
> COPY COMMAND.COM B:
> COPY ANSI.SYS B:

4. Put the CALIPSO distribution diskette in drive A.

5. Enter the DOS command

> COPY *.* B:

6. Move the Program Diskette from drive B to drive A.

7. Restart the computer by holding down Ctrl and Alt and pressing Del.

To establish the right working environment, the program disk should contain a file named CONFIG.SYS containing the following commands:

> FILES = 20
> BUFFERS = 12
> DEVICE = ANSI.SYS

Such a CONFIG.SYS file is contained on the distribution diskette. You may want to have other statements in your CONFIG.SYS file to run other programs on your computer; however, it is important that the three commands just given be included in the file.

The CALIPSO program is started by typing CALIPSO. If you want the program to start automatically when you boot the computer from the program diskette, create a file with the name AUTOEXEC.BAT containing the single statement:

> CALIPSO

An AUTOEXEC.BAT file containing this statement is on the distribution diskette.

Setup File

We use a file named SETUP to describe the configuration of your computer to various programs. If you have a color monitor, it contains your choices of colors. It always contains the location of various files that programs will be looking for. You can change the SETUP file by typing the letter s when the first screen appears after starting CALIPSO or by selecting the option to change the SETUP file from the main menu of CALIPSO. You will be asked to respond to several questions; appropriate responses are described for each configuration below. Note that if you have an IBM System II computer with a monochrome screen, you should respond to the question concerning the type of monitor with a c (as though you had a color monitor) and then select white as the color for characters and black as the background color.

If you are using a very old DOS machine (made before the early 1970s) that was not manufactured by IBM, the screen memory may be stored in a memory segment other than B000h or B8000h. You might be able to make the programs work on such a computer by changing the last number in the SETUP file to the *decimal* equivalent of your computer's screen memory segment. The SETUP files created by CALIPSO have the last five digits 45056 (for monochrome) or 47104 (for color).

One of the questions you are asked when making a SETUP file concerns the amount of memory your computer has. You must have at least 256K. If you don't know how much memory your computer has, choose 256K. All of the programs will run well in 256K; however, the limitation on the number of variables the SOLVER program can handle will be 250 instead of 500 if you select 256K. Since problems with more than 250 variables are fairly unusual, the limitation will probably not trouble you. The most memory that the programs can use is 384K.

If a screen that is supposed to be visible does not appear, it probably means that your SETUP file contains the wrong choice of monitor. Start the CALIPSO program again, type s (to make a new SETUP file) when the first screen appears, and enter the correct choice for your monitor.

If you are using one of IBM's new PS2 computers, select a color monitor whether or not you actually have a color monitor.

Configuration 1: Using One Fixed Drive

The program diskette should remain in the drive during execution. The SETUP file should direct the programs to seek all files on drive A. If it is necessary to store programs on a diskette other than the program diskette, you will need to remove the program diskette and replace it with the other diskette when you are reading or writing problem data. You must remember to insert the program diskette before attempting to execute any of the programs in CALIPSO.

Configuration 2: Using Two Fixed Drives

The program diskette should be kept in drive A, and drive A should be the default drive. If problem files are also stored on the program diskette, operations are conducted exactly as for Configuration 1. If problem files are stored on a separate diskette, this diskette should be kept in drive B. The SETUP file should direct the programs to seek program files on drive A and problem files on drive B.

Configuration 3: Hard Disk

If you have a hard disk, you can copy all of the programs to it. You will probably want to store them in a special directory created for the purpose. Specify its path for all program and data files when you make your SETUP file. You will still need to make a CONFIG.SYS file similar to the one described above and place it in the root directory of your hard disk.

Configuration 4: RAM Disk

If you have DOS version 3.1 or higher or other software capable of making a RAM disk, you certainly want to make use of it as described here. Using it will load programs faster and save wear on disk drives.

Creation of a RAM disk varies with the software you have. We will describe use of the VDISK command of DOS 3.1 and higher. You need to create a RAM disk of 300K. Since 256K is required to execute our programs, you will need 640K of memory to use a RAM disk to full potential. If you have less than 640K, you might still consider putting the more commonly used programs on a smaller RAM. Note that if you have only a 640K machine, do not try to make a RAM disk larger than 300K. If you do so, there will not be enough memory left for the SOLVER program. (Even though 256K + 300K is less than 640K, some room must be reserved for the DOS command processor.)

Commands to create RAMs are always placed in the CONFIG.SYS file. This file is executed every time the computer is started. For DOS 3.1, put the following command in it.

$$device = vdisk.sys\ size = 300\ size = 512\ entries = 16$$

Next, you will probably want to create a batch file to copy the programs into the RAM. It should contain the following statements. The letter X has been substituted for the drive letter on which the programs will be found. If you have a hard disk, you will probably replace X by C. If the programs are in a subdirectory of your hard disk, you will need to replace X by a path. If you plan to keep the programs on a diskette, you will replace X by A. You must replace the letter Y by the drive letter for your RAM disk.

```
COPY X:CALIPSO.EXE Y:
COPY X:TUTOR.EXE Y:
COPY X:MODELER.EXE Y:
COPY X:SOLVER.EXE Y:
COPY X:HELP.LP Y:
COPY X:SETUP Y:
Y:
CALIPSO
```

RAM.BAT would be a good name for this batch file. When starting a session, you would type RAM to copy the files to the RAM disk. Throughout the session, the RAM disk Y should be the default disk.

When you make your SETUP file, specify the RAM disk letter Y for the location of the program files. You will probably want to keep data files on a hard or fixed disk, because otherwise they will be lost when you shut down.

(See page xii for the order form for CALIPSO.)

Appendix B

ANSWERS FOR ODD-NUMBERED EXERCISES

EXERCISES 1.1

1. Let x_1 be the number of tons of shoestring potatoes and x_2 the number of tons of nuggets processed each day.

Maximize: Subject to:

$$z = 60x_1 + 45x_2 \qquad x_1 + x_2 \leq 8 \qquad \text{(Peeling)}$$

$$\frac{5}{2}x_1 + \frac{3}{2}x_2 \leq 16 \qquad \text{(Slicing)}$$

$$x_1, x_2 \geq 0$$

3. Let x_1, x_2, x_3 be the amounts invested in low-risk, medium-risk, and high-risk stocks, respectively.

Maximize: Subject to:

$$z = 0.07x_1 + 0.09x_2 + 0.11x_3 \qquad x_1 - x_2 \qquad \leq 2000$$

$$x_3 \leq 8000$$

$$x_2 + x_3 \leq 14{,}000$$

$$x_1 + x_2 + x_3 = 18{,}000$$

$$x_i \geq 0, \qquad 1 \leq i \leq 3$$

5. Let L be acres in lawn, and let $T, B, P, C_1,$ and C_2 be acres of tomatoes, beans, peas, corn, and carrots, respectively.

Maximize: $z = 750T - 150T$ (Tomatoes)

$+ 150B - 40B$ (Beans)

$+ 160P - 30P$ (Peas)

$+ 50C_1 - 25C_1$ (Corn)

$+ 180C_2 - 55C_2$ (Carrots)

$- 15L$ (Lawn)

$$\text{Subject to:} \quad 90T + 18B + 27P + 10C_1 + 30C_2 + 6L \leq 300 \quad \text{(Labor)}$$
$$150T + 40B + 30P + 25C_1 + 55C_2 + 15L \leq 400 \quad \text{(Capital)}$$
$$T + B + P + C_1 + C_2 + L = 4 \quad \text{(Land)}$$
$$T, B, P, C_1, C_2, L \geq 0$$

7. Let a, b, c, d, e be the respective amounts of products A, B, C, D, E that are produced. Measure time in minutes.

(a) Maximize: $z =$
$$12a + 11b + 12c + \frac{21}{2}d + 6e \quad \text{(Selling price)}$$
$$-2a - 2b - 2c - d - e \quad \text{(Materials cost)}$$
$$-\frac{135}{60}a - \frac{72}{60}b - \frac{72}{60}c - \frac{108}{60}d - \frac{81}{60}e \quad (M_1 \text{ cost})$$
$$-\frac{72}{60}a - \frac{90}{60}b - \frac{108}{60}c - \frac{36}{60}d - \frac{54}{60}e \quad (M_2 \text{ cost})$$
$$-\frac{72}{60}a - \frac{108}{60}b - \frac{120}{60}c - \frac{144}{60}d \quad (M_3 \text{ cost})$$

(b) $c \geq 20$ (Minimum order)

$d \geq 30$

$15a + 8b + 8c + 12d + 9e \leq 4800 \quad (M_1)$

$8a + 10b + 12c + 4d + 6e \leq 4800 \quad (M_2)$

$6a + 9b + 10c + 12d \leq 4800 \quad (M_3)$

$a, b, c, d, e \geq 0$

9. Let a, b, c, d, e be the respective amounts of products A, B, C, D, E. Measure time in minutes. Let c_1 be the number of units of C in excess of 20, and let d_1 be the number of units of D in excess of 30.

Maximize:

$$z = 12a + 11b + 12c + \frac{21}{2}d + 6e + 9c_1 + 8d_1$$
$$-2a - 2b - 2c - d - e - 2c_1 - d_1$$
$$-\frac{135}{60}a - \frac{72}{60}b - \frac{72}{60}c - \frac{108}{60}d - \frac{81}{60}e - \frac{72}{60}c_1 - \frac{108}{60}d_1 \quad \text{(Extra sales)}$$
$$-\frac{72}{60}a - \frac{90}{60}b - \frac{108}{60}c - \frac{36}{60}d - \frac{54}{60}e - \frac{108}{60}c_1 - \frac{36}{60}d_1 \quad \text{(Extra sales)}$$
$$-\frac{72}{60}a - \frac{108}{60}b - \frac{120}{60}c - \frac{144}{60}d - \frac{120}{60}c_1 - \frac{144}{60}d_1 \quad \text{(Extra sales)}$$

Subject to:

$c = 20$ (or c could be replaced throughout by 20)

$d = 30$ (or d could be replaced throughout by 30)

$15a + 8b + 8c + 12d + 9e + 8c_1 + 12d_1 \leq 4800$

$8a + 10b + 12c + 4d + 6e + 12c_1 + 4d_1 \leq 4800$

$6a + 9b + 10c + 12d + 10c_1 + 12d_1 \leq 4800$

$a, b, c, d, e, c_1, d_1 \geq 0$

11. Since the cost coefficients of x_1, x_2, x_3 are in descending order, we should try to make x_1, x_2, x_3 large in that order. Since $x_1 + x_2 \leq x_3$, it follows that $x_1 + x_2 \leq 50{,}000$ (if $x_1 + x_2 > 50{,}000$, then $x_1 + x_2 + x_3 > 100{,}000$). We obtain the largest contribution to the objective function by setting $x_1 = (50{,}000)/3$ and $x_2 = (\frac{2}{3})(50{,}000)$. It follows that $x_3 = 50{,}000$.

13. Let m_1, m_2, m_3 be the number of pounds of each mineral in a 100-lb batch.

Minimize: $\quad z = \dfrac{35}{10} m_1 + \dfrac{25}{10} m_2 + 3m_3$

Subject to: $\quad \dfrac{3}{100} m_1 + \dfrac{7}{100} m_2 + \dfrac{9}{100} m_3 \geq 4 \quad$ (A)

$$\dfrac{5}{100} m_1 + \dfrac{8}{100} m_2 + \dfrac{1}{100} m_3 \geq 3 \quad \text{(B)}$$

$$\dfrac{35}{100} m_1 + \dfrac{32}{100} m_2 + \dfrac{27}{100} m_3 \geq 30 \quad \text{(C)}$$

$$\dfrac{24}{100} m_1 + \dfrac{12}{100} m_2 + \dfrac{15}{100} m_3 \geq 16 \quad \text{(D)}$$

$$-\dfrac{1}{100} m_1 + \dfrac{99}{100} m_2 - \dfrac{1}{100} m_3 \leq 0$$

$$m_1 + m_2 + m_3 = 100$$

$$m_i \geq 0, \quad 1 \leq i \leq 3$$

Note that the fifth inequality is equivalent to the constraint

$$m_2 \leq 0.01(m_1 + m_2 + m_3)$$

EXERCISES 1.2

1. Minimize:

$$\begin{aligned} z = \quad & 10x_{11} + 6x_{12} + 12x_{13} + 15x_{14} \\ + \; & 8x_{21} + 4x_{22} + 8x_{23} + 10x_{24} \\ + \; & 11x_{31} + 8x_{32} + 5x_{33} + 6x_{34} \end{aligned}$$

Subject to:

$$x_{11} + x_{12} + x_{13} + x_{14} \leq 50{,}000$$
$$x_{21} + x_{22} + x_{23} + x_{24} \leq 50{,}000$$
$$x_{31} + x_{32} + x_{33} + x_{34} \leq 30{,}000$$
$$x_{11} + x_{21} + x_{31} = 35{,}000$$
$$x_{12} + x_{22} + x_{32} = 25{,}000$$
$$x_{13} + x_{23} + x_{33} = 40{,}000$$
$$x_{14} + x_{24} + x_{34} = 30{,}000$$
$$x_{ij} \geq 0 \text{ and integer}, \; 1 \leq i \leq 3, \; 1 \leq j \leq 4$$

3. (We let $\infty = 99$.)

Minimize:

$$z = \begin{aligned} &4x_{11} + 99x_{12} + 6x_{13} + 12x_{14} + 21x_{15} \\ &+ 5x_{21} + 5x_{22} + 12x_{23} + 6x_{24} + 99x_{25} \\ &+ 10x_{31} + 7x_{32} + 15x_{33} + 12x_{34} + 10x_{35} \\ &+ 99x_{41} + 8x_{42} + 10x_{43} + 10x_{44} + 12x_{45} \end{aligned}$$

Subject to:

$$x_{11} + x_{12} + x_{13} + x_{14} + x_{15} \leq 250$$
$$x_{21} + x_{22} + x_{23} + x_{24} + x_{25} \leq 100$$
$$x_{31} + x_{32} + x_{33} + x_{34} + x_{35} \leq 300$$
$$x_{41} + x_{42} + x_{43} + x_{44} + x_{45} \leq 150$$
$$x_{11} + x_{21} + x_{31} + x_{41} = 100$$
$$x_{12} + x_{22} + x_{32} + x_{42} = 200$$
$$x_{13} + x_{23} + x_{33} + x_{43} = 250$$
$$x_{14} + x_{24} + x_{34} + x_{44} = 150$$
$$x_{15} + x_{25} + x_{35} + x_{45} = 100$$
$$x_{ij} \geq 0 \text{ and integer, } 1 \leq i \leq 4, 1 \leq j \leq 5$$

5. The cost table is

	Jan	Apr	Jul	Oct	Supply
Jan	25	35	45	55	300
Apr	∞	40	50	60	900
Jul	∞	∞	30	40	220
Oct	∞	∞	∞	35	60
Demand:	150	200	600	400	

(We let $\infty = 999$.)

Minimize:

$$z = \begin{aligned} &25x_{11} + 35x_{12} + 45x_{13} + 55x_{14} \\ &+ 999x_{21} + 40x_{22} + 50x_{23} + 60x_{24} \\ &+ 999x_{31} + 999x_{32} + 30x_{33} + 40x_{34} \\ &+ 999x_{41} + 999x_{42} + 999x_{43} + 35x_{44} \end{aligned}$$

Subject to:

$$x_{11} + x_{12} + x_{13} + x_{14} \leq 300$$
$$x_{21} + x_{22} + x_{23} + x_{24} \leq 900$$
$$x_{31} + x_{32} + x_{33} + x_{34} \leq 220$$
$$x_{41} + x_{42} + x_{43} + x_{44} \leq 60$$
$$x_{11} + x_{21} + x_{31} + x_{41} = 150$$
$$x_{12} + x_{22} + x_{32} + x_{42} = 200$$
$$x_{13} + x_{23} + x_{33} + x_{43} = 600$$
$$x_{14} + x_{24} + x_{34} + x_{44} = 400$$
$$x_{ij} \geq 0 \text{ and integer, } \quad 1 \leq i, j \leq 4$$

7. From the statement of the problem, it is natural to assume that the supply of new napkins is large enough to meet any demand. In our solution we set the supply of new napkins to be 500.

The values of i are 1 for new, 2 for Monday's dirty napkins, 3 for Tuesday's, 4 for Wednesday's, and 5 for Thursday's. The values of j are 1 for Monday, 2 for Tuesday, 3 for Wednesday, 4 for Thursday, and 5 for Friday. The cost table is

	M	T	W	Th	F	Supply
New	35	35	35	35	35	500
M	∞	8	5	3	3	95
T	∞	∞	8	5	3	75
W	∞	∞	∞	8	5	108
Th	∞	∞	∞	∞	8	90
Demand:	95	75	108	90	120	

We use 99 as a high value to act as infinite cost.

Minimize:

$$
\begin{aligned}
z = \quad & 35x_{11} + 35x_{12} + 35x_{13} + 35x_{14} + 35x_{15} \\
+ & 99x_{21} + 8x_{22} + 5x_{23} + 3x_{24} + 3x_{25} \\
+ & 99x_{31} + 99x_{32} + 8x_{33} + 5x_{34} + 3x_{35} \\
+ & 99x_{41} + 99x_{42} + 99x_{43} + 8x_{44} + 5x_{45} \\
+ & 99x_{51} + 99x_{52} + 99x_{53} + 99x_{54} + 8x_{55}
\end{aligned}
$$

Subject to:

$$
\left.
\begin{aligned}
x_{11} + x_{12} + x_{13} + x_{14} + x_{15} &\le 500 \\
x_{21} + x_{22} + x_{23} + x_{24} + x_{25} &\le 95 \\
x_{31} + x_{32} + x_{33} + x_{34} + x_{35} &\le 75 \\
x_{41} + x_{42} + x_{43} + x_{44} + x_{45} &\le 108 \\
x_{51} + x_{52} + x_{53} + x_{54} + x_{55} &\le 90
\end{aligned}
\right\} \text{ (Supply equations)}
$$

$$
\left.
\begin{aligned}
x_{11} + x_{21} + x_{31} + x_{41} + x_{51} &= 95 \\
x_{12} + x_{22} + x_{32} + x_{42} + x_{52} &= 75 \\
x_{13} + x_{23} + x_{33} + x_{43} + x_{53} &= 108 \\
x_{14} + x_{24} + x_{34} + x_{44} + x_{54} &= 90 \\
x_{15} + x_{25} + x_{35} + x_{45} + x_{55} &= 120
\end{aligned}
\right\} \text{ (Demand equations)}
$$

$$x_{ij} \ge 0 \text{ and integer}, \quad 1 \le i, j \le 5$$

9. Minimize:

$$
\begin{aligned}
z = \quad & 8x_{11} + 10x_{12} + 6x_{13} + 9x_{14} \\
+ & 10x_{21} + 8x_{22} + 8x_{23} + 11x_{24} \\
+ & 12x_{31} + 9x_{32} + 10x_{33} + 9x_{34} \\
+ & 10x_{41} + 10x_{42} + 9x_{43} + 14x_{44}
\end{aligned}
$$

Subject to:

$$\Sigma_{j=1}^{4} x_{ij} = 1, \; 1 \le i \le 4$$

$$\Sigma_{i=1}^{4} x_{ij} = 1, \; 1 \le j \le 4$$

$$x_{ij} = 0 \quad \text{or} \quad 1, \qquad 1 \le i, j \le 4$$

11. Minimize:

$$
\begin{aligned}
z = \quad & 8x_{11} + 10x_{12} + 6x_{13} + 9x_{14} \\
+ & 10x_{21} + 8x_{22} + 8x_{23} + 11x_{24} \\
+ & 0x_{31} + 0x_{32} + 0x_{33} + 0x_{34} \\
+ & 10x_{41} + 10x_{42} + 9x_{43} + 14x_{44}
\end{aligned}
$$

Subject to:

$$\Sigma_{j=1}^{4} x_{ij} = 1; \qquad 1 \le i \le 4$$

$$\Sigma_{i=1}^{4} x_{ij} = 1; \qquad 1 \le j \le 4$$

$$x_{ij} = 0 \quad \text{or} \quad 1; \qquad 1 \le i, j \le 4$$

13. Minimize:

$$z = \quad 6x_{11} + 5x_{12} + 4x_{13} + 3x_{14} + 2x_{15} + 1x_{16}$$
$$+ 1x_{21} + 2x_{22} + 3x_{23} + 4x_{24} + 5x_{25} + 6x_{26}$$
$$+ 5x_{31} + 1x_{32} + 2x_{33} + 6x_{34} + 4x_{35} + 3x_{36}$$
$$+ 5x_{41} + 1x_{42} + 2x_{43} + 6x_{44} + 4x_{45} + 3x_{46}$$
$$+ 6x_{51} + 4x_{52} + 2x_{53} + 3x_{54} + 5x_{55} + 1x_{56}$$
$$+ 1x_{61} + 3x_{62} + 4x_{63} + 5x_{64} + 2x_{65} + 6x_{66}$$

Subject to:

$$\sum_{j=1}^{6} x_{ij} = 1; \qquad 1 \le i \le 6$$
$$\sum_{i=1}^{6} x_{ij} = 1; \qquad 1 \le j \le 6$$
$$x_{ij} = 0 \quad \text{or} \quad 1; \qquad 1 \le i,j \le 6$$

EXERCISES 1.4

1. Minimize:

$$z = \quad 8U_1 + 8U_2 + 8U_3 + \quad 8U_4 \qquad \text{(Hiring costs)}$$
$$+ 3D_1 + 3D_2 + 3D_3 + \quad 3D_4 \qquad \text{(Firing costs. Note that } D_1 \text{ may not be}$$
$$\qquad\qquad\qquad\qquad\qquad\qquad\qquad \text{zero since } P_0 \ne 0)$$
$$+ 2I_1 + \; 2I_2 + \; 2I_3 + 2 \cdot 500 \qquad \text{(Storage costs; 1000 for period 4)}$$

Subject to:

$$\left. \begin{array}{l} U_1 - D_1 = P_1 \\ U_2 - D_2 = P_2 - P_1 \\ U_3 - D_3 = P_3 - P_2 \\ U_4 - D_4 = P_4 - P_3 \end{array} \right\} \quad \text{(Hiring–firing constraints)}$$

$$\left. \begin{array}{l} 300 + P_1 = 500 + I_1 \\ I_1 + P_2 = 900 + I_2 \\ I_2 + P_3 = 400 + I_3 \\ I_3 + P_4 = 800 + 500 \end{array} \right\} \quad \text{(Production constraints)}$$

$$U_i, D_i, P_i \ge 0, \qquad 1 \le i \le 4; \qquad I_i \ge 0, \qquad 1 \le i \le 3$$

3. 50 decision variables: U_i $(1 \le i \le 12)$; I_i $(0 \le i \le 12)$; P_i $(0 \le i \le 12)$; D_i $(1 \le i \le 12)$

5. (a) Minimize: $\quad z = \quad 25U_1 + 25U_2 + 25U_3 + \quad 25U_4$
$$+ 10D_1 + 10D_2 + 10D_3 + \quad 10D_4$$
$$+ \quad 5I_1 + \quad 5I_2 + \quad 5I_3 + 5 \cdot 500$$

(b) $\left. \begin{array}{l} U_1 - D_1 = P_1 - 4000 \\ U_2 - D_2 = P_2 - P_1 \\ U_3 - D_3 = P_3 - P_2 \\ U_4 - D_4 = 3000 - P_3 \end{array} \right\} \quad \text{(Hiring–firing constraints)}$

$\left. \begin{array}{l} P_1 + 750 = 4000 + I_1 \\ P_2 + I_1 = 5000 + I_2 \\ P_3 + I_2 = 3000 + I_3 \\ 3000 + I_3 = 2500 + 500 \end{array} \right\} \quad \text{(Production constraints)}$

$$U_i, D_i \ge 0, \qquad 1 \le i \le 4; \qquad P_i \ge 0, \qquad 1 \le i \le 3; \qquad I_i \ge 0, \qquad 1 \le i \le 3$$

7. Add $P_1 \le 4000$ and $P_2 \le 4000$ to the constraints for Exercise 5.

9. The profit contribution to the objective function for May is

$$2.0x_5 - 2.2y_5 + 2.3(x_5 - y_5) = 4.3x_5 - 0.1y_5$$

Therefore, it is advantageous to use as little water as possible for irrigation ($y_5 = 1$) in May.

11. The January and February constraints are equivalent to $x_1 \geq -3$ and $x_1 + x_2 \geq -1$, respectively. Since $x_1, x_2 \geq 0$, these constraints are always satisfied. The constraints $y_1, y_2, y_3, y_{11}, y_{12} \geq 0$ occur twice.

13. Minimize:

$$z = x_1 + x_2 + b_1 + 50,000$$

Subject to:

$$0.08x_1 + 0.06x_2 + 1.05b_1 - b_2 = 50,000 \quad \text{(Year 2)}$$
$$0.08x_1 + 0.06x_2 + 1.05b_2 - b_3 = 50,000 \quad \text{(Year 3)}$$
$$0.08x_1 + 0.06x_2 + 1.05b_3 - b_4 = 50,000 \quad \text{(Year 4)}$$
$$0.08x_1 + 0.06x_2 + 1.05b_4 - b_5 = 50,000 \quad \text{(Year 5)}$$
$$0.08x_1 + 0.06x_2 + 1.05b_5 - b_6 = 50,000 \quad \text{(Year 6)}$$
$$1.18x_1 + 0.06x_2 + 1.05b_6 - b_7 = 50,000 \quad \text{(Year 7)}$$
$$1.31x_2 + 1.05b_7 - b_8 = 50,000 \quad \text{(Year 8)}$$
$$1.05b_8 - b_9 = 50,000 \quad \text{(Year 9)}$$
$$1.05b_9 = 50,000 \quad \text{(Year 10)}$$
$$x_1, x_2 \geq 0; \quad b_i \geq 0, \quad 1 \leq i \leq 9$$

In Chapter 3 you will learn to solve this problem and discover that the minimal cost to SL is \$369,595.48.

EXERCISES 2.1

1.

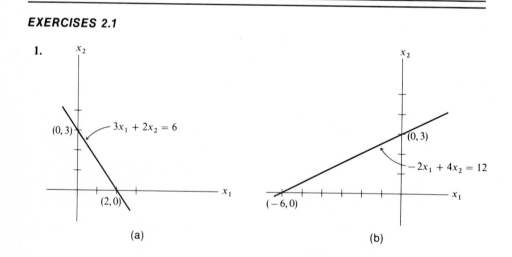

(a)

(b)

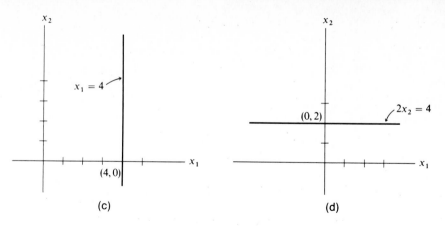

(c)

(d)

3.

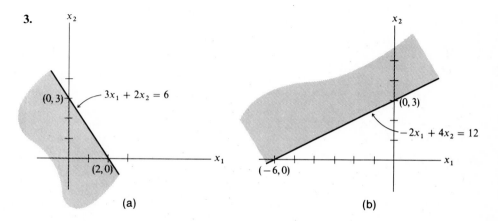

(a)

(b)

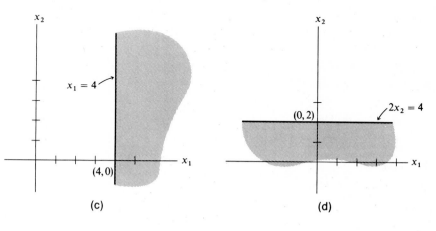

(c)

(d)

5.

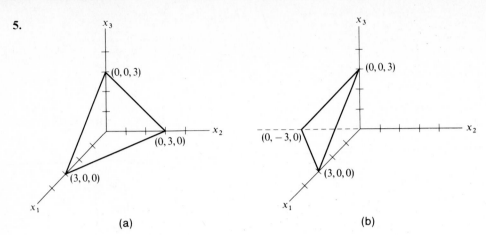

(a)   (b)

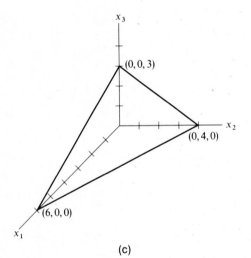
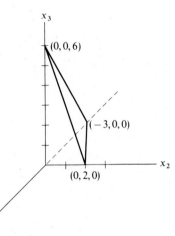

(c)   (d)

7.

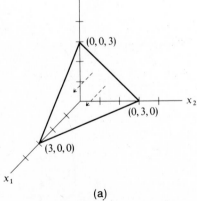

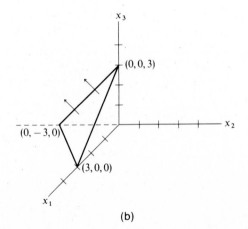

(a)   (b)

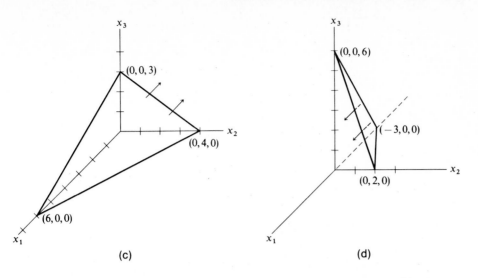

(c)   (d)

9. (a) and **(c)**   **11. (a)** and **(b)**   **13. (a), (b), (f), (h)**   **15. (a)** Yes   **(b)** no   **(c)** yes

17. $(\frac{3}{2}, 3)$ **19.** $(\frac{5}{4}, \frac{5}{2})$ **21.** Let $t = t_2$, and observe that $t_1 = 1 - t$.

23. We must show that if

$$\mathbf{p} = t_1\mathbf{p}_1 + t_2\mathbf{p}_2 + \cdots + t_k\mathbf{p}_k$$

and

$$\mathbf{q} = s_1\mathbf{p}_1 + s_2\mathbf{p}_2 + \cdots + s_k\mathbf{p}_k$$

are convex combinations of $\mathbf{p}_1, \mathbf{p}_2, \ldots, \mathbf{p}_k$, then so is $(1 - t)\mathbf{p} + t\mathbf{q}$ for each t, $1 \leq t \leq 1$. Note that

$$(1 - t)\mathbf{p} + t\mathbf{q} = (1 - t)t_1\mathbf{p}_1 + (1 - t)t_2\mathbf{p}_2 + \cdots + (1 - t)t_k\mathbf{p}_k$$
$$+ ts_1\mathbf{p}_1 + ts_2\mathbf{p}_2 + \cdots + ts_k\mathbf{p}_k$$
$$= [(1 - t)t_1 + ts_1]\mathbf{p}_1 + [(1 - t)t_2 + ts_2]\mathbf{p}_2 + \cdots + [(1 - t)t_k + ts_k]\mathbf{p}_k$$

Moreover, the coefficients of the points $\mathbf{p}_i$ are nonnegative, and

$$[(1 - t)t_1 + ts_1] + [(1 - t)t_2 + ts_2] + \cdots + [(1 - t)t_k + ts_k]$$
$$= (1 - t)(t_1 + t_2 + \cdots + t_k) + t(s_1 + s_2 + \cdots + s_k)$$
$$= (1 - t) + t = 1$$

EXERCISES 2.2

1. Optimal solution:
$x_1 = \frac{3}{2}, x_2 = \frac{5}{2}$
$z = 430$

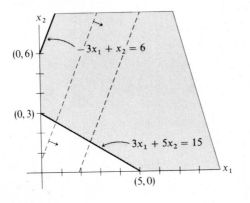

3. There is no optimal solution
because the feasible region
is unbounded.

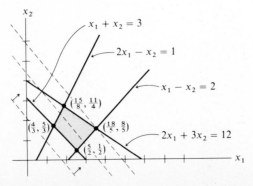

5. Optimal solution:
$x_1 = \frac{18}{5}, x_2 = \frac{8}{5}$
$z = \frac{166}{5}$

7. Optimal solution:
$x_1 = 6, x_2 = 0$
$z = 30$

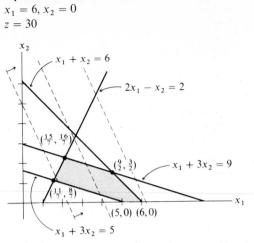

9. Optimal solution:
$x_1 = 4, x_2 = 0$
$z = -16$

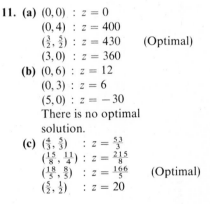

11. (a) $(0,0) \;:\; z = 0$
$(0,4) \;:\; z = 400$
$\left(\frac{3}{2}, \frac{5}{2}\right) \;:\; z = 430$ (Optimal)
$(3,0) \;:\; z = 360$

(b) $(0,6) \;:\; z = 12$
$(0,3) \;:\; z = 6$
$(5,0) \;:\; z = -30$
There is no optimal solution.

(c) $\left(\frac{4}{3}, \frac{5}{3}\right) \;:\; z = \frac{53}{3}$
$\left(\frac{15}{8}, \frac{11}{4}\right) \;:\; z = \frac{215}{8}$
$\left(\frac{18}{5}, \frac{8}{5}\right) \;:\; z = \frac{166}{5}$ (Optimal)
$\left(\frac{5}{2}, \frac{1}{2}\right) \;:\; z = 20$

13. Any point on the line segment connecting the points $(2, 1)$ and $(3, 0)$ corresponds to an optimal solution.

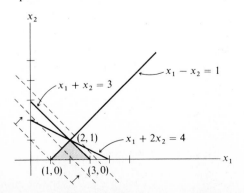

15. (1, 7, 8) : $z = 31$ (1, 13, 4) : $z = 35$
    (1, 13, 8) : $z = 43$ (4, 13, 6) : $z = 42$
    (4, 13, 8) : $z = 46$ (Optimal)   (4, 5, 4) : $z = 22$
    (4, 7, 8) : $z = 34$ (4, 10, 4) : $z = 32$
    (1, 5, 4) : $z = 19$ (2, 13, 4) : $z = 36$

17. $1 \le c_1/c_2 \le \frac{5}{3}$ **19.** $-1 \le c_1/c_2 \le \frac{2}{3}$

21. Each intersection is counted twice. For instance, the pair of lines L_1 and L_2 and the pair of lines L_2 and L_1 correspond to the same intersection point.

23. The planes P_1, P_2, and P_3 occur in the triples (P_1, P_2, P_3), (P_1, P_3, P_2), (P_2, P_1, P_3), (P_2, P_3, P_1), (P_3, P_1, P_2), and (P_3, P_2, P_1).

25. (a) 35   **(b)** 252   **(c)** 184,756   **(d)** 8008

EXERCISES 2.3

1. (a) $\mathbf{k}_1 = (1, 1)$, $\mathbf{k}_5 = (4, 3)$ **(b)** $(\frac{9}{2}, 2)$ **(c)** $f(\mathbf{k}_1) = 5$, $f(\mathbf{p}) = \frac{25}{3}$, $f(\mathbf{q}) = 15$, $f(\mathbf{k}_5) = 17$

3. $b_1\mathbf{p}_1 + b_2\mathbf{p}_2 + b_3\mathbf{p}_3 = 2(1, 2, 3) + 4(2, 2, 4) + 5(1, 0, 2) = (15, 12, 32)$
    $f(b_1\mathbf{p}_1 + b_2\mathbf{p}_2 + b_3\mathbf{p}_3) = f(15, 12, 32) = 114$
    $b_1f(\mathbf{p}_1) + b_2f(\mathbf{p}_2) + b_3f(\mathbf{p}_3) = 18 + 56 + 40 = 114$

5. (a) You need to solve the system of equations

$$\frac{3}{2} = t_1 + 2t_2$$

$$\frac{5}{2} = 2t_1 + 3t_2 + 4t_3$$

$$1 = t_1 + t_2 + t_3$$

The solution is $t_1 = \frac{1}{2}$, $t_2 = \frac{1}{2}$, $t_3 = 0$; therefore,

$$\left(\frac{3}{2}, \frac{5}{2}\right) = \frac{1}{2}\mathbf{k}_1 + \frac{1}{2}\mathbf{k}_2 + (0)\mathbf{k}_3$$

(b) $(2, 2) = (\frac{6}{11})\mathbf{k}_1 + (\frac{4}{11})\mathbf{k}_2 + (\frac{1}{11})\mathbf{k}_3$
(c) There is no solution to the system of equations

$$\frac{5}{2} = t_1 + 3t_2 + t_3$$

$$\frac{5}{2} = 2t_1 + 4t_2 + t_3$$

$$1 = t_1 + t_2 + t_3 \qquad (t_i \ge 0, \qquad 1 \le i \le 3)$$

and thus you cannot express $\mathbf{p}$ as a convex combination of $\mathbf{k}_1, \mathbf{k}_2, \mathbf{k}_3$.

7. (a) $(\frac{3}{2}, \frac{3}{2}) = (\frac{1}{2})(1, 1) + 0(1, 2) + (\frac{1}{2})(2, 2) + 0(2, 1)$. You can also use any other solution of the system of equations

$$\frac{3}{2} = t_1 + t_2 + 2t_3 + 2t_4$$

$$\frac{3}{2} = t_1 + 2t_2 + 2t_3 + t_4$$

$$1 = t_1 + t_2 + t_3 + t_4 \qquad (t_i \geq 0, \qquad 1 \leq i \leq 4)$$

as the coefficients of the points $(1, 1), (1, 2), (2, 2)$, and $(2, 1)$.

(b) $(\frac{5}{4}, \frac{9}{8}) = (\frac{5}{8})(1, 1) + (\frac{1}{8})(1, 2) + 0(2, 2) + (\frac{1}{4})(2, 1)$. As in part (a), there are many other answers as well.

9. Note that any point **p** in K lies in one of the triangles in Figure 9, say, the triangle with vertices $\mathbf{k}_1, \mathbf{k}_m, \mathbf{k}_{m+1}$. Then, by Exercise 4, it follows that

$$\mathbf{p} = t_1\mathbf{k}_1 + 0\mathbf{k}_2 + \cdots + 0\mathbf{k}_{m-1} + t_m\mathbf{k}_m + t_{m+1}\mathbf{k}_{m+1} + 0\mathbf{k}_{m+2} + \cdots + 0\mathbf{k}_n,$$

where $t_1 + t_m + t_{m+1} = 1$ and $t_1, t_m, t_{m+1} \geq 0$.

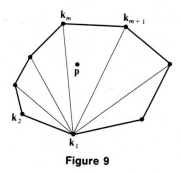

Figure 9

11. Let $\mathbf{k}_1 = (x_{11}, x_{12}, \ldots, x_{1n}), \mathbf{k}_2 = (x_{21}, x_{22}, \ldots, x_{2n}), \ldots, \mathbf{k}_m = (x_{m1}, x_{m2}, \ldots, x_{mn})$. Then,

$$f(b_1\mathbf{k}_1 + b_2\mathbf{k}_2 + \cdots + b_m\mathbf{k}_m)$$

$$= f(b_1 x_{11} + b_2 x_{21} + \cdots + b_m x_{m1}, b_1 x_{12} + b_2 x_{22}$$

$$+ \cdots + b_m x_{m2}, \ldots, b_1 x_{1n} + b_2 x_{2n} + \cdots + b_m x_{mn})$$

$$= c_1(b_1 x_{11} + b_2 x_{21} + \cdots + b_m x_{m1}) + c_2(b_1 x_{12}$$

$$+ b_2 x_{22} + \cdots + b_m x_{m2}) + \cdots$$

$$+ c_n(b_1 x_{1n} + b_2 x_{2n} + \cdots + b_m x_{mn})$$

$$= b_1(c_1 x_{11} + c_2 x_{12} + \cdots + c_n x_{1n})$$

$$+ b_2(c_1 x_{21} + c_2 x_{22} + \cdots + c_n x_{2n}) + \cdots$$

$$+ b_m(c_1 x_{m1} + c_2 x_{m2} + \cdots + c_n x_{mn})$$

$$= b_1 f(\mathbf{k}_1) + b_2 f(\mathbf{k}_2) + \cdots + b_m f(\mathbf{k}_m)$$

13. Note that $\partial f/\partial x_1 = c_1, \partial f/\partial x_2 = c_2, \ldots, \partial f/\partial x_n = c_n$. At a maximum or minimum interior point, all of these partial derivatives will be equal to 0.

EXERCISES 2.4

1. **(b)** is not linear because of the x^2 term.   **(c)** is not linear because of the xy term.

3. **(a)** $2x_1 + 3x_2 - 4x_3 + 5x_4 \leq -5$ **(b)** $-3x - 2y = 3$ **(c)** $-x + 4y \leq 0$

5. **(a)** $(0, 0, 0)$ and $(1, 2, -1)$. There are many other answers.
   (b) $(1, 1, 11)$ and $(2, 1, \frac{19}{2})$. There are many other answers.

7. Substitute the given values for x_1, x_2, x_3 to obtain

$$2 + 3(1) - 2(-3) = 11 \quad \text{and} \quad -3(2) + 5(1) - 4(-3) = 11$$

9. $\begin{aligned} x_1 + 2x_2 \qquad\quad + x_4 + x_5 \qquad\qquad &= 20 \\ x_1 - x_2 + x_3 \qquad\qquad + x_6 \quad &= 10 \\ -x_1 + 4x_2 + 2x_3 - 3x_4 \qquad\quad + x_7 &= 40 \end{aligned}$

   Basic variables: x_5, x_6, x_7

11. **(a)** Not in canonical form because no basic variables are identified.
    (b) Not in canonical form because no basic variables are identified.
    (c) x_1, x_4, x_6 are basic variables as are x_1, x_2, x_4.

13. **(c)** $x_1 = 1, x_4 = 2, x_6 = 3$, or $x_1 = 1, x_2 = 3, x_4 = 2$

15. If x_1, x_2, x_3 is a solution to System 1, multiply the first equation of System 1 by k to obtain $ka_{11}x_1 + ka_{12}x_2 + ka_{13}x_3 = kb_1$. Then add this equation to the second equation of System 1 to obtain the second equation of System 2. If x_1, x_2, x_3 is a solution to System 2, multiply the first equation of System 2 by k, and then subtract it from the second equation of System 2. The difference is the second equation of System 1.

17. Answers will vary for different choices of basic variables.

(a) $\begin{bmatrix} 1 & 1 & -2 & 3 & 2 \\ -2 & 0 & ① & 0 & 2 \end{bmatrix}$

$\begin{bmatrix} -3 & 1 & 0 & 3 & 6 \\ -2 & 0 & 1 & 0 & 2 \end{bmatrix}$

$\begin{aligned} -3x_1 + x_2 \qquad + 3x_4 &= 6 \\ -2x_1 \qquad + x_3 \qquad &= 2 \end{aligned}$

(b) $\begin{aligned} x_1 &= \frac{201}{33} \\ x_2 &= -\frac{51}{33} \\ x_3 &= \frac{1}{11} \end{aligned}$

19. $\begin{bmatrix} ① & 2 & 0 & 1 & 20 \\ 2 & 1 & 1 & 0 & 10 \\ -1 & 4 & -2 & 3 & 40 \end{bmatrix}$

$\begin{bmatrix} 1 & 2 & 0 & 1 & 20 \\ 0 & ⊖3 & 1 & -2 & -30 \\ 0 & 6 & -2 & 4 & 60 \end{bmatrix}$

$\begin{bmatrix} 1 & 0 & \frac{2}{3} & -\frac{1}{3} & 0 \\ 0 & 1 & -\frac{1}{3} & \frac{2}{3} & 10 \\ 0 & 0 & 0 & 0 & 0 \end{bmatrix}$

One or more rows will eventually contain all zero entries.

EXERCISES 2.5

3. (a) a, c, j, g, i
(c) a, c, j, g and a, i, g

(b)

Letter	x_1	x_2	$z = 120x_1 + 100x_2$
a	0	0	0
c	2	0	240
j	$\frac{21}{8}$	$\frac{5}{8}$	$\frac{755}{2}$
g	$\frac{3}{2}$	$\frac{5}{2}$	430
i	0	4	400

5.

x_1	x_2	$z = x_1 + x_2$
0	0	0
3	0	3
3	2	5
2	1	3
2	4	6
1	3	4
1	4	5
0	3	3

The answer is no.

7. The canonical system

$$\frac{2}{5}x_2 + s - \frac{1}{5}t = 1$$

$$x_1 + \frac{3}{5}x_2 \quad + \frac{1}{5}t = 3$$

determines the feasible intersection point $(3, 0)$. The canonical system

$$-\frac{2}{3}x_1 \quad + s - \frac{1}{3}t = -1$$

$$\frac{5}{3}x_1 + x_2 \quad + \frac{1}{3}t = \quad 5$$

determines the infeasible intersection point $(0, 5)$. The canonical system

$$x_1 + x_2 + s \quad = 4$$

$$2x_1 \quad - 3s + t = 3$$

determines the feasible intersection point $(0, 4)$. The canonical system

$$x_1 + x_2 + s \quad = \quad 4$$

$$-2x_2 - 5s + t = -5$$

determines the infeasible intersection point $(4, 0)$. The canonical system

$$x_1 \quad -\frac{3}{2}s + \frac{1}{2}t = \frac{3}{2}$$

$$x_2 + \frac{5}{2}s - \frac{1}{2}t = \frac{5}{2}$$

determines the feasible intersection point $(\frac{3}{2}, \frac{5}{2})$.

9. Of the ratios $b_1/a_{11}, b_2/a_{21}, b_3/a_{31}$, calculate those having a nonzero denominator. If all the ratios you calculated are negative, then no choice of pivot will work. Otherwise, pivot on the x_1 term corresponding to the smallest of the positive ratios. We will develop this technique fully in the next section.

11. Let $u_2 = -x_2$.

Maximize:

$z' = -z = -2x_1 - 3u_2$

Subject to:

$5x_1 + 2u_2 \leq 3$
$-2x_1 - 3u_2 \leq 5$
$x_1, u_2 \geq 0$

13. Let $x_2 = u_2 - v_2$.

Maximize:

$z = 2x_1 - 3u_2 + 3v_2$

Subject to:

$-x_1 + 2u_2 - 2v_2 \leq 5$
$x_1 - 3u_2 + 3v_2 \leq 6$
$x_1, u_2, v_2 \geq 0$

15. Let $u_1 = -x_1, x_2 = u_2 - v_2$.

Maximize: Subject to:

$z' = -z = 2u_1 + 3u_2 - 3v_2$ $-2u_1 - 5u_2 + 5v_2 \leq 12$

                              $3u_1 + 2u_2 - 2v_2 \leq 15$

                                  $u_1, u_2, v_2 \geq 0$

EXERCISES 3.1

1. (a) x_1 (b) x_1 (c) Either x_1 or x_2

3. (a) x_8 (b) x_5 (c) x_8 (d) x_5

5. Solve the first constraint for x_3 and the second constraint for x_4. Substitute for these variables in the given objective function to obtain $z = 3x_1 - 5x_2 + 17$.

7. The pivot elements are circled.

(a)

$$x_1 + \quad x_2 + \quad s_1 \qquad\qquad = \quad 4$$
$$\boxed{5x_1} + \quad 3x_2 \qquad\qquad + \quad s_2 = \quad 15$$
$$z - 120x_1 - 100x_2 \qquad\qquad\qquad = \quad 0$$

$$\boxed{\frac{2}{5}x_2} + \quad s_1 - \frac{1}{5}s_2 = \quad 1$$
$$x_1 + \quad \frac{3}{5}x_2 \qquad\quad + \frac{1}{5}s_2 = \quad 3$$
$$z \qquad\quad - 28x_2 \qquad + 24s_2 = 360$$

$$x_2 + \frac{5}{2}s_1 - \frac{1}{2}s_2 = \quad \frac{5}{2}$$
$$x_1 \qquad\quad - \frac{3}{2}s_1 + \frac{1}{2}s_2 = \quad \frac{3}{2}$$
$$z \qquad\qquad + 70s_1 + 10s_2 = 430$$

Optimal solution: $(\frac{3}{2}, \frac{5}{2}, 0, 0)$; $z = 430$
(b) Optimal solution: $(50, 100, 0, 0)$; $z = 1000$
(c) Optimal solution: $(3, 12, 0, 27, 0, 0)$; $z = 174$
(d) The objective function is unbounded on the feasible region.

9. $x_1 = 2x_2 + 4$
$x_5 = 3x_2 + 6$
$x_4 = \quad x_2 + 3$
$z = 4x_2 + 14$

11. The answer is to build $\frac{45}{2}$ sofas and $\frac{45}{2}$ chairs for a total selling price of \$6750. This is not a very practical answer because it calls for constructing half a chair and half a sofa. We will learn to deal with problems that require integer solutions in Chapter 8.

13. (b) $b_1' = b_1 - a_{11}(b_2/a_{21}) = a_{11}(b_1/a_{11} - b_2/a_{21})$
$b_2' = b_2/a_{21}$
$b_3' = b_3 - a_{31}(b_2/a_{21}) = a_{31}(b_3/a_{31} - b_2/a_{21})$
The condition is $b_2/a_{21} \le b_1/a_{11}$ and $b_2/a_{21} \le b_3/a_{31}$.
(c) The condition is $b_3/a_{31} \le b_1/a_{11}$ and $b_3/a_{31} \le b_2/a_{21}$.
(d) The departing basic variable must correspond to the row with smallest positive ratio.

15. We add a slack variable to each inequality to obtain

$$x_1 + \quad x_2 + s_1 \qquad = \quad 4$$
$$5x_1 + 3x_2 \qquad + s_2 = 15$$

The Greedy Entering Basic Variable Rule picks x_1 as the entering basic variable.

Pivoting on $5x_1$ gives the equivalent system:

$$\frac{2}{5}x_2 + s_1 - \frac{1}{5}s_2 = 1$$

$$x_1 + \frac{3}{5}x_2 \quad\quad + \frac{1}{5}s_2 = 3$$

The basic solution is $x_1 = 3$, $s_1 = 1$. The corresponding value of the objective function is $z = 100 \cdot 3 = 300$.

If we select x_2 as the entering basic variable and pivot on x_2 in the first equation, we obtain the equivalent system

$$x_1 + x_2 + s_1 \quad\quad = 4$$

$$2x_1 \quad\quad - 3s_1 + s_2 = 3$$

The basic feasible solution is $x_1 = 0$, $x_2 = 4$, and the value of the objective function is $z = 100 \cdot 0 + 90 \cdot 4 = 360$.

17. (a) Since $10x_5 + 10x_6 = 0$, $x_5 \geq 0$, and $x_6 \geq 0$, we have $x_5 = 0$ and $x_6 = 0$.
    (b) You can assign any value to x_2 that does not make $x_1 = 2 - \frac{5}{4}x_2 < 0$. For instance, setting $x_2 = 1$ gives the alternative optimal solution

$$(x_1, x_2, x_3, x_4, x_5, x_6) = (\tfrac{3}{4}, 1, 10, 26, 0, 0)$$

(c) See part b. You can find an alternative optimal basic solution by assigning any value to x_2 satisfying $0 \leq x_2 \leq \frac{8}{5}$.

EXERCISES 3.2

1. (c) $$\frac{2}{5}x_1 \quad\quad + s_1 - \frac{1}{5}s_2 = \quad 1$$

$$\frac{3}{5}x_1 + x_2 \quad\quad + \frac{1}{5}s_2 = \quad 3$$

$$z - 28x_1 \quad\quad\quad + 24s_2 = 360$$

(d) $s_1 = 1$, $x_2 = 3$.

(e) $z = 100 \cdot 0 + 120 \cdot 3 = 360$

(g) $$x_1 \quad\quad + \frac{5}{2}s_1 - \frac{1}{2}s_2 = \quad \frac{5}{2}$$

$$x_2 - \frac{3}{2}s_1 + \frac{1}{2}s_2 = \quad \frac{3}{2}$$

$$z \quad\quad\quad + 70s_1 + 10s_2 = 430$$

$$z = 430 = 100(\tfrac{5}{2}) + 120(\tfrac{3}{2})$$

3. The optimal solution is $(0, 25, \frac{35}{3}, 0, 0)$; $z = \frac{170}{3}$.

5. The optimal solution is $(8, 18, 0, 0, 0, 25)$; $z = 70$.

7. The minimal solution is $(0, \frac{3}{5}, \frac{13}{10}, 0, \frac{61}{10}, 0)$; $z = -\frac{77}{10}$.

9. Let $u_1 = -x_1$ and $v_1 - v_2 = x_2$. The optimal tableau is

	u_1	v_1	v_2	s_1	s_2	
v_2	$-\frac{2}{3}$	-1	1	$\frac{1}{3}$	0	4
s_2	-1	0	0	1	1	20
	$\frac{4}{3}$	0	0	$\frac{1}{3}$	0	4

The solution is $x_1 = 0$, $x_2 = -4$; $z = 4$.

11. (a)
$$x_2 - \frac{3}{5}x_3 + \frac{2}{5}x_4 = \frac{23}{5}$$
$$x_1 \qquad - \frac{4}{5}x_3 + \frac{1}{5}x_4 = \frac{4}{5}$$
$$- \frac{2}{5}x_3 + \frac{3}{5}x_4 + x_5 = \frac{72}{5}$$
$$z \qquad - 2x_3 + x_4 = 10$$

(c)
$$x_2 = \frac{3}{5}x_3 + \frac{23}{5}$$
$$x_1 = \frac{4}{5}x_3 + \frac{4}{5}$$
$$x_5 = \frac{2}{5}x_3 + \frac{72}{5}$$
$$z = 2x_3 + 10$$

13. (a) The optimal tableau is

	x_1	x_2	x_3	x_4	x_5	
x_2	0	1	$-\frac{3}{4}$	$\frac{5}{4}$	$-\frac{1}{2}$	$\frac{5}{2}$
x_1	1	0	$-\frac{1}{4}$	$-\frac{3}{4}$	$\frac{1}{2}$	$\frac{3}{2}$
	0	0	0	35	10	430

(c) If $x_4 = 0$ and $x_5 = 0$, the system of equations corresponding to the final tableau in (a) is

$$x_1 = \frac{1}{4}x_3 + \frac{3}{2}$$

$$x_2 = \frac{3}{4}x_3 + \frac{5}{2}$$

$$z = 0x_3 + 430$$

Thus, we obtain infinitely many solutions for x_1 and x_2, all giving the same value for z.

15. Examine the reduced costs of the nonbasic variables in the final tableau. If all of these reduced costs are negative, then the optimal solution is unique. If the reduced cost of any nonbasic variable is zero, then there are alternative optima provided there is at least one positive coefficient in the column corresponding to this nonbasic variable. You can find an alternative basic solution by pivoting on such a positive coefficient.

17. Optimal solution: $(\frac{1}{5}, 0, \frac{8}{5}, 0, 0, 4)$; $z = \frac{27}{5}$

19. Optimal solution: $(1, 0, 0, 3, 0, \frac{1}{2}, 0)$; $z = \frac{11}{3}$

21. Optimal solution: $(0, \frac{44}{15}, \frac{7}{3}, \frac{53}{15}, 0, \frac{66}{5}, 0, 0)$; $z = \frac{1190}{3}$

23. A tableau indicating that the objective function is unbounded on the feasible region is

	x_1	x_2	x_3	x_4	s_1	s_2	s_3	
x_3	2	-6	1	0	0	$-\frac{1}{2}$	$\frac{1}{2}$	25
x_4	6	-13	0	1	0	$-\frac{3}{2}$	$\frac{1}{2}$	15
s_1	12	-21	0	0	1	$-\frac{7}{2}$	$\frac{1}{2}$	15
	6	-10	0	0	0	$-\frac{3}{2}$	$\frac{5}{2}$	135

EXERCISES 3.3

1. One of the possible optimal tableaus is

	x_1	x_2	x_3	s_1	s_2	s_3	
s_1	0	0	2	1	0	0	1
x_1	1	0	17	0	$\frac{3}{2}$	2	$\frac{17}{2}$
x_2	0	1	7	0	$\frac{1}{2}$	1	$\frac{7}{2}$
	0	0	19	0	$\frac{5}{2}$	3	$\frac{27}{2}$

Optimal solution: $(\frac{17}{2}, \frac{7}{2}, 0, 1, 0, 0); z = \frac{27}{2}$

3. $x_1 = 1, x_2 = 0, x_3 = 1, x_4 = 0; z = \frac{5}{4}$

5. We first divide the ith row by a_{ij}, obtaining b_i/a_{ij} in the resource column. To replace $-c_j$ in the objective row by 0, we add c_j times the pivot row to the objective row. Thus, the objective function value becomes $v^* = v + c_j(b_i/a_{ij})$; $v^* > v$, since $c_j > 0$, $b_i > 0$, and $a_{ij} > 0$.

7. Set all nonbasic variables except the entering basic variable x_1 to 0 to obtain the system of equations

$$x_4 = 3x_1 + 60$$
$$x_5 = x_1 + 10$$
$$x_6 = 2x_1 + 50$$
$$z = 2x_1 + 0$$

We can assign any nonnegative value to x_1 and, thereby, make z as large as we please.

EXERCISES 3.4

1.

	x_1	x_2	x_3	y_1	y_2	
y_1	3	2	-4	1	0	7
y_2	1	-1	3	0	1	2
(z)	2	2	-5	0	0	0
(w)	-4	-1	1	0	0	-9

3.

	x_1	x_2	x_3	s_1	s_2	s_3	y_1	y_2	y_3	
y_1	3	-1	4	-1	0	0	1	0	0	5
y_2	-1	2	5	0	-1	0	0	1	0	6
y_3	2	1	-1	0	0	-1	0	0	1	2
(z)	1	2	-1	0	0	0	0	0	0	0
(w)	-4	-2	-8	1	1	1	0	0	0	-13

5. $(\frac{4}{3}, \frac{1}{3}, 0)$. There are other possible answers.

7. $(0, \frac{140}{51}, \frac{46}{17}, \frac{110}{51})$. There are other possible answers.   **9.** $(\frac{8}{7}, \frac{2}{7})$

11. $(0, 0, 3)$. There are other possible answers.   **13.** $(5, 0)$; $z = 150$ **15.** $(3, 5)$; $z = 59$

17. $(\frac{87}{5}, \frac{33}{5}, 0)$; $z = \frac{3666}{5}$ **19.** The objective function is unbounded on the feasible region.

23. The final tableau for phase 1 is

	x_1	x_2	x_3	x_4	x_5	y_1	y_2	
x_1	1	0	$\frac{1}{7}$	0	$\frac{4}{7}$	0	$-\frac{4}{7}$	$\frac{1}{7}$
y_1	0	0	$-\frac{3}{7}$	-1	$-\frac{5}{7}$	1	$\frac{5}{7}$	$\frac{4}{7}$
x_2	0	1	$\frac{1}{7}$	0	$-\frac{3}{7}$	0	$\frac{3}{7}$	$\frac{8}{7}$
(z)	0	0	$\frac{5}{7}$	0	$\frac{6}{7}$	0	$-\frac{6}{7}$	$\frac{19}{7}$
(w)	0	0	$\frac{3}{7}$	1	$\frac{5}{7}$	0	$\frac{2}{7}$	$-\frac{4}{7}$

25. The final tableau for phase 1 is

	x_1	x_2	x_3	s_1	s_2	s_3	y_1	y_2	
x_2	$\frac{3}{4}$	1	$\frac{1}{4}$	$\frac{1}{4}$	0	0	0	0	$\frac{5}{4}$
y_1	$-\frac{17}{4}$	0	$-\frac{19}{4}$	$-\frac{3}{4}$	-1	0	1	0	$\frac{13}{4}$
y_2	$\frac{17}{4}$	0	$\frac{19}{4}$	$-\frac{1}{4}$	0	-1	0	1	$\frac{3}{4}$
(z)	$\frac{17}{4}$	0	$-\frac{17}{4}$	$\frac{3}{4}$	0	0	0	0	$\frac{15}{4}$
(w)	0	0	0	1	1	1	0	0	-4

27. This is a transportation problem with cost table

		Demand points		
		Idaho Falls	Pocatello	
Supply points		1	2	Supply
Burley 1		6	3	45
Twin Falls 2		9	5	40
Demand		25	30	

Minimize:
$$z = 6x_{11} + 3x_{12} + 9x_{21} + 5x_{22}$$

Subject to:
$$x_{11} + x_{21} = 25$$
$$x_{12} + x_{22} = 30$$
$$x_{11} + x_{12} \leq 45$$
$$x_{21} + x_{22} \leq 40$$
$$x_{ij} \geq 0, \qquad 1 \leq i, j \leq 2$$

The optimal solution is $x_{11} = 25$, $x_{12} = 20$, $x_{21} = 0$, $x_{22} = 10$; $z = 260$.

29. The answers are given above.

31. (low-risk) 4000; (medium-risk) 6000; (high-risk) 8000; $z = 1700$

33. The optimal objective function value for both problems is $\frac{2875}{3}$. The second constraint in part (b) should be multiplied by -1 to give $6x_1 - 9x_2 \leq 2$.

EXERCISES 3.5

1. (a) Pivot on the entry $\frac{5}{2}$. If you do so, you will discover that the problem is infeasible.
   (b) Pivot on the entry $\frac{5}{2}$ (or $-\frac{23}{6}$). If you do so, you will discover that the problem is feasible and has one redundant constraint.

3. The final tableau for phase 1 is

	x_1	x_2	x_3	x_4	y_1	y_2	y_3	
x_3	$\frac{1}{3}$	$-\frac{2}{3}$	1	0	$\frac{1}{3}$	0	0	$\frac{1}{3}$
x_4	$\frac{2}{3}$	$-\frac{1}{3}$	0	1	$\frac{2}{3}$	$\frac{1}{2}$	0	$\frac{5}{3}$
y_3	0	0	0	0	-3	-2	1	0
(z)	$\frac{5}{3}$	$\frac{5}{3}$	0	0	$\frac{2}{3}$	0	0	$\frac{2}{3}$
(w)	0	0	0	0	4	3	0	0

The third constraint is redundant because it contains zeros in the columns corresponding to decision variables. Notice that this is also the final tableau for phase 2.

5. The final tableau for phase 1 is

	x_1	x_2	x_3	s_1	s_2	y_1	y_2	
x_1	1	$\frac{1}{3}$	0	$\frac{1}{3}$	0	0	$\frac{1}{3}$	$\frac{5}{3}$
y_1	0	$-\frac{7}{3}$	0	$-\frac{4}{3}$	-1	1	$-\frac{1}{3}$	$\frac{4}{3}$
x_3	0	$-\frac{5}{3}$	1	$-\frac{2}{3}$	0	0	$\frac{1}{3}$	$\frac{2}{3}$
(z)	0	$\frac{7}{3}$	0	$\frac{1}{3}$	0	0	$-\frac{2}{3}$	$-\frac{7}{3}$
(w)	0	$\frac{7}{3}$	0	$\frac{4}{3}$	1	0	$\frac{4}{3}$	$-\frac{4}{3}$

This tableau shows that the problem is infeasible because the minimum value of $y_1 + y_2$ is $\frac{4}{3}$.

7. (a) $2x_1 + 3x_2 + 4x_3 \leq 1$ **(b)** $(a_{11} + 1)y_1 + (a_{12} + 2)y_2 + (a_{13} + 3)y_3 \leq 1$
    $3x_1 + 4x_2 + 5x_3 \leq 2$ $(a_{21} + 1)y_1 + (a_{22} + 2)y_2 + (a_{23} + 3)y_3 \leq 1$
    $4x_1 + 5x_2 + 6x_3 \leq 3$
   (c) $x_1 - x_2 + x_3 - x_4 \geq 1$
    $-x_1 + x_2 - x_3 + x_4 \geq 2$
    $x_1 - x_2 + x_3 - x_4 \geq 3$

9. (a) Let $\quad a_{11} = 1, a_{12} = 3, a_{13} = -4, b_1 = 3$
   $a_{21} = 1, a_{22} = 0, a_{23} = 2, b_2 = 5$
   $\sum_{j=1}^{3} a_{ij}x_j = b_i, \qquad 1 \leq i \leq 2$
   (b) Let $\quad a_{11} = 2, a_{12} = -3, a_{13} = 5, b_1 = 3$
   $a_{21} = 3, a_{22} = -1, a_{23} = 5, b_2 = -1$
   $a_{31} = 1, a_{32} = 0, a_{33} = 1, b_3 = 1$
   $\sum_{j=1}^{3} a_{ij}x_i \leq b_i, \qquad 1 \leq i \leq 3$

11. It is not possible for the artificial objective function to be unbounded because we are minimizing $w = y_1 + y_2$ and $y_1 + y_2 \geq 0$.

13. The type (5) constraints require three slack variables, the type (6) constraints require five artificial variables, and the type (7) constraints require six slack variables and six artificial variables. Thus, the phase 1 problem has

$$8 + 3 + 5 + 12 = 28$$

variables.

EXERCISES 3.6

1. The answers are those given for exercises in Section 3.4.

3. The final tableau with $M = 100$ is

	x_1	x_2	x_3	x_4	y_1	y_2	
x_4	0	8	0	1	6	-6	2
x_1	1	$-\frac{1}{2}$	0	0	-1	2	3
x_3	0	$\frac{1}{2}$	1	0	-1	1	0
	0	3	0	0	102	95	-9

The Big-M method does not encounter any difficulty.

5. The final tableau with $M = 100$ is

	x_1	x_2	s_1	s_2	y_1	s_3	y_2	
x_1	1	0	$\frac{1}{7}$	0	0	$\frac{4}{7}$	$-\frac{4}{7}$	$\frac{1}{7}$
y_1	0	0	$-\frac{3}{7}$	-1	1	$-\frac{5}{7}$	$\frac{5}{7}$	$\frac{4}{7}$
x_2	0	1	$\frac{1}{7}$	0	0	$-\frac{3}{7}$	$\frac{3}{7}$	$\frac{8}{7}$
	0	0	$\frac{305}{7}$	100	0	$\frac{506}{7}$	$\frac{194}{7}$	$-\frac{381}{7}$

7. The final tableau with $M = 100$ is

	x_1	x_2	x_3	s_1	s_2	y_1	y_2	
x_1	1	$\frac{1}{3}$	0	$\frac{1}{3}$	0	0	$\frac{1}{3}$	$\frac{5}{3}$
y_1	0	$-\frac{7}{3}$	0	$-\frac{4}{3}$	-1	1	$-\frac{1}{3}$	$\frac{4}{3}$
x_3	0	$-\frac{5}{3}$	1	$-\frac{2}{3}$	0	0	$\frac{1}{3}$	$\frac{2}{3}$
	0	$\frac{707}{3}$	0	$\frac{401}{3}$	100	0	$\frac{398}{3}$	$-\frac{407}{3}$

Because $y_1 \neq 0$, the problem is infeasible.

9. The answers are with those for Section 3.4.

11. The answer is with those for Section 3.4.

13. The final tableau with $M = 100$ is

	x_1	x_2	x_3	x_4	y_1	y_2	y_3	
x_3	$\frac{1}{3}$	$-\frac{2}{3}$	1	0	$\frac{1}{3}$	0	0	$\frac{1}{3}$
x_4	$\frac{2}{3}$	$-\frac{1}{3}$	0	1	$\frac{2}{3}$	$\frac{1}{2}$	0	$\frac{5}{3}$
y_3	0	0	0	0	-3	-2	1	0
	$\frac{5}{3}$	$\frac{5}{3}$	0	0	$\frac{1202}{3}$	300	0	$\frac{2}{3}$

If an artificial variable remains in the basis with value 0, and if the coefficients of all decision variables for the row corresponding to this artificial variable are 0, then the problem has a redundant constraint.

EXERCISES 3.7

1. (a) Peanuts: $341\frac{2}{3}$ pounds; almonds: 0 pounds; cashews: 0 pounds   (b) $74.81

3. Because it takes $\frac{1}{12}$ hour to produce 1000 ft of rope on Machine A, and the labor cost is $10 per hour, the total labor cost on machine A is $10(\frac{1}{12})(A_1 + A_2)$. By similar reasoning, the labor cost on machine B is $14(\frac{1}{9})(B_2 + B_3)$, and the labor cost on machine C is $16(\frac{2}{15})(C_2 + C_3)$.

The objective function is

$$\text{Maximize:} \quad z = 15A_1 + 18A_2 - \frac{5}{6}(A_1 + A_2)$$

$$+ 18B_2 + 20B_3 - \frac{14}{9}(B_2 + B_3)$$

$$+ 18C_2 + 20C_3 - \frac{32}{15}(C_2 + C_3)$$

$$= \frac{85}{6}A_1 + \frac{103}{6}A_2 + \frac{148}{9}B_2 + \frac{166}{9}B_3$$

$$+ \frac{238}{15}C_2 + \frac{268}{15}C_3$$

The constraints are

$$\left.\begin{array}{c} \dfrac{1}{12}A_1 + \dfrac{1}{12}A_2 \le 30 \\[2mm] \dfrac{1}{9}B_2 + \dfrac{1}{9}B_3 \le 35 \\[2mm] \dfrac{2}{15}C_2 + \dfrac{2}{15}C_3 \le 40 \end{array}\right\} \quad \text{(Limits on machine time)}$$

$$A_1 + A_2 + B_2 + B_3 + C_2 + C_3 \le 600 \quad \text{(Limit on storage)}$$

$$\left.\begin{array}{c} A_1 \ge 200 \\ A_2 + B_2 + C_2 \ge 115 \\ B_3 + C_3 \ge 110 \end{array}\right\} \quad \text{(Production requirements)}$$

$$A_1, A_2, B_2, B_3, C_2, C_3 \ge 0$$

It is better to enter the constants in the CALIPSO program as fractions than it is to convert them to decimal approximations because the program will carry more accuracy through the calculations if it converts the fractions to decimals itself. This problem can be solved with the TUTOR program. You can enter the problem as inequalities with the MODELER program or as a tableau with the tableau editor of TUTOR. The solution is $A_1 = 200$, $A_2 = 115$, $B_3 = 285$, and the rest of the decision variables are 0. The maximal profit is \$10,064.17.

5. (a) Tomatoes: 2.33 acres; peas: 1.67 acres; $z = 1616.67$ (b) $A = 71.11$; $C = 320$; $D = 97.78$; $z = 2440$ (c) $A = 84.65$; $B = 376.30$; $C = 20$; $D = 30$; $z = 2387.14$
(d) $m_1 = 48.25$; $m_2 = 1.00$; $m_3 = 50.75$; $z = 323.625$

7. [Note that there may be alternative solutions to these problems.]
(a) $x_{11} = 25{,}000$; $x_{12} = 25{,}000$; $x_{21} = 10{,}000$; $x_{23} = 40{,}000$; $x_{33} = 0$; $x_{34} = 30{,}000$; $z = 980{,}000$ (b) $x_{11} = 100$; $x_{13} = 150$; $x_{24} = 100$; $x_{32} = 200$; $x_{35} = 100$; $x_{43} = 100$; $x_{44} = 50$; $z = 5800$ (c) $x_{11} = 150$; $x_{13} = 150$; $x_{22} = 200$; $x_{23} = 230$; $x_{24} = 340$; $x_{33} = 220$; $x_{44} = 60$; $z = 59{,}100$ (d) $x_{11} = 95$; $x_{12} = 25$; $x_{22} = 50$; $x_{23} = 45$; $x_{33} = 63$; $x_{34} = 12$; $x_{44} = 78$; $x_{45} = 30$; $x_{55} = 90$; $z = 6883$

9. (a) $x_{13} = 1$; $x_{22} = 1$; $x_{34} = 1$; $x_{41} = 1$; $z = 33$ (b) $x_{13} = 1$; $x_{22} = 1$; $x_{34} = 1$; $x_{41} = 1$; $z = 24$ (c) $x_{16} = 1$; $x_{21} = 1$; $x_{33} = 1$; $x_{42} = 1$; $x_{54} = 1$: $x_{65} = 1$; $z = 10$

11. (a) $I_2 = 100$; $I_3 = 500$; $P_1 = 200$; $P_2 = 800$; $P_3 = 800$; $P_4 = 800$; $U_1 = 200$; $U_2 = 600$; $z = 8600$ (b) $D_4 = 0$; $I_1 = 875$; $P_1 = 4125$; $P_2 = 4125$; $P_3 = 3000$; $U_1 = 125$; $D_3 = 1125$; $I_3 = 0$; $z = 21{,}250$ (c) $D_3 = 1125$; $I_1 = 875$; $P_1 = 4125$; $P_2 = 1425$; $P_3 = 3000$; $P_4 = 3000$; $U_1 = 125$; $z = 18{,}750$

13. $b_1 = 161{,}830.73$; $b_2 = 132{,}543.45$; $b_3 = 101{,}791.80$; $b_4 = 69{,}502.57$; $b_5 = 35{,}598.88$; $b_7 = 136{,}162.40$; $b_8 = 92{,}970.52$; $b_9 = 47{,}619.05$; $x_1 = 157{,}764.75$; $z = 369{,}595.48$

EXERCISES 4.1

1. (a) $A + B = \begin{bmatrix} 5 & 4 & 2 \\ 1 & 3 & 1 \end{bmatrix}$ (b) $A - B = \begin{bmatrix} -3 & 2 & -4 \\ 3 & -1 & -1 \end{bmatrix}$

(c) $2A = \begin{bmatrix} 2 & 6 & -2 \\ 4 & 2 & 0 \end{bmatrix}$ (d) $2A + 3B = \begin{bmatrix} 14 & 9 & 7 \\ 1 & 8 & 3 \end{bmatrix}$

3. (a) $AB = \begin{bmatrix} 4 & 4 \\ 9 & 0 \end{bmatrix}$

(b) $AC = \begin{bmatrix} 0 & -1 & 4 \\ 2 & -1 & 6 \end{bmatrix}$

(c) $CB = \begin{bmatrix} 12 & 0 \\ 1 & 2 \\ 4 & 4 \end{bmatrix}$

(d) $CC = \begin{bmatrix} 4 & 7 & 0 \\ 0 & 1 & 0 \\ 0 & -1 & 4 \end{bmatrix}$

5. $AB = \begin{bmatrix} 8 & 9 \\ 18 & 19 \end{bmatrix}$ $BA = \begin{bmatrix} 5 & 8 \\ 15 & 22 \end{bmatrix}$

7. $AB = \begin{bmatrix} -3 & -1 & -2 & 1 \\ 6 & 2 & 4 & -2 \\ 3 & 1 & 2 & -1 \\ 9 & 3 & 6 & -3 \end{bmatrix}$ $BA = [-2]$

9. $\begin{bmatrix} -4 & -2 & -2 & -2 \\ 0 & 1 & 5 & 2 \\ 2 & 6 & 6 & 5 \\ 7 & -1 & 5 & -1 \end{bmatrix}$

11. $\begin{bmatrix} 6 & 5 & 6 \\ 3 & 0 & 5 \end{bmatrix}$

13. The inverse is $\begin{bmatrix} \frac{1}{3} & -\frac{2}{3} \\ \frac{1}{3} & \frac{1}{3} \end{bmatrix}$

15. $A^T = \begin{bmatrix} 1 & 0 \\ 2 & 1 \end{bmatrix}$ $B^T = \begin{bmatrix} 1 & 1 & 3 \\ 2 & 0 & 1 \end{bmatrix}$ $C^T = \begin{bmatrix} 1 & 0 & 1 \\ 2 & 1 & 3 \\ 1 & 1 & 1 \end{bmatrix}$

$D^T = \begin{bmatrix} 1 \\ 2 \\ 1 \\ 3 \end{bmatrix}$ $E^T = [2 \quad 1 \quad -1 \quad 4]$ $F^T = \begin{bmatrix} 1 & 6 \\ 2 & 7 \\ 3 & 8 \\ 4 & 9 \\ 5 & 10 \end{bmatrix}$

21. Since $\alpha \neq 0$, $1/\alpha$ is defined. Let $B = (1/\alpha)A^{-1}$. Then

$$(\alpha A)B = (\alpha A)(1/\alpha)A^{-1} = (\alpha)(1/\alpha)AA^{-1} = 1AA^{-1} = AA^{-1} = I$$

23. $(BB^T)^T = (B^T)^T B^T = BB^T$

EXERCISES 4.2

1. The elementary matrices are a, b, and c.

3. $E = \begin{bmatrix} 0 & 0 & 1 \\ 0 & 1 & 0 \\ 1 & 0 & 0 \end{bmatrix}$

5. $E = \begin{bmatrix} 1 & 0 & 0 & 2 \\ 0 & 1 & 0 & 0 \\ 0 & 0 & 1 & 0 \\ 0 & 0 & 0 & 1 \end{bmatrix}$

7. You should premultiply by the matrices

(a) $\begin{bmatrix} 1 & 0 & 0 \\ 0 & 3 & 0 \\ 0 & 0 & 1 \end{bmatrix}$ (b) $\begin{bmatrix} 1 & 0 & 0 \\ 2 & 1 & 0 \\ 0 & 0 & 1 \end{bmatrix}$ (c) $\begin{bmatrix} 1 & -1 & 0 \\ 0 & 1 & 0 \\ 0 & 0 & 1 \end{bmatrix}$

9. The inverse is   **11.** The inverse is   **13.** The inverse is

$\begin{bmatrix} \frac{1}{3} & -\frac{1}{3} \\ \frac{1}{3} & \frac{2}{3} \end{bmatrix}$ $\begin{bmatrix} \frac{1}{2} & -\frac{1}{2} \\ \frac{1}{2} & \frac{1}{2} \end{bmatrix}$ $\begin{bmatrix} 1 & 0 & 0 \\ -\frac{1}{2} & \frac{1}{2} & 0 \\ 0 & -\frac{1}{3} & \frac{1}{3} \end{bmatrix}$

15. (a) $\begin{bmatrix} 1 & -2 \\ 0 & 1 \end{bmatrix}$ (b) $\begin{bmatrix} 1 & 0 & -\frac{2}{3} \\ 0 & 1 & -\frac{4}{3} \\ 0 & 0 & \frac{1}{3} \end{bmatrix}$

17. (a) $\begin{bmatrix} 2 & 0 & 0 \\ -8 & 1 & 0 \\ -3 & 0 & 1 \end{bmatrix}$ (b) $\begin{bmatrix} 1 & 0 & \frac{3}{2} \\ 0 & 1 & -1 \\ 0 & 0 & \frac{1}{2} \end{bmatrix}$

19. Let $P_1, P_2, \ldots, P_r$ be the pivoting matrices corresponding to the sequence of pivots performed on matrix A. By Theorem 1, we can write the matrix resulting from the first pivot as $P_1 A$. Again by Theorem 1, we can perform the second pivot by premultiplying $P_1 A$ by P_2 to obtain $P_2 P_1 A$. Continuing in this manner, we can write the matrix resulting from r pivots as $P_r P_{r-1} \cdots P_2 P_1 A$. The matrix D of the corollary is $D = P_r P_{r-1} \cdots P_2 P_1$. By the corollary to Theorem 2 of Section 4.1, the product of invertible matrices is invertible. Each P_i is a product of elementary matrices (which are invertible) and, hence, D is a product of invertible matrices.

21. The matrix D is the product of r pivoting matrices:

$$D = P_r P_{r-1} \cdots P_2 P_1$$

For $1 \le i \le r$, each P_i is obtained from the identity matrix I_m by replacing a column j $(1 \le j \le m - 1)$ by the column L given in Theorem 3. Thus, the mth column of P_i is always a unit column $\mathbf{e}_m$ that has a 1 in the mth row and zeros elsewhere. Theorem 4 follows from the observation that the product of any two matrices whose last column is $\mathbf{e}_m$ also has its last column equal to $\mathbf{e}_m$.

EXERCISES 4.3

1. *Tableau 1*
$h(1) = 3, h(2) = 5; \mathbf{x}_B = [x_3 \quad x_5]^T;$
$\mathbf{c}_B = [0 \quad 0]; z^* = 0; B^{-1} = I_2; B = I_2$

Tableau 2
$h(1) = 3, h(2) = 1; \mathbf{x}_B = [x_3 \quad x_1]^T;$
$\mathbf{c}_B = [0 \quad 1]; z^* = \frac{4}{3};$

$B^{-1} = \begin{bmatrix} 1 & -\frac{1}{3} \\ 0 & \frac{1}{3} \end{bmatrix}; \quad B = \begin{bmatrix} 1 & 1 \\ 0 & 3 \end{bmatrix}$

Tableau 3
$h(1) = 2, h(2) = 1; \mathbf{x}_B = [x_2 \quad x_1]^T; \mathbf{c}_B = [2 \quad 1]; z^* = \frac{28}{5};$

$B^{-1} = \begin{bmatrix} \frac{3}{5} & -\frac{1}{5} \\ \frac{2}{5} & \frac{1}{5} \end{bmatrix}; \quad B = \begin{bmatrix} 1 & 1 \\ -2 & 3 \end{bmatrix}$

3. $h(1) = 3, h(2) = 2, h(3) = 7; \mathbf{x}_B = [x_3 \ x_2 \ x_7]^T; \mathbf{c}_B = [-9 \ -57 \ 0]; z^* = 0;$

$$B^{-1} = \begin{bmatrix} -\frac{3}{2} & \frac{11}{2} & 0 \\ \frac{1}{2} & -\frac{5}{2} & 0 \\ 0 & 0 & 1 \end{bmatrix}; \qquad B = \begin{bmatrix} -\frac{5}{2} & -\frac{11}{2} & 0 \\ -\frac{1}{2} & -\frac{3}{2} & 0 \\ 0 & 0 & 1 \end{bmatrix}$$

5. $h(1) = 4, h(2) = 2, h(3) = 1; \mathbf{x}_B = [x_4 \ x_2 \ x_1]^T; \mathbf{c}_B = [0 \ -1 \ -3]; z^* = -9;$

$$B^{-1} = \begin{bmatrix} 1 & 22 & -22 \\ 0 & -2 & 2 \\ 0 & -2 & 3 \end{bmatrix}; \qquad B = \begin{bmatrix} 1 & 11 & 0 \\ 0 & -\frac{3}{2} & 1 \\ 0 & -1 & 1 \end{bmatrix}$$

7.

	x_1	x_2	x_3	x_4	x_5	x_6	
x_4	$-\frac{1}{12}$	0	$\frac{5}{24}$	1	0	$\frac{19}{24}$	$\frac{55}{12}$
x_2	$\frac{1}{4}$	1	$-\frac{1}{8}$	0	0	$\frac{1}{8}$	$\frac{5}{4}$
x_5	$-\frac{2}{3}$	0	$\frac{1}{6}$	0	1	$\frac{5}{6}$	$\frac{2}{3}$
	$\frac{3}{2}$	0	$\frac{1}{4}$	0	0	$\frac{19}{4}$	$\frac{75}{2}$

Tableau T^*

9.

	x_1	x_2	x_3	x_4	x_5	x_6	
x_4	$\frac{1}{3}$	$\frac{2}{3}$	0	1	0	$\frac{1}{3}$	4
x_5	$\frac{1}{3}$	$\frac{5}{3}$	0	0	1	$-\frac{2}{3}$	2
x_3	$-\frac{2}{3}$	$-\frac{4}{3}$	1	0	0	$\frac{1}{3}$	0
	0	1	0	0	0	1	13

Tableau T^*

11. $B^{-1} = \begin{bmatrix} \frac{1}{4} & \frac{1}{8} \\ -\frac{1}{4} & \frac{3}{8} \end{bmatrix} \quad (x_1 \ x_2 \ x_3) = (1 \ 0 \ 0); z^* = 2$

13. (a) There are n ways to choose $h(1)$. After $h(1)$ is chosen, there are $n - 1$ ways to choose $h(2)$. Continue in this manner until m values have been chosen. The total number of ways of assigning values to h is $n(n - 1)\cdots(n - m + 1)$.

(b) From part (a), there are $n(n - 1)\cdots(n - m + 1) = n!/(n - m)!$ ways of choosing *ordered* m-tuples to assign to h. A subset of m elements can be arranged in $m!$ distinct orderings. Therefore, we must divide the number of ordered m-tuples by $m!$ to obtain the number of subsets of m variables.

(c) Each set of basic variables can be reordered in $m!$ ways.

EXERCISES 4.4

1. $\begin{bmatrix} -\frac{3}{5} & \frac{2}{5} & 0 & 0 \\ -\frac{4}{5} & \frac{1}{5} & 0 & 0 \\ -\frac{2}{5} & \frac{3}{5} & 1 & 0 \\ -2 & 1 & 0 & 1 \end{bmatrix}$

3. $\begin{bmatrix} 0 & \frac{2}{7} & 0 & \frac{1}{7} & 0 & 0 \\ 0 & -\frac{3}{7} & 0 & \frac{2}{7} & 0 & 0 \\ 0 & -\frac{5}{7} & 1 & -\frac{6}{7} & 0 & 0 \\ 1 & -\frac{5}{7} & 0 & \frac{1}{7} & 0 & 0 \\ 0 & 0 & 1 & 0 & 1 & 0 \\ 0 & -\frac{1}{7} & 0 & \frac{10}{7} & 0 & 1 \end{bmatrix}$

5. (a)
$$\begin{bmatrix} 1 & 0 & -\frac{19}{4} & \frac{7}{4} & 0 \\ 0 & 1 & -\frac{3}{4} & -\frac{1}{4} & 0 \\ 0 & 0 & \frac{5}{4} & -\frac{1}{4} & 0 \\ 0 & 0 & -\frac{1}{4} & \frac{1}{4} & 0 \\ 0 & 0 & 1 & 1 & 1 \end{bmatrix}$$
(b)
$$\begin{bmatrix} -\frac{3}{10} & \frac{19}{10} & 0 & 0 & 0 \\ -\frac{1}{10} & \frac{3}{10} & 0 & 1 & 0 \\ \frac{1}{5} & \frac{2}{5} & 1 & 0 & 0 \\ \frac{1}{10} & \frac{7}{10} & 0 & 0 & 0 \\ \frac{2}{5} & \frac{9}{5} & 0 & 0 & 1 \end{bmatrix}$$

7. $R: m \times m$; $\mathbf{0}: m \times 1$; $\mathbf{s}: 1 \times m$; $\mathbf{1}: 1 \times 1$; $B: m \times m$; $Q: m \times (n - m)$; $\mathbf{b}: m \times 1$;
$-\mathbf{c}_B: 1 \times m$; $\mathbf{q}: 1 \times (n - m)$; $z_0: 1 \times 1$;

9. (a) $\mathbf{x}_B = [x_1 \ \ x_5 \ \ x_2]^T$; $\quad \mathbf{c}_B = [0 \ \ 4 \ \ 2]$

(b)
$$B^{-1} = \begin{bmatrix} 1 & \frac{13}{5} & \frac{1}{5} \\ 0 & \frac{3}{5} & \frac{1}{5} \\ 0 & -\frac{2}{5} & \frac{1}{5} \end{bmatrix}; \quad B = \begin{bmatrix} 1 & -3 & 2 \\ 0 & 1 & -1 \\ 0 & 2 & 3 \end{bmatrix}$$

(c)

$$S^* = \begin{bmatrix} 1 & 0 & 0 & \frac{1}{5} & \frac{43}{5} & \frac{13}{5} & \frac{99}{5} & 46 \\ 0 & 1 & 0 & \frac{1}{5} & \frac{13}{5} & \frac{3}{5} & \frac{19}{5} & 12 \\ 0 & 0 & 1 & \frac{1}{5} & -\frac{7}{5} & -\frac{2}{5} & -\frac{11}{5} & 0 \\ 0 & 0 & 0 & \frac{6}{5} & \frac{53}{5} & \frac{8}{5} & \frac{64}{5} & 102 \end{bmatrix}$$

(d) Observe that $DS = D^*$, where

$$D = \left[\begin{array}{ccc|c} 1 & \frac{13}{5} & \frac{1}{5} & 0 \\ 0 & \frac{3}{5} & \frac{1}{5} & 0 \\ 0 & -\frac{2}{5} & \frac{1}{5} & 0 \\ 0 & \frac{8}{5} & \frac{6}{5} & 1 \end{array} \right]$$

11. The new solution is $(x_1 \ \ x_2 \ \ x_3) = (\frac{7}{3} \ \ 0 \ \ \frac{5}{3})$; $z^* = \frac{1325}{3}$

EXERCISES 4.5

Note that transportation problems often have alternative optimal solutions. Therefore, your answers may not agree with those given, except, of course, your optimal objective values should be the same as ours.

1. $x_{11} = 25{,}000$, $x_{12} = 25{,}000$, $x_{21} = 10{,}000$, $x_{23} = 40{,}000$, $x_{34} = 30{,}000$; $z^* = 980{,}000$

3. $x_{11} = 100$; $x_{13} = 150$; $x_{24} = 100$, $x_{32} = 200$, $x_{35} = 100$, $x_{43} = 100$; $z^* = 5800$

5. $x_{11} = 150, x_{12} = 30, x_{14} = 120, x_{22} = 170, x_{33} = 600, x_{34} = 220, x_{44} = 60;$
   $z^* = 59,100$

7. $x_{11} = 95, x_{12} = 25, x_{22} = 50, x_{23} = 45, x_{33} = 63, x_{34} = 12, x_{44} = 78, x_{45} = 30,$
   $x_{55} = 90, s_1 = 380; z^* = 6883$

9. $x_{13} = 1, x_{22} = 1, x_{34} = 1, x_{41} = 1; z^* = 33$

11. $x_{13} = 1, x_{22} = 1, x_{34} = 1, x_{41} = 1; z^* = 24$

13. $x_{16} = 1, x_{21} = 1, x_{33} = 1, x_{42} = 1, x_{54} = 1, x_{65} = 1; z^* = 10$

15. The answer is given in Example 3.

17. $x_{16} = 100, x_{25} = 75, x_{26} = 10, x_{32} = 5, x_{37} = 85; z^* = 6560$

19. $x_{15} = 1, x_{26} = 1, x_{57} = 1; z^* = 1525$

EXERCISES 5.1

1. Tableau 1: Since this is the initial tableau, there is no associated pivoting matrix.
   $h(1) = 4, h(2) = 5, h(3) = 6; \mathbf{c}_B = [0 \quad 0 \quad 0]; \mathbf{b}^* = [20 \quad 9 \quad 30]^T$

   Tableau 2:

   $$P = \begin{bmatrix} 1 & -\frac{5}{3} & 0 \\ 0 & \frac{1}{3} & 0 \\ 0 & -\frac{2}{3} & 1 \end{bmatrix} \quad \begin{matrix} h(1) = 4, h(2) = 2, h(3) = 6 \\ \mathbf{c}_B = [0 \quad 8 \quad 0]; \mathbf{b}^* = [5 \quad 3 \quad 24]^T \end{matrix}$$

   Tableau 3:

   $$P = \begin{bmatrix} \frac{3}{4} & 0 & 0 \\ -\frac{1}{4} & 1 & 0 \\ -4 & 6 & 1 \end{bmatrix} \quad \begin{matrix} h(1) = 1, h(2) = 2, h(3) = 6 \\ \mathbf{c}_B = [6 \quad 8 \quad 0]; \mathbf{b}^* = [\frac{15}{4} \quad \frac{7}{4} \quad 4]^T \end{matrix}$$

   Tableau 4:

   $$P = \begin{bmatrix} 1 & 0 & \frac{5}{24} \\ 0 & 1 & -\frac{1}{8} \\ 0 & 0 & \frac{1}{6} \end{bmatrix} \quad \begin{matrix} h(1) = 1, h(2) = 2, h(3) = 5 \\ \mathbf{c}_B = [6 \quad 8 \quad 0]; \mathbf{b}^* = [\frac{55}{12} \quad \frac{5}{4} \quad \frac{2}{3}]^T \end{matrix}$$

3. **(a)** Column 3
   (b) Row 3

   (c) $P_3 = \begin{bmatrix} 1 & 0 & 0 \\ 0 & 1 & -6 \\ 0 & 0 & 3 \end{bmatrix}$

   (d) $\mathbf{b}^* = \begin{bmatrix} 2 \\ 9 \\ 3 \end{bmatrix}$

   (e) $\mathbf{w} = [\frac{1}{3} \quad 0 \quad \frac{7}{3}]$
   (f) $z^* = 12$

5. **(a)** Column 3
   (b) Row 1

   (c) $P_3 = \begin{bmatrix} \frac{6}{5} & 0 & 0 & 0 \\ -3 & 1 & 0 & 0 \\ -\frac{9}{5} & 0 & 1 & 0 \\ \frac{54}{5} & 0 & 0 & 1 \end{bmatrix}$

   (d) $\mathbf{b}^* = \begin{bmatrix} \frac{6}{5} \\ 9 \\ \frac{16}{5} \\ \frac{84}{5} \end{bmatrix}$

   (e) $\mathbf{w} = [36 \quad 0 \quad 0 \quad 0]$
   (f) $z^* = 216$

7. Yes, provided the initial tableau is in canonical form. Then the objective row entry in the column corresponding to the jth column of the B^{-1} matrix is $\mathbf{c}_B B^{-1} \mathbf{e}_j - 0$, and $\mathbf{c}_B B^{-1} \mathbf{e}_j$ is the jth term in $\mathbf{c}_B B^{-1}$.

13. Theorem 3 of Section 4.2 implies that a pivot can be performed on any column of a matrix S by premultiplying S by the corresponding pivoting matrix P. If we take S to be the A matrix with the column $\mathbf{b}^*$ adjoined as the rightmost column of S, then premultiplication of $\mathbf{b}^*$ by P performs the pivot on $\mathbf{b}^*$.

EXERCISES 5.2

1. (a) $4 \cdot 5 \cdot 7 = 140$ **(b)** $3 \cdot 1 \cdot 4 = 12$

3. If $R = UV$, then $(UV)W = RW$. R is of size $r \times t$, and we can calculate R in rst multiplications. We can calculate RW in rtq multiplications. Thus, altogether, we can calculate (UV) in $rst + rtq$ multiplications. Calculating $U(VW)$ requires $rsq + stq$ multiplications.

5. We use Theorem 2 to write the product as

$$\begin{bmatrix} 0 + 2 \cdot 2 \\ -1 \cdot 2 \\ 1 + 3 \cdot 2 \\ 4 + 2 \cdot 2 \\ 1 - 2 \cdot 2 \\ 3 + 1 \cdot 2 \end{bmatrix} = \begin{bmatrix} 4 \\ -2 \\ 7 \\ 8 \\ -3 \\ 5 \end{bmatrix}$$

7. We use Theorem 3 to calculate the product as

$$[0 \quad (0 \cdot 2 + 2 \cdot (-1) + 1 \cdot 3 + 4 \cdot 2 + 1 \cdot (-2) + 3 \cdot 1) \quad 1 \quad 4 \quad 1 \quad 3]$$
$$= [0 \quad 10 \quad 1 \quad 4 \quad 1 \quad 3]$$

9. We use the corollary to Theorem 3 to calculate the product as

$$\begin{bmatrix} 1 \cdot 2 + 2 \cdot 3 + 3 \cdot 1 + 1 \cdot 4 & 2 & 3 & 1 \\ 4 \cdot 2 + 5 \cdot 3 + 6 \cdot 1 + 2 \cdot 4 & 5 & 6 & 2 \\ 3 \cdot 2 + 2 \cdot 3 + 1 \cdot 1 + 1 \cdot 4 & 2 & 1 & 1 \\ 2 \cdot 2 + 0 \cdot 3 + 1 \cdot 1 + 2 \cdot 4 & 0 & 1 & 2 \end{bmatrix} = \begin{bmatrix} 15 & 2 & 3 & 1 \\ 37 & 5 & 6 & 2 \\ 17 & 2 & 1 & 1 \\ 13 & 0 & 1 & 2 \end{bmatrix}$$

11. (a)

Column	1	2	3	4	5	6
	↓	↓	↓	↓	↓	↓

entry → $\begin{bmatrix} -1 \\ 4 \end{bmatrix}$ $\begin{bmatrix} 2 \\ 2 \end{bmatrix}$ $\begin{bmatrix} 1 \\ 2 \end{bmatrix}$ $\begin{bmatrix} 9 \\ 1 \end{bmatrix}$ $\begin{bmatrix} -2 \\ 1 \end{bmatrix}$ $\begin{bmatrix} 1 \\ 2 \end{bmatrix}$
row →
pointer →

entry → $\begin{bmatrix} 1 \\ 4 \end{bmatrix}$ $\begin{bmatrix} 1 \\ 3 \end{bmatrix}$ $\begin{bmatrix} 1 \\ 2 \end{bmatrix}$
row →
pointer →

entry → $\begin{bmatrix} 1 \\ 4 \end{bmatrix}$
row →
pointer →

(b)

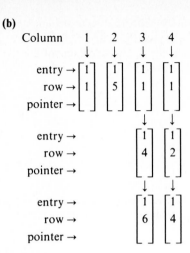

Column 1 2 3 4

13. Tableau 3

Column 1 2 3

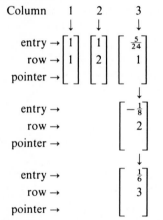

15. (a) A has density $\frac{9}{12} = \frac{3}{4}$, and B has density $\frac{11}{12}$. **(b)** $\frac{2}{3}$

17. The A matrix has n^2 entries. Each variable occurs with coefficient 1 in two rows of each column for a total of $2n$ nonzero entries. Therefore, the density is $2n/n^2 = 2/n$.

EXERCISES 5.3

1. 5050

3.

	Ordinary simplex method	Revised simplex method
Space	4131	807.15
Time	104,525	8250

5.

	Ordinary simplex method	Revised simplex method
Space	501,501	55,150.1
Time	50,200,100	1,770,000

EXERCISES 5.4

1. (a) The objective function is the sum of the net profits for each division. All of the constraints are given except for the one for total investment, which is

$$\sum_{i=1}^{12} x_i \leq 400$$

(b) There are 12 decision variables and 8 slack variables for a total of 20 variables. The size of the A matrix is $8 \cdot 12 = 96$. There are 39 nonzero coefficients. Therefore, the density is $\frac{39}{96} = 0.406$.

(c) $x_1 = 5$; $x_6 = 2.778$; $x_7 = 2.593$; $x_{10} = 11.667$; $x_{11} = 25.000$; $z = 250.741$

3. (a) Minimize L subject to the constraints given in Example 2.

(b) The minimum value of L is \$219,909.

5. (a) The LP problem is given with the answers for Section 1.2.
The solution is $x_{11} = 100$, $x_{13} = 150$, $x_{24} = 100$, $x_{32} = 200$, $x_{35} = 100$, $x_{43} = 100$, $x_{44} = 50$, and the rest of the variables are 0. The minimum transportation cost is \$5800.

(b) The LP problem is given with the answers for Section 1.2. The solution is given with the answers for Section 3.5, Exercise 5. The problem can be entered with the inequality editor as follows:

```
assign ( &c_ij, i = 1 to 5, j = 1 to 5,
         35, 35, 35, 35, 35,
         99,  8,  5,  3,  3,
         99, 99,  8,  5,  3,
         99, 99, 99,  8,  5,
         99, 99, 99, 99,  8);
assign ( &a_i, i = 1 to 5, 500, 95, 75, 108, 90);
assign ( &b_j, j = 1 to 5, 95, 75, 108, 90, 120);
for i = 1 to 5 do add ( j = 1 to 5, − &c_ij * x_ij);
for i = 1 to 5 do add ( j = 1 to 5, x_ij) <= &a_i;
for j = 1 to 5 do add ( i = 1 to 5, x_ij) = &b_j;
```

7. The LP problems are given with the answers for Section 1.4. The solutions are given with the answers for Section 3.7, Exercises 11 and 13.

EXERCISES 6.1

1. Suppose the plant manufactures x_1, x_2, x_3, x_4 units of products F_1, F_2, F_3, F_4, respectively. The answer is

$$(x_1, x_2, x_3, x_4, s_1, s_2, s_3) = \left(0, \frac{28}{3}, 0, \frac{136}{3}, \frac{10}{3}, 0, 0\right)$$

The maximum profit is \$733.33.

3. The value realized from the sale of B_3 is

$$3u_1^* + 3u_2^* + 4u_3^* = \frac{1}{2} + \frac{23}{4} = \frac{25}{4}$$

Since the profit of \$4 per unit of B_3 is less than the value of B_3, ACI would not make blend B_3. The value of B_4 is

$$7u_1^* + 2u_2^* + u_3^* = 7\left(\frac{1}{6}\right) + 2\left(\frac{23}{12}\right) = 5$$

Since the per-unit profit for blend B_4 is also \$5, it is not to ACI's advantage to produce blend B_4.

5. u_1^* represents the value of a unit increase in parts manufacturing time. Since more time is already available than could be profitably used, there is no advantage in increasing the available time.

7. If we let p_1, p_2, p_3 be the hours that we run process P_1, P_2, P_3, respectively, then the primal problem is

Minimize: | Subject to:
$z = 220p_1 + 480p_2 + 375p_3$ | $4p_1 + 7p_2 + 7p_3 \geq 4000$
| $5p_1 + 7p_2 + 4p_3 \geq 2700$
| $3p_1 + 8p_2 + 5p_3 \geq 3000$
| $p_i \geq 0, \quad 1 \leq i \leq 3$

If we let a, b, c, be the cost of producing a ton of rebar of types A, B, C, respectively, the dual problem is

Maximize: | Subject to:
$y = 4000a + 2700b + 3000c$ | $4a + 5b + 3c \leq 220$
| $7a + 7b + 8c \leq 480$
| $7a + 4b + 5c \leq 375$
| $a, b, c \geq 0$

9. The solution to the primal problem is

$$p_1 = 105.5556, \qquad p_2 = 42.5926, \qquad p_3 = 468.5185; \qquad z = 219361.111$$

The solution to the dual problem is

$$a = 28.61111, \qquad b = 0.2778, \qquad c = 34.72222; \qquad y = 219361.1111$$

11. If u_1 is the value per hour of the peeling machine, and u_2 is the value per hour of the slicing machine, then the dual problem is

$$\text{Minimize:} \qquad \text{Subject to:}$$

$$y = 8u_1 + 16u_2 \qquad u_1 + \frac{5}{2}u_2 \geq 60$$

$$u_1 + \frac{3}{2}u_2 \geq 45$$

$$u_1, u_2 \geq 0$$

EXERCISES 6.2

1. Minimize:

$y = 12u_1 + 6u_2 + 18u_3$

Subject to:

$4u_1 + 2u_2 + 5u_3 \geq 24$

$3u_1 - u_2 + 2u_3 \geq 32$

$u_i \geq 0, \qquad 1 \leq i \leq 3$

3. Minimize:

$y = 12u_1 + 21u_2 + 5u_3$

Subject to:

$3u_1 + 2u_2 \qquad\quad \geq -27$

$2u_1 - 3u_2 + u_3 \geq \quad 24$

$4u_1 + u_2 + 5u_3 \geq \quad 15$

$u_i \geq 0, \qquad 1 \leq i \leq 3$

5. The dual problem is

$$\text{Minimize:} \qquad \text{Subject to:}$$

$$y = 4u_1 + 3u_2 \qquad u_1 + u_2 \geq \quad 2$$

$$u_1 - 2u_2 \geq -3$$

$$u_1, u_2 \geq 0$$

If we replace the second constraint by the equivalent inequality $-u_1 + 2u_2 \leq 3$, we can use the two-phase method to obtain the solution $u_1 = \frac{1}{3}, u_2 = \frac{5}{3}; y = \frac{19}{3}$.

7.

Corner point for (7)	Value of z	Corner point for (8)	Value of z
$(0,0)$	0	$(120,0)$	480
$(3,0)$	360	$(70,10)$	430
$(0,4)$	400	$(0,\frac{100}{3})$	500
$(\frac{3}{2},\frac{5}{2})$	430		

Observe that each value of z for LP problem (7) is less than or equal to every value of y for LP problem (8).

9. One of many possible examples is the pair of dual problems

Maximize:

$z = 2x_1 - x_2$

Subject to:

$-x_1 + x_2 \leq -2$

$x_1 - x_2 \leq \quad 1$

$x_1, x_2 \geq 0$

Minimize:

$y = -2u_1 + u_2$

Subject to:

$-u_1 + u_2 \geq \quad 2$

$u_1 - u_2 \geq -1$

$u_1, u_2 \geq 0$

11. $z^* = 0.15b_1 + 1.22b_2 + 0b_3 + 0b_4 + 0b_5 + 1.07b_6$
$= 0.15 \cdot 0 + 1.22 \cdot 0 + 1.07 \cdot 200 = 214$
(The above calculation does not agree exactly with the answer given in Tableau 4, because numbers in that tableau were rounded to 2 decimal places.)
(a) The new value of z is $1.07 \cdot 210 = 224.7$, a change of 10.7.
(b) They are all 0.

13. The dual prices are $(u_1, u_2, u_3) = (0, 0, 2)$.
(a) The new value is $2 \cdot 50 + 100 = 200$, a decrease of 10.
(b) The reduced costs of x_1, x_2, x_3 are $0, 0, -3$, respectively. If we set all nonbasic variables except x_3 equal to 0, and if we set $x_3 = 20$, then

$$x_1 = (\tfrac{7}{3})20 + \tfrac{5}{4}$$
$$x_5 = (\tfrac{5}{2})20 + \tfrac{45}{2}$$
$$x_7 = (\tfrac{5}{2})20 + \tfrac{15}{2}$$

is a feasible solution with $z = 160$.

EXERCISES 6.3

1. Let $\mathbf{c} = [80 \quad 100]$

$A_1 = \begin{bmatrix} 2 & -3 \\ 3 & 2 \end{bmatrix} \quad \mathbf{b}_1 = \begin{bmatrix} 4 \\ 3 \end{bmatrix}$

$A_2 = \begin{bmatrix} 4 & 5 \\ 2 & 8 \end{bmatrix} \quad \mathbf{b}_2 = \begin{bmatrix} 12 \\ 8 \end{bmatrix}$

$\mathbf{x} = \begin{bmatrix} x_1 \\ x_2 \end{bmatrix}$

(b) Let $\mathbf{u} = \begin{bmatrix} u_1 \\ u_2 \end{bmatrix} \quad \mathbf{w} = \begin{bmatrix} w_1 \\ w_2 \end{bmatrix}$.
The dual problem is

Minimize:
$y = \mathbf{b}_1^T \mathbf{u} + \mathbf{b}_2^T \mathbf{w}$

Subject to:
$A_1^T \mathbf{u} + A_2^T \mathbf{w} \geq \mathbf{c}^T$
$\mathbf{u} \geq 0, \mathbf{w} \leq 0$

(d) Let $u_3 = -w_1$ and $u_4 = -w_2$.

Maximize:
$y = -4u_1 - 3u_2 + 12u_3 + 8u_4$

(a) The LP problem is

Maximize: Subject to:
$z = \mathbf{cx}$ $A_1\mathbf{x} \leq \mathbf{b}_1$
                $A_2\mathbf{x} \geq \mathbf{b}_2$
                $\mathbf{x} \geq 0$

(c) Minimize:
$y = 4u_1 + 3u_2 + 12w_1 + 8w_2$
Subject to:
$2u_1 + 3u_2 + 4w_1 + 2w_2 \geq 80$
$-3u_1 + 2u_2 + 5w_1 + 8w_2 \geq 100$
$u_1, u_2 \geq 0; \quad w_1, w_2 \leq 0$

Subject to:
$2u_1 + 3u_2 - 4u_3 - 2u_4 \geq 80$
$-3u_1 + 2u_2 - 5u_3 - 8u_4 \geq 100$
$u_i \geq 0, \quad 1 \leq i \leq 4$

3. (a) The LP problem is

Maximize:

$z = \mathbf{cx}$

Subject to:

$A_1\mathbf{x} \leq \mathbf{b}_1$

$A_2\mathbf{x} = \mathbf{b}_2$

$A_3\mathbf{x} \geq \mathbf{b}_3$

$\mathbf{x} \geq 0$

(b) Let $\mathbf{u} = [u_1]$, $\mathbf{v} = [v_1]$, $\mathbf{w} = \begin{bmatrix} w_1 \\ w_2 \end{bmatrix}$.

The dual is

Minimize:

$y = \mathbf{b}_1^T\mathbf{u} + \mathbf{b}_2^T\mathbf{v} + \mathbf{b}_3^T\mathbf{w}$

Subject to:

$A_1^T\mathbf{u} + A_2^T\mathbf{v} + A_3^T\mathbf{w} \geq \mathbf{c}^T$

$\mathbf{u} \geq 0;$ $\mathbf{v}$ unrestricted; $\mathbf{w} \leq 0$

(c) Minimize:

$y = 12u_1 + 3v_1 + 8w_1 + 2w_2$

Subject to:

$3u_1 + v_1 + 4w_1 + 3w_2 \geq 2$

$2u_1 + v_1 + 3w_1 + 2w_2 \geq 3$

$-u_1 + v_1 + w_1 - w_2 \geq -4$

$u_1 \geq 0;$ v_1 unrestricted; $w_1, w_2 \leq 0$

(d) Let $u_2 - u_3 = v_1, u_4 = -w_1, u_5 = -w_2$.

Maximize:

$y = -12u_1 - 3u_2 + 3u_3 + 8u_4 + 2u_5$

Subject to:

$3u_1 + u_2 - u_3 - 4u_4 - 3u_5 \geq 2$

$2u_1 + u_2 - u_3 - 3u_4 - 2u_5 \geq 3$

$u_1 - u_2 + u_3 + u_4 - u_5 \leq 4$

$u_i \geq 0,$ $1 \leq i \leq 5$

5. (a) Maximize:

$z = \mathbf{cx}$

Subject to:

$A_1\mathbf{x} \leq \mathbf{b}_1$

$A_2\mathbf{x} = \mathbf{b}_2$

$A_3\mathbf{x} \geq \mathbf{b}_3$

$\mathbf{x} \geq 0$

(b) Let $\mathbf{u} = \begin{bmatrix} u_1 \\ u_2 \end{bmatrix}$ $\mathbf{v} = \begin{bmatrix} v_1 \\ v_2 \end{bmatrix}$ $\mathbf{w} = \begin{bmatrix} w_1 \\ w_2 \end{bmatrix}$.

Minimize:

$y = \mathbf{b}_1^T\mathbf{u} + \mathbf{b}_2^T\mathbf{v} + \mathbf{b}_3^T\mathbf{w}$

Subject to:

$A_1^T\mathbf{u} + A_2^T\mathbf{v} + A_3^T\mathbf{w} \geq \mathbf{c}^T$

$\mathbf{u} \geq 0;$ $\mathbf{v}$ unrestricted; $\mathbf{w} \leq 0$

(c) Minimize:

$y = 8u_1 + 12u_2 + 9v_1 + 6v_2 + 8w_1 + 15w_2$

Subject to:

$u_1 + 2u_2 + v_1 + 2v_2 + w_1 + 2w_2 \geq 2$

$3u_1 + 3u_2 + v_1 - 3v_2 + 3w_1 + 9w_2 \geq 3$

$-2u_1 - 4u_2 + 3v_1 + 2v_2 + 5w_1 + 4w_2 \geq 5$

$4u_1 + u_2 + v_1 - v_2 + 7w_1 + 8w_2 \geq 7$

$u_1, u_2 \geq 0;$ v_1, v_2 unrestricted; $w_1, w_2 \leq 0$

(d) Let $v_1 = r_1 - t_1; v_2 = r_2 - t_2; p_1 = -w_1; p_2 = -w_2$.

Maximize:

$y = -8u_1 - 12u_2 - 9r_1 + 9t_1 - 6r_2 + 6t_2 + 8p_1 + 15p_2$

Subject to:

$$u_1 + 2u_2 + r_1 - t_1 + 2r_2 - 2t_2 - p_1 - 2p_2 \geq 2$$
$$3u_1 + 3u_2 + r_1 - t_1 - 3r_2 + 3t_2 - 3p_1 - 9p_2 \geq 3$$
$$-2u_1 - 4u_2 + 3r_1 - 3t_2 + 2r_2 - 2t_2 - 5p_1 - 4p_2 \geq 5$$
$$4u_1 + u_2 + r_1 - t_2 - r_2 + t_2 - 7p_1 - 8p_2 \geq 17$$
$$u_1, u_2, r_1, r_2, t_1, t_2, p_1, p_2 \geq 0$$

7. **(a)** Maximize: **(b)** Let $\mathbf{u} = \begin{bmatrix} u_1 \\ u_2 \end{bmatrix}$ $\mathbf{w} = [w_1]$.

$z = \mathbf{cx}$

Subject to: Minimize:
$A_1\mathbf{x} \geq \mathbf{b}_1$ $y = \mathbf{b}_1^{\mathsf{T}}\mathbf{u} + \mathbf{b}_2^{\mathsf{T}}\mathbf{w}$
$A_2\mathbf{x} \leq \mathbf{b}_2$
$\mathbf{x} \geq \mathbf{0}$ Subject to:
                        $A_1^{\mathsf{T}}\mathbf{u} + A_2^{\mathsf{T}}\mathbf{w} \geq \mathbf{c}^{\mathsf{T}}$
                        $\mathbf{u} \leq \mathbf{0}, \qquad \mathbf{w} \geq \mathbf{0}$

(c) Minimize: **(d)** Let $p_1 = -u_1$ and $p_2 = -u_2$.

$y = 4u_1 + 5u_2 + 3w_1$ Maximize:

Subject to: $y = -4p_1 - 5p_2 + 3w_1$
$3u_1 + 2u_2 + w_1 \geq -6$
$-2u_1 + 3u_2 + w_1 \geq 3$ Subject to:
$u_1, u_2 \leq 0; \qquad w_1 \geq 0$ $3p_1 + 2p_2 - w_1 \leq 6$
                            $2p_1 - 3p_2 + w_1 \geq 3$
                            $p_1, p_2, w_1 \geq 0$

9. The answers to (a), (b), and (c) are straightforward. In part (d) you should change the objective function to

$$\text{Maximize:} \quad z = -6x_1 + 3x_2$$

before using Table 1.

11. We first write the LP problem in the standard form of a primal problem by letting $x_1 = t_1 - t_2$ and $t_4 = -x_4$ to obtain

Maximize: The dual problem is

$z = 3t_1 - 3t_2 - 4x_2 + 9x_3 - t_4$ Minimize:

Subject to: $z = 0u_1 - 10u_2$
$-t_1 + t_2 + 5x_2 - x_3 \qquad \leq \quad 0$ Subject to:
$-3t_1 + 3t_2 - 8x_2 \qquad - t_4 \leq -10$
$t_1, t_2, x_2, x_3, t_4 \geq 0$ $-u_1 - 3u_2 \geq \quad 3$
                                $u_1 + 3u_2 \geq -3$
                                $5u_1 - 8u_2 \geq -4$
                                $-u_1 \qquad \geq \quad 9$
                                $-u_2 \geq -1$
                                $u_1, u_2 \geq 0$

We can combine the first two inequalities of the dual to obtain an equality and rearrange the other constraints to obtain

Maximize: Subject to:

$z = 10u_2$ $u_1 + 3u_2 = -3$

 $-5u_1 + 8u_2 \leq 4$

 $u_1 \leq -9$

 $ u_2 \leq 1$

 $u_1, u_2 \geq 0$

It would be a good exercise for you to form the dual of this dual to verify that you obtain the primal given in this problem.

13. After restating LP problem (8) of Section 6.1 as a maximization problem, we obtain the dual.

Minimize: Subject to:

$y = 8w_1 + 6w_2 + 4w_3 + 5w_4$ $2w_1 + 4w_2 + 3w_3 + 7w_4 \geq -8000$

 $4w_1 + 5w_2 + 3w_3 + 2w_4 \geq -6400$

 $4w_1 + w_2 + 4w_3 + w_4 \geq -6000$

 $w_i \leq 0, \qquad 1 \leq i \leq 4$

If we let $x_i = -w_i$, $1 \leq i \leq 4$, change to a maximization problem, and change these $\geq$ inequalities to $\leq$ inequalities by multiplying by -1, we obtain the primal problem (1) of Section 6.1.

15. The answers are given in Tableaus (2) and (9) of Section 6.1.

17. One possibility is $\mathbf{v}^+ = \begin{bmatrix} 7 & 12 & 8 \end{bmatrix}^T$ and $\mathbf{v}^- = \begin{bmatrix} 4 & 16 & 14 \end{bmatrix}^T$.

19. (a) The solution is $x_1 = 1$, $x_3 = 6$. The dual prices are $u_1 = \frac{9}{4}$ and $u_2 = \frac{3}{4}$ and are found in columns 4 and 5, respectively.

   (b) $z^* = 0 + 18 = 18$

   The dual objective value is $(\frac{9}{4})7 + (\frac{3}{4})3 = 18$.

21. (a) The solution is $x_1 = 19$, $x_2 = 8$. The dual prices are $u_1 = -1$ and $u_2 = -1$ and are found in columns 2 and 5, respectively.

   (b) We first write the primal problem as

   Maximize: $z = -2x_1 + 3x_2 - x_3 - x_4$

   Then $z^* = -2 \cdot 19 + 3 \cdot 8 = -14$ and the dual objective value is $(-1)3 + (-1)11 = -14$.

23. The primal problems have the same solution because the constraints $2x_1 + x_2 \leq 2$ and $-2x_1 - x_2 \geq -2$ have the same solution set. The dual problems are

Minimize: Minimize:

$y = 3u_1 + 2u_2$ $y = 3u_1 - 2u_2$

Subject to: Subject to:

$u_1 + 2u_2 \geq 2$ $u_1 - 2u_2 \geq 2$

$2u_1 + u_2 \geq 2$ $2u_1 - u_2 \geq 2$

$u_1, u_2 \geq 0$ $u_1 \geq 0, \qquad u_2 \leq 0$

You can easily solve these problems geometrically. The solution to the first problem is $(u_1, u_2) = (\frac{2}{3}, \frac{2}{3})$, and the solution to the second problem is $(u_1, u_2) = (\frac{2}{3}, -\frac{2}{3})$.

EXERCISES 6.4

3.

Row	Dual price	Slack
1	$v_1 = \frac{1}{2}$ (column 6)	No slack variable
2	$u_1 = 1$ (column 4)	$s_1 = 0$
3	$w_1 = 0$ (column 7)	$s_2 = \frac{19}{3}$

Observe that the product of each dual price and its corresponding slack variable is 0.

5. The dual prices are found in the artificial variable columns. The solution to the dual problem is $(\frac{7}{11}, \frac{6}{11}, \frac{16}{11})$.

9. After multiplying by $\mathbf{x}^{*T}$ as suggested, we have

$$\mathbf{x}^{*T}A_1^T\mathbf{u}^* + \mathbf{x}^{*T}A_2^T\mathbf{v}^* + \mathbf{x}^{*T}A_3^T\mathbf{w}^* - \mathbf{x}^{*T}\mathbf{t}^* = \mathbf{x}^{*T}\mathbf{c}^T$$

We take transposes to obtain

$$(\mathbf{u}^{*T}A_1 + \mathbf{v}^{*T}A_2 + \mathbf{w}^{*T}A_3)\mathbf{x}^* - \mathbf{t}^{*T}\mathbf{x}^* = \mathbf{c}\mathbf{x}^* \tag{1}$$

We use the constraints of the primal and the fact that $\mathbf{u}^* \geq 0$, $\mathbf{w}^* \leq 0$ to obtain

$$A_1\mathbf{x}^* \leq \mathbf{b}_1, \quad \text{which implies} \quad \mathbf{u}^{*T}A_1\mathbf{x}^* \leq \mathbf{u}^{*T}\mathbf{b}_1 \tag{2}$$

$$A_2\mathbf{x}^* = \mathbf{b}_2, \quad \text{which implies} \quad \mathbf{v}^{*T}A_2\mathbf{x}^* = \mathbf{v}^{*T}\mathbf{b}_2 \tag{3}$$

$$A_3\mathbf{x}^* \geq \mathbf{b}_3, \quad \text{which implies} \quad \mathbf{w}^{*T}A_3\mathbf{x}^* \leq \mathbf{w}^{*T}\mathbf{b}_3 \tag{4}$$

We add the expressions on the right of (2), (3), and (4) and use (1) to obtain

$$(\mathbf{u}^{*T}\mathbf{b}_1 + \mathbf{v}^{*T}\mathbf{b}_2 + \mathbf{w}^{*T}\mathbf{b}_3)\mathbf{x}^* - \mathbf{t}^{*T}\mathbf{x}^* \geq \mathbf{c}\mathbf{x}^* \tag{5}$$

By the Fundamental Principle of Duality [Theorem 1, Part (d)],

$$\mathbf{u}^{*T}\mathbf{b}_1 + \mathbf{v}^{*T}\mathbf{b}_2 + \mathbf{w}^{*T}\mathbf{b}_3 = \mathbf{c}\mathbf{x}^*$$

so from (5) we have

$$-\mathbf{t}^{*T}\mathbf{x}^* \geq 0 \quad \text{or} \quad \mathbf{t}^{*T}\mathbf{x}^* \leq 0$$

Since the entries in $\mathbf{t}^*$ and $\mathbf{x}^*$ are nonnegative, we have

$$\mathbf{t}^{*T}\mathbf{x}^* = 0 \qquad\qquad\blacksquare$$

11. (a) Since $\mathbf{q}^*$ is a feasible solution of (27), we have

$$A\mathbf{x}^* + I_m\mathbf{s}^* = \mathbf{b}^*; \qquad \mathbf{x}^* \geq 0, \qquad \mathbf{s}^* \geq 0$$

Therefore, $A\mathbf{x}^* \leq \mathbf{b}^*$.

(b) The objective-row entries in the columns of the final tableau corresponding to the slack variable columns are

$$\mathbf{u}^* = \mathbf{c}_B B^{-1} I_m - 0 = \mathbf{c}_B B^{-1}$$

All of the objective-row entries in the final tableau are nonnegative. Therefore, the objective-row entries corresponding to the decision variables satisfy

$$c_B B^{-1} A - c = u^* A - c \geq 0$$

Thus,

$$u^* A \geq c$$

Taking the transpose, we have $A^T u^{*T} \geq c^T$.

(c) and (d) By the weak duality principle for "twins" (Exercise 10), it is sufficient to show that $cx^* = b^T u^{*T}$. We have

$$cx^* = c_B B^{-1} b = u^* b = b^T u^{*T}$$

13. Here is one of many possible examples.

Maximize:

$$z = x_1 - \frac{1}{2}x_2$$

Subject to:

$$x_1 - x_2 \leq 1$$
$$x_1 - x_2 \geq 2$$
$$x_1, x_2 \geq 0$$

Minimize:

$$y = u_1 + 2u_2$$

Subject to:

$$u_1 + u_2 \geq 1$$
$$-u_1 - u_2 \geq -\frac{1}{2}$$
$$u_1 \geq 0, \qquad u_2 \leq 0$$

EXERCISES 6.5

1.

	x_1	x_2	x_3	x_4	x_5	x_6	
x_4	$-\frac{3}{2}$	0	0	1	$-\frac{7}{2}$	15	$\frac{7}{2}$
x_2	$\frac{1}{2}$	1	0	0	$\frac{5}{2}$	-6	$\frac{3}{2}$
x_3	$-\frac{1}{2}$	0	1	0	$-\frac{1}{2}$	3	$\frac{3}{2}$
	$\frac{1}{2}$	0	0	0	$\frac{9}{2}$	3	$\frac{39}{2}$

3.

	x_1	x_2	x_3	x_4	x_5	x_6	
x_6	-2	0	0	-6	-4	1	2
x_2	$\frac{1}{2}$	1	0	1	$\frac{5}{2}$	0	2
x_3	$-\frac{1}{2}$	0	1	-1	$-\frac{1}{2}$	0	2
	$\frac{3}{2}$	0	0	4	$\frac{7}{2}$	0	10

5. The problem is infeasible.

7. $(x_1, x_2, s_1, s_2) = (70, 10, 0, 0); z^* = 430$

9. $(x_1, x_2, x_3, x_4, s_1, s_2, s_3) = (0, \frac{3}{2}, 0, \frac{5}{2}, 0, 7, 0); z^* = \frac{23}{2}$

11. The problem is infeasible. The final tableau is

	x_1	x_2	x_3	s_1	s_2	s_3	
x_1	1	-4	0	-2	-1	0	2
x_3	0	-2	1	-1	-1	0	3
s_3	0	7	0	4	3	1	-5
	0	7	0	3	2	0	-5

13. The objective row of the initial tableau has nonnegative entries. Equality constraints can be replaced by two $\leq$ constraints, and $\geq$ constraints can be changed to $\leq$ constraints by multiplying them by -1.

15. When we divide the 1st row by a_{1j}, the entry in the resource column becomes b_1/a_{1j}. To obtain a 0 in the jth column of the objective row, we multiply the 1st row by $-d_j$ to obtain $-d_j b_1/a_{1j}$. We then add this value to the 0 in the objective row to obtain $-d_j b_1/a_{1j}$.

17. If the row having the properties described in the problem is the ith row, then the corresponding constraint is

$$a_{i1}x_1 + a_{i2}x_2 + \cdots + a_{in}x_n = b_i$$

The left-hand side of this equation is nonnegative since $a_{ij} \geq 0$ and $x_j \geq 0$ for $1 \leq j \leq n$. The right-hand side is negative; and therefore, the constraint is infeasible.

19. (a) $(x_1, x_2) = (1.429, 1.714)$ **(b)** $(x_1, x_2) = (3, 0.667)$
    (c) $(x_1, x_2) = (1, 2)$ **(d)** $(x_1, x_2) = (3.143, 0.571)$

EXERCISES 6.6

1. If we let k represent a unit change in the mixing time, then [as in expression (3)] k must satisfy

$$4u_1 + (3 - k)u_2 + u_3 < 15$$

in order that the current solution remain optimal. It follows that $k > \frac{1}{2}$. Thus, if the mixing time is reduced by more than $\frac{1}{2}$ hr, then the plant should produce some units of F_3.

3. Since the reduced cost of F_1 is 5, each unit produced will cost \$5. To find the largest number of units of F_3 that could be produced without making the solution infeasible, we consider the constraints from the final tableau with all nonbasic variables except x_1 set to 0.

$$s_1 = x_1 + \frac{10}{3}$$

$$x_2 = -\frac{2}{5}x_1 + \frac{28}{3}$$

$$x_4 = -\frac{4}{5}x_1 + \frac{136}{3}$$

Since s_1, x_2, and x_3 must be nonnegative, the constraints on x_1 are

$$-\frac{2}{5}x_1 + \frac{28}{3} \geq 0$$

$$-\frac{4}{5}x_1 + \frac{136}{3} \geq 0$$

Therefore, x_1 cannot exceed $\frac{70}{3}$.

5. You can use SOLVER to show the problem is infeasible.

7. $\hat{z} = 0(b_1 + \Delta b_1) + \frac{10}{3}(b_2 + \Delta b_2) + \frac{20}{3}(b_3 + \Delta b_3)$

$= \frac{10}{3}(120 + \Delta b_2) + \frac{20}{3}(50 + \Delta b_3)$

$\approx 733.33 + 3.33\,\Delta b_2 + 6.67\,\Delta b_3$

9. If the call goes first to Paris or Tokyo, then it cannot go to London.

11. (a) It is easy to make a mistake in modeling this problem by allowing trips around disconnected loops to occur. We give a specialized solution here and consider this sort of problem in generality in Chapter 9 under the topic "the traveling salesman problem." Here is a correct model:

Minimize:

$z = \quad 60x_{12} + 35x_{13} + 70x_{25} + 70x_{52} + 20x_{54}$
$\quad\quad + 20x_{46} + 40x_{74} + 60x_{73} + 75x_{37} + 75x_{32}$

Subject to:

$x_{12} + x_{13} = 1$	(Leave Moscow)
$x_{12} + x_{32} + x_{52} = x_{25}$	(Leave Salmon without returning to Moscow)
$x_{25} = x_{52} + x_{54}$	(Enter and leave Twin Falls at most once)
$x_{54} + x_{74} = x_{46}$	(Enter and leave Sun Valley at most once)
$x_{46} = 1$	(Stop in Pocatello)
$x_{73} + x_{13} = x_{37}$	(Leave Lewiston and don't go back to Moscow)

All variables are binary.

(b) The solution is $x_{46} = 1$, $x_{12} = 1$, $x_{25} = 1$, $x_{54} = 1$, and the remaining variables are 0.

(c) The reduced cost of x_{73} is 135. If the traveler is forced to go from Boise to Lewiston, then she must complete a loop back to Boise at a cost of $60 + 75 = 135$. The reduced cost of x_{52} is 140. If the traveler must go from Twin Falls to Salmon, then she must complete the loop back to Twin Falls, thereby incurring a cost of $70 + 70 = 140$. The reduced cost of x_{32} is 125. This case is more interesting because forcing her to take a trip from Lewiston to Salmon results in an infeasible solution. We can see that this is so from the 5th and 6th constraints in the tableau in part (b). They are

$$x_{74} + x_{73} - x_{37} + x_{32} = 0$$

$$x_{73} - x_{37} = 0$$

The solution of these two constraints with $x_{32} = 1$ causes x_{74} to be -1.

EXERCISES 7.1

1. $(x_1, x_2, x_3, s_1, s_2, s_3) = (0, 142.5, 405, 0, 0, 1120)$; $z = 511.875$

3. $(x_1, x_2, x_3, s_1, s_2, s_3) = (0, 0, 340, 960, 0, 2300)$; $z = 340$

5. $(x_1, x_2, x_3, s_1, s_2, s_3) = (0, 455, 30, 0, 0, 1120)$; $z = 365.25$

7. $(x_1, x_2, x_3, s_1, s_2, s_3) = (0, 0, 333.33, 0, 1213.33, 2333.33)$; $z = 500$

9. $(x_1, x_2, x_3, s_1, s_2, s_3) = (0, 455, 30, 0, 0, 1120)$; $z = 438$

11. $(x_1, x_2, x_3, x_4, s_1, s_2, s_3) = (0, 460, 0, 20, 0, 0, 1200)$; $z = 373$. Yes, it is more profitable.

13. The value of b_2 can decrease until the line $3x_1 + 6x_2 = b_2$ passes through P_4. The value of b_2 when this line passes through P_4 is $b_2 = 3(\frac{3}{2}) + 6(2) = 16.5$. The value of b_2 can increase indefinitely. Thus the range is $16.5 \le b_2 < \infty$.

15. The final tableau is

	x_1	x_2	x_3	s_1	s_2	s_3	
x_3	$-\frac{1}{8}$	0	1	$\frac{3}{8}$	$-\frac{1}{4}$	0	30
x_2	$\frac{15}{16}$	1	0	$-\frac{5}{16}$	$\frac{3}{8}$	0	455
s_3	3	0	0	0	-1	1	1120
	$\frac{5}{64}$	0	0	$\frac{9}{64}$	$\frac{1}{32}$	0	$\frac{1485}{4}$

If **b** is the column of initial resource values, and **b*** is the column of final resource values, then $\mathbf{b^*} = B^{-1}\mathbf{b}$. Also, changing the values of **b** only changes **b***. Since the third column of B^{-1} is the unit column $\mathbf{e}_3$, the only effect of increasing b_3 is to increase b_3^*.

17. (a) When the objective function $z = c_1x_1 + c_2x_2$ has a positive slope and $c_1 < 0$, the direction of increasing z for the level lines for z is to the northwest. Therefore, when a level line of $z = c_1x_1 + c_2x_2$ passes through P, the range over which its slope can vary without changing the solution is bounded by the slope of $2x_1 + x_2 = 10$ and a vertical line through P. Thus $-\infty < -c_1/c_2 \le -2$.
    (b) $-2 \le -c_1/c_2 \le 1$ (c) $4 \le b_1 \le 16$ (d) $2 \le b_2 \le 20$ (e) $1 \le b_3 < \infty$

EXERCISES 7.2

1. $1920 \le b_1 \le 3456$; $1666.67 \le b_2 \le 3000$; $2880 \le b_3 < \infty$

3.
$$\mathbf{b^*} = B^{-1}\mathbf{b} = \begin{bmatrix} \frac{3}{8} & -\frac{1}{4} & 0 \\ -\frac{5}{16} & \frac{3}{8} & 0 \\ 0 & -1 & 1 \end{bmatrix} \begin{bmatrix} 3500 \\ 2880 \\ 4000 \end{bmatrix} = \begin{bmatrix} 592.5 \\ -13.75 \\ 1120 \end{bmatrix}$$

$$z^* = \mathbf{c}_B\mathbf{b^*} + z_0 = \begin{bmatrix} 1 & \frac{3}{4} & 0 \end{bmatrix} \begin{bmatrix} 592.5 \\ -13.75 \\ 1120 \end{bmatrix} = 582.1875$$

The optimal solution is $(0, 0, 576, 44, 0, 1120)$; $z = 576$.

5. If $\Delta b_2 = 0$ and $\Delta b_3 = 0$, we obtain $-3\,\Delta b_1 \leq 240$ and $5\,\Delta b_1 \leq 7280$. This is the same as $-80 \leq \Delta b_1 \leq 1456$.

If $\Delta b_1 = 0$ and $\Delta b_2 = 0$, we obtain $-\Delta b_3 \leq 1120$. This is the same as $-1120 \leq \Delta b_3 < \infty$.

If $\Delta b_1 = 0$ and $\Delta b_3 = 0$, we obtain $2\,\Delta b_2 \leq 240$ and $-6\,\Delta b_2 \leq 7280$. This is the same as $-\frac{3640}{3} \leq \Delta b_2 \leq 120$.

7. If we set $\Delta b_3 = 200$ and $\Delta b_2 = -100$ in (17), we obtain

$$-3\,\Delta b_1 \leq 440$$
$$5\,\Delta b_1 \leq 6680$$
$$-300 \leq 1120$$

and so, $-\frac{440}{3} \leq \Delta b_1 \leq 1336$.

11.
$$\lambda_1 = \frac{700}{1456} = 0.4808 \quad \text{and} \quad \lambda_2 = -\frac{-600}{\frac{3640}{3}} = \frac{1800}{3640} = 0.4945$$

Since $\lambda_1 + \lambda_2 \leq 1$, the 100% rule applies.

13. Since these three perturbations are within the allowable ranges (13), the column matrices

$$\mathbf{q}_1 = \begin{bmatrix} 1400 \\ 0 \\ 0 \end{bmatrix} \quad \mathbf{q}_2 = \begin{bmatrix} 0 \\ 100 \\ 0 \end{bmatrix} \quad \mathbf{q}_3 = \begin{bmatrix} 0 \\ 0 \\ -1000 \end{bmatrix}$$

satisfy the hypothesis of Theorem 3. If we let

$$\lambda_1 = 0.4, \qquad \lambda_2 = 0.5, \qquad \lambda_3 = 0.1$$

then $\lambda_1 + \lambda_2 + \lambda_3 = 1$. By Theorem 3, the perturbation

$$\mathbf{q} = \lambda_1 \mathbf{q}_1 + \lambda_2 \mathbf{q}_2 + \lambda_3 \mathbf{q}_3 = \begin{bmatrix} 560 \\ 50 \\ -100 \end{bmatrix}$$

leaves the set of basic variables unchanged.

15. $-4 \leq \Delta b_1 \leq \frac{4}{7}$; $-1.5 \leq \Delta b_2 < \infty$; $-1.6 \leq \Delta b_3 \leq 8$

17. The answers obtained from examining Tableau 5 of Example 1 of Section 3.3 are $-10 \leq \Delta b_1 < \infty$; $-14 \leq \Delta b_2 < \infty$; $-10 \leq \Delta b_3 < \infty$; $-21 \leq \Delta b_4 \leq \frac{35}{3}$; $-14 \leq \Delta b_5 \leq \frac{21}{5}$.

19. You must retain the artificial variable columns during phase 2. The resource ranges are $-1 \leq \Delta b_1 < \infty$; $-\infty < \Delta b_2 \leq 3$.

21. The ranges obtained from Tableau 4 of this example are $-\infty < \Delta b_1 \leq 2$; $-2 \leq \Delta b_2 < \infty$; $-1 \leq \Delta b_3 \leq 2$.

23. (a)
$$\frac{1}{8}\Delta b_1 - \frac{3}{16}\Delta b_3 \leq \frac{3}{2}$$

$$-\frac{1}{2}\Delta b_1 + \frac{1}{4}\Delta b_3 \leq 2$$

$$\frac{21}{8}\Delta b_1 - \Delta b_2 - \frac{15}{16}\Delta b_3 \leq \frac{3}{2}$$

(b)
$$-\frac{1}{7}\Delta b_4 + \frac{2}{7}\Delta b_5 \leq 3$$

$$-\frac{2}{7}\Delta b_4 - \frac{3}{7}\Delta b_5 \leq 10$$

$$-\Delta b_3 + \frac{6}{7}\Delta b_4 - \frac{5}{7}\Delta b_5 \leq 10$$

$$-\Delta b_1 \qquad -\frac{1}{7}\Delta b_4 - \frac{5}{7}\Delta b_5 \leq 10$$

$$-\Delta b_2 \qquad -\;\Delta b_5 \leq 14$$

(c) $-3\,\Delta b_1 + \Delta b_2 \leq 3$ **(d)** $\qquad -\;\Delta b_3 \leq 4$

$\qquad -\Delta b_1 \qquad \leq 5$ $\qquad\qquad -\Delta b_2 + \;\Delta b_3 \leq 2$

$$\Delta b_1 \qquad -2\,\Delta b_3 \leq 2$$

25.
$$2 + \;\Delta b_1 - \frac{2}{3}\Delta b_2 \geq 0$$

$$\frac{3}{2} - \frac{1}{2}\Delta b_1 + \frac{1}{2}\Delta b_2 \geq 0$$

27. 9 $\qquad + \;\Delta b_2 - 6\,\Delta b_3 \geq 0$

1 $\qquad\qquad + \;\Delta b_3 \geq 0$

8 + $\;\Delta b_1 \qquad - 4\,\Delta b_3 \geq 0$

29. $(x_1, x_2, x_3, s_1, s_2, s_3) = (3.5, 0.5, 0, 1.5, 0, 0); \; z = 11.5$

$$-1.5 \leq \Delta b_1 < \infty$$

$$-3 \leq \Delta b_2 \leq 1$$

$$-1 \leq \Delta b_3 \leq 0.6$$

31. Let x_1, x_2, x_3, x_4 be the respective numbers of bikes of type A, B, C, D produced by MBM.

(a) The solution is

$$(x_1, x_2, x_3, x_4, s_1, s_2, s_3, s_4, s_5, s_6) = (24, 11, 16, 6, 0, 5, 0, 0, 2820, 0)$$

The allowable individual perturbations of the resource ranges are

$$-7.5 \leq \Delta b_1 \leq 16.5 \qquad -3 \qquad \leq \Delta b_4 \leq 6.6$$

$$-5 \quad \leq \Delta b_2 < \infty \qquad -2820 \leq \Delta b_5 < \infty$$

$$-3.75 \leq \Delta b_3 \leq 8.25 \qquad -33 \quad \leq \Delta b_6 \leq 15$$

(b) 2820. Yes, some of the values of x_1, x_2, x_3, x_4 will change.

(c) 33, since $\Delta b_3 \geq -33$

(d) Yes, Δb_2 cannot decrease by more than 5 and still have x_2 in the basis.

(e) No, $24 - 7.5 = 16.5$. Therefore, they do not need to change until it drops to 16.

EXERCISES 7.3

1. Basic variables: $-\frac{3}{4} \le \Delta c_2 \le \frac{1}{2}$; $-1 \le \Delta c_4 \le 1$; $-\frac{1}{2} \le \Delta c_7 < \infty$
   Nonbasic variables: $-\infty < \Delta c_1 \le 5$; $-\infty < \Delta c_3 \le 2$; $-\infty < \Delta c_5 \le 4$;
   $-\infty < \Delta c_6 \le 2$; $-\infty < \Delta c_8 \le 3$

3. Basic variables: $-\frac{2}{3} \le \Delta c_5 \le \frac{9}{2}$; $-3 \le \Delta c_2 \le 1$; $-5 \le \Delta c_7 \le \frac{2}{3}$
   Nonbasic variables: $-\infty < \Delta c_1 \le 2$; $-\infty < \Delta c_3 \le 32$; $-\infty < \Delta c_4 \le 9$;
   $-\infty < \Delta c_6 \le 15$; $-\infty < \Delta c_8 \le 12$

5. $(x_3, x_6, x_7) = (24, 12, 46)$; $z = 120$

7. $(x_3, x_4, x_7) = (15, 3, 64)$; $z = 110$

9. $z = \mathbf{c_B b^*} = \begin{bmatrix} 5 & 6 & 0 \end{bmatrix} \begin{bmatrix} 12 & 12 & 34 \end{bmatrix}^T = 132$

11. $z = \mathbf{c_B b^*} = \begin{bmatrix} 1 & 6 & \frac{1}{2} \end{bmatrix} \begin{bmatrix} 12 & 12 & 34 \end{bmatrix}^T = 101$

13. $(x_3, x_2, x_7) = (12, 12, 34)$; $z = 36$

15. (a) See Exercise 13.
    (b) If $\mathbf{q} = [-2, -4, 0, -2, 0, 0, 0] = [\lambda_1 \Delta c_1, \lambda_2 \Delta c_2, \lambda_3 \Delta c_3, \lambda_4 \Delta c_4, \lambda_5 \Delta c_5, \lambda_6 \Delta c_6,$
    $\lambda_7 \Delta c_7]$, then $-4 = \lambda_2 \Delta c_2$. Together with $\Delta c_2 \ge -\frac{7}{2}$, this implies $-4/\lambda_2 \ge -\frac{7}{2}$.
    Therefore, $\lambda_2 \ge \frac{8}{7} > 1$, which contradicts the fact that $\lambda_1 + \lambda_2 + \cdots + \lambda_7 = 1$
    and $\lambda_i \ge 0$, $1 \le i \le 7$.

17. Let x_1 be the number of small dragons and x_2 be the number of large dragons
    produced each week. The optimal solution is $x_1 = 31.54$, $x_2 = 7.24$; $z = 3026.87$.
    (b) Increase the selling price of small dragons by at least \$10.54.
    (c) Reduce the selling price of large dragons by at least \$63.25.

19. (a) You can obtain the given sensitivity ranges from SOLVER's sensitivity analysis
    report.
    (b) If we let the slack variables for the first and third constraints of Problem A be s_1
    and s_2, respectively, and let the slack variables for the constraints in Problem B be
    t_1, t_2, t_3, t_4, in the order of the constraints, then the solutions to the problems
    given in part (a) are
    Problem A:
    $(x_1, x_2, x_3, s_1, s_2) = (0, 4.2, 7.8, 0, 32.4)$; $z = 47.4$
    Problem B:
    $(x_1, x_2, x_3, t_1, t_2, t_3, t_4) = (0, 4.2, 7.8, 0, 0, 0, 32.4)$; $z = 47.4$
    (c) The values of the decision variables and the value of the objective function are the
    same in both solutions given in part (b). However, the solutions (which include the
    values of the slack variables) are different. In problem B, $\Delta c_2 > 3$ or $-4 \le \Delta c_3 <$
    -3 switches some of the slack variables in the basis without affecting the values of
    the decision variables.

EXERCISES 7.4

1. (a) Basic variables: $-\frac{18}{5} \le \Delta c_1 \le 2$; $\frac{32}{5} \le c_1 \le 10$; $-1 \le \Delta c_4 \le 23$; $4 \le c_4 \le 28$
   Nonbasic variables: $-\infty < \Delta c_2 \le \frac{17}{4}$; $-\infty < c_2 \le \frac{41}{4}$; $-\infty < \Delta c_3 \le \frac{9}{4}$;
   $-\infty < c_3 \le \frac{25}{4}$

(b) From Tableau (2) of Section 6.1, the dual prices are $(u_1, u_2, u_3) = (\frac{1}{6}, \frac{23}{12}, 0)$. The condition on c_2 obtained from the second constraint of the dual problem is

$$4\left(\frac{1}{6}\right) + 5\left(\frac{23}{12}\right) + 0 = \frac{41}{4} \geq c_2$$

By similar reasoning, the condition on c_3 is

$$3\left(\frac{1}{6}\right) + 3\left(\frac{23}{12}\right) + 0 = \frac{25}{4} \geq c_3$$

(c) Since x_2 is nonbasic, we use the technique explained in Example 3 to obtain the condition

$$a_{12}u_1 + a_{22}u_2 + a_{32}u_3 = 3\left(\frac{1}{6}\right) + 6\left(\frac{23}{12}\right) + 4(0) = 12$$

Since $12 > 8$, the optimal basis does not change.

(d) Since x_3 is nonbasic, the new coefficients must satisfy

$$2\left(\frac{1}{6}\right) + 2\left(\frac{23}{12}\right) + 4(0) + 3(0) = \frac{25}{6} \geq 5$$

Since this inequality is false, the optimal basis will change.

(e) From the final tableau for the dual problem, we obtain

$$\frac{9}{4} + \frac{5}{8}\Delta c_1 + \quad 0\Delta c_2 + (-1)\Delta c_3 + \frac{1}{4}\Delta c_4 \geq 0$$

$$\frac{23}{12} + \frac{7}{24}\Delta c_1 + \quad 0\Delta c_2 + \quad 0\Delta c_3 - \frac{1}{12}\Delta c_4 \geq 0$$

$$\frac{17}{4} + \frac{9}{8}\Delta c_1 + (-1)\Delta c_2 + \quad 0\Delta c_3 + \frac{1}{4}\Delta c_4 \geq 0$$

$$\frac{1}{6} - \frac{1}{12}\Delta c_1 + \quad 0\Delta c_2 + \quad 0\Delta c_3 + \frac{1}{6}\Delta c_4 \geq 0$$

(f) If $\Delta c_2 = \Delta c_3 = \Delta c_4 = 0$, then Δc_1 must satisfy

$$\Delta c_1 \geq \max\left\{-\frac{18}{5}, -\frac{46}{7}, -\frac{32}{9}\right\} = -\frac{18}{5} \quad \text{and} \quad \Delta c_1 \leq 2$$

If $\Delta c_1 = \Delta c_3 = \Delta c_4 = 0$, then Δc_2 must satisfy

$$\Delta c_2 \leq \frac{34}{8} = \frac{17}{4}$$

If $\Delta c_1 = \Delta c_2 = \Delta c_4 = 0$, then Δc_3 must satisfy

$$\Delta c_3 \leq \frac{18}{8} = \frac{9}{4}$$

If $\Delta c_1 = \Delta c_2 = \Delta c_3 = 0$, then Δc_4 must satisfy

$$\Delta c_4 \geq \max\left\{-\frac{18}{2}, -\frac{34}{2}, -\frac{2}{2}\right\} = -1 \quad \text{and} \quad \Delta c_4 \leq \frac{46}{2} = 23$$

(g) We use the technique introduced in Example 6. The addition of x_5 adds the constraint

$$2u_1 - 2u_2 + 4u_3 \geq 3$$

to the dual problem. If the dual prices satisfy this constraint, the optimal solution is unchanged. Substituting these prices gives

$$2\left(\frac{1}{6}\right) - 2\left(\frac{23}{12}\right) + 4(0) = -\frac{7}{2}$$

Since $-\frac{7}{2} < 3$, the optimal solution will change.

3. From the optimal tableau for the primal problem, the dual prices are $[\frac{19}{5} \quad \frac{3}{5} \quad 0 \quad 0]$.
(a) From the first constraint for the dual problem, we obtain the condition

$$\frac{19}{5} + 2\left(\frac{3}{5}\right) - 2(0) + 0 \geq c_1$$

and so $c_1 \leq \frac{25}{5} = 5$. This agrees with the obvious limit $\Delta c_1 \leq 4$ that we can read directly from the tableau.
(b) To solve the dual problem by the simplex method, we must form an equivalent problem with nonnegative variables. We replaced the negative dual variables corresponding to the third and fourth constraints of the primal by $-u_3$ and $-u_4$, respectively.
(c) The B^{-1} matrix is found in the artificial variable columns. The desired system of inequalities is

$$\frac{19}{5} + 0\Delta c_1 + \frac{2}{5}\Delta c_2 + \frac{3}{5}\Delta c_3 \geq 0$$

$$4 - \Delta c_1 + 0\Delta c_2 + \Delta c_3 \geq 0$$

$$\frac{3}{5} + 0\Delta c_1 - \frac{1}{5}\Delta c_2 + \frac{1}{5}\Delta c_3 \geq 0$$

(d) $-\infty < \Delta c_1 \leq 4$
$-9.5 \leq \Delta c_2 \leq 3$
$-3 \leq \Delta c_3 < \infty$

5. $3 + 0\Delta c_1 + 0\Delta c_2 + 3\Delta c_3 + 0\Delta c_4 \geq 0$
$14 + 0\Delta c_1 + 4\Delta c_2 - 1\Delta c_3 - 1\Delta c_4 \geq 0$
$7 - 1\Delta c_1 + 2\Delta c_2 + 4\Delta c_3 + 0\Delta c_4 \geq 0$
$5 + 0\Delta c_1 + 1\Delta c_2 - 1\Delta c_3 + 0\Delta c_4 \geq 0$

7. The dual prices are $[31 \quad 0 \quad 11]$. The constraint in the dual problem corresponding to the new column is

$$3u_1 + 2u_2 - u_3 \geq 3$$

Since $3 \cdot 31 + 2 \cdot 0 - 1 \cdot 11 > 3$, the tableau remains optimal.

EXERCISES 7.5

1. The new column of the A matrix is

$$
B^{-1}A_7 = \begin{bmatrix} \frac{3}{5} & -\frac{1}{5} & 0 \\ -\frac{1}{5} & \frac{2}{5} & 0 \\ -\frac{2}{5} & -\frac{1}{5} & 1 \end{bmatrix} \begin{bmatrix} 1 \\ 1 \\ -1 \end{bmatrix} = \begin{bmatrix} \frac{2}{5} \\ \frac{1}{5} \\ -\frac{8}{5} \end{bmatrix}
$$

The new cost coefficient is

$$
-c_7^* = \mathbf{c}_B A_7^* - c_7 = \begin{bmatrix} 4 & 3 & 0 \end{bmatrix} \begin{bmatrix} \frac{2}{5} \\ \frac{1}{5} \\ -\frac{8}{5} \end{bmatrix} - 5 = -\frac{14}{5}
$$

Since $-c_7^* < 0$, the old solution is no longer optimal. One pivot of the simplex method shows that the new LP problem is unbounded.

3. One pivot of the simplex method yields the new basic solution

$$
\begin{bmatrix} x_8 & x_1 & x_6 \end{bmatrix} = \begin{bmatrix} \frac{6}{7} & \frac{16}{7} & \frac{41}{7} \end{bmatrix}; z = 12
$$

5. (a) The new column of the A matrix is

$$
B^{-1}A_8 = \begin{bmatrix} 3 & -1 & 0 \\ 0 & 1 & 0 \\ -4 & -1 & 1 \end{bmatrix} \begin{bmatrix} 4 \\ 4 \\ 5 \end{bmatrix} = \begin{bmatrix} 8 \\ 4 \\ -15 \end{bmatrix}
$$

The new cost coefficient is

$$
-c_8^* = \mathbf{c}_B A_8^* - c_8 = \begin{bmatrix} 1 & 6 & 0 \end{bmatrix} \begin{bmatrix} 8 \\ 4 \\ -15 \end{bmatrix} - \frac{3}{2} = \frac{61}{2}
$$

Since $-c_8^* \geq 0$, the old solution is still optimal.

(b) The new column of the A matrix is

$$
B^{-1}A_8 = \begin{bmatrix} 3 & -1 & 0 \\ 0 & 1 & 0 \\ -4 & -1 & 1 \end{bmatrix} \begin{bmatrix} -2 \\ 1 \\ 1 \end{bmatrix} = \begin{bmatrix} -7 \\ 1 \\ 8 \end{bmatrix}
$$

The new cost coefficient is

$$
-c_8^* = \mathbf{c}_B A_8^* - c_8 = \begin{bmatrix} 1 & 6 & 0 \end{bmatrix} \begin{bmatrix} -7 \\ 1 \\ 8 \end{bmatrix} - \frac{1}{2} = -\frac{3}{2}
$$

Since $-c_8^* < 0$, the old solution is no longer optimal. One pivot of the simplex method yields the new solution

$$
\begin{bmatrix} x_3 & x_2 & x_8 \end{bmatrix} = \begin{bmatrix} \frac{167}{4} & \frac{31}{4} & \frac{17}{4} \end{bmatrix}; z = \frac{723}{8} .
$$

7. The dual prices are $[3 \quad 5 \quad 0]$.

(a) Since $3 \cdot 4 + 5 \cdot 4 + 0 \cdot 5 = 32 > \frac{3}{2}$, the optimal solution is unchanged.

(b) Since $3 \cdot (-2) + 5 \cdot 1 + 0 \cdot 1 = -1 < \frac{3}{2}$, the optimal solution is changed.

EXERCISES 7.6

1. By Table 2, a change of $\Delta b_1 = -64$ does not alter the mix of processes. By using the B^{-1} matrix from (17), we find that the new solution is

$$x_B = \begin{bmatrix} x_2 \\ s_2 \\ s_3 \\ x_3 \\ x_1 \end{bmatrix} = B^{-1} \begin{bmatrix} 1136 \\ 200 \\ 600 \\ 2000 \\ 1800 \end{bmatrix} = \begin{bmatrix} 17.01 \\ 55.80 \\ 3.68 \\ 4.52 \\ 22.39 \end{bmatrix}$$

From Table 2 we see that the dual price for the labor constraint is 14.46. Therefore, the cost increases by

$$64(14.46) = 925.44$$

and the new total cost of the week's production is

$$28{,}825.27 + 925.44 = 29{,}750.71$$

3. From (17) we see that the changes $\Delta b_1 = -40$ and $\Delta b_4 = 100$ must satisfy five inequalities. The second of these is

$$0.0062(-40) - 0.0677(100) \geq -56.1990$$

Since this inequality is not satisfied, the mix of processes will change. For the reasons given in the answer to Question 2, MOM elects to solve the problem again. The solution is

$$(x_1, x_2, x_3, x_4, s_1, s_2, s_3, s_4, s_5)$$
$$= (9.29, 21.07, 10.40, 4.93, 0, 18.21, 0, 0, 0)$$
$$z = 32{,}551.82$$

5. The increased costs contribute

$$5(15x_1 + 10x_2 + 20x_3 + 8.5x_4) + 5(3x_1 + 3.2x_2 + 5x_3 + 7x_4)$$
$$= 90x_1 + 66x_2 + 125x_3 + 77.5x_4$$

Since we are solving a maximization problem, $\Delta c_1 = -90$, $\Delta c_2 = -66$, $\Delta c_3 = -125$, and $\Delta c_4 = -77.5$. We see that these changes satisfy (21) and conclude that the solution will not change. We use Table 2 to compute the new cost to be

$$28{,}825.27 + 90(17.62) + 66(22.99) + 125(3.48) + 77.5(0) = \$32{,}363.41$$

7. The raise of $1.50 per hour contributes

$$1.5(24x_1 + 32x_2 + 12x_3 + 28x_4) = 36x_1 + 48x_2 + 18x_3 + 42x_4$$

to the cost function. We see that $\Delta c_1 = -36$, $\Delta c_2 = -48$, $\Delta c_3 = -18$, and $\Delta c_4 = -42$ satisfy (21) and conclude that the solution is not changed. We use Table 2 to compute the new cost to be

$$28,825.27 + 36(17.62) + 48(22.99) + 18(3.48) + 42(0) = \$30,625.75$$

9. From Table 2 we see that 56.2 tons of white clay could be mined in excess of that used for producing slurry; therefore, we do not need to change our production schedule. Since it costs \$70 to mine a ton of white clay, and we can sell a ton for \$85, the profit is $15 \cdot 50 = \$750$.

EXERCISES 8.1

1.

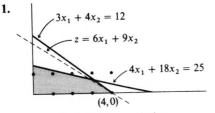

3. Let x_i, $1 \leq i \leq 6$, be the number of times pattern i is used.

Minimize:
$$z = x_1 + 4x_2 + x_3 + x_4 + 4x_5 + x_6$$

Subject to:

$$
\begin{array}{llll}
4x_1 + 2x_2 + 2x_3 + x_4 & & \geq 200 & \text{(6-ft sections)} \\
x_2 & + 2x_4 + x_5 & \geq 250 & \text{(9-ft sections)} \\
x_3 & + x_5 + 2x_6 & \geq 300 & \text{(12-ft sections)} \\
\end{array}
$$
$$x_i \geq 0 \text{ and integer}, \quad 1 \leq i \leq 6$$

5. If we set $x_2 = x_3 = 0$ in the first constraint, then x_1 can attain its largest value; however, x_1 must satisfy $2x_1 \leq 24$. Therefore, $x_1 \leq 12$. Similarly, we set $x_1 = x_3 = 0$ to conclude that $3x_2 \leq 24$ and hence that $x_2 \leq 8$; and we set $x_1 = x_2 = 0$ to conclude that $4x_3 \leq 24$ and hence that $x_3 \leq 6$.

7. (a) Since $x_1, x_2, x_3 \geq 0$, $-x_1 - x_2 - x_3 \leq 0$, and hence $-x_1 - x_2 - x_3 + 3 \leq 3$.
   (b) From $x_1 + 5x_2 + 8x_3 \leq 40$, we conclude that $x_1 \leq 40$, $x_2 \leq 8$, and $x_3 \leq 5$.
   Therefore,

$$2x_1 + 3x_2 + 4x_3 - 12 \leq 2 \cdot 40 + 3 \cdot 8 + 4 \cdot 5 - 12 = 112$$

   (c) The condition is

$$2x_1 + 3x_2 + 4x_3 - 12 \leq 112y$$
$$-x_1 - x_2 - x_3 + 3 \leq 3(1 - y)$$
$$y \quad \text{binary}$$

9. (a) $M = 3 \cdot 5 + 2 \cdot 6 + 6 \cdot 8 - 24 = 51$ (b) $M = 2 \cdot 5 + 3 \cdot 6 + 5 \cdot 8 - 4 = 64$
   (c) For the negative terms $-2x_2$ and $-3x_3$, use $x_2 \geq 0$ and $x_3 \geq 0$ to obtain
   $M = 3 \cdot 5 - 2 \cdot 0 - 3 \cdot 0 - 10 = 5$
   (d) $M = 6 \cdot 5 - 3 \cdot 0 - 4 \cdot 0 + 5 = 35$

11. From the constraint $x_1 + 5x_2 + 4x_3 \leq 15$, conclude by setting pairs of variables equal to 0 that

$$x_1 \leq 15, \qquad x_2 \leq 3, \qquad x_3 \leq \frac{15}{4}$$

Conclude that

$$x_1 + 2x_2 + 3x_3 - 11 \leq \frac{85}{4}$$

$$3x_1 + 2x_2 + 4x_3 - 15 \leq 51$$

$$4x_1 + 6x_2 + x_3 - 18 \leq \frac{255}{4}$$

$$x_1 + 5x_2 + 2x_3 - 13 \leq \frac{49}{2}$$

The condition is

$$x_1 + 2x_2 + 3x_3 - 11 \leq \frac{85}{4} y_1$$

$$3x_1 + 2x_2 + 4x_3 - 15 \leq 51y_2$$

$$4x_1 + 6x_2 + x_3 - 18 \leq \frac{255}{4} y_3$$

$$x_1 + 5x_2 + 2x_3 - 13 \leq \frac{49}{2} y_3$$

$$y_1 + y_2 + y_3 \leq 2$$

$$y_i \text{ binary}, \qquad 1 \leq i \leq 3$$

(Note that a computer program will usually find an integer solution faster when fractional bounds are rounded to the next higher integer.)

13. (a) By setting pairs of variables equal to 0 in the first constraint, we obtain $x_1 \leq 6, x_2 \leq 4, x_3 \leq 3$. By setting pairs of variables in the second constraint equal to 0, we obtain $x_1 \leq 4, x_2 \leq 4, x_3 \leq 4$. In either event, we have $x_1 \leq 6$, $x_2 \leq 4, x_3 \leq 4$.
   (b) Since we always have

$$2x_1 + 3x_2 + 4x_3 - 12 \leq 2 \cdot 6 + 3 \cdot 4 + 4 \cdot 4 - 12 = 28$$

$$x_1 + x_2 + x_3 - 4 \leq 6 + 4 + 4 - 4 = 10$$

the condition is

$$2x_1 + 3x_2 + 4x_3 \leq 12 + 28y$$

$$x_1 + x_2 + x_3 \leq 4 + 10(1 - y)$$

$$y \text{ binary}$$

15. (a) We have $a_1x_1 + a_2x_2 + a_3x_3 - 10 \leq 20$ and $-b_1x_1 - b_2x_2 - b_3x_3 + 12 \leq 0 + 0 + 0 + 12$. The condition is

$$a_1x_1 + a_2x_2 + a_3x_3 - 10 \leq 20y$$

$$-b_1x_1 - b_2x_2 - b_3x_3 + 12 \leq 12(1 - y)$$

$$y \text{ binary}$$

17. From constraint 1, $x_1 \leq 60$, $x_2 \leq 60$, and $x_3 \leq 15$. If in the notation of Example 5, we let $f(x_1, x_2, x_3) = x_2 + x_3$ and $g(x_1, x_2, x_3) = x_1 + x_3 - 54$, then $f(x_1, x_2, x_3) \leq 75$ and $g(x_1, x_2, x_3) \leq 21$. The "If ..., then ..." condition is equivalent to

$$x_2 + x_3 \quad \leq 75y$$

$$x_1 \quad + x_3 - 54 \leq 21(1 - y)$$

$$y \text{ binary}$$

19. (a) $u + v \geq 1$ **(b)** $u + v = 1$ **(c)** $u + v \leq 1$ **(d)** $u \leq v$ **(e)** $u \leq v$

EXERCISES 8.2

1. Branch at node 3 with either ($x_1 \leq 12$ and $x_1 \geq 13$) or ($x_2 \leq 5$ and $x_2 \geq 6$)

3. The solution at node 2 is maximal.

5. $(x_1, x_2) = (3, 0)$; $z = 9$ **7.** $(x_1, x_2) = (2, 2)$; $z = 16$

9. Make six units of product A and five units of product B for a maximum profit of $2600.

11. $(x_1, x_2, x_3) = (0, 3, 6)$; $z = 21$ **13.** $(x_1, x_2, x_3) = (11.5, 0, 6.5)$; $z = 42.5$

15. $(x_1, x_2, x_3) = (4.5, 1, 0)$; $z = 12.5$

17. $b_1 = 4818$, $b_2 = 7182$, $g_1 = 9818$, $g_2 = 2182$, $f_1 = 12000$, and $z = 2253.6$, where a subscript 1 refers to inside cost and a 2 refers to outside cost.

19. $(x_1, x_2, x_3) = (3.545, 2.818, 0)$; $z = 15.545$

21. (a) $(x_1, x_2, x_3, x_4, x_5, x_6) = (15, 0, 8, 125, 0, 146)$; $z = 294$
(b) $(x_1, x_2, x_3) = (0, 0, 6.5)$; $z = 32.5$
(c) $(x_1, x_2, x_3) = (0, 60, 0)$; $z = 240$

23. For $1 \leq i \leq 3$ and $1 \leq j \leq 2$, let x_{ij} be the number of barrels produced by plant i in period j; let $g_{ij} = 1$ if plant i *must* be started in period j; and let $r_{ij} = 1$ if plant i is operated in period j. The cost function to be minimized is

$$z = 5x_{11} + 5x_{12} + 4x_{21} + 4x_{22} + 6x_{31} + 6x_{32}$$
$$+ 700r_{11} + 700r_{12} + 800r_{21} + 800r_{22} + 900r_{31} + 900r_{32}$$
$$+ 3000g_{11} + 3000g_{12} + 2400g_{21} + 2400g_{22} + 1800g_{31} + 1800g_{32}$$

We express the demand constraints as

$$x_{11} + x_{21} + x_{31} \geq 2950 \quad \text{(Equality is ensured by the objective function)}$$
$$x_{12} + x_{22} + x_{32} \geq 4000$$

For each period, $j = 1$ and 2, we express the limits on maximum production for each plant and the requirement that $r_{ij} = 1$ if $x_{ij} > 0$ by

$$x_{1j} \leq 2100 r_{1j}, \qquad x_{2j} \leq 1900 r_{2j}, \qquad x_{3j} \leq 3000 r_{3j}$$

For each plant i, we express the logical conditions governing startup costs as

$$\text{If } r_{i1} = 1, \text{ then } g_{i1} = 1 \tag{9}$$

$$\text{If } g_{i1} = 0, \text{ then if } r_{i2} = 1, \text{ then } g_{i2} = 1 \tag{10}$$

Note that if g_{ij} is not set to 1 by condition (9) or (10), it is automatically set to 0 in an attempt to minimize z. Condition (9) is realized by the constraints

$$r_{11} \leq g_{11}; \qquad r_{21} \leq g_{21}; \qquad r_{31} \leq g_{31} \tag{11}$$

Condition (10) is equivalent to

$$\text{If } g_{i1} = 0, \text{ then } r_{i2} \leq g_{i2} \qquad (1 \leq i \leq 3) \tag{12}$$

Condition (12) is equivalent to

$$r_{i2} - g_{i2} \leq g_{i1} \qquad (1 \leq i \leq 3) \tag{13}$$

The admissible ranges for the variables are

$$x_{ij} \geq 0, \qquad r_{ij} \text{ and } g_{ij} \text{ binary}$$

The solution found by CALIPSO is

$$x_{11} = 1050, x_{12} = 2100, x_{21} = 1900, x_{22} = 1900$$

$$r_{11} = 1, r_{12} = 1, r_{21} = 1, r_{22} = 1,$$

$$g_{11} = 1, g_{21} = 1$$

$$z = 39{,}350$$

You can reduce the number of variables used to solve this problem by capitalizing on the fact that there are only two periods of operation. However, such solutions do not generalize to the situation when there are more than two periods. (See Exercise 24.)

25. $x_{11} = 2, x_{14} = 2.5, x_{15} = 3.5, x_{21} = 1, x_{23} = 2.5, x_{24} = 3.5, x_{32} = 2, x_{35} = 6.5; t = 7.5$

EXERCISES 8.3

1. $(x_1, x_2, x_3) = (1, 1, 0); z = 4$ **3.** $(x_1, x_2, x_3, x_4) = (1, 0, 0, 1); z = 4$

5. For $1 \leq i \leq 6$, let $x_i = 1$ if a fire station is built in City i; otherwise, let $x_i = 0$. Since City 1 can only be serviced by a fire station in City 1 (0 miles) or City 2 (20 miles), we have the constraint $x_1 + x_2 \geq 1$. Constraints for the other five cities are formed in a similar manner. The optimal solution is $(x_1, x_2, x_3, x_4, x_5, x_6) = (0, 1, 0, 1, 0, 0); z = 90$

7. Let $x_i = 1$ if project P_i is undertaken; otherwise, let $x_i = 0$. The condition that P_1 cannot be undertaken unless P_4 is undertaken is modeled by the constraint $x_1 \leq x_4$.

The condition that P_2 cannot be done if both P_4 and P_5 are done can be phrased as

$$\text{If } x_4 + x_5 = 2, \text{ then } x_2 = 0 \tag{3}$$

This condition is implied by the constraint

$$x_2 \leq 2 - (x_4 + x_5)$$

which is equivalent to

$$x_2 + x_4 + x_5 \leq 2 \tag{4}$$

You should take care that introducing a constraint like (4) does not introduce a spurious constraint on x_4 and x_5. In this case, it does not because setting $x_2 = 1$ (its only other possible value) in (4) yields $x_4 + x_5 \leq 1$. This is a valid constraint because the contrapositive of (3) is

$$\text{If } x_2 \neq 0, \text{ then } x_4 + x_5 \neq 2$$

which in view of the binary values of x_1, x_4, and x_5 is equivalent to

$$\text{If } x_2 = 1, \text{ then } x_4 + x_5 < 2$$

The situation would be quite different if the variables could take on integer values larger than 1.

The condition that P_3 cannot be done if either P_4 or P_8 is done is equivalent to the condition

$$\text{If } x_4 = 1 \text{ or } x_8 = 1, \text{ then } x_3 = 0 \tag{5}$$

The contrapositive of (5) is

$$\text{If } x_3 = 1, \text{ then } x_4 = 0 \text{ and } x_8 = 0 \tag{6}$$

Condition (6) is implied by the constraint

$$x_4 + x_8 \leq 2 - 2x_3$$

The standard form LP problem is

Minimize:
$$z = 12x_1 + 14x_2 + 15x_3 + 20x_4 + 21x_5 + 22x_6 + 25x_7 + 27x_8$$
Subject to:
$$
\begin{aligned}
x_1 + x_2 + \; x_3 + x_4 + x_5 + x_6 + x_7 + x_8 &\geq 4 \\
-x_1 \qquad\qquad\quad + x_4 \qquad\qquad\qquad &\geq 0 \\
-x_2 \qquad - x_4 - x_5 \qquad\qquad &\geq -2 \\
- 2x_3 - x_4 \qquad\qquad - x_8 &\geq -2 \\
x_i \text{ binary}, \qquad 1 \leq i \leq 8 &
\end{aligned}
$$

The optimal solution is

$$(x_1, x_2, x_3, x_4, x_5, x_6, x_7, x_8) = (1, 1, 0, 1, 0, 1, 0, 0); \; z = 68$$

9. **(a)** The sum on the left side of the ith constraint would be evaluated by adding (in order, starting with $j = 1$) those coefficients a_{ij} that correspond to variables x_j that

are 1 in the current partial solution. You should stop adding coefficients when the sum first exceeds b_i. In computer jargon,

$$s = 0; j = 1;$$

While $(s < b_i$ and $j \le n)$ do begin

If $x_j = 1$ then $s := s + a_{ij};$

$j := j + 1;$

end;

(b) If the value of j at the end of part (a) is less than n, continue adding any *positive* coefficients a_{ik} (for $k > j$) that may occur to see if the sum can be made greater than or equal to b_i. In computer jargon

While $(s < b_i$ and $j \le n)$ do begin

If $a_{ij} > 0$, then $s := s + a_{ij};$

$j := j + 1;$

end;

EXERCISES 8.4

1. We multiply the objective function by -1, and let $y_1 = 1 - x_1, y_2 = x_3, y_3 = 1 - x_2$ to obtain the standard form problem

Minimize: Subject to:

$z' = 2y_1 + 3y_2 + 4y_3 - 6$

$$y_1 - y_2 - 2y_3 \ge -6$$
$$-2y_1 - y_2 - y_3 \ge -1$$
$$y_1 - 2y_2 + y_3 \ge -2$$
$$y_i \text{ binary}, \quad 1 \le i \le 3$$

It is obvious that $(0, 0, 0)$ is a feasible solution, and, hence, it is the minimal solution. Therefore, the solution to the original problem is $(x_1, x_2, x_3) = (1, 1, 0)$, and the maximum value of z is 6.

3. We substitute $y_1 = 1 - x_3, y_2 = x_1, y_3 = x_2$ for the variables in the problem to obtain the standard form problem

Minimize: Subject to:

$z' = y_1 + 2y_2 + 3y_3 - 1$

$$-y_1 + y_2 + y_3 \ge 0$$
$$-y_1 + 2y_2 - y_3 \ge 1$$
$$-2y_1 + 3y_2 - y_3 \ge -\frac{1}{2}$$
$$y_i \text{ binary}, \quad 1 \le i \le 3$$

The solution is $(y_1, y_2, y_3) = (0, 1, 0)$. The solution to the original problem is $(x_1, x_2, x_3) = (1, 0, 1)$ and $z = 1$.

5. From the second constraint we conclude that $x \le 3$, and from the first constraint we conclude that $y \le 4$. Therefore, we let

$$x = u_0 + 2u_1$$
$$y = v_0 + 2v_1 + 4v_2$$

Setting $x_1 = 1 - v_0$, $x_2 = 1 - u_0$, $x_3 = 1 - v_1$, $x_4 = 1 - u_1$, and $x_5 = 1 - v_2$, and multiplying both constraints by -1 we obtain the standard form problem

Minimize:
$$z' = 100x_1 + 120x_2 + 200x_3 + 240x_4 + 400x_5 - 1060$$

Subject to:
$$x_1 + x_2 + 2x_3 + 2x_4 + 4x_5 \ge 6$$
$$3x_1 + 5x_2 + 3x_3 + 10x_4 + 12x_5 \ge 18$$
$$x_i \text{ binary}, \qquad 1 \le i \le 5$$

The solution is $(x_1, x_2, x_3, x_4, x_5) = (0, 0, 0, 1, 1)$, and the minimal value of z' is -420. Thus,

$$(u_0, u_1, v_0, v_1, v_2) = (1, 0, 1, 1, 0)$$
$$(x, y) = (1, 3); \qquad z = 420$$

7. By considering the first constraint, we see that $x \le 3$ and $y \le 2$. Let $x = u_0 + 2u_1$ and $y = v_0 + 2v_1$.

$$\text{Setting } x_1 = 1 - u_0, \qquad x_2 = 1 - v_0, \qquad x_3 = 1 - u_1, \qquad x_4 = 1 - v_1$$

we obtain the standard form problem

Minimize:
$$z' = 40x_1 + 50x_2 + 80x_3 + 100x_4 - 270$$

Subject to:
$$2x_1 + 3x_2 + 4x_3 + 6x_4 \ge 9$$
$$4x_1 + x_2 + 8x_3 + 2x_4 \ge 11$$
$$x_i \text{ binary}, \qquad 1 \le i \le 4$$

The solution is

$$(x_1, x_2, x_3, x_4) = (1, 1, 1, 0) \quad \text{or} \quad (u_0, u_1, v_0, v_1) = (0, 0, 0, 1)$$

The solution to the original problem is

$$x = u_0 + 2u_1 = 0, \qquad y = v_0 + 2v_1 = 2, \qquad z = 100$$

9. Since $(0, 0, 6)$ is a feasible solution, the optimal solution (x_1^*, x_2^*, x_3^*) satisfies

$$12x_1^* + 14x_2^* + 5x_3^* \le 12 \cdot 0 + 14 \cdot 0 + 6 \cdot 6 = 36$$

Therefore, we can restrict attention to variables satisfying

$$12x_1 \le 36, \qquad 14x_2 \le 36, \qquad 5x_3 \le 36$$

Since all variables are integer valued this means that

$$x_1 \le 3, \qquad x_2 \le 2, \qquad x_3 \le 7$$

We set

$$x_1 = u_0 + 2u_1$$
$$x_2 = v_0 + 2v_1$$
$$x_3 = w_0 + 2w_1 + 4w_2$$

to obtain the binary problem

Minimize:
$$z = 12u_0 + 24u_1 + 14v_0 + 28v_1 + 6w_0 + 12w_1 + 24w_2$$

Subject to:
$$2u_0 + 4u_1 + 3v_0 + 6v_1 + 4w_0 + 8w_1 + 16w_2 \geq 12$$
$$4u_0 + 8u_1 + 3v_0 + 6v_1 + 2w_0 + 4w_1 + 8w_2 \geq 12$$
$$u_0 + 2u_1 + v_0 + 2v_1 + w_0 + 2w_1 + 4w_2 \geq 5$$
$$u_0, u_1, v_0, v_1, w_0, w_1, w_2 \text{ binary}$$

To arrange the cost coefficients of the objective function in ascending order, let

$$y_1 = w_0, \qquad y_2 = u_0, \qquad y_3 = w_1, \qquad y_4 = v_0, \qquad y_5 = w_2, \qquad y_6 = v_1$$

and write the expressions of the problem in ascending order of these subscripts.

11. Add twice the first constraint to the second constraint to obtain $3x_1 - 3x_3 \geq 5$. Clearly, this constraint is infeasible.

13. The objective function is given in (12), the constraints guaranteeing that each city is entered exactly once are given in (13), and the constraints guaranteeing that each city is left exactly once are given in (14). The constraints of type (19) are

$$t_i - t_j + 6x_{ij} \leq 5, \qquad 1 \leq i \leq 6, \qquad 1 \leq j \leq 6, \qquad i \neq j$$

15. You could use the general model given in Example 4, the general model of Exercise 12, or the special model for a tour of five or fewer locations given in Exercise 14. We let $x_{ij} = 1$ if the drill moves from hole i to j, and $x_{ij} = 0$ if it does not. The optimal solution is

$$x_{23} = 1, \qquad x_{12} = 1, \qquad x_{31} = 1$$

and the other variables are 0.

17. Bill should take the following things: sleeping bag, book, fishing gear, rubber raft, camera, beer, food, and either a flashlight or more food. If he decides to take the flashlight, then he needs to carry only 38 pounds. To take more food, he must carry 40 pounds.

EXERCISES 8.5

1. The answers are given as the sum of the integer and fractional parts.
   (a) $\frac{3}{2} = 1 + \frac{1}{2}$ (b) $2 = 2 + 0$ (c) $-2 = -2 + 0$ (d) $-\frac{3}{2} = -2 + \frac{1}{2}$
   (e) $1.7 = 1 + 0.7$ (f) $-1.7 = -2 + 0.3$ (g) $0 = 0 + 0$ (h) $0.2 = 0 + 0.2$
   (i) $-0.2 = -1 + 0.8$ (j) $-13.7 = -14 + 0.3$

3. **(a)** $-(\frac{3}{4})x_1 - (\frac{1}{2})x_2 - (\frac{1}{4})x_4 \le -\frac{1}{4}$ **(b)** $-(\frac{3}{4})x_1 - (\frac{1}{2})x_2 - (\frac{3}{4})x_4 \le -\frac{1}{4}$
   (c) $-0.8x_1 - 0.5x_2 - 0.7x_4 \le -0.7$ **(d)** $-0.5x_1 - 0.4x_2 - 0.2x_4 \le -0.6$

5. The optimal solution is $(x_1, x_2, x_3) = (2, 0, 5)$; $z = 7$

7. The optimal solution is $(x_1, x_2, x_3) = (5, 0, 8)$; $z = 37$

9. The optimal solution is $(x_1, x_2) = (4, 0)$; $z = 28$

11. $x_1 = 2, x_2 = 2; z = 16$ **13.** $x_1 = 0, x_2 = 4, x_3 = 2; z = 18$

EXERCISES 8.6

1. $f(x) = 100z_2 + 250z_3 + 450z_4$
   $z_1 + z_2 + z_3 + z_4 = 1$
   $\quad y_2 + y_3 + y_4 = 1$
   $z_1 \le y_2$
   $z_2 \le y_2 + y_3$
   $z_3 \le y_3 + y_4$
   $z_4 \le y_4$
   $z_i \ge 0, \quad 1 \le i \le 4; \quad y_i \text{ binary}, \quad 2 \le i \le 4$

3. $f(x) = 50z_1 + 100z_2 + 250z_3 + 450z_4$
   The other constraints are the same as those given in the answer to Exercise 1.

5. If $a_1 < x \le a_2$, then z_1 must be 1 to satisfy the first constraint. Since $x - a_i < 0$ for $2 \le i \le n$, the other z_i are set to 0 in a minimization of the objective function. Thus, $f(x) = b_1$. The other cases are similar.

7. For any $x, 0 < x \le 50$, the LP model is

   Minimize: Subject to:
   $z = 2z_1 + 3z_2 + 2.5z_3$ $x - 40 \le 50z_1$
                            $x - 10 \le 50z_2$
                            $x - 30 \le 50z_2$
                            $z_1, z_2 \text{ binary}$

9.

Permutation	Constraints	
(123)	$x_{13} + \frac{1}{2} \le x_{23} + By_{123};$	$x_{23} + 1 \le x_{33} + By_{123}$
(132)	$x_{13} + \frac{1}{2} \le x_{33} + By_{132};$	$x_{33} + 2 \le x_{23} + By_{132}$
(213)	$x_{23} + 1 \le x_{13} + By_{213};$	$x_{13} + \frac{1}{2} \le x_{33} + By_{213}$
(231)	$x_{23} + 1 \le x_{33} + By_{231};$	$x_{33} + 2 \le x_{13} + By_{231}$
(312)	$x_{33} + 2 \le x_{13} + By_{312};$	$x_{13} + \frac{1}{2} \le x_{23} + By_{312}$
(321)	$x_{33} + 2 \le x_{23} + By_{321};$	$x_{23} + 1 \le x_{13} + By_{321}$

$y_{123} + y_{132} + y_{213} + y_{231} + y_{312} + y_{321} \ge 5; \quad y_{ijk} \text{ binary}$

11. (a) Let x_i, $1 \le i \le n$, be the number of times pattern i is used. We do not need to know what n is. We maximize

$$z = -x_1 - x_2 - \cdots - x_n$$

Suppose pattern 1 cuts one 3-ft board from a 20-ft piece of stock and wastes the rest, pattern 2 cuts one 6-ft board and wastes the rest, and pattern 3 cuts one 9-ft board and wastes the rest. The initial basic matrix is $B = I_3$. (Note that there are several ways to solve this problem because of ties in choosing the entering and departing variables. Most ways take only three pivots.)

(b) *Iteration 0*

$$\mathbf{c}_B B^{-1} \begin{bmatrix} a_1 \\ a_2 \\ a_3 \end{bmatrix} - (-1) = -a_1 - a_2 - a_3 + 1$$

We find the entering basic variable by solving

Minimize: Subject to:
$w = -a_1 - a_2 - a_3 + 1$ $3a_1 + 6a_2 + 9a_3 \le 20$
                              $a_i \ge 0$ and integer, $1 \le i \le 3$

One optimal solution is

$$\begin{bmatrix} a_1 \\ a_2 \\ a_3 \end{bmatrix} = \begin{bmatrix} 6 \\ 0 \\ 0 \end{bmatrix}; w = -5$$

To find the departing basic variable we calculate

$$\min \left\{ \frac{25}{6}, -, - \right\} = \frac{25}{6}$$

and conclude that the variable for row 1 departs. The pivoting matrix is

$$P_1 = \begin{bmatrix} \frac{1}{6} & 0 & 0 \\ 0 & 1 & 0 \\ 0 & 0 & 1 \end{bmatrix}$$

The next inverse of the basic matrix and resource column are

$$B_1^{-1} = \begin{bmatrix} \frac{1}{6} & 0 & 0 \\ 0 & 1 & 0 \\ 0 & 0 & 1 \end{bmatrix} \qquad \mathbf{b}_1 = \begin{bmatrix} \frac{25}{6} \\ 50 \\ 40 \end{bmatrix}$$

(c) *Iteration 1*

The entering pattern is $\begin{bmatrix} 0 \\ 3 \\ 0 \end{bmatrix}$

The departing variable is in row 2. The pivoting matrix is

$$P_2 = \begin{bmatrix} 1 & 0 & 0 \\ 0 & \frac{1}{3} & 0 \\ 0 & 0 & 1 \end{bmatrix}$$

The next inverse of the basic matrix and resource column are

$$B_2^{-1} = \begin{bmatrix} \frac{1}{6} & 0 & 0 \\ 0 & \frac{1}{3} & 0 \\ 0 & 0 & 1 \end{bmatrix} \qquad b_2 = \begin{bmatrix} \frac{25}{6} \\ \frac{50}{3} \\ 40 \end{bmatrix}$$

(d) In two more iterations of the method, you can find that a solution is to use the pattern

$$\begin{bmatrix} 6 \\ 0 \\ 0 \end{bmatrix}$$

$\frac{25}{6}$ times, the pattern

$$\begin{bmatrix} 0 \\ 3 \\ 0 \end{bmatrix}$$

$\frac{50}{3}$ times, and the pattern

$$\begin{bmatrix} 0 \\ 0 \\ 2 \end{bmatrix}$$

20 times. The minimal number of pieces of stock to cut is

$$\frac{25}{6} + \frac{50}{3} + 20 = \frac{245}{6} = 40\frac{5}{6}$$

If we round $\frac{25}{6}$ to 5, $\frac{50}{3}$ to 17, and $z = 40\frac{5}{6}$ to 41, we obtain the optimal solution (to see why rounding gives the optimal solution, see Exercise 13 of this section).

(e) There are five practical patterns given in the columns labeled $x_1 - x_5$ in the following tableau and three "unit" patterns which we introduce to obtain a starting basic feasible solution given in the columns labeled $y_1 - y_3$ in the following tableau. If we write the objective function in terms of the nonbasic variables, the initial tableau is

	x_1	x_2	x_3	x_4	x_5	y_1	y_2	y_3	s_1	s_2	s_3	
y_1	6	0	0	1	3	1	0	0	-1	0	0	25
y_2	0	3	0	1	0	0	1	0	0	-1	0	50
y_3	0	0	2	1	1	0	0	1	0	0	-1	40
	-5	-2	-1	-2	-3	0	0	0	1	1	1	-115

(f) The optimal solution found by SOLVER is

$$(x_1, x_2, x_3) = \left(\frac{25}{6}, \frac{50}{3}, 20\right); \qquad z = \frac{245}{6}$$

13. Let $\hat{z}^* = (a_1 + d_1) + (a_2 + d_2) + \cdots + (a_n + d_n)$
    $= (a_1 + a_2 + \cdots + a_n) + (d_1 + d_2 + \cdots + d_n)$
    $< z^* + 1$

    Then $z^* \leq \hat{z}^* < z^* + 1$ and $\hat{z}^*$ is an integer. Therefore, $\hat{z}^*$ is the smallest integer $\geq z^*$. Since the integer solution of the LP problem cannot be $< z^*$ (otherwise, z^* would not have been minimal), the integer solution must be $\hat{z}^*$.

15. If we round $\frac{75}{4}$ to 19, then

$$(x_1, x_2, x_3, x_4, x_5, x_6) = (19, 2, 0, 125, 0, 150)$$

is a feasible solution of the LP problem and has a corresponding z value of 294. Since 294 is the next integer greater than the optimal z value of 293.75, it must be the integer solution because the integer solution could not be less than 293.75.

EXERCISES 9.1

1. Add a dummy variable D^* with demand $d^* = 190$, and set all shipping costs into D^* to be 0.

3. Add a dummy supply point S^* that supplies $s^* = 100$ units. Set all shipping costs from S^* to be ∞.

5. You can verify that the given values satisfy the constraints by substituting the values into the equations, or you can argue as follows.

$$x_{11} + x_{12} + x_{13} = (s_1/s)(d_1 + d_2 + d_3) = (s_1/s)s = s_1$$

since

$$s = s_1 + s_2 + s_3 = d_1 + d_2 + d_3$$

Similarly,

$$\sum_{j=1}^{3} x_{2j} = (s_2/s) \sum_{j=1}^{3} d_j = (s_2/s)s = s_2$$

$$\sum_{i=1}^{2} x_{ij} = (d_j/s) \sum_{i=1}^{2} s_i = (d_j/s)s = d_j, \qquad 1 \leq j \leq 3$$

7. We pivot to bring x_{23} into the basis and take x_{22} out to obtain the basic feasible solution

$$x_{13} = 20, \qquad x_{23} = 5, \qquad x_{21} = 15, \qquad x_{12} = 10$$

9. $c_{11} = 10, c_{12} = 9, c_{21} = 3, c_{33} = -8$

11. $c_{11} = 9, c_{12} = 9, c_{14} = 7, c_{21} = 2, c_{24} = 2, c_{31} = 2, c_{33} = -5, c_{34} = 2, c_{43} = -7$

13. $c_{11} = 1, c_{12} = 8, c_{21} = -4, c_{33} = 0, c_{42} = 8, c_{43} = 6$

15. $c_{11} = -9, c_{12} = -4, c_{13} = -8, c_{21} = 0, c_{25} = 2, c_{31} = 1, c_{33} = 0, c_{34} = 2, c_{35} = 5,$
$c_{43} = 1, c_{44} = -2, c_{45} = 5$

EXERCISES 9.2

1. The departing basic variable is x_{14}. The new table is

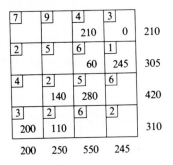

				Supply
[7]	[9]	[4] 210	[3] 0	210
[2]	[5]	[6] 60	[1] 245	305
[4]	[2] 140	[5] 280	[6]	420
[3] 200	[2] 110	[6]	[2]	310
200	250	550	245	

3. The departing basic variable is x_{23}. The new table is

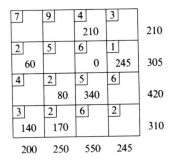

				Supply
[7]	[9]	[4] 210	[3]	210
[2] 60	[5]	[6] 0	[1] 245	305
[4]	[2] 80	[5] 340	[6]	420
[3] 140	[2] 170	[6]	[2]	310
200	250	550	245	

5. The departing basic variable is x_{13}. The new table is

				Supply
[7]	[9]	[4] 0	[3] 210	210
[2]	[5] 140	[6] 130	[1] 35	305
[4]	[2] 250	[5] 420	[6]	420
[3] 310	[2] 0	[6]	[2]	310
60	390	550	245	

7. The departing basic variable is x_{31}. The new table is

			Supply
[7] 70	[9]	[4] 80	150
[3]	[2] 50	[5] 120	170
[6] 0	[1] 250	[4]	250
[2] 110	[5]	[6]	110
180	300	200	

9. The departing basic variable is x_{31}. The new table is

					Supply
[2] 40	[8]	[4]	[5] 5	[7] 155...	
[2] 40	[8]	[4]	[5] 155	[7] 250	450
[8]	[3] 130	[2] 195	[6]	[8]	325
[5] 0	[1] 220	[9]	[4]	[6]	220
[2] 110	[2]	[7]	[7]	[7]	110
150	350	200	155	250	

11. The departing basic variable is x_{42}. The new table is

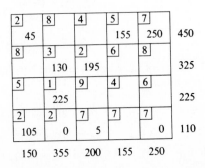

					Supply
[2] 45	[8]	[4]	[5] 155	[7] 250	450
[8]	[3] 130	[2] 195	[6]	[8]	325
[5]	[1] 225	[9]	[4]	[6]	225
[2] 105	[2] 0	[7] 5	[7]	[7] 0	110
150	355	200	155	250	

13. The answer is given with the answer for Exercise 3. The minimum transportation cost is

$$z = 4 \cdot 210 + 2 \cdot 60 + 2 \cdot 80 + 3 \cdot 140 + 2 \cdot 170 + 5 \cdot 340 + 1 \cdot 245$$
$$= 3825$$

15. The cost table corresponding to the given basic feasible solution is

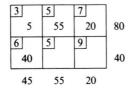

The optimal basic solution is $x_{11} = 45$, $x_{12} = 15$, $x_{13} = 20$, $x_{22} = 40$; $z = 550$.

17. An optimal solution is $x_{13} = 150$, $x_{21} = 250$, $x_{31} = 100$, $x_{32} = 200$, $x_{33} = 0$; $z = 2450$.

19. $x_{12} = 600$, $x_{21} = 100$, $x_{23} = 300$, $x_{32} = 200$, $x_{33} = 300$; $z = 4900$.

EXERCISES 9.3

Note that your answers may differ from those given and still be correct.

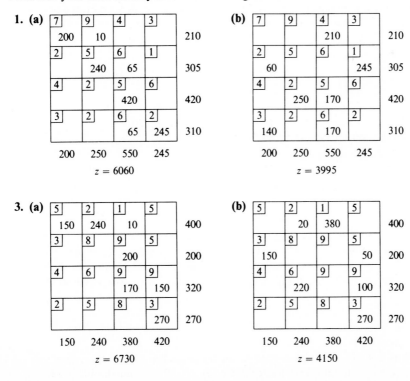

5. (a)

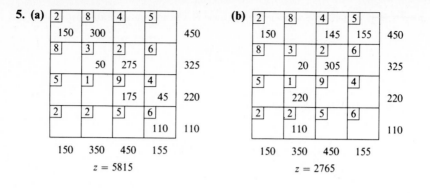

$z = 5815$

(b)

$z = 2765$

EXERCISES 10.1

1. Suppose that x_5 is the entering basic variable.

   (a) The Departing Basic Variable Rule gives

   $$m_1 = \min\left\{\frac{b_1^*}{2}, \frac{b_2^*}{4}, -, -\right\}$$

   (b) Since x_5 is a bounded variable, $m_2 = B_5$.

   (c) Since there are bounded basic variables, we write the equations corresponding to the rows of Tableau 1 with all nonbasic variables except x_5 set equal to 0.

   $$x_3 + 2x_5 = b_1^*$$
   $$x_1' + 4x_5 = b_2^*$$
   $$x_2 - 5x_5 = b_3^*$$
   $$x_7' + 0x_5 = b_4^*$$

   The bounds on x_1, x_3, and x_7 imply that

   $$x_3 = -2x_5 + b_1^* \leq B_3 \tag{1}$$
   $$x_1' = -4x_5 + b_2^* \leq B_1 \tag{2}$$
   $$x_2 = 5x_5 + b_3^* \tag{3}$$
   $$x_7' = b_4^* \leq B_7 \tag{4}$$

   Since none of expressions (1)–(4) impose an upper bound on x_5, we have $m_3 = \infty$. If $m_1 \leq m_2$, then we use the ordinary Departing Basic Variable Rule to choose the departing basic variable. If $m_2 < m_1$, then we let $x_5' = B_5 - x_5$ and adjust the tableau accordingly.

3. (a) $m_1 = \min\{\frac{2}{2}, -, -, -\} = 1$

   (b) $m_2 = 4$

   (c) $m_3 = \min\{-, (7 - 6)/2, (4 - 3)/1, -\} = \frac{1}{2}$

   Since m_3 is smallest, the basic variable for the second row, x_1', is the departing basic variable. In Tableau 2, we replace x_1' by $x_1 = 7 - x_1'$ and multiply the

ANSWERS FOR ODD-NUMBERED EXERCISES   **643**

second row by -1 to obtain Tableau 2a.

Tableau 2a

	x_1	x_2	x_3	x_4	x_5	x_6	x_7'	
x_3	0	0	1	2	2	0	0	2
x_1	1	0	0	2	4	2	0	1
x_2	0	1	0	-1	-4	-3	0	3
x_7'	0	0	0	-3	0	1	1	4
	0	0	0	-3	-4	-5	0	27

5.

	x_1'	x_2	x_3	x_4	x_5	x_6	x_7'	
x_5	0	0	$\frac{1}{2}$	1	1	0	0	1
x_1'	1	0	2	2	0	-2	0	10
x_2	0	1	2	3	0	-3	0	7
x_7'	0	0	0	-3	0	1	1	4
	0	0	2	1	0	-5	0	31

7.

	x_1	x_2	x_3	x_4	x_5	x_6	x_7'	
x_3	0	0	1	2	2	0	0	2
x_1	1	0	0	2	4	2	0	6
x_2	0	1	0	-1	-4	-3	0	3
x_7'	0	0	0	-3	0	1	1	4
	0	0	0	-3	-4	-5	0	27

9. $[x_1 \quad x_2 \quad x_3] = [2 \quad 1.2 \quad 2.3]; z = 22.8$

====

EXERCISES 10.2

1. The original problem in matrix form is

Maximize: Subject to:

$z = \mathbf{c}_1\mathbf{v}_1 + \mathbf{c}_2\mathbf{v}_2 + \mathbf{c}_3\mathbf{v}_3$ 　　$F_1\mathbf{v}_1 + F_2\mathbf{v}_2 + F_3\mathbf{v}_3 \leq \mathbf{b}_0$

$$D_1\mathbf{v}_1 \qquad\qquad\quad \leq \mathbf{b}_1$$
$$D_2\mathbf{v}_2 \qquad\quad \leq \mathbf{b}_2$$
$$D_3\mathbf{v}_3 \leq \mathbf{b}_3$$
$$\mathbf{v}_i \geq \mathbf{0}, \qquad 1 \leq i \leq 3$$

$$\mathbf{c}_1 = [1 \quad 2] \qquad \mathbf{c}_2 = [3 \quad -1] \qquad \mathbf{c}_3 = [1 \quad 1]$$

$$\mathbf{v}_1 = \begin{bmatrix} x_1 \\ x_2 \end{bmatrix} \qquad \mathbf{v}_2 = \begin{bmatrix} x_3 \\ x_4 \end{bmatrix} \qquad \mathbf{v}_3 = \begin{bmatrix} x_5 \\ x_6 \end{bmatrix}$$

$$F_1 = \begin{bmatrix} 1 & -1 \\ 2 & 1 \end{bmatrix} \quad F_2 = \begin{bmatrix} 2 & 1 \\ -1 & 3 \end{bmatrix} \quad F_2 = \begin{bmatrix} -1 & 0 \\ 1 & 2 \end{bmatrix}$$

$$b_0 = \begin{bmatrix} 12 \\ 10 \end{bmatrix}$$

$$D_1 = \begin{bmatrix} 1 & 1 \\ 5 & 3 \end{bmatrix} \quad b_1 = \begin{bmatrix} 4 \\ 15 \end{bmatrix}$$

$$D_2 = \begin{bmatrix} 1 & -3 \\ 1 & 2 \end{bmatrix} \quad b_2 = \begin{bmatrix} 4 \\ 6 \end{bmatrix}$$

$$D_3 = \begin{bmatrix} 4 & 3 \\ 3 & 4 \\ 5 & 2 \end{bmatrix} \quad b_3 = \begin{bmatrix} 12 \\ 12 \\ 10 \end{bmatrix}$$

The master problem is

Maximize:
$$z = \quad (c_1 P_1)\alpha_1 + (c_1 P_2)\alpha_2 + \cdots + (c_1 P_r)\alpha_r$$
$$+ (c_2 Q_1)\beta_1 + (c_2 Q_2)\beta_2 + \cdots + (c_2 Q_s)\beta_s$$
$$+ (c_3 R_1)\gamma_1 + (c_3 R_2)\gamma_2 + \cdots + (c_3 R_t)\gamma_t$$

Subject to:
$$(F_1 P_1)\alpha_1 + (F_1 P_2)\alpha_2 + \cdots + (F_1 P_r)\alpha_r$$
$$+ (F_2 Q_1)\beta_1 + (F_2 Q_2)\beta_2 + \cdots + (F_2 Q_s)\beta_s$$
$$+ (F_3 R_1)\gamma_1 + (F_3 R_2)\gamma_2 + \cdots + (F_3 R_t)\gamma_t + S = b_0$$
$$\alpha_1 + \alpha_2 + \cdots + \alpha_r \qquad\qquad\qquad = 1$$
$$\beta_1 + \beta_2 + \cdots + \beta_s \qquad = 1$$
$$\gamma_1 + \gamma_2 + \cdots + \gamma_t = 1$$
$$\alpha_i \geq 0, \quad 1 \leq i \leq r; \quad \beta_i \geq 0, \quad 1 \leq i \leq s; \quad \gamma_i \geq 0, \quad 1 \leq i \leq t; \quad S \geq 0$$

3. The original problem in matrix form is

Maximize: \qquad\qquad Subject to:
$$z = c_1 v_1 + c_2 v_2 + c_3 v_3$$

$$F_1 v_1 + F_2 v_2 + F_3 v_3 + \begin{bmatrix} -s_1 \\ s_2 \\ 0 \end{bmatrix} = b_0$$

$$D_1 v_1 \qquad\qquad + \begin{bmatrix} s_3 \\ s_4 \end{bmatrix} = b_1$$

$$D_2 v_2 \qquad + \begin{bmatrix} s_5 \\ -s_6 \end{bmatrix} = b_2$$

$$D_3 v_3 + \begin{bmatrix} 0 \\ s_7 \end{bmatrix} = b_3$$

$$v_i \geq 0, \quad 1 \leq i \leq 3$$

$$c_1 = [1 \quad 2] \qquad c_2 = [1 \quad -1] \qquad c_3 = [2 \quad 1 \quad 3]$$

$$v_1 = \begin{bmatrix} x_1 \\ x_2 \end{bmatrix} \qquad v_2 = \begin{bmatrix} x_3 \\ x_4 \end{bmatrix} \qquad v_3 = \begin{bmatrix} x_5 \\ x_6 \\ x_7 \end{bmatrix}$$

$$F_1 = \begin{bmatrix} 2 & 3 \\ 1 & 1 \\ 1 & 0 \end{bmatrix} \quad F_2 = \begin{bmatrix} 1 & -1 \\ 2 & 1 \\ -1 & 0 \end{bmatrix} \quad F_3 = \begin{bmatrix} 1 & 2 & 1 \\ -1 & 1 & -1 \\ 1 & 0 & -1 \end{bmatrix}$$

$$\mathbf{b}_0 = \begin{bmatrix} 6 \\ 12 \\ 4 \end{bmatrix}$$

$$D_1 = \begin{bmatrix} 1 & 1 \\ 5 & 3 \end{bmatrix} \quad \mathbf{b}_1 = \begin{bmatrix} 4 \\ 15 \end{bmatrix}$$

$$D_2 = \begin{bmatrix} 1 & 1 \\ 1 & 1 \end{bmatrix} \quad \mathbf{b}_2 = \begin{bmatrix} 5 \\ 2 \end{bmatrix}$$

$$D_3 = \begin{bmatrix} 1 & -1 & 2 \\ 2 & 3 & 1 \end{bmatrix} \quad \mathbf{b}_3 = \begin{bmatrix} 3 \\ 12 \end{bmatrix}$$

The auxiliary master problem is

Maximize:

$$w = -y_1 - y_2 - y_3 - y_4 - y_5$$

Subject to:

$$(F_1 P_1)\alpha_1 + (F_1 P_2)\alpha_2 + \cdots + (F_1 P_r)\alpha_r$$
$$+ (F_2 Q_1)\beta_1 + (F_2 Q_2)\beta_2 + \cdots + (F_2 Q_s)\beta_s$$
$$+ (F_3 R_1)\gamma_1 + (F_3 R_2)\gamma_2 + \cdots + (F_3 R_t)\gamma_t + \begin{bmatrix} -s_1 + y_1 \\ s_2 \\ y_2 \end{bmatrix} = \begin{bmatrix} 6 \\ 12 \\ 4 \end{bmatrix}$$

$$\alpha_1 + \alpha_2 + \cdots + \alpha_r \qquad + y_3 \qquad = 1$$
$$\beta_1 + \beta_2 + \cdots + \beta_s \qquad + y_4 \qquad = 1$$
$$\gamma_1 + \gamma_2 + \cdots + \gamma_t + y_5 \qquad = 1$$

$$0 \le \alpha_i \le 1, \quad 1 \le i \le r; \quad 0 \le \beta_i \le 1, \quad 1 \le i \le s; \quad 0 \le \gamma_i \le 1, \quad 1 \le i \le t;$$
$$s_1, s_2 \ge 0; \quad y_i \ge 0, \quad 1 \le i \le 5$$

To express w in terms of nonbasic variables, we let

$$\mathbf{e}_1 = [1 \quad 0 \quad 0] \qquad \mathbf{e}_3 = [0 \quad 0 \quad 1]$$

and solve the constraints for y_i to obtain

$$-y_1 = \quad (\mathbf{e}_1 F_1 P_1)\alpha_1 + (\mathbf{e}_1 F_1 P_2)\alpha_2 + \cdots + (\mathbf{e}_1 F_1 P_r)\alpha_r$$
$$+ (\mathbf{e}_1 F_2 Q_1)\beta_1 + (\mathbf{e}_1 F_2 Q_2)\beta_2 + \cdots + (\mathbf{e}_1 F_2 Q_s)\beta_s$$
$$+ (\mathbf{e}_1 F_3 R_1)\gamma_1 + (\mathbf{e}_1 F_3 R_2)\gamma_2 + \cdots + (\mathbf{e}_1 F_3 R_t)\gamma_t - s_1 - 6$$
$$-y_2 = \quad (\mathbf{e}_3 F_1 P_1)\alpha_1 + (\mathbf{e}_3 F_1 P_2)\alpha_2 + \cdots + (\mathbf{e}_3 F_1 P_r)\alpha_r$$
$$+ (\mathbf{e}_3 F_2 Q_1)\beta_1 + (\mathbf{e}_3 F_2 Q_2)\beta_2 + \cdots + (\mathbf{e}_3 F_2 Q_s)\beta_s$$
$$+ (\mathbf{e}_3 F_3 R_1)\gamma_1 + (\mathbf{e}_3 F_3 R_2)\gamma_2 + \cdots + (\mathbf{e}_3 F_3 R_t)\gamma_t - 4$$
$$-y_3 = \alpha_1 + \alpha_2 + \cdots + \alpha_r - 1$$
$$-y_4 = \beta_1 + \beta_2 + \cdots + \beta_s - 1$$
$$-y_5 = \gamma_1 + \gamma_2 + \cdots + \gamma_t - 1$$

5. By setting three variables at a time equal to 0 in the first constraint, we obtain

$$x_1 \leq 6, \qquad x_2 \leq 3, \qquad x_3 \leq 4, \qquad x_4 \leq 6$$

The feasible region for the first block of independent constraints is bounded if either of the constraints

$$x_1 \leq 6 \quad \text{or} \quad x_2 \leq 3$$

is added to the block. The feasible region for the second block of independent constraints is bounded if both

$$x_3 \leq 4 \quad \text{and} \quad x_4 \leq 6$$

are added to the block.

Maximize: Subject to:

$$w = \beta_1 + \beta_2 + \cdots + \beta_s - 1$$

$$(F_1 P_1)\alpha_1 + (F_1 P_2)\alpha_2 + \cdots + (F_1 P_r)\alpha_r$$

$$+ (F_2 Q_1)\beta_1 + (F_2 Q_2)\beta_2 + \cdots + (F_2 Q_s)\beta_s + \begin{bmatrix} s_1 \\ s_2 \end{bmatrix} = \begin{bmatrix} 12 \\ 3 \end{bmatrix}$$

$$\alpha_1 + \qquad \alpha_2 + \cdots + \qquad \alpha_r \qquad = 1$$

$$\beta_1 + \qquad \beta_2 + \cdots + \qquad \beta_s + y_1 \quad = 1$$

All variables are nonnegative.

EXERCISES 10.3

1. The corner points of the feasible region for the first set of independent constraints are

$$P_1 = \begin{bmatrix} 0 \\ 0 \end{bmatrix} \qquad P_2 = \begin{bmatrix} 4 \\ 0 \end{bmatrix} \qquad P_3 = \begin{bmatrix} 2 \\ 4 \end{bmatrix} \qquad P_4 = \begin{bmatrix} 0 \\ 2 \end{bmatrix}$$

The corner points corresponding to the second set of independent constraints are

$$Q_1 = \begin{bmatrix} 0 \\ 0 \end{bmatrix} \qquad Q_2 = \begin{bmatrix} 3 \\ 3 \end{bmatrix} \qquad Q_3 = \begin{bmatrix} 1 \\ 0 \end{bmatrix} \qquad Q_4 = \begin{bmatrix} 0 \\ 4 \end{bmatrix}$$

The master problem is

Maximize:
$$z = 12\alpha_2 + 10\alpha_3 + 2\alpha_4 + 9\beta_2 + 2\beta_3 + 4\beta_4$$

Subject to:
$$8\alpha_2 + 16\alpha_3 + 6\alpha_4 \qquad + 9\beta_2 + \beta_3 + 8\beta_4 + s_1 = 10$$
$$4\alpha_2 + 2\alpha_3 \qquad + 3\beta_2 - 2\beta_3 + 12\beta_4 + s_2 = 8$$
$$\alpha_1 + \alpha_2 + \alpha_3 + \alpha_4 \qquad = 1$$
$$\beta_1 + \beta_2 + \beta_3 + \beta_4 = 1$$
$$\alpha_i, \beta_i \geq 0, \qquad 1 \leq i \leq 4; \qquad s_1, s_2 \geq 0$$

The initial tableau is

	α_1	α_2	α_3	α_4	β_1	β_2	β_3	β_4	s_1	s_2	
s_1	0	8	16	6	0	9	1	8	1	0	10
s_2	0	4	2	0	0	3	-2	12	0	1	8
α_1	1	1	1	1	0	0	0	0	0	0	1
β_1	0	0	0	0	1	1	1	1	0	0	1
	0	-12	-10	-2	0	-9	-2	-4	0	0	0

3. The corner points corresponding to the first set of independent constraints are

$$P_1 = \begin{bmatrix} 2 \\ 4 \end{bmatrix} \quad P_2 = \begin{bmatrix} 0 \\ 2 \end{bmatrix} \quad P_3 = \begin{bmatrix} 0 \\ 8 \end{bmatrix}$$

The corner points corresponding to the second set of independent constraints are

$$Q_1 = \begin{bmatrix} 12 \\ 0 \end{bmatrix} \quad Q_2 = \begin{bmatrix} 1 \\ 0 \end{bmatrix} \quad Q_3 = \begin{bmatrix} 3 \\ 3 \end{bmatrix}$$

The auxiliary master problem is

Maximize:
$$w = 17\alpha_1 + 7\alpha_2 + 25\alpha_3 + 13\beta_1 + 2\beta_2 + 10\beta_3 - s_1 - 12$$

Subject to:
$$\begin{aligned}
16\alpha_1 + 6\alpha_2 + 24\alpha_3 + 12\beta_1 + \beta_2 + 9\beta_3 - s_1 + \qquad\quad y_1 \qquad\qquad\qquad &= 10 \\
2\alpha_1 \qquad\qquad\qquad\qquad\quad - 24\beta_1 - 2\beta_2 + 3\beta_3 \qquad + s_2 \qquad\qquad\qquad &= 8 \\
\alpha_1 + \alpha_2 + \alpha_3 \qquad\qquad\qquad\qquad\qquad\qquad\qquad + y_2 \qquad &= 1 \\
\beta_1 + \beta_2 + \beta_3 \qquad\qquad\qquad\qquad + y_3 &= 1
\end{aligned}$$
$$\alpha_i, \beta_i, y_i \geq 0, \quad 1 \leq i \leq 3; \quad s_1, s_2 \geq 0$$

The tableau for the auxiliary master problem is

	α_1	α_2	α_3	β_1	β_2	β_3	s_1	s_2	y_1	y_2	y_3	
y_1	16	6	24	12	1	9	-1	0	1	0	0	10
s_2	2	0	0	-24	-2	3	0	1	0	0	0	8
y_2	1	1	1	0	0	0	0	0	0	1	0	1
y_3	0	0	0	1	1	1	0	0	0	0	1	1
(z)	-10	-2	-8	-24	-2	-9	0	0	0	0	0	0
(w)	-17	-7	-25	-13	-2	-10	1	0	0	0	0	-12

5. $(x_1, x_2, x_3, x_4, x_5, x_6) = (0, 4, 5.2, 0.4, 1.14, 2.14); \ z = 26.49$

7. $(x_1, x_2, x_3, x_4, x_5, x_6) = (1.5, 2.5, 1.7, 0.3, 4.2, 1.2); \ z = 17.5$

BIBLIOGRAPHY

[1] E. M. L. Beale, *Naval Research Quarterly*, Vol 2, 1955   (3.3)

[2] Bernard Kolman and Robert Beck, *Elementary Linear*   (8.5)
     Programming with Applications, New York: Academic Press,
     1980, 246–249

[3] R. G. Bland, "New Finite Pivoting Rules for the Simplex   (3.3)
     Method," *Mathematics of Operations Research*, 2, May 1977,
     103–110

[4] A. Charnes, "Optimality and Degeneracy in Linear   (3.3)
     Programs," *Econometrica* (20), 1952

[5] Vasek Chvatal, *Linear Programming*, New York: Freeman,   (3.3)
     1980, 37–38

[6] R. E. Gomory, "Essentials of an Algorithm for Integer   (8.5)
     Solutions to Linear Programs," *Bulletin of AMS* (64), 1958

[7] H. W. Kuhn, "The Hungarian Method for the Assignment   (9.4)
     Problem," *Naval Research Quarterly*, Vol 2, 83–97, 1955 and
     Vol 3, 253–258, 1956

[8] V. Klee and G. J. Minty, "How Good Is the Simplex   (10.4)
     Algorithm?", *Inequalities-III*, O. Shisha, ed, New York:
     Academic Press, 159–175

[9] Katta Murty, *Linear Programming*, New York: Wiley, 1983,   (10.1)
     Ch 11

[10] Ibid., Ch 13   (9.2)

[11] Gilbert Strang, *Introduction to Applied Mathematics*,   (10.4)
     Wellesley-Cambridge Press, 1986, Section 8.2
     Wayne L. Winston, *Operations Research: Applications and*
     Algorithms, Duxbury, 1987

[12] Steven Roman, *Linear Algebra with Applications*, San Diego: (4.1)
     Harcourt Brace Jovanovich, 1988

[13] H. Taha, *Integer Programming: Theory, Applications, and*   (8.4)
     Computations, New York: Academic Press, 1975

[14] Jan Van Tiel, *Convex Analysis*, New York: Wiley, 1984   (2.3)

INDEX

A 8
B 9
C 0
D 1
E 2
F 3
G 4
H 5
I 6
J 7